Technology of Metalcasting

Frederick P. Schleg
Principal Author

Frederick H. Kohloff
Jeet Radia, Dan Oman, Jim Schifo
Michael Gwyn
Contributing Authors

An American Foundry Society Publication

Published and Distributed

by the

American Foundry Society
Des Plaines, Illinois 60016-8399
www.afsinc.org

ISBN 0-87433-257-5

Printed in the United States of America

This book, *Technology of Metalcasting,* is largely based on the book, *Cast Metals Technology,* originally published in 1972, but has been revised and expanded to cover updated foundry operations and equipment, waste management, emission control and the establishment of foundry operation controls.

Acknowledgments

American Foundry Society gratefully acknowledges the Foundry Education Foundation (FEF) for its support toward the making of this book. AFS also acknowledges J. Gerin Sylvia for his book, *Cast Metals Technology,* which contributed so much to metalcasting education.

A very special thank you to John Deere Co., Inc. for its generous contribution, which enabled the ongoing efforts of various authors and editors to complete this publication.

AFS also thanks the following companies for their help in making this book possible:

ASM, International
Penton Publishing
McGraw-Hill Education
Hitchiner Manufacturing
Copper Development Association Inc.
Eirich
Crucible Institute, Inc.
Metallurgical Services, Surrey, England
Steel Founders' Society of America, SFSA
QIT, now Rio Tinto
NADCA (North American Die Casting Association)
Butterworth/Heinemann, Elsevier Science

In Appreciation

I would like to first of all acknowledge the encouragement and patience of my wife, Alice, who stuck with me during this literary effort and who held up well after hearing the following many times over... "I'm going below to work on the book."

Next I would like to thank Sue Thomas and Karen Frink of the AFS Special Publications Department for their questions and review of the material presented in this textbook. They kept me on my toes!

A great debt of gratitude has to go to my colleagues at AFS and CMI who, over the many years, shared with me their technical know-how and answered my many questions regarding the metalcasting process. These folks along with the many instructors with whom I have had the privilege of working in CMI courses are responsible for much of the information contained in this textbook. I, as well as the metalcasting industry, owe these men and women who have served and are now serving as instructors in CMI courses many, many thanks!

FPS

Preface

The foundation for this book was laid back in 1965 when J. Gerin Sylvia was contracted by AFS to write a textbook for use in junior colleges and area technical schools where metalcasting was taught. That book was *Cast Metals Technology.* Although never updated or rewritten, the textbook had been reprinted several times by AFS and Cast Metals Institute.

Since 1965, a lot of molten metal has gone over the spout of the ladles in the metalcasting industry and, with this, many new and exciting technological advances have been made in the process of producing metal castings. It is the objective of this textbook, *Technology of Metalcasting,* to expand and delve a little deeper into the technology of how metal castings are produced. This industry is not stagnant but vibrant and robust.

The material contained in this textbook has been gleaned from many technical articles, books and the author's personal notes, taken while teaching at the Cast Metals Institute for 25 years. The technical articles have appeared in the industry's magazines, research reports, and technical literature prepared by the many different suppliers of equipment and materials to the metalcasting industry.

The books include textbooks written over the years by many well-known educators and technologists and also those published by the American Foundry Society. These latter books were written by foundry men and women and suppliers who are members of various AFS Technical Committees, and deal with specific subjects such as cupola melting, cleaning room operation, aluminum casting technology, and copper-base alloys.

It was the author's intent to widen the scope and raise the technology level of this textbook, but not to a level such that it would overwhelm the reader. Every chapter in this textbook could be a complete textbook in its own right with additional information presented. This textbook has not been written to *impress* but to *express* and to help those who are new to the wonderful world of metalcasting to become better acquainted and knowledgeable in its many processes and how they all come together to produce quality metal castings.

Some may feel that the material contained in this textbook is too basic. If one were to really study the metalcasting process, he or she would find that what has really changed the most over the years have been the methods in which new technology has been applied. There have been many new molding, coremaking and melting materials developed; however, the basics, which affect their success or failure, haven't changed over the years. Many of these basics are covered in this textbook and will more than likely be covered in future textbooks. We still have to learn to walk before we can run.

It is hoped that, as students or nonstudents reads this textbook, they will find it to be interesting and informative reading. It is also hoped that this textbook will serve as the launching pad toward a long and successful career in the wonderful and exciting world of metalcasting.

Remember one key basic point of information... *"the cope is the top and the drag is the bottom"!*

Frederick P. Schleg, Retired
Cast Metals Institute
September 2002

Table of Contents

1. Introduction to Metalcasting 1–6

Brief History 1
Metalcasting Today 4
Foundry Types 5
Foundry Departments 5

2. Molding and Casting Processes 7–26

Nonpermanent Molding Processes 7
Sand Molding Processes 7
Green Sand Molding 7; Skin Dried Molding 8; Dried Sand Molding 8; Shell Molding 8; Coldbox Molding 9; Nobake Molding 10
Lost-Foam Casting/Molding Process 11
Vacuum (V-Process) Molding 12
Cosworth Molding Process 12
Vacuum-Assisted (Countergravity) Casting/Molding Process (For Sand or Permanent Molding) 12
Plaster Molding Process 16
Ceramic Molding Process 16
Investment Casting/Molding Processes 16
Solid Investment Casting/Molding 17; Shell Investment Casting/Molding 17
Permanent Molding Processes 18
Gravity Permanent Molding Process 21
Static Pour Process 21; Tilt Pour Process 21; Low-Pressure Permanent Molding (LPPM) Process 22; Diecasting Process 23; Graphite Molding Process 24; Centrifugal Casting/Molding Process 24; Continuous Casting Process 25
Innovative Molding/Casting Processes 25
Squeeze Casting Process 25
FM Process 25
Rheocasting and Thixomolding Processes (Semi-Solid Metal Casting) 26

3. Patternmaking 27–38

Equipment and Materials 27
Building a Pattern 28
Types of Patterns/Tooling 30
Loose Patterns (Solid or Split) 30; Matchplate Patterns 31; Cope-and Drag Matchplate Patterns 31; Shell Molding Patterns 31; Special Patterns or Pattern Devices 31; Expandable Polystyrene (EPS) Patterns 32; Wax or Plastic Patterns 33; Permanent Mold Tooling 33; Diecasting Mold Tooling 34; Centrifugal Casting Mold Tooling 35
Corebox Construction 35
Rapid Prototyping 36
Computer-Aided Design 36

4. Molding Sands 39–54

Types of Sand 39
Silica Sand 39
Zircon Sand 40
Olivine Sand 40
Chromite Sand 41
Physical Characteristics of Sand 41
Selecting Sand 41
Green Sand Mold Mixtures 42
Green Sand Mold Components 42
Clay 43; Water 45; Cereal 46; Cellulose 46; Iron Oxide 47; Carbonaceous Material 47; Polymers and Chemical Additives 47; Preblends 48
Green Sand Preparation 48
Mulling 48
Mixing 49
Quality Control of Green Sand Components 49
Green Sand Testing 50
Moisture 50; Compactability 51; Specimen Weight 51; Permeability 51; Green Compressive Strength 51; Methylene Blue Clay Content 52; AFS Grain Fineness 52; Mold Hardness Testing 52; Loss-On-Ignition (LOI) 53; Automatic and Miscellaneous Testing 53

5. Green Sand Mold Production 55–62

Hand Molding 55
Machine Molding 56
Jolt Molding 56
Squeeze Molding 57
Jolt-Squeeze Molding 58
Sand Slinger Molding 58
Impact/Impulse Molding 58
Vacuum-Squeeze Molding 59
Additional Molding Methods 60
Stack Molding 60
Flaskless Molding 60
High-Density Green Sand Molding 60
Green Sand Handling System 60

6. Coremaking 63–80

Selecting a Core Binder System 63
Core Sand 63
Core Sand Additives 64
Chemical Binder Systems 64
Heat-Activated Systems 65
Core Oil Process 65; Shell Process 65; Hotbox Systems 66; Warmbox Systems 66
Coldbox Systems 66
Sodium Silicate/CO_2 66; Phenolic Urethane/Amine (PUA) 67; Epoxy Acrylic/SO_2 68; Phenolic/Methyl Formate 68; Phenolic/CO_2 68; Furan/SO_2 68; Acrylic/SO_2 68

Nobake (Air-Set) Systems 69
Silicate/Ester Nobake 70; Phosphate/Metal Oxide Nobake 70; Alkyd Oil/Urethane Nobake 70; Phenolic/Urethane Nobake (PUNB) 71; Polyol/Urethane Nobake 71; Ester-Cured/Phenolic Nobake 71; Furan/Acid Nobake 71; Phenolic/Acid Nobake 71

Mixing and Corebox Filling Equipment 71
Heat-Activated System Equipment 72
Core Oil Process Equipment 72; Shell Process Equipment 72
Coldbox System Equipment 73
Nobake (Air-Set) System Equipment 74

Testing Chemically Bonded Sands 74
Tensile Strength Testing 75
Transverse Strength Testing 75
Gas Evolution Testing 75
Acid Demand Value (ADV) Testing 75
Shell Core and Mold Testing 75

Reclamation and Reuse of Spent Sand 76
Mechanical Reclamation and Binder Compatibility 76
Thermal Reclamation 77
Wet Reclamation 77
Reuse of Spent Sand 77

Refractory Coatings 77
Venting 78
Effect of Relative Humidity 78
Shakeout 79

7. Metal Alloys Cast in the Foundry 81–84
Classification of Casting Alloys 81
Properties of Metals and Alloys 82

8. Family of Ferrous Alloys 85–96
Cast Irons 85
Gray Cast Iron 85
Chilled Cast Iron 87
White Cast Iron 87
Malleable Cast Iron 88
Ductile Cast Iron 88
Compacted Graphite Iron 91
Austempered Ductile Iron 92
Other Cast Irons 93

Cast Steels 93
Plain Carbon Steels 94
High- and Low-Alloy Steels 96

9. Family of Nonferrous Alloys 97–112
Aluminum-Base Alloys 97
Copper-Base Alloys 103
Coppers 105
High-Copper Alloys 105
Brasses 105
Red and Semi-Red Brasses, Unleaded and Leaded 105; Yellow Brasses 105; High-Strength and Leaded High-Strength Yellow Brasses 105; Silicon Brasses/Bronzes 105
Bronzes 106
Tin Bronzes 106; Aluminum Bronzes 106; Copper-Nickel Alloy 106; Nickel Silvers 106; Leaded Coppers 106

Magnesium-Base Alloys 107
Titanium-Base Alloys 109
Cobalt-Base Alloys 109
Nickel-Base Alloys 111
Zinc-Base Alloys 111
Aluminum Composites 111

10. Melt Furnaces 113–134
Furnace Selection 113
Metallurgical Factors 114
Types of Furnaces 115
Cupola Furnaces 115
Crucible Furnaces 120
Electric/Direct-Arc Furnaces 121
Induction Furnaces 122
Channel Induction Furnace 123; Coreless Induction Furnace 125; Lift-Coil and Push-Out Crucible Coreless Induction Furnaces 126
Induction Furnace Refractory Practice 127
Channel Induction Furnace Linings 128; Coreless Induction Furnace Linings 128
Electric Resistance Furnaces 130
Electric Globar Furnaces 130
Reverberatory Furnaces 130
Direct-Fired Reverberatory Furnace 130; Sloping Dry-Hearth Reverberatory Furnace 131; Stack-Melting Reverberatory Furnace 131; Wet-Bath Reverberatory Furnace 132; Electric Radiant Reverberatory Furnace 132; Front-Charging Reverberatory Furnace 132
Dual-Energy Reverberatory Furnace 133
Regenerative and Recuperative Burner Systems 133

11. Ferrous Melting Practice 135–170
Cupola Melting of Ferrous Alloys 135
Cupola Bottom Installation 135
Charging and Feeder Charging Mechanisms 136
Charge Buckets 136; Feeder Charging 138
Coke Bed Fundamentals 138
Foundry Coke 138; Coke Bed Preparation 138; Coke Bed Performance 141
Cupola Blast Air System 141
Blast Air Blowers and Controls 141; Blast Air Measurement 142; Blast Air Duct Systems 142; Blast Air Conditioning 143
Charge Materials and Makeup 145
Cupola Combustion 147
Cupola Zones 147
Well Zone 147; Oxidation (Combustion) Zones 148; Reduction Zone, Melting Zone, Preheating Zone 148
Slagging and Tapping 149
Slagging, Tapping 149; Slag Collecting and Disposal 150
Dropping Bottom 152
Cold Bottom Dropping, Hot Bottom Dropping 152
Computer-Assisted Operation 153

Electric Furnace Melting of Ferrous Alloys 153
Induction Furnace Melting 153
Coreless Induction Melting (Cast Iron Alloys) 153; Coreless Induction Melting (Carbon Steels) 156; Coreless Induction Melting (High-Alloy Steels) 156; Channel Induction Melting 157
Electric/Direct-Arc Melting (General) 158

Electric Arc Melting (Carbon Steel) 160
Acid Melting Practice 160; Basic Melting Practice 162
Electric Arc Melting (Alloy Steel) 163
AOD Refining 163; Desulfurization 164
Electric Arc Melting (Cast Iron) 164
Charge Materials 164; Melting Procedure 165; CE and Silicon Analysis 165; Desulfurization 165
Testing and Analysis 166
Metal Chemistry Control 166
Eutectometer 166; Chill Test 167; Fluidity Spiral 167
Chemical Analysis 167
Temperature Measurement 167
Optical Pyrometer 167; Radiation Pyrometer 168; Immersion Thermocouple (Pyrometer) 168

12. Nonferrous Melting Practice 171–186
Aluminum Alloys 171
Charge Materials 172
Melting Safety 172
Tools and Their Maintenance 172
Temperature Control and Measurement 173
Automatic Ladling Devices 173
Furnaces Used for Melting Aluminum Alloys 174
Crucible Furnaces and Crucibles 174; Induction Furnaces 179; Electric Resistance Furnaces 180; Reverberatory Furnaces 180
Handling Molten Aluminum 181
Magnesium Alloys 181
Melting Procedures for Magnesium Alloys 181
Degassing Procedure 182
Handling Molten Magnesium 182
Copper Alloys 182
Charge Materials 182
Furnaces Used for Melting Copper Alloys 183
Gas-Fired Crucibles 183; Induction Furnaces 183
Melting Considerations 183
Metal Loss Control 183; Gases and Degassing 184
Basics of Good Copper Alloy Melting 185
Titanium Alloys 186

13. Microstructure/Ferrous Alloys 187–212
Ferrous and Nonferrous Microstructure 187
Atomic Arrangement 187
Phase Diagrams 188
Cooling Curves 189; Lever Arm Principle 190; Component Solubility 190
Ferrous Alloy Microstructure 192
Gray Cast Iron Microstructure 192
Gray Iron Graphitization 193; Effect of Gray Iron Inoculation 197; Effect of Heat Treatment on gray Iron 198; Effect of Alloying on Gray Iron 198
Malleable Iron Microstructure 199
Effect of Heat Treatment on Malleable Iron (Annealing or Malleablization) 200
Ductile Iron Microstructure 201
Effect of Desulfurization on Ductile Iron 202; Other Elemental Effects 202; Spheroidal Graphite Formation in Ductile Iron 203; Effect of Inoculation on Ductile Iron 205; Process Control of Ductile Iron 206; Effect of Heat Treatment on Ductile Iron 206
Compacted Graphite Iron (CGI) Microstructure 207
Austempered Ductile Iron (ADI) Microstructure 208
Carbon Steel Microstructure 208
Austenite Transformation of Carbon Steel 209; Segregation in Carbon Steel 210; Effect of Heat Treatment on Carbon Steel 210
High- and Low-Alloy Steel Microstructure 211

14. Microstructure/Nonferrous Alloys 213–224
Nonferrous Phase Diagrams 213
Nonferrous Alloy Microstructure 213
Aluminum-Silicon Alloy Microstructure 213
Effect of Grain Refinement 213; Effect of Structural Modification 214; Effect of Heat Treatment 217
Other Aluminum Alloy Microstructure 218
Aluminum-Copper Alloys 218; Aluminum-Magnesium Alloys 218; Aluminum Composites 218
Magnesium Alloy Microstructure 219
Copper Alloy Microstructure 220
Solidification (Freezing Ranges) 220; Effect of Grain Refinement 222

15. Gating Practice 225–246
Fluid Flow of Molten Metals 225
Bernoulli's Theorem 225
Law of Continuity 227
Velocity 227
Momentum 228
Frictional Forces 228
Reynolds Number 228
Fluid Life (Fluidity) 229
Gating System Requirements 230
Pouring Practice 230
Manual Pouring 231; Automatic Pouring 231
Gating system Design 232
Horizontal Gating System 233
Pouring Basin 233; Pouring Cup or Sprue Cup 233; Sprue 234; Strainer/Choke Core 234; Sprue Well (Base) 235; Runners 236
Vertical Gating System 237
Tile Gating System 238
Pressurized and Nonpressurized Gating Systems 238
Pressurized Gating System 239; Nonpressurized Gating System 240
Gating System Aids 241
Filters/Screens 241
Filter Size 243; Filter Placement 243; Direct-Pour System Filtering 243; Advantages of Filtering 244
Gating for Aluminum Composites 244
Gating System Choke Calculations 245

16. Solidification of Metals/Alloys 247–258
Heat Transfer 247
Measurement of Heat 247
Measurement of Thermal Gradient 248
Methods of Heat Transfer 248
Areas of Heat Transfer 249
Tapping 249; Pouring 249; Molds 250
Control of Heat Transfer 251; Thermal Deception 251

Solidification (Freezing) 251
Grain Growth/Solidification (Freezing) Ranges 252
Narrow (Short) Freezing Range 252; Wide (Long) Freezing Range 253; Medium (Intermediate) Freezing Range 254
Progressive and Directional Solidification 255; Solidification of Corners 255
Solidification Modeling 356
Summary 356

17. Risering Practice 259–274
Shrinkage 259
Riser Design 260
Determining Riser Placement 260
Grouping Castings 260
Determining Riser Shape 260
Determining Riser Size 261; Calculation Methods 262
Determining Riser Connection 264
Riser-Altering Aids 265
Riser Sleeves 265
Radiation Shields 265
Backpouring 266
Riser Feeding 266
Feeding Distance 266
Control of End Effect 269
Chills 269
Types and Purposes of Chills 270; Chill Material Selection 271; Choosing Chill Size 271; Chill Preparation 271; Chill Placement 272; Problems Associated With Chills 272; Chill Aggregates 273

18. Cleaning and Inspection 275–296
Cooling 275
Shakeout 276
Shakeout Methods 276
Dump 276; Punch/Blow 276; Hammering 276; Flask Removal 276
Cleaning and Finishing Operations 277
Cutting 277
Knockoff (Flogging) 277; Abrasive Cutoff 277; Flame Cutting 278; Plasma Arc Cutting 278; Gouging and Pad Reduction 278; Powder Cutting 278; Air-Carbon Arc Process 279; Bandsawing 279; Friction Sawing 280; Trim Press Operations 280; Shearing 280; Coining and Broaching 281; Waterjet Cutting 281; Chipping 282
Blasting 282
Grit and Shot 283; Abrasive Blasting Equipment 283; Blast Equipment Selection 283; Air Blasting 286; Abrasives Selection 286; Water (Hydraulic) Blasting 287; Mechanical Blasting 287
Grinding 288
Grinding Wheels 288; Grinding Machine Types 289; Abrasive Belts and Discs 289; Pneumatic Tools 291
Post-Cleaning Options 291
Impregnation 291; Hot Isostatic Pressing (HIP) 292; Value-Added Operations 292
Inspection and Testing 293
Standards and Specifications 293
Inspection and Testing Methods 293
Visual Inspection 293; Dimensional Inspection 294; Dye Penetrant and Fluorescent Powder Testing 294; Magnetic Particle Inspection 294; Ultrasonic Testing 294; Radiographic Inspection 294; Eddy Current Inspection 295; Pressure (Leak) Testing 295; Mechanical and Physical Property Tests 295
Final Comments 295

19. Environmental Protection and Safety Control 297–308
Air Quality Control 297
EPA Requirements 297
Emission Control 297
Particulate Matter Control, Organic Compound Control 297
Water Pollution Control 298
Discharged Foundry Waters 298
Water Treatment Options 298
Government Control 298
Legislation 301
Solid Waste Management 302
Solid Waste Generation 302
Hazardous Wastes 302
Beneficial Reuse 302
Legislation 303
Safety and Health Program 303
Management Commitment/Employee Involvement 303
Worksite Analysis 304
Hazard Prevention and Control 304
Safety and Health Training 304
OSHA Compliance Checklist 305
OSHA Training Requirements 307

20. Casting Design 309–318
Structural Geometry 309
Design Parameters 310
Fluid Life 310
Solidification Shrinkage 311
Liquid Shrinkage 311; Liquid-to-Solid Shrinkage 311; Patternmaker's Contraction 314
Pouring Temperature 314
Slag/Dross 314
Section Modulus 314
Area Moment of Inertia 314
Modulus of Elasticity 315
Castability Vs. Geometry 316
Junction Design 316
Postcasting Considerations 316
Geometric Dimensioning & tolerancing 317
Tolerance Capabilities 317

Glossary 319

Index 339

Introduction to Metalcasting

1

What is a metal casting? A metal casting may be defined as a metal object produced by pouring molten metal into a mold containing a cavity that has the desired shape of the casting, and by allowing the molten metal to solidify in the cavity. If sand molds are used, the molds are destroyed upon solidification of the metal. If a permanent-type mold is used, it is merely separated to remove the casting, and the mold can be used again.

BRIEF HISTORY

Historical data indicates that metalcasting had its beginning some 4000 years before the Christian era. There seems little question that metal was first used in the part of the world known as the Eurasiatic steppe belt (the Russian Black Sea area) **(Fig. 1-1).**

It is only natural that gold was the first metal to gain attention. It must surely have been evident in streams and waterways. Early man undoubtedly picked up pure nuggets and admired their excellent surfaces. It was soon discovered that gold was malleable, and could be flattened without splitting. At about the same time, early man discovered copper beads that had come from copper-bearing ore, which had been used to bank fires. Heating and hammering (forging) could shape copper, and thus—because it suited the purpose of early man—copper became the material used to produce castable articles.

When the people of the Black Sea area swept down into Mesopotamia some 4000 years before Christ, conquering as they traveled, their victories were due primarily to their forged weapons. It was in their new habitat that some early foundrymen invented a high temperature forge fire and produced a cast object from molten metal. Thus, before the wondering eyes of one early foundryman, the art of casting metal was born.

Copper was the first metal to be cast. Later, it was noted that sometimes copper contained certain other substances that made it harder; thus, bronze was discovered. The material that had been accidentally combined with copper was tin.

Melting was first done in clay-lined holes in the ground. Gradually, as time went by, artisans developed more-permanent melting media, which eventually evolved into melting furnaces. Then, artisans began to notice that the amount of metal produced was somewhat proportional to the air draft, which produced a hotter fire.

Eventually, an upright shell, lined with clay, was charged with alternate layers of copper ore and wood. Hollow wood blowers (tuyeres) were introduced into the furnace at the level of the firebox. These blowers were crude bellows, covered with goatskin at the blower end. With the aid of spring poles tied to each bellows, the operator stepped on each bellows, forcing air into the furnace. When he removed his foot, the bellows snapped back into normal position and air flowed back into them. This provided a steady stream of air to the furnace, thus creating a blast furnace.

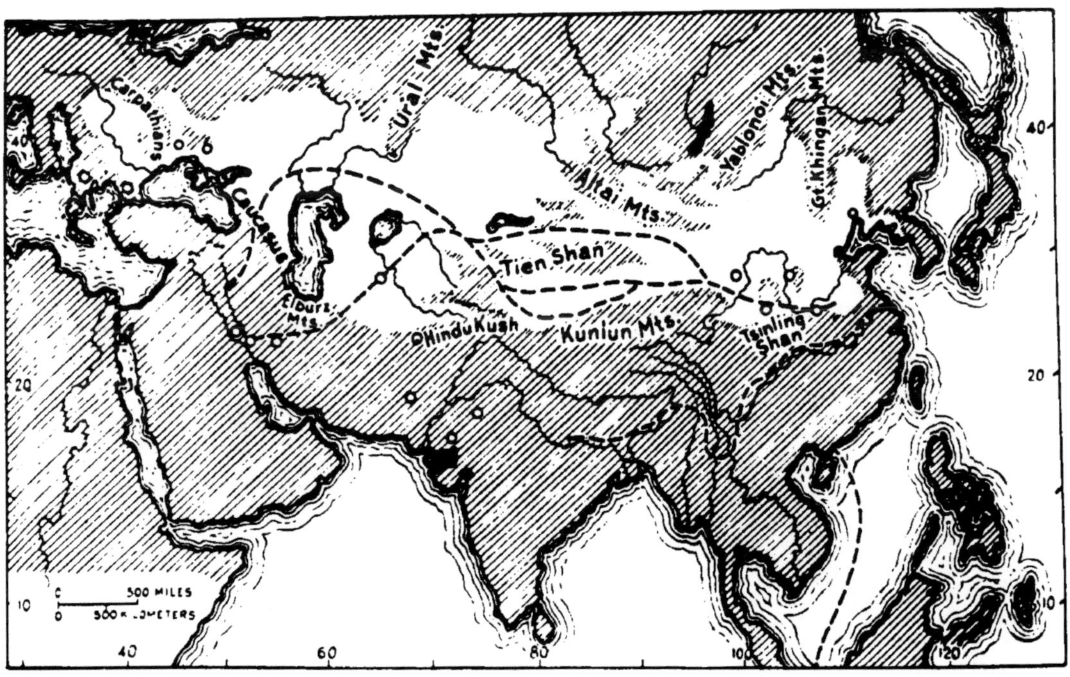

Fig. 1-1. The birthplace of metals has been traced with some accuracy to the steppe corridor of Eurasia, to the area north of the Black Sea in the Carpathian Mountains of Russia. [B.L. Simpson, 1948.]

Many of the earliest molds, into which the molten metal was poured, were open half-molds, made of sand. Later, the molds were cut in stone or fireclay **(Fig. 1-2)**. Simple shapes were formed to produce spearheads, ax heads and simple agricultural implements.

The early masses of people were migratory. Hence, the casting process moved eastward into the Orient. It continued to be developed there until about the 10th century. The intricate and delicate work of the early Chinese foundrymen indicates the use of lost-wax techniques, closed stone molds and sectional loam molds. The Chinese achieved a mastery of bronze casting and an advanced knowledge of metallurgy in earlier times **(Fig. 1-3)**.

Iron had been discovered as far back as 2000 B.C. Early man first shaped it by reducing the ore, melting it, and puddling it with slag into a ball, then forging it as wrought iron. The Chinese are known to have made castings of iron about 600 B.C. In India, cast crucible steel was first produced about 500 A.D., but the process disappeared and was lost until rediscovered later by Benjamin Huntsman, in England, about 1750.

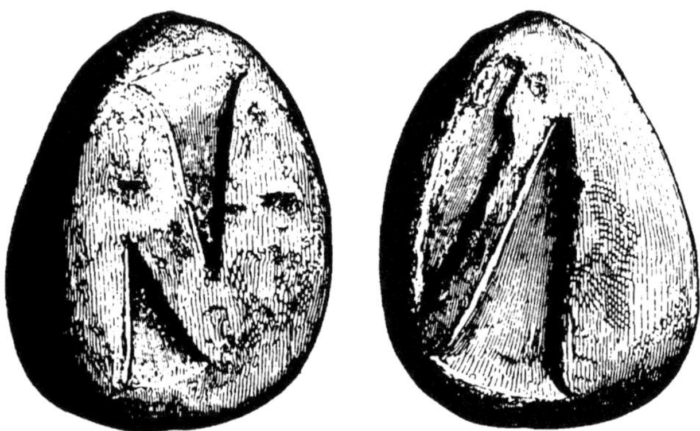

Fig. 1-2. Early mold for a spearhead, cut in fire clay.

Fig. 1-3. A ceremonial bronze elephant cast in two parts and joined. This is an example of the workmanship of early Chinese foundrymen. Probably from the Chou dynasty (1122–255 B.C.) or earlier. [B.L. Simpson, 1948]

After spreading eastward, the casting process spread westward into the Near East, the Mediterranean basin and the rest of Europe. The Egyptians improved the already well-developed casting techniques of the Orient. They are also credited with the discovery of the lost-wax process of casting metal. The use of cope and drag, and of core molding, appeared early in the Egyptian artisans' development of metalcasting **(Fig. 1-4)**.

Iron was not generally cast in Europe until about the 14th century. Previously, parts were shaped by forging. The high temperature needed for pouring iron castings was obtained in furnaces resembling small blast furnaces. During melting, the air was supplied by large bellows operated by hand, foot or water power. Iron ore was reduced in these furnaces, and when it was molten, it flowed directly into molds.

At the beginning of the 13th century, the chief interest of foundrymen was the casting of bells for the large cathedrals then being built in Europe. The molds for these bells were very often made in the churchyard or somewhere else near the cathedral. The furnace for melting the metal was constructed alongside the mold. The bells were cast using the loam molding or sweep molding process **(Fig. 1-5)**.

According to history, a monk in the city of Ghent cast the first cannon in bronze in the year 1313. Benvenuto Cellini, who used loam molding combined with the lost-wax process, made artistic articles and statues in Italy. Leonardo da Vinci is also credited with the casting of many fine works of art in metal. It is also interesting to note that during wartime, bells were melted down to produce cannons because metal supplies were so limited.

Fig. 1-4. Cast bronze cat (made with a removable core). Cast in Egypt, probably in Sakkarah, in the seventh century B.C.

Vannoccio Biringuccio (1480–1539) became head of the papal foundry in Rome in 1538. He is the first man known to have set down foundry practice in writing, in detail. His clarity of analysis, together with a common-sense practical approach, marks him as an accomplished artisan. Biringuccio's *Pirotechnia* undoubtedly covers all that was known of metallurgy in the 16th century. Even today, his statements of the three most important principles of making castings go unchallenged. According to him, these principles are "making and arranging the molds well, smelting and liquefying the materials of the metals well, and making the composition of their associations according to the results you wish to have." The exacting art of bell casting bears out a point made by Biringuccio—that the art of casting is a "necessary means to very many ends."

Following the Renaissance, trade and commerce gradually revived, and the rise of a "free" industry brought into being the craft guilds. The guilds exercised complete control over all skilled workers, and exerted a great influence on foundry operations. The guilds accomplished a great deal of good by establishing principles of good workmanship, quality and honesty. However, because of their monopolistic power and the fact that they often carried rules to extremes, the guilds finally deteriorated.

In 1730, in England, a man named Abraham Darby, of Coalbrookdale, initiated the use of coke as a fuel. Iron could now be produced at about two thirds the earlier cost, and thus coke became one of the principal tools of the iron foundries. It was also at this time that an energy crisis (which began in England) swept the world, causing foundries to search out a new source of fuel for their furnaces. In 1794, there appeared the first metal-clad cupola, similar in appearance to those of today, invented by John Wilkinson of England. To provide the air blast needed, he used for the first time the steam engine invented by James Watts in 1765. Naturally enough, after the invention of the steam engine, the need for iron castings greatly expanded.

The first American foundry was established in 1642 near Lynn, Massachusetts, on the Saugus River. It was known as the Saugus Iron Works. The first American casting, the famous "Saugus pot," is the treasured property of the city of Lynn **(Fig. 1-6)**. The Saugus-area bog ore proved suitable for the start of an industry that eventually was to include more than 5000 plants in the United States. In quick succession, plants were established along the eastern seacoast as far south as Virginia.

No history of the American foundry industry would be complete without a reference to the iron plantations, great estates that existed principally in eastern Pennsylvania in the 18th century. Mount Joy Forge, later known as Valley Forge (of Revolutionary War fame) was one of many that started in 1742.

Paul Revere, the Revolutionary War patriot who made the ride from Boston to Lexington on the night of April 18, 1775 to warn of the approach of the British, was a foundryman by trade. He operated a bell-and-fittings foundry in Boston. His success in metallurgy is well known today and a great American company bearing his name—Revere Copper and Brass Company—is the direct descendant of Revere's original enterprise. Also of interest is the fact that seven of the signers of our country's Declaration of Independence were foundrymen. These men were protesting against England's prohibition of manufacturing within the colonies.

An important development of the 19th century involved the introduction of chilled-iron car wheels. This was, of course, of great importance to the railroad industry. In 1847, Asa Whitney of Philadelphia obtained a patent on a process for annealing chilled-iron car wheels cast with chilled tread and flange. Cold-blast charcoal iron was first used. Later, a small amount of ferromanganese introduced directly into the ladle produced a chilled-iron car wheel that performed excellently. This made possible long hauls and heavier railroad freight loads.

Fig. 1-5. Sweeping cope of bell mold in pit. Note complete core and cope, at left. Vertical sweeping, coupled with pit molding, was usually employed in molding bells. [B.L. Simpson, 1948]

Fig. 1-6. The first American casting. The iron pot known as the "Saugus pot" was made at the Saugus Iron Works, Saugus, Massachusetts.

In order to have a complete description of iron, there needs to be a mention of American "blackheart" malleable iron, as contrasted with European "whiteheart" malleable iron. Seth Boyden of Newark, New Jersey is credited with much of the progress and growth of the American malleable-iron industry. In attempting to duplicate European whiteheart malleable iron, Boyden experimented with an iron containing a larger percentage of silicon than was available in the European product. He produced a strong iron that could be machined easily. In the process, he managed to shorten the time of anneal to between six and 10 days.

Steel castings appear to have been made first in India, and may have been poured in England as early 1609. Inadequate equipment, however, held back the manufacture of steel castings in large quantities. The introduction of the converter, the open-hearth furnace and, finally, the electric furnace made it possible to produce steel commercially in great quantity, and to do so economically. Development continued, and in 1831, William Vickers of England was able to make cast steel from wrought-iron scrap by combining manganese-oxide and carbon. Cast-steel guns were made by the Krupp Works in Germany in 1847.

In the United States, cast steel was produced by the crucible process in 1818 at the historic Valley Forge foundry. It wasn't until 1831 that William Garrard of Cincinnati, Ohio, utilizing the excellent refractory clays of Cumberland, West Virginia, established the first commercial crucible-steel operation in this country.

In 1851, William Kelly of Kentucky invented a converter that enabled him to produce rather soft steel. Sir Henry Bessemer, in England, developed the process that bears his name. He succeeded in purifying the metal and assuring the presence of enough carbon to make steel. The first Bessemer converters in the United States were installed in Troy, New York in 1865.

With the development of the open-hearth furnace in 1845 and the perfection (in 1857) of the regenerative open-hearth furnace with its great heat, the steel industry was given the tonnage capacity required for successful operation. The first open-hearth furnace in the United States was installed in 1870. Thus, a vital tool for the growth of a nation came into being.

METALCASTING TODAY

One could continue tracing the technological advances of the metalcasting industry through the remainder of the 19th century, the entire 20th century and now into the 21st century; however, there would not be time or space to discuss the technologies involved in producing metalcastings. For instance, today, ductile iron has replaced malleable iron to the point that there is very little cast iron alloy poured. Now, ductile iron has spawned the use of austempered ductile iron and an equal ground between gray and ductile iron; compacted graphite iron has also been developed.

The nonferrous alloys have seen changes in the various alloys that have been developed over the years. This is true of the aluminum, copper-base, magnesium and zinc-based alloys. Superalloys, such as titanium, cobalt, etc., now entered the industry. The ferrous and nonferrous alloys will be discussed later in this book.

Moldmaking has also advanced over the years, from early versions crafted from stone to bonded sand, graphite and metal molds. Today, much of the manual labor used in moldmaking, especially in sand molds, has been eliminated and high-speed molding machines are used. The science and technology of preparing and maintaining a molding sand system has also kept pace with the other advances being made within the metalcasting industry.

Probably the one area in the metalcasting process that has seen the greatest change is making sand cores. Years ago, sand cores were made by hand and then baked (hardened) to offer rigidity. Today, core sands are blown into coreboxes and "cured" (for rigidity) within seconds. For instance, today's engine block wall thickness' between cylinders were unheard of 50 years ago.

Other molding processes that do not use sand as the molding refractory have also seen great changes in their technologies. These processes include diecasting and permanent mold casting, which have found great success in nonferrous casting applications.

The ways and means of melting various foundry alloys has seen considerable change since the early foundry days. The use of electric energy and natural gas has changed the way in which many foundries melt alloys. Here, too, mechanization and automation has brought this area of the metalcasting process into areas unknown many years ago.

Patternmaking, including the making of coreboxes, has kept pace with the other processes. This progress includes the use of new pattern and corebox materials, CNC machining and other computer-controlled equipment. Rapid prototyping has also played a vital role in helping the patternmaker keep current.

Finally, the cleaning, finishing and inspection areas have also seen improvement and advancement over the years. The improvements included taking out much of the manual labor in the cleaning room, installing improved mechanical handling systems, automating the equipment used to clean castings and much needed improvements with the addition of computers to track castings. New and improved testing and quality inspection techniques and equipment are now common in foundries.

The words "technological revolution" might be used to adequately describe the tremendous mechanization, automation and system controls that have grown with and within the metalcasting industry to develop it into one of the largest manufacturing industries in the United States today.

FOUNDRY TYPES

Foundries or metalcasting plants can be divided into several types. For instance, foundries can be ferrous or nonferrous based on the alloy poured. Some foundries pour both ferrous and nonferrous alloys. The specific type of alloy poured, such as ductile iron, aluminum, carbon steel, copper-base alloys, can further typify these foundries.

Foundries can also be classified as jobbing or captive. Jobbing foundries (often independently owned) contract for outside casting jobs to fit their capabilities and usually have lower production rates when compared to captive foundries. Captive foundries usually produce castings for use within their parent organization. Today, some captive foundries also produce castings for outside firms.

Next, the type of molding process employed also classifies the foundry. In some cases, more than one molding process is used within the foundry. For example, foundries that use sand as the refractory aggregate in moldmaking are called sand casting foundries. Various molding processes are covered in Chapter 2.

Foundry Departments

Foundries are departmentalized, most of them having molding, coremaking, melting, cleaning, quality control, maintenance, patternmaking and engineering departments. **Figure 1-7** is a flowchart of a typical foundry operation. Not all foundries follow all steps outlined.

Molding—The molding department is concerned primarily with the making of molds, and may use any one or a combination of the following methods: green sand, dry sand, bench, floor, machine, shell, pit or loam molding. In permanent-mold and diecasting foundries, metal molds are used, and investment-casting foundries use ceramic molds.

Coremaking—Cores are sand shapes inserted into molds to form internal cavities. Coremaking may be done by hand ramming, machine ramming, or air blowing. Cores may be made of oil-bonded sand, silicate-bonded sand, thermosetting resin-coated sand, or by the air-setting or the hotbox method. Some cores require baking and a storage area, while others are made and used within a short period. Permanent molds can use cores made of sand or metal inserts, whereas diecasting requires metal-insert cores. Investment-casting cores can either be made using the shell material or ceramic.

Melting—The melting department is concerned with one or more of the following types of furnaces: cupola, electric furnaces and crucible furnaces. Gas, oil, coke or electricity are fuels used in melting.

Cleaning—The cleaning department is concerned with the removal of gates and risers, and the chipping, grinding, welding, shotblasting, or tumbling of the castings, so that all sand clinging to the surface is removed. During the shotblasting and tumbling operations, some, if not all, of the sand cores are also removed. Additional surface cleaning techniques may be employed to ensure quality levels are maintained. Surface condition is, of course, of paramount importance and great care is taken to give the casting a clean appearance.

Quality Control—The quality-control department (QC) is responsible for the control of metal analysis and castability, so that certain standards of physical and mechanical properties are maintained. Control of the mold and coremaking materials is necessary so that mold and coremaking operations yield quality castings. Control of the casting dimensions, as well as surface finishes that are sound and free of any voids or defects, is the goal of the QC Department. To yield the highest casting quality, nondestructive testing techniques are utilized, since closer tolerances and uniform quality are constantly being demanded by the foundry's customers.

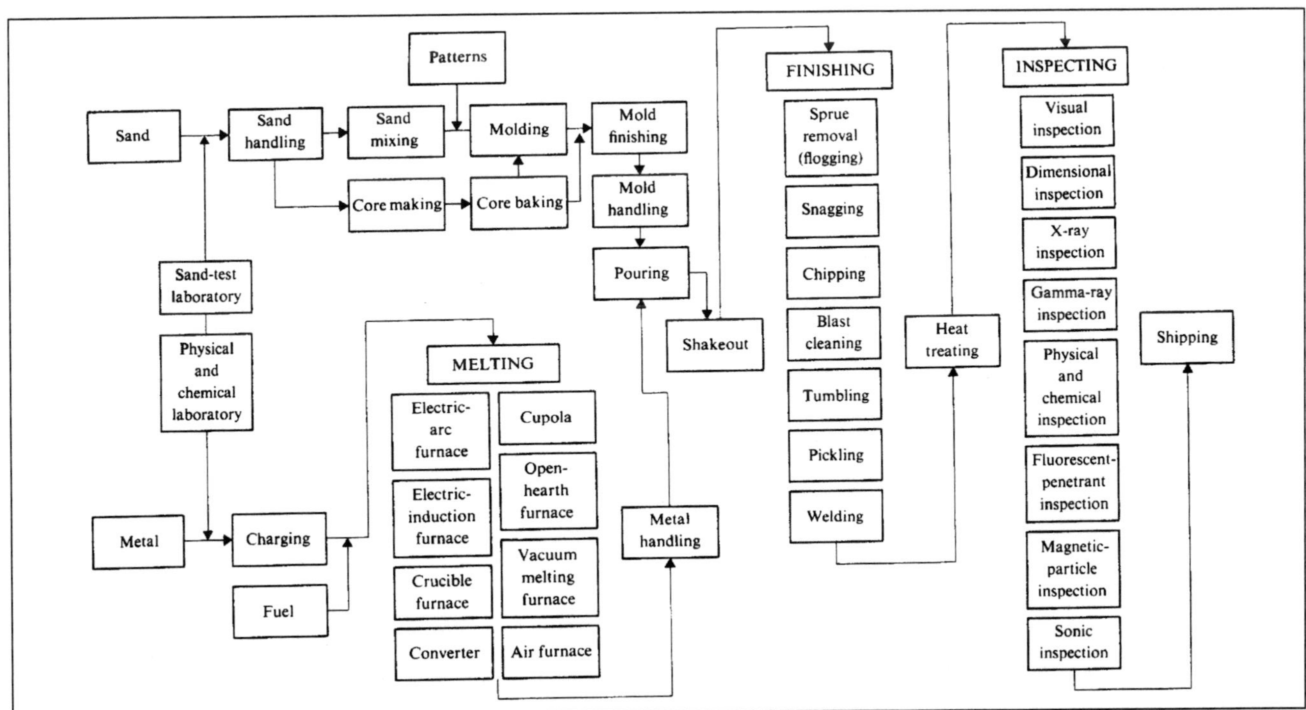

Fig. 1-7. Flowchart of a typical foundry operation.

Maintenance—The maintenance department plays a very essential role in producing quality castings. With the advent of today's highly sophisticated foundry equipment, the role of the maintenance department has been upgraded. Maintenance personnel must be skilled in the use of the latest diagnostic equipment, electronics, pneumatics and hydraulics. Preventive maintenance is the key in keeping today's foundries running smoothly.

Patternmaking—The patternmaking departments in most foundries perform mainly pattern and corebox maintenance. Some foundries still use their pattern shops to build new patterns and coreboxes. It is this department's responsibility to maintain the patterns and coreboxes in peak condition so that quality molds and cores can be made.

Engineering—Foundry engineering departments can usually be divided into several areas. The process engineers work to improve the various foundry processes. Production and plant engineers work with the equipment and equipment layout in the foundry. Industrial engineering is also employed in many foundries to study ways and means to make it more conducive to perform assigned tasks. Today's foundries also must be concerned with the environment inside as well as outside the foundry. To meet these challenges, many foundries employ environmental engineers. Environmental departments must keep up with the many changes in state and federal regulations and apply them, where applicable, into the everyday foundry operations.

Many foundries also employ technical-school graduates and degreed engineers in the area of pattern engineering. These people many times will work with foundry customers in the design and redesign of castings. They will also make use of the latest computer techniques, such as solidification modeling and fluid-flow modeling to improve the quality of the castings, and decrease the extensive lead-time in producing castings. Stereolithography and Fast Freeform Fabrication methods are only a few of the emerging rapid prototyping and patternmaking tools.

BIBLIOGRAPHY

Aitchison, L., *A History of Metals*, Vol. 1, Interscience Publishers, New York, New York, 1960.

Ekey, D.C. and W.P. Winter, *Introduction to Foundry Technology*, McGraw-Hill, New York, New York, 1958.

Simpson, B.L., *Development of the Metal Casting Industry*, American Foundrymen's Society, Des Plaines, Illinois, 1948.

Smith, C.S. and M.T. Gnudi, translators for *The Pirotechnia of Vannoccio Biringuccio*, The American Institute of Mining and Metallurgical Engineers, New York, New York, 1942.

Molding and Casting Processes

2

There are literally dozens of molding and casting processes available today that can be used make molds for producing metalcastings. Each process offers distinct advantages and benefits when matched with the proper alloy and application. When selecting a molding or casting process, some areas to be considered include the following:

- required quality of the metalcasting surface;
- required dimensional accuracy of the metalcasting;
- number of metalcastings required per order;
- type of pattern equipment required;
- moldmaking costs;
- effect of selected molding process on the casting design;
- effect of secondary operation(s) performed on the casting.

High-quality metalcastings can only be produced from high-quality molds and correct molding processes. The commercial success of a metalcasting process may reflect speed of production, dimensional accuracy, surface finish, replication of detail, metallurgical considerations, or some other particular feature inherent to a molding process. There are many different molding processes available from which to choose, and all have advantages and disadvantages.

Molding processes can be simply broken down into two main categories:

1. Nonpermanent
2. Permanent

In the nonpermanent molding processes, the mold is destroyed when the metalcasting is removed from the mold. The opposite is true for permanent molds. In this case, the mold can be reused, once the metalcasting has been removed. There are also some new, innovative molding/casting processes, which will be discussed later in this chapter.

Fig. 2-1. Workers perform final checks on green sand molds prior to assembly.

NONPERMANENT MOLDING PROCESSES

The discussion of molding processes will begin with the nonpermanent method. One way of classifying metalcastings is according to the materials used in making molds. The highest total weight metalcastings are produced in molds where sand is used as the molding material. The term "sandcasting" is used to describe the process of making metalcastings in sand molds. There are several different types of sand molding processes that can be used.

Sand Molding Processes

For "dry" sand molding (other than "green" sand molding), the terms "refractory," "binder" and "catalyst" will be used in this chapter. Refractory materials are those able to withstand high temperatures without fusing (i.e., sand). Therefore, these materials need a binder to hold them together when shaping the mold. A catalyst is used to harden (or cure) the binder and assure that the mold retains its shape during the pouring process. The catalyst can be in the form of heat or chemical compound (liquid or gas). More on this in Chapter 6, Coremaking.

Green Sand Molding

The most common process for making molds for metalcastings is the green sand molding process **(Fig. 2-1).** In this process, a granular refractory mineral, usually silica sand, is coated with a mixture of bentonite clay, water and, in some cases, other additives. When these coated grains of sand are compacted around the pattern, they are held together by the "glue" composed of clay and water. When the pattern is removed from the mold, the mold cavity retains the shape of the pattern surface around which the molding sand has been compacted.

The following are some important points to review when considering the green sand molding process:

- For many metal alloy applications, green sand molding processes are the most cost-effective of all metal-forming methods.
- This process readily lends itself to automated systems for high-volume work, as well as short runs and prototype work.
- In the case of hand molding, sand slinging, manual jolt or squeeze molding, wood or plastic pattern materials can be used.
- When high-pressure, high-density molding methods are used, metal and plastic patterns will be required.
- High-pressure, high-density molding normally produces well-compacted molds, which, in turn, yield better surface finishes and details, casting dimensions and tolerances.
- The properties of green sand are adjustable within a wide range, which makes it possible to use this process with all types of green sand molding equipment and for the majority of alloys poured in the metalcasting industry.

- Titanium cannot be poured into green sand molds made of silica sand; however, all other metal alloys can be poured into these molds.
- Small, medium and relatively large metalcastings can be produced using the green sand molding process.

A more detailed discussion concerning the preparation of green sand for molding operations and the actual methods for making the molds will be discussed in Chapter 5.

Skin Dried Molding

A type of green sand molding, called skin dried molding, is normally used when large ferrous metalcastings are to be made. Points to consider when using this process are itimized in the next section on Dried Sand Molding.

As stated earlier, one of the ingredients in green sand molding is water. When large quantities of molten ferrous metals are poured into green sand molds, surface and other defects can occur on the metalcasting. Since only the water content near the surface of the mold cavity or the parting line is affected by the heat of the molten metal, it should be removed to help reduce or eliminate these defects. Depending on the size of the metalcasting and the pouring temperature of the molten metal, the water can be removed from 0.25 in. (6.35 mm) to 1.0 in. (25.4 mm) of the parting line surface. The water on and near the surface of the mold cavity can be removed by torching, with natural gas, hot air or infrared heating elements **(Fig. 2-2).**

To further improve the strength and heat-resistant characteristics of the mold parting line, special refractory coatings can be brushed or sprayed onto these surfaces. These refractory coatings are usually a mixture of water, finely crushed silica, zircon, chromite or mullite. In cast iron castings, the refractory coating may be graphite, which can be dusted or brushed onto the surface of the sand. Alcohol and other volatile liquids may be used to replace about 90% of the water. When this is done, the surface of the molds may be ignited, producing enough heat to drive the water away from the surface of the mold cavity and develop the necessary dry strength and eliminate the need for other drying methods. To prevent the formation of gas, all the solvent should be allowed to burn off before closing and pouring the mold. Extreme care should be used when contemplating the use of alcohol or other solvents.

Fig. 2-2. Skin drying of drag molds.

Dried Sand Molding

Another variation of green sand molding is dried sand molding. This method is similar to skin dried molding, except that the mold is placed in an oven and the water is baked out of the entire mold. At times, a petroleum binder is used, resulting in a very high-strength mold after oven curing or drying.

This molding process is used for larger ferrous metalcastings, and the mold size is limited by the size of the mold drying ovens. A refractory coating can be applied to the parting line surface for the same reasons as mentioned in skin dried molding.

Some points to consider when using either the skin dried or dried sand molding processes are as follows:

- It can eliminate some defects on the surface of large ferrous metalcastings when the molds are made of green sand.
- It produces a very hard and strong mold-cavity surface, which can produce better dimensional tolerances and smoother metalcasting surfaces.
- It improves the separation of the molding sand from the metalcasting surface.
- It increases the time needed to make the mold.
- It adds to the cost of producing the mold.
- These two molding processes are normally used for ferrous alloys.

Shell Molding

The less common sand molding processes use bonding materials *other than clay and water* to hold the sand grains together. As with the green sand molding process, the mold is destroyed when the metalcasting is removed. An example of this process is seen in **Fig. 2-3.**

Shortly after World War II, a process requiring heat to cure the mold was brought to this country. As will be learned in Chapter 6, this process can also be used to make cores. In the shell molding process, the sand grains are coated with a thermosetting plastic resin. The pattern used for this molding process requires heating up to 475F (246C). Because the pattern has to be heated to this

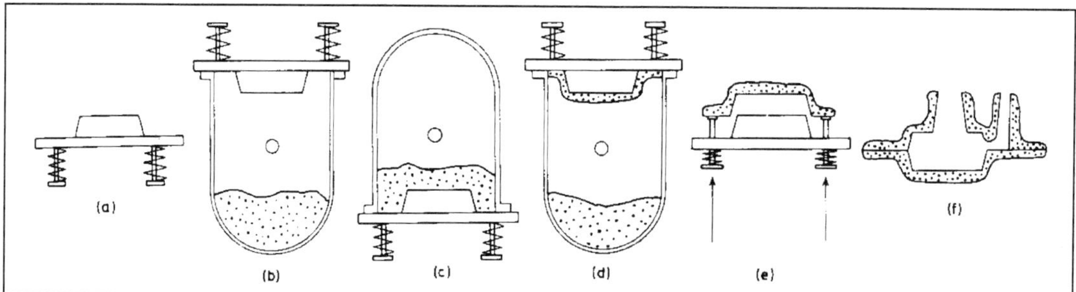

Fig. 2-3. Basic production sequence in shell molding process: (a) pattern plate; (b) pattern plate mounted on dump box; (c) and (d) investment; (e) ejection; (f) shell assembly.

temperature, the pattern must be metal. The pattern material most commonly used is cast iron. The patterns are normally heated with natural gas burners located near the back of the pattern.

The thermosetting resin-coated sand is dumped or blown on top of the pattern, which will produce the surface of the mold cavity. The coated sand is allowed to lay on the pattern surface until a "shell" is produced to the desired thickness. This is called the invest cycle or set time. When this is achieved, the remaining uncured resin-coated sand is removed by a dumping action and can be used to make another mold. The shell remains on the pattern and is subjected to heat until it is thoroughly cured. The mold is made in two halves and can have either a horizontal or vertical parting line **(Fig. 2-4).**

The two halves are then glued, clamped or clipped together after the cores have been set. They are then ready for the pouring operation. The molds are set in beds of sand or, in the case of vertically parted molds, placed into a container. After the mold has been placed in the container, the container is filled with some type of backing material. This material can be used foundry sand or, in some cases, metal shot.

Some considerations, when selecting the shell molding process, include the following:

- Patterns must be made from metal and can be expensive.
- This process lends itself best to high-production rate casting quantities.
- The surfaces of the mold cavities are very smooth and can produce high-quality surface finishes on metalcastings.
- Because of their rigidity, shell molds can produce good dimensional accuracy.
- The cooling rate of the metalcasting can be controlled by using the appropriate backing material around the shell mold.
- The process lends itself well to the use of refractory materials other than silica sand.
- Shell molds can be stored indefinitely and improve just-in-time delivery of castings.
- Sizes are usually limited to small or medium.
- The shell molding process is energy-intensive and thus can be more expensive than green sand.
- All metal alloys, with the exception of titanium, can be poured into shell molds.

Fig. 2-4. Set of shells forming shell mold.

Coldbox Molding

Sodium Silicate-CO_2 Molding—The sodium silicate-CO_2 (carbon dioxide) molding process is normally referred to as CO_2 molding. It is a molding process in which liquid sodium silicate is mixed with sand. Other additives can be used to aid the shakeout process and also the collapsibility of the mold as the casting solidifies.

The sand mixture is placed around the pattern and compacted. The compaction can be done manually with the aid of pneumatic rammers or squeezers. After compaction, CO_2 gas is passed through the mold **(Fig. 2-5).** The CO_2 gas acts as a catalyst and causes the sodium silicate to cure or harden, locking the sand grains in place. After the mold has been cured, the pattern is removed, cores are set and the mold is closed, ready for the pouring operation.

There are several less-frequently used variations of this molding process that are self-hardened without the use of CO_2 gas. These methods require the use of ferrosilicon fines, Portland cement, or an ester to act as the catalyst.

From the metalcaster's point of view, the CO_2 process is one of the most environmentally acceptable of all the chemical molding processes available. Despite this, the process presents some production difficulties that have limited its use. Chief among these is short bench life. In other words, once the molding sand has been mixed, it must be used as quickly as possible. The reason for this is that the binder is very prone to picking up moisture (hydroscopic), which weakens the binder. This is true even after the binder has been cured. Once the mold has been closed, it should be poured as soon as possible.

A sodium silicate binder creates a very hard and rigid mold. This can cause poor collapsibility of the mold and set up undo stresses or even hot tears in the metalcasting. Also, shakeout can be a problem, which can slow down the production rate. Despite these factors, work does continue to overcome these shortcomings because the molding process offers a variety of significant benefits, such as the following:

- A very hard and rigid mold is typical of the process, which gives good metalcasting dimensional tolerances.
- Good metalcasting surfaces are readily obtainable.
- Wooden and plastic patterns can be used. However, for high-production runs, metal is preferred.
- A wide range of mold sizes can be produced.

Fig. 2-5. Molding machine uses sodium silicate bonded sand cured by CO_2.

- The binder used is probably the least gas-generating of the molding processes, reducing the possibility of gas-related defects.

Amine and SO_2—There are two other gas-cured binder systems that can be used to make molds: the amine process and the SO_2 (sulfur dioxide) process. The difference between the two lies in the gas catalyst used to cure the binder. The binder in both cases is a phenolic urethane resin that is mixed with the sand. In the amine process, an amine gas is used to catalyze the binder. In the SO_2 process, sulfur dioxide gas is used as the catalyst. Both of these processes are being used extensively in producing cores. They are mentioned here because they can also be used to make molds.

Both the amine and SO_2 sand molding processes offer several benefits, as follows:

- The dimensional accuracy of metalcastings produced is very good.
- Both processes are especially adaptable to high-production runs, since the production cycles are very short.
- The casting's external surface finishes are excellent.
- The molds have excellent shelf (or storage) life.

Nobake Molding

Nobake molding is very similar to the coldbox molding processes in that the binder coating the sand is catalyzed. The difference in this case is that the catalyst is a liquid, not a gas. There are various binder systems available from which to choose, such as furan, alkyd-oil and phenolic urethane. In each case, a liquid catalyst is used to cure the binder. Another term commonly used in place of nobake is "air-set." This denotes the fact that the cure takes place under ambient conditions without the aid of gas or heat.

The mixing of the sand, binder and catalyst is done in a high-speed, high-intensity continuous mixer **(Fig. 2-6).** Once the catalyst makes contact with the binder, the "cure" cycle begins and continues until the cure is completed. Once the molding sand has been mixed, it should be used as soon as possible in order to get well-compacted molds. The time available for using the mixed molding sand is called "work time." Different binder systems have different work times.

To make the mold, the pattern is placed in a flask, which can be made of wood or metal. The mixed molding sand leaves the continuous mixer and falls into the flask. The molding sand can be hand tucked into places where compaction can be difficult and can also be compacted with the use of a pneumatic rammer. Final compaction is usually done on a vibratory compaction table.

The next step is to strip the pattern from the mold. The time before this can be done is called "strip time." Strip time can be controlled by the molder. Once the pattern is removed from the mold, the mold is allowed to cure. This cycle can also be controlled.

The nobake molding process also lends itself to core molding. Core molding exists when the mold is made up entirely of nobake cores. Very intricate molds can be made using this process. By using this process, you could conceivably have outside surfaces of the metalcasting with negative draft, if required. Although normally used for making large metalcastings, it can also be used for medium-size castings.

The nobake process does not require metal patterns. Wood and plastic patterns can be used. One precaution to take, if plastic patterns are used, is to be sure that the plastic used for the pattern is compatible with the catalyst used in the molding sand. Plastic pattern surfaces can be softened by a catalyst if it is not compatible with the plastic material.

The nobake process is not a high-production-rate molding process. Attempts have been made to speed up the strip time and cure cycle; however, it is still considered to be a low-production, jobbing type of molding process.

Although all alloys, with the exception of titanium, can be poured into nobake molds, one precaution should be taken when pouring the lower-pouring-temperature alloys. Normally, these alloys are the nonferrous alloys. Due to the rigidity of nobake molds, hot tears can become a problem. Nobake binder suppliers have developed binders that will "burn" out faster at the lower pouring temperatures and alleviate this problem.

Some of the advantages and disadvantages of nobake molding are as follows:

- Nobake molds are very rigid and lend themselves to very good dimensional tolerances.
- Wood and plastic patterns can be used.
- Metalcasting surface finishes are normally very good.
- Most of the nobake molding systems have very good shakeout performance.
- Molds can be stored indefinitely.
- Nobake systems are not used for high-speed, high- production runs.

Pit Molding—Pit molds, which are used to produce castings too large for a flask, may be made in a pit by a bedding-in method **(Fig. 2-7)**. The pattern is set in a pit in the position in which the casting is to be poured. Usually, nobake sand is rammed or compacted around the sides of the pattern to a predetermined thickness and allowed to harden. The remaining empty space in the mold will then be filled with heap sand and then compacted. The cope for the complete mold may rest on the drag at or above floor level, and may be bolted down to prevent runout at the parting plane.

Fig. 2-6. Nobake mold filling box from a continuous mixer is shown. Nobake binder systems can be used to produce either cores, core molds or casting molds.

Some foundries have a concrete-lined pit equivalent to the size of the mold that they customarily produce. The mold may be made of a layer of nobake sand backed up with heap sand. A cope is then made and placed on top of the drag, which is in the pit. At times, when the design of the casting is such that a pattern cannot be drawn out of the mold, the entire mold cavity may be constructed with cores.

To prevent the forming of excessive internal stresses, large castings should cool slowly. Thus, it may be days after these castings are poured before they can be subjected to air cooling.

The advantage of pit molding is that it allows the pouring of very large castings, some weighing in excess of 50 tons (45.4 metric tons). Normally, only the ferrous alloys are poured in pit molds.

Lost Foam Casting/Molding Process

The lost foam casting (LFC) process, also called the expendable pattern casting (EPC) process, is a molding and casting procedure using gasifiable, expandable polystyrene (EPS) patterns. With this process, the pattern consists either of one piece or of several pieces glued together. Loose unbonded sand is vibrated around the EPS pattern in a flask. The EPS pattern remains in the flask and, as the molten metal enters the mold, the EPS pattern vaporizes. The molten metal displaces the pattern, thus forming the metalcasting.

The principle of "full mold" casting (the original name for this casting process) was disclosed in 1958 by an American patent published by H.F. Shroyer. To begin with, the technique was used for the development of art castings. It was not until 1962 that EPS patterns were utilized commercially to produce castings with a high degree of precision.

The full mold patterns are cut from large "boards" of EPS as one piece or several pieces that are glued together. The gating system and risers are made of the same material and glued to the pattern. The early metalcastings were large in size and consisted of stamping dies, valves, machine-tool bases, etc. A layer of chemically bonded sand was used to face the pattern in order to avoid any possible deformation of the pattern during subsequent backup with unbonded dry sand.

Fig. 2-7. Floor and pit molding techniques are used when the casting is too large to be made in a conventional jolt-squeeze or automatic molding machine.

The use of cores in the process can be eliminated, as the pattern need not be withdrawn. Any pockets, undercuts or cavities may simply be filled with the chemically bonded molding sand. The pattern can be formed in such a manner that it may serve as a corebox at the same time.

This material may also be utilized as an addition to an existing wooden or metal pattern. The expandable polystyrene is vaporized and eliminated, thus leaving metal where metal is desired.

Another advantage of this process is being able to use spherical risers. This is made possible because the riser pattern does not have to be removed from the mold. More advantages of this spherical-shaped riser will be discussed in Chapter 17, Risering Practice.

In the 1980s, a resurgence took place in the use of the lost foam casting process. Foundries no longer had to pay a license for using the process, making it more appealing to the metalcaster. Along with that, the technology of producing the EPS beads had advanced tremendously. Now, smaller-size metalcastings can be produced, and in high volumes.

Patterns are produced in high-production molding machines. Aluminum dies are placed in the molding machines and EPS beads are blown into the dies. Steam is injected into the dies, and the beads are expanded and fused together, forming the pattern. The die is cooled and the pattern ejected from the die. Patterns can be designed to be used immediately or after a cure period. Patterns can be made in one piece or several pieces and glued together. Cored passageways become part of the pattern.

After the pattern has been assembled, if needed, the gating and risering system is attached to the pattern. The next step is to coat the pattern assembly with a refractory wash. After the wash is applied and dried, the pattern assembly is suspended in a flask, and loose, unbonded sand is poured into the flask. As the sand is placed in the flask, the flask is vibrated and the unbonded sand is compacted around and into the internal passageways of the pattern. Thus, the mold and cores are formed at the same time. The mold is then ready for pouring. Shakeout consists of dumping the mold after the casting has solidified. The unbonded sand simply drains out of the cored passageways. **Figure 2-8** shows an example of an EPS pattern and the resulting casting.

The advantages and disadvantages of the LFC process are the following:

- The process lends itself to high or low production rates.
- Dimensional accuracy is said to be very good.

Fig. 2-8. Example of EPS pattern and lost foam casting.

- The molten metal never touches the molding sand and thus assumes the surface finish of the refractory coating.
- All alloys can be poured using the lost foam process.
- There are no limitations to the size of the metalcastings.
- In most cases, no cores are needed; thus, there are no core fins or parting lines.
- With glued patterns, there is the possibility of "glue" marks on the casting.
- A die is required to produce the patterns, which can add to the cost of the castings.
- The patterns have to be handled with care.
- The refractory coating has to be applied and maintained properly.

There are other variations of the LFC process. These processes also require an EPS pattern.

In one case, a nobake mold is made around the pattern. After the mold has been cured, the pattern is burned out and cores are set. This variation is normally used for large ferrous castings.

Another process (Replicast) was developed in Great Britain to produce carbon steel metalcastings. In this process, the pattern is invested in a ceramic shell similar to the investment casting process. The pattern is then burned out of the ceramic shell. In this case the ceramic slurry flows into the cored passageways in the pattern and forms the cores. This process was developed because, when the pattern vaporizes, carbon soot is given off, which can be absorbed by the carbon steel. Recently, however, a manufacturer of the EPS beads has developed a formulation that gives off very little, if any, carbon.

Vacuum (V-Process) Molding

Another name for vacuum molding the is V-process. This process was first developed in Japan in 1972. The Japanese used it to reproduce detailed works of art in aluminum and bronze, including sculptures, statues and ornamental grillwork. With the V-process, castings of all sizes and shapes, from a few ounces to tons, can be cast in most alloys, including aluminum, gray and ductile iron, and steel. While the V-process lends itself to prototyping, it is ideal for short- and medium-production jobs as well.

The primary differences between the V-process and conventionally bonded sandcasting are:

1) the V-process uses a thin plastic film that is heated to the deformation point and then vacuum-formed over a pattern, which is mounted on a hollow carrier plate;
2) the V-process uses dry, free-flowing, unbonded sand to fill the special flask over the pattern.

Since the molding sand is unbonded, permeability of the molding sand is not a concern; much finer sand can be used. A slight vibration quickly compacts the fine-grained sand to its maximum bulk density. The cope half of the mold is then covered with a second sheet of plastic film. The vacuum is then drawn through the sides of the flask and on the sand, which becomes very rigid. Releasing the vacuum on the pattern permits easy stripping of the mold half from the pattern. The drag half of the mold is made the same manner.

Cores, if required, can be set in the drag half of the mold before the mold is closed. With the mold closed, the mold cavity has a plastic film lining. The molding sand maintains its hardness by holding the vacuum within the mold halves at 300–600 mm Hg. As the molten metal is poured into the mold, the plastic film melts and is replaced immediately by the molten metal. After the metal has solidified and cooled, the vacuum is released and the sand falls away. This sequence is illustrated in **Fig. 2-9.**

The patterns must be vented or drilled to allow the plastic film to be drawn over the top, which typically eliminates the use of metal patterns. Use of wood patterns is possible but not recommended. While the patterns are touched by the molding sand, they are still subject to changes in temperature and humidity. The more-progressive foundries use cast epoxy or a machinable plastic component for patterns.

Some of the advantages and disadvantages of the V-process are as follows:

- The process produces excellent surface finishes.
- The castings have excellent dimensional accuracy.
- Zero draft is possible, which offers a wide range of advantages.
- There is unlimited pattern life, since the pattern never touches the sand.
- Design/drafting is less complex because calculations and depictions related to draft are eliminated.
- There is excellent reproduction of details.
- Maintenance of molding equipment requires strict attention, especially the vacuum equipment.
- Pattern equipment can, at times, become more costly than normal pattern equipment.

Cosworth Molding Process

The Cosworth process was developed in Great Britain to produce aluminum castings for the automotive industry. More recently it was brought to North America by the Ford Motor Company, which now owns the process.

The main aim of this process was to reduce and eliminate the entrapment of aluminum oxide in aluminum alloy castings. This is done by the use of an electromagnetic pump that pumps molten aluminum, from under the surface of the melt bath, into the mold cavity. By pumping the molten metal from under the surface of the bath and through a tube, the risks of forming aluminum oxide are tremendously reduced. The molds are made of sand and can be nobake or green sand types. Aluminum alloy castings produced using this process were used in the famous "Cosworth" racing engine **(Fig. 2-10).**

Vacuum-Assisted (Countergravity) Casting/Molding Process (For Sand or Permanent Molding)

In these molding processes, the molten metal is moved into the mold with the application of a vacuum in the mold. The pressure differential between the atmospheric pressure pushing on the melt and the reduced pressure inside the mold causes the metal to be pushed up into the mold cavity. The flow of metal into the mold cavity is controlled by the rate and amount of the vacuum application. This process is very similar to the way in which liquid is sucked through a straw, out of a glass or bottle. Another term used when discussing these processes is "countergravity" pouring of the molds.

The first of these processes was developed by Aurora Metals, Aurora, Illinois, to produce bronze impellers. In this process, a

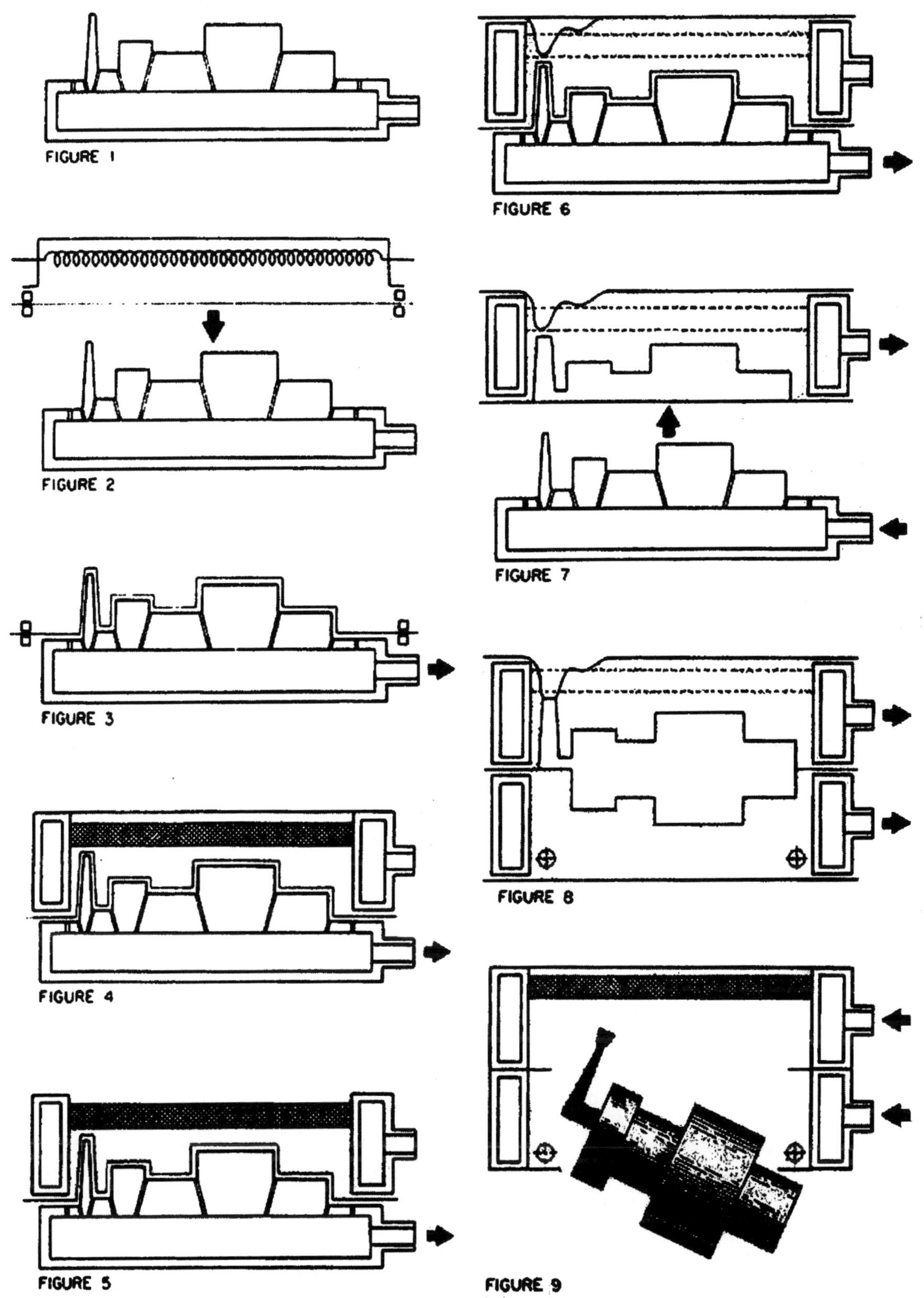

Fig. 2-9. How the V-process works: 1) pattern, with vent holes, is placed on hollow carrier plate; 2) heater softens the 0.003–0.007-in. plastic film; 3) film drapes over pattern with 300–600 Hg vacuum acting through pattern vents to draw it tightly around pattern; 4) flask is placed on pattern; 5) flask is filled with sand and vibration compacts sand to maximum bulk density; 6) sprue cup is formed, mold surface is leveled, and back of mold is covered with unheated plastic film; 7) vacuum is applied to flask, atmospheric pressure hardens sand, vacuum is released on pattern plate, mold strips easily; 8) cope and drag assembly form a plastic-lined cavity, and molds are kept under vacuum during pouring; 9) casting is cooled, vacuum is released, sand drops away and casting is released.

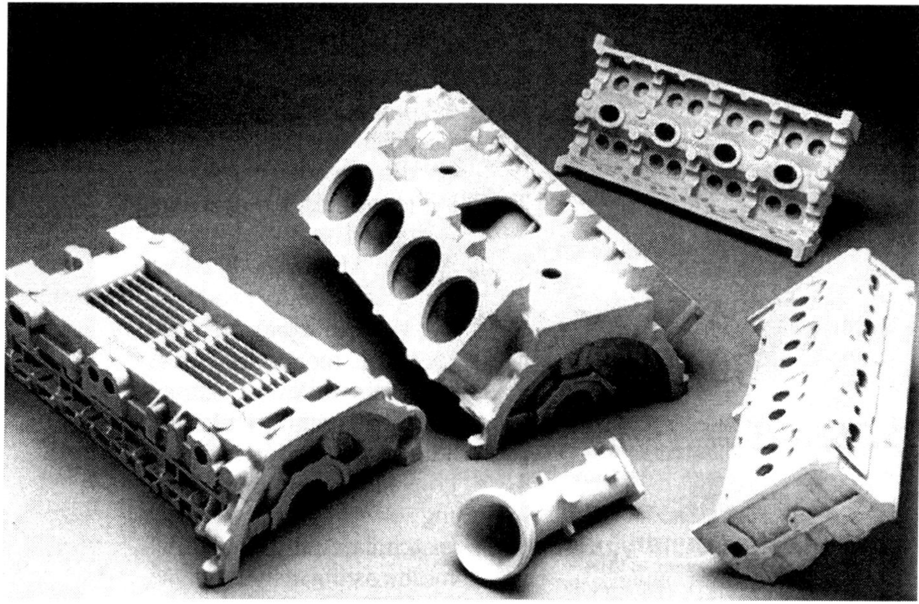

Fig. 2-10. Examples of castings produced by the Cosworth process.

silicon carbide fill tube is attached to the drag surface of a semi permanent or permanent mold. The mold, with the end of the tube attached to it, is placed in a sealed chamber.

The container with the mold in it is placed over a furnace containing the molten metal. The end of the silicon carbide tube that is not attached to the mold is lowered into the bath of molten metal in the furnace. A vacuum is created in the container, and the molten metal is moved up into the mold cavity. The vacuum is maintained until the casting has sufficiently solidified. When the vacuum is released, the container and the silicon carbide tube are withdrawn from the bath of molten metal and the remaining molten metal drains back into the furnace.

The next process is the CLAS process (Countergravity Low-pressure Air-Sand) as shown in **Fig. 2-11.** This process was developed by Hitchiner Manufacturing, Milford, New Hampshire and first published in 1983. Briefly, it consists of attaching a resin-bonded sand mold (either shell or nobake) that has gates through the drag half to an open-bottom vacuum chamber.

The machine then places the mold bottom into a bath of molten metal. A small vacuum is applied to the vacuum chamber, and molten metal is moved into the mold cavities. After the gates are allowed to solidify, the mold is removed from the melt. The mold is then released from the vacuum chamber and the castings shaken out from the mold. This process permits the making of thin-wall steel castings with 0.60-in. (1.5-mm) wall thickness.

The sand molds used in this process are limited in size by the size of the metal bath or furnace size, and this limits the size and number of castings that can be produced at one time. The CLAS process, also known as the VAC (Vacuum Assisted Casting) process is licensed by Hitchiner for the General Motors Corporation.

Due to limitations on size of castings, castings per mold and the size of the mold itself, a newer version of the CLAS process was developed by Hitchiner. The Loose Sand Vacuum Assisted Casting (LSVAC) process was developed to use thin-section sand molds made by the Croning Shell process or the nobake molding process. In the LSVAC process, molds up to 30 in.^{2} (76.2 cm^2) tall can be cast. The molds have gate or sprue openings along the bottom edge of the mold.

The molds are placed into a casting chamber and placed on a sheet of aluminum foil in a contoured base. Several molds can be cast at the same time, depending on how many molds will fit in the chamber. Loose, dry, unbonded sand is used to fill the space between the thin molds and casting chamber. This loose sand replaces the costly bonded sand that would have to be used in the CLAS or VAC process. The casting chamber is then attached to a machine, and a vacuum is applied to the chamber to rigidize the unbonded sand. The casting chamber and molds are then moved to the source of molten metal and lowered partially into the melt to submerge the gate and sprue openings in the bottom of the molds. The vacuum is held with the molds in the metal until the castings or gates solidify. With the vacuum already applied, the molten metal is rapidly moved upward to fill the mold cavity(s). Camshafts and thin-wall (2 to 3 mm wall) stainless steel exhaust manifolds have been cast using this process, along with a variety of other metalcastings. This process is also licensed by Hitchiner Manufacturing.

Some of the advantages of these vacuum-assisted or countergravity–poured metalcasting processes are as follows:

- The flow rate of molten metal into the mold cavity(ies) is accurately controlled by the amount of vacuum drawn, improving overall metalcasting soundness.
- The metal can be moved into the mold cavities more quickly than gravity pouring, resulting in the fillout of thinner casting sections.
- Only clean molten metal is moved into the mold cavity(ies), reducing the potential for slag or inclusions in the metalcastings.
- Lower and more consistent metalcasting temperatures can be used because there is no ladle transfer of the metal before casting the molds. This results in a reduced grain size and better mechanical properties.
- Metalcastings are said to have good surface finishes.
- Dimensional tolerances are said to be excellent.

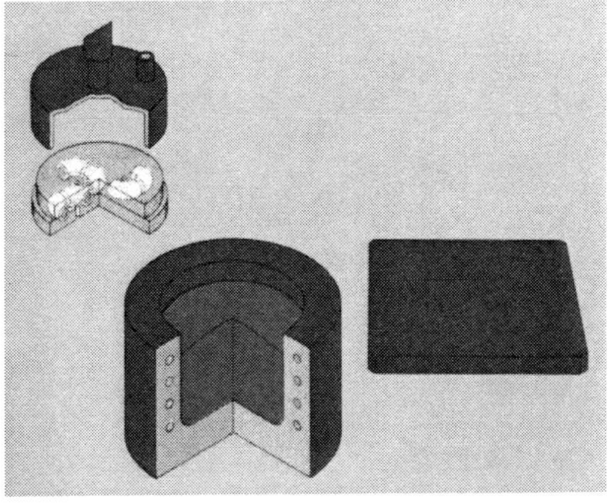

1. Cross section of resin-bonded sand mold before mold chamber is aligned and picked up for transfer into molten metal.

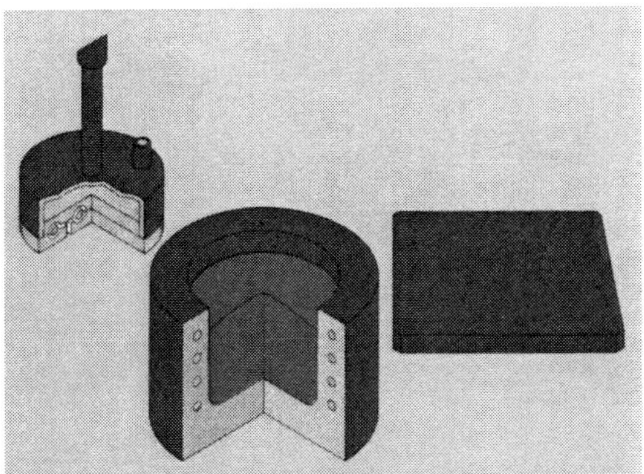

2. Mold chamber is mated and clamped.

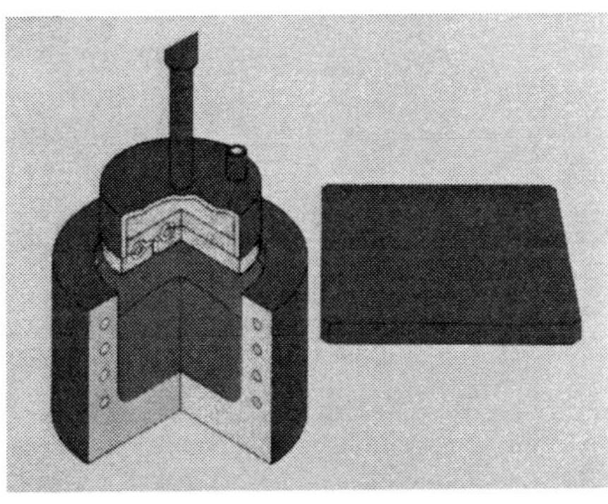

3. Chamber is transferred to pour.

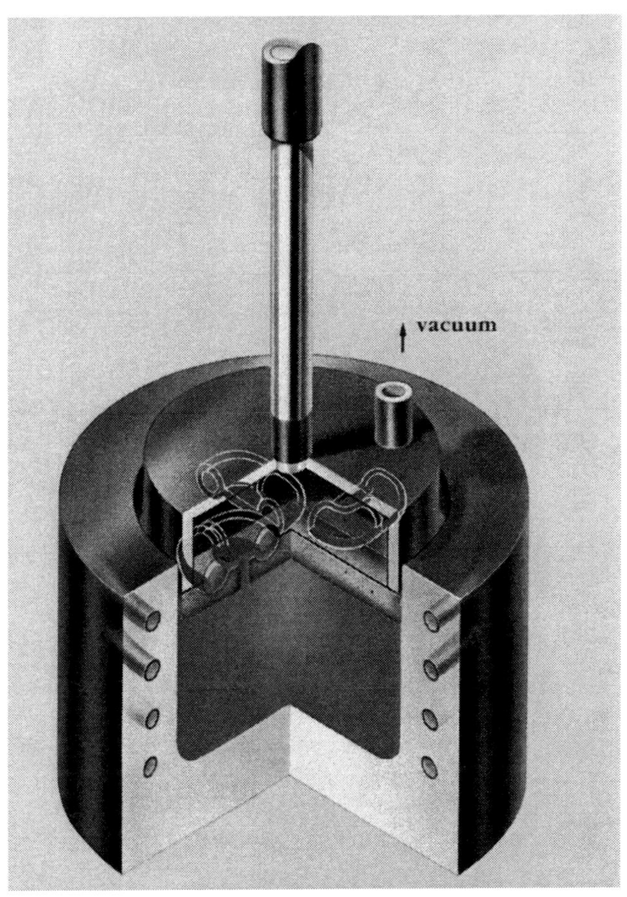

4. Mold chamber is lowered into melt and vacuum is applied.

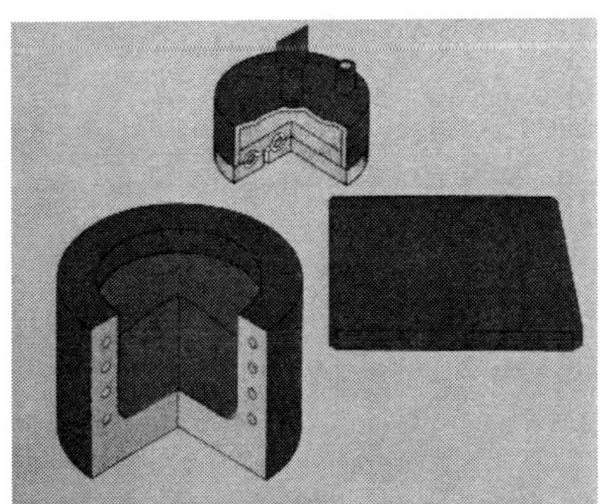

5. Mold is extracted from melt.

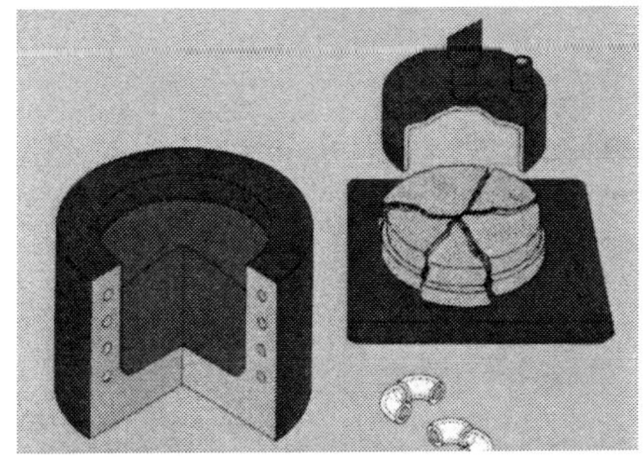

6. Finished castings are separated from sand mold on shakeout tray. Gating and risering are minimal.

Fig. 2-11. The countergravity low-pressure air sand (CLAS) process. [Courtesy of Hitchiner Mfg.]

Plaster Molding Process

Plaster molding can be used to produce metalcastings in which the lower melting temperature alloys, such as aluminum, are poured. The four generally recognized plaster molding processes are 1) conventional plaster molding, 2) matchplate pattern plaster molding, 3) Antioch process and 4) foamed plaster process.

In the conventional plaster molding process, slurry containing calcium sulfate (also called gypsum) is poured into a flask that contains the pattern. After the slurry has set, the pattern and flask are removed, and the drying cycle to remove moisture from the mold is begun. After the mold sections have been allowed to cool **(Fig. 2-12),** the mold and core sections can be assembled into a finished mold. Most molds are preheated to about 250F (121C) before pouring. These molds have poor permeability; and, in many cases, vacuum or pressure assist is required during pouring.

Matchplates, used as the pattern tooling in green sand molding, can be sometimes cast to size using the matchplate pattern plaster molding process. This process is the same as the conventional process, but the parting line of the mold is separated by a small amount (typically 6-9 mm) to form the plate of the matchplate. The outer edges of the matchplate are defined by a frame that is placed on the parting line of the plaster mold.

The Antioch process was developed to increase the permeability of the plaster mold. The process treats the plaster molds with steam to increase the mold permeability 15 times higher than conventional plaster molds.

The foamed plaster process also creates a high permeability plaster mold but uses a foaming agent in the plaster. Air is whipped into the plaster to form a dispersion of tiny air bubbles that are stabilized by the foaming agent.

The plaster molding processes are especially suited for short run and prototype work with lower-temperature alloys. Among the advantages and disadvantages of the plaster molding process are the following:

- Castings have exceptionally smooth surfaces, and intricate designs and details are readily obtainable.
- Dimensional accuracy of the metalcastings is very good.
- Because of the insulating mold material and vacuum assist, thinner-wall metalcastings can be produced.
- Slow cooling of plaster molds minimizes warping and promotes uniformity of microstructure and mechanical properties in the metalcasting.
- Only the lower-melting-temperature alloys can be poured into plaster molds.
- Patterns used in the process are best made from materials other than wood.

Fig. 2-12. Foundry employee touches up the drag section of a plaster mold.

Ceramic Molding Process

Another method of producing nonpermanent molds is called ceramic molding. (This process and its offshoots are also known as the Shaw process, the Unicast process, the Osborn-Shaw process and the Ceramicast process.) These processes are essentially the same, with slight variations. Generally, these processes employ a mixture of graded refractory fillers; in most cases, hydrolyzed ethyl silicate binder and a liquid catalyst that are blended to a slurry consistency. Various refractory materials can be used as filler material. The pattern is placed into a container, and the pattern is coated with a high-grade slurry to form the surface of the mold cavity. The container is then filled with a coarser backup slurry.

After the slurry binder has gelled, the pattern is stripped from the mold. The cope and drag mold sections and cores made in this manner are assembled into a completed mold. The mold is then heated to an elevated temperature to cure the ceramic. Molten metal is then poured into the mold, with or without preheating the mold. The ceramic molding processes have proved to be very effective with smaller-size castings in low-volume runs. They also have the following advantages:

- They provide excellent surface finishes.
- There are good-to-excellent dimensional tolerances in the metalcastings.
- Depending on the refractory material used, these processes lend themselves to all alloys.

Investment Casting/Molding Processes

The investment casting process, also known as the lost wax process, is one of the oldest molding processes used to produce metalcastings. It was originally used to produce art objects and jewelry. It still is probably the most commonly used molding process to produce these types of castings. However, today it is also used to produce metalcastings that can be found in aircraft, space vehicles, machinery, automotive, dental and many other industries.

There are two types of molds used in this process: solid and shell. Today, the predominant type of mold used is the shell mold. The solid mold method is similar to the ceramic molding processes discussed earlier. The major difference is that in investment casting, the wax or plastic pattern is melted out of the mold. In ceramic molding, a permanent pattern is used.

Solid Investment Casting/Molding

The basic steps of the solid investment casting process are as follows:

1. Heat-disposable wax or plastic patterns are made.
2. The patterns are attached to a gating system.
3. The pattern assembly (sometimes called a "tree") is placed in a metal can or flask and "invested" or covered with a ceramic slurry and refractory material to produce a solid monolithic ceramic mold.
4. The wax or plastic pattern is melted out to leave a precise cavity in the ceramic mold that replicates the shape of the pattern.
5. The ceramic mold is then fired to remove the last traces of the pattern material, to develop the high-temperature bond and to preheat the mold, making it ready for casting.
6. The mold is then poured.

These basic process steps can be seen in **Fig. 2-13.** Patterns for production investment casting are made in dies. For prototype castings, stereolithography or some other rapid prototyping process can be used to make a pattern. The wax for investment casting patterns is a specially prepared wax for use in the investment casting process. The dies are placed in an injection molding machine and the wax is injected into the die in the liquid state or extruded into the die in a pasty consistency.

The dies can have multiple side actions and pulls to create holes and undercuts in the wax pattern. When the part has core geometry that cannot be formed by the die, the core is usually made of soluble wax or ceramic materials. In the case of soluble wax cores, they are removed from the pattern before the pattern is invested. Ceramic cores, on the other hand, stay in the mold during the entire casting process and are removed during the cleaning operation. Extremely complex castings can be made by this process.

Shell Investment Casting/Molding

In the shell mold process, the ceramic mold is b the tree assembly into a ceramic slurry. After eac stucco of fused silica sand or zircon sand is rai slurry. Another method of applying the refractory the slurry-coated tree assembly into a fluidized be material. After each dipping cycle and stuccoing entire assembly is allowed to dry thoroughly before the next dipping and stuccoing cycles begin. Thus, a ceramic shell is built up around the tree assembly and its thickness or number of layers required is determined by the size of the metalcasting and the temperature of the alloy to be poured.

After the ceramic shell process has been completed, the entire assembly is placed upside-down in an autoclave or a flash-fire furnace, which is at a high temperature. This operation removes the bulk of the wax from the mold. The shell is then heated to about 1800F (982C) to burn out the pattern material and any residue of this material that may still be in the mold. At these elevated temperatures, a high-temperature bond is formed in the shell mold **(Fig. 2-14).**

The shell molds can be stored for future use or immediately poured while still at these elevated temperatures. If the shells are to be stored for any length of time, they will require preheating before molten metal is poured into them.

Investment casting shells can be gravity poured in air or under vacuum for superalloy castings. Investment casting shells can also be cast by countergravity processes. These countergravity processes are licensed by Hitchiner Manufacturing. Two countergravity casting processes are CLA (Countergravity Low-pressure Air-melt) and CLV (Countergravity Low-pressure Vacuum-melt). In the original investment casting process, the molds were gravity poured in the conventional manner. These two processes (CLA and CLV) make use of a vacuum in the mold to move the molten metal

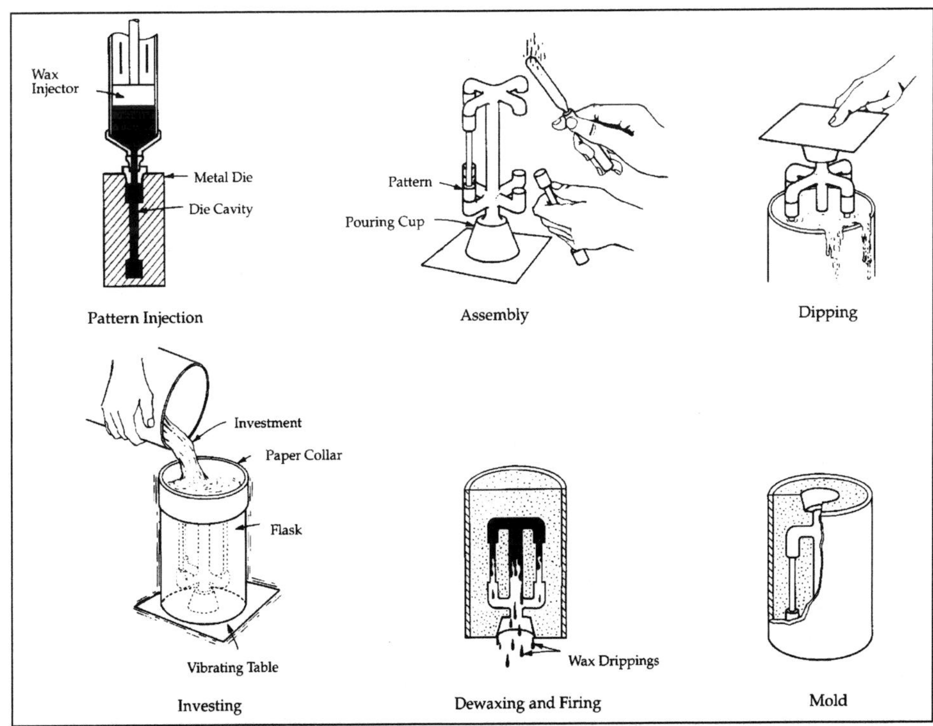

Fig. 2-13. Solid mold investment casting process.

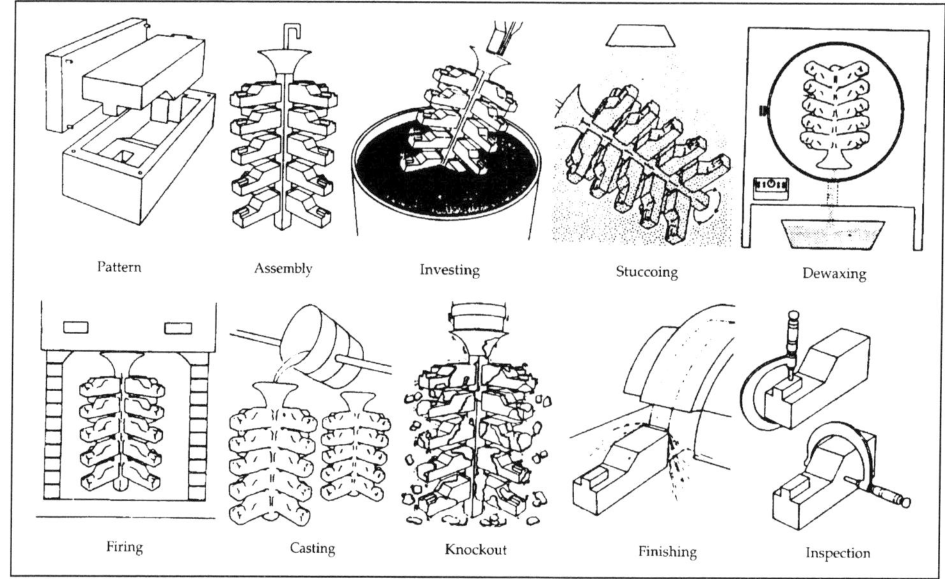

Fig. 2-14. Schematic illustration of shell investment casting.

up into the mold cavity(ies). Controlled, fast filling of the mold cavities results in high-quality thin-wall castings. As with the countergravity sand castings, the processes also have the advantages of good metal cleanliness and low casting temperatures, leading to small grain size and improved mechanical properties.

CLA Process—In this process **(Fig. 2-15),** the mold with a ceramic fill pipe is placed into a chamber with the fill pipe extending through a hole in the chamber. The open end of the fill pipe is lowered into a bath of molten metal. A vacuum is applied in the chamber, and the molten metal is moved up into the mold. The vacuum is held long enough for the metalcastings to solidify but short enough for the gating system to be still liquid and able to flow back into the bath of molten metal after the vacuum is released. The result is metalcastings with a minimum amount of solidified gate metal. The parts do not need to be cut off of a central "tree," so automated gate cutoff methods can be used. The CLA process has made it possible to manufacture investment castings in very high volume—as high as 100,000 parts per day.

CLV Process—This process **(Fig. 2-16)** operates in the same manner as the CLA process; however, the major difference is the atmosphere in which the metal is melted and cast. The CLV process involves the melting of the metal in a vacuum chamber. The hot mold is placed into a separate chamber and a vacuum is created. Both chambers are then back-filled with argon. The valve is opened and the melt is raised until the opening in the ceramic fill pipe enters the molten metal. The pressure in the mold chamber is lowered respective to the melt chamber, to move the molten metal up into the mold. The vacuum is then released after the parts and gates have solidified. The CLV method is used to cast aerospace parts from superalloys that must be melted and cast under vacuum or a protective atmosphere. The fillout abilities of the countergravity casting process have allowed the manufacture of superalloy castings with wall thickness as thin as 0.3 mm.

Most investment castings weigh less than 5 lb (2.3 kg), but there is a trend toward producing larger investment castings in the range of 10–30 lb (4.5–13.6 kg). This is largely due to the use of robots for the dipping and mold-handling operations. Investment castings weighing up to 800 lb (362.9 kg) have been poured.

Some of the advantages and disadvantages of the investment casting process are as follows:

- The process produces excellent surface finishes.
- Very tight dimensional tolerances can be held.
- The investment casting process is normally considered a high-production-oriented process.
- This process lends itself very well to the production of metalcastings using all alloys including nickel-base, cobalt, carbon, stainless and high-alloy steels, titanium, copper base and aluminum alloys.
- Due to this process's ability to reproduce excellent detail and maintain close tolerances, many machining operations can be eliminated.
- Although efforts are being made to reduce the time to produce the molds, the process still has a longer lead time than do other molding processes.
- The size of the metalcasting is still somewhat limited.
- When compared to other molding processes, it still is more expensive. However, reductions in postcasting operations, such as machining and surface finishing, may overcome this disadvantage.

PERMANENT MOLDING PROCESSES

The next series of molding processes will be those in which the molds can be used many times. In most cases, the molds can be used for thousands of metalcastings before they have to be repaired. The mold material used for the majority of these processes is ferrous alloys. In some cases, graphite is used for the mold material.

There are several variations of permanent molding. These variations are in the mold materials and the methods by which the molten metal enters the mold cavity.

Permanent mold castings can range in size from ounces to more than 100 lb (45.4 kg). Various complex shapes can be cast out of aluminum alloys, copper-base alloys, gray iron, zinc-aluminum alloys and carbon steel. These metalcastings can also be cored, if necessary.

The molds used in permanent molding can be made of gray iron, high-alloy iron, steel and graphite. The metal molds are capable of producing 10,000 to 120,000 or more metalcastings. However, for some special designs, metal molds or dies may be good for only 200 to 500 metalcastings. Graphite molds are used to cast carbon steel and the zinc-aluminum alloys.

Gray and high-alloy iron metal molds can be cast to shape and then finish-machined or machined out of billets to final shape and dimensions. Steel molds, on the other hand, are machined out of steel billets. Some graphite molds are cast as a graphite slurry over a pattern, and cured. The next step, if necessary, is to finish-machine the mold cavity. The mold can also be machined out of solid blocks of graphite. In all cases, if required, the gating and risering systems are machined into the mold. Not all permanent molds require a gating system.

Removal of the metalcastings from the mold can be performed manually or by some mechanical means. In the case of some metal molds, ejector pins are used to remove the castings from the mold.

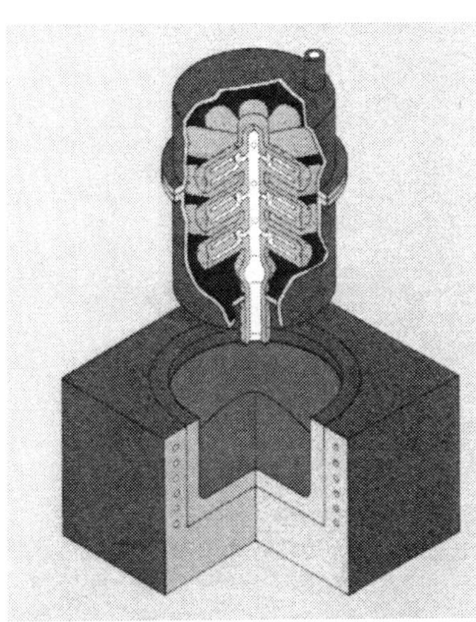

1. Cross section of mold in chamber, open end down.

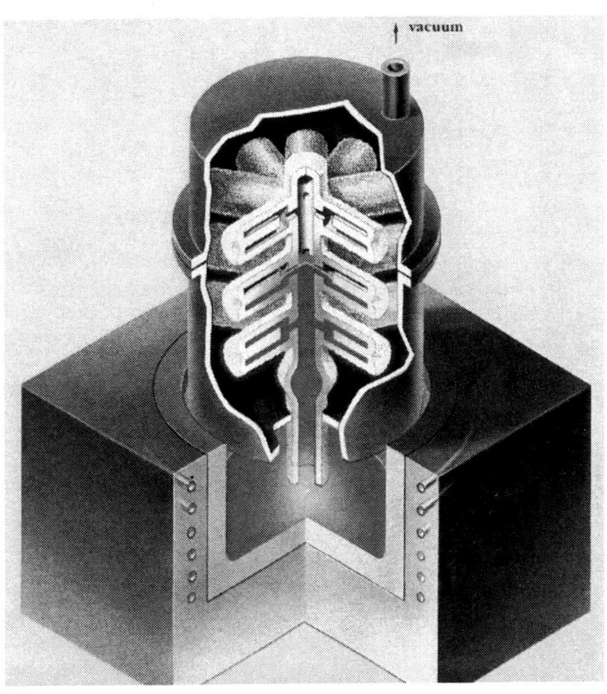

2. Open end is immersed in furnace and molten metal drawn up into the mold by vacuum.

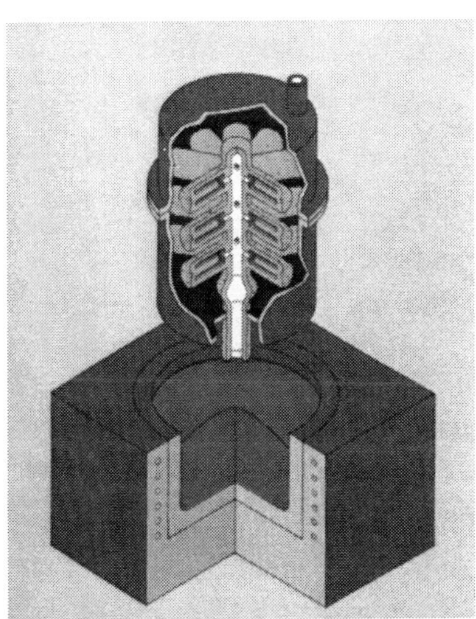

3. Vacuum is time-released when castings solidify. Excess melt returns to furnace.

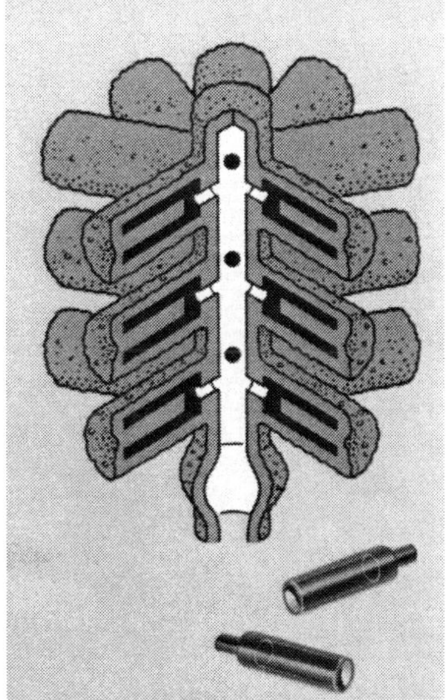

4. There is no center stick. Ceramic shell is then removed. Loose castings with minimal residual gate are ready for final machining.

Fig. 2-15. The countergravity low-pressure air (CLA) process. [Courtesy of Hitchiner Mfg.]

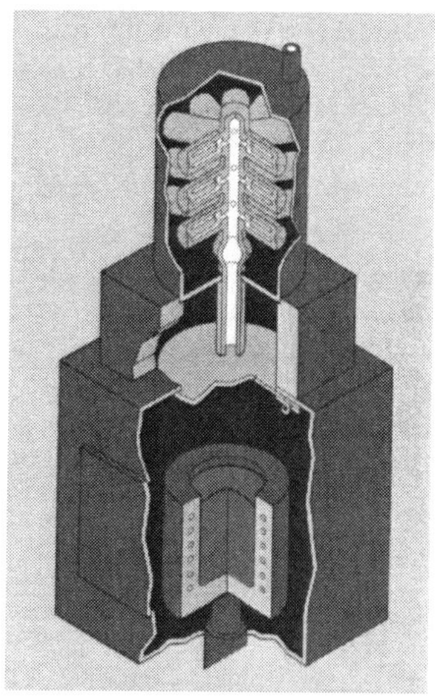

1. Cross section of mold and metal in vacuum chamber.

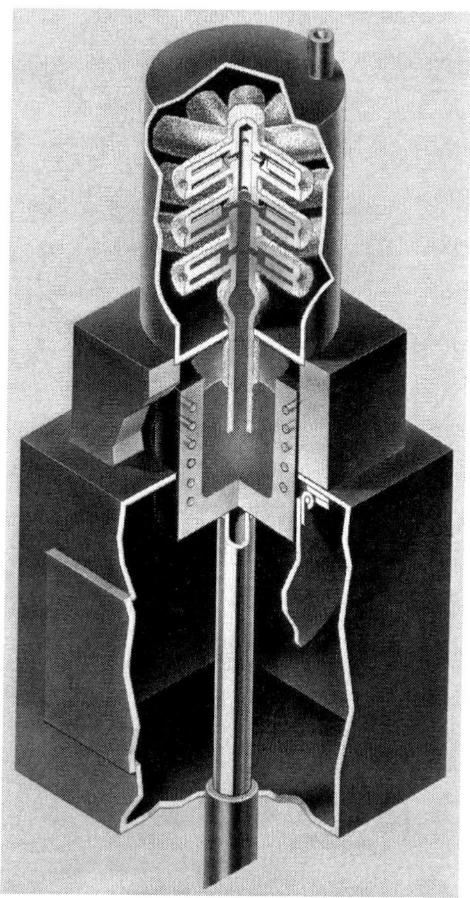

2. Both chambers are filled with argon and the melt is raised to the open end of the mold. Additional vacuum is applied and molten metal is drawn into the mold.

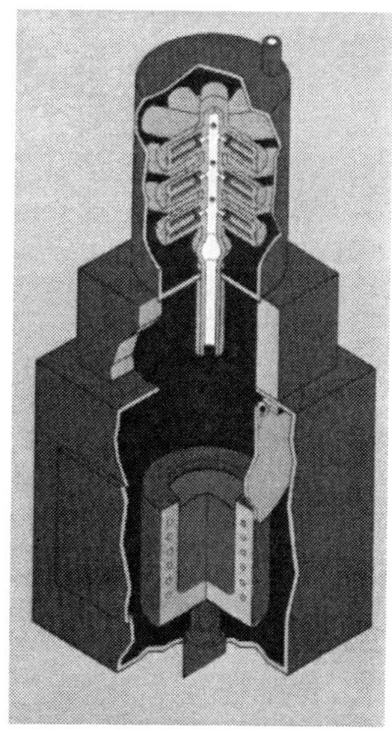

3. Excess metal returns to the melt.

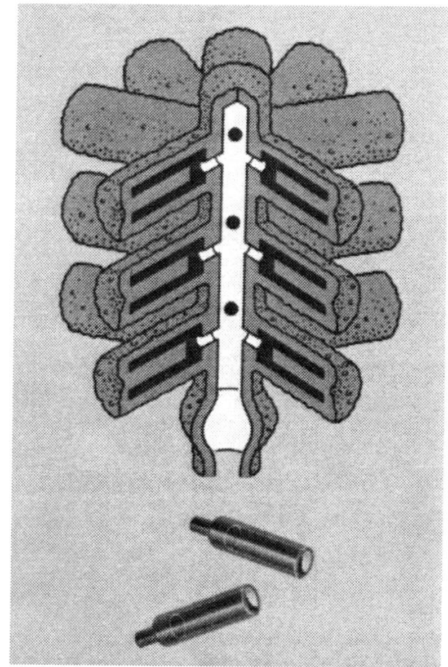

4. There is no center stick. Ceramic shell is then removed. Loose castings with minimal residual gate are ready for final machining.

Fig. 2-16. The countergravity low-pressure vacuum (CLV) process. [Courtesy of Hitchiner Mfg.]

Regardless of the mold material, alloyed poured cores can be used in the permanent mold processes. Although preferred, metal cores do not have to be used. If they are used, the design of the mold becomes more complex, as does the method or mechanism used to move the core insert(s) into and out of the mold. On the other hand, in the case of metal molds, sand cores can be used. When sand cores are used, the term "semipermanent molding" is used. Most of the usual sand-core processes can be used.

Since the materials used to make these molds are not permeable, some means of venting the molds has to be provided. This can be done in several ways. One method is the use of parting line vents **(Fig. 2-17).** These would be very small scratch-type vents machined in the parting line on one half of the mold. Another method is to is to machine a very shallow flat spot on the side of the ejector pins, which are usually round. The holes through which the ejector pins move are also round, so the flat surface provides a vent for air, mold and core gases to escape. The design of the mold cavity may also be such that as the mold is filling, pockets of air can be trapped. In these cases, vents will have to be provided into these areas.

A very important part of the permanent molding process is the maintenance of the mold cavity surfaces. In most cases, a refractory coating of some sort has to be applied to the surfaces of the mold . These coatings can serve several functions, as follows:

1. In all cases, they protect the mold surfaces from damage from the molten metal.
2. In the case of metal molds, these refractory coatings can act as an insulator or a chill.
3. Again, in metal molds, the refractory coating can act as a lubricant to help in the ejection of the metalcasting from the mold.
4. The surface of the refractory coating can serve as a vent.

Fig. 2-17. Open permanent mold shows the gating and venting systems.

In any case, the operator of the molding machine has to pay close attention to the condition of the refractory coating.

It should also be mentioned here that, in most cases, the metal molds are preheated to 250–500F (121–260C). The preheating is done prior to the coating of the mold and before it is placed in service. Although the heated molds still act as a chill, there are times when certain parts of the casting may require a faster or slower cooling rate. This can be done by varying the refractory coating or applying external cooling or heating to these areas.

Gravity Permanent Molding Process

There are two ways in which metalcastings are produced in permanent molds: static pouring and tilt pouring.

Static Pour Process

In the static pour process, the molds (normally metal) are clamped into a device or machine and poured with the mold's parting line in the vertical position. After the metalcastings have solidified, they are removed from the mold. One of the inherent problems is that, depending on the design of the gating system, air can get trapped in the mold as the molten metal enters. With alloys such as aluminum, the drop of the molten metal causes dross formation, which can be trapped in the metalcasting as inclusions.

Advantages and disadvantages of the static pour process include the following:

- The molds, although heated, still act as a chill, which improves mechanical properties of the metalcasting.
- The surface finish of these metalcastings is normally very good.
- The cost of the molds is high; thus, this process is normally used for high-production runs.
- Gray iron can be poured using the static method.
- The inherent problems caused by the turbulent flow of molten metal can cause dross inclusions.

Tilt Pour Process

The tilt pour process was developed to overcome the problems associated with static pour. In this case, the molds are held in an automated tilt pour molding machine **(Fig. 2-18).** The cycle begins with the mold's parting line in the horizontal position. On the upper

Fig. 2-18. Typical tilt pour casting machine.

end of the mold is a pouring cup(s), which is attached to the mold. This cup(s) will hold enough molten metal so that at the end of the pouring cycle, there will be sufficient molten metal to completely fill the mold cavity and gating system.

After the pouring cup(s) is filled with molten metal, the operator presses a button on the machine, and it begins to tilt the mold from a horizontal position to a vertical position. As this happens, the molten metal runs out of the pouring cup(s) and down the gating system, and begins to fill the mold cavity. The speed at which this part of the casting cycle proceeds can be preset and controlled automatically by the machine. For instance, it can begin tilting the mold slowly and then gradually speed up as the mold begins to reach the vertical position. At the end of the pouring cycle, the mold's parting line is in the vertical position.

After the casting(s) has solidified, it is automatically ejected from the mold by ejector pins as the mold is opened. The mold coating is checked and repaired, if necessary, and the mold is then lowered into the horizontal position. Cores are then set if needed, the mold is closed and the casting cycle begins all over again.

By automatically controlling the tilting of the mold to the vertical position, the velocity of the molten metal as it goes down the sprue(s) can be controlled. Slowing this velocity can help reduce the turbulence in the gating system and thus reduce the formation of dross. Also, since the gating system is not completely full of molten metal, air inside the mold cavity can escape by way of the gating system. In other words, as the molten metal is going down, air can be going up the sprue(s).

Some advantages and disadvantages of the tilt pour process are as follows:

- Aluminum castings produced in permanent molds have finer dendrite arm spacing, grain structure and strength properties than do similar castings produced in sand molds.
- Aluminum castings produced in permanent molds using the tilt pour method are less prone to entrapping gas.
- The tilt pour process is used primarily for aluminum alloys; however, copper-base alloys have been cast in tilt pour machines.
- The filling of the mold can be controlled automatically by adjusting the tilting speed.
- Due to the cost of the molds, this casting process is used mainly for high-quantity production rates.
- Extremely complex or irregular shapes may be difficult or even impossible to cast using permanent molds.
- There are limitations as to the size of metalcasting that can be produced.

Low-Pressure Permanent Molding (LPPM) Process

In the low-pressure process, a permanent mold, made of metal, is mounted over an airtight furnace containing molten metal (**Fig. 2-19**). A feeder tube, or stalk, is mated or joined to the nozzle at the bottom of the mold and extends into the molten metal in the furnace. Heat is usually applied externally to the nozzle. With the mold closed, low-pressure air, normally between 2 and 5 psi (1406 and 3515 kg/m^2), is admitted into the furnace. The air pressure forces the surface of the melt downward, and causes molten metal to rise up the stalk and enter the mold cavity.

The casting is fed by pressure on the molten metal in the nozzle opening. This eliminates the need for risers because the mold cavity is filled and fed from the bottom. This simplifies the gating system and greatly increases the yield. After enough time has elapsed for the casting to solidify, the pressure is released and the still-molten metal in the stalk runs back into the furnace. The mold is opened, and the casting is extracted. Because the molten metal enters the mold slowly, air is not entrapped.

Advantages and disadvantages of LPPM can include the following:

- By controlling the amount of air pressure exerted on the surface of the bath of molten metal, the fill rate of the mold can be controlled.
- Very thin sections can be successfully cast in suitable low-pressure process dies with aid of vacuum.
- The pressure exerted on the bath of molten metal remedies the need for risers.
- Again, like all permanent mold processes using metal molds, the cost of producing these molds limits this process to high-production-rate metalcastings.
- Aluminum alloys are currently the only alloys being cast at this time.

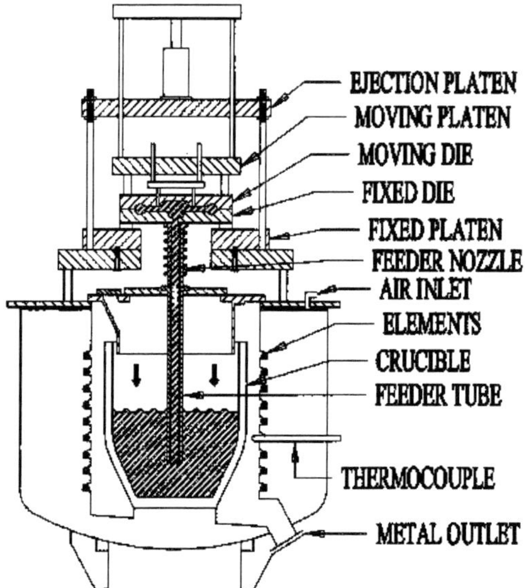

Fig. 2-19. Cross section of a low-pressure machine

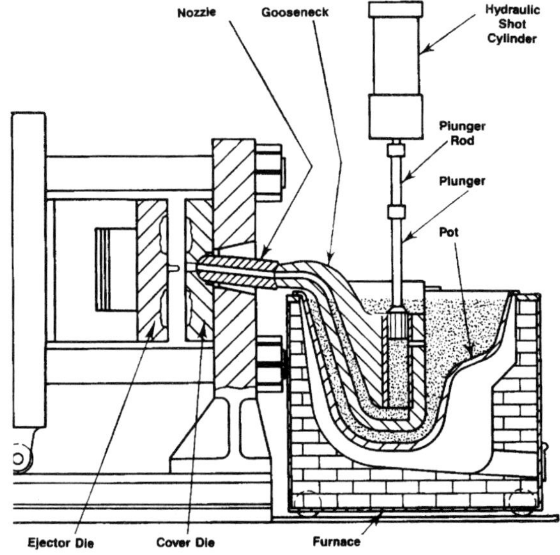

Fig. 2-20. Diagram of hot chamber diecasting machine.

Diecasting Process

Diecasting is a process involving the injection of molten metal at high pressures (as opposed to casting by gravity pressure) into metal dies. This process is believed to have begun sometime during the middle of the 19th century. What is known as diecasting in the United States is called pressure diecasting in Europe.

The first step in diecasting is to produce a steel mold, or die, capable of producing tens of thousands of castings in rapid succession. Some foundries are equipped with a moldmaking department whereas other foundries purchase molds from a reputable supplier (i.e., pattern/mold shop, machine shop). The die must be made in two sections to permit removal of the castings. These sections are mounted on a machine and are arranged so that one is stationary (fixed die half) while the other is movable (ejector die half).

To begin the casting cycle, the diecasting machine clamps the two die halves tightly together. Molten metal is then injected into the die cavity under very high pressure, where it solidifies very quickly. The die halves are drawn apart and the casting(s) is ejected. The dies used in diecasting can be simple or complex, having movable slides, cores or other sections, depending on the complexity of the metalcasting. The complete cycle of the diecasting process is by far the fastest known for producing precise nonferrous parts.

Diecasting machines, regardless of the type of machine used, have one main objective, and that is to hold the two die halves, cores and/or other movable sections securely locked in place during the casting cycle. Most machines use toggle-type mechanisms actuated by hydraulic cylinders (sometimes air pressure) to achieve locking.

Diecasting machines, large or small, vary fundamentally only in the method used to inject molten metal into the die cavity. These are classified and described as either hot chamber or cold chamber machines.

Hot chamber machines **(Fig. 2-20)** are used primarily for zinc and low-melting-point alloys that do not readily attack and erode metal pots, cylinders and plungers. Advanced technology and development of new, higher-temperature materials have extended the use of this equipment to magnesium alloys.

In the hot chamber machine, the injection mechanism is immersed in molten metal in a furnace attached to the machine. As the plunger is raised, a port opens, allowing molten metal to fill the cylinder. As the plunger moves downward sealing the port, it forces the molten metal through the gooseneck and nozzle into the die. After the metal has solidified, the plunger is withdrawn, the die opens and the resulting metalcasting is ejected.

Hot chamber diecasting machines are rapid in operation. Cycle times vary from less than one second for small castings weighing less than one ounce, to 30 seconds for a metalcasting weighing several pounds. Dies fill very quickly, normally between five and 40 milliseconds, and the molten metal is injected at high pressures—1500 to over 4500 psi (1,054.65–3,163,950 kg/m^2). Nevertheless, modern technology gives close control over these values.

Cold chamber diecasting machines **(Fig. 2-21)** differ from hot chamber diecasting machines primarily in one respect: The injection plunger and cylinder are not submerged in molten metal. The molten metal is poured into a "cold chamber" through a port or pouring slot by a hand ladle or an automatic ladle. A hydraulically operated plunger, advancing forward, seals the port, forcing molten metal into the locked die at high pressure. Injection pressures range from 3000 to over 10,000 psi (2,109,300–7,031,000 kg/m^2) for both aluminum and magnesium alloys, and from 6000 to over 15,000 psi (4,218,600–10,546,500 kg/m^2) for copper-base alloys.

Dies can be classified as single-cavity, multiple-cavity, combination and unit dies **(see Fig. 2-22)**. A single-cavity die requires no explanation. Multiple-cavity dies have several cavities that are identical. If a die has cavities of different shapes, it's called a combination or family die. A combination die is used to produce several parts for an assembly.

For simple parts, unit dies might be used to effect tooling and production economies. Several parts for an assembly, or for different customers, might be cast at the same time with unit dies. One or more unit dies are assembled in a common holder and connected by runners to a common opening or sprue hole. This permits simultaneous filling of the die cavities.

The alloys cast in the diecasting process fall into the nonferrous family. These alloys include aluminum, zinc, zinc-aluminum, magnesium, copper, and lead and tin alloys.

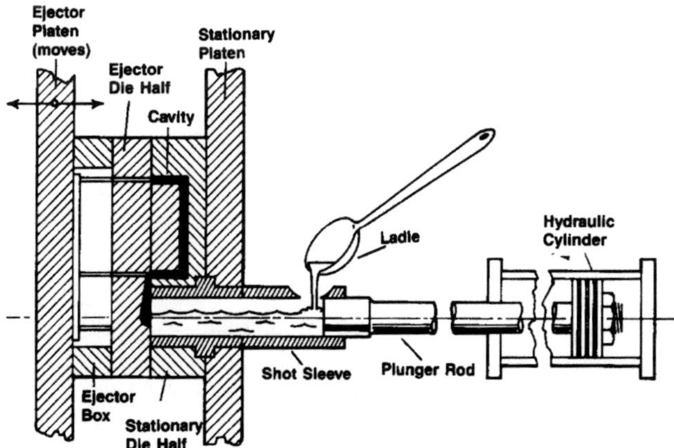

Fig. 2-21. Diagram of cold chamber diecasting machine.

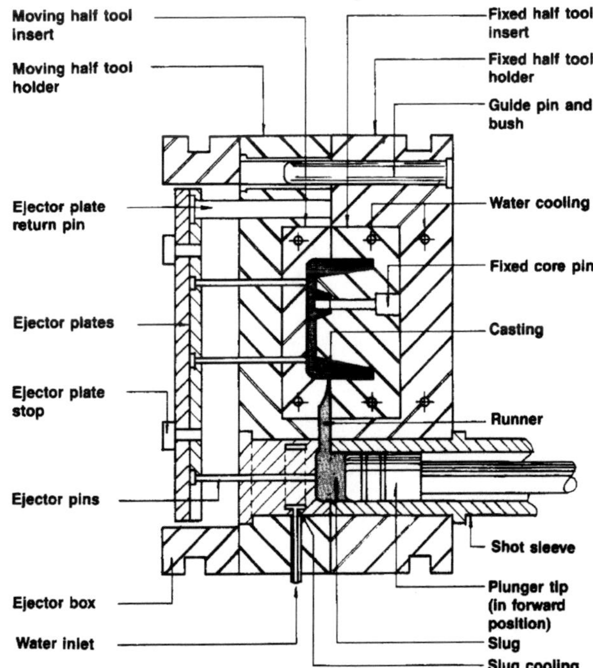

Fig. 2-22. Diagram of a typical die used in diecasting.

Some of the advantages and disadvantages of diecasting are as follows:

- Diecasting provides complex shapes within closer tolerances when compared to other mass production processes.
- Diecastings are produced at high rates of production and little or no machining may be required.
- Diecastings can be produced with thinner walls than those attainable by other casting processes—and they are stronger than plastic injection parts with the same dimensions.
- Diecasting dies can produce thousands of identical metalcastings within specific tolerances before additional tooling is required.
- Diecastings can be produced with surfaces simulating a wide variety of textures.
- Inserts of other metals and some nonmetals can be cast in place.
- Diecasting is a high-production process for very light- to medium-weight metalcastings.
- The metalcasting process, at this time, is limited to nonferrous alloys.
- Due to the high rate of velocity with which molten metal enters the die cavity(ies), there will be severe turbulence, which can cause oxide inclusions in the metalcastings. (Research is being conducted to decrease this potential problem.)
- Cores, when used in diecasting, are made of metal.
- The movement of these cores, mechanically, can lead to complex dies.

Graphite Molding Process

The use of graphite to make permanent molds has seen a resurgence with the advent of the zinc-aluminum alloys. In the 1950s, research was ongoing regarding the casting of aluminum alloys in graphite permanent molds. The main obstacle to its success was the short life span of the molds. However, today, carbon steel railroad wheels are produced in graphite molds using the Griffin process.

Griffin Process—This process is very similar to the LPPM process, in that air pressure is used to force the molten metal up into the mold. In this case, a ladle of molten steel is placed in a pit, in the floor, that can be sealed. The graphite mold is placed over the pit, and a silicon carbide tube is put in place. The tube extends down into the ladle of molten steel and connects with the sprue in the mold. Air pressure is applied to the pit, and the molten steel goes up the tube into the mold cavity. The air pressure is maintained until the casting has solidified and then is released. The molten metal in the tube falls back into the ladle, and the tube is removed from the mold.

The mold moves on and the casting is removed from the mold. A coating is then applied to the surface of the mold cavity. If the mold cavity requires "resurfacing," the mold is sent to the machine shop, where a fly cutter in a vertical mill produces a new mold-cavity surface. The fly cutter is ground to the shape and contours of the surfaces of a railroad wheel.

In the case of the zinc-aluminum alloys, graphite molds that have been cast from a slurry or machined out of a graphite block are used. Since zinc-aluminum alloys are poured at lower temperatures than the aluminum alloys, the life of the molds has been greatly extended. These molds are poured statically, usually in the vertical position. After the castings have solidified, they are removed from the mold using ejector pins. A very thin coating of a graphite aerosol is used as a mold coating.

Some of the advantages and disadvantages of graphite permanent molds are as follows:

- Titanium can be poured into graphite molds.
- The graphite molds act as chills, thus improving the mechanical properties of the metalcastings.
- Graphite is easy to machine; thus, the cost of producing the molds is reduced.
- When compared to metal molds and pouring zinc-aluminum alloys, graphite molds have a shorter life.
- The zinc-aluminum alloys have a size and weight limitation using graphite molds.

Centrifugal Casting/Molding Process

Centrifugal casting can be considered a permanent molding process because some of the molds used are reusable. Centrifugal casting has been used for years as an economical method for producing cylinders and tubes. With centrifugal casting, the mold revolves at high rates of speed in a horizontal, vertical, or inclined position as the molten metal is poured into it.

Centrifugally cast pipe and tubing can be cast in almost any required length, thickness and diameter. Because the mold forms only the outside surface and length of the metalcasting, different wall thicknesses can be produced using the same mold. The centrifugal force of this process keeps the casting hollow, eliminating the need for cores. Pipe and tubing are normally cast in horizontal or inclined casting machines.

Metalcastings other than cylinders and tubes can also be produced in vertical centrifugal casting machines **(Fig. 2-23).** An example of this type of vertical centrifugal casting is a controllable pitch propeller hub for ships.

Molds for centrifugal castings are generally divided into three classifications, according to how the refractory is applied: 1) permanent, 2) rammed and 3) spun.

1. *Permanent.* One type of mold is a permanent mold made of cast iron, steel or graphite. These molds are usually coated on the inside surface with a thin refractory wash to increase mold life.
2. *Rammed.* The second type of mold is a rammed mold. It consists of a metal flask, usually steel, lined with a layer of refractory molding mix, which has been rammed into place. The surface of the mold cavity is coated with a refractory wash, and then baked and dried to a very hard consistency.
3. *Spun.* A third type of mold is the spun or centrifugally cast mold. It consists of a metal flask into which is poured a predetermined weight of refractory material in slurry form. The flask is rotated rapidly, and the refractory material is centrifuged onto the walls of the flask. The flask rotation is stopped, and the liquid part of the slurry is drained off. This leaves a mold with a refractory coating, which is then baked dry prior to use.

In the centrifugal casting process, molten metal is poured into a rotating mold, where it is accelerated to mold speed. Centrifugal force causes the molten metal to spread over and cover the inside mold surface. Continued pouring of the molten metal increases the thickness to the intended dimension. Rotational speeds vary but sometimes develop more than 150 times the force of gravity on the outside surface of the metalcasting.

Once the molten metal has been distributed over the mold surface, solidification begins immediately. Most of the heat in the molten metal is extracted through the mold. This induces progressive solidification. During solidification, the molten head of metal feeds the solid-liquid interface as it progresses toward the bore. This, combined with the centrifugal pressure being applied, results in a sound, dense structure across the wall, with impurities generally being confined near the inside surface. The inside layer of metal and inclusions can be removed by boring if an internal machine surface is required.

Some advantages of centrifugal casting are as follows:

- Any alloy common to static pouring can be cast centrifugally.
- Mechanical properties of centrifugal castings are excellent.
- A cleaner, denser metal is found on the outside of the metalcasting, and the impurities are on the inside, where they can be bored out.

Continuous Casting Process

Continuous casting is used to produce wrought shapes such as sheet, strip, wire, tubing and rods or bars of various cast shapes. Approximately 90% of the annual copper-base production consists of wrought shapes produced by continuous casting. Ferrous alloys, such as steel and gray and ductile iron, are also continuous-cast into wrought shapes.

The technology of the continuous casting process is well developed, but is neither widely understood nor applied in typical foundries **(Fig. 2-24).** The basic equipment for this process consists of a tundish for holding the molten metal, which is attached directly to a machined graphite mold of the desired cavity configuration. The mold is fitted with an additional graphite mandrel if hollow cast sections are made. The lower end of the mold is fitted with a water jacket for heat extraction and extraction of the base for closing the lower end of the mold during start-up operations, and an extraction mechanism (usually in the form of power-driven rollers).

The extraction mechanism serves to draw the solidified shape from the mold at the rate the casting solidifies by water cooling. Additional water cooling, below the mold, can be used to accelerate solidification and increase the casting rate.

Continuous casting can be done either vertically or horizontally. There is a distinct improvement of mechanical properties of materials cast using the continuous casting process. Also, the intrinsic metallurgical advantage of wrought shapes plus a high yield of 98% have made this a widely exploited casting process.

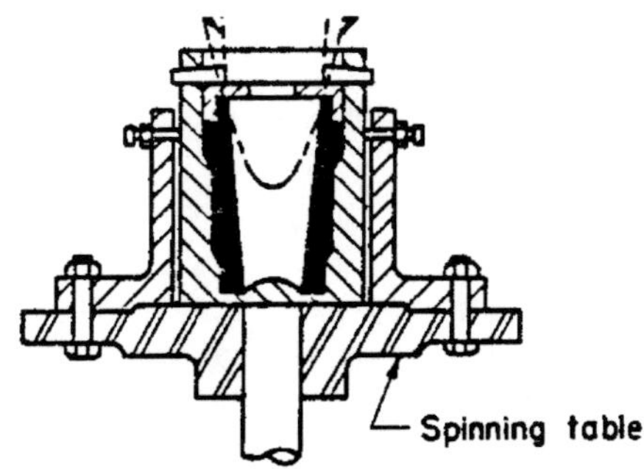

Fig. 2-23. Diagram of vertical centrifugal casting mold.

INNOVATIVE CASTING/MOLDING PROCESSES

The casting and molding processes making up this group are termed innovative because they are unique and most are relatively new.

Squeeze Casting Process

Squeeze casting is a technique in which molten metal is metered into a permanent mold die cavity and, as the molten metal solidifies, pressure is applied to it **(Fig. 2-25).** Pressure as low as 8000 psi (5,624,800 kg/m^2) is required for forming in this process.

Generally, the squeeze-casting process is used for high-production runs in aluminum alloys. If a particular application calls for relatively small castings with well-defined geometries, squeeze casting provides many advantages.

FM Process

The FM process has been developed specifically to produce thin-wall iron and steel metalcastings. It was developed by the French firm Pont-a-Mousson and is said to mean "fonte mince," or thin iron. The FM process is a mold-filling technique, versus pouring, that uses a controlled differential pressure to fill molds with high-melting alloys, including the superalloys.

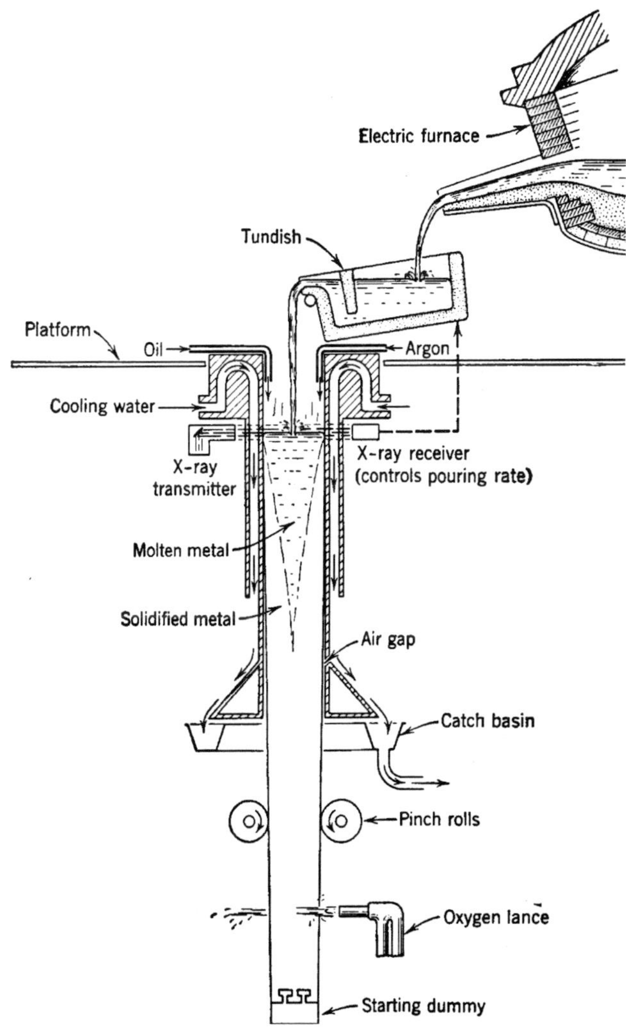

Fig. 2-24. Schematic of continuous casting operation.

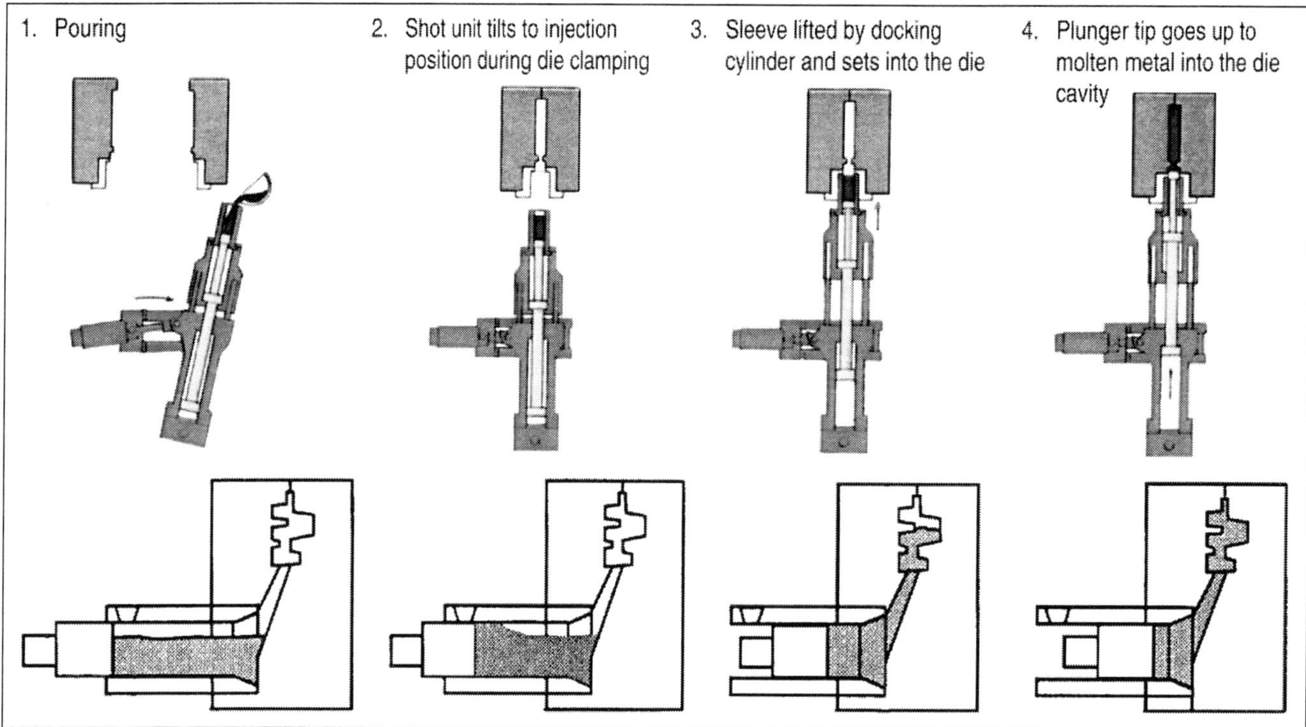

Fig. 2-25. Squeeze casting injection systems. Top row is vertical, bottom row is horizontal.

The FM process uses bottom filling of the mold that is controlled yet allows for rapid filling of the mold. Any type of mold can be used—green sand, metal, nobake, shell, etc. The rapid fill rates are obtained by exerting a low pressure on the molten metal in the ladle or furnace and a negative pressure in the mold. Evacuation of mold and core gases is also achieved during mold filling.

Casting wall thicknesses of 0.10–0.12 in. (2.5–3.0 mm) have been obtained in gray, ductile and alloyed cast irons, and in low-carbon steels. The properties of nickel-base superalloys and high-chrome steels are improved in this process.

Rheocasting and Thixomolding Processes (Semi-Solid Metal Casting)

In 1976, a new metal-forming process was developed at the Massachusetts Institute of Technology and was called rheocasting. This process uses a phenomenon called thixotropy, which involves the vigorous agitation of a semisolid metal to produce a highly fluid, diecastable alloy. Advantages of this process are reportedly longer die and chamber life, finer-grained metalcastings with fewer defects and greater economy (less loss) of molten metal fed to the diecasting machine.

A consortium of entrepreneurs has moved rheocasting research into the area of thixomolding. The major difference between these two casting processes is that in thixomolding, the metal alloy is heated only to its mushy state between the solidus and liquidus lines on the phase diagram. The thixotropic material enters the die as one mass and does not entrap air. Lower porosity levels also aid in the heat-treating process, thus enhancing the mechanical properties of the heat-treated metalcastings. At this time, magnesium alloys are being used in these two processes, but it is predicted that zinc and aluminum alloys will also be cast in the future.

CONCLUSION

An attempt has been made to give a brief description of the various types of molding and metalcasting processes that are available and used successfully today. Space constraints have limited the depth and detail needed in discussing these processes. Some metalcasting processes may not have been mentioned, but those covered are the ones most commonly used today. For further information on these or other metalcasting processes, please contact the American Foundry Society, Inc., at 800/537-4237 or visit our website at www.afsinc.org

BIBLIOGRAPHY

Aluminum Casting Technology, second edition, American Foundrymen's Society, Inc., Des Plaines, Illinois, 1993.

Aluminum Permanent Mold Handbook, AFS, Des Plaines, IL, 2001.

Bex, T., "Thixomolding Promises Savings," *Modern Casting,* Aug, 1992.

Binder System for Cold Box Processes, American Colloid Co., Skokie, IL.

Bralower, P.M., "The Metalcasting Process, Parts 5 & 6," *Modern Casting,* May and June, 1989.

Casting Copper-Base Alloys, AFS., Des Plaines, Illinois, 1984.

Ekey, D.C. and W.P. Winter, *Introduction to Foundry Technology,* McGraw-Hill, New York, NY, 1958.

Handbook of Sand Technology, Hill and Griffith Co., Cincinnati, OH.

Heine, R., C.R. Loper, Jr. and P. Rosenthal, *Principles of Metalcasting,* McGraw-Hill, New York, NY, 1967.

High Pressure Molding, first edition, AFS, Inc., Des Plaines, IL, 1973.

Imagination in Metallurgy, Hitchner Mfg. Co., Inc., Milford, NH.

Introduction to Die Casting, NADCA, Rosemont, IL, 1997.

Mihaichuk, W., "Graphite Mold Casting the Zinc Foundry Alloys," *Modern Casting,* July, 1981.

Molding Methods and Materials, AFS, Des Plaines, IL, 1962.

Shell Process Foundry Practice, 2nd edition, AFS, Des Plaines, IL, 1973.

Technical Data Report, Hill and Griffith Co., Cincinnati, OH.

Patternmaking 3

A pattern may be defined as a three-dimensional model used to establish the mold cavity into which molten metal is poured to produce a casting. Therefore, it is necessary to have a pattern, whether a single casting or a number of castings are to be made. A pattern can also be thought of as the "tooling" that brings repeatability and reproducibility to all metalcasting processes. A corebox, on the other hand, is used to produce a sand core that, when placed in the mold, blocks where the metal can travel thus establishing desired passageways inside the casting. It is important to have a quality pattern and corebox(es) to produce a quality casting. The fundamental principles of patterns and coreboxes will be discussed so that the reader will have a basic knowledge of their construction and use.

The patternmaker must have both skill and vision to interpret the casting drawings and see the finished product. In order to communicate with the designer, the metalcaster and, if required, the machine shop that will machine the casting, the patternmaker must understand the molding and coremaking processes that will be used to make the casting. In some cases, he/she may also be required to design the gating and risering system. Patternmakers must have a strong background and working knowledge of mechanical drawing, computer-aided design (CAD) and of plane and solid geometry, as well as knowledge of wood, metal, other materials and finishes required in order to transmit an idea into a casting.

EQUIPMENT AND MATERIALS

The number of castings to be made from a pattern, the molding process and the molding equipment used to make the mold are items to be considered when selecting the material from which the pattern is to be made. Typical materials used to make patterns are wood, metal, plastic, expandable polystyrene (EPS) and wax. The patternmaker must have knowledge of the molding process used to make the mold and knowledge of the equipment on which the mold is to be made. The number of castings to be made from the pattern is also a very important consideration.

The first thing to consider is the material from which the pattern is constructed. When the molds are made from bonded sand the pattern in turn must possess high wear resistance. Sand is an abrasive material, and when compacted around the pattern, will cause wear. In the case where only a few castings are to be made from a pattern, a soft pine can be used. For a longer pattern life and a sharper impression in the sand, hardwood may be used. Although molding processes will be discussed later, in the case of green sand molds, wood patterns are used mainly for hand-molding operations. Care must also be taken when using wood patterns to keep them dry, to prevent warpage from taking place. It should be mentioned that wood patterns are used a great deal for making chemically bonded molds. **Figure 3-1** shows a patternmaker carving a wood pattern.

Fig. 3-1. Patternmaker working to close-shape tolerances, carving the wood pattern of a special core plate pattern.

Today, most green sand foundries are using some form of machine molding. This is true for short as well as long production runs. These machines can exert extreme pressures when compacting the molding sand around the pattern. Under these conditions, wood patterns would not hold up and would soon lose their shape and dimensions. Metal patterns and plastic patterns, while more expensive to make, are dimensionally more stable and wear-resistant, and will last considerably longer than wood in this type of molding operation. Metal patterns can be machined from solid cast iron or aluminum billets; however, cast iron and aluminum patterns can first be cast to shape and then machined to finished dimensions. Since metal patterns are used for long production runs, their surfaces can be coated with a wear-resistant material, which reduces the wear caused by the sand.

Polyurethane elastomers (urethanes) and epoxies are one of the most common types of plastics used for patterns, and are used in a number of plastic tool-construction techniques. They can be used as solids, such as thermosetting polyurethane foundry board available for computer numerical control (CNC) machining or as castable liquids. Plastic patterns can give manufacturing flexibility, quick response time and shorter lead times between prototype and production castings.

Wood, metal or plastic patterns are used for making most chemically bonded molds. Wood will normally hold up well in the making of chemically bonded molds because the molding sand is not compacted under extreme pressures around the pattern. Chemically bonded molding sands are also said to have good flowability, which can reduce wear on the pattern surfaces. If plastic patterns are to be used to make chemically bonded molds, precautions must be taken to ensure that the plastic is compatible with the catalyst used in the chemically bonded sand mixture. Some plastic patterns have been known to soften on the surface when used in making these types of molds. In some cases, chemically bonded foundries will use metal patterns to reduce the amount of wear and tear of the pattern. *The importance of proper maintenance of the surface of chemically bonded molding patterns cannot be emphasized enough.* Nicks and scratches in the surface of the pattern can result in very difficult removal of the pattern from the mold.

Fig. 3-2. Shell pattern and mold half.

So far, the previously mentioned patterns have not been subjected to heat when making the sand mold. There is a molding process, called shell molding, in which the pattern is heated to temperatures around 475F (246C). At these temperatures, only metal patterns will hold up. In addition, this process is used mainly for high-production runs of castings; thus, metal is the chosen pattern material. Steel **(Fig. 3-2)** and cast iron are the metals normally used for these patterns, with cast iron being used most often.

Up to now, patterns that will be *reused* have been discussed. There are several molding processes in which the pattern is destroyed. Expandable polystyrene (EPS) is used for patterns to be utilized in the full mold and lost foam processes. Each pattern is used only once, and multiple patterns must be made for multiple castings. These patterns are vaporized as the molten metal enters the mold or can be vaporized before the molten metal is poured into the mold.

In the case of investment casting, the pattern is made of wax or plastic. In either case, the patterns are burned out of the mold before molten metal is poured into the mold. These wax patterns are made in metal dies.

BUILDING A PATTERN

Before the pattern is actually laid out and made, the patternmaker must determine a parting line. The parting line separates one half of the pattern from the other half. In a horizontally parted mold, the top half is called the cope and the bottom half the drag. The mold may also be parted vertically. To the patternmaker, the parting line is the base, the largest cross section, from which the pattern and corebox are developed and drafted. When possible, this is a flat plane but can also be irregular.

Parting-line location often presents options that must be agreed upon by the foundry and the customer. Parting lines are ideal places to locate gates and risers, which after removal will require some grinding, and this may not be acceptable to the customer. Also, in most molding processes, the parting line will leave a line on the casting, which again may be unacceptable to the customer at that location.

Patterns are true three-dimensional models of the final casting to be produced. However, the pattern has been modified to take into account some of the problems that will occur during the making of the sand mold and the solidification of the metal. The first modification is draft or taper, which is added to the walls of the pattern, perpendicular to the parting line. Draft is added so that the pattern can be removed from the mold without tearing or damaging the mold surfaces. The amount of draft to be added is dependent on how well the molding sand is compacted and on the sand-molding process. Normally, outside surfaces require less draft than internal pockets. With today's modern molding machines and chemically bonded molding, some patterns may require only one, or less, degree of draft on outside surfaces. Remember that normally the deeper the draw the more draft that is needed **(Fig. 3-3).**

Foremost, fillets and radii are used on the pattern to aid in making sand molds. Fillets are used in internal corners and radii are placed on external corners. These items aid in the compaction of the mold material and in withdrawing the pattern from the mold. Another term used for radii is "breaking" the edges. Besides being used as an aid to molding, these radiused edges also help prevent chill cracks in iron castings from forming on the edges of the casting. Meanwhile, the fillets also aid in removing heat from internal corners of the casting, which, if retarded, can cause casting defects in these corners **(Fig. 3-4).**

Shrinkage allowance on patterns is a correction for the metal's contraction once it has turned solid and is cooling to ambient temperature. The total contraction is volumetric, but the correction for it is usually expressed linearly. Thus, the solid-shrinkage allowance refers to the amount the pattern must be made larger than the casting dimensions to provide for this shrinkage. The linear allowances in **Table 3-1** are representative of values for castings in sand molds. Special conditions prevail with some alloys, and the patternmaker must consider these when determining the amount of solid-shrinkage allowance to add to the pattern.

The patternmaker uses a shrink rule to allow solid-shrinkage allowance to be added to the pattern. For example, on a 1/8-inch (3.175-mm) shrink rule, each foot is 1/8 inch (3.175 mm) longer, and each graduation is proportionately longer than the conventional length. Sometimes, if a pattern is initially made from wood and then from some other metal (as one does in making master patterns) double allowances are made. A cast aluminum pattern to be made from a wood master pattern might require an allowance of 1/4 in. (6.35 mm) per foot. The total allowance on the original wood master pattern would then provide for solid shrinkage of the aluminum pattern casting and of gray iron castings made using the aluminum pattern. Shrinkage allowances are also expressed as percentages; 1/8 in. per foot is approximately 1%.

When constructing patterns for castings where various points on the casting's surface must be machined, one should provide enough excess metal for all machined surfaces. That allowance, commonly called finish, depends on the metal used, the shape of the part, the size of the part, the tendency to warp, and the machining method and setup. It may be necessary to locate surfaces to be machined in the drag side of the mold. This matter should be discussed between the foundry and customer. It is important that surfaces to be machined are free of any inclusions, which may cause hard spots that are detrimental to machining operations. **Table 3-2** gives data that one may use as a guide to machine-finish allowances for castings

Table 3-1. Pattern Shrinkage Allowances

	Pattern Dimensions In. (mm)	Construction	Shrinkage In. per Ft Fract. (Dec.)	Multiplier
Aluminum	0–48 (1220)	Solid	5/32 (0.156)	1.0130
	49–72 (1830)	Solid	9/64 (0.141)	1.0118
	Over 72 (1830	Solid	1/8 (0.125)	1.0104
	0–24 (610)	Cored	5/32 (0.156)	1.0130
	25–48 (1220)	Cored	9/64 (0.141)	1.0118
			to 1/8 (0.125)	1.0104
	Over 48 (1220)	Cored	1/8 (0.125)	1.0104
			to 1/16 (0.063)	1.0053
Brass	—	—	3/16 (0.188)	1.0157
Bronze	—	—	1/8 (0.125)	1.0104
	—	—	to 1/4 (0.250)	1.0208
Gray Cast Iron	0–24 (610)	Solid	1/8 (0.125)	1.0104
	25–48 (1220)	Solid	1/10 (0.100)	0.0083
	Over 48 (1220)	Solid	1/12 (0.083)	1.0069
	0–24 (610)	Cored	1/10 (0.100)	1.0083
	25–36 (915)	Cored	1/12 (0.083)	1.0069
	Over 36 (915)	Cored	1/8 (0.125)	1.0104
Cast Steel	0–24 (610)	Solid	1/4 (0.250)	1.0208
	25–72 (1830)	Solid	3/16 (0.188)	1.0157
	Over 72 (1830)	Solid	5/32 (0.156)	1.0130
	0–18 (480)	Cored	1/4 (0.250)	1.0208
	19–48 (1220)	Cored	3/16 (0.188)	1.0157
	49–66 (1675)	Cored	5/32 (0.156)	1.0130
	Over 66 (1675)	Cored	1/8 (0.125)	1.0104
Magnesium	0–48 (1220)	Solid	1/16 (0.063)	1.0053
	Over 48 (1220)	Solid	5/32 (0.156)	1.0130
	0–24 (600)	Cored	5/32 (0.156)	1.0130
	Over 24 (600)	Cored	5/32 (0.156)	1.0130
Ductile Iron				
Pearlitic	0–24 (610)	Solid	1/8 (0.125)	1.0104
Ferritic	0–12 (505)	Solid	0	1.0000
White Irons	—	Solid	5/32 (0.156)	1.0130
High Alloy Irons	—	Solid	1/8 (0.125)	1.0104

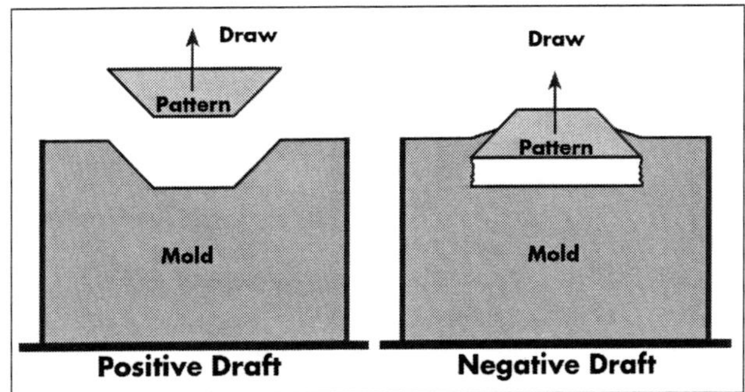

Fig. 3-3. Pattern having negative draft (right) will damage the mold when withdrawn. Pattern with positive draft (left) wil not damage mold when withdrawn.

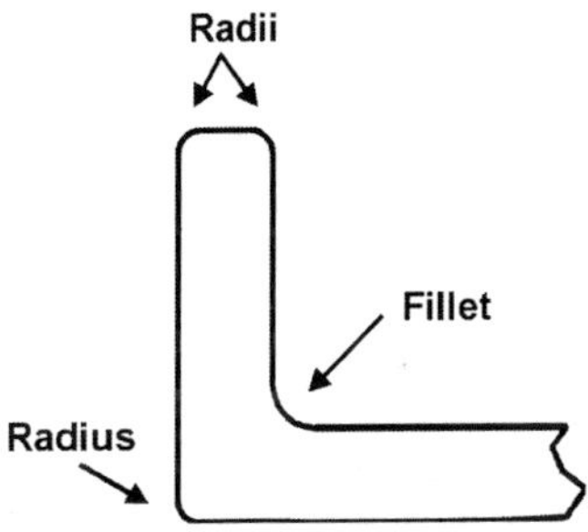

Fig. 3-4. Drawing of typical fillet showing radius and radii.

Table 3-2. Machine Finish Allowances

	Pattern Size In. (mm)	Bore In. (mm)	Surface In. (mm)	Cope Side In. (mm)
Cast Iron	0–6 (150)	1/8 (3)	3/32 (2)	3/16 (5)
	6–12 (300)	1/8 (3)	1/8 (3)	1/4 (6)
	12–20 (500)	3/16 (5)	5/32 (4)	1/4 (6)
	20–36 (900)	1/4 (6)	3/16 (5)	1/4 (6)
	36–60 (1500)	5/16 (8)	3/16 (5)	5/16 (8)
Cast Steel	0–6 (150)	1/4 (6)	1/8 (3)	1/4 (6)
	6–12 (300)	1/4 (6)	3/16 (5)	1/4 (6)
	12–20 (500)	1/4 (6)	1/4 (6)	5/16 (8)
	20–36 (900)	9/32 (7)	1/4 (6)	3/8 (10)
	36–60 (1500)	5/16 (8)	1/4 (6)	1/4 (6)
Nonferrous	0–3 (75)	1/16 (1.5)	1/16 (1.5)	1/16 (1.5)
	3–8 (200)	3/32 (2)	1/16 (1.5)	3/32 (2)
	6–12 (300)	3/32 (2)	1/16 (1.5)	1/8 (3)
	12–20 (500)	1/8 (3)	3/32 (2)	1/8 (3)
	20–36 (900)	1/8 (3)	1/8 (3)	5/32 (4)
	36–60 (1500)	5/32 (4)	1/8 (3)	3/16 (5)

produced in sand molds. Other casting processes may require different finish allowances to be used.

At times, it may be necessary to include such items as lettering, numbers and identification marks on castings. The location and size of these items are largely dependent on the ability of the molten metal to flow and fill them in. Alloys such as cast iron and aluminum have good fluid life and can easily fill in these items. On the other hand, carbon steels have poor fluid life, and larger letters or numbers will have to be used. In the latter case, these items may have to be placed in the drag.

When cores are to be used in producing the casting, some form of location, support and restraint must be provided in the mold. Coreprints, which become extensions of the pattern, fulfill these requirements. These extensions of the pattern will produce coreprints (sockets) in the mold into which the coreprints of the core will fit.

During layout, the patternmaker should allow a slight coreprint clearance on small patterns, and increasingly, allow more clearance with large patterns to completely close the mold. The more accurate the pattern, not only in the diameter or size of the coreprints but also in their alignment, the less clearance is required. Individual foundry coremaking processes should also be taken into account when laying out coreprints **(Fig. 3-5).**

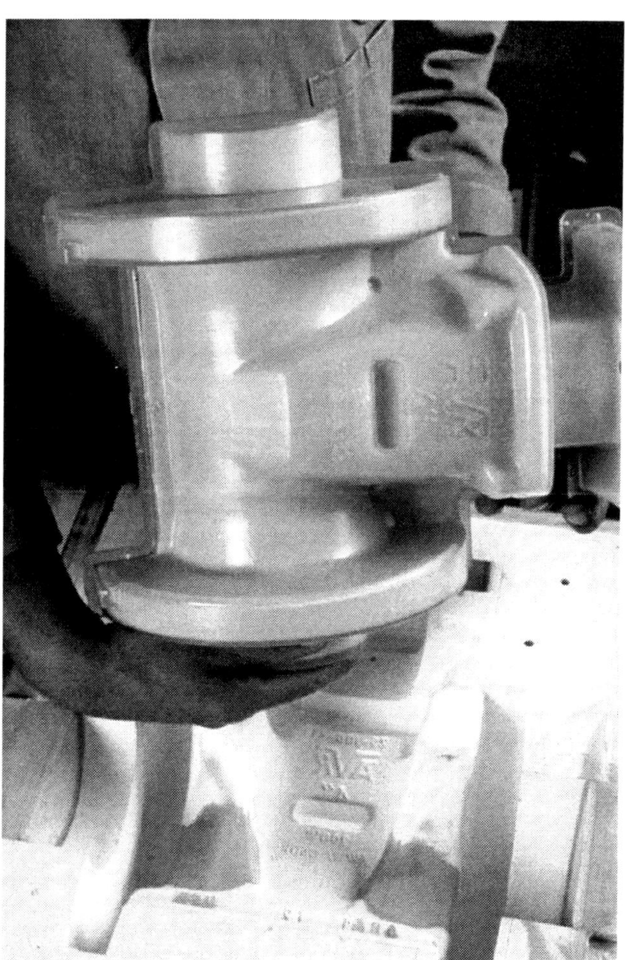

Fig. 3-5. Urethane pattern with core prints.

Types of Patterns/Tooling

There are 10 types of patterns used in the production of castings. They are as follows:

1. Loose (solid or split)
2. Matchplate
3. Cope-and-drag matchplate
4. Shell-molding
5. Special patterns or pattern devices
6. Expendable polystyrene (EPS)
7. Wax or plastic
8. Permanent mold
9. Diecasting mold
10. Centrifugal casting mold tooling

Loose Patterns (Solid or Split)

Loose patterns are single copies of the casting to be made. However, they incorporate the allowances and coreprints necessary to produce the casting. They are generally made from wood but may be made from plastic or any other suitable material **(Fig. 3-6).** Hand molding is usually practiced with loose patterns; thus, the process is slow and labor-intensive. At times, a follow board or pattern bed **(Fig. 3-7)** is used to support the solid pattern during the molding of the drag half of the mold. In this case, the molder cuts a parting line. The pattern can be split into cope and drag halves. In the case of the split pattern, the parting line is formed where the pattern is split. Gates and risers are usually cut by hand, but these can also be constructed as loose pieces and molded with the pattern. Drawing the pattern from the sand is also done by hand, with a draw pin, after the pattern is rapped to loosen it from the sand. Consequently, the dimensions of the casting may vary.

Fig. 3-6. Loose piece pattern.

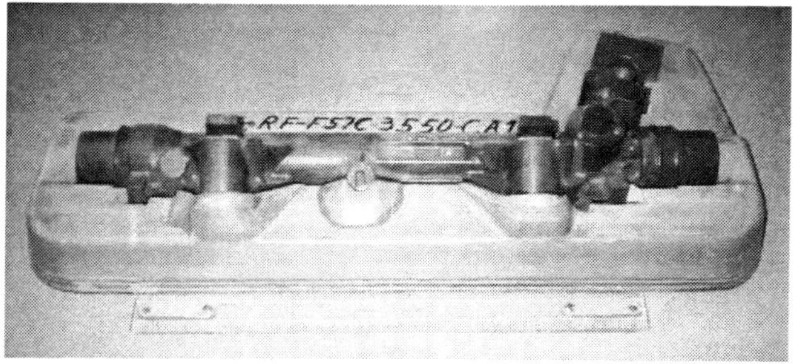

Fig. 3-7. Pattern with follow board.

Special Patterns or Pattern Devices

Special devices are used with loose patterns having an irregular parting line. A follow board or pattern bed (match) is often used in this case. The pattern bed serves to support the loose pattern during the molding of the drag half of the mold, and establishes the parting surface when the match is removed. A hard sand bed (match) supports the pattern. The term hard-sand bed originates in the method of construction of the mold. A flask is first rammed with sand and rolled over. The pattern is then carefully embedded in (cut into) the sand, with care being taken to cut and smooth the parting-line sand. This is the pattern bed. A drag half flask is rammed over this bed. The mold is inverted, the pattern bed is lifted off and set aside for reuse and the cope half of the flask is rammed over the drag. Then normal procedure for loose pattern molding is followed. The bed or match can be used a number of times. Plaster beds have a much longer expected life than sand beds before deterioration of the plaster interferes with efficiency.

Hollowed-out boards that perform the same function as pattern beds are known as follow boards **(Fig. 3-7).** They support the pattern in the first molding operations and form the natural parting line of the castings. The molding sequence with a follow board is quite similar to the molding sequence used with a pattern bed. The drag half of the flask is rammed around the pattern and follow board. The mold is then inverted, the follow board is removed and the cope half is rammed. The mold is then finished in the normal molding procedure.

It is desirable to have solid patterns that are designed to draw readily out of the sand toward the parting line. Drawbacks, loose pieces and cores are often used to form internal or external cavities or projections on patterns that perhaps would not "draw."

Sometimes a loose piece deep in the mold cavity is hard to withdraw. To make it easier to withdraw the pattern, one may use a loose boss on the end, dovetailed to facilitate removal. A simpler way to make the angle box is to use a core. This eliminates the loose piece; it also eliminates the possibility of disturbing the mold while one is withdrawing a pattern piece from a deep draw. The loose boss can be redesigned so the face of the boss extends to the parting line and has adequate draft. This makes it easier to withdraw the pattern without a loose piece.

Certain types of castings may call for "sweeps" instead of three-dimensional patterns to make a mold. Ordinarily, these special sweeps are used when the shape to be molded may be formed by the rotation of a curved line element about an axis, as seen back in **Fig. 1-5.** The use of a simple sweep eliminates the necessity for construction of a large, expensive three-dimensional pattern. An example of castings made using this method is church bells, especially large ones. These sweeps are normally made of wood. Occasionally, a straight sweep is used to form a plane surface.

Matchplate Patterns

For high-production runs, a matchplate pattern is used. The matchplate pattern is usually limited to small- to medium-size castings. The cope and drag portions of the pattern are mounted on opposite sides of a metal, plastic or wood plate, which forms the parting line between the cope and the drag. The metal matchplate and pattern can be pressure-cast as one piece in a plaster mold or simply cast in a sand mold **(Fig. 3-8).** Today, it is common practice to have the gating and risering systems cast or mounted on the plate. Matchplate patterns can be used with manually operated and automated green sand molding machines. The register between the cope and drag halves of the pattern has to be perfect, or defective castings will result. The improved production rate, together with the increase of dimensional accuracy of the casting, justifies the increased cost.

Matchplate patterns can be used to make just one casting per mold or a number of castings per mold. The term "multiple cavity matchplate" is sometimes used to describe a matchplate that is used to produce more than one casting cavity per mold. The cavities can be used exclusively for one particular casting or for more than one type of casting. In the case of very small castings, such as those used in plumbing systems, the number of castings can exceed 50, depending on the size of the molding machine and matchplate size for that machine.

Cope-and-Drag Matchplate Patterns

When larger castings are desired, the pattern halves are mounted on two plates. The cope half of the pattern is mounted on a cope plate and the drag half is mounted on a drag plate. As with matchplates, the plates can be made of metal, wood or plastic. The patterns themselves can be cast directly with the metal plate or mounted on the plate.

The two halves of the mold are then made separately. In high-production operations, two automated molding machines are normally used. For medium-to-low production runs, the pattern plates can be mounted in manually controlled molding machines. Again, perfect register must be the rule with this type of operation.

Shell-Molding Patterns

Shell molding requires patterns that will have to be heated up to about 475F (246C). Since heat is required, patterns used to make shell molds are generally made of gray cast iron or steel. Aluminum has been used, but gray cast iron and steel seem to be the metals of choice. These patterns are either attached to a plate of metal or cast to shape along with the plate. The pattern is then finished to shape and size by machining and hand tooling. Shell-molding pattern plates can also be machined out of blocks of metal **(see Fig. 3-2).** Often, the two halves of the pattern are on one plate, and if not, then two separate plates are used, each having half of the pattern on it. Pattern plates have a flat back against which the gas flames impinge to heat the pattern plate.

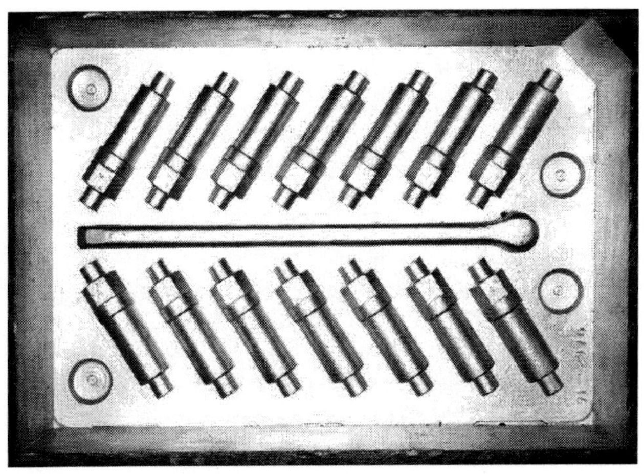

Fig. 3-8. Matchplate for green sand molding.

Shell molds are made by joining the two halves of the mold together. Depending on the orientation of the parting line, there can be either horizontally parted molds or molds that are parted vertically. The pattern walls should be kept to an even thickness throughout the pattern for proper heating. The gating and risering system becomes part of a pattern plate half or can be partially on both halves of the pattern plate. Contraction of the alloy to be poured into the mold has to be taken into account when laying out the pattern.

Some form of ejection mechanism has to be designed to remove the mold halves from their respective pattern plates. The ejection mechanism is usually metal pins, which are evenly distributed over the surface of the shell mold half so that it is not damaged during ejection. The burners that supply the heat for the patterns are part of the machine in which the patterns are mounted.

Expandable Polystyrene Patterns

Patterns that are made of expandable polystyrene (EPS) are used in either the full mold or the lost foam casting process. One method of producing these patterns is by fabrication, or simply cutting the pattern out of expandable polystyrene board. In the first case, various parts of the pattern can be machined using simple woodworking machines and tools, and then gluing the parts together to form the finished pattern. This material can also be cut using a hot wire or a sharp knife. No draft is needed on the pattern, as it is not withdrawn from the molding sand after pouring. Small or irregular pattern parts are often formed and attached to a wood or metal pattern, and these parts are left in the mold to vaporize. The gating and risering system is also made out of EPS and is glued to the pattern. The pattern is coated with a permeable refractory coating before the mold is made.

For high-production EPS patterns, another method of production is used. These patterns are made from expandable polystyrene beads injected into a preheated, 220F (40C), aluminum mold. Steam is then injected into the mold, and the beads expand and bond together to form the pattern. The mold is then cooled with cold water and the pattern ejected from the mold. At this point, if the pattern is complete, the pattern moves on to the next operation. In some designs, the ejected part may have to be glued to another part(s) to form the completed pattern **(Fig. 3-9).**

One of the advantages of this casting process is that separately made cores can be eliminated. Should the casting require "cored" passageways, the pattern can be made to form them. These passageways are formed in the pattern by incorporating removable inserts and/or pullbacks in the mold. The molds are normally made of aluminum because of its excellent heat-transfer characteristics, which can result in shorter cycles. A mold-wall thickness of 1/2 inch is normally recommended and maintained as closely as possible over the entire mold. If the pattern has a thick section, the mold wall in that area can be made thinner to permit faster bead expansion. Side walls and back plates can be an inch or more, for added strength.

Should inserts be necessary to make "cored" holes in the pattern, they can be either loose pieces or pullbacks. If they are pullbacks, they should be made of common brass because closely fitted aluminum will gall against aluminum. If brass is used, it should be thinner because it has less heat-transfer ability.

An important thing to remember in making these molds is that they do require proper venting. Since the EPS beads are blown into the mold, some means has to be provided to allow the "blow air" to leave the mold. If this is not done properly, defective patterns will result.

Expandable polystyrene patterns will usually continue to expand slightly after being ejected from the mold. After this, they will begin to contract at a determined rate, based on section thickness, bead density, time and atmospheric conditions (heat and humidity). Therefore, the mold designer and builder must have a good working knowledge of the customers' equipment and know the actual pattern storage times.

Another key point to remember in the mold design is that sharp internal corners of the mold cavity itself should be avoided, not only because of the potential to crack but also because it produces a sharp external corner on the pattern. These sharp external pattern corners, when coated, are coated more thinly and less uniformly. This problem can cause castings to have poor edges. On the other hand, sharp internal corners on the pattern can cause excessive coating buildup, which will be difficult to dry. Since these refractory coatings are usually water-based, any remaining moisture can cause casting defects in both ferrous and nonferrous castings.

After these patterns have been completed, the gating system and risers are glued to them. The next step will be to apply a permeable refractory coating to the pattern, gating system and risers.

Wax or Plastic Patterns

Patterns used for investment castings can be made of wax or plastic. In the case of wax patterns, liquid wax is injected into an aluminum die to produce a pattern that is the replica of the part to be produced. Another method of putting the wax into the die is by extrusion. In this case, a semisolid wax is used. The die may be made of aluminum, with provisions for opening and removing the wax pattern. The die cavity must be exact in detail and extremely smooth to provide quality wax patterns. A wax pattern is required for each casting to be produced.

Fig. 3-9. Complex EPS pattern made from multiple pieces.

When plastic is used, it too is injected into a die. However, it can be machined from solid blocks of plastic or formed by the new rapid prototyping processes.

If cores are required to make internal passageways in the casting, several methods can be used. First, if it is possible, the pattern can be designed so that it will allow a core to be formed of the mold material. Another method is to use preformed ceramic cores, which are inserted into the die before the wax is injected or extruded into the die. Another method is to use soluble wax cores, which become part of the wax pattern but are removed prior to the pattern being coated.

The pattern(s) is attached to a wax or plastic sprue **(Fig. 3-10).** This sprue with the pattern(s) attached to it can be called a cluster or tree. The cluster is then ready for the next step in the investment casting process.

Fig. 3-10. Assembly of wax patterns for investment casting.

Permanent Mold Tooling

The permanent mold process does not use patterns in the way they have been discussed up to now. In this process, the molten metal is poured into either a metal or a graphite mold. The mold can be used over and over again—thus, the term *permanent mold.* To produce the permanent mold, two methods of manufacturing can be used. In one case, the mold can be machined out of metal or graphite billets. In the other case, a pattern is made and then the mold is cast to shape in metal. After the mold has been cast, it is finished by machining and hand tooling.

There are several different types of permanent-molding processes. One can be called the static-pouring method **(Fig. 3-11)** and the other the tilt-pour method. Yet, another is called low-pressure permanent molding (LPPM). These processes were discussed in more detail in the previous chapter.

Since the molds used in the first two methods are very similar, they will be discussed first. The most accurate molds are made by machining from metal billets. The two metal alloys most commonly used for the molds are steel and gray cast iron. As with the patterns made for sand molding, the design of the mold includes allowances for solid shrinkage and draft. Wherever possible, the thickness of the mold is kept as uniform as possible. Appropriate changes in mold thicknesses can be made should the design of the casting require help in solidifying properly.

Fig. 3-11. Permanent mold for static pouring.

The molds are made in two halves. One half is called the male and the other, the female. Unlike sand molding, some form of removing the casting from the mold is required. A mechanical ejection system is designed and is usually attached to the male half of the mold. This ejection system can be hydraulically operated, or a simple mechanical spring-actuated system can be used. Hydraulic-actuated ejector systems are preferred because they will return the ejector pins flush with the mold surface. In the case of some static-poured permanent molds, the casting is stripped out of the mold by hand. Since permanent molds are not permeable, some form of venting the air out of the mold is required. One way of doing this is to grind a flat surface on the round ejector pins. A maximum flat of 0.008 in. (0.2 mm) can be used.

Cores used in the permanent molding process can be made of core sand or metal. If the cores are to be made of sand, the coreprints have to be part of the mold design. On the other hand, metal cores, sometimes called inserts, have to be mechanically moved in and out of the mold. Again, this can be done with a rack-and-pinion system or hydraulically. If metal inserts are to be used as cores, they should be coated with high-temperature grease. High-temperature greases containing graphite are very popular. If the casting requires cast-in inserts, the inserts can be held in place by pins or pockets.

The gating and risering systems are either cast to shape and then finished to size or simply machined into the mold when the mold cavity is machined. In static pouring, the molten metal drops to the bottom of the mold and then fills from the bottom up. In tilt pouring, the molten metal runs to the bottom of the mold and then fills the gating system and mold cavity as the mold is tilted to the vertical position. Where possible, risers are located in the top portion of the mold.

To help the casting cool and solidify properly, selective cooling and heating of the mold cavity can be done. When required, heavy sections of the mold can be cooled by installing a steel insert directly below that area and passing water through the insert. Oil can also be used in place of water for either heating or cooling. Air is often used for cooling.

Another type of permanent molding is the low-pressure permanent-molding (LPPM) process. As with the previous permanent-mold process, metal molds are used—the difference being the manner in which the molten metal enters the mold. This difference was explained in the previous chapter.

The metal alloys used for the molds are usually gray cast iron or H-13 tool steel. The type of mold material chosen is usually based on how long the mold will be used and the complexity of the design. Such items as metal sections or long projections in the mold would be better if built from tool steel because they are less fragile, and easier to repair. Complex castings with erratic parting lines can be cast using the LPPM process, since only one entry of molten metal into the casting is usually used and the mold can be built with five moving parts quite easily. Allowances have to be made in the mold design for the solid shrinkage that will take place in the casting.

Again, sand cores or metal cores (inserts) can be used to form internal passageways in the casting. The usual practice used for the other permanent-mold processes using metal molds is followed. The mechanism used to seat the sand core in the mold is somewhat dependent on the complexity of the mold design. On the other hand, metal cores can become part of the mold itself or may require some form of mechanical means to be removed from the casting before the casting is ejected from the mold.

Gating is usually nonexistent, as the casting itself becomes the gating system. The molten metal enters the casting at only one point. Normally, this point is the heaviest section of the casting. However, if necessary, a normal gating system can be used. The big difference in the LPPM process is that risers are not used. Solidification shrinkage is fed from the "feeding" of the casting under pressure, promoting directional solidification back to the point where the molten metal enters the casting. Directional solidification can be aided with the use of water-cooled steel inserts in the mold. If necessary, these inserts can be water cooled.

One major precaution has to be followed when making the molds. There can be no machine or tool marks on the surface of the mold that encounters molten metal. Since the molten metal enters the mold under pressure, it can be forced into these marks and they will appear on the surface of the casting.

Another permanent-molding process uses graphite molds. One such use of graphite molds is the "Griffin Process," discussed in Chapter 2, which is used to make steel railroad wheels. In this process, the mold cavity, including the risers, is machined into large solid blocks of graphite. Since the molten steel is forced into the mold cavity at only one point, no gating system is required.

Other types of castings can be produced in graphite molds that are machined out of solid graphite blocks. Again, solid shrinkage has to be taken into account when designing the mold. In most of these cases, the gating system and risers also have to be machined into the mold.

Diecasting Mold Tooling

Diecasting is another form of permanent molding geared to the high production rate of small-to-medium size castings, although, equipment to produce large castings is being developed. In this process, the molten metal will be "shot" into the mold at extremely high pressures. Thus, the mold material used is steel. As with the permanent molds previously discussed, these molds may also be cast to shape, and then finished to size or machined out of solid blocks of metal **(Fig. 3-12).** Since the molten metal will be entering the mold cavity under pressure, it is important that the cavity surfaces be smooth and have no tool marks on them.

Again, solid shrinkage has to be taken into account when dimensioning the mold cavity. The molten metal is normally sent through a gating system, which is part of the mold. No risers are used in this process since most of the castings produced have thin sections. If necessary to aid in solidification or for some other reason, the mold may be water-jacketed. Some method of mechanically ejecting the casting from the mold also has to be manufactured and has to become part of one half of the mold. The molds can produce one casting per shot or many castings, based on their size and overall size of the mold **(Fig. 3-13).**

Centrifugal Casting Mold Tooling

Centrifugal casting is done by rotating a mold about a vertical or horizontal axis while the molten metal is being poured into it. The mold into which the molten metal is poured is made out of metal or graphite, and thus can be reused.

The predominant types of castings poured in these molds are cylindrical in shape and hollow. Pipe and tubing are two examples

Fig. 3-12. Large automotive die. [Courtesy of NADCA]

of centrifugally cast parts; thus, the design of the mold is rather straightforward. On the other hand, castings of varying geometries can be centrifugally cast and require molds that are more complex to produce. Since the molten metal is poured directly into the mold cavity, no gating or risering systems are required.

COREBOX CONSTRUCTION

Just as patterns are needed to make molds, coreboxes are needed to make the sand cores used to create the internal passageways in the casting. It should be said here that cores could also be used to make up a mold or part of a mold.

Corebox construction begins with understanding the various coremaking processes available. With this understanding, an educated judgment can be made in selecting the material to be used to build the corebox. As with pattern materials, wood, plastic and metals can be used. The selection of the material to be used for building a corebox is related to the coremaking process, the method of filling the corebox and the quantity of cores that is required. Coreboxes can be used to make one core or several cores at one time. The size of the cores will determine how many cores can be made at one time along with the size limitations of the corebox and the core machine.

The patternmaker has to know what core process will be used to make the cores. Once this is known, a material can be selected from which the corebox can be made. If the core process requires a heated corebox, then metal coreboxes will be needed. The material generally chosen is gray cast iron. As with metal patterns, the corebox can be completely machined or it can be cast to shape and machined to size.

Should a chemically bonded process be selected, wood or plastic materials can be used, especially if small quantities of cores will be required. However, should the core sand be blown into the corebox and large quantities of cores are required then metal will be used. If plastic materials are selected for the corebox, care should be taken that they are compatible with the catalyst used to cure the core.

Fig. 3-13. Section of automotive die. [Courtesy of NADCA]

How the core sand enters the corebox is also critical in the selection of a corebox material and construction. If the core sand is blown into the corebox, it is done with high-pressure air. Since the core sand is abrasive, material used for making the corebox must be able to withstand severe abrasion. In most cases, this will require that metal be used to help withstand the abrasion. However, some contend that there are plastic materials available that will withstand abrasion. It is important that the construction of coreboxes into which core sand is to be blown includes sufficient reinforcement, to prevent the corebox from distorting. On the other hand, if the core sand is simply to be dumped into the corebox from a mixer or by hand, then wood or plastic will do the job.

When the core sand is blown into the corebox, a special plate is built containing blowtubes. It is through these tubes that the core sand will be blown into the box. The number and placement of these blowtubes is of great importance in producing uniformly good, dense cores. In the case of heated coreboxes, these blowtubes may require cooling. The blowtube plate is clamped to the corebox by the coreblowing machine. The corebox itself must be fitted with vents, through which the air in the closed corebox and the air used to transport the core sand can escape. The number and location of these vents is also very critical in producing uniformly good, dense cores.

As will be discussed in Chapter 6, Coremaking, there are many different coremaking processes that utilize gas to cure (or harden) the core. In order to do this, gassing tubes are an integral part of the corebox construction. Again, the patternmaker must be familiar with the core process and the proper location of the tubes. It is critical that these tubes be properly placed in the corebox so that the gas is properly distributed in the core(s) to be gassed. Normally, these tubes are placed in one half of the corebox. In the other half of the corebox, vents are located to carry unused gas out of the corebox. These same tubes are used to pass purging air into and away from the cores after the gassing cycle is complete.

The number of cores to be produced by the corebox also plays a role in the selection of a corebox material. It stands to reason that the more cores that are to be made in a corebox, the more abrasion the core-forming surfaces will have to withstand. Where the chance of severe abrasion exists, the choice of materials would be either metal or plastic. On the other hand, where a few cores are required and the core process allows, wood can be used to make the corebox.

Coreboxes are normally built in two halves and can be either horizontally or vertically placed in a coremaking machine. In the case of handmade cores, the cores are usually made in halves, stripped from the corebox and, when cured, glued together. When the cores are made in automated coremaking machines and cured in the corebox, some type of ejection mechanism is designed to strip the core out of the corebox. Normally, when this method is used, the cores are in one piece and not in two halves.

There are times when many of the same small cores will be needed. An example of this would be making small plumbing fittings. A gang corebox is designed and made in which maybe up to 50 cores can be made at one time. The number of cores is dependent on the size corebox that the coremaking machine can accommodate.

When placing the cores in the mold, some method of locating and holding the cores in place is required. This is accomplished by using coreprints. Coreprints are extensions that are part of the core. Earlier in this chapter, coreprints were discussed, as they apply to patternmaking. Coreprints have to be included in the construction of coreboxes. The number of coreprints required will depend on the shape of the core, and normally at least two coreprints are required for each core.

RAPID PROTOTYPING

Today's foundry customers are demanding shorter and shorter lead times from the conceptual design of the casting to its production and delivery. Rapid prototyping is a method in which these lead times can be significantly shortened. The parts made by rapid prototyping can be used as models or in some cases as patterns from which actual metal castings will be made.

Current technologies such as desktop manufacturing, rapid prototyping, functional prototyping, rapid manufacturing, solid free-form fabrication or conceptual modeling allow the patternmaker and foundry engineer to make 3-D models, prototypes, patterns, tooling and production parts directly from CAD data very quickly. Practical systems are available and some of the systems are as follows:

- *Laminated Object Manufacturing*—Sheets of material, such as paper, plastic or composites, are laminated with a thermally activated bonding agent and a CO_2 laser scans the contour of each layer. An outline of that layer of excess material is later removed. Materials are relatively inexpensive and buildup is rapid. Post-processing is required.
- *Selective Laser Sintering*—Models are constructed one layer at a time by a modulated 50-W CO_2 laser that sinters a thin layer of powdered material in an area defined by the part geometry.
- *Stereolithography*—Depending on the system used, a laser cures a liquid polymer (layer by layer) on a platform that is submerged in a vat of liquid polymer. Vertical support structure members are needed for extreme horizontal overhangs. Post curing processing is a function of finish and detail. A special version of stereolithoraphy has been developed to help build plastic patterns for direct shell investment casting.
- *Fused Deposition Modeling*—A continuous filament of thermoplastic polymer or wax, at a temperature just above the material's flow point, is deposited by an extruding head. Models are constructed layer by layer. Horizontal overhangs require no supporting structures.
- *Solid Ground Curing*—Incremental layers of liquid photopolymer that are covered by a photomask are cured with a 2-kW ultraviolet lamp. No support structure or post curing is necessary.
- *Direct Shell Production Casting*—Ceramic shell, complete with integral cores, is made automatically and directly from a CAD file without tooling or patterns. This process creates the equivalent of ceramic shells made for the investment casting process. There is no need for tooling or setup.

Rapid prototyping greatly reduces prototyping costs and design time, and the ability to achieve part geometry in one operation that would otherwise require multiple steps to produce. These processes allow participants in simultaneous engineering projects to evaluate designs at an early stage. In a matter of hours, changes can be made to correct design errors, improve manufacturability, cut costs and optimize performance. The patternmaker's skill and knowledge of the many metalcasting processes is a valuable tool in using these rapid prototyping methods.

COMPUTER-AIDED DESIGN

Today's patternmaker should be well-trained in Computer Aided Design (CAD) and Computer Aided Manufacturing (CAM). It is not uncommon today to use computers to design the casting and go directly from the computer to the automated machine tool that will produce the pattern and coreboxes. If the patternmaker is going to design the gating and risering systems, he/she should also be able to use the solidification modeling and fluid-flow computer programs.

The integrated patternmaking approach begins with a geometric database of the part to be cast. The database may be constructed from drawings. The pattern designer adds a parting line, draft and machining allowances at the CAD workstation. The part dimensions can be automatically scaled up to the designed pattern dimensions that included the shrinkage allowance by inputting the appropriate scale factor. With the parting line and draft added by the pattern designer, the appropriate gates and risers can then be added by using one of a number of software packages that are available. A software pattern geometry database that can be used as-is or modified will assist the pattern designer **(Fig. 3-14).**

Solidification modeling software can be used to model the solidification of the final resultant casting from the geometry data before prototype patterns are made. Pattern dimensions, gating and risering design can all be evaluated and modified by adjusting the pattern geometry database.

Numerical control (NC) machining of patterns to very close tolerance can be readily performed using the pattern geometry database. The NC part programmer can retrieve the pattern-geometry model and use that model to construct the appropriate cutter paths for pattern machining. Changes can be made to the pattern database, and the NC output will modify the appropriate pattern dimension. Numerical control machining can be used to machine

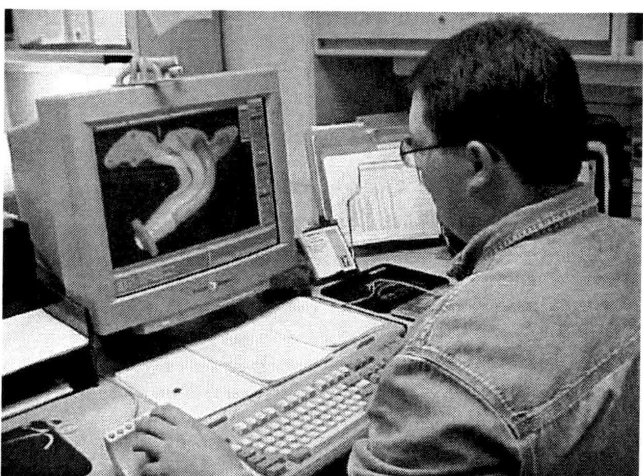

Fig. 3-14. Pattern designer using computer-aided design application.

master-pattern equipment from which production patterns and coreboxes are made. The verification of pattern and casting dimensions can be performed using an automated coordinate-measuring machine (CMM).

BIBLIOGRAPHY

Cast Metals Handbook, fourth edition, American Foundrymen's Society, Inc., Des Plaines, IL, 1957.

Heine, R., C.R. Loper, Jr., P. Rosenthal, *Principles of Metal Casting,* McGraw-Hill, New York, NY, 1967.

Marek, C.T., *Fundamentals in the Design and Production of Castings,* John Wiley, New York, NY, 1950.

Metals Handbook, ninth edition, Vol 15, ASM International, 1988.

"Pattern Making Allowance-Distortion," *Modern Casting,* April, July, August, 1978.

Patternmakers' Manual, second edition, American Foundrymen's Society, Inc., Des Plaines, IL, 1950.

Patternmaking Guide, second edition, American Foundrymen's Society, Inc., Des Plaines, IL, 1990.

Rogers, H.S., "Laminated Plastic Equipment," *AFS Transactions,* Vol 75, pp 292-293, 1967.

Sicha, W.E., "Aluminum Pattern Castings," *AFS Transactions,* Vol 69, pp 479-482, 1961.

"The Metal Casting Process," *Modern Casting,* January through December, 1989.

Weiser, P.F., *Steel Castings Handbook,* fifth edition, Steel Founders Society of America, 1980.

Molding Sands 4

The first part of this chapter will cover types of sand, the physical characteristics of sand and things to consider when selecting sand for moldmaking. The balance of the chapter will cover the aspects and testing of *clay bonded* sand mixtures, which are used in green sand systems.

There are sand molding processes commonly used in foundries where the sand grains are *chemically* bonded together. These processes include shell molding and nobake molding. Since these processes are also used for making cores, they will discussed in Chapter 6, Coremaking.

As molding sands are discussed, the reader should keep in mind that much of what will be said in this chapter concerning raw sand will also apply to coremaking. Sand can be defined as granular particles resulting from the disintegration or crushing of rocks. The following AFS definition of sand illustrates the degree of variance that can be obtained when ordering silica sand: *Sand is mineral matter, irrespective of chemical composition, from 2.0–0.05 mm; 1/12–1/500 in.; 10–250 mesh.*

Sand also denotes a class of several minerals, rather than just one mineral such as silica or quartz. Zircon, olivine, chromite and ground ceramic minerals are also classified as sand when they are in this size range and used in the foundry.

TYPES OF SAND

The types of sand discussed in this section will be as follows:

- Silica
- Zircon
- Olivine
- Chromite

Silica Sand

Silica sand is essentially SiO_2 (silicon dioxide) and is found in sand dunes, river deposits, lakes and other large bodies of water. In many cases, underground streams that are no longer in existence have left large deposits of silica sand. However, where the sand is mined can change the chemical composition of the silica sand. Silica sand is used more extensively in foundries than any other sand substance—the two main reasons being, 1) it is very plentiful in the United States and 2) it is relatively low in cost.

When first introduced into the mold as silicon dioxide, SiO_2, it is in the alpha-quartz phase and goes through a transformation to beta-quartz when heated by the molten metal. The beta-quartz reverts to the alpha phase upon cooling. The major effect on the sand grains during this cycle is one of expansion. Casting defects, such as mold wall movement, buckles, scabs and rattails, can occur if this expansion is not accommodated. Measurable expansion in silica sand is in the order of 0.018 in. per in. at 2000F (1093C). The refractory quality of silica sand is generally considered good. The melting point is approximately 3250F (1787C), which is normally above the pouring temperature of most alloys poured in the foundry industry. **Tables 4-1 and 4-2** contain information on chemical composition, melting point and physical properties of sand.

Thermal conductivity of silica sand is satisfactory with a high thermal stability. However, silica sand is also a good insulator in that, once it reaches its specific heat content it has a tendency to retain the heat. Silica sand is wetted easily by molten metals and fluxes with iron at approximately 2240F (1226C). Wetted means that the sand surface of the mold will be covered with the liquid or molten metal. Use of mold washes and facing can help reduce the wetting tendency.

The silica level of the sand has a direct bearing on the sand quality and the resulting castings. Too high a silica level limits the ability to dissipate heat; on the other hand, too low a silica level causes the sand to lose its refractoriness. In either case, sand-related casting defects may occur.

Silica sand can be further classified as bank sand, lake sand and *pure* silica sand (pure SiO_2).

Bank sands are those found in dried-up riverbeds or from the bed or banks of active rivers. Lake sands, on the other hand, are mined in the dune areas around the Great Lakes in the Midwestern United States—Lake Michigan in particular. When these two sands (bank and lake) are chemically analyzed, it is found that the sand grains contain entrapped materials and thus are *not pure* silica sands. These materials are iron oxide, rootile (roots from trees), calcium carbonate and others. These foreign materials can cause problems in making good molds and cores, especially when chemical binders are used, such as in nobake molding and coremaking. Silica sand on the other hand is considered *pure* silica and is mined from the St. Peter deposits found in Illinois and Missouri.

Table 4-1.
Chemical Composition (%) and Melting Point of Sand

Chemical	Silica	Olivine	Chromite	Zircon
SiO_2	98.82	41.2	1.34	33.50
MgO	0.031	49.4	8.75	—
Cr_2O_3	—	—	45.80	—
ZrO_2	—	—	—	65.00
Al_2O_3	0.049	1.8	21.34	1.00
Fe_2O_3	0.019	7.1	19.50	0.03
CaO	0.0016	0.2	0.94	—
TiO_2	0.012	—	0.03	0.19
Melt Point F(C)	3110(1710)	3400(1875)	3800(2093)	4600(2538)

Table 4-2.
Physical Properties of Sand

Property	Silica	Olivine	Chromite	Zircon
Color	White/Brown	Green	Black	White
Spec. Gravity	2.65–2.67	3.27–3.37	4.3–4.5	4.6–4.7
Bulk Density	95–97	98–103	156–165	152–183
Thermal Expan.	0.018	0.0083	0.0045	0.0037
Temp. Reaction	Acidic	Basic	Basic/Neut.	Slightly Acidic
Shape	Varied	Angular	Angular	Rounded

These three silica sands can be further classified by their grain shape: rounded, subangular, angular and compound **(Fig. 4-1).** Silica sand is said to be rounded and the other two, lake and bank, are subangular to angular. The shape of the sand grains has an effect on the amount of binder or bonding material required to hold the sand grains together, the ability to compact the sand grains together and the permeability of the mold. Permeability is a measure of the mold to allow gases to pass through it. Rounded sand grains will require less binder than subangular with angular requiring the most. Also the ability to compact the sand grains together is either helped or hindered by the grain shape. Rounded sand grains compact very well, whereas angular grains are difficult to compact. The properties of subangular sands fall between the other two shapes.

Subangular and angular sand grains, when compacted to the same density, will offer better mold permeability than rounded sand grains. Subangular and angular sand grains are better equipped to accommodate the sand grain expansion that will take place when the sand is heated by the molten metal.

The surface area of each sand grain also will determine how much binder material will be needed to hold the sand grains together. Rounded sand grains will require the least amount and angular the most amount of binder. Subangular, again, falls somewhere between the rounded and angular.

It should be pointed out here that both angular and subangular sand grains are constantly undergoing a change in shape as they are reused. In the case of angular sand grains, if they are not transported with care, the sharp projection on the grains can be broken off and produce "fines," which can cause problems. Fines can be compared to dust or very, very fine particles. Even the abrasion of angular sand grains together in the mixing and mulling processes can cause these fines to be generated. Subangular and angular sands have small "gullies" on their surface which will fill with bonding materials and, after repeated rebonding, can become more rounded in shape.

Compound sand grains are agglomerates of small grains bonded to form a large grain. This bonding together of the grains is normally not strong enough to prevent the compound grains from breaking apart during normal foundry processes. What results are excessive fines in the sand system that can lead to sand-related casting defects.

There are several other types of sand used in the foundry. These sands are sometimes called "specialty sands," and are zircon, olivine and chromite sand.

Zircon Sand

Zircon sand is known as zirconium silicate or $ZrSiO_4$. It is found in Australia, Florida and California. Zircon sand has a lower thermal expansion rate than that of silica sand. It has high refractoriness, high specific heat content, low thermal expansion rate and resistance to wetting, thus making it desirable as a molding or mold facing material. The zircon sand grain shape is elliptical; however, based on the grain shapes previously discussed, it will be labeled as rounded grains **(see Fig. 4-1).**

The low expansion and high refractory value permit use of fine grains, thereby producing smoother surfaces. It is not easily wetted by molten metal, and it can help reduce some sand-related casting defects. Because of its high cost, it can be best used as a mold facing sand. Its useful properties can be appreciated with savings in the cleaning room, especially in areas of the mold where there will be large amounts of heat due to the design of the casting. Zircon sand is also used in mold and core washes and coatings.

Olivine Sand

Olivine sand is an ortho-silicate of magnesium and iron, occurring in nature as forsterite and fayalite. Only the sand with 90% forsterite is considered useful as a foundry sand. Olivine is found in large masses that must be crushed to produce sand grains that are mostly angular in shape **(see Fig. 4-1).** Olivine exhibits somewhat lower thermal expansion characteristics than does silica.

The melting point of olivine is approximately 3400F (1871C), slightly higher than silica. The real value of olivine comes from a thermal behavior unlike that of silica. It exhibits low thermal expansion and low contraction, and thus has good stability when compared to silica sand. The angular grain shape makes it harder to produce good rammed densities but can also provide for a better control of rammed density. It is more expensive than silica because it is less abundant. Olivine sand is used primarily in the nonferrous segment of the foundry industry; however, it can also be used in cast iron and steel foundries. For instance, manganese steel foundries use olivine sand to overcome the unfavorable chemical reaction that takes place if this alloy is poured into molds using silica sand.

Original Magnification 20X
Angular Sand Grains

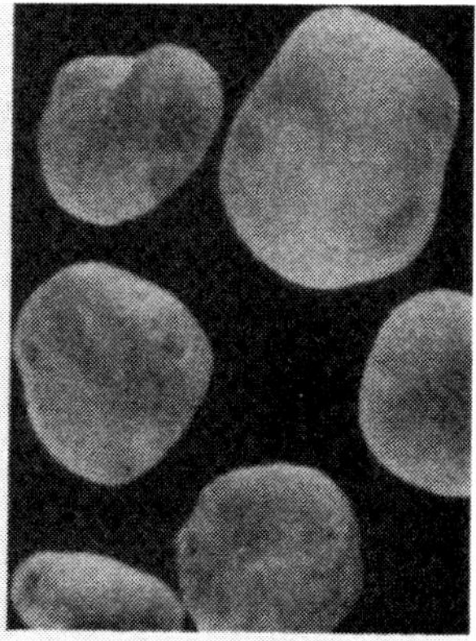
Original Magnification 20X
Rounded Sand Grains

Original Magnification 20X
Sub-angular Sand Grains

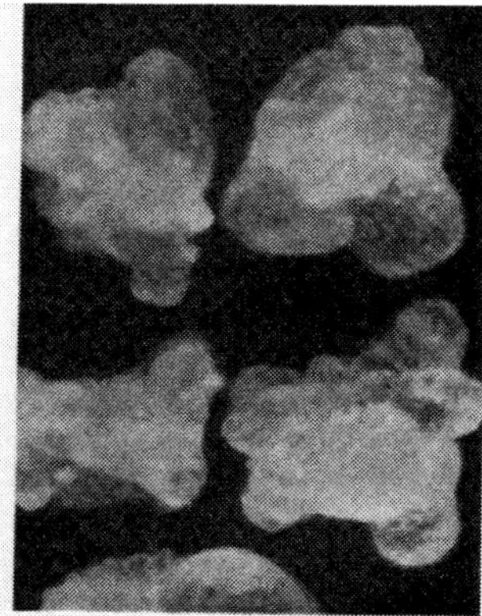
Original Magnification 40X
Compound Sand Grains

Fig. 4-1. Four shapes of sand grains.

Chromite Sand

Chromite sand is mined in Africa; thus, most of its high cost comes from the shipping charges. It is an angular sand **(see Fig. 4-1),** which exhibits good chilling characteristics, a higher refractoriness than silica sand, and less thermal expansion than silica sand. As molding sand, it is used mainly in steel foundries. When ground into finer material, it is also used in mold and core coatings. As with zircon sand, chromite sand can be used in parts of molds or cores where extreme amounts of heat may be present. This is especially true in very large castings in that there is the opportunity to do some hand molding or coremaking.

PHYSICAL CHARACTERISTICS OF SAND

There are some differences between the physical characteristics of silica sand and the specialty sands (zircon, olivine and chromite).

As mentioned earlier, all of the specialty sands have less thermal expansion than does silica sand. Silica sand has a thermal expansion rate of 0.018 in./in. (0.457 mm/mm); olivine sand's thermal expansion is 0.008 in./in. (0.203 mm/mm); chromite sand's thermal expansion is 0.004 in./in. (0.101 mm/mm) and finally zircon sand's thermal expansion is 0.003 in./in. (0.076 mm/mm). This comparison of thermal expansion characteristics is shown in **Fig. 4-2.** Thus, if the foundry should be having major problems with sand expansion-type defects, the specialty sands, among other things, may be able to help solve the problem.

The three specialty sands also have a higher specific gravity or density and volumetric specific heat content than silica sand. These physical characteristics are what help them become a chilling material. That is, they can absorb, hold and transfer more heat than silica sand. **Table 4-2** lists the physical properties just discussed.

When specialty sands are used, precautions should be taken when preparing the mix formulations for molding sand. The bulk densities of the specialty sands are greater than that of silica sand. What this means is, that for a given weight there will be less sand grains to be coated with the binder in the specialty sands than in the silica sand. This can also come into play when specialty sands are used in parts of a mold or core and end up blending into the silica molding sand system at the shakeout. For instance, if we took one cubic foot of silica sand, which may weigh 98 lb (44.5 kg) and compared it to the same volume of olivine, it would weigh 104 lb (47.2 kg), zircon 170 lb (77.1 kg), and chromite 160 lb (72.6 kg).

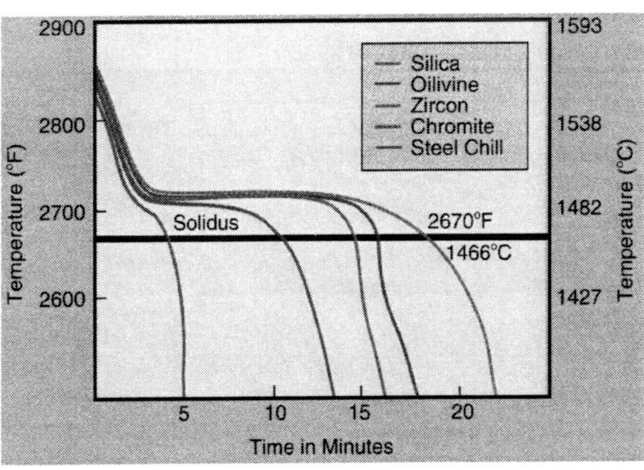

Fig. 4-2. Comparison of thermal expansion characteristics.

SELECTING SAND

The main ingredient found in the formulation for making sand molds or cores is sand! It is also very evident then that selection of the proper sand has a large effect on the quality of the metal castings to be produced. **Table 4-3** lists guidelines that may be used to help select the right sand for the job to be done.

First, consider the alloy to be poured and the size of the casting(s) to be produced. The main thing to be concerned about is the alloy's pouring temperature. Another item to consider is whether there is any thing in the chemical composition of the alloy that may react with the sand selected. The weight of the metal casting to be made is also an important consideration, along with the number of metal castings per mold or, as they say in the foundry, "mold weight." In this regard, the term "metal-to-sand ratio" comes into play. This term refers to the weight of all the molten metal in the mold, as compared to the weight of all the sand in the mold, not including the cores. The higher this ratio the more punishment the sand will have to withstand.

The next three items—grain fineness, particle size distribution and grain shape—have to do with the sand grains themselves. First, the AFS grain fineness number (AFS gfn) refers to how fine or coarse (small or large) the average sand grain is in a given sand sample. Remember, this is an average particle size and sometimes can be deceiving. A finer sand will generally provide a smoother surface finish, but if too fine, it will inhibit the passing of gases generated during the casting process and can lead to defects in the metal casting. The higher the AFS gfn, the finer (smaller) the average grain particle will be. On the other hand, a low AFS gfn will produce a rougher casting surface and will have larger interstices between the sand grains. Foundry grades run from 25 to 170 AFS grain fineness number.

The AFS gfn is important. However, the particle size distribution is also just as important to keep in mind when selecting the sand to be used. Particle size distribution is also called screen distribution and can have a significant affect on the mold and casting. A blend of marbles and dust may give an acceptable AFS gfn, but it is easy to imagine what the casting surface would look like and the other casting defects that may result. The sand used for sand molding should at least have a three-screen distribution and can range as high as a five-screen distribution. This will be discussed later during the sand sieve test, which is run to determine the AFS gfn and screen distribution and other foundry sand tests.

Table 4-3.
Sand Selection Checklist

- Alloy type
- Melting temperature
- Metal:sand ratio
- AFS gfn
- Particle size distribution
- Grain shape (structure)
- Sand type (silica, olivine, chromite or zircon)
- Chemical composition (affects fusion point, pH and ADV)
- Cleanliness (AFS clay content)
- Transportation cost
- Sand cost
- Supplier technical support
- Dependable delivery
- Adequate reserves
- Consistency of product

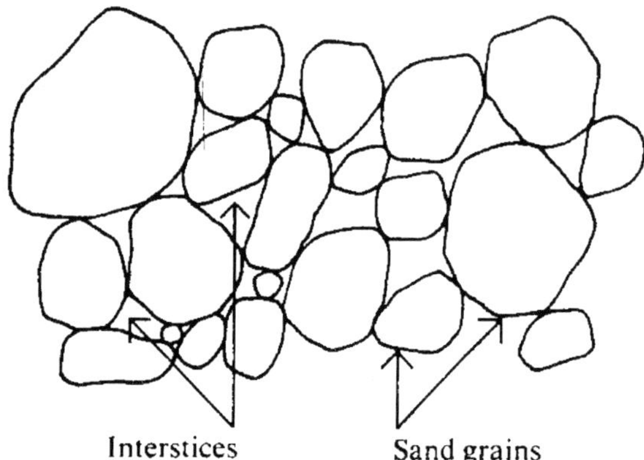

Fig. 4-3. Schematic illustration of sand grains as they may appear under magnification.

Grain shape is determined by viewing a sand sample under a microscope. Earlier, the pros and cons of the grain shapes available were discussed. It is safe to say that the higher pouring temperature alloys and larger castings poured from these alloys will require an angular shaped sand grain. This does not mean that subangular and rounded sand grains can't be used with adjustments made to the molding sand formulation. However, lower pouring temperature alloys can be poured in molds using subangular sands and in some cases even rounded sand grains. It should be pointed out, here, that some steel and nonferrous foundries do use olivine sand for their molds and, as is known, olivine is angular sand.

Selecting the sand type to be used is, again, determined primarily by understanding any special needs that will have to be met by the mold as the molten metal is poured into it and then solidifies. The question is, which sand should be selected... silica or one of the specialty sands.

The chemical composition of the sand comes into play more in the area of nobake molding and coremaking than in green sand molding. Things to keep in mind are the fusion point, pH and acid demand value (ADV). Fusion point would have to be considered with the higher pouring temperature alloys than the lower pouring temperature alloys. This would be true for both green sand molds and nobake molds and cores. The pH and ADV come into play more importantly in the nobake processes. In the case of the specialty sands, the purity of the sand affects its ability to provide the desired properties for which it is used.

When ordering foundry sand it is always a good idea to order washed and dried sand. The cleanliness of the sand is a very important consideration, as it can affect the ability of the sand to be properly coated with the bonding material. There may also be other organic materials attached to the sand grains, which could cause problems. The sand should also be dried after washing to get rid of the excess water, which may still be attached to the sand grains. Also, water will be added in measured amounts to the sand when it is being mulled and any excess water that is unaccounted for can cause severe problems in making good green sand molds.

The remaining items listed in **Table 4-3** are self-explanatory. One should choose a foundry sand supplier that consistently meets the foundry's needs for quality sand. Of course, the foundry should provide the sand supplier with the technical specifications of the sand that detail acceptable or unacceptable sand characteristics. It is also a good practice to test the incoming sand to see if it meets the required specifications. It is advantageous to form a good working relationship with the foundry sand supplier.

The compactness of the sand grains affects the permeability of the mold because of its effect on the interstitial structure. The moisture content in the molding sand also affects the permeability, because excess moisture tends to collect in the interstices. A schematic illustration **(Fig. 4-3)** shows how this may appear when looked at under low-magnification. The bond content of the molding sand also affects the permeability in a similar manner.

Before leaving the topic of selecting foundry sands, "natural-bonded sands" should be briefly discussed. The earliest iron castings produced in the United States were cast in natural-bonded sand. This type of molding sand is mined with the sand grains naturally coated with clay. Natural-bonded sands can have anywhere from 8–50% of weak, naturally occurring clay attached to the grains. When water is added, it becomes the bonding agent, or glue, that holds the sand grains together when they are compacted. Other additives can be added to this molding sand to help it meet certain requirements. The deposits of good natural-bonded sands are not as plentiful as they once were; thus, it is used in very few foundries. Today, nonferrous foundries are the primary users of this molding sand. Often, it is used in school foundry programs, since it is rather easy to maintain.

GREEN SAND MOLD MIXTURES

The largest tonnage or total weight of metalcastings produced in the United States are made using the green sand molding process. For that reason, some time will be spent discussing this molding process. The question usually asked by the non-foundry community is, "Why is it called green sand when it isn't green?" The answer is that the molding sand is bonded with a mixture of clay and water, and the water is not driven out or fired, before its use. This is very similar to the greenware in ceramics, where the term *green* means that the ceramic has not been fired or dried in a kiln.

As was discussed, natural-bonded sand can be called green sand. However, as the foundry industry progressed, more and more demands were being made of the green sand molds from a physical and production standpoint. Then, along came what is today called "synthetic sand." Synthetic sand is a mixture of "raw" sand, water, clay and other materials. The following sections will take a closer look at the materials that make up synthetic sand or green sand.

Of course, the most abundant material found in green sand molds is sand. These sand grains have to be glued or bonded together so that the molds can be handled and molten metal can be poured into them, without damage. The main materials used for this bonding purpose are clay (bond) and water. Other materials (additives) will be added to this mixture to help the green sand mold meet the many physical demands that will be placed on it during pouring of the molten metal.

Green Sand Mold Components

In foundry terminology, clay is considered the bond and sand is the aggregate or refractory. The other materials added to the green sand for molding purposes are called additives. Among these are water, cereal, cellulose, iron oxide, carbonaceous materials, polymers and chemical additives.

Clay

Since clay is the next most abundant material (percentage-wise) found in the green sand mold, it will be discussed first.

The main two geological classifications of clays used in green sand molds are *montmorillonite* and *kaolinite.* These two classifications can be further broken down into names used more commonly in the foundry. The two types of montmorillonites used in the foundry are *southern (calcium)* and *western (sodium) bentonites*. The reader should be made aware of the fact that today's technical literature concerning these two bentonites refers to them by their chemical designation, that is, sodium and calcium bentonite. Kaolinite is also known as fireclay. **Table 4-4** gives an overview of the clays most commonly used in green sand. These three clays will be discussed further, near the end of this section on Clay.

As a rule, green sand can contain between 5 and 12% clay, based on the weight of the sand. The amount of clay or combination of clays needed to determine the percentage is based on several factors. These factors include:

1. Pouring temperature of the alloy
2. Size and section thickness of the casting(s) to be made in the mold
3. Shakeout characteristics of the mold.

The first two factors were discussed previously in the sand section.

Shakeout characteristics are the ease with which the solidified casting can be removed from the mold. Also included with this would be how well the molding sand breaks down into a size that can be handled easily and reconstituted for reuse in making other molds. However, one essential area that many times is forgotten is the "collapsibility" of the mold. As the casting begins its solid contraction or shrinkage, it gets smaller in size. If, for any reason, this contraction is restricted or hindered, it can cause casting defects such as hot tears, and severe internal stresses can be set up in the casting. In other words, the mold has to begin to collapse as the casting begins its solid contraction. The type and amount of clay used can help or hinder this contraction of the casting as well as help or hinder the collapsibility of the mold.

These three factors also come into play when discussing "green strength" and "hot strength" of a green sand mold. Simply put: green strength is the ability of the mold to stay together until it is filled with molten metal; hot strength is the ability of the mold to hold the shape and contours of its internal cavity until the casting has turned completely solid. The manner in which we withdraw the pattern from the mold, and then further handle the mold, will determine how much green strength the mold will need. The rougher the completed mold is handled the more green strength it will need.

As for the hot strength the mold will need, the higher the pouring temperature, the heavier the casting and casting section thickness, the more hot strength the mold will need. Remember, however, that the mold has to "give" and allow the casting to contract as it cools to the solid state. The ability of the mold to break down at shakeout is also affected by the amount of sodium bentonite in the molding sand. Too much sodium bentonite will cause poor shakeout properties.

Table 4-5 lists the physical properties of the three commonly used clays in green sand molding. Research into the chemical and physical characteristics of these clays has shown that both composition and structure have a great influence on the working properties of clays. The distinctive property of clay is its plasticity, which, in the presence of moisture, is imparted to a mass of sand.

Bonding Mechanism—These clays work to hold the sand grains together in a mold. The bonding forces involved in holding particles of clay together (and thus the sand grains) may be accounted for by several theories; electrostatic bonding, surface tension forces, and interparticle friction bond.

The mechanism of *electrostatic bonding* of clays may be described as a network of dipolar forces operating at the sand-clay and clay-clay interfaces. This network of forces is initiated by the preferential adsorption of positive ions and negative ions on combined water and clay (hydrated) surfaces. **Figure 4-4** shows a micelle (pronounced *my-sel)* or hydrated clay particle.

When water is added to dry clay, the negative ions are adsorbed on the nuclei of the clay atoms and form an integral part of the

Table 4-4.
Clay Minerals Used for Bonding Molding Sands

Clay Type	Composition Type	Base Exchange	Refractoriness (Softening Point)	Swelling Due to Water	Shrinkage Due to Loss of Water	Particle Size and Shape
Montmorillonite Class IA, Western Bentonite (Wyo., S.Dak., Utah)	$(OH)_4Al_4Si_8O_{20}{\cdot}nH_2O$ (90% montmorillonite 10% quartz, feldspar mica, etc.)	High. Na is adsorbed ion. pH = 8–10	2100–2450F 1148–1342C	Very high, gel-forming.	Very high.	Flake size less than 0.00001 inch
Montmorillonite Class IB, Southern Bentonite (Mississippi)	$(OH)_4Al_4Si_8O_{20}{\cdot}nH_2O$ (85% montmorillonite, 15% quarts, limonite, etc.)	High. Ca is adsorbed ion. pH = 4–6.50	1800F + 982C +	Slight, little tendency to gel.	Very high.	Flake size less than 0.00001 inch
Kaolinite Class IV, Fireclay (Ill., Ohio)	$(OH)_8Al_4Si_4O_{10}$ (60% kaolinite, 30% illite, 10% quartz, etc.)	Very low.	3000–3100F 1647–1703C	Very low, non-gel-forming.	Low.	Fire clays often ground; therefore, may be relatively coarse, or ground to a flour.

Table 4-5.
Physical Properties of Clay Types

Property	Sodium Bentonite	Calcium Bentonite	Fireclay
% Clay	5.0	5.0	12.0
% Moisture	2.5	2.5	3.0
Green compression, psi	11.2	12.4	10.2
Dry compression, psi	101	69	71.5
Hot compression (psi) at:			
1000F (537.7C)	110	55	75
1500F (815.5C)	210	103	145
1850F (1010C)	520	150	170
2000F (1093.3C)	345	72	510
2500F (1371.1C)	8	4	27

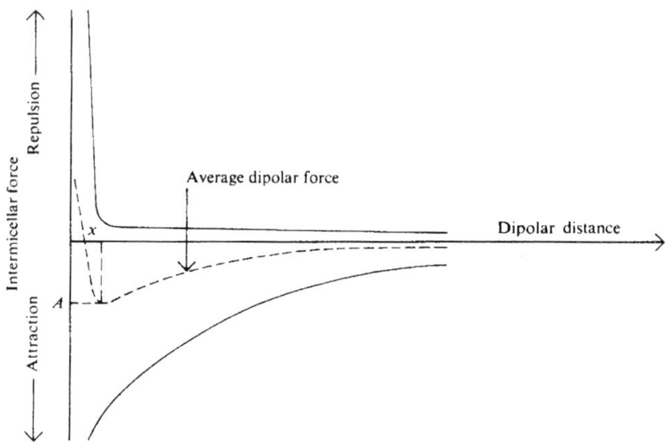

Fig. 4-5. Forces of attraction and repulsion as a function of dipolar distance.

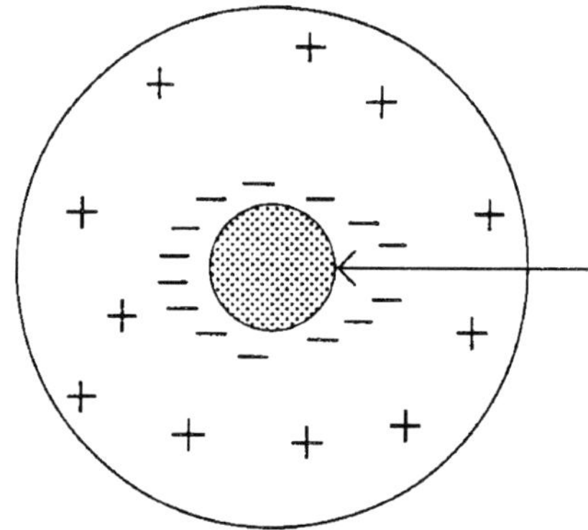

Fig. 4-4. Hydrated clay particle (micelle). Arrow points to the nucleus of an atom in a clay molocule.

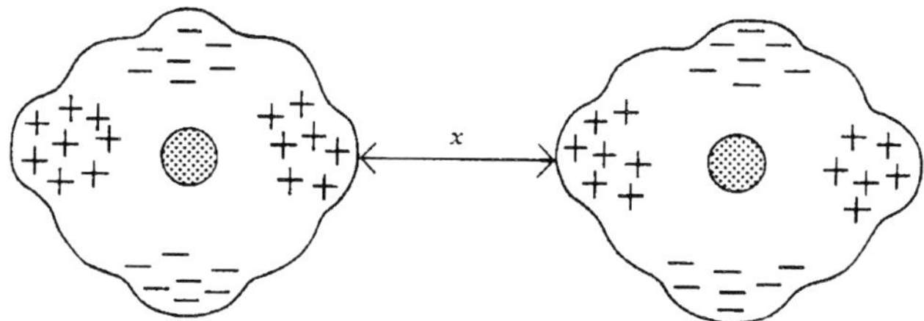

Fig. 4-6. Micellular dipoles.

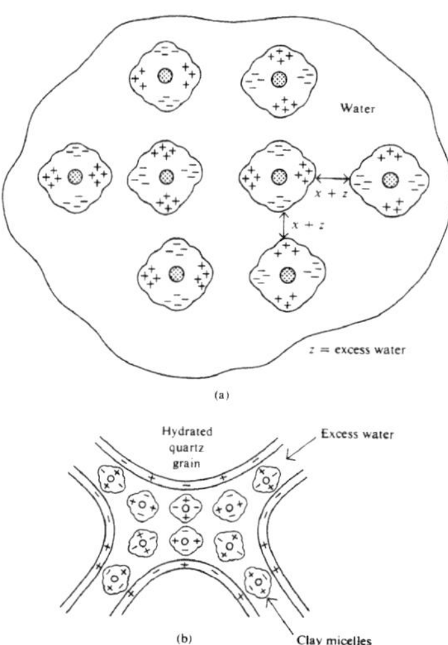

Fig. 4-7. Diopole alignment of hydrated clay particles (a). Schematic of disposition of clay and quartz dipoles (b). In green sand, the intermicellular voids are filled with water.

crystal. The positive ions are attracted by the negative ions, but repelled by the nuclei of the clay atoms, with the result that the positive ions take up equilibrium positions. The water forms neutralized clay micelles whose kinetic energy caused them to move toward one another. There is then a force of attraction between positive ions themselves and the nuclei of the clay particle **(Fig. 4-5).** As the distance between the clay micelles increases, the force of attraction increases and that of repulsion decreases, resulting in a net intermicellular force.

The drawing together of two micelles causes the orientation of unlike ions, forming a clay dipole **(Fig. 4-6),** and a maximum attractive force is at an optimum distance of separation, x. There are many such dipoles in a clay-water medium. Depending on the type of clay, a maximum degree of hydration is necessary to develop a dipole completely. An important point to remember is that the strengths of clay-bonded sands increase with increasing amounts of water, up to a maximum value.

As the amount of water is further increased, water enters the spaces between the dipoles to a distance greater than x **(Fig. 4-7),** resulting in a decrease in the net intermicellular force.

Surface tension of the water surrounding the clay and clay-sand particles provides another possible source of bond strength. The interstices of the clay particles are filled with water. The surface layers of water act on a stretched membrane of hydrated clay, forcing the clay particles together. When the amount of water is reduced by drying, the force holding the particles together increases.

The geometry of the aggregate can provide another force, adding to the strength of the bond between particles. The theory of *interparticle friction,* or the so-called block-and-wedge theory, involves materials under pressure. When molding sand is rammed inside a flask, the particles are jammed against one another. Sand that is rammed to a certain shape and that resists deformation is held together through interparticle friction. A favorable orientation of sand grains causes a packing action, which supports the sand grains within the flask. When various sizes and shapes of sand grains are used, the strength properties of sand mixtures can be changed, which indicates the existence of an interlocking or frictional force.

Bentonites and Fireclay—There are various reasons why bentonites or fireclays are used in green sand molds.

Sodium or western bentonite clay provides a very high hot strength and is used mainly in ferrous alloy castings where the metal temperatures and pressures require this higher hot strength in the mold. In some cases, nonferrous foundries may also use some sodium bentonite, especially when producing large, heavy-section castings.

Sodium bentonite has the capability to hold large quantities of water and can swell 14–21 times its size when exposed to sufficient moisture. This is one reason it is used to seal cracks in basement walls. This water retention phenomenon aids in reducing sand-related defects that are caused by the evaporation of the moisture from the metal-mold interface and its re-condensation away from this interface.

Sodium bentonite will provide green strength; however, it takes a longer mulling time to generate the same green strength that calcium bentonite can give in a shorter mulling time. The reason for this is that it is harder to get the water into sodium bentonite particles than into calcium bentonite particles. It is not uncommon to have steel foundries using only sodium bentonite clay in their green sand. Most cast iron foundries will use a combination of the two bentonites. Of course, for large heavy-section cast iron castings, the bentonite ratio will favor sodium over calcium bentonite. As mentioned earlier, nonferrous foundries may use some sodium bentonite, in very low amounts when pouring heavy-section, large castings.

Calcium or southern bentonite is similar to sodium bentonite, but due to changes in the chemistry of the clay, it is classified as a "non-swelling" bentonite and thus provides much lower hot strengths in a green sand mold. This lower hot strength provides excellent shakeout properties and slightly higher green strengths than sodium bentonite. It also requires less mulling energy to get the water into the clay particles, and evenly distributed, and develops the clay coating for the sand grains.

Calcium bentonite is the choice of clay for the majority of nonferrous foundries. In most cases, these foundries will use a straight calcium bentonite bonded green sand. However, as mentioned earlier, it can be used in combination with sodium bentonite, if necessary. The calcium-to-sodium bentonite ratio, in this case, would favor calcium bentonite. Calcium and sodium bentonite combinations are used in most cast iron foundries. The ratio of calcium to sodium bentonite is, once again, determined by the pouring temperature, metal-to-mold ratio and section size. Shakeout is also another big factor to be considered.

Fireclay or kaolinite clay, is very different from the bentonite clays. The clay particles are much larger, and it requires a significantly higher amount of fireclay to provide green strength similar to the bentonites. For instance, 12% fireclay is required to provide the same green strength as 5% of either of the two bentonites. However, at the 12% level, fireclay will provide the same hot strengths similar to sodium (western) bentonite. Combinations of fireclay and sodium bentonite provide hot strengths higher than either individual clay. Fireclay would be used in making green sand molds for large, heavy-section ferrous castings.

Table 4-6 shows the different hot strengths that can be obtained by blending clays at the proportions indicated.

Precautions—Clay varieties used in green sand molding should be used with caution. The clay at or close to the metal-mold interface will be "burned," which means that its ability to be used again to bond or glue sand grains together is lost. It has become dead clay and is not reusable. The amount of this dead clay (after shakeout) is dependent on the pouring temperature of the molten metal and the metal-to-mold ratio. The higher the pouring temperature and the higher the metal-to-mold ratio, the more dead clay will be generated. This means that clay additions will have to be made to the system sand or green sand for reuse. A test that is used to determine the amount of "live" clay in the molding sand is briefly mentioned later in this chapter.

Additions of clay to the used sand will require an increase in the moisture to activate these clays. If the water and clay levels are both increased excessively, the molding sand can become heavy and will become difficult to feed through hoppers into molds. This heavy, wet sand will resist compaction and may cause casting defects such as mold wall movement, oversized castings, a rough wire wool appearance on the casting surface, penetration, difficult shakeout and hot tears. If the clay is increased but the water remains at its previous level, there will not be sufficient water to activate all the clay present. This type of dry and friable (easily crumbled) molding sand can result in other casting defects such as washes, cuts, sand inclusions and runouts.

Water

Water is the most abused ingredient. Probably the two essential reasons are that it is the cheapest and requires the least amount of physical effort to introduce it into the green sand system. Water affects almost every physical and mechanical property requiring control, if quality green sand molds and castings are to be made.

Water or moisture in the green system sand is as essential as the clay substance itself. This water exists in two forms in the mold:

1) as free moisture, which can be removed by drying the mold

or

2) as combined or absorbed moisture in the clay particles, which can be removed only by heating the clay to high temperatures.

Water, which is present in amounts of about 1.5 to 8%, activates the clay in the sand and causes the aggregate to develop plasticity

Table 4-6.
Hot Strength (in psi) of Various Clay Blends

Blend	1000F 537.2C	1500F 814.7C	1850F 1004C	2000F 1092.2C	2500F 1369.7C
75%SB/25%CB	70	185	395	227	3
50%SB/50%CB	67	185	280	155	3
25%SB/75%CB	60	110	150	105	2
75%SB/25%FC	215	350	575	545	10
50%SB/50%FC	190	300	535	470	11
25%SB/75%FC	270	350	775	910	24
75%CB/25%FC	76	140	255	210	8
50%CB/50%FC	100	110	265	325	21
25%CB/75%FC	120	155	240	400	18

SB = Sodium Bentonite; CB = Calcium Bentonite; FC = Fireclay

All clays were adjusted to provide a typical proportion for each of the respective clays (i.e., the 50% SB/50%FC mixture would contain 2.5% SB and 6.0 FC for a total clay addition of 8.5%.

and strength. Water in molding sand is often referred to as *tempering water*. This water (up to a limited amount) is absorbed and held rigidly by the clay. Any excess of that which can be absorbed by the clay exists as *free water*. Only the tempering water appears to be effective in developing strength. The rigid clay coatings of the sand grains may be forced together, causing a wedging action and, thus developing strength. Free water can act as a lubricant, making the sand more plastic and moldable, though it may lower the strength. Thus, it is evident that control of the water percentage in molding sand is very important.

Since 1996, several foundries in the United States have been using an advanced oxidation (AO) technique in their green sand systems. The AO clarifier *black water* systems allow foundries to segregate active or live clays and seacoal from undesirable silica fines so that these clays and seacoal can return to the molding sand system while the dead clays and sand fines can be disposed of. The system uses an AO process that generates reactive radicals and highly oxidized species such as ozone, which, when combined with the black water system, can degrade, break down and/or mineralize organics at low temperatures.

What this does is clean the recycled molding sand and allow the mold additives to better adhere to the sand grains. These systems have also effectively removed active clay from the dust collected by air pollution control devices such as wet collectors and baghouses. The active clay is returned to the green sand system via the black water, reducing the amount of new sand and additives that have to be added to the molding sand system. The foundries using this process report that the use of new sand and additives has been appreciably reduced.

In addition to the three basic ingredients of molding sand (sand, clay and water), other materials may be present. They are usually added to improve certain properties, and are often referred to as additives. They are divided into two classes: organic and inorganic additives. The cereals, resins, proteins and oils are organic additives. Cement, silicates and some esters are inorganic additives.

Cereal

Cereals are generally added to green sand to increase the plasticity or deformation of the molding sand. The cereal (corn flour, milo flour, wheat flour or rye flour) makes a paste (like Elmer's glue) with the water present and provides plasticity in the molding sand. This plasticity reduces the brittleness of the molding sand, often allowing the molder or molding machine to draw deep pockets that could not be drawn without the cereal addition. The paste, when allowed to dry, aids in reducing the friable edges of the molds, which can lead to sand inclusions in the castings.

When cereal is exposed to water and then heat, the soluble starch and sugars are carried off by the moisture to the mold's surface. The water can then evaporate, leaving the starch and sugar as a bonding agent on the surface of the mold. This deposition of the starch and sugars can aid in holding the sand grains together, thereby decreasing the chances of sand expansion-related defects on the castings. In other words, in molding systems with hot sand, the addition of cereal can aid in decreasing loss of moisture and reducing sand inclusions related to the dislodging of the sand grains due to moisture loss at the mold's surface.

A cereal additive, as used in the foundry, is finely ground and frequently gelatinized. Cereals are used in the range of 0.25–2.00% to increase the green and dry strength of the green sand mold. Cereals burn out at approximately 500–700F (260-371C). Therefore, they maintain their usefulness during molding, and do not disintegrate until they meet the molten metal.

Precautions—There are some precautions to follow when adding cereal to green sand. Because cereal is an organic material, it will decompose when exposed to the heat of the molten metal and will generate the gases associated with this decomposition. Excessive amounts of cereal in the molding sand can cause gas-related casting defects. Cereal absorbs water very quickly and takes water away from the clay, reducing the bond strength developed from the clay. If a small amount of cereal is placed on a microscope and water is added, the cereal will rapidly absorb the water and expand, similar to popcorn when it is exposed to heat. This quick absorption ties up water that may be needed for the clay's softening and distribution on the sand grain's surface. If excess amounts of cereal are present and additional water is added, the bonding caused by the combination of clay and cereal may result in difficult shakeout and poor collapsibility of the mold, causing hot tears or excessive internal stresses in the castings. Cereals also affect the flowability of the green sand; excessive amounts of cereal can lead to poorly compacted molds and related defects in castings.

Cellulose

Cellulose additives include materials such as wood flour, oat hulls, rice hulls and ground nutshells. Initial additions of cellulose can be anywhere from 1–2% and after that 0.1–0.2% based on the weight of the sand mixture. Cellulose materials tie up excessive water in the green sand and are generally added to enhance the flowability of the molding sand during the fill sequence and to aid in promoting a better breakdown of the molding sand during the shakeout cycle. The particles of cellulose fiber hold the sand grains apart during the compaction phase of making the mold, and are thermally degraded by the casting during pouring and solidification. The reduction in size of the cellulose particles, when exposed to the molten metal, allows the sand grains room to expand and fill the newly created voids. This degradation of the cellulose particles promotes a better shakeout and the removal of the sand from the casting.

Cellulose is a fractured or ground organic product with a very porous or open structure. These products quickly absorb water and give the molding sand greater flowability by inhibiting a portion of the water present from reacting with and being absorbed by the clay platelets. A lower water-to-clay ratio can result in dry, friable molds, since the cellulose materials do not actually bond the sand grains; thus, the porous structure surrenders its moisture through evaporation into the air.

Precautions—All organic additives decompose when exposed to the temperatures present during the pouring of molten metal, and the gases, if not adequately vented out of the mold, may result in gas defects in the castings. Since these materials do not form starch or sugar, they do not contribute to the plasticity of the molding sand and do not form the "glue" described in the section on cereals. Excessive use of cellulose can cause premature shakeout, cuts, washes, sand inclusions, mold wall movement and oversize castings.

Iron Oxide

Iron oxide is not normally added to the green sand system but is generally introduced into the system by the decomposition of cores containing iron oxide. When iron oxide is added directly into the green sand system, it is usually to introduce fines that give a greater degree of rigidity to the mold by increasing the grain-to-grain contact. The lower melting temperature of the iron oxide can give some thermal cushion to the molding sand, reducing expansion-type defects. This thermal expansion benefit is only helpful if the metal being poured has a higher fusion point than the melting temperature of the iron oxide being used.

Iron oxides are a finely ground, inorganic material that will fill the voids between the sand grains, thus reducing the permeability of the molding sand. This reduction in permeability may result in gas-related defects in the castings. The fine particles of iron oxide increase the contact points between the sand grains and can provide higher compression strength. This higher value can lead to a reduction in clay additions, which can lead to erosion-type defects.

Precautions—The fineness of the iron oxide provides more surface area, which will take up water, reducing the amount available for absorption by the clay. The increase in the compression by increasing the contact points between the sand grains, the filling of the voids between the sand grains, and the reduction in water available for the clay creates a brittle molding sand that is difficult to draw from pattern pockets. Also, excessive levels of iron oxide lower the refractoriness of the molding sand and can result in burn-on, with the molding sand fused to the casting surface. This defect is usually first noticed in the ingate areas and in heavy sections of the casting.

Carbonaceous Materials

Foundries pouring cast irons will add carbonaceous material to their green sand. In addition, some copper-base foundries pouring large, heavy-section castings will also add carbonaceous materials to their green sand. The main reason these materials are added to the green sand is to help give the castings a good surface finish. In this regard, the term "peel" is used, which describes the way the molding sand peels away from the casting at shakeout. These materials would not be used in carbon steel foundries because the molten metal may pick up the carbon available and thus change the carbon content of the carbon steel. The carbonaceous materials used in green sand include seacoal, asphalt, Gilsonite, lignite (causticized) and petroleum distillates. All of these materials are primarily composed of carbon.

The most commonly used of these carbonaceous materials is *seacoal.* Seacoal is a finely ground, bituminous coal. It got its name during colonial times, because it was shipped from Great Britain, after mining under the Sea of Wales. Today, this bituminous coal is mined here in the United States and prepared for use as seacoal in the green sand. Seacoal is added to the green sand at a rate of 2% to 8%, based on the weight of the green sand mixture.

There are several theories as to just how seacoal helps to give the cast iron castings a better surface finish. The first is that, as the molten metal enters the mold cavity, the heat radiated to the mold surfaces causes the seacoal to burn. As the seacoal burns it depletes the amount of oxygen in the mold atmosphere, thus causing a reducing atmosphere to form. This hinders the chances of Fe_3O_4 (iron oxide) and other oxides from forming, which can create slag. Second, the gasses generated by the burning seacoal recondense when they contact the sand grains and coat them with lustrous carbon. The lustrous carbon fills the voids between the sand grains, which enhances the casting finish. Third is the concept that the seacoal expands in the reducing atmosphere, and fills the voids between the sand grains, stopping penetration by extruding the carbon in to the casting area, and forming a carbon interface between the mold surface and the molten metal.

The carbon supplements, *asphalt, Gilsonite, lignite and petroleum distillates,* usually contain higher levels of volatile carbons that increase carbon peel. One of the rules of thumb applied to these supplements is: for every 1% of these supplements used in a preblend, 2% of seacoal can be replaced.

When these carbonaceous materials are used, their sulfur content should be monitored, especially if ductile iron is poured into the green sand molds. This is especially true if the carbon supplements are used.

Since these carbonaceous materials are added to produce a reducing atmosphere in the mold cavity, and to produce lustrous carbon, they produce volumes of smoke and gas. For this reason, the green sand molds should be properly vented. Excessive carbon in the mold cavity can displace the molten metal and produce a defect called kish. Kish forms a mottled surface on the cope surface of cast iron castings. The castings may also have a bluish tint to them, especially in thin sections.

Precautions—Excessive quantities of asphalt or Gilsonite can waterproof the sand and clay particles, inhibiting the development of the sand/clay bond. In addition, the introduction of excessive amounts of fine seacoal and the other carbon additives may fill the voids between the sand grains, lowering the mold's permeability while increasing the gases that must be vented.

Polymers and Chemical Additives

Polymers are rather new additives used in green sand molding. These additives are used to lower the surface tension of the water to make the clay platelets easier to wet and soften. The introduction of hydrocarbons and residual core sands into the return sand system often produces sand and clay particles that do not readily absorb water. These polymers are usually used in foundries that have problems with hot return sand, as well as with insufficient mulling or mixing and sand cooling capacities.

Additions of chemicals, such as soda ash, are made to increase or control the pH of the molding sand. The excessive use of cereals or the introduction of residual acids from acid-catalyzed chemically bonded cores can lower the pH of a green sand molding system, thus depressing the performance of the bentonite. The clays like to work in an eight to nine pH level. The basic categories of polymers and chemicals are; soda ash, wetting agents, chemical modifiers and organic polymers.

Since these additives are introduced to activate clay, which is usually latent or dormant, the first indications of their over-use will be an increase in the muller amperage, an increase in the green compressive strength, and a stiffness and compaction resistance in the molding sand. The normal response is to reduce the clay level and bring the system back to its typical performance levels. It is up to the individual foundry to determine if this is the correct response.

The increased strength and mulling efficiency is the result of the use of clay that is present, but has not been wetted and developed as bonding clay. This latent clay provides a cushion for normal variations, like temperature or moisture level, in the return molding sand.

Precautions—It is important to stress the need to keep the system as simple as possible. In other words, do not put something into the sand system without first studying the effect it will have on the performance of the green sand molding system. Next, if something new is put into the system, can anything be taken out of the system? Remember, the more material there is in the green sand molding system the more there is to be controlled, and the more chances for something to go wrong.

Preblends

The term preblend is used when the clay and additives used in the green sand molding system are blended together before adding to the system. In other words, the clay and additives are not added separately. In this case, the foundry notifies the foundry supplier what percentage of each ingredient is required in the preblend and the foundry supplier prepares it accordingly. Preblending does not include sand.

GREEN SAND PREPARATION

Now it is time to put all of the ingredients together to make green sand molds. The preparation of green sand can be done in two different ways: one is called "mulling" and the other "mixing."

Mulling

Mulling can be defined as the application of work forces to cause kneading, smearing, compression and shear. This action disperses the water and clay additions throughout the sand; and converting the water and clay into clay/water bonding entities, which are uniformly distributed over the sand grain surface. Mullers are machines in which plows use wheel faces to blend compressed and uncompressed sand for recompression. There are two types of mullers: vertical wheel and horizontal wheel. Another name for the horizontal wheel muller is "speed muller." **Figure 4-8** shows a typical slow-speed muller.

When preparing a new batch of green sand, there is a preferred procedure to follow. First, all of the sand and water are added to the muller. These two ingredients are mixed until the water is evenly distributed over the sand grains. Second, the clay and additives are added to the muller. Today, many foundries order their clay(s) and additives preblended, so that the proper amount of each ingredient is added to the muller. The clay is attracted to the wet sand grains, absorbs the water and becomes a putty-like clay glue. The additives, such as seacoal and cereal, become imbedded in the clay glue. The compressive forces exerted by the mulling wheels cause the sliding of the sand, which, in turn causes shearing, spreading and, ultimately, coating of the grains by the clay glue.

This all sounds very simple and easy to do; however, it is a Herculean task to ask these machines to do it in a few minutes and seconds. For instance, one pound of typical molding sand can vary from five million to 15 million sand grains. Alternatively, a mulling machine having a two-ton capacity is expected to properly distribute the water and clay among 40 billion sand grains. It cannot be expected that all sand grains will receive identical treatment or that perfect sand grain coatings will be realized. A specific mulling machine must be equated to a specific procedure so that the given mulling time results in the input of the needed mulling energy.

The next step is to calculate the time and energy required to develop the clay into a glue. Imagine that the bentonite particles are in the form of a deck of cards, with the water being absorbed between the cards or platelets. As the water enters the spaces between the platelets, the clay is softened and can then be distributed on the sand's surface. Since the clay platelets are small (sodium bentonite: 0.2 microns, calcium bentonite: 0.5 microns and fireclay: 20 microns) the small openings between the platelets require a great amount of time to fully absorb the water and to soften them for distribution over the sand grain's surface. (Just as a reference, a human hair is 80–100 microns in diameter.)

Regarding mulling, there are two terms used when discussing mulling procedures: mulling cycle and mulling time. These are two separate terms and should not be considered the same thing. In batch mulling, cycle time means from the time the sand (new or returned) is placed into the muller until the batch is discharged. Mulling time, on the other hand, is the time from when the clay, water and additives are placed in the muller until some period just before discharge. The manufacturer of the muller can give you the specific recommendations for the correct cycle and mulling times.

When mulling "return" sands (green sand returned to the system after shakeout) a new problem rears its ugly head, and that is hot sand. This is especially true in high-production foundries pouring ferrous alloys and copper-base alloys. A rule of thumb is that any return sand over 120F (49C) is to be considered "hot sand." This sand should be cooled before it is placed in the muller. There are varieties of sand coolers available for cooling the hot sand before it

Fig. 4-8. Typical slow-speed muller for mixing molding sand.

reaches the muller. Many times, the foundry will use the muller to cool the hot sand and to try to keep up with production demands and reduce the mulling time. This can lead to serious problems. The proper place to begin cooling hot sand is at the shakeout.

To meet increased production demands, continuous mullers have become very popular in high-production foundries. A continuous muller consists of two batch-type mullers that are connected together. The sand to be mulled enters at one end of the machine and discharges out the other end. In the case of continuous mulling, some sand grains go through the muller faster than others, due to sand grain movement. One advantage of continuous mulling is that there is less temperature variation in the prepared green sand.

Many times, the question asked is: "What is sufficient mulling?" Simply put: Sufficient mulling is the least amount of mulling resulting in the satisfactory quantity and quality of castings.

Mixing

Another way to prepare a batch of molding sand is by mixing. There are two types of mixers used to produce green sand: high-intensity and vacuum cooling. These machines coat the sand with the bond in a different way than do the mullers.

The first of these two machines is called a high-intensity mixer. This machine has either one or two high-speed turbine agitators, depending on their size. The inside pan of the mixer also rotates at a high rate of speed. The agitators cause the sand to rotate in one direction while the pan rotates it in another direction. This action not only mixes the sand, clay, water and additives but also causes these materials to bond to the sand grains by impacting the sand grains with the bonding material. **Figure 4-9** is an illustration of the mixing action taking place inside a high-intensity mixer. These mixers can be used for either batch or continuous preparation of green sand.

Also available is a vacuum cooling system, which can be used to cool the green sand as it is being mixed. This system works on the principle that when a liquid, in this case water, is exposed to a pressure lower than its vapor pressure, it begins to boil. Part of the liquid vaporizes whereas the remaining liquid cools down. The cooling effect occurs because the required energy for vaporization is taken from its surroundings, in this case, the green sand. This process, known as flash cooling, continues until the temperature corresponds with the vapor pressure or the pressure inside the mixer.

Maintenance of mullers and mixers is a very essential part of properly prepared green sand. Included in this maintenance program should be the water-metering devices. The manufacturer's maintenance recommendations should be followed. Time should be given to perform the necessary maintenance procedures at the proper time. Poor maintenance can only lead to poor green sand molds and, ultimately, poor-quality castings.

Fig. 4-9. Mixing action in a high-intensity mixer. [Courtesy of Eirich]

QUALITY CONTROL OF GREEN SAND COMPONENTS

The place to begin controlling the green sand is at the receiving dock! It is at this point—where the foundry receives the materials to be used in their green sand system—that quality control begins. These materials should be checked to see that they meet the specifications set by the foundry. These specifications should be set based on the performance and properties of the green sand required to produce quality castings.

Some recommended tests to be run on incoming sand are sieve analysis (AFS gfn), grain distribution, LOI (loss-on-ignition) and moisture. Of course, a microscope should be used to check the grain shape and look for foreign material. It should be noted, at this point, that all of the test procedures, and the tests discussed here, can be found in the *AFS Mold & Core Test Handbook, 3rd. Edition.* The clays can be sized by comparing them with the commonly used size of processed clay, being 60–92% passing through a USA sieve 200 (USA sieve No. ASTM E-11). Carbonaceous materials should be checked for volatile combustible matter (VCM), fixed carbon and ash. Sulfur should also be checked, especially if ductile iron is to be poured into the green sand molds. Other additives should also be checked, especially for sizing. In the case of preblended materials, a good working relationship should be built up with the company supplying the material. The supplier should be able to furnish information regarding the particular application.

Another incoming material that bears watching is water. Many times, little thought is given to water. The number, concentration and type of impurities, including mineral matter, organic matter and gases, will vary, especially in different geographical locations. In addition, the pH value of water will depend on types and amounts of impurities.

Green Sand Testing

As part of a good quality control system, the foundry must have a program to test the various properties of their green sand. Once the green sand properties are such that consistent good quality castings are being produced, some means of keeping a check on these properties is required. This requirement is met by a good green sand, or molding sand, testing program. With this program in place, the foundry should be able to produce green sand molds capable of turning out quality castings consistently. The people assigned to conduct these tests should be well trained in the proper procedure to run the tests and have a working knowledge of green sand molding.

Remember, this section is about green sand systems. There are also tests for chemically bonded sands in Chapter 6, Coremaking.

The basic tests commonly used as controls for green sand systems include sand temperature, moisture, compactability, density and specimen weight, permeability, green strength, AFS/25 micron clay, AFS grain fineness, methylene blue clay and loss-on-ignition (LOI). These tests relate to the control of the four basic variables in green sand: water, bonding agent, sand and carbonaceous materials. Other additives may be added to this list.

These basic tests can be broken down into two classifications: daily tests and weekly tests. **Table 4-7** shows the breakdown of the tests, as to whether they are to be run daily or weekly. It should be noted here that there is not total agreement between green sand "experts" when it comes to how often some of these tests are to be run. The sand technician in the foundry may find it necessary to run certain tests more or less frequently than indicated in **Table 4-7** to produce quality green sand molds.

Sampling is an important consideration before conducting the tests. Samples of molding sand for routine testing should ideally be taken at the molding machine. If more than one molding machine is used, it is best if the sample is taken at the machine farthest from the muller or mixer that supplies the molding sand. When evaluating the effectiveness of the sand system control program, samples can be taken at various points such as at the muller or mixer, at the molding machine, at shakeout and before and after cooling.

When a sample of molding sand is taken for testing, the sample should not be carried in an open container. Sealed containers with lids are recommended to keep the sample from picking up or losing moisture, etc. The green sand to be tested must not be packed into the containers, but should be of the same loose consistency as found at the point of sampling.

Riddling the molding sand through a coarse screen into the container, before testing, helps to improve repeatability of results. In this way, large core butts, pieces of tramp metal and other foreign materials are removed. Again, the container must have a lid and the sample should be kept covered at all times.

If the sand temperature is to be taken, a suitable thermometer can be inserted into the molding sand in the container at the time of sampling. The temperature can be read as soon as the thermometer stabilizes. The daily tests should then immediately be run following the sand temperature reading. The weekly tests, which use dried samples of sand can be performed after the daily tests are run.

The following text briefly discusses the basic sand tests. More complete information may be found in the *AFS Mold & Core Test Handbook, 3rd. Edition.*

Table 4-7. Suggested Green Sand Testing Schedule

Daily Tests

- Moisture content
- Compactability
- 2x2-in. Specimen weight
- Permeability
- Green compressive strength
- Methylene blue clay

Weekly Tests

- AFS grain fineness
- Loss-on-ignition/volatiles

Moisture

The moisture test should be run first. This test is simply a quantitative measure of the amount of water in the molding sand. The test is run by placing a sample of the molding sand into a specially designed pan and forcing hot air through it. The air temperature should be between 220F and 230F (104C and 110C). Within this temperature range, constant weight can be reached in about five minutes, without the loss of volatile organic materials such as seacoal. The weight loss upon drying is used to calculate the percent moisture. **Figure 4-10** shows a drier that can be used to dry the molding sand sample for this test.

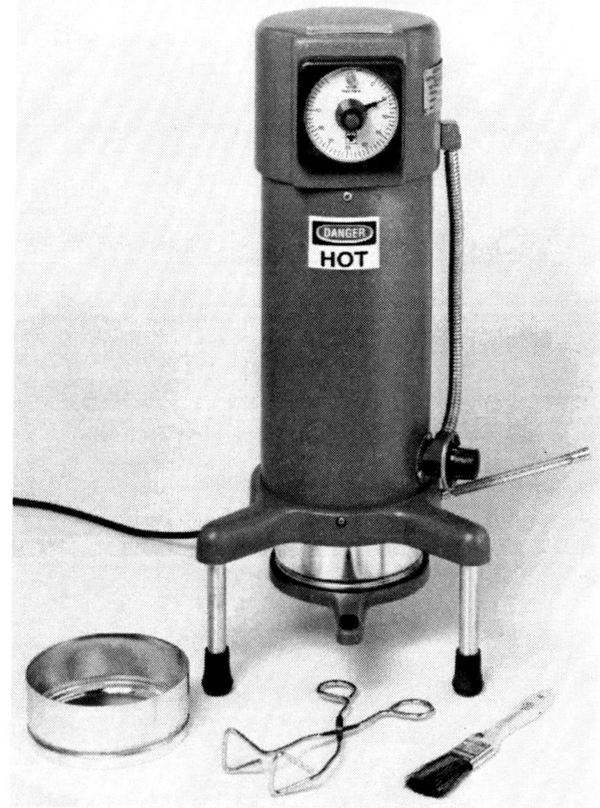

Fig. 4-10. Forced hot air drying apparatus and sample pan with 500 mesh filter cloth bottom.

Compactability

The compactability test is a measure of bulk density of the molding sand, which can be used to control the moisture in the molding sand. The higher the moisture the higher the compactability and vice versa. This test can also be used to control the bond/water ratio. **Figure 4-11** shows the test equipment used to perform the compactability test.

Specimen Weight

Before proceeding with other green sand tests, which require a standard test specimen, the specimen weight must be determined. Compacted density can be determined simultaneously with the compactability test. This test is used to produce the AFS standard test specimen, which is a cylindrical specimen that is two inches (50.8 mm) in diameter and two inches (50.8 mm) high, after three rams with the standard rammer, which has an indicator that can be used to make this determination. **Figure 4-12** shows this piece of testing equipment. This test can indicate changes in the green sand composition (such as silica sand content), increased additives or that dead clay and ash have accumulated.

Permeability

Permeability is a measure of the green sand's ability to allow the air in the mold cavity and the gasses formed as the molten metal is poured into the mold to pass between the sand grains and exit the mold. AFS permeability is the rate at which 2000 cc of air passes through an AFS standard specimen with a head pressure of four inches (10 cm) of water. Many factors can affect permeability, but the compaction and sand grain fineness are the major variables. This test provides an important relative measure of the venting characteristics of the green sand. An attachment is available that allows this test to be made on actual green sand molds ready for pouring. An electric permmeter is seen in **Fig. 4-13.**

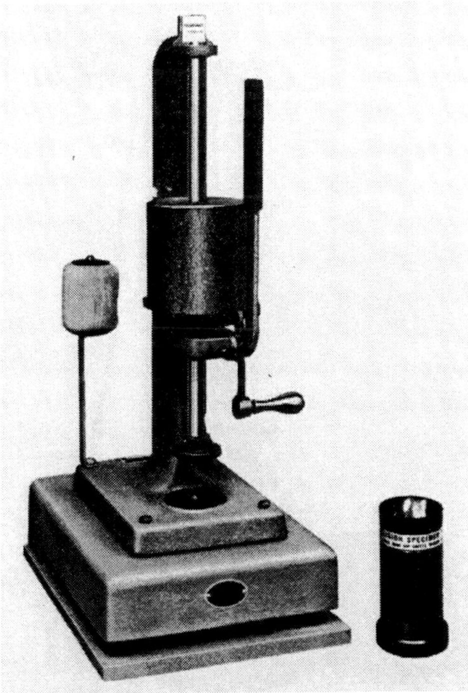

Fig. 4-11. Standardized ramming instrument with precision specimen tube.

Green Compressive Strength

AFS standard green compressive tests are typically used to control the strength characteristics of the green sand. In this test, an AFS standard specimen is loaded in compression and the maximum load to failure is recorded. This test is commonly used to relate to the molding and handling properties of the green sand. Dry strength is sometimes also used to compliment the green strength data, by relating to how well molds hold up to the heat and metallostatic pressure of the molten metal and the mold's shakeout properties. An electronic universal sand strength machine is seen in **Fig. 4-14a,** whereas **Fig. 4-14b** shows a computerized strength machine.

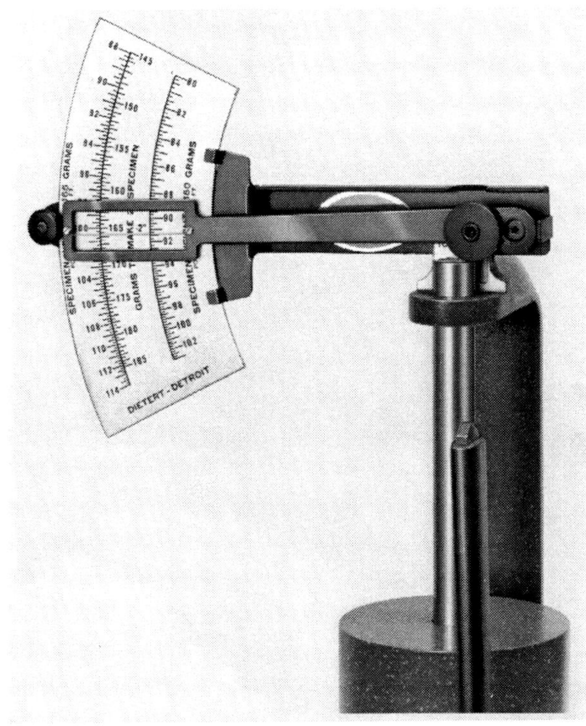

Fig. 4-12. Density and specimen weight indicator accessory for sand rammer.

Fig. 4-13. Electric permeability meter.

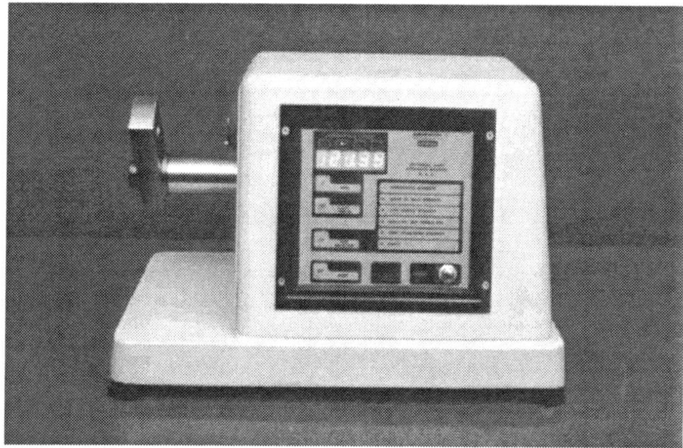

Fig. 4-14a. Electronic universal sand strength machine.

Fig. 4-14b. Computerized sand strength machine.

Methylene Blue Clay Content

The methylene blue clay test measures the "live" clay in the green sand. Live clay is capable of acting as bonding material. Return or used green sand also contains "dead" clay, which has been so close to the mold-metal interface that it is exposed to high temperatures. These temperatures can be high enough to destroy the clay to the point that it will not absorb water and form a bonding material. In controlling foundry green sand, it is important to know how much of the total clay in the green sand system is live clay. The methylene blue clay test is used for this determination. **Figure 4-15** shows some of the equipment necessary to perform the methylene blue clay test.

AFS Grain Fineness

The screen analysis is used to determine the AFS grain fineness number (gfn) and the grain size distribution of the sand. The fineness of the sand grains in the green sand system can change as it is used repeatedly. Weak sand grains break down and generate fines. Dust collection systems can remove fines and cause the sand to become coarser. Core sand that gets into the return sand can cause a change in the fineness of the sand. Many factors can cause the sand fineness to change. The screen analysis test provides a tool for monitoring these changes.

In the screen analysis test, testing sieves are arranged in a stack from the coarser, on the top, to finer, on the bottom **(Fig. 4-16).** The USA Sieve's numbers are six, the coarsest, to 270, the finest, and finally pan. A sample of the sand to be tested is weighed out and placed in the top sieve. The sieve stack is then placed in a sieve shaking device for 15 minutes. The sieve-shaking device simulates rotary and tapping motion of hand sieving, but in a more consistent mechanical manner.

After agitating, the weight of the sand retained on each of the sieves and pan material is recorded. With this information, the AFS grain fineness and grain distribution can be calculated. As mentioned earlier, the AFS gfn is an average. Thus, sands with widely different grain distributions can have the same AFS gfn. Some foundries go one step further with this test. The sand that has been tested is placed in test tubes with numbers that correspond to the sieve number and pan. These test tubes are then placed in a rack in numerical order and this then gives a visual picture of what is happening with the grain distribution.

There are some other tests that can be run on the green sand; however, the ones mentioned here are the ones normally run in the foundry on a daily or weekly basis. There are two more tests that should be discussed, as they can be used to help control the green sand and molding machine in which it will be used to make a mold.

Mold Hardness Testing

Mold surface hardness, as determined by this test, is the resistance offered by the surface of a green sand mold to penetration by a loaded plunger. **Figure 4-17** shows an instrument for determining the hardness of a green sand surface. The principle of this test is similar to that of the Brinell hardness test: The softer the surface of the mold, the greater the penetration of the ball into it. This

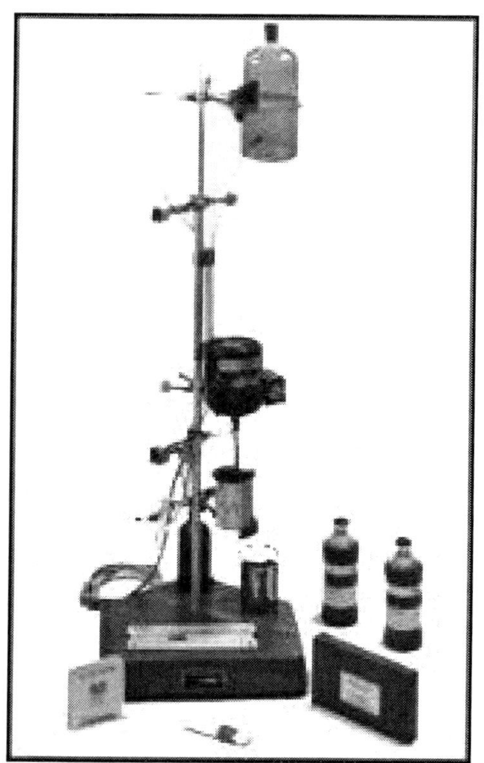

Fig. 4-15. Equipment for methylene blue clay test (tester only).

Fig. 4-16. A nest of standard foundry sieves.

Fig. 4-17. B-scale electronic mold hardness tester.

instrument measures mold hardness using the AFS "B" scale. The one-inch diameter indenting ball is spring-loaded. When it is pressed against a flat hard surface, the force exerted by the ball is 980 grams or 10 Newtons, and the dial reads 100 hardness. The scale on the instrument is uniform from 0 to 100 and the ball moves 1/10 of an inch from one end to the other.

The hardness of a green-sand mold should be in the following ranges:

Type of Mold	*Hardness*
Very soft rammed mold	20–40
Soft rammed mold	40–50
Medium rammed mold	50–70
Hard rammed mold	70–85
Very hard rammed mold	85–100

The practical advantage of using the mold hardness test lies in the fact that, by using this device, one can standardize the degree of ramming. Hardness may be regulated by proper adjustment of the molding machine and the green sand. The maximum hardness to which a mold may be rammed can be determined by the mold-hardness tester. There may be a variation in hardness in different parts of the same mold due to the shape of the pattern and the way the molding machine compacts the molding sand.

Loss-On-Ignition (LOI)

Organic materials are sometimes used in green sand for producing copper-base and ferrous metal castings. In addition, the return sand from the shakeout can contain core sands that contain organic materials. This particular test, sometimes also called the combustibles test, is an important tool to help control the amount of these materials in the green sand. In this test, the amount of organic materials in the green sand is determined by measuring the weight loss in a weighed amount of green sand—after it has been fired and the organic materials are burned out.

Automatic and Miscellaneous Testing

Other green sand tests that can be run are; Friability Test, Cone Jolt Toughness Test and the Wet Tensile Test. The average foundry does not run these tests. These tests are normally used in research laboratories. For further information about these tests, refer to the *AFS Mold and Core Test Handbook, 3rd Edition.*

Today's foundry using the green sand molding process has the option to use automatic and continuous on-line testing equipment to examine the prepared green sand before it is used. These pieces of equipment can be located just about anywhere in the green sand system to help control the green sand properties. Some of the properties this equipment can test for are compactability, compressive strength, shear strength or any combination of these. The results of these tests are then automatically communicated to the devices used to control the ingredients going into the muller or mixer so that they can add the proper amounts of each ingredient.

BIBLIOGRAPHY

Application of Additives to Clay-Bonded Sand Systems, American Foundrymen's Society, Inc., Des Plaines, IL, 1979.

Bralower, P.M., Granlund, M.J., "Controlling Properties of Molding Sands," *Modern Casting,* April, 1989.

Drews, B., "Simultaneous Mixing and Cooling of Moulding Sand Under Vacuum," *Casting Plant Technology International,* February, 1996.

Eirich Intensive-Action Mixer Type R, Eirich Machines, Inc, Gurnee, IL

Foundry Sand Handbook, seventh edition, AFS, Des Plaines, IL, 1963.

Furness, Jr, J.C.; Cannon, F.S.; Voight, R.C.; Neill, D.A.; Bigge, R.H.; "Effects of Advanced Oxidants on Green Sand System Performance in a Black Water System," *AFS Transactions,* 2001.

Green Sand Additives, second edition, American Foundrymen's Society, Inc., Des Plaines, IL, 2000.

Heine, R.W., Loper, C.R., and Rosenthal, P.C., *Principles of Metal Casting,* McGraw-Hill, New York, 1982.

Hoyt, D.F., "Back to the Basics of Silica Sand," *Modern Casting,* September, 1987.

Hoyt, D.F., Sand Technology for Foundry Applications, (Course Handout), Wedron Silica Company, Marseilles, Illinois, April, 1997.

Hoyt, D.F., "If It's Black, Why Do They Call It Green Sand?" Wedron Silica Co., Marseilles, IL.

Introduction to Metalcasting, (CMI Class Handout), Cast Metals Institute, Des Plaines, IL.

Krysiak. M.B., "Reducing Casting Defects—A Basic Green Sand Control Program," Georg Fischer DISA Inc., Holly, MI January, 1994.

LaFay, V.S., "Additives Improve Sand Molding," *Modern Casting,* July, 1992.

Mizzi-Krysiak, M.B., Pedicini, L.J., Sand Testing Design, 3 parts, *Modern Casting,* February, March, April, 1990.

Molding Methods and Materials, first edition, American Foundrymen's Society, Inc., Des Plaines, IL, 1962.

Operation and Application Guide for Simpson Multi-Mull, National Engineering Company, Chicago, IL, 1975.

Schleg, F.P., Cast Metals Institute (Personal Class Notes) 1965–1992.

Simpson/Gerosa Product Guide, Simpson Tech. Corp., Aurora, IL.

Green Sand Mold Production

5

Once the green sand has been prepared, tested, and is ready for use, the various ways to make a green sand mold can be explored. The green sand molding process is probably the most diversified of the molding processes, as these molds can be made in a variety of ways on different types of molding machines or made by hand. The smaller, repetitive, medium and medium-to-heavy castings can be made in green sand molds. High-quality metal castings with excellent surface finish and tolerances of less than ±0.015in./in. (0.038cm/cm) can be produced when the green sand process is properly employed. The key to high-quality green sand molds is to obtain uniform mold properties throughout the mold.

A few terms the reader shoud be acquainted with are: Cope, Drag, Flask and Mold Weights:

- *Cope*—The cope is the top half of the mold.
- *Drag*—The drag is the bottom half of the mold.
- *Flask*—The flask is the container into which the molding sand is placed and compacted around the pattern. The flask plays an important role in the production of quality metal castings. Flasks used for mold production, especially those made in molding machines, must be strong. In most cases, flasks are made of steel.

  Flask maintenance is critical. The pins and bushings, which align the cope and drag, should be inspected on a regular basis and replaced when worn.

  There are two types of flasks: permanent (tight) and removable (snap) **(Figs. 5-1 and 5-2).** A tight flask is used to make the mold and contains the mold until shakeout. A snap flask is used to make the mold and is removed from the mold before the pouring operation. Sometimes a "jacket" is placed around the mold, at the parting line, to stabilize the mold during pouring until it is ready for shakeout. The jacket is a metal box, open at the top and bottom.
- *Mold Weights*—Mold weights are usually used on horizontally parted molds, either flaskless or those in tight flasks. The mold weights are used to prevent the cope from "lifting" off of the drag during pouring due to the buoyancy of the molten metal. Mold weights can be placed on the cope by hand or mechanical means. The mold weights are removed from the cope after the metal has been poured into the mold and has begun to solidify **(can be seen in Fig. 5-3).**

The text in this chapter relates to *green sand mold production,* where the mold material is held together by clay, water and additives. Mold production using mold materials that are held together by chemical binders is covered in Chapter 6, Coremaking. Other casting and molding processes are covered in Chapter 2.

HAND MOLDING

The first green sand molds were made by hand. The molder had complete control over filling of the flask and compaction of the green sand. The molder may also have had the responsibility of "cutting in" the gating and risering system.

Figure 5-3 shows the steps in making a green sand mold using a loose-piece pattern. First, the drag half of the loose-piece pattern is placed on the molding board, along with the drag half of the flask. The pattern is then coated with a parting compound, which helps to keep the green sand from adhering to the pattern when it is drawn from the mold.

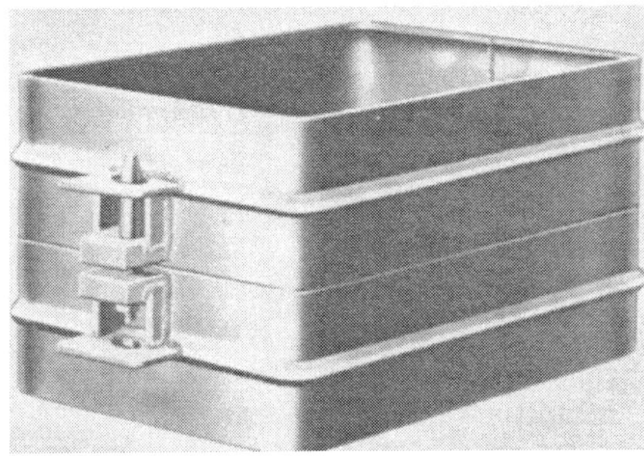

Fig. 5-1. Permanent (tight) flask. [From Principles of Metalcasting]

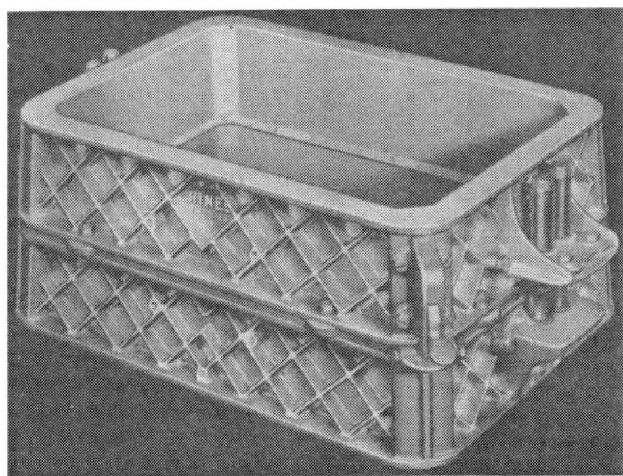

Fig. 5-2. Removable (snap) flask. [From Principles of Metalcasting]

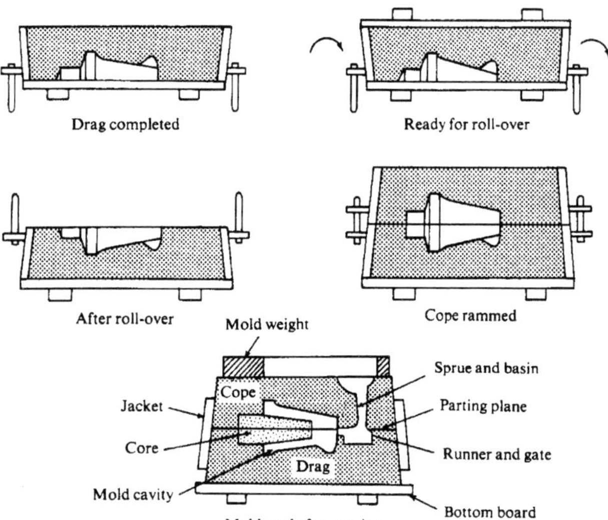

Fig. 5-3. Steps in the construction of a mold made by hand-ramming of sand.

Next, the flask is filled with green sand that is compacted around the pattern by the molder with a hand rammer. As the green sand is compacted, hardness and strength are developed, making the sand firm and rigid. After the pattern is drawn from the mold, this hardness and strength will help the mold cavity retain its shape and not erode during filling. After the drag half of the mold has been completed, it is rolled over.

Next, the cope part of the flask is placed over the drag half of the mold and the cope half of the pattern put in place. Again, a parting compound is spread over the pattern. The flask is then filled with green sand, which is then rammed or compacted by the molder. In some cases the molder may make the cope half in the same manner in which the drag half was made using a separate molding board.

The cope and drag halves of the mold are then separated and the pieces of the pattern are drawn from the mold. If the gating system was not cut in during the making of the drag half of the mold it is done at this time. If cores are required they are now set in place in the mold and the mold is then closed and ready for pouring. The two halves of the mold can be clamped together or weighted down to prevent the cope from lifting off the drag during the pouring operation.

When a matchplate is used, the drag half is made first, followed by the cope. In this operation, no molding board is used because the pattern is mounted on the matchplate. After the drag half has been made, a bottom board is put on top of the just-completed drag and the drag is rolled over. The cope half of the flask is then put in place and the cope half of the mold is compacted. With the two halves of the mold made, the cope half is lifted from the drag half and the matchplate is removed from the mold. If a gated matchplate was used, there will be no need to cut in a gating system or risers; if not, the gating and risering system would have to be cut in. Next, cores, if required, are set in place, the mold is closed and it is ready for pouring.

This method of hand-making green sand molds is not well suited for production quantities requiring the same mold properties and quality. The human factor plays a large role in the consistency of making quality molds by hand. At the start of the work shift, the molder is well rested but grows fatigued as the work shift continues. The molds made early in the shift do not always have the same properties as those made at the end of the shift. In addition, today, it is difficult to find true hand molders.

As time progressed and the technology became available, the molder used a pneumatic hand rammer, which, if operated correctly, could yield more consistent mold properties. Rammer usage decreased the amount of physical force needed to ram or compact the green sand during mold making. However, even with this help, the molder could not keep pace with the demand for increased production.

MACHINE MOLDING

There are six ways in which machines can be used to compact the green sand around the pattern: jolt molding, squeeze molding, jolt/squeeze molding, impact or impulse molding, sand slinger molding and vacuum squeeze molding. Each of these compaction methods use machines for compacting the molds and will be briefly discussed.

The first three methods—jolt, squeeze, jolt/squeeze—are depicted in **Fig. 5-4.**

Jolt Molding

The first method of machine molding used to make green sand molds was the jolt-molding machine **(Fig. 5-5).** The basic principle behind this compaction method can be explained through settling of the sand. For example: a box of breakfast cereal is filled to the top at the processing plant. As the cereal box is transported to the supermarket, it is jolted during handling. This jolting causes the cereal in the box to compact, or settle, and thus there is less box volume used for the same weight of cereal.

This same principle explains the jolt-molding machine for the compaction of the green sand. With this machine, cope and drag pattern plates are used because the cope and drag are made separately.

Initially the drag pattern plate is placed in the drag half of the flask, which is filled with green sand. The table of the molding machine is lifted, either hydraulically or pneumatically, and then dropped. This "jolting" action causes the green sand to compact around the pattern. The jolting action continues until the green sand is well compacted. The number of "jolts" will depend on the compactability properties of the green sand, the depth of the flask half and the configuration of the pattern. The same sequence is repeated for the cope half of the mold. Two jolt molding machines could be used simultaneously, one compacting the cope and the other compacting the drag.

Once the two halves of the mold have been made, the pattern plates are drawn out and the cores, if needed, can be set in the drag half of the mold. The cope is then placed over the drag and the mold is ready for pouring. The only manual labor required is placing the flask halves on the machine, putting the pattern plate into the flask and then removing the completed half of the mold from the machine. Mechanical devices have relieved the molder from performing this work manually, especially for large molds.

Although the jolt-molding machine made the job of making molds easier and faster, there still was a problem developing uniform mold properties. A study of soil mechanics shows that soil can be easily compacted downward but not sideways. The same holds true for green sand. On the jolt machine all of the compaction force is in a downward direction and there is very little, if any, sideways compaction taking place.

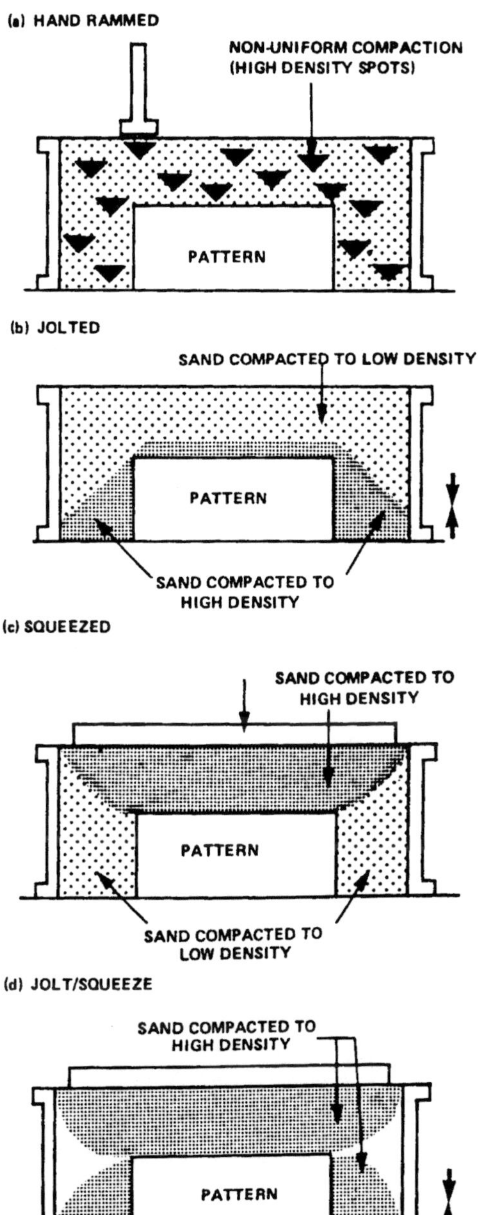

Fig. 5-4. Compaction of molding sands using various molding methods.

Squeeze Molding

In the squeeze molding process, hydraulic or pneumatic pressure is used to compact the green sand around the pattern. A plus for this type of molding machine is that both halves of the flask can be used along with a matchplate pattern. First, the drag half of the mold is squeezed. The next step is to roll over the flask and squeeze the cope. The squeezing of the green sand is done between a squeeze head (or plate), which is attached to the arm at the top of the machine, and the table of the machine on which the mold sits. The squeeze plate has dimensions that allow it to fit into the flask. The table is raised by either hydraulic or pneumatic power and the green sand in the flask is squeezed or compacted around the pattern. On some squeeze molding machines, the table remains fixed and the squeeze plate is forced down into the flask filled with green sand.

Since the object of sand compaction is to obtain uniform mold density in all parts of the mold, various means had to be devised to overcome this problem. The first efforts were to design contour squeeze plates. As the squeeze pressure was applied, the extensions (contours) exerted the pressure deeper into the mold and around the pattern. This action did produce more uniform mold hardnesses on the mold cavity surfaces.

One molding machine manufacturer used a rubberized bladder-type of squeeze plate that exerted air pressure on the bladder, which somewhat followed the configuration of the top surface of the pattern. Another manufacturer used a squeeze plate equipped with many individual compensating heads. Each of these heads could individually be set at different predetermined squeeze pressures. The latter molding machines can be found in foundries producing large molds at high production rates, such as automotive cast iron foundries.

Although it is assumed that the green sand has been dropped into the flask before compaction, there are squeeze-molding machines that first blow the green sand into the flask and then squeeze. This blowing of the green sand into the flask does cause some compaction to take place; however, the squeezing action causes the greater amount of compaction. Squeeze molding machines can be operated either manually or automatically.

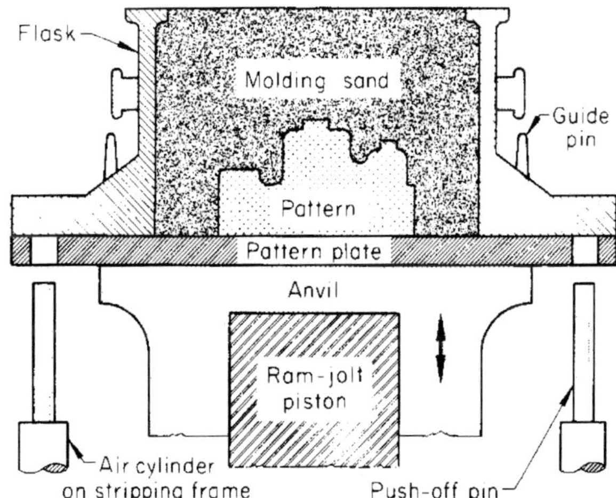

Fig. 5-5. Schematic of jolt-molding machine. (From Metals Handbook]

So far, only molding machines making horizontally parted molds have been discussed. There are also molding machines that produce vertically parted molds. In this case, the green sand is blown into the flask in which the pattern halves are located. In the case of vertical molding, the two halves of the mold are called the ram half and the swing half. Next, squeeze pressure is exerted on half of the mold. After compaction of the ram half, the swing plate is removed from that half of the mold and cores are set. The pattern, in this case, is mounted on the swing plate, which acts as a match plate. The ram, with its half of the pattern still in the mold, pushes the just-completed half of the mold out of the machine, pushing it up against the previously completed mold half. The ram is then drawn back, the swing plate moves into place and the compaction process begins again. **Figure 5-6** is an illustration of how these molds are made. Vertically parted molding machines are run automatically.

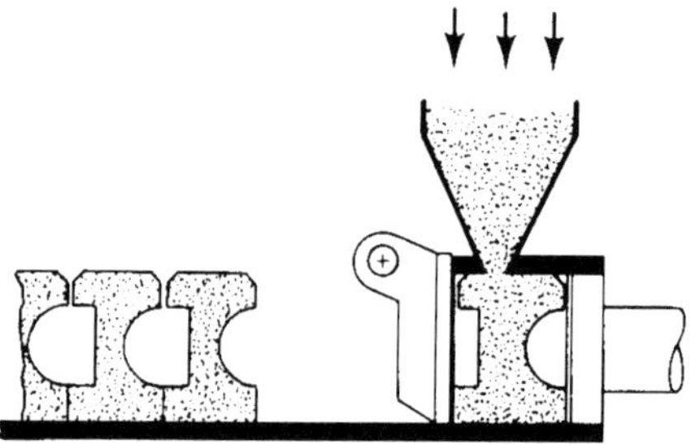

(5-6a) mold chamber is filled with sand

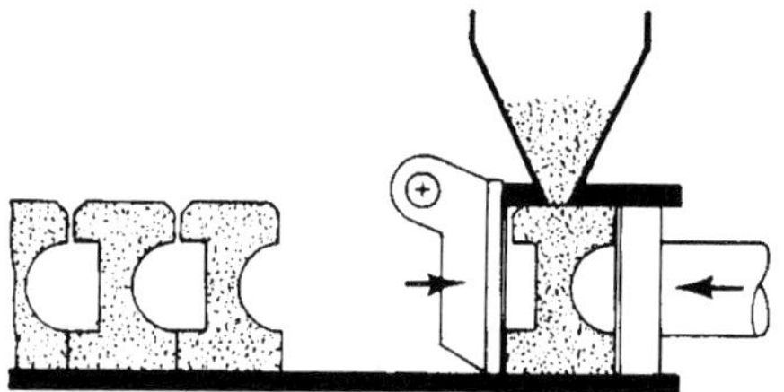

(5-6b) squeeze pressure applied

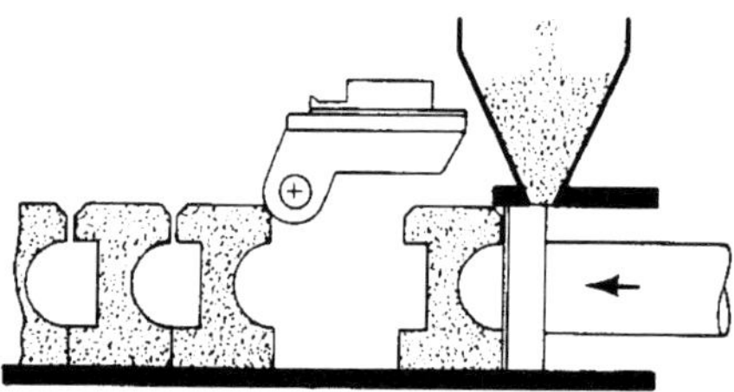

(5-6c) finished mold ejected from chamber

Fig. 5-6. Vertically parted molds being produced in a squeeze molding machine. [From Metals Handbook]

Jolt-Squeeze Molding

There are molding machines that combine both jolting and squeezing to compact the molds. **Figures 5-7 and 5-8** show the steps in making a complete mold on a jolt-squeeze machine. In this case, a matchplate pattern is being used. The compaction of the green sand can be accomplished by jolting and squeezing. In some foundries, two jolt-squeeze molding machines work in tandem; one produces the drag half of the mold and the other the cope half. After the pattern has been stripped from the drag, cores (if needed) are set. Next, the drag and cope halves of the mold are moved to the close-over station where the cope is placed on the drag. The latter type of jolt-squeeze operation is normally found in larger, high-production foundries such as those producing cast iron automotive and agricultural castings. In these cases, the operation of the machine is automated so that a machine operator oversees the operation of the molding machine. Cores can be set individually or with the use of a coresetting fixture. The molds made in such an operation are usually very large. If only one machine is used to make the mold and the matchplate is not removed from the flask, the drag is made by jolting and squeezing and the cope is squeezed.

Sand Slinger Molding

The sand slinger is a molding machine that, as the name implies, slings the sand onto the pattern and continues to fill the flask, in layers, until full **(Fig. 5-9).** The slinging action of the green sand, first against the pattern and then the continuing layers of sand, provides the compaction. The green sand is fed by a conveyor belt to the impeller head, in which an impeller, rotating at a very high rate of speed, slings the green sand into the flask. The choice of the diameter of the impeller head and the horsepower of the motor driving the impeller are determined by the mold hardness required. This type of molding equipment comes in various sizes, with some large machines actually being moved about on a rail system on the floor of the foundry. The sand slinger impeller can be operated by hand or "joy stick" to obtain a properly compacted green sand mold. Larger machines are operated by a person equipped with joystick style controls located adjacent to the impeller head above the flask.

The sand slinger is not a high-production molding operation. It is usually found in ferrous foundries pouring large castings. This operation is normally a floor-molding operation, where the molds are made and poured in one location on the foundry floor. If the mold is to be moved, it usually requires an overhead crane. This type of molding machine is not normally automated.

Impact/Impulse Molding

In impact/impulse molding, the green sand is blown into the flask and compacts over the pattern surface. The operating pressure can be increased to improve molding sand compaction. After the flask has been filled with green sand, an impact/impulse pressure, typically 40–75 psi (2.72–5.1 atmospheres), is applied, further compacting the molding sand. Due to the permeability of the molding sand, this pressure is distributed throughout the mold and yields a more uniformly compacted mold.

The impact/impulse pressure can be produced two different ways. The first method injects a controlled amount of combustible gas into a chamber above the mold, then agitates the mold. The gas is then ignited and pressure caused by the ignition of the gas is transmitted onto and into the mold. This is called gas impact. The second method fills the pressure vessel above the mold with

Fig. 5-7. Jolt-squeeze-rollover operation.

compressed air. When the valve is opened, the air pressure is released and within milliseconds, the sand is compacted.

With this type of molding machine, the cope and drag are made separately, transferred to a core-setting station and then closed over. The mold is then moved to the pouring area. It has been shown that this type of compaction yields uniform compaction properties throughout the mold. These molding machines are automated.

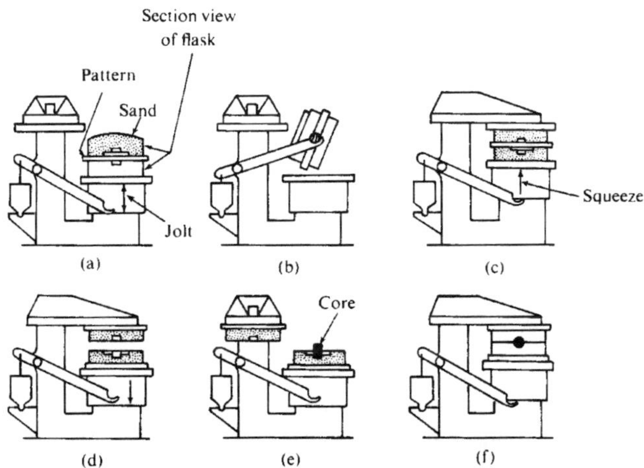

Fig. 5-8. Steps in making a mold on a jolt-squeeze-rollover machine. (a) Jolting the sand on the pattern in the inverted drag. (b) Rolling over the flask to fill the cope. (c) Squeezing to pack the sand in the cope. (d) Raising the cope to remove the pattern. (e) Placing a core. (f) Replacing the cope on the completed mold.

Vacuum-Squeeze Molding

In opposition to impact/impulse molding, another green sand molding process establishes a partial vacuum in the flask, before filling with molding sand. This creates a pressure differential to accelerate the movement of green sand from the hopper into the direction of the pattern, which causes some compaction of the green sand on the pattern surface. Final compaction is carried out by a hydraulic squeeze action.

Typically, this process, along with the impact/impulse method, produces the highest compaction values on the pattern face and parting line areas. This type of compaction produces uniform mold strength throughout the pattern. Mold permeability increases with distance from the pattern face, resulting in better mold venting characteristics.

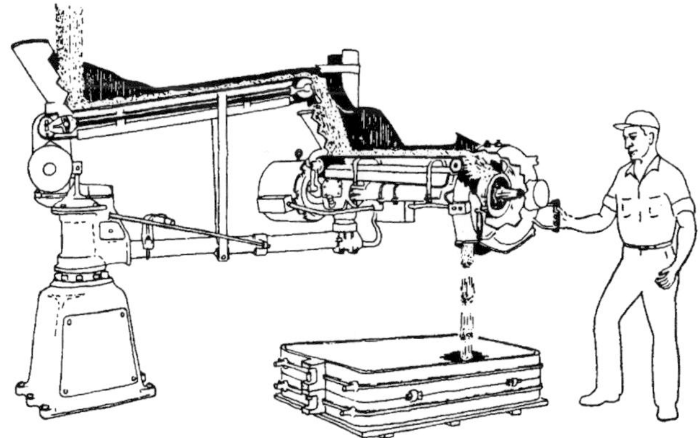

Fig. 5-9. Filling a flask with a sand slinger.

ADDITIONAL MOLDING METHODS

Stack Molding

Stack molding is another form of high-production green sand molding. (Other molding sands such as shell and nobake can also be used.) The principle of stack molding is economical because it is possible to produce one complete mold per flask section. A series of flasks are stacked to make a complete stack mold. In this process, pattern impressions are made in both sides of the mold. The bottom of a given flask in a stack provides the cope of the flask below and the drag of the next layer up and so on **(Fig. 5-10).**

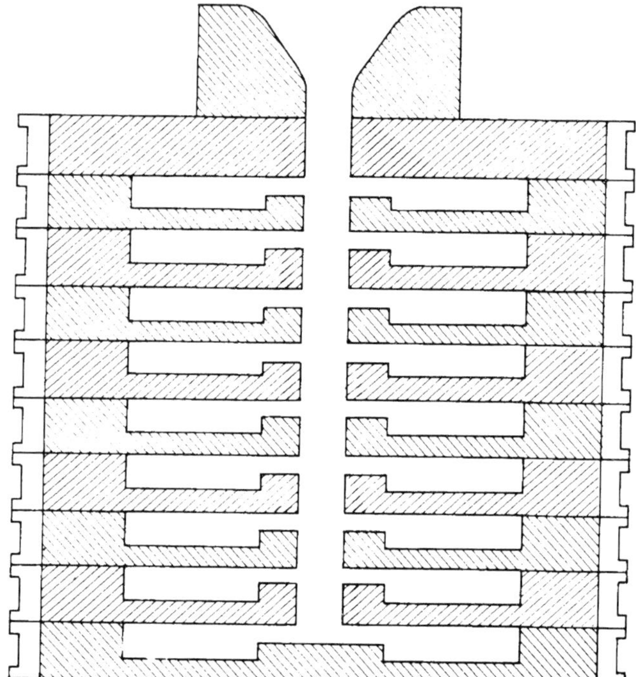

Fig. 5-10. Schematic of arrangement of stacked molds displaying gating system and both faces of molded parts to be cast.

Flaskless Molding

Flaskless molding is done in automated horizontally and vertically parted green sand molding machines. In these cases, the flask is an integral part of the molding machine. After the mold halves have been compacted, cores are set and the mold is closed and removed from the machine. The complete molds are placed on a mold conveyor and transported to the pouring area.

High-Density Green Sand Molding

Another term used many times to describe this type of green sand molding is "high-pressure molding." The preferred term is "high-density." The AFS Green Sand Molding Committee (4M) defines high-density molding as follows: "High-density molding is any type of impact/impulse molding and/or molding process where the pressure on the sand, as it is molded, is equal to or greater than 100 psi (6.8 atmospheres). Molds can be made to various mold harnesses and measured with a mold hardness tester. (See Mold Hardness Testing in Chapter 4.)

High-density green sand molding is said to start in the very hard rammed mold or a range of 85 to 100 on the mold hardness tester. The mold hardness tester can only read to 100 (100 is comparable to glass for hardness). Molds made to the 85–100 range can only be made in high density molding machines and these machines are usually automated.

GREEN SAND HANDLING SYSTEM

After the green sand mixture is prepared, it has to be conveyed to the molding station. In most foundries, the green sand is transported from the preparation area (the muller or mixer) to the molding station via bucket elevators and conveyor belts. One large molding machine may have its own exclusive sand handling system, whereas several smaller molding machines may share a molding sand handling system. In either case, the molding sand is placed by the conveyor into a hopper or bin directly above the molding machine. From the hopper, it moves by gravity into the molding machine. Normally, this whole operation is fully automated and various sensing devices are used to control its operation. This can all be done under the watchful eye of the sand technician.

Several precautions should be taken regarding heat, moisture and compaction, when using the molding sand handling system. Usually, this system is located above the foundry floor where temperatures can be very high. Such a system is called an "over-head' sand distribution system. When the green sand is being transported through these hot areas, moisture can evaporate. If green sand moisture samples are to be taken, it is a good idea to take the sample from the hopper above the last molding machine to be fed. In addition, the green sand can become slightly compacted as it is transported to the molding machines. The compaction problem can be alleviated by locating an aerating device where the sand is discharged from the hopper into the molding machine. This can aid the flowability of the green sand and make for better-compacted molds.

After the mold has been poured, cooled and shaken out, the green sand proceeds to the return sand system. It eventually will end up in the return sand bin or hopper above the mulling or mixing machine. The return sand is moved again by a series of bucket elevators and conveyor belts.

At shakeout, the returned green sand can be prepared for reuse. Return sand cooling should begin at shakeout. First, lumps of green sand should be broken up so that the heat contained within the lumps can be released. Several shakeout machines can perform this operation during or after shakeout. Sand coolers can do this job after shakeout. There is also a green sand mixer that has this operation incorporated into the mixer.

Return green sand cooling is very important because it can affect the control of properties that need to be maintained within the green sand system. The temperature to which the green sand can rise is dependent on the size and section thickness of the castings and the pouring temperatures.

There have been many technical papers written on the effect of hot sand on the quality of green sand molds. These papers are available from the American Foundry Society, Inc.

BIBLIOGRAPHY

Bosworth, T.J., Heine, R.W., Parker, J.J., King, E.H. and Schumacher, J.S., "Sand Movement and Compaction in Green Sand Molding," 59-32, *AFS Transactions,* 1959.

Guide to Selecting Casting Processes, Casting World, Summer 1994, American Foundrymen's Society, Inc., Des Plaines, IL.

Metal Casting Processes, 12 articles January-December 1989, *Modern Casting,* American Foundrymen's Society, Inc., Des Plaines, IL.

Operating Guide to High-Density Molding, 1st Edition, 1993, American Foundrymen's Society, Inc., Des Plaines, IL.

Coremaking 6

A core may be defined as a preformed mass of bonded sand or some other suitable material that is used to make the internal passageways of a casting. The entire mold may be constructed of cores, which, in that case, is called "core molding." It is in the coremaking area that the most recent developments have occurred in metalcasting technology over the past 50 years. **Table 6-1** is a listing of coremaking processes that were developed over this period.

Like castings, cores are made in a mold (called a corebox). Unlike castings, which are made of metal, cores are made of sand. In coremaking the sand is combined with other materials, so it is important to carefully select the proper core binder system appropriate for the casting.

Table 6-1.
Approximate Time of Commercial Introduction of Core and Mold Processes in North America

Date	Process
pre 1940	core oil
1950	shell: liquid and flake
1952	silicate/CO_2
1953	airset oils
1958	phenolic acid-catalyzed nobake
1960	furan hotbox
1962	phenolic hotbox
1965	oil urethane nobake
1968	phenolic/urethane/amine coldbox
1968	silicate ester-catalyzed nobake
1970	phenolic urethane nobake
1974	alumina phosphate nobake
1977	furan SO_2
1978	polyol urethane nobake
1978	warmbox
1982	FRC SO_2
1983	epoxy SO_2
1984	phenolic ester nobake
1985	phenolic ester coldbox
1992	phosphate/metal oxide

SELECTING A CORE BINDER SYSTEM

The selection of a core binder system should be based on an evaluation of many factors. Some of the system factors to be considered are production rate, capital outlay, operating costs, quality level, environmental impact, labor skills involved, existing equipment (coreboxes and core machines) and available sand.

Some core requirements that should be considered are:

- *The strength of the core.* The binder used must develop sufficient strength at room temperature and at molten metal temperatures—room-temperature strengths for handling the cores and molten metal-temperature strength for the core to hold its shape when surrounded by molten metal.
- *The production rate and size of the core.* High production rates require rapid strength development, hence curing rates that are fast as well as controllable. Large cores require that the curing rate be slow enough to allow the corebox to be filled with the core sand mixture prior to the formation of the bond.
- *The rate of thermal decomposition of the core binder.* It should be compatible with the molding methods and production rates used.
- *The collapsibility of the core.* It should be such that it does not hinder the solid contraction of the casting.

An important consideration in the selection of a binder system is the reclaimability of the bonded sand. The system must be economical and ecologically safe to use. *The environmental impact in the foundry and community is always an important parameter in the final decision.* (Reclamation will be discussed later in this chapter.)

In regard to "strength" and "collapsibility" mentioned above, when considering the job a core has to do, it seems to be a Herculean task. First, it has to be strong enough to withstand the handling it will be subjected to in the foundry. Second, it has to be as strong enough to withstand the pressures and temperature to which it will be subjected and still hold its shape. Then, at the proper time, as the casting is shrinking, it has to collapse, and come out of the casting at shakeout as loose sand. The collapsibility characteristics of the core process have to be such that the core does not restrict shrinkage and cause excessive internal stresses to be set up in the casting. Poor collapsibility can also cause hot tears in castings. The collapsibility of a core is often overlooked and causes many unnecessary problems in producing quality castings.

CORE SAND

In Chapter 4 on Molding Sands, it was said that much of the material would also apply to core sands. However, since sand makes up almost 99% or more of a core, it is worth repeating some of that material, especially as it applies to making cores.

Each of the binder systems is affected differently by alkaline or acidic impurities in the sand, moisture content and the overall type and purity of the sand. Care should be taken to select sand that will produce the best results, based on binder consumption, casting surface requirements and workability of the core sand mixture. It should be pointed out again that the density of the sand could vary between silica sands and the specialty sands.

Knowledge of how impurities in the sand affect the performance of each binder system is the first step to good sand selection and, eventually, quality cores. The source of this information is the binder manufacturer or supplier. Such items as acid demand value (ADV), pH, percentage of fines, moisture content, loss-on-ignition (LOI), values of new and reclaimed sand and the purity are basic information that should be known.

The following text will discuss some of the sands or aggregates that are used in coremaking.

Silica sand is available in rounded, subangular, compound and angular sand grain shapes. The available AFS grain fineness number (AFS gfn) may vary from approximately 30 to 180. This sand may be used with all binder systems. Some impurities found in some of the silica sands can affect their performance adversely. These impurities can include clays, calcium carbonates and other materials that affect the catalyst or binder, reducing their effectiveness. The effects of impurities can result in longer cure or set times and reduced strengths. The ADV and pH tests can be used as indicators of impurities in the sand. Silica sand is the aggregate most commonly used in the foundry industry. A wide choice of silica sands is available, varying in grain size, shape, distribution and purity. The silica sand supplier can be of valuable assistance when selecting the type of silica sand to be used in coremaking operations.

Olivine sand is angular sand. The AFS gfn is available from 60 to 180. This sand cannot be used with acid-catalyzed binder systems due to its inherent alkalinity.

Chromite sand is angular sand used for its chilling tendency and resistance to metal penetration. The available AFS gfn range is between 50 and 80. Chromite sand is compatible with all binder systems.

Zircon sand is rounded (elliptical) grain sand, generally used in the same manner as chromite sand. Available AFS gfn may vary from 65 to 130, depending on the sand source.

Table 6-2 summarizes the compatibility of these sands with the various core binder systems. Grain shape, fineness and screen distribution are three other areas to consider when selecting core sand.

Grain shape affects the amount of binder required and the compaction of the bonded sand in the corebox. Generally, rounded sand grains require less binder than subangular or angular sand grains. Also, the true round sand grains, silica sand, are 99% pure silica and the purest of the other silica sands, bank and lake.

Fineness—Higher amounts of fineness require additional binder, or a decrease in core properties may be realized.

Screen Distribution—Sands of similar AFS fineness with narrow screen distribution (1 or 2 screens) require less binder than sands with broad screen distribution (4 or 5 screens), due to the difference in surface area to be covered by the binder. The grain size distribution is an important variable in core sand. The density of the compacted core sand increases as the screen distribution is broadened. Sands with at least three to four major (+10%) screens should be used with chemical binder systems. Segregation of sand grains, by screen size, can occur at any time in the sand handling system and must be minimized with antisegregation devices. According to one very knowledgeable binder supplier, sand segregation is the cause of most of the problems in making quality cores consistently. An excellent video on controlling sand segregation is available from sand suppliers. Sand storage and handling facilities should be designed to minimize contamination by moisture, foreign materials, segregation and extreme temperature changes.

It should be mentioned here that some of the chemical binder systems used to make cores are also used to make nobake molds. Thus, the same items just discussed on core sand selection and handling applies to sands used in making nobake molds.

Table 6-2.
Compatibility of Various Binder Systems With Typical Foundry Sands

	Silica	Chromite	Olivine	Zircon
Furan	Yes	Yes	No	Yes
Alkyd Urethane	Yes	Yes	Yes	Yes
Phenolic Urethane	Yes	Yes	Yes*	Yes
Phenolic Acid Set	Yes	Yes	No	Yes
Phenolic Ester Cured	Yes	Yes	Yes*	Yes
Phosphate	Yes	Yes	Yes	Yes**
Sodium Silicate	Yes	Yes	Yes	Yes**

*Requires more binder to maintain performance properties.
**Shakeout characteristics poor.

Core Sand Additives

Additives can be added to the core sand mixture to improve surface finish, shakeout and to control certain defects. Kaolin-type clay, fine zircon, carbonates, complex carbohydrates and sugar/clay blends are included for better surface finish. Potassium fluoroborate (KBF_4) is added to minimize gas defects. Lithium carbonate and calcium carbonate also may be used to improve humidity resistance. All of these additives improve shakeout.

Iron oxides are also used as additives. Iron oxides can be used to control lustrous carbon when cast irons are poured, and to control veining and/or nitrogen pinholing. Magnetite (black) iron oxide and hematite (red) iron oxide can be used. However, if aluminum alloys are to be poured, iron oxide should not be used as an additive.

As a rule, the use of additives should be limited to a "when needed" basis. Not all additives are compatible with all chemical binder systems. In addition, it should be kept in mind that when additives are added to the core sand mixture, more fines are added, and additional binder will be required.

CHEMICAL BINDER SYSTEMS

Before getting into the various chemical binder systems available, green sand cores should be mentioned. Green sand cores are made using the same bonding mechanism discussed in Chapter 4, Molding Sands. In many cases, the green sand core is formed as part of the mold. An example of this would be the boltholes found in the flange of a valve body. Several years ago, there was a foundry still using green sand cores to produce pipe elbows and there may be many other applications of green sand cores. However, for the majority of cores produced, some type of chemical binder system is usually used to produce the core.

Inorganic binder systems are those that do not contain carbon in the binder molecules and produce very little or no gas when raised to elevated temperatures. Unfortunately, inorganic systems are the least reactive, achieve the least strength and require the most attention during the coremaking, moldmaking and casting processes. These systems are based on silicate and phosphate/metal oxide technology. All of the other systems are organic.

Systems also can be based on how they are chemically catalyzed **(Table 6-3)**. Acid-type binder systems having a pH greater than seven are cured by adding alkaline material. Basic-type binder systems have a pH less than seven and are cured by adding acidic material. Other systems are cured by various chemical mechanisms that are independent of a distinct pH shift.

Binders also can be based on the way that the core and mold sand physically hardens (cures) **(Table 6-4)**. Resins cure in several ways: by being in contact with a vaporized catalyst (coldbox); by being exposed to various forms of heat (heat cured); and by being mixed with a liquid catalyst that reacts with the binder at ambient temperatures (nobake). When a binder has fully reacted (polymerized) and reached its final cure stage, it has become a thermally stable, infusible material called a thermoset. In addition to activating binders, heat can be a way to accelerate the cure rates of all binder systems. Shell, core oil, hotbox and furan warmbox binder systems are cured by applying heat to the resin-coated sand. Although various forms of catalysts and/or accelerators also may be added to the mixture, heat is the primary curing agent in these systems.

Chemical binder systems are classified into three types: heat activated, coldbox and nobake (air-set). These systems will be covered in the following text.

Heat-Activated Systems

Heat-activated systems, just as the name implies, require heat to cure the binder. There are four processes associated with this system: core oil, shell, hotbox and warmbox.

Core Oil Process

The oldest of all known chemically bonded sand cores are those bonded with core oils. They are referred to in the foundry industry as "oil sand" cores. This process begins with cereal and other dry additives being mixed with sand. Water is added to "activate" the cereal (to give green strength for handling). Kerosene-type oil is then added to the sand mixture, near the end of the coating cycle, to improve flowability and release.

After coating, the core sand is blown or rammed into a suitable corebox. The green strength property of the mix allows the wet, pliable core to be removed easily from the corebox. The core is then placed onto a suitable support (dryer) for subsequent oven drying. To achieve maximum productivity without burning thin sections of cores, the oven must have a drying atmosphere that is low in humidity and high in convection air volume. The best temperature in the core oven is 350F (177C) for core oil, but many ovens operate at about 500F (260).

The core oil process produces cores quickly, even with the poorest tooling (coreboxes). There are some limitations in terms of tight dimensional tolerances because of handling and drying of the pliable core shape.

Shell Process

The shell core process was discovered after World War II, in Germany, where it was used to make German hand grenades and other castings. At that time, the process was called the Croning or "C" process after its inventor Dr. Johannes Croning. The process remains one of the most convenient forms of core production. Foundries can use storable, precoated sand that can be supplied by

Table 6-3.
Binder Systems Classified by Catalyst Curing Mechanism

Acidic	Basic (Alkaline)	Other
silicate CO_2 plus dehydration	phenolic ester nobake	shell (neutral)—hexa addition
warmbox	phenolic ester coldbox (addition)	silicate nobake—saponification plus dehydration
hotbox	oil urethane nobake plus oxidation	core oil (neutral)—oxidation
SO_2 furan	phenolic urethane nobake and coldbox	phosphate/metal oxide
SO_2 acrylic epoxy and free radical cure	polyol urethane nobake	

Table 6-4.
Categories of Resin Core/Mold Curing Processes

Coldbox	Nobake	Heat-Activated
phenolic/urethane/amine	furan/acid	shell
silicate/CO_2	phenolic/acid	core oil
furan/SO_2	phenolic/ester	phenolic hotbox
acrylic/epoxy/SO_2	oil urethane	furan hotbox
(acrylic) FRC/SO_2	silicate/ester	urea formaldehyde hotbox
phenolic/ester (methyl formate)	phenolic urethane	warmbox
phenolic/CO_2	phosphate/metal oxide	
	polyol urethane	

commercial coaters, or they can coat their own sand. A number of foundries use the shell process for making molds in addition to coremaking.

Shell sand can be coated using one of two different processes:

- *Hot Coating*—In the hot coating process, a flaked phenolic novolak resin is mixed with sand heated to about 350F (177C). The resin melts and coats the sand. The hot (275F/135C) resin-coated sand is quenched with a water solution of hexamethylene tetramine and mulled until the sand mass breaks down. Sand is discharged then screened and aerated to further cool and particulate it. The hot coating process is used to produce the majority of shell process coated sand.
- *Warm Coating*—In the warm coating process, calcium stearate, hexa-powder and water/alcohol solution of novolak resin is added to sand heated to about 200F (93C). When the coated sand mass breaks down after the carrier evaporates, the coated sand is discharged to an aerator to be cooled and particulated. Warm-coated sand builds up the sand shell more quickly, potentially reducing the core or molding machine cycle.

To produce a shell core requires a heated corebox. The corebox is heated up to 500–550F (260–287C) and the coated sand is dumped and/or blown into the corebox. The coated sand remains in the corebox until the required shell thickness has been achieved (dwell time). Next, the corebox is maneuvered into a position so that the coated sand unaffected by the heat drains out of the inside of the core. The shell that was just produced remains in the corebox until it is cured (cure cycle) and then ejected from the corebox. It should be mentioned that not all shell cores are hollow. The design of the shell core may be such that the uncured sand cannot be drained out of the core.

The shell process offers several distinct economic advantages. The coated sand is said to have an indefinite bench life and the shell cores have an indefinite shelf life. Bench life means that the coated sand can be stored indefinitely and shelf life means that shell cores can be stored indefinitely, before use. Producing a hollow core can aid the shakeout of the core from the casting. Shell cores can also produce smooth cored surfaces in the casting due to the better flowability and density of the coated sand

Hotbox Systems

The hotbox systems were developed in the early 1960s in response to the auto industry's need to cast lighter engine blocks. The use of these binder systems allowed the casting of thinner wall sections in the engine blocks, because the hotbox cores were stronger and had good shakeout characteristics.

There are three distinct hotbox binder systems, based on three types of resins: urea formaldehyde, furan and phenolic. Although the resins can be used individually (along with the proper catalyst), the resins are sometimes blended to optimize properties.

One significant drawback to hotbox systems is that formaldehyde is given off at several points in the process, which makes careful venting and fume control necessary. Under certain circumstances, detectable formaldehyde might be measured in the areas around the sand coating operation, around the coremaking machine while the core is cured at elevated temperatures in the corebox, and around the core after it has been removed from the corebox and is cooling to ambient temperature. Formaldehyde can also be released as core coatings are dried in a drying oven. To help minimize these problems, formaldehyde scavengers can be added to the core sand mixture.

Urea Formaldehyde—This system is simply made up of a water solution of urea and formaldehyde. It is often modified with up to 30% furan to improve cure speed and strength. This heat-cured system is ideally suited for dielectric oven curing. Productivity of this system is quite fast, but every step of the coremaking and casting process liberates formaldehyde into the workplace.

Furan and Phenolic—Furan and phenolic hotbox systems are catalyzed with a water solution of acid salt, based primarily on ammonium nitrate and ammonium chloride. The catalyst, upon being exposed to elevated corebox temperatures, decomposes to form ammonia and a mild acid that cures the binder.

It is important to realize that the hotbox core continues to cure after being ejected from the heated corebox. The residual heat in the sand plus a mild exothermic reaction (which causes an increase in temperature of the core) from the resin polymerization continues to make the catalyst decompose and cure the sand until the temperature drops to less than 125F (52C).

The bench life of the coated sand will vary from 1 to 4 hours, depending on the type of resin and the temperature. Furan is said to have a much longer bench life than the phenolics.

Warmbox Systems

The warmbox system is composed of furan resin catalyzed with a copper salt/water/methanol solution of aromatic sulfonic acid. An advantage to using the warmbox rather than the hotbox system is that the warmbox system will effectively cure at a processing temperature about 100°F (56°C) lower than the hotbox systems. Warmbox systems with very low free formaldehyde have been developed. These systems are also phenol-free. These two aspects are distinct environmental advantages compared to the hotbox systems.

Another advantage of the warmbox system is that there is less heat required, thus saving on heat energy. In addition, it has been reported that plastic coreboxes can be used due to the lower box temperatures.

Coldbox Systems

Coldbox systems are used in most foundries today, especially those requiring a high volume of cores. These systems require a vaporized (or gas) catalyst to cure the resin-coated sand. The coldbox classification includes sodium silicate/CO_2; phenolic urethane/amine; epoxy acrylic/SO_2; phenolic/methyl formate; phenolic/CO_2; furan/SO_2; and acrylic/SO_2.

Sodium Silcate/CO_2

This is the inorganic binder system that is cured with carbon dioxide gas. The chemistry of the sodium silicate/CO_2 process was first described in an 1898 British patent. The sodium silicate/CO_2 system was originally used as an oven-heat-cured system. Its earliest use as a true coldbox system was in a Czechoslovakian foundry at the end of World War II. Sodium silicate/CO_2 was introduced into the U.S. as America's first true coldbox foundry binder system around 1952.

Because carbon dioxide gas is odor-free and nonirritating, it does not have to be collected for disposal. The inorganic classification of the binder and catalyst (CO_2) help to make this system the least gas generating of all the coremaking processes when surrounded by molten metal.

The properties of the finished core depend on the ratio of silicate to soda. The available ratios range from 1.90–3.22 parts of SiO_2 to 1 part Na_2O. The lower ratios (2.0–2.2) result in higher final core strengths. The higher ratios (>2.7) give lower final strengths, but are stronger initially. For optimum properties, most formulators select a ratio of about 2.5:1.

The binder's strength and the ability to store the core or mold is based on a specific amount of carbon dioxide gas reaching the binder-coated sand surface **(Fig. 6-1)**. The complex curing mechanism involves first gelation and then dehydration.

It is best to over-cure cores and molds if they are going to be used the very same day they are made because they have no humidity resistance. In other words, they are very hydroscopic. Alternatively, one could say they have very poor shelf life.

The cores that are under-gassed and stored continue to cure, rather than to lose strength in storage. In most cases, it is preferable to under-gas a sodium silicate/CO_2 core or mold. Special care must be taken when a core has various section thicknesses. Thin sections will cure rapidly and the thicker sections will cure slowly. A chemical called phenolthalein can be added to the core sand mixture. When the CO_2 gas encounters the phenolthalein, it will turn the core pink. Thus, any section of the core that the CO_2 gas does contact will not turn pink. In addition, studying the various shades of pink will give the metalcaster some idea of gas flow pattern through the core.

Sodium silicate coated sands are very sensitive to cold temperatures, from the standpoint of coating and reactivity. **Figure 6-2** illustrates how the strength of the coated sand changes as it is subjected to temperature changes during the metalcasting process.

The sodium silicate/CO_2 system is normally considered to have poor shakeout characteristics. Additions of various additives to the formulation can help improve shakeout. These additions can be helpful when ferrous alloys are poured, but not so when pouring aluminum alloys. Another problem with using additives is that they are mostly organic materials and, thus, add a gas-generating material to the cores and mold.

Reclamation of the sodium silicate binder is often cited as one of the negative aspects of this system. If the coating is vitrified by exposure to high metal-pouring temperatures, the glass-like sodium silicate coating does not lend itself to mechanical or thermal reclamation. However, it is possible to reclaim the coated sand using wet reclamation techniques.

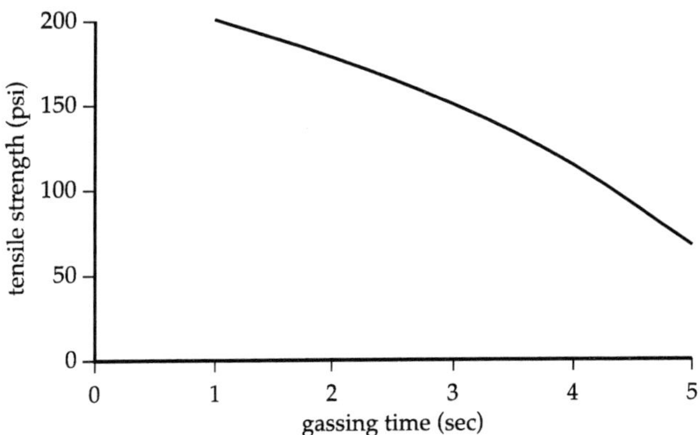

Fig. 6-1. The 48-hour tensile strength of 2.5/1 SiO_2 silicate binder at various CO_2 gassing times.

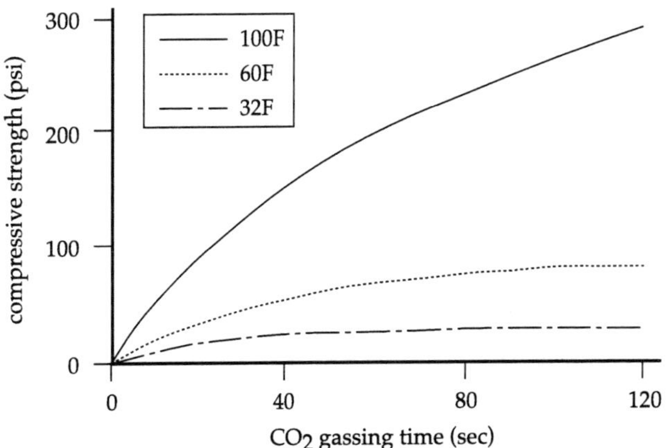

Fig. 6-2. Effects of sand temperature on compressive strength of CO_2 silicate cores with 4% binder.

Phenolic Urethane/Amine (PUA)

This phenolic urethane coldbox system is comprised of two resin parts. Part 1 is a phenolic resin dissolved in an organic solvent, and part 2 is an isocyanate component dissolved in a similar solvent. Parts 1 and 2 are mixed with clean sand. Either silica, zircon or olivine sand can be used. After the sand has been coated, it is blown into the corebox through entry ports, called blow tubes or blow slots, located in one half of the corebox. The resin-coated sand is then cured by a vaporized amine catalyst, which is introduced into the corebox through cope vents, and through the core sand entry ports.

The placement and sizing of these cope vents is very critical to ensure that the amine catalyst is distributed throughout the coated sand in the corebox. The open area of the exhaust vents is limited to about 80% of the cross-sectional area of the cope vents. The pressurization causes the amine catalyst to travel to all parts of the core before it exits through the exhaust vents in the opposite half of the corebox.

After the amine catalyst has been introduced into the corebox, a very dry "purge air" is introduced. This volume of air travels through the core and out the exhaust vents, distributing the amine catalyst to all cavities of the corebox. This flushing of the amine catalyst also helps to remove the residual odors of the amine catalyst from the core. This purge air can also be heated to increase its effectiveness. The purge air is then piped to an exhaust scrubber, which neutralizes the amine and forms an acid salt.

There are two types of amines that are commonly used as the catalyst: triethylamine (TEA) and dimethylethylamine (DMEA). DMEA is the more effective catalyst, but, because of its stronger odor, higher cost and only marginal improvement in total core machine cycle time, most foundries use TEA.

All polymeric resin binders shrink as they cure, and PUA is no exception **(Fig. 6-3)**. When the core is removed from the corebox, it is about 70–80% cured. The cure is completed as the solvent evaporates from the binder coating on the sand. The amount of shrinkage depends on the density of the core, the type of binder and aggregate, and the age of the core.

The keys to a successful PUA coldbox process are:

- using the resin coated sand as soon after coating as possible,
- correct rigging of the corebox.

Most productivity problems can be traced to using improperly rigged coreboxes, which promotes sticking and poor cycles, or to using coated sand that has sat too long in a hopper after coating before it was used for coremaking.

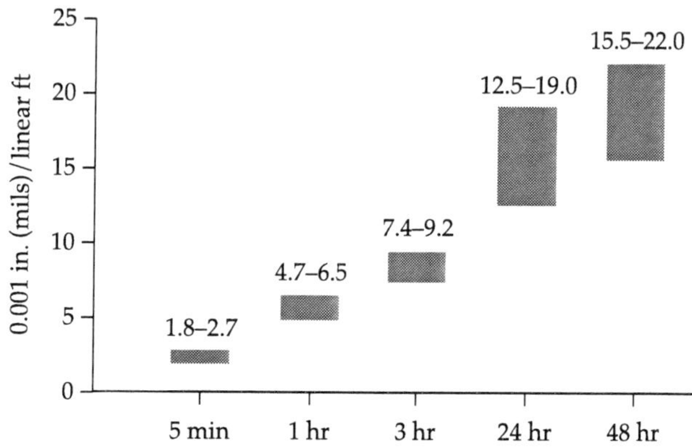

Fig. 6-3. Amount of linear shrinkage in a PUA coldbox core with 1.5% binder level and part 1:part 2 ratio of 55:45.

Epoxy Acrylic/SO_2

Resin formulators recognized the fact that acrylic binders offered outstanding cure speed but poor hot strength, and that epoxy binders had good superior strength but needed long gassing cycles because of their slow cure rates. Therefore, the epoxy/acrylic/SO_2 system, using a blend of epoxy and acrylic, was developed to optimize both the handling and casting properties. This system was introduced to the foundry industry in the 1980s.

This SO_2-gassed resin system usually uses a ratio of 65 parts of epoxy to 35 parts of acrylic, with peroxide added to the epoxy component. However, because the shakeout properties of this system can be difficult for aluminum operations, a 50/50 blend is used to improve the shakeout characteristics.

Phenolic/Methyl Formate

This system was introduced to the foundry industry in 1984 and is based on a high-water-content phenolic resole resin that reacts with vaporized methyl formate coreactant. The sweet smelling, flammable methyl formate typically is used in amounts of 30–40 parts per 100 parts of resin (by weight).

Because the methyl formate is a coreactant and not a catalyst, the resin binder must be exposed to and react with the proper amount of the vaporized liquid material for a complete chemical reaction to occur. This factor sometimes leads to overgassing to make sure that a complete cure takes place.

This water-based system offers water cleanup and is low in smoke and odor compared to other organic coldbox systems. Because of the low strength of the cores produced using this system, it is not used where delicate cores are needed. However, the lower strength is adequate in heavier section cores. There is a significant increase in strength of the coated sand when it is heated by the molten metal, and can cause shakeout problems especially with aluminum castings.

Phenolic/CO_2

The phenolic/CO_2 system is based on the same type of high-water-content phenolic resole resin used in the phenolic/methyl formate system. The difference is that carbon dioxide gas replaces the methyl formate coreactant.

The CO_2 gas does not have to be collected and disposed of, which is one advantage. The same advantages and disadvantages found in the phenolic/methyl formate system are apparent in this system, with the obvious difference that the less offensive carbon dioxide gas is used.

Furan/SO_2

The furan/SO_2 system originated in Europe around the middle 1970s. This system is comprised of three components: 1) a standard-type furan resin, 2) a peroxide oxidizer and 3) silane, which is added during the mixing cycle to give better binder adhesion and higher strengths. The mix, which has an outstanding bench life, is either blown into the corebox for gassing or can be hand rammed into a corebox and cured under a gassing hood.

In this system, when the SO_2 gas comes in contact with the organic peroxide, it forms an acidic material that cures the furan resin. The reaction is slow, and it benefits from the addition of heated purge air and from warming the corebox surface, against which the core sand will rest, to temperatures above 120F (49C). As might be expected, cold sand and cold operating conditions can negatively affect the furan/SO_2 process curing.

Although a sulfur-type catalyst is present in this system, the color-staining problem associated with the liquid sulfonic acid-cured furan nobake resin system is not a problem. Outstanding bench life and extraordinary shakeout properties are plusses for this coremaking system.

When coreboxes are clean, the core is strong and uniformly dense, and most importantly, the furan/SO_2 system shakes out better than any other unmodified binder system **(Fig 6-4)**. Aluminum foundries use this system to make castings because of its shakeout characteristics.

Acrylic/SO_2

Improvements to the furan/SO_2 process led to the development of the foundry industry's fastest-curing binder system: the acrylic/SO_2 system. Some of the good characteristics of this system are rapid cure, outstanding bench life and good shakeout properties. Again, this system would be ideal for use in an aluminum foundry. However, the system has relatively poor hot strength, so its use eventually faded in all but a few aluminum foundries where small castings are produced.

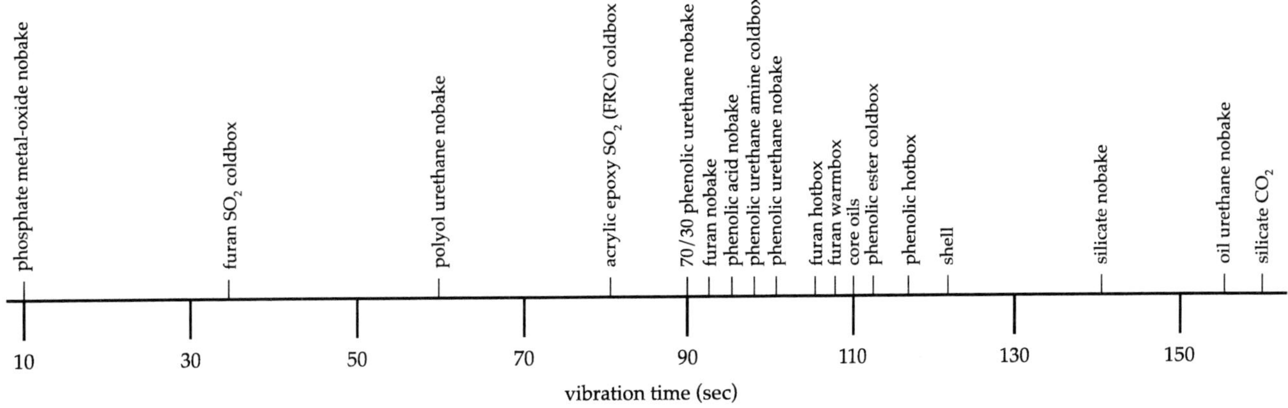

Fig. 6-4. Comparison of shakeout times for chemical binder systems used to cast Alloy 356.

Nobake (Air-Set) Systems

Chemically bonded core systems that need heat or a gas catalyst to cure the coated sand have been discussed. Here, the discussion shifts to chemically bonded systems in which a *liquid catalyst* is used to cure the binder at *ambient temperature,* over a period of time. These systems are not noted for their speed of cure and, thus, are not usually used for high production rates for cores and molds, as are the heat-cured and gas-catalyzed systems. Very large cores, as well as small cores, can be made using these systems. Also, molds of various sizes can be made, and are typically used in jobbing or low production rate foundries.

Any nobake process is a binder system that will cure at ambient temperature with a predictable and controllable "work time," "strip time" and "cure time." *Work time* is the amount of time, after coating the sand with a reactive resin catalyst, that the coated sand remains workable, flowable and capable of being made denser by compacting. Of these three terms, work time is the most important operator-controllable element in nobake core and mold production. *Strip time* means the amount of time before the core is strong enough to be removed from the corebox and retains its shape. On the other hand, the pattern can be removed from the mold and the mold will hold its shape. *Cure time* is the time necessary for the binder coating to be totally cured throughout the core and/or mold. The cure rate is based on the type of resin, the amount and type of catalyst used, and especially the temperature and humidity.

Some binders, such as the acid-cured furans, actually generate water as they cure. This water must evaporate to allow the resin to cure thoroughly. High ambient relative humidity slows the rate of evaporation and extends the strip time. Because under-cured cores and residual solvents produce excessive amounts of hydrogen gas, care must be taken to optimize the time between stripping and pouring to minimize gas-related defects in the castings and to maintain dimensional integrity.

Temperature is also an important controllable factor in nobake systems. Cold sand, cold coreboxes, patterns and low ambient temperatures slow every part of the core and moldmaking operations. Conversely, heat can speed up each of the processes. Temperatures under 50F (10C) prevent effective sand coating and necessitate special adjustments to core and mold production. Ideally, process temperature operating parameters for nobake cores and molds should be in the 75–85F (24–29C) range. Lower and upper temperatures are 50F (10C) and 110F (43C), respectively.

A good rule to keep in mind is the, "10-degree Celsius or 18-degree Fahrenheit Rule." This rule states that, for each 18°F (10°C) increase or decrease in temperature, the speed of the chemical reaction takes twice as long to occur (if cooler) or happens twice as fast (if warmer). Temperature is one of the primary things that controls such things as work time, bench life and strip time. To overcome this potential problem, foundries have installed sand heaters/coolers and binder heaters/coolers.

The effects of humidity and temperature control are as important to a nobake binder system as the type and amount of catalyst. In many ways, they are even more important than the effects of the chemicals used in the process.

For example, when working with "Bondo" or other products containing a resin and hardener, these products work the same way; the only difference is that, in cores and molds, sand grains are bonded together and the hardener is called a catalyst.

It should be mentioned that the nobake cores and molds could be made stronger by placing them in an oven for a period after curing **(Fig. 6-5)**. This action can also help drive out any moisture that may have been absorbed by the cores or molds for a long storage period.

The nobake systems include silicate/ester, phosphate/metal oxide, alkyd oil/urethane, phenolic urethane (PUNB), polyol urethane, ester-cured phenolic, furan/acid and phenolic/acid.

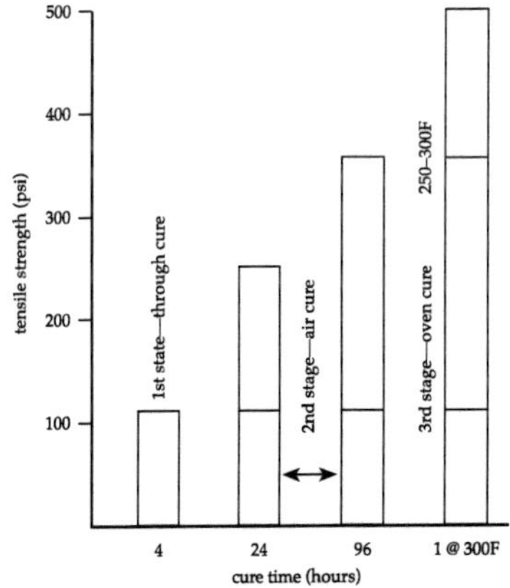

Fig. 6-5. Effect of baking at 300F (149C) on oil-urethane system tensile strength development, even after up to 96 hours of room-temperature curing. 1.2% (based on sand) alkyd resin (with 0.6 phr catalyst) + 20% (based on resin) isocyanate —> cured resin.

Silicate/Ester Nobake

Ester-cured silicate nobake binder systems were introduced in 1954. Silicate binder, when used in the nobake system, uses a liquid, ester-type catalyst or hardening agent to about 10 parts of resin (by weight). The liquid catalyst eliminates the effects of under/over-gassing, which can be a problem in sodium silicate/CO_2 systems. Thus, the nobake silicates have better humidity resistance and core storage capabilities than the gassed (coldbox) version.

As with sodium silicate/CO_2 systems, shakeout problems normally associated with this system in ferrous casting production may not be a concern in aluminum foundries due to the lower pouring temperatures. The hot strength and shakeout properties of silicate binders can be altered by the use of various additives mentioned in the section on sodium silicate/CO_2 systems.

Relatively poor productivity, humidity resistance problems, burn-in, and difficult shakeout and reclamation are potential problems with this system. However, silicate nobakes are successfully used with all types of sand to make cores and molds in many aluminum foundries.

Phosphate/Metal Oxide Nobake

This inorganic binder system was introduced in 1992 consisting of an acidic, water soluble, liquid phosphate binder and a powdered metal oxide hardener.

Like other inorganics, this binder suffers from slow cure, low strengths and poor humidity resistance. However, it exhibits excellent veining resistance and, like all inorganics, virtually no smoke or odor and very little, if any, gas generation when subjected to molten metal. The system offers better shakeout qualities than any other nobake system, and is easily reclaimed using mechanical reclamation equipment.

Alkyd Oil/Urethane Nobake

This system dates back to about 1965, and is the first of the modern nobake systems. The alkyd oil/urethane nobake system is the one preferred for producing very large cores and molds because of its long work time and excellent stripping characteristics.

This system is comprised of three parts: A, B and C. Part A is an alkyd oil-type resin used in the range of 0.8–1.2%, based on the sand. Part B is a liquid amine/metallic catalyst adjusted for speed in the range of 2–10% by weight of part A. Part C is a polymeric MDI-type isocyanate (the urethane component), which must be used at a ratio of 20–100 parts, based on the weight of part A.

This three-part system can be supplied as a two-part system by blending part A (resin) with part B (catalyst). The unique three-stage curing mechanism **(Fig. 6-5)** illustrates how the strength of the system can be controlled in the process. These stages are detailed as follows:

- Stage 1 of the curing process begins when part A reacts with part C to form a urethane-type polymer. This stage produces enough strength to strip the core from the corebox or pattern from the mold. Because the sand is plastic at this early stage, the stripping characteristics are outstanding.
- Stage 2 is a cross-linking of the unsaturated double bonds in the alkyd-type oil. This is the so-called oxidation stage, where oxygen reacts with the binder to produce additional strength for up to 72 hours.
- Stage 3 is an optional part of the process where the core or mold is oven baked to provide maximum physical properties and to minimize gas. This system is affected negatively by moisture in the sand.

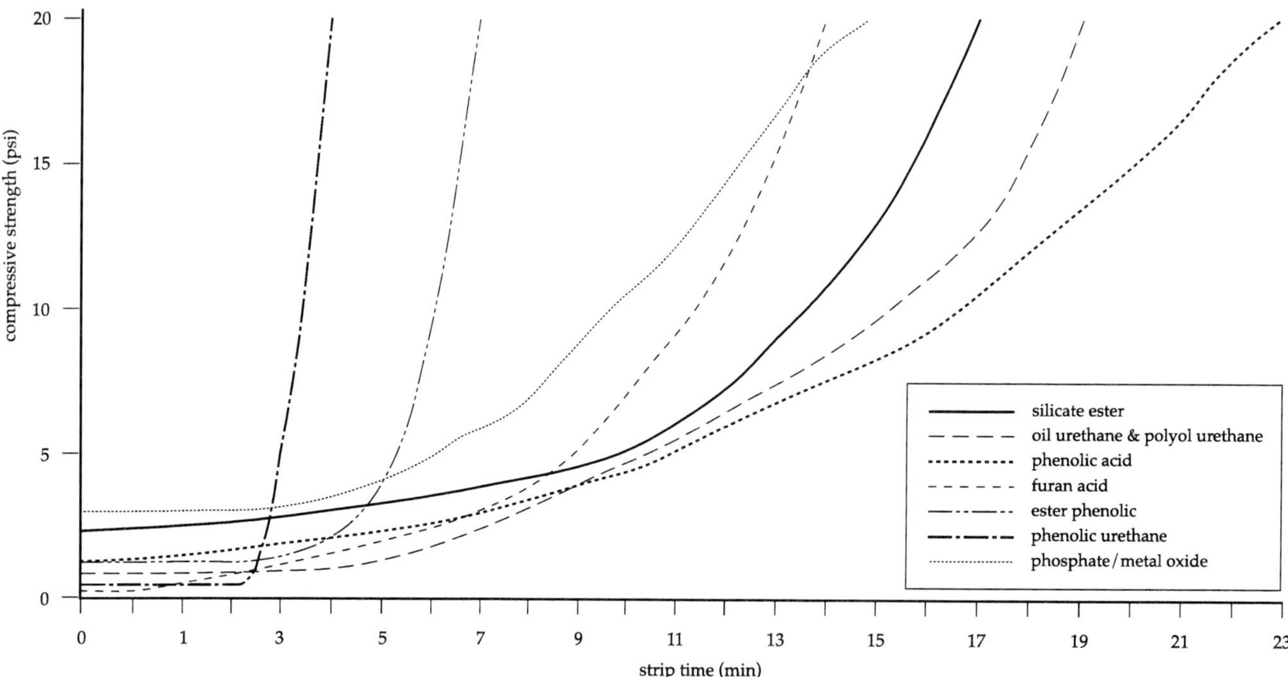

Fig. 6-6. Compressive strength development of nobake binder systems.

Phenolic/Urethane Nobake (PUNB)

This system is standard for making small- to medium-sized cores and molds, is highly productive and offers precise, variable work and strip times. **Figure 6-6** compares the typical rates of strength development for the nobake systems and illustrates the high productiveness of the phenolic urethane system.

The PUNB system was introduced in 1970. The chemical components of the system are similar to the PUA coldbox system mentioned earlier, with the exception of a liquid catalyst being substituted for the vaporized amine. An advantage of the nobake over the coldbox version is that this system has much better humidity degradation resistance.

One notable chemical improvement to the PUNB system has resulted in a unique version that uses a part 1:part 2 ratio of 70:30. This enhancement offers a system that requires much less binder to achieve a given strength. This results in less gas, better flowability, less smoke during pouring and shakeout, and greatly improved shakeout characteristics. Typical binder levels with this system are 0.8–1.0%.

Polyol/Urethane Nobake

This is a three-part system that is similar to the phenolic urethane system. The main difference between these two systems is that, in the polyol urethane nobake system, the resin is formulated to contain reacted polyol, which gives less hot strength than the phenolic.

The decrease in hot strength helps shakeout problems, especially in aluminum alloy castings, which can be a problem with the urethane nobakes. This system is an excellent alternative to both alkyd oil/urethane and phenolic urethane systems in the lower pouring temperature alloys.

Ester-Cured/Phenolic Nobake

The ester-cured phenolic nobake binder system, introduced in the U.S. in 1984, is an alkaline-phenolic resin used with an ester coreactant that controls the rates of cure and strip time. The resin is mixed with a variety of ester coreactants at 20–25%, based on the weight of the binder. Strip times vary from 2 to 80 minutes, depending on which coreactant is selected.

This system has a number of processing advantages: no sulfur, low odor, excellent erosion resistance, no lustrous carbon, low gas evolution and good expansion defect resistance. There is very little smoke during pouring, and the binder cleanup can be done with water. When compared to silicates, this system has better flowability, higher strength and improved handling properties.

The ester-cured phenolic nobake system does not contain furfuryl alcohol and is not compatible with acid-cured systems. It typically is used at a binder level of 1.5–2.0%, based on the weight of the sand, which is somewhat higher than other organic systems. Acid demand value is not especially important with this system, and it runs well with olivine sand.

During the casting operation and using this system, the strength of the core or mold increases dramatically, which can lead to potential shakeout problems. These problems are not as serious in mold shakeout as they are in core collapsibility or shakeout. Another drawback of this system is that the resin's viscosity increases during storage. This can be a problem during hot weather.

Weak strength has led to handling problems with delicate cores, and shakeout problems have limited its use in large-sized core applications. This system has gained acceptance in Europe, and its use, it is said, will increase in North American aluminum foundries.

Furan/Acid Nobake

The acid-cured furan nobake binders were introduced in the U.S. market around 1958. Various types of acid catalysts are used to cure the resin **(Table 6-5)**. The type and amount of the acid determines the strip and cure rates. High acid demand value sand, moisture from any source and low temperatures significantly impair the reaction.

The exothermic (heat producing) chemical reaction between furan and the acid produces water as a by-product. This water slows the cure rate and makes the core or mold curing process appear to proceed from the side exposed to the air toward the side facing the corebox or pattern surface. This explains the so-called "from the outside in" cure characteristics of furan nobakes.

This system is noted for its satisfactory reactivity with all types of acid catalysts, excellent erosion resistance, good hot strength and outstanding shakeout characteristics.

Phenolic/Acid Nobake

The acid-cured phenolic binders were introduced in the early 1960s. Because the phenolic resin is less reactive than the furan, it requires the stronger sulfonic-type acid catalyst for curing. Phenolic nobakes offer a low-cost alternative to furan nobakes, but the system is slower to cure, and the resin has limited storage.

MIXING AND COREBOX FILLING EQUIPMENT

The mixing equipment and coremaking equipment depends somewhat on the type of coremaking system chosen. This equipment can also be selected based on three of the main designations for core binder systems, namely: heat-activated systems, coldbox systems and nobake systems.

The hotbox and warmbox system sands can be mixed in a muller or mixer. A mixer merely stirs or mixes; a muller smears or kneads the mixture while stirring.

These two core processes use the core blowing method of getting the core sand into the corebox. This operation is similar to that discussed earlier, with one main exception: the cores produced

Table 6-5.
Acid-Cured Nobake Catalysts

Type of Catalyst	Solvent	Flashpoint	App. wt/gal
BSA	water/alcohol	50/>200	10.5
XSA	—	—	9.7
TSA	—	—	10.1
85% Phosphate	water	>200	13.8
75% Phosphate	—	>200	13.5

are solid cores and the sand blown into the corebox cavity remains there. Curing of these cores takes place usually in the corebox or can be continued once the core is removed from the corebox, in the case of hotbox cores. The corebox temperatures are much higher for the hotbox cores than the warmbox cores, as previously discussed.

Heat-Activated System Equipment

Core Oil Process Equipment

The core oil process sand mixture can be produced in a muller. Since this process requires heat for curing, it has a relatively good bench life. Depending on the number of cores required, a batch-type muller or continuous-type muller could be used.

The prepared core sand is placed into the corebox, either by hand or by a core machine. When making the cores by hand, the core sand is rammed into the corebox by using either a hand-powered rammer or a pneumatic rammer.

There are two types of core machines used: a core blower or a core shooter.

Core Blower—A core blower fills the corebox with core sand by using a high-pressure air stream. The core sand hopper is above the machine and has an opening at the bottom, which allows the core sand to fall into a reservoir. The high-pressure air is blown through the reservoir, which allows the air and core sand to mix and enter the corebox cavity. The air pressure is applied to the sand reservoir through a hand-actuated or automatic valve. As the core sand and air enter the corebox, the core sand is compacted and the air is vented through vent screens located at suitable places in the corebox. The blowing action gives a very rapid filling of the corebox within a matter of seconds **(Fig. 6-7).**

This rapid filling makes venting of the corebox very critical. Not only does the air in the corebox cavity have to exit ahead of the core sand, but also the air mixed in with the core sand must exit. If the venting of this air is impeded in any way, there will be problems producing quality cores. When using a core blower, a critical part of quality coremaking is proper placement of the vents and proper maintenance of the vents.

Core Shooter—A core shooter operates differently than a core blower. The main difference between these two machines is how the high-pressure air is used to move the core sand into the corebox cavity. In the core shooter, a measured amount of core sand is dropped into a cylindrical-shaped chamber, directly below the hopper. The chamber is then swung into place over the corebox and under a plunger. High-pressure air forces the plunger down, "shooting" the core sand in the chamber into the corebox, and the core sand becomes compacted in the corebox cavity. Again, venting of the corebox is very critical. The air in the corebox must exit ahead of the incoming core sand to get a uniformly compacted core. Core shooters are not as widely used today as they were years ago.

After compaction, the oil sand core is removed from the corebox and placed on a dryer, if necessary. The dryer is used to help the core maintain its shape until cured. The core and dryer, if used, are placed in an oven to cure the core. There are several types of ovens used for curing oil sand cores. The smaller ovens are similar in size to a household oven, whereas, there are some ovens large enough to require a utility vehicle to place the cores in the oven. Tunnel ovens as the name implies, have a conveyor that transports the cores through the tunnel oven. There are tower ovens and, as the name implies, are built vertically. This type of oven has trays that are attached to a conveyor, which lifts the trays and cores up to the top of the oven and then back down to be removed from the oven.

All of these ovens can be heated using natural gas, electricity or infrared heating. Microwaves and high-frequency dielectric equipment can be used to provide the heat to cure the cores as well.

Shell Process Equipment

The shell core process, as described earlier in the section on Heat-Activated Systems, uses resin-coated sand. Depending on the amount of coated sand required, it can be purchased as coated or the foundry can create its own resin-coated sand. The coating process was described earlier in this chapter. The following text will briefly discuss the various methods used to fill the corebox.

Dump Method—The first and oldest method of filling the corebox cavity is the dump box method. In this method, the core sand is placed in a box that is then attached to the pattern plate or corebox. (For the sake of simplicity, from here on, the corebox and pattern will be referred to as simply "pattern equipment.") The box and pattern equipment are inverted so that the shell sand falls out of the box into the corebox cavity or over the pattern surface. After the dwell time has been reached, the machine returns the box and pattern equipment back to their original position. After the cure cycle is complete, the core or mold is ejected from the pattern equipment and the machine is ready for the next cycle. This core machine can be operated manually or run in a semi-automatic or automatic mode. **Figures 6-8 and 6-9** show two different types of shell molding machines.

Dump/Blow Method—This machine uses a dump-blow method to transport the core sand to the pattern equipment. This method operates similar to the dump box method with the exception that the air is blown into the coated sand as it falls into or onto the pattern equipment. The same sequence as the dump box method takes place and the core or mold is removed from the pattern equipment.

Blow Method—The final method of getting the coated sand into or onto the pattern equipment is the blowing method. The equipment to do this operation and steps necessary were discussed earlier in this section. The only difference is that the entire process of curing the core or mold is accomplished in the machine. Again, it must be stressed that vents be appropriately placed and maintained, especially in the corebox. **Figure 6-10** shows an automated shell core machine.

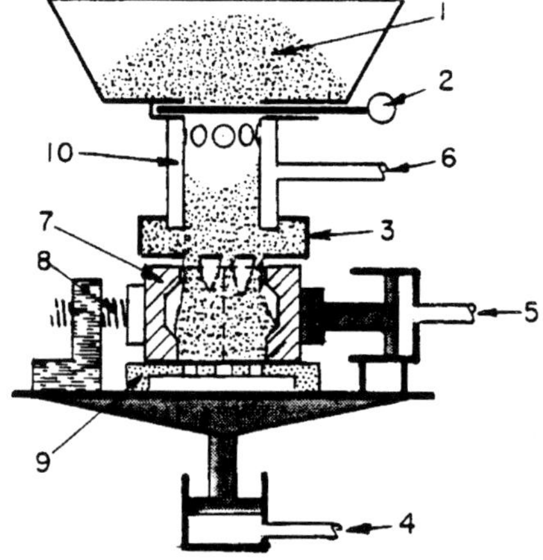

Fig. 6-7. Illustration of a core blower machine: mixed sand (1), gate valve (2), blowing head (3), air cylinder (4), air clamp (5), air gun (6), open top of corebox (7), box location repeater (8), bottom plate (9), feed chute (10).

Coldbox System Equipment

For the sodium silicate/CO_2 system, either continuous or batch-type mixers can be used to prepare the core sand using metering pumps to add the sodium silicate. Over-mixing should be avoided, because there may be a reduction in the work time of the sand mixture. Mixtures normally have a good bench life, but it becomes shorter when higher ratio silicates are used. Containers of prepared sand should be covered with a plastic sheet or damp sacking material to prevent premature hardening of the surface (crusting).

Both continuous and batch-type mixers can be used with the phenolic/urethane/amine (PUA) coremaking system. Mixers with low heat buildup are preferred. The furan/SO_2, epoxy acrylic/SO_2 and acrylic/SO_2 sand systems can be mixed in almost any type of mixer. Because of the exceptional usable bench life of the mixed sand (8–24 hours for furan, and weeks for epoxy), the normal less-expensive batch-type mullers and slow-speed single auger mixers are just as good as the newer generation of high-speed mixers. Once the chemicals are mixed with the sand, no chemical reaction takes place, thus the exceptional bench life.

In the case of the sodium silicate/CO_2 systems, the prepared sand can be placed into the corebox or over the pattern by hand and compacted by hand. A coremaking machine can also be used, where the prepared sand is blown into the corebox. As with all coreblowing machines, proper venting of the corebox is very critical. In addition, these vents can aid the distribution of CO_2 throughout the core. These coremaking machines can be fully automated or semi-automatic.

All of the other coldbox systems use coremaking machines that are either automatic or semi-automatic in their operation. These sand mixtures are blown into the corebox cavity and compacted; thus, proper venting of the corebox becomes very critical. Next, the gas or vaporized catalyst is released into the core, after which air is used to purge any of the remaining catalyst out of the core. These two procedures also require proper placement of the gassing tubes. The purge air is then piped to a chemical scrubber that makes it suitable for release to the atmosphere.

Fig. 6-9. Single-station, fully automatic shell molding machine, capable of producing accurate molds in short cycle times. [Photo courtesy of Dependable Foundry Equipment Co.]

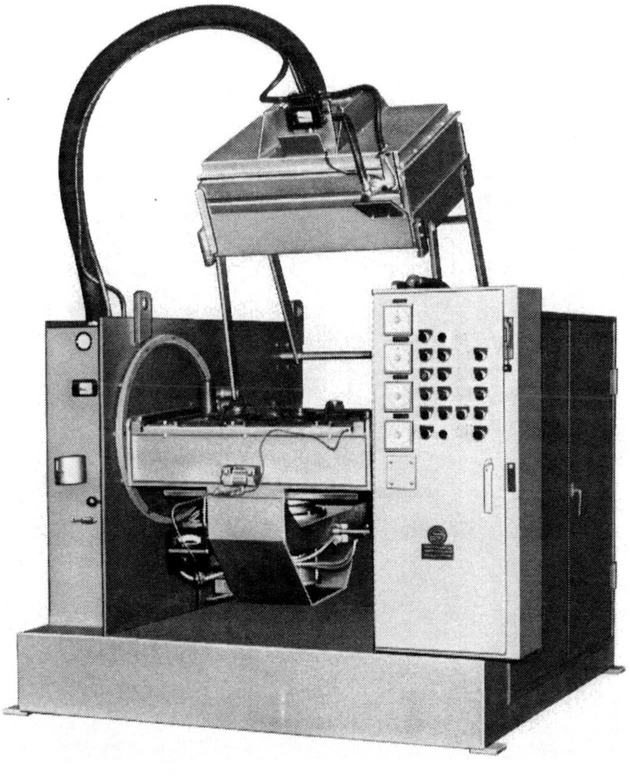

Fig. 6-8. This type of automatic shell molding machine is widely used throughout the world. Pattern size: two 20.5x15.5-in. plates or one 20.5x31.5-in. plate.

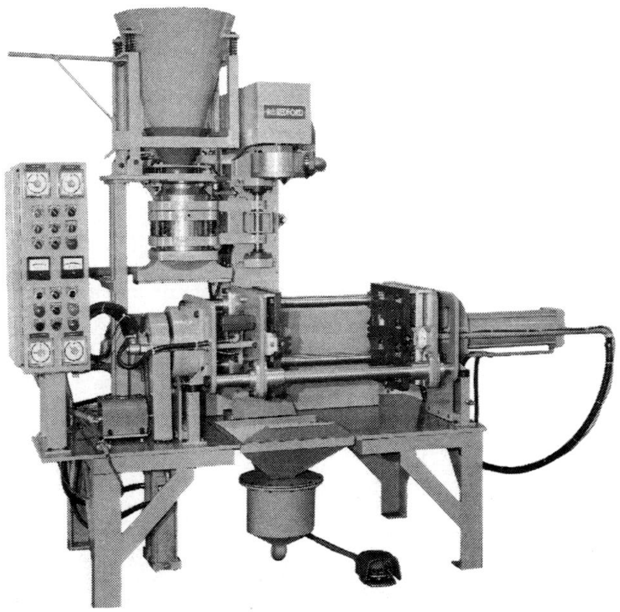

Fig. 6-10. This automatic shell core machine has a blow capacity of 40 pounds. [Photo courtesy of Dependable Foundry Equipment Co.]

Nobake (Air-Set) System Equipment

The silicate/ester nobake system can use batch mixers or high-speed intensive mixers to prepare the core sand. It is recommended that the ester hardener (catalyst) be added to the sand first and thoroughly dispersed before adding the silicate binder. Care is necessary with batch mixers not to over-mix the sand, which reduces the work time. On the other hand, high-speed intensive mixers may cause dehydration of the silicate binder, which impairs bond strengths. Over-mixing in batch mixers produces a dry, powdery mixture, which should be discarded to prevent the production of weak, friable molds and cores.

Longer cycles on batch mixers or mullers result in phosphate/metal oxide sand mixtures with higher green strengths at discharge. Excessive blending should be avoided when using batch mixers, as this will shorten work life. When a high-speed muller is used, the mulling cycles should be shortened. The best mixing in continuous mixers is usually accomplished by minor adjustments in blade angles and the reduction of sand flow below that normally used for more flowable organic binder systems. In all mixers or mullers, it is advisable to disperse the catalyst/hardening agent first, before the addition of the resin component.

Alkyd oil/urethane nobake system sand can be prepared in a muller or continuous mixer. Oxides, if required, are added first to the sand; next, parts A and B, which have been premixed, are added; and, finally, part C is added. The amount of time required for proper mixing or mulling of each component can be obtained from the manufacturer of the system's chemical components. In continuous mixers, these additions are in the same order, with distances of 2–12 inches separating the various additions. Equipment manufacturers can also advise the proper positioning.

The phenolic urethane nobake (PUNB) system sand can be prepared in most foundry batch-type mullers, auger-type continuous mixers and the newer generation of high-speed intensity mixers. Part 1 and the proper catalyst/activator are added to the sand first. Because of the small amount of catalyst activator used, it is recommended that it be preblended with part 1 before it is added to the sand. This will ensure efficient distribution and optimal performance.

Mixing times for batch mixers are best determined at the point of use; over-mixing of the sand and resin may result in lower tensile strength, due to the partial curing of the mix. However, under-mixing may result in poor distribution and lower tensile. Properly mixed sand will be discharged from the mullers and mixers as a free-flowing mass that can be distributed and compacted easily, since there is essentially no green strength.

Ester phenolic nobake system sand can be mixed in any of the mixers discussed. The resin portion is slightly higher in viscosity than other binders, thus requiring more intensive action to distribute the resin uniformly. The resin is usually added to the sand first, then the esters; however, adding the ester first is acceptable.

Furan/acid nobake system sands are prepared in batch mullers and continuous mixers. Batch mulling is used when strip times are relatively long. Catalyst is added first and mixed with the sand for two minutes. Next, the resin is added and mixed for an additional two minutes. In continuous mixers, the catalyst is again first and the resin second.

Batch mulling is an acceptable method of preparing phenolic/acid nobake system sands when the desired strip times are long. When making this decision, consideration should be given to mulling time, handling time to filling station, and filling/compacting time of core or mold. The total time of these operations should not exceed one-third the strip time, or cured sand may be encountered in the core or mold.

In trough-type continuous mixers, the acid catalyst should be added first and then the resin at a distance of 2–8 in. (5.08–20.32 cm) further down the trough. In the high-speed turbo-head mixers, the resin and catalyst can be added at the same time because of the high mixing speed and short retention time.

When preparing any of the nobake system sands, it must be remembered that, as soon as the catalyst meets the resin, the cure begins. In other words, when the mixed sand leaves the muller or mixer, it has already begun to cure. When mixed correctly, and the work time is watched carefully, these sands are said to have good flowability. That is, they will flow very readily over and around the pattern, in the case of molds. In the case of cores, it will flow very readily into the corebox. Some hand tucking may be required in areas such as deep pockets and areas where the sand flow is restricted. Final compaction is done on a compaction table, which vibrates the mold in the flask or core in the corebox.

All of this equipment—mixer or muller, stripping machine and compaction table—can be placed into a loop, straight line or circular production line. In some cases, a tunnel oven may even be used. The production line is usually made up of roller conveyor sections. For larger, floor molds and very large cores, the continuous mixer may be a mobile mixer that moves on rails to the flask or corebox filling stations. Again, final compaction takes place on a compaction table.

TESTING CHEMICALLY BONDED SANDS

When compared to clay bonded sand, there is relatively little testing done with the chemically bonded sands. The procedures to test for specific core and mold properties are well established and can be found in the *AFS Mold and Core Test Handbook,* 3rd Edition, 2000.

The type of test and the upper and lower control limits that are used to measure core and mold quality should vary according to job size, complexity, required handling and storage considerations. A meaningful testing procedure should be used as a value for adjusting core and mold strength to compensate for outside factors, such as temperature and humidity, which affect performance.

The following tests (discussed in Chapter 4, Molding Sands) are used for chemically bonded sands: AFS Grain Fineness Number; Loss-On-Ignition (LOI); Sand Density; Permeability; and Screen Distribution.

Other tests used for the chemically-bonded sands include Tensile Strength, Transverse Strength, Gas Evolution, Acid Demand Value (ADV) and Shell Core and Mold Testing. Each will be discussed in the following text.

Tensile Strength Testing

Essentially every foundry using chemically bonded sands uses tensile strength as the primary control procedure. The test is simple and fast, but its values often are too narrowly and strictly interpreted.

The tensile strength value of core sand should be just high enough to prevent breakage during handling. Molds should be strong enough to prevent runouts. It is a mistake to believe that most defects can be solved simply by making everything "rock hard." This approach will be likely to create other defects, and should not be considered a realistic alternative.

Transverse Strength Testing

Transverse strength testing of resin-bonded sand is an excellent alternative to simple tensile strength testing. There are a few times a core or mold is subjected to tensile stress in the casting process. The transverse testing apparatus **(Fig. 6-11)** uses a 1x1x8-in. (2.5x2.5x20-cm) specimen.

Fig. 6-11. Transverse strength testing apparatus.

Tensile and transverse tests using a 1-in. (2.5-cm.) thick specimen provide an overall strength measurement that relates to heavy-section cores. Strength readings obtained on 1-in. (2.5-cm.) thick cross-section specimens can be misleading if the interior has high strength but the surface is weak.

An alternative transverse test uses a thin-disk specimen, 0.312 in. thick x 1.95 in. diameter (8 mm thick x 50 mm dia). The thin-disc specimen more accurately reflects the properties of the core surface. It also relates better to the strength properties of today's complex core configurations with thin sections.

Gas Evolution Testing

The rate and amount of gas that evolves from resin-coated sand during the casting operation is one of the most important aspects of the core and mold's ability to produce a quality casting. The gas evolution test measures those aspects.

The test is performed with a 0.25-gram resin-coated core or mold section that is inserted into a furnace preheated to 1850F (1010C). As gas is generated from the decomposing coating, its quantity is measured at specific time intervals. It should be noted that the standard procedure for determining gas evolution has a predrying step, the results of which can be misleading. In this procedure, the sample is dried for about an hour at 220–235F (104–113C) before the sample is inserted into the heated furnace. Consequently, any moisture or solvent-based material that exists on the sand will be driven off during this predrying step.

Acid Demand Value (ADV) Testing

The acid demand value test is used to determine the acid-soluble elements that are present in the raw sand. It is also an indicator of impurities in the sand. The ADV test is especially useful in predicting how sand will work with the catalyst materials of the various binder systems. High ADV sand minimizes the bench life of phenolic urethane coldbox (PUCB) systems. High ADV sands require neutralization of the alkaline materials by the acid catalyst in furan and phenolic nobakes.

ADV is a good test to run on incoming sand as soon as it is unloaded. This could give an indication of some incompatible material that might have contaminated the shipping container during the previous load.

Shell Core and Mold Testing

There are several tests run on shell sands that are not run on other core or mold sands. Several of these tests are run for shell sand production control and the others as laboratory control tests on the coated sand mix itself.

The *production control tests* include *cold and hot tensile strength*; and *stick point* or *melt point*. In the tensile strength tests, a sample of shell sand is made in the form of a dog bone and removed from the mold while hot, or is allowed to cool and then pulled. The dog-bone shaped samples are made in the tensile testing machine itself. **Figure 6-12** shows this testing machine.

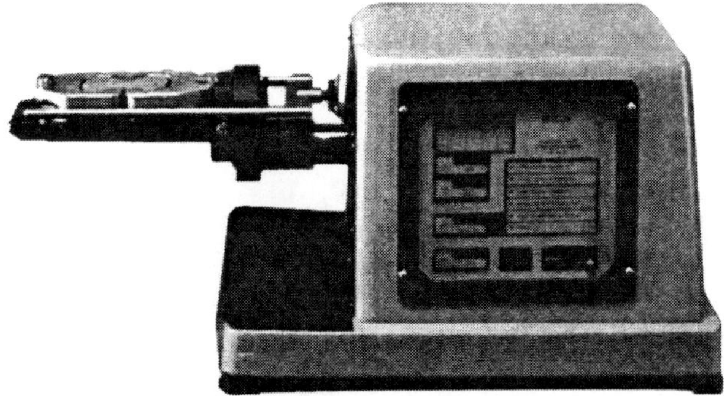

Fig. 6-12. Shell tensile strength tester.

Table 6-6.
Compatibility of Chemical Binder Systems Used on High-Purity Silica Sand (No Additives) After Mechanical Reclamation
(0 = incompatible, 10 = compatible)

Binder system used for recoating	Original binder system used to coat sand: Shell	Core Oil	Furan & Phenolic Hotbox[D]	Warmbox	Furan & Phenolic Nobake[E]	Oil Urethane Nobake	Phenolic Urethane Nobakes & Coldbox[D]	Epoxy/ Acrylic SO_2	Phenolic Resole Nobake & Coldbox[F]	Phenolic Resole/ CO_2	Silicate CO_2 & Nobake[C]	Inorganic Phosphate Nobake
Shell	6–10	10	8–10	7–10	5–8	10	10	10	2–8	8	0–5	7
Core Oil	10	10	10	10	10	10	10	10	10	10	10	10
Furan & Phenolic Hotbox	10	10	10	10	10	9	8	10	0	8	5	7
Warmbox	10	10	10	10	10	9	7	10	0	7	1	2
Furan & Phenolic Nobake	5–10	10	10	10	10	9	6	10	0–4	6	5	7
Oil Urethane Nobake	10	10	2–5[A]	1–5	2–5	10	10	10	0–10	8	6	8
Phenolic Urethane Nobakes & Coldbox	7–10	10	2	1–5	1–5	10	10	9	0	8	5	7
Epoxy/Acrylic SO_2	10	10	10	10	10	10	9	10	0–4	6	1–4	3
Phenolic Resole Nobake & Coldbox	10	10	8–10	8–10	10	10	10	10	7–10[B]	8	2	2
Phenolic Resole/CO_2	10	10	4	2	1	10	10	10	0	6	2	2
Silicate CO_2 & Nobake	10	10	2	0	0	10	10	10	0	8	0–8[C]	7
Inorganic Phosphate Nobake	10	10	10	10	10	10	10	10	0	6	7	10

A—Needs extra catalyst B—Needs to have a high percentage of new sand added and has third part additive to help reclamation C—May be difficult to particulate. Recycling accumulates soda and changes SiO_2/Na_2O ratio
D—Might build up nitrogen E—Might build up sulfur F—Will build up potassium and shorten work time

The stick point or melt point test is run on a gradient heated metal bar. A sample of the coated sand is placed along the length of the bar. After a period, a small camel's-hair brush or air at a very low pressure is used to brush or blow the coated sand off the bar. A point at where the coated sand begins to stick to the bar and can't be removed is said to be the stick or melt point. The temperature of the bar is taken at this point and becomes the stick point or melt point of the coated sand.

Tests on the prepared mix that require more than five minutes to complete are called *laboratory tests*. These tests include shell tensile strength, or shell transverse strength and deflection, shell permeability and burn-off or total loss-on-ignition.

RECLAMATION AND REUSE OF SPENT SAND

Sand reclamation is the removal of the chemical binder coating from the surface of the spent core and mold sand. It should be mentioned here that spent green sand could also be reclaimed, if necessary. Reclamation can be performed by using mechanical, thermal or water-type (wet) processes.

Any serious sand reclamation project should begin with a study of the sand, binder(s), sand:metal ratio and reclamation process alternatives. After a tentative decision of the type of reclamation method to be used is made, various vendors of this method should be consulted. Only after a representative sample of spent coated sand has gone through the entire reclamation process, and has been recoated with resin at the appropriate binder and catalyst levels a number of times, should a final decision be made.

Good sand reclamation cannot start at the waste sand pile. It must be an integral part of the foundry process, based on consultation with all parties involved—the operating management, the equipment suppliers, the engineering firm, the environmental consultants and the relevant permitting agencies.

Mechanical Reclamation and Binder Compatibility

When core and mold sands coated with different binder systems are mixed within the same mechanical reclaimer, unexpected results are likely to occur when the reclaimed sand is recoated with a chemical binder. The chemistry of a mechanically reclaimed sand surface is no longer that of the original sand. The surface chemistry of the reclaimed sand grains after mechanical reclamation is much like the binder system that the reclaimer tried to remove, and the original physical and chemical nature of the sand has probably been altered dramatically.

Table 6-6 can be used as a general guide to predict the binder system compatibility when recoating mechanically reclaimed sand with the same or different type of binder system. Note that the table is very general. A single chart cannot take into account all variables of mechanical reclamation systems.

What is mechanical reclamation or, as it is sometimes called, mechanical scrubbing? Mechanical reclamation takes place in a machine where the spent sand is literally thrown against a "target" (composed of hard, wear-resistant cast iron) at a very high rate of speed. When the spent sand impacts the target, the resulting impingement causes the binder to crack and break away from the sand surface. This impact also causes the sand grains to rebound back into the oncoming stream of sand. The action causes a "scrubbing" between the sand grains, which also helps remove the binder from the sand grain. The spent sand, which is usually in lump form, is first particulated into individual sand grains before being placed into the mechanical reclamation unit.

When castings are poured, the extreme high temperatures of the molten metal cause embrittlement of the molding sand grains. Mechanical reclamation benefits from the "thermal embrittlement" of the coated surface of the sand. However, alloys with low pouring temperatures, such as aluminum, will make it more difficult for the mechanical reclaimer to do its job.

Most system sands will tolerate a minimum amount of incoming sand having a completely incompatible binder system. Some systems do not experience a serious loss of properties in a single coating cycle, but may suffer drastically after multiple passes. Also, depending on the physical nature of the base sand, some mechanically reclaimed sands will get stronger, and some will not have the resin-bonded strength of the original sand.

Another point to keep in mind is the shape of the sand grains and their screen distribution. As the sand grains are "scrubbed," their shape can be greatly changed. If an angular sand was used, it will, after reclamation, tend to be become more rounded. The sharp points of the sand grains will be broken off and become fines, thus affecting the screen distribution. When the sand grains are subjected to very high pouring temperatures, they can begin to thermally crack. This can weaken the sand grain and cause it to split apart in the reclaimer, thus changing the shape and screen distribution of the reclaimed sand.

Thermal Reclamation

There are fewer problems with chemical interactions when thermal reclamation is used. Complete thermal reclamation requires about 1,000,000 Btu/ton of spent sand to thoroughly remove binder from silica sand. Organic binders contribute energy to the calcining process because they burn. Calcination is a heat treatment process performed on materials (in this case, sand) to remove volatile substances. Based on resin content and binder type, 500,000–900,000 Btu/ton of spent sand will be obtained from the organic resin-bonded sand systems.

Inorganic binders do not burn. They, unlike organic binders, contribute nothing of fuel value for burning the coating off the spent sand. In addition, any type of inorganic material present in the sand's coating tends to build up on the reclaimed sand, often resulting in curing problems and a significant loss in strength.

Some evidence of thermal reclamation can be seen at shakeout. At times, the resin-coated sand on the surface of the core or mold can reach temperatures at which thermal reclamation can begin. The affected sand will be white in color and fall away like salt, indicating that the binder has been burned away. An aluminum permanent mold foundry uses shell cores in its casting process. Rather than removing the shell cores from the castings, the castings with the cores still inside are put into heat treat furnaces. As the castings are heat treated, the shell cores are burned out and somewhat thermally reclaimed.

Some foundries will use a combination of thermal and mechanical reclamation. They feel that this combination will give them even better results than either method individually.

Wet Reclamation

Wet reclamation is generally not used because the wastewater disposal process is extremely expensive. It requires a few hundred gallons of water to wet-reclaim a ton of sand that has been coated with a water-soluble binder like silicate. Thus, if wet reclamation is used, it will be with a silicate-bonded system.

Reuse of Spent Sand

In the past, after being used, resin-coated sands were 1) metered into the clay-bonded molding sand system; 2) treated via some reclamation process and recoated with resin; or 3) relegated to a dumpsite.

A fourth option is now available. Current studies show that resin-bonded core/mold sands are nonhazardous by U.S. government standards. Therefore, they can be used as a raw material in diverse applications such as sanitary landfill covers, roadbeds and road fills. These sands also can be used as one of the ingredients for making bricks and concrete blocks, or as filler for cement, concrete or asphalt.

REFRACTORY COATINGS

Refractory coatings can be applied to resin-bonded cores and molds. However, before applying a coating (or wash, as it is sometimes called), it is important to realize that any coating placed on a core or mold that has a very poor surface finish will not fulfill its intended purpose. A coating will only be as good as the substrate material to which it is applied. Thus, using a coating to cover up poor coremaking practices is not a good idea. The cores should be made to meet the highest quality requirements possible and then, if necessary, a coating can be used.

Why are coatings used? There are several reasons: 1) to improve the surface finish of the cored passageways in the casting or, in the case of molds, the casting's external surfaces; 2) to control the heat transfer characteristics and thus the microstructure in the casting; 3) to improve the venting of a core; and 4) to prevent certain defects in the casting.

Improved surface finish—If the cored passageway or casting surface calls for a smoother surface than previous specifications, a coating can help improve surface finish, if applied properly.

Heat transfer—If there are known areas in a casting where the heat-transfer action from the casting to the core or mold is hindered, a coating that has "chilling" characteristics can be used. In other words, the coating will absorb heat very readily, causing faster solidification. This can prevent some of the core sand- or mold sand-related defects in areas where heat can be trapped. Should there be an area where a harder surface is required, a chill-type of coating can be used. The steeper thermal gradient produced by the chill coating will cause rapid solidification and a harder surface.

Venting—Gas generated by the core, as it is heated by the molten metal, will follow the path of least resistance to escape the core. This path may be through the surface of the core. By "sealing" the surface of the core with a coating, the gas can be forced to leave through passageways that have been provided, called vents.

Defects—In certain cases, the core or mold may be subjected to severe turbulence from the flowing molten metal. A coating can be used in those areas to prevent erosion. In very large castings, especially the ferrous castings, cores and mold surfaces in the drag can be subjected to tremendous metallostatic pressures, which can cause the molten metal to penetrate between the sand grains. A coating can help prevent this type of defect from occurring.

Coatings can be applied by 1) dipping certain core sections or the entire core into the coating mix, 2) brushing on the coating and 3) spraying very large cores with the coating. In the case of molds, the coating is usually brushed or sprayed onto the mold cavity surface. The application of the coating is very important and should be closely monitored.

Coatings consist of several ingredients: refractory material, carrying agent, suspension agent and, sometimes, clay.

The *refractory material* can be silica, olivine, zircon or chromite. The latter two are used primarily for chilling characteristics. Whichever refractory is used, it has to be "carried" to the core or mold surface and actually penetrate between the sand grains. The *carrying agents* most commonly used are water, chlorinated solvent or alcohol light-off (methyl or ethyl alcohol). The latter two carriers will more than likely be phased out, due to government regulations. Thus, coatings in the future will be water-based coatings.

Some type of a *suspension agent* is used to help keep the refractory in suspension. If this is not done, and there is no agitation in the storage container from which the coating is applied, the refractory will settle out of solution and to the bottom of the container. Some coatings also contain some *clay*, which gives the coating refractory material some adhesiveness when it encounters the sand grain.

When water-based coatings are used, they should not be applied to heat-activated binder systems (hotbox, warmbox) until the cores or molds are cooled to just above room temperature. In coldbox systems, the coatings should be applied as soon as the core is removed from the corebox, and the coating should be formulated to contain little (if any) carrying or "penetrating" agent so that there is little penetration into the core or mold surface. Nobake systems should not be coated until the cure has been completed. It is best to warm the nobakes before applying the coating. Under no circumstances should a coating be applied to cold cores or molds.

In all cases, the applied coating should be 1–10 mils (0.001–0.0010 in.) thick over the surface **(Fig. 6-13).** The refractory coating must be formulated at the proper Baumé, with just enough wetting agent for it to penetrate no further than 3–4 sand grains into the resin-bonded sand surface. The wetting agent aids in coating adsorption. This level of coating penetration anchors the coating and prevents spalling. When multiple layers of coating are to be applied, each layer must be dry before the next coat is applied.

VENTING

It is impossible to overvent. In some cases, venting is a lost technology in foundry practice. Good venting practice not only allows core and mold gases to escape, it also allows the air in the mold cavity to escape in front of the incoming molten metal.

Molds should have vent holes drilled into the cope and scratch vents etched on the parting line. Cores should have vent holes drilled or produced by arbors, or vent wire placed in the core prints, wherever possible.

EFFECT OF RELATIVE HUMIDITY

All binder systems lose strength when they are exposed to high relative humidity (RH). How much strength is lost is based on the type of binder and the relative humidity. The higher the RH and the higher the temperature, the greater the effect of RH on the core or mold.

Humidity degradation becomes a greater problem as the temperature increases in climates where the RH is high. This factor is true because the air can hold more moisture as the temperature increases and, therefore, can release more moisture as condensation when the air temperature drops. This condensation can settle onto core and mold surfaces.

If the resin film is thin, as is the case with low resin levels, the moisture may soften and penetrate through the binder film. When the moisture reaches the area between the resin and the sand grain surfaces, it can cause adhesive failure. This phenomenon begins on the core or mold surface and progresses inward, partially explaining why thin cores are so sensitive to humidity degradation.

Humidity degradation reverses somewhat if the RH decreases, or if the temperature increases significantly. When this occurs, the cores and molds will regain some of their strength. When cores are used in green sand molds, they are often sent through a small drying oven just prior to being set in the mold. This practice can help eliminate the possibility of water from the green sand condensing on the cooler core surface.

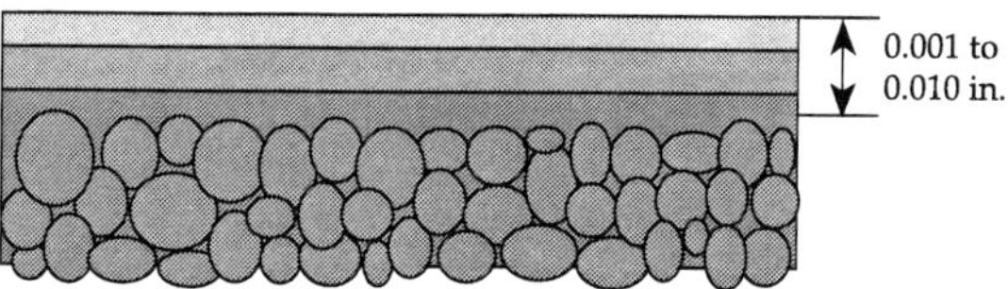

Fig. 6-13. Multiple layers of coatings should be of contrasting colors. A coating layer must be completely dry before the next layer is applied.

The effect of humidity degradation becomes more severe when air pressure is increased, as it is in the blow magazine of a core machine. When using a core blower, "bone dry" blow air must be used to produce strong, storable cores. It is advantageous to use some type of air dryer in the air system.

SHAKEOUT

Shakeout in some of the lower pouring temperature alloys can present a problem. To help shakeout, optimize casting properties and satisfy environmental requirements, the resin-bonded sand and the casting process should satisfy the following requirements in all alloys:

- Hot strength of the binder should be high enough to prevent expansion defects but low enough to promote adequate shakeout.
- Low pouring temperature alloy "versions" of binder systems should be used when available.
- The binder system should generate a minimum amount of gas, and should be able to make an acceptable core or mold with minimal binder catalyst.
- Shakeout additives should be used when necessary.
- Shakeout should be performed before the sand has cooled.

BIBLIOGRAPHY

AFS Mold and Core Test Handbook, 3rd Ed., American Foundry Society, Des Plaines, IL, 2000.

Aluminum Casting Technology, 2nd Edition, The American Foundrymen's Society, Inc., Des Plaines, IL (1993).

Carey, P., "The Effects of Humidity on Modern Core Practices," *modern casting,* American Foundrymen's Society, Inc., p. 52 (Sep 1981).

Chemically Bonded Cores and Molds, The American Foundrymen's Society, Inc., Des Plaines, IL (1987).

No-Bake Cores and Molds, The American Foundrymen's Society, Inc., Des Plaines, IL (1980).

Metal Alloys Cast in the Foundry

7

Since the pattern equipment, mold and core(s) have been discussed, it is now time to choose the appropriate alloy for the casting application. The casting designers may make this choice because of their familiarity with the casting's end use. The material plays an important role in the casting design. However, casting designers don't always understand the details of this selection process and will consult the foundry for advice or refer to one of many handbooks available.

Very few foundries pour pure metal; what they pour is actually a metal alloy. For instance, aluminum, copper, iron, magnesium and titanium are just a few of the metals that can be cast in foundries. A metal alloy is a combination of chemical elements, of which at least one has to be a metal. Aluminum alloy, Alloy 535.0, is an alloy in which, by percentage, aluminum is the major alloying element. This alloy also contains smaller percentages of the following metals: magnesium, iron, copper, titanium, manganese and silicon (which is chemically classified as a nonmetal but metallurgically as a metal).

Another alloy is gray cast iron. The predominant chemical element in gray iron is iron, which is a metal. A typical gray iron may also have smaller percentages of silicon, manganese, phosphorous and sulfur. Again, silicon can be considered a metal, metallurgically.

CLASSIFICATION OF CASTING ALLOYS

Metal alloys can be divided into two major families: Ferrous and Nonferrous. Ferrous alloys, which will be covered in Chapter 8, are iron-based alloys. Nonferrous alloys, which will be covered in Chapter 9, contain very little, if any, iron.

Following is a list of the subdivisions under the two alloy families:

Ferrous Alloys

A. Carbon Steel
   1. Plain carbon steel
      a. Low-carbon Steel
      b. Medium-carbon Steel
      c. High-carbon Steel
   2. Low-alloy steel
   3. High-alloy steel

B. Cast Iron
   1. Gray cast iron
   2. White cast iron
   3. Malleable iron
   4. Ductile or nodular cast iron
   5. Compacted graphite cast iron
   6. Austempered ductile cast iron

Nonferrous Alloys

A. Light Alloys
   1. Aluminum-base
   2. Magnesium-base

B. Heavy alloys
   1. Copper-base
   2. Lead-base
   3. Nickel-base
   4. Tin-base
   5. Zinc-base

C. Others
   1. Titanium
   2. Cobalt

It is impossible to provide a complete listing of all the castable alloys that do not fall into either the ferrous or the nonferrous families. However, some are worthy of mention here:

High Temperature Alloys—Although parts using complex, high-temperature alloys can be fabricated by other processes, they can be cast using precision-casting methods. Discussing these various alloy types goes beyond the intent of this book. However, elements such as cobalt, chromium, columbium (niobium), tantalum, tungsten, nickel and titanium can be used in various combinations to produce these alloys.

Nuclear Power Alloys—There are various special alloys that are well suited for nuclear power applications.

Dental Alloys—These alloys are used to make dental crowns and other dental appliances. Dental alloys compare with the high-temperature alloys in cost, base elements used and complexity of composition. These alloys are cast using the precision casting processes.

Precious Metal Alloys—Castings made of silver, gold or platinum are used where the need for high corrosion and oxidation resistance or aesthetic qualities are required.

The various alloys mentioned can be further subdivided into a number of different variations of alloys. Some of these variations can be very extensive. The final selection of the alloy to be cast can include one or more of the following factors:

- Cost
- Corrosion resistance
- Toughness
- Weight
- Tradition
- Strength
- Casting properties (castability)
- Appearance
- High or low temperature properties
- Electrical properties
- Susceptibility to heat-treatment
- Wear resistance and machinability
- Personal preference

PROPERTIES OF METALS AND ALLOYS

The physical and mechanical properties of various metals and their alloys come into play when choosing which metal or metal alloy will best meet the casting's end use and the environment.

Physical properties are those properties determined by static tests and include 1) density, 2) electrical and thermal conductivity, 3) corrosion resistance, 4) expansion and 5) specific heat.

Mechanical properties reveal the elastic or inelastic reaction of the metal or metal alloy when force is applied. Mechanical properties also involve relationships between stress and strain; for example, modulus of elasticity, tensile strength, fatigue limit and impact strength. These latter properties have often been designated physical properties, but the term mechanical properties is more appropriate.

Table 7-1 lists some of the physical properties of particular chemical elements. A comparison of the values listed will show the large differences in properties that exist between some metals—properties that can indicate rejection or acceptability for a particular application. For instance, aluminum and copper are found superior to iron in electrical conductivity (reciprocal of resistivity). The resistivity of aluminum is somewhat higher than that of copper-base alloys on volume. However, if it were based on weight, aluminum would be found to have an electrical conductivity superior to that of copper.

The modulus of elasticity is important in determining the metal to use for a given function, For instance, **Table 7-1** shows that iron is about 2.5 times as stiff as aluminum. This means that it would take 2.5 times the load to deflect a bar of iron, compared with an aluminum bar of the same dimensions. However, if the weight factor is considered and the bars are of equal weight or unit length (but differing in cross-sectional area), the difference would not be so marked. This fact would be more valid if the added volume of the aluminum bar was properly distributed to stiffen the bar, relative to the applied load.

Note the various melting temperatures of the different metals. These melting temperatures may be significant in the use of a metal at elevated temperatures, such as in heat treat furnaces, furnaces in general, and even in selecting a molding process. For instance, iron cannot be poured into plaster molds, however, aluminum can. This is because aluminum has a lower melting point.

These are only a few of the factors that need to be considered when selecting a metal for a particular application. In addition to the physical data for pure metals, other data on alloys are useful and are tabulated in **Table 7-2**.

BIBLIOGRAPHY

Aluminum Casting Technology, 2nd. Edition, The American Foundrymen's Society, Inc., Des Plaines, IL.

Casting Copper-Base Alloys, The American Foundrymen's Society, Inc., Des Plaines, IL (1984).

Principles of Metal Casting, 2nd. Edition, Heine, Loper and Rosenthal, McGraw-Hill, Inc., New York, NY (1967).

Steel Casting Metallurgy, Svoboda and Monroe, Steel Founders' Society of America, Des Plaines, IL (1984).

Table 7-1.
Some Physical Properties of Elements

Element	Aluminum	Carbon (graphite)	Copper	Iron	Magnesium	Manganese	Nickel	Tin	Zinc	Lead
Symbol	Al	C	Cu	Fe	Mg	Mn	Ni	Sn	Zn	Pb
Atomic no.	13	6	29	26	12	25	28	50	30	82
Atomic weight (1947)	26.98	12.011	63.54	55.85	24.32	54.93	58.71	118.70	65.38	207.21
Density at 20 C (68 F), g/cu cm	2.699	2.25	8.96	7.87	1.74	7.43	8.902	7.298	7.138[a]	11.36
Density at 68 F (20 C), lb/cu in.	0.09751	0.081	0.324	0.284	0.0628	0.270	0.322	0.2637	0.258[a]	0.4097
Atomic volume, cu cm/g atom	9.996	5.33	7.09	7.10	14.0	7.39	6.59	16.26	9.17	18.27
Melting point, °C	660	3727	1083.0 ± 0.1	1536.5	650 ± 2	1245	1453	231.9 ± 0.1	419.46	327.4
°F	1220	6740	1981.4 ± 0.2	2797.7 ± 2	1202 ± 4	2273	2647	449.4 ± 0.2	787.11	621.4
Boiling point, °C	2450	4830	2595	3000 ± 150	1110	2150	2730	2270	906	1725
°F	4442	8730	4703	5430 ± 270	2025 ± 20	3900	4950	4120	1663	3137
Specific heat at 20 C, cal/g/°C	0.215	0.165	0.092	0.11	0.245	0.115	0.105	0.054	0.0915	0.031
Heat of fusion, cal/g	94.5	....	50.6	65.5	89 ± 2	63.7	73.8	14.5	24.09	6.26
Btu/lb	170	....	91.1	117.9	160	114.7	132.8	26.1	43.36	11.27
Coefficient of linear thermal expansion:										
Near 20 C, μin./°C	23.6	0.6–4.3	16.5	11.76	27.1	22	13.3	23	39.7[b]	29.3
Near 68 F, μin./°F	13.1	0.3–2.4	9.2	6.53	15.05	12.22	7.39	13	22.0	16.3
Thermal conductivity near 20 C, cal/(sq cm/cm)/°C/sec	0.53	0.057	0.94	0.18	0.367	....	0.22	0.15	0.27	0.083
Electrical resistivity, μohm-cm	2.655 (20 C)	1375 (0 C)	1.673 (20 C)	9.71 (20 C)	4.45 (20 C)	185 (20 C)	6.84 (20 C)	11.0 (20 C)	5.916 (20 C)[c]	20.65 (20 C)
Modulus of elasticity in tension, 10^6 psi	9	0.7	16	28.5 ± 0.5	6.35	23	30	6–6.5	[d]	2.0
Crystal structure	Face-centered cubic	Hexagonal[e]	Face-centered cubic	Body-centered cubic	Close-packed hexagonal	Cubic (complex)[e]	Face-centered cubic	Body-centered tetragonal	Close-packed hexagonal	Face-centered cubic

* Adapted from the ASM "Metals Handbook," 8th ed., vol. 1, 1961.
[a] At 40 C (104 F).
[b] For polycrystalline zinc; in single crystals, varies from 61.5 (parallel to hexagonal axis) to 15 (perpendicular to hexagonal axis).
[c] For polycrystalline zinc; in single crystals, varies from 6.16 (parallel to hexagonal axis) to 5.89 (perpendicular to hexagonal axis).
[d] Pure zinc has no clearly defined modulus of elasticity.
[e] Ordinary form; other modifications known or probable.

Table 7-2.
Casting Properties of Foundry Products

Metal or alloy	Nominal composition, %	Solidification range, F	Shrinkage, in./ft	Specific gravity	Wt., lb/cu in.
Steels:					
Carbon cast steel	Less than 0.20 C; 0.50–1.00 Mn; 0.20–0.75 Si; 0.05 P max; 0.06 S max	2730–2615	1/8–1/4	7.86	0.284
	0.20–0.40 C; 0.50–1.00 Mn; 0.20–0.75 Si; 0.05 P max; 0.06 S max	2695–2590	1/8–1/4	7.86	0.284
	More than 0.40 C; 0.50–100 Mn; 0.20–0.75 Si; 0.05 P max; 0.06 S max	2670–2160	1/8–1/4	7.80–7.86	0.282–0.284
Alloy cast steel	....	2730–2280	Up to 5/8	7.50–8.10	0.271–0.293
Cast irons:					
Gray iron	....	2400–2000	1/8	7.00–7.50	0.252–0.271
White iron	....	2550–2065	3/16–1/4	7.70	0.277
Malleable iron	....	2550–2065	3/16	7.20–7.45	0.259–0.268
Aluminum alloys:					
Aluminum	99+ Al	1215 M.P.	5/32+	2.70	0.098
Aluminum-copper alloy	92 Al; 8 Cu	1165–975	5/32	2.83	0.102
Aluminum-copper-silicon alloy	4 Cu; Si; 93 Al	1170–970	5/32	2.75	0.099
Aluminum-magnesium alloy	4.0 Mg; 96.0 Al	1185–1075	5/32	2.63	0.095
Aluminum-manganese alloy	2.0 Mn; 98.0 Al	1255–1215	5/32	2.73	0.099
Aluminum-silicon alloy	12.0 Si; 88.0 Al	1150–1070	5/32	2.65	0.096
Aluminum-zinc alloy	2.5 Cu; 1.3 Fe; 11.0 Zn; 85.2 Al	1165–980	5/32	2.94	0.106
Copper alloys:					
Copper	99+ Cu	1980 M.P.	3/16	8.80	0.317
Copper-nickel alloy	45 Ni; 55 Cu	2325–2235	1/4	8.60	0.310
Cupronickel	30 Ni; 70 Cu	2250–2140	3/16	8.80	0.317
Bell metal	20 Sn; 80 Cu	1600–1450	3/16	8.70	0.313
Nickel brass (nickel silver)	20 Ni; 15 Zn; 65 Cu	2060 approx	3/16	8.50	0.316
Red brass	5 Zn; 5 Pb; 5 Sn; 85 Cu	1775 approx	3/16	8.75	0.314
Medium red brass	77 Cu; 10 Zn; 10 Pb; 3 Sn	1775 approx	3/16	8.68	0.312
Gun metal	8 Sn; 4 Zn; 88 Cu	1825–1775	1/8	8.30–8.60	0.300–0.310
Aluminum bronze	10 Al; 90 Cu	1920 approx	7/32	7.50	0.271
Manganese bronze	55–60 Cu; 38–42 Zn; 3.5 max Mn	1675 approx	7/32	8.40	0.303
Leaded bearing bronze	80 Cu; 10 Pb; 10 Sn	1725 approx	3/16	8.90	0.324
Silicon bronze	94 Cu; 5 Si; 1 Mn	1830 approx	3/16	8.20	0.297
Lead alloys:					
Lead-base bearing alloy	75 Pb; 15 Sb; 10 Sn; 0.50 max Cu; 0.20 As	515–465	....	9.73	0.350
Lead-antimony alloys	90 Pb; 10 Sb	500–475	....	10.67	0.386
Lead-tin-antimony alloys	80 Pb; 5 Sn; 15 Sb	500–460	....	10.04	0.363
Magnesium-base alloy	3.5–10 Al; 3.5 max. Zn; remainder Mg	1150 approx	5/32–1/8	1.83 max	0.066
Nickel alloys:					
High nickel-copper alloy	70 Ni; 30 Cu	2500–2400	1/4	8.80	0.317
High nickel-copper-silicon alloy	65 Ni; 30 Cu; 3–5 Si	2400–2300	1/4	8.65	0.312
Tin alloy, babbitt	90 Sn; 5 Cu; 5 Sb	690–430	....	7.34	0.265
Zinc-base alloy (die casting)	0.10 max Cu; 3.5–4.3 Al; remainder Zn	720 approx	5/32	6.60–6.70	0.238–0.242

* Adapted from *Metals and Alloys*, April, 1949.

Family of Ferrous Alloys

8

Ferrous alloys are made up of predominantly iron (Fe) plus carbon (C). The amount of carbon will determine whether it is an "iron" or a steel. Iron contains greater than 2% carbon, whereas steel contains less than 2% carbon.

CAST IRONS

Of all the metal alloys—ferrous and nonferrous—poured in the metalcasting industry, the largest tonnage is poured from the cast iron family of ferrous alloys. Cast iron metalcasting foundries send castings into every conceivable industry. The demand for these castings is based on the nature of cast irons: they are good engineering materials and are economically advantageous. Cast iron offers a tremendous range of metallic properties: strength, hardness, machinability, wear resistance, abrasion resistance, corrosion resistance and others. In addition, the metalcasting properties of cast iron, such as yield, fluidity, shrinkage, casting soundness, ease of production, make this material highly desirable for casting purposes. Whichever way it is looked at, the family of cast irons offers a variety of engineering properties that ensure continued and widespread use. Since many cast irons of different properties are employed, it is desirable that the student obtain an overall picture of the entire field. This section will present some of the simpler and more fundamental differences between members of the cast iron family.

As mentioned earlier, *cast iron* is a generic term that refers to a family of materials differing widely in their properties. An alloy of iron, carbon (up to approximately 4.0%) and silicon (up to approximately 3.50%) that is not usually malleable, as cast, is known as cast iron. Malleable means the material has properties that enable it to be mechanically deformed under compression, without breaking. Another way of putting it is that it can be "worked." Other chemical elements, such as manganese, sulfur and phosphorus, all in small amounts, are also found in cast iron **(Table 8-1).** When alloyed cast iron is required, any of the following chemical elements—nickel, copper, chromium and/or molybdenum—may be added to the iron.

There are many variations of cast iron commonly poured today. Chemical composition, solidification process, cooling rate and microstructure control the nature of these irons. The following discussions present some of the simpler and more fundamental differences among the cast irons.

Table 8-1.
Chemical Composition Range of Typical Unalloyed Cast Irons

Iron Type	%C	%Si	%Mn	%S	%P
White	1.8–3.6	0.5–1.9	0.25–0.8	0.06–0.2	0.06–0.2
Malleable (Cast White)	2.2–2.9	0.9–1.9	0.15–1.2	0.02–0.2	0.02–0.2
Gray	2.5–4.0	1.0–3.0	0.2–1.0	0.02–0.25	0.02–1.0
Ductile	3.0–4.0	1.8–2.8	0.1–1.0	0.01–0.03	0.01–0.1
CG	2.5–4.0	1.0–3.0	0.2–1.0	0.01–0.03	0.01–0.1

Gray Cast Iron

This cast iron has a chemical composition such that, when solidified, a large portion of its carbon is randomly distributed throughout the casting as graphite in "flake" form. **Figure 8-1** shows a photomicrograph of gray iron (note the black graphite flakes). **Figure 8-2** indicates graphite flake structure. Following are some typical chemical compositions of gray cast iron: (Remember, iron is the main chemical element and metal in any of the cast irons.)

Typical Chemical Compositions of Gray Cast Iron

Type	*C %*	*Si %*	*Mn %*	*S %*	*P %*
A	3.30–3.60	2.30–2.60	0.50–0.80	0.20 max	0.30 max
B	3.10–3.50	1.90–2.30	0.60–0.90	0.125 max	0.12–0.18
C	3.50–3.90	2.20–3.10	0.40–0.80	0.10	0.30–0.80
D	2.90–3.20	0.90–1.10	0.65–0.90	0.05–0.12	0.20 max
E	2.60–2.80	2.20–2.50	0.90–1.00	0.08 max	0.08 max

To illustrate how the chemical composition can vary the use of gray cast iron, "A" is a general-purpose use gray cast iron, "B" is used for motor blocks, "C" is used for piston rings, "D" can be used for heavy (2000–10,000 lb / 907–4536 kg) machine tool bases, and "E" is a high-strength gray cast iron for diesel engine blocks and crank shafts. In other words, the chemical composition can affect the mechanical and physical characteristics of the gray cast iron and, thus, the end use. In addition to iron, carbon and silicon, gray iron may also contain small percentages of manganese, sulfur and phosphorus, which may be present because of the blast-furnace process that produces pig iron. Pig iron is sometimes used as a charge material when melting cast iron. When alloyed cast irons are desired, nickel, copper, chromium and molybdenum may be added.

Fig. 8-1. Flake graphite in thin section of gray cast iron, unetched, 100X.

Table 8-2.
Relationship Between Structure and Tensile Strength and Brinell Hardness Ratio, Listed by Composition Range

Carbon equivalent, %	Tensile strength divided by Bhn	Structure
3.45–3.65	210 and over	Smallest cell, normal graphite
	190–210	Small cell, normal graphite
	180–190	Medium cell, some type D† graphite
	170–180	Large cell, some type D—medium cell, completely type D
	160–170	Large cell, partial type D
	160 and below	Large cell, complete type D
3.65–3.85	210 and over‡	Smallest cell, normal graphite
	190–210	Smallest cell, normal graphite
	180–190	Medium cell, normal graphite or small cell, partial type D
	170–180	Large to medium cell with partial type D graphite
	160–170	Large cell, type D graphite or free ferrite
3.85–4.20	190–210	Medium cell, normal graphite
	180–190	Medium cell, large normal graphite
	170–180	Medium or large cell, some type D
	160–170	Large cell, type D graphite
	160 or below	Free ferrite, type D graphite

* From T. E. Barlow and C. H. Lorig.[12]
† Type D graphite—AFA-ASTM graphite flake-size chart—also called modified, eutectiform, dendritic, pseudo-eutectic, etc. Undesirable for wear resistance; low deflection, low toughness values, and poor transverse properties.
‡ Very few irons in this range.

Fig. 8-2. Flake graphite structure of gray iron under scanning electron microscope.

This is true for the family of cast irons; more information may be found in the many metalcasting handbooks that have been published.

The solidification of the alloy also plays an important role in the properties that can be achieved. This and the various structures that are formed will be discussed in Chapter 16, which discusses the solidification of cast irons.

From a metallurgical-engineering standpoint, gray iron may be viewed as a microstructurally sensitive alloy. In other words, when looking at a solidified and prepared sample of a gray iron casting under a microscope, the "structure" can bne seen. This structure, which includes the graphite flakes and the material surrounding it, gives the gray iron its mechanical properties. The carbon and silicon are the most important composition factors influencing mechanical properties. The carbon equivalent (CE) has been related to a number of mechanical properties and can be calculated as follows:

$$CE = \%C + 1/3\ (\%Si + \%P)$$

If the amount of phosphorus is negligible, many metallurgists will not use it when calculating the CE. **Figure 8-3** shows the relationship of CE to the tensile strength. As this figure indicates, decreasing CE results in an increasing tensile strength.

The relationship of hardness (the resistance to surface penetration) to tensile strength is subject to substantial variations due to the influence of the various flake graphite types. The various flake graphite types will be discussed in Chapter 13 on Microstructure of Ferrous Alloys. **Table 8-2** lists the relationships between tensile strength and Brinell hardness (Bhn or HB). Note how the various flake graphite types affect the tensile strength:HB ratio. When the more normal, small graphite flakes exist, the tensile strengths are higher than when the larger flakes are present and the tensile strengths are lower.

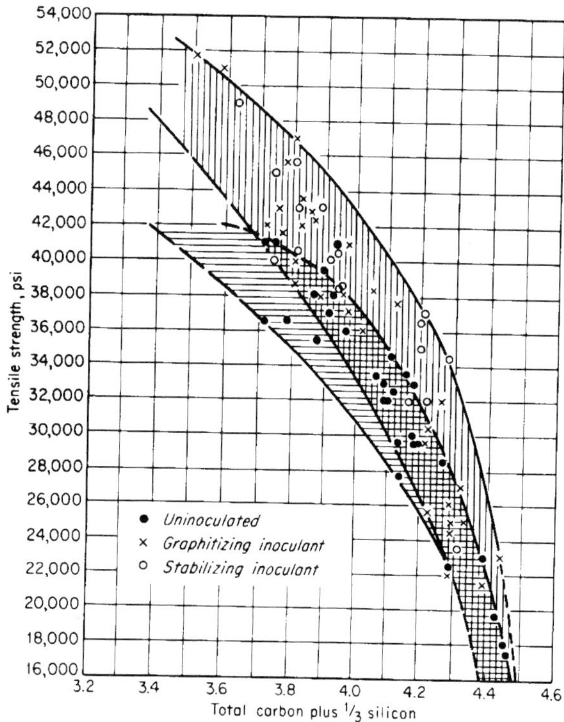

Fig. 8-3. Tensile strength of 1.20-in. dia gray iron bars as affected by carbon equivalent. [From T.E. Barlow and C.H. Lorig, AFS Transactions, vol 54, 1946]

As with most alloys, gray iron is "section sensitive." The speed at which a section of the casting solidifies is normally determined by the section thickness. The thinner the section, the faster it will solidify; the thicker the section, the longer it will take to solidify. The speed at which the section solidifies has a profound effect on the microstructure. It is safe to say that fast solidification in thin sections results in higher hardness and higher tensile strengths. The opposite is true for heavy sections. This topic will be discussed in Chapter 16, Solidification.

Following are some of the other properties of gray iron:

- Gray iron has outstanding compressive strength; usually 3–5 times it's tensile strength.
- The torsion strength of gray iron is about 1.20–1.40 times its strength in tension.
- Modulus of elasticity in tension varies from 12 to 22 million psi, depending upon the microstructure.
- The shear strength of gray iron is about 1.0–1.60 times the tensile strength.
- Gray iron's endurance limit is about 35–50% of its tensile strength.
- Hardness may vary in much the same way as its strength.
- The wear resistance of gray iron is outstanding in its resistance to sliding-friction type wear, especially when lubricated.
- Gray iron has excellent machinability—the best among the ferrous alloys.
- Heat resistance to scaling and retention of moderate strength at elevated temperatures are desirable properties of gray iron.
- The damping capacity of gray iron is excellent. Damping capacity is the ability to absorb energy due to vibrations and thus dampen the vibrations. For example, when gray iron is struck with a hammer it will not ring.
- The corrosion resistance of gray iron is important in some applications.
- Gray iron's electrical resistance is sufficiently high so that it is used extensively for resistance grids.
- Gray iron may be heat treated to change its hardness or temper.

Having this range of properties, one can correctly expect to find gray iron used in many different end uses. **Figure 8-4** shows some gray iron castings. Gray iron is used for engine blocks (both gasoline and diesel), cylinder liners, piston rings, machine tool bases, valves, pump manifolds, bathtubs and sinks, small engines, ship anchors, architectural grillwork, agricultural equipment...and the list can go on.

Fig. 8-4. Example of brake disc, cast in gray iron to withstand heat without warping.

Chilled Cast Iron

Chilled cast iron can solidify as a gray iron; however, by rapid cooling (chilling), white iron will be produced. Both white and gray iron can be seen on a fractured surface in **Fig. 8-5**. The white iron is the result of rapid cooling and the gray iron is the result of a normal cooling rate. The chemical composition of chilled-iron castings has the carbon and silicon content balanced so that the portions of the casting cast against a metal chill will solidify white and the remainder of the casting, not chilled, will solidify gray.

Chilled cast irons were used for railroad freight car wheels. Today, this cast iron alloy is used for grain-mill rolls, rolls for crushing ores and rolling metals, etc.

The chilled zone has high hardness; however, it is also brittle. This brittleness can be reduced with little loss of hardness, by heat treating. The combination of abrasion-resistant surfaces and a gray iron core makes these castings suitable for use as sludge-pump liners, jaw-crusher plates, grinding mill liners, camshafts, grinding balls, abrasive materials-handling equipment, and other such uses.

White Cast Iron

White cast iron has a chemical composition such that, after solidification, its carbon is present in a chemically combined form called *cementite*. Cementite is a compound of iron and carbon also known as iron carbide, which has an approximate chemical formula of Fe_3C. Cementite is a very hard material. Disregarding cooling rates and any other variables, silicon in amounts of less than 1.5% contribute to the formation of white iron. When the percentage of silicon is increased, the iron is changed from white to mottled or gray. The transitional area between the gray and the white iron is called mottled iron. Thus, carbon and silicon may be varied to produce a white or gray iron, as desired. White iron may be produced in one of two ways: 1) by cooling the casting rapidly enough to prevent the removal of carbon from chemical combination, or 2) by proper adjustment of the composition.

Chilled iron structure —>

Mottled iron structure —>

Gray iron structure —>

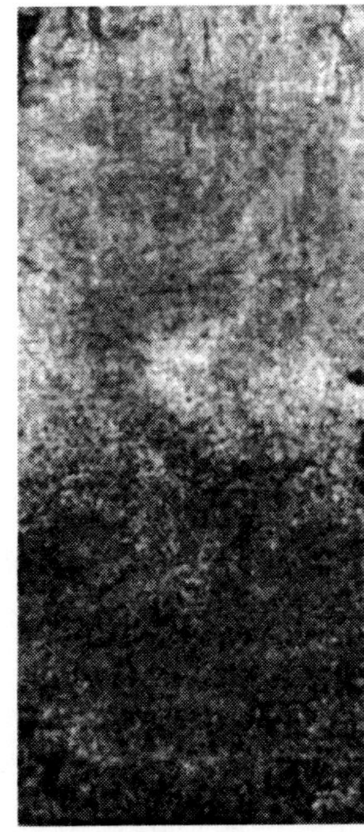

Fig. 8-5. Fractured surface of chilled iron casting. Upper surface on the figure was chilled by a metal block. [From AFS]

Because of rapid loss of heat, castings containing thin sections have a greater tendency to become white iron than do those containing thick sections. Another influential factor is the heat conductivity of the mold material. A metal mold is more likely to produce a casting of white structure than is a green sand mold. A piece of metal can be deliberately inserted in a certain area of the mold, to act as a chill and produce a finer-grained structure in that area of the casting.

The principal use of white iron is to produce malleable iron. Some white iron castings are used as such because of the hardness of the material. Hardness of 350–550 HB and over may occur in white irons. The abrasion resistance accompanying this hardness is used to advantage in tumbling mill jack-stars, pulverizing mill plates, and similar applications. The more common use of the abrasion-resistant type white iron microstructure is made in chilled iron castings.

Malleable Cast Iron

American malleable iron is truly a product born of the American foundrymen's inventiveness. Foundrymen were looking for a cast iron that had the ductility that gray iron couldn't offer. Seth Boyden in Newark, New Jersey developed the first "blackheart" malleable iron castings in 1826. Boyden's work resulted in the growth of the American, or blackheart, malleable foundry industry.

Malleable iron offers to the metalcasting designer and user the following desirable properties: ease of machinability, toughness and ductility, corrosion resistance in certain applications, strength adequate for widespread use, magnetic properties, and uniformity resulting from 100% heat-treatment of all castings produced.

The properties of malleable iron are mainly related to its metallographic structure (microstructure). This cast iron can be described microstructurally as a ferrous alloy composed of temper carbon in a matrix of ferrite (and sometimes pearlite) containing dissolved silicon. **Figure 8-6** shows two pearlitic malleable castings.

Table 8-3.
Typical Chemical Composition of White Irons Heat-Treatable to Malleable Iron

	ASTM No. 32510	ASTM No. 35018	Cupola-malleable
% C	2.30–2.65	2.00–2.45	2.80–3.30
% Si	0.9–1.40	0.90–1.30	0.60–1.10
% Mn	0.25–0.55	0.21–0.55	Less than 0.65
% P	0.18	Less than 0.18	Less than 0.20
% S	0.05–0.18	0.05–0.18	Less than 0.25

To define a few metallurgical terms:

- *Temper carbon* is the free or graphitic carbon that precipitates from solution, usually in the form of rounded or equiaxed nodules in the structure, during the graphitizing or malleablizing of white cast iron. In other words, malleable iron, when first solidified in the mold, comes out as white iron.
- *The matrix* is the material that surrounds the graphite.
- *Ferrite* is a solid solution of iron and silicon.
- *Pearlite* is alternate layers of ferrite and cementite that can be formed in cast irons and carbon steels. Pearlite supposedly got its name because, under the microscope, it looked like "mother of pearl."

Thus, malleable iron can be either pearlitic malleable iron or ferritic malleable iron or a combination of the two. **Table 8-3** lists typical chemical compositions of malleable iron; **Fig. 8-7** shows a microstructure of white cast iron before heat treatment; and **Fig. 8-8** after heat treatment to malleable iron.

It is the heat treatment after shakeout that converts the white cast iron to malleable iron. The annealing heat treatment is often called "malleablization," since it converts the hard, brittle, white cast iron into malleable iron. This heat treatment, in some instances, can take several days to complete, which creates very long lead times to get malleable iron castings produced.

Some of the principal industries using castings made of malleable iron are the railroad, agricultural machinery, marine and mining equipment, and pipe fitting. With the onset of ductile iron, the use of malleable iron castings declined considerably.

Ductile Cast Iron

Ductile cast iron was first announced as a new engineering material at the 1948 American Foundrymen's Society's Casting Congress and Exposition held in Philadelphia, Pennsylvania. It was at this meeting that two independent organizations announced their success in producing this revolutionary cast iron. First, to make the announcement was the British Cast Iron Research Association (BCIRA). The second announcement at the same technical session was made by the International Nickel Company (INCO).

The BCIRA process consists of the addition of cerium (a rare earth) to molten hypereutectic cast irons of essentially the same analysis of gray iron. The cerium removes the sulfur from the gray iron and, with about 0.02% residual cerium, produces graphite spheroids instead of flakes.

Fig. 8-6. A differential gear carrier and a hydrauic clutch torque plate made of pearlitic malleable iron.

The INCO process uses magnesium additions to either hypo- or hypereutectic cast irons to achieve the same results. After initial exposure to these processes, the foundry industry soon realized the greater potential and economy of the magnesium process, which is now almost universally practiced. Ductile cast iron is also known as spheroidal graphite iron (SG iron), and nodular iron.

Essentially, ductile iron consists of graphite spheroids dispersed in a matrix similar to steel **(Fig. 8-9)**. The only significant difference between gray cast iron and ductile cast iron is in the shape of graphite phase; the matrices (matrix) can be similar. **Figure 8-10** shows a graphite nodule with the matrix etched away.

The ease with which ductile iron can be processed and cast into complex shapes is very dependent on a carbon content (or carbon equivalent). During solidification, most of the carbon forms as graphite spheroids, or nodules, which exert only a minor influence on the mechanical properties in contrast to the influence that the graphite flakes have in gray iron. The matrix structure has the greatest effect on the properties of the iron. Ductile cast irons, thus,

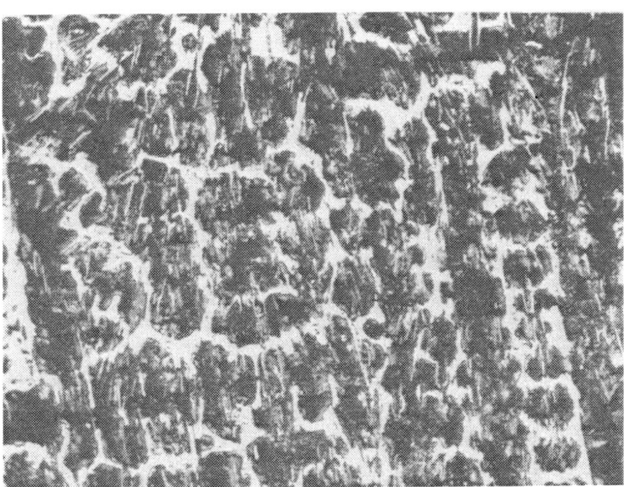

Fig. 8-7. Microstructure of white cast iron: white, massive carbide areas; dark, pearlite areas. Nital etched, 150X.

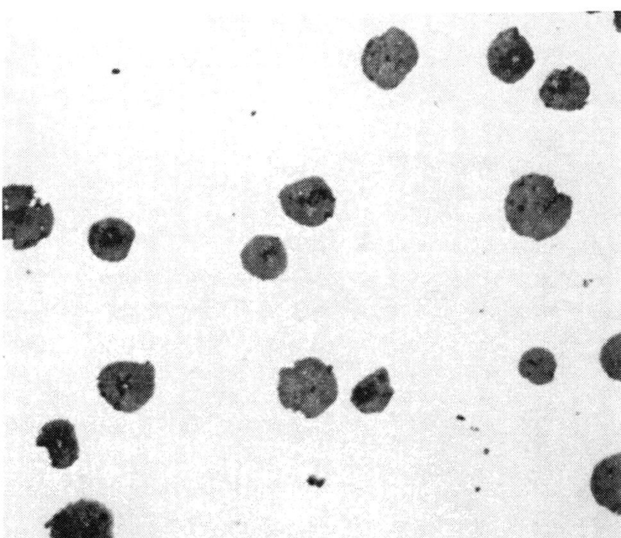

Fig. 8-8. Microsamples of the heats tapped on April 12, 1943 showed that the graphite had taken on a spheroidal shape ...and ductile iron was born.

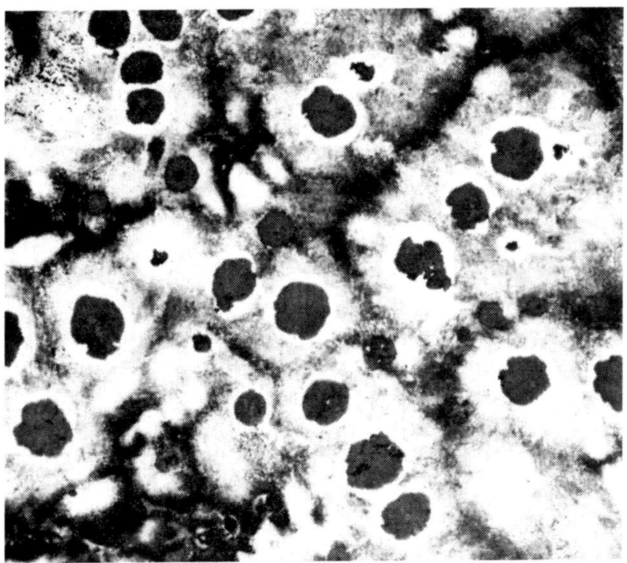

Fig. 8-9. General features of as-cast structure before annealing, etched in picral; nodules of graphite with surrounding ferrite and the remainder of the matrix consisting of pearlite, 100X.

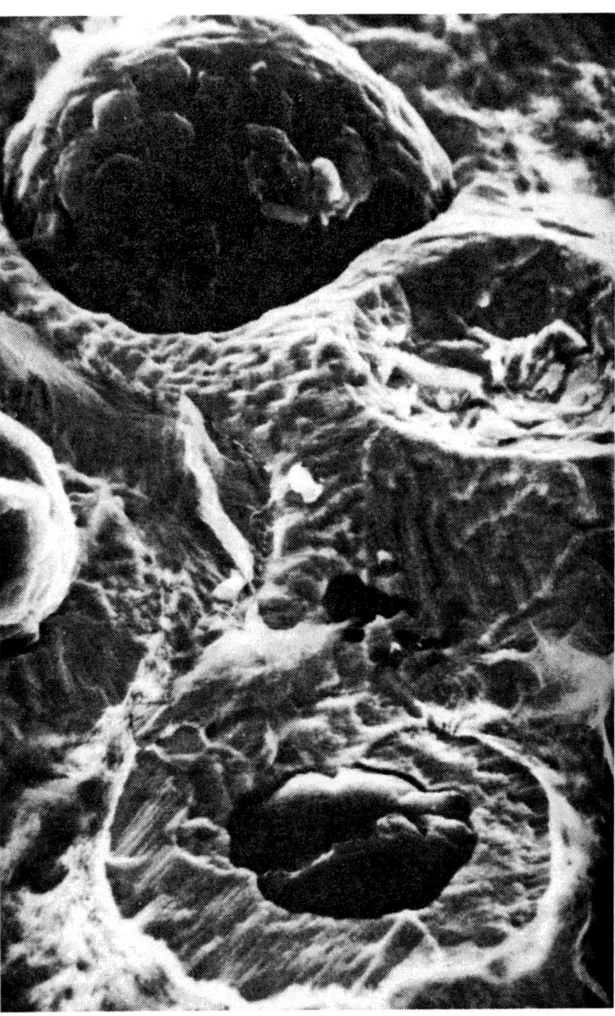

Fig. 8-10. Nodular graphite in ductile iron under a scanning electron microscope.

are a family of cast alloys that combine the principal advantages of gray iron with the engineering advantages of steel.

The matrix of ductile iron can be controlled by the base composition, by foundry practice and/or by heat treatment. Ductile iron castings with a minimum tensile strength of 60,000 psi with over 25% elongation, or up to 150,000-psi minimum tensile strength, yet having 1–4% elongation, can be produced. **Table 8-4** shows a summary of the principal types of ductile iron, their characteristics and some casting applications.

It is said that ductile iron, as an engineering material, has seen the most phenomenal growth in tonnage produced of any other casting material, to date. The December 1998 issue of *modern casting,* published by the American Foundry Society, reported that in 1997 there were 4,128,000 metric tons of ductile iron produced in the United States; worldwide, approximately 13,212,440 metric tons were produced. These figures ranked it second, behind gray iron, as the metal alloy most commonly cast, tonnage wise, in the United States and worldwide. **Figure 8-11** shows some of the different types of ductile iron castings produced.

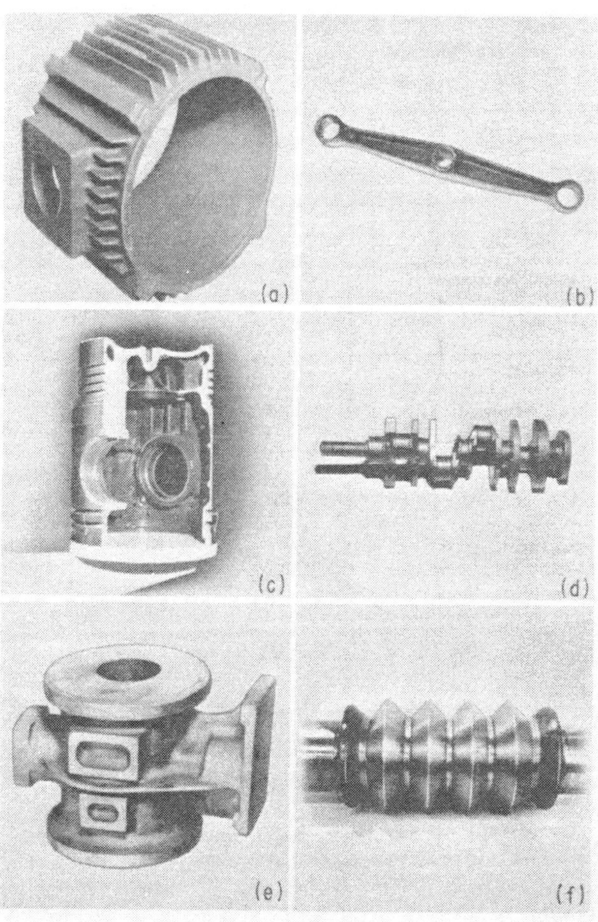

Fig. 8-11. Examples of the application of ductile iron castings: [a to d, courtesy of International Nickel Co., Inc.; e and f, courtesy of Climax Molybdenum Company] (1) motor frame for U.S. Navy made of 60-40-15 DI, because of its high resistance to explosion impact; (b) axle equalizer beam for heavy-duty truck produced in normalized DI to replace a steel forging; (c) sectioned diesel engine piston produced in DI; (d) shell molded DI automotive crankshaft; (e) rough casting of hot air valve for Caravelle plane, produced in DI containing Ni, Cr and Mo, for necessary hot strength; (f) alloyed DI used for 2.5x2.5-in. angle finishing roll.

Table 8-4.
Principle Types of Ductile Iron

Type No.†	Brinell hardness no.	Characteristics	Applications
80-60-03	200-270	Essentially pearlitic matrix, high-strength as-cast. Responds readily to flame or induction hardening	Heavy-duty machinery, gears, dies, rolls for wear resistance, and strength
60-45-10	140-200	Essentially ferritic matrix, excellent machinability and good ductility	Pressure castings, valve and pump bodies, shock-resisting parts
60-40-15	140-190	Fully ferritic matrix, maximum ductility and low transition temperature (has analysis limitations)	Navy shipboard and other uses requiring shock resistance
100-70-03	240-300	Uniformly fine pearlitic matrix, normalized and tempered or alloyed. Excellent combination of strength, wear resistance, and ductility	Pinions, gears, crankshafts, cams, guides, track rollers
120-90-02	270-350	Matrix of tempered martensite. May be alloyed to provide hardenability. Maximum strength and wear resistance	

*Courtesy Gray and Ductile Iron Founder's Society.
† The type numbers indicate the minimum tensile strength, yield strength, and per cent of elongation. The 80-60-03 type has a minimum of 80,000 psi tensile, 60,000 psi yield, and 3 per cent elongation in 2 in.

The effect of section size on the properties of ductile iron is not as great as it is on gray iron. The number and size of graphite nodules does affect the properties due to their ability to influence the remaining structure of the iron. A carbidic matrix and/or deteriorated graphite shapes generally accompany low nodule counts. Carbidic matrix means that there are carbides, which are very hard, present in the matrix. On the other hand, if the number of nodules is increased, the amount of ferrite in the as-cast structure increases. The effect of section size on nodule number and size is shown in **Figs. 8-12 and 8-13.**

Ductile iron holds a unique position among the engineering materials because of its unusual combination of castability and mechanical properties when compared to other ferrous alloys. Castability simply means the ability to be cast.

Some of the engineering advantages of ductile iron are:

- The machinability of ductile iron is superior to that of gray iron at equivalent hardnesses and better than that of steel at equivalent strength levels.
- Corrosion resistance of ductile iron to corrosive media has been shown to be equal to that of gray iron and generally better than that of steel.
- When it comes to wear resistance, ductile iron has demonstrated that it is equal to that of the best grades of gray iron and superior to that of carbon steel.
- The thermal shock resistance of ferritic ductile iron was shown to be good. In one experiment, ferritic ductile iron was heated to temperatures above 1300F (704C) and then drastically quenched in cold water, without cracking.

Of course, ductility is the one engineering property that ductile iron has more of than gray iron. It also has the high strength, toughness, hot workability and hardenability of steel.

In general, it may be said that ductile iron combines the processing advantages of gray iron with the engineering advantages of steel. More detailed information on the properties of ductile iron is available from the American Foundry Society, and the Ductile Iron Society.

Compacted Graphite Iron

The first observations and documentation of compacted graphite iron (CGI) were made during the early work on ductile iron in the late 1940s. Researchers found that when the magnesium (or cerium) was not able to stabilize a fully spheroidal graphite structure (nodule), an intermediate form of compacted (vermicular) graphite appeared **(Fig. 8-14)**. It was not until the mid 1960s, when the first CGI patent was granted, that CGI began to be recognized as a new type of cast iron that combined the strength and stiffness of ductile iron with the thermal conductivity and castability approaching that of gray cast iron.

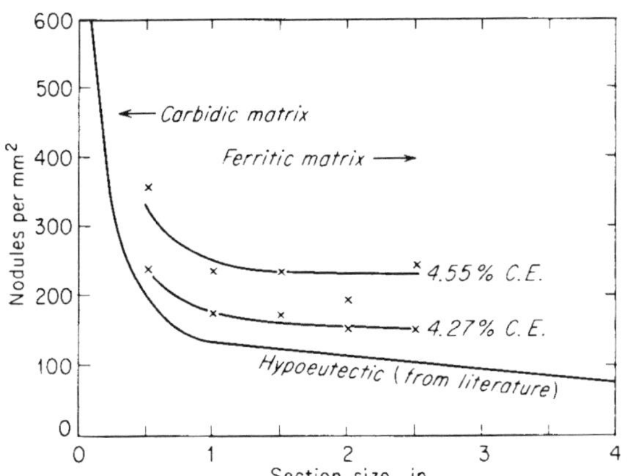

Fig. 8-12. Effect of section size of castings on the number of nodules produced from a standardized treatment procedure.

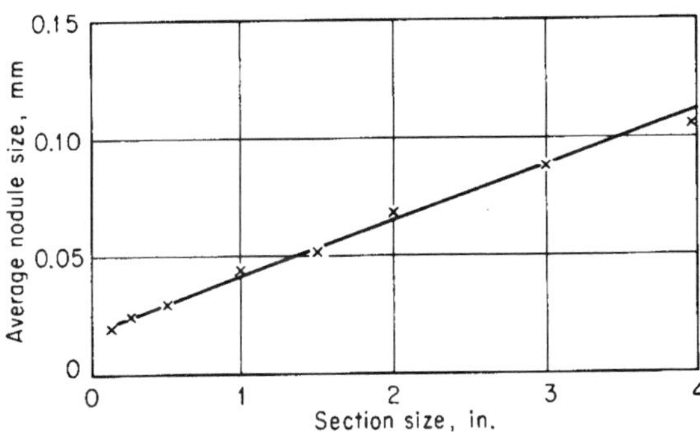

Fig. 8-13. Effect of section size on the average nodule size of the same castings as in Fig. 8-18.

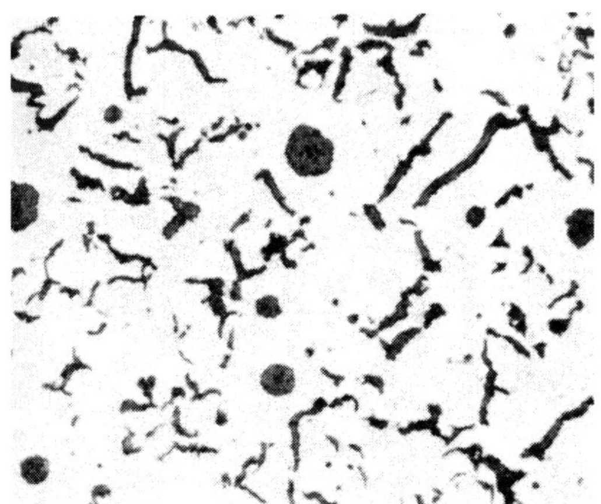

Fig. 8-14. Compacted graphite iron with compacted graphite, unetched, 100X.

The graphite in CGI is in the form of interconnected flakes, as in gray cast iron. This enables the production of sound castings, especially those of complex shapes or with intricately cored passages. The deep-etched scanning electron microscope (SEM) micrograph seen in **Fig. 8-15** shows that the compacted graphite particles are shorter and thicker, and have rounded edges and roughened surfaces, which is different than the graphite flakes found in gray cast iron. The irregular shape of the compacted graphite clusters and the roughened surface improve the adhesion with the iron matrix and minimize the initiation and propagation of cracks. This results in substantially increased tensile strength, fatigue strength and elastic modulus relative to gray cast iron

The form of the graphite in CGI also provides for improved strength, some ductility and a better machine finish than gray cast iron. With the interconnected graphite, a higher thermal conductivity, more damping capacity and better machinability than with ductile iron is obtained. Close metallurgical control is necessary to obtain successive castings with consistent properties. The nodularity (or number of nodules that are formed) should be restricted to a maximum of 20% in all performance-critical sections, in order to optimize castability, thermal conductivity, damping capacity and machinability, together with the mechanical properties. Along with close metallurgical control and some rare earth element additions,

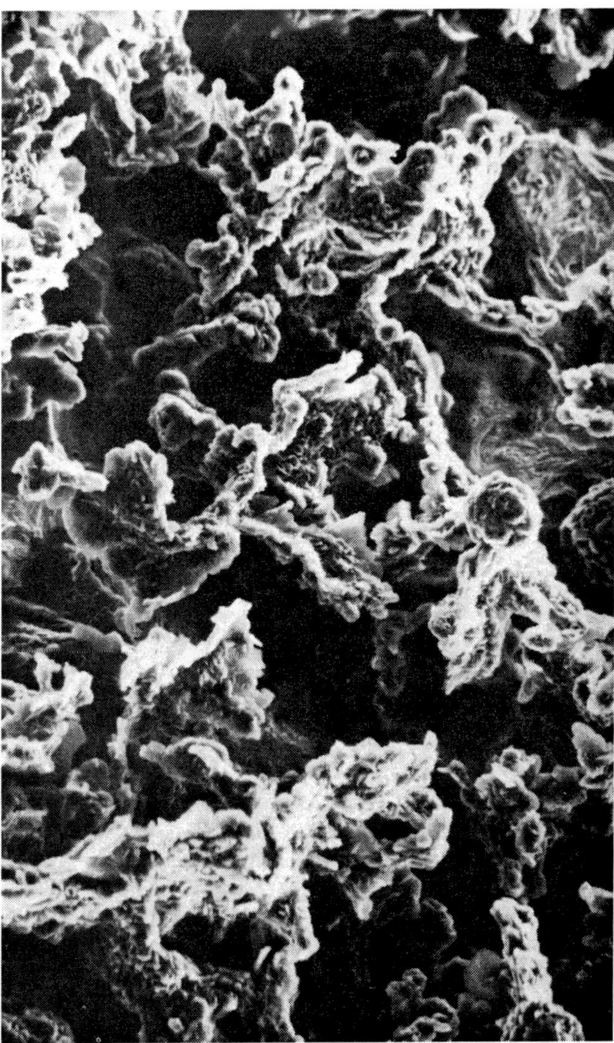

Fig. 8-15. Compacted graphite iron under a scanning electron microscope.

it also requires an alloying element, such as titanium, to minimize the formation of spheroidal graphite. Alloying or heat treatment can adjust its matrix structure. **Table 8-5** shows typical mechanical and physical properties of CGI.

The tensile properties and Brinell hardness of CGI are dependent upon microstructure of the iron; in particular, the larger the quantity and distribution of the graphite particles approaching spherical shape, the higher the relative tensile strength. A typical stress analysis indicates that CGI does not deviate significantly from linear elastic behavior until a stress level of 30,000 psi (206.85 MPa) is reached.

Section sensitivity is an item that has to be considered when using CGI. Thinner sections, with high cooling rates, will promote the formation of graphite nodules, whereas thicker sections will have lower nodularity. This phenomenon can be used to advantage by the foundry engineer and the casting designer. The natural tendency of CGI toward higher nodularity in thin sections provides the opportunity to place thermally efficient low-nodularity CGI in thicker sections. Under conditions of consistent graphite nodularity, the strength of CGI decreases as section size increases. Like gray cast iron, this phenomenon is due to the coarsening of the graphite particles, although the rate of loss with CGI is less than that of gray cast iron.

Compacted graphite irons have been proven very successful in some applications where ductile iron cannot be used because of its lower thermal conductivity or its tendency to warp. Because of continued and increased demands for reduced-weight metal castings with improved performance, many manufacturing sectors have investigated and adopted CGI as an alternative material. This is particularly true in the automotive industry for cylinder blocks and heads. Up until this time, however, production of CGI castings have been limited to components such as exhaust manifolds, hydraulic housings and brackets, and large castings such as ingot molds, stationary diesel engine frames and other special parts **(Figs. 8-16 and 8-17)**.

Table 8-5.
Typical Mechanical Properties of CG Irons in 1.2-in. (30 mm) Test Bars

Property	Type 1	Type 2	Type 3
Brinell Hardness	217-270	163-241	130-179
Min. Tensile Strength			
psi	65,000	50,000	40,000
MPa	*448*	*345*	*276*
Min. Yield Strength 0.2% Offset			
psi	55,000	40,000	28,000
MPa	*379*	*276*	*193*
Min. Elongation	1%	1%	3-5%
Percent Pearlite	90% Min.	10-90%	10% Max.

Fig. 8-16. Compacted graphite iron engine block.

Fig. 8-17. Compacted graphite iron hay baler knotter.

Austempered Ductile Iron

In the late 1970s, a new type of cast iron was added to the cast iron family: austempered ductile iron (ADI). This heat-treated ductile iron gave the foundry industry another material to challenge forgings.

ADI's microstructure consists of acicular ferrite (needle-shaped) in a high-carbon austenite matrix, called ausferrite. This microstructure is shown in **Fig. 8-18**. While some captive foundries that examined ADI in the late 1970s experienced only limited success, today's technology and market conditions are propelling wider use of the material. As heat-treating technology needed to produce ADI has matured, foundries have also become more advanced in producing high-quality ductile irons. Without high-quality ductile iron, there is very little chance of producing high-quality ADI. Chapter 13 will detail how ductile iron is converted to ADI through heat treating.

A wide choice of mechanical properties for ADI is available to the designer and user. At this point, higher austempering temperatures produces ADI with lower strength and hardness and higher ductility and toughness compared to lower austempering temperatures. **Table 8-6** lists the five ASTM standard grades of ADI and the property combinations available. These specifications give the minimum tensile and impact levels along with the typical Brinell

Table 8-6.
Five ASTM Standard ADI Grades (ASTM 897-90)

Grade	Tensile Strength (KSI)*	Yield Strength (KSI)*	Elongation (%)*	Impact Energy (Ft-Lb)**	Typical Hardness (BHN)
1	125	80	10	75	269-321
2	150	100	7	60	302-363
3	175	125	4	45	341-444
4	200	155	1	25	388-477
5	230	185	N/A	N/A	444-555

* Minimum Values
**Un-notched Charpy Bars Tested at 72 7F

hardness values. Typical properties of ADI produced in foundries as a function of Brinell hardness are given in **Fig. 8-19.**

For a given level of ductility, ADI provides twice the strength of conventional ductile iron. The strength of ADI is comparable to a variety of steels. The modulus of ADI, however, is about 20% less than that of steel and must be considered in the early stages of designing the ADI casting. Component weight reductions of greater than 10% can be realized by using ADI in place of steel forgings. If relative weight per unit of yield strength is considered, ADI will perform remarkably well. In most instances aluminum, which is perceived as a lightweight engineering material, is unable to match the weight per unit of yield strength of ADI.

In addition, when compared to steel forgings, ADI has been shown to exhibit fatigue strengths that are equal to or greater than forged steel. ADI also offers excellent abrasion resistance. The damping capacity of ADI is better than that of steel due to the presence of graphite in the microstructure.

Some concern has been raised over the machinability of ADI. Machining requirements for the higher grades of ADI can be completed before austempering because the part growth relationships during heat treatment are predictable. Grades 1 and 2 are easier to machine than steels of an equivalent hardness, due to the graphite in the microstructure.

ADI castings are found in numerous automotive, truck, agricultural, industrial, railroad, construction and military applications. Automotive and truck applications include suspension components, camshafts, ring and pinion gears, timing gears, CV joints, engine mounts, differential housings and wheel hubs. The wear resistance of ADI is taken advantage of in agricultural and construction castings such as digger teeth, plow points, grader blades, pavement breakers and fertilizer knives.

Conveyor components, pump components, dies and wear plates are some examples of ADI's use in industrial applications. The wear resistance and high static and fatigue strength make ADI a good material for railroad castings such as wear shoes, nipper hooks, shock absorbers and engine parts. The military use of ADI includes projectiles, armor, rocket bodies, track shoes, track guides, engine rotors and struts. **Figure 8-20** shows a typical ADI casting.

Fig. 8-18. This photomicrograph of Grade 1 ADI (etched with 5% nital) illustrates the ausferrite matrix. Originally 500X.

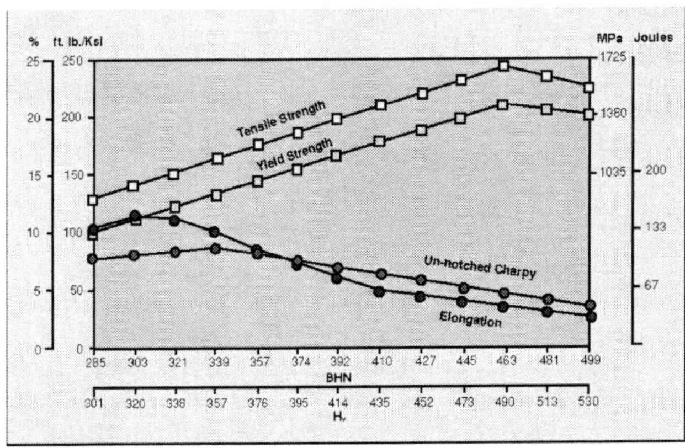

Fig. 8-19. This chart shows the typical properties of ADI as a function of Brinell hardness.

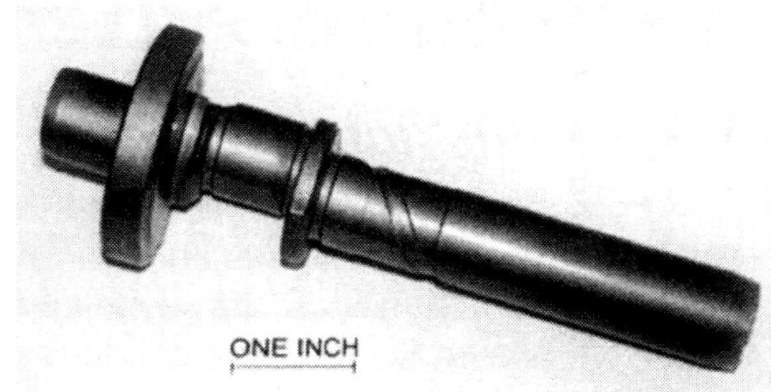

Fig. 8-20. Example of a ductile iron application; a 1-lb crankshaft for a hermetically sealed refrigerator compressor.

Other Cast Irons

There are two other cast iron materials that will be mentioned briefly: Ni-Hard and Meehanite irons.

Ni-Hard is a cast iron that contains about 4.5% nickel (Ni) and 1.5% chromium (Cr). This material is a hard, white iron, offering very good wear resistance. This material incidentally led to the discovery of ductile iron by Keith Millis at the International Nickel Company. This company was promoting Ni-Hard but, due to World War II, was afraid that its supply of Cr would be endangered since it came from Africa. Millis was given the task of finding a substitute for Cr and, in turn, when he tried the use of magnesium (Mg), ductile iron, as we know it, was born.

Meehanite iron is a proprietary material. Alkaline earth silicides, such as calcium silicide and magnesium silicide, are used to control the uniformity of the matrix and graphite distribution. This material can have either flake or nodular graphite. In most respects, this material is no different in properties than the normal gray or ductile iron.

CAST STEELS

Steel is essentially an alloy of mostly iron containing carbon. The smaller amounts of the other alloying elements in steel will be discussed later. It is the presence of carbon with the iron that gives carbon steel its remarkable properties. When carbon is missing, iron is quite soft and weak. However, adding just 0.2–0.3% carbon to the iron appreciably raises the strength; the ductility, although reduced, is still appreciable. The result is that carbon steel exhibits versatility not found in any other metal or alloy.

The properties of carbon steel castings are uniform, regardless of the direction in which the castings are tested. This "isotropic" behavior is absent in structural shapes produced by forging and rolling. Shapes made by forging or rolling develop directional properties during the working process. Thus, cast carbon steel, which does not have this directionality, is better suited in many applications.

Cast carbon steel is strong, with tensile strengths ranging from 60,000 to 280,000 psi (413.7–1930.6 MPa). Tensile strength is the maximum tension load that a material will withstand prior to fracture. This alloy also has ductility. Ductility is the property in a material that permits permanent deformation without rupture caused by stress in tension. In simple terms, carbon steel can be bent without tearing apart. This combination of strength and ductility adds up to give carbon steel great toughness and resistance to shock.

Controlling the composition, specifically the carbon content, can control carbon steel properties within rather wide limits.

Carbon steel castings may be easily welded with no serious property changes. Because of this, the cast-weld process has evolved (also known as composite design) in which large or irregular shapes are cast individually and then welded together to make a complex steel shape **(Fig. 8-21)**.

Heat treating steel castings improves their physical properties. Metallurgists requiring castings to meet certain physical and mechanical characteristics specify castings of carbon steel alloys. The properties (such as hardness and resistance to abrasion) of low-alloy carbon steels are controlled largely by the response of the steel to heat treatment.

High-alloy steels take advantage of some specific alloy property, such as a resistance to corrosion or heat, or some property other than hardenability. Examples are the outstanding wear resistance of 14% manganese steel or the corrosion resistance of an 18% chromium stainless steel.

Cast steel has many desirable properties, but the designer, metallurgist and foundry worker must all understand the casting requirements. Steel must be poured at high temperatures (2900–3000F / 1593–1649C), and thus requires special consideration from the standpoint of refractories, molding sands, metal handling, and pouring and molding techniques. Steel has a high rate of solidification shrinkage, which produces the need for many risers that are normally much larger than those used for casting other alloys. Measures must be taken to control carbon levels in the carbon steel during the melting process to assure that levels are where they should be.

Carbon steel is made up of the same six basic elements as cast iron: iron, carbon, silicon, manganese, phosphorus and sulfur; however, the carbon content must be less than 2%. The carbon content controls the tensile strength, hardness and toughness properties of carbon steels as well as the heat treatment requirements. The alloying elements in carbon steels are usually in the following amounts:

Carbon: less than 2%; Silicon: 0.2–0.8%
Manganese: 0.5–1.0%; Phosphorus, max.: 0.05%
Sulfur, max.: 0.06%

Figure 8-22 shows a comparison of some of the various ferrous alloys, based on the carbon and silicon content.

Plain carbon steels have three commercial classifications determined by their carbon content.

1. Low-carbon steel: carbon less than 0.20%
2. Medium-carbon steel: carbon between 0.20 and 0.50%
3. High-carbon steel: carbon more than 0.50%.

High-alloy and low alloy carbon steels have two commercial classifications determined by their alloy content.

1. Low-alloy steels: alloy content less than 8%.
2. High-alloy steels: alloy content totaling more than 8%.

Plain Carbon Steels

Low-carbon steels are those that have a carbon content less than 0.20%. The elements present in low-carbon cast steel, other than carbon, are manganese (0.50–1.00%), silicon (0.25–0.80%), and sulfur and phosphorus (up to 0.05% max. in each). Residual alloys, such as nickel, chromium, copper and molybdenum, are present in small amounts. Cast carbon steels that contain less than 0.20% carbon are used in electrical and magnetic equipment. They are usually given a full annealing heat treatment.

Medium-carbon steels have a carbon content of 0.20–0.50%. Much of the medium-carbon cast steels are heat treated by normalizing, which consists of cooling the castings in air from approximately 100°F above the upper critical temperature to the ambient temperature. In this chapter, carbon steel castings will be referred to as "steel castings."

High-carbon steels contain over 0.50% carbon. In these cast steels, there are 0.50–1.00% manganese, 0.30–0.80% silicon and 0.05% maximum each of phosphorus and sulfur.

Fig. 8-21. Composite structure by welding. [Courtesy of SFSA]

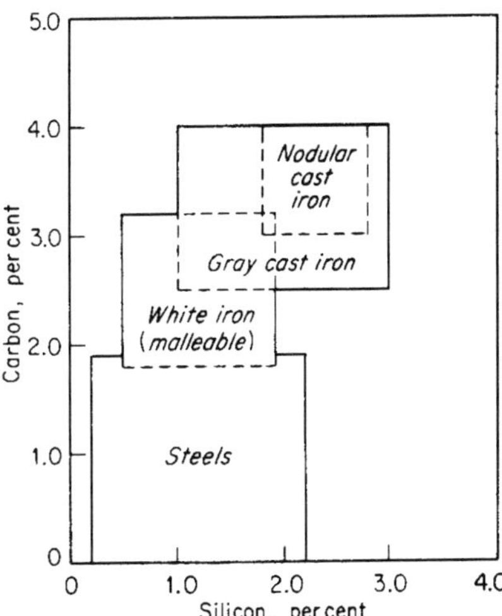

Fig. 8-22. Carbon and silicon percentage ranges present in cast irons. Note the overlapping compositions of the various grades.

Table 8-7.
Principal Effects of Major Alloying Elements in Steel

Element	Percentage	Primary function
Manganese	0.25–0.40	Combines with sulfur to prevent brittleness
	>1%	Increases hardenability by lowering transformation points and causing transformations to be sluggish
Sulfur	0.08–0.15	Contributes to ease of machining metal
Nickel	2–5	Acts as a toughener
	12–20	Improves corrosion resistance
Chromium	0.5–2	Increases hardenability
	4–18	Improves corrosion resistance
Molybdenum	0.2–5	Forms stable carbides; inhibits grain growth
Vanadium	0.15	Forms stable carbides; increases strength while retaining ductility; promotes fine grain size
Boron	0.001–0.003	Acts as powerful hardening agent
Tungsten		Enhances hardness at high temperatures
Silicon	0.2–0.7	Increases strength of spring steels
	2	
	Higher percentages	Improves magnetic properties
Copper	0.1–0.4	Improves corrosion resistance
Aluminum	Small	Makes a good alloying element in nitriding steels
Titanium		Fixes carbon in inert particles
		Reduces martensitic hardness in chromium steels

Table 8-7 lists some of the elements used as alloys in cast steel, and the primary function of each. Usually, manganese and silicon are already present in steel as residuals from the refining process. When steel is cooled from an elevated temperature, manganese and silicon increase its hardness. However, when sulfur is present in amounts of more than 0.06%, it has a harmful effect on both ductility and toughness. Excess phosphorus causes steel to become brittle at low temperatures, which can cause cold shortness in castings. Manganese, silicon and phosphorus are soluble in iron, and thus do not reveal themselves in the microstructure (structure of the solidified metal sample, after preparation, as seen under a microscope).

The strength and hardening characteristics of cast steel are directly related to the percentage of carbon contained within. Steel castings can be readily annealed, normalized, quenched and tempered. Special surface hardening treatments can be applied. Steel castings are available in a wide range of mechanical properties, depending upon the composition and heat treatment. Properties within the following ranges can be obtained at normal temperatures:

Tensile strength, ksi (MPa)	60–300 (414–2068)
Yield strength, ksi (MPa)	30–250 (207–1724)
Elongation, %	50–4
Reduction in area, %	65–5
Brinell hardness, HB	120 to 623
Charpy V-notch impact, ft-lb	200–5 to 271–6.8

There is no difference between the transverse and longitudinal mechanical properties of steel castings.

Steel castings can be found abundantly in the automotive, railroad, and earth-moving industries, in electrical machinery, and in all kinds of manufacturing. Cast steels have played an important role in the development of rockets, jet aircraft, and nuclear power. These industries have a steadily increasing need for metal, which has good strength characteristics and a high degree of resistance to corrosion and creep at high temperatures. Steels with special properties can also be made to fit any segment of the wide range of properties employed in the engineering design of steel parts. Steel castings have kept pace with their needs. **Figures 8-23 and 8-24** show some examples of carbon steel castings.

Fig. 8-23. Carbon steel pump casting.

Fig. 8-24. Carbon steel steam generator housing.

High- and Low-Alloy Steels

Low-alloy cast steels usually have a carbon content under 0.45%, along with small amounts of alloying elements to produce certain definite properties. The cast low-alloy steels are more resistant to atmospheric corrosion than are cast carbon steels **(Figs. 8-25 and 8-26).**

High-alloy cast steels are commonly referred to as cast stainless steels. However, this classification of steels can have one of many different names. These names include heat-resistant high-alloy steels, corrosion-resistant high-alloy steels, austenitics, ferritics, martensitics, duplex stainless steels, super ferritics, austenitic age-hardenable steels, martensitic age-hardenable steels, stabilized stainless steels and H grades.

The primary components are typically iron, chromium, nickel and carbon. However, the ranges and ratios vary widely and thus increase the variety of these steel alloys. A minimum of 12% chromium addition will impart corrosion and oxidation resistance to steel. Stainless steels typically contain one or more alloying elements in addition to chromium: Copper, molybdenum, or small amounts of other elements, such as columbium and nitrogen, when added, can significantly improve strength or corrosion resistance.

This wide range of alloys gives the user many options with regard to heat or corrosion resistance and cost. Additionally, the ranges for strength and toughness achievable with this class of materials cover almost as broad a spectrum as is possible. Subtle changes in composition and heat treatment may have a profound effect on microstructure, strength, pitting resistance, toughness, weldability, magnetic properties, machinability, hardness or other engineering properties. The proper use of metallurgical principles is essential for getting the desired properties while avoiding disastrous results.

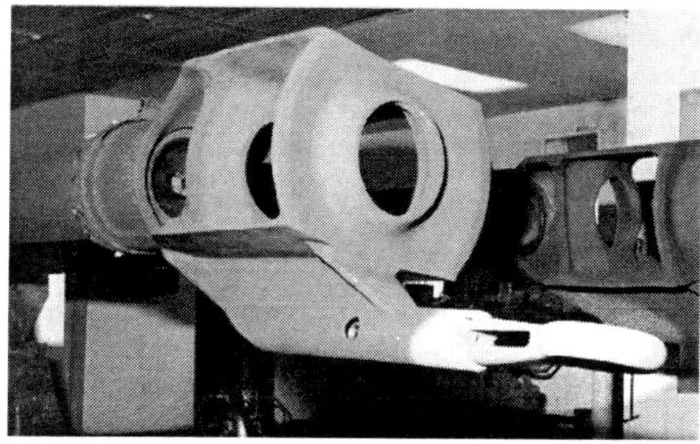

Fig. 8-25. Howitzer made of low-alloy steel.

Fig. 8-26. Tank tread made of low-alloy steel.

Many tests have verified that the properties achieved in cast alloys are very similar to those found in wrought counterparts. However, whether wrought or cast material is used, the chance of misapplication exists. Since these alloys are relatively costly, high-performance materials, they are chosen for conditions of temperature, environment or stress, which have the potential of being very destructive if proper precautions are not taken. Examples are pumps, impellers and propellers in seawater, or molten metal service, furnace tubes, chemical equipment, boiler components, refinery valves and pumps, food processing equipment, etc. **Figures 8-27 and 8-28** show some high-alloy steel castings.

BIBLIOGRAPHY

Bex, T., Ductile Iron One of the Century's Metallurgical Triumphs, American Foundrymen's Society, Inc., *modern casting,* pp 45-48 (Feb 1991).

Hayrynen, K.L., ADI: Another Avenue for Ductile Iron, American Foundrymen's Society, Inc., *modern casting* (Oct 1998).

Heine, R.W., Loper, C.R., Rosenthal, P.C., *Principles of Metal Casting,* Second edition, McGraw-Hill, New York (1967).

Iron Castings Handbook, Iron Castings Society, Inc., 1981.

Kovacs, Sr., B.V., Austempered Ductile Iron Fact and Fiction, American Foundrymen's Society, Inc., *modern casting,* pp 38-41 (Mar 1990).

Metalcasting Dictionary, AFS, Des Plaines, IL 1968.

Millis, K.D., The Father of Ductile Iron, AFS, *modern casting* (Oct 1998).

Samans, C.A., *Engineering Metals and Their Alloys,* Macmillan, New York (1957).

Schleg, F.P., Personal Class Notes from American Foundrymen's Society, Inc., Cast Metals Institute (CMI) Ferrous and Nonferrous Metallurgy Courses 1965-1992.

Steel Casting Metallurgy, Steel Founders' Society of America, Des Plaines, IL (1984).

For additional information on the unveiling of ductile iron in the USA since 1948, refer to *modern casting* magazine (Oct 1998), American Foundry Society, Inc, Des Plaines, IL.

Fig. 8-27. Stainless steel ball separator.

Fig. 8-28. Stainless steel ice cutter.

Family of Nonferrous Alloys

9

Nonferrous alloys—that is, alloys other than those based largely on iron (i.e., cast iron, carbon steel)—are classified according to the principal base elements of which they are composed. The base elements used commercially are mainly aluminum, copper, magnesium, lead, tin, zinc, nickel and titanium. Most commercially cast nonferrous alloys are not pure metals; they are composed of the principal base element plus one or more primary alloying elements, with perhaps two or three additional elements in smaller amounts. To obtain the desired physical and mechanical properties, one must be able to manipulate the amounts of the principal and trace alloying elements. Binary alloys consist of a base element plus one primary alloying element. When a third element (of more than a few percentage points) is included, the alloy becomes a ternary alloy.

Nonferrous alloys are cast into many shapes and end products. **Figures 9-1 through 9-4** show a variety of nonferrous castiangs. Different casting processes are used, depending on the size, shape, production quantity and quality of casting desired. When selecting an alloy for a specific application, a variety of characteristics must be considered:

- Physical properties including density, coefficient of thermal expansion, thermal conductivity, electrical resistivity and magnetic properties
- Mechanical properties, including strength, ductility, hardness and stiffness
- Corrosion resistance
- Wear and abrasion resistance
- Machinability
- Weldability
- Heat and electrical conductivity

Physical and mechanical properties—usually are considered first because they often form the basics in material selection (e.g., a component needs to meet certain weight and strength criteria). The most commonly specified mechanical properties are tensile strength, yield strength, percent elongation, modulus of elasticity, Brinell hardness, fatigue strength and impact strength.

Corrosion resistance—Corrosion is the chemical reaction between a material and its environment. The environment may be liquid or gas. The metal ion may leave the metallic structure to enter the environment as a free ion or it may react to form a compound at the surface of the metal that reduces further corrosion action.

Wear and abrasion resistance—There is no one simple number to express wear resistance of a material. Wear occurs due to intimate contact and relative movement between two materials, resulting in shearing or fracture of the softer surface and/or welding of the two surfaces and subsequent rupture of the weld. Wear is not simply a material characteristic. With the continuous use of an oil film, for instance, contact between surfaces can be prevented and thus wear is minimized.

Machinability—is difficult to rate or characterize relative to a material. To a machine shop, the type of chip generated during machining, the life of the cutting tool, the tendency of the material to build up on the cutting tool, the need for lubricant or coolant and the ability to generate a desired surface finish are all important measures of machinability. Any given material can, therefore, rate very well relative to one such measure and poorly relative to another.

Weldability—of a material depends on the ease with which it can be welded without cracking, generally, directly a function of its chemistry and solidification mode or hot shortness. (Hot shortness is the brittleness of an alloy at elevated temperatures.)

Heat and electrical conductivity—A physical characteristic also to be considered when selecting a nonferrous alloy is its ability to conduct heat or electricity. Some nonferrous alloys are chosen for use in electrical fittings due to their ability to conduct electricity.

ALUMINUM-BASE ALLOYS

Before studying the properties of aluminum and its alloys, a brief look at the infancy of these alloys is in order. History chronicles that Colonel William Frishmuth produced the first aluminum castings in his foundry in Philadelphia, Pennsylvania in 1876. These first aluminum castings were small parts used to make an engineer's transit. At that time, Frishmuth's aluminum foundry was the only one in the industry. The parts were cast from a chemically produced aluminum alloy, which cost $1/oz ($1/28.35 g).

Aluminum was not widely used at the time. Charles Martin Hall, a 22-year-old student at Oberlin College, didn't discover his method of producing pure aluminum until 1886, using an electric

Fig. 9-1. Compressor wheel casting has thin blades made possible by the good fluidity and excellent mechanical properties of aluminum alloys.

current.)Frishmuth's most crowning achievement was the casting of the aluminum cap for the Washington Monument in Washington, D.C. The casting weighed more than 6 lb (2.7 kg) with walls 0.5 in. (1.27 cm) thick. The pyramid was 9 in. (22.8 cm) tall and almost 6 in. (15.24 cm) on a side at the base. The alloy composition was 1.7% iron (Fe), 0.55% silicon (Si) and the balance aluminum (Al). The metal was poured into cast iron molds. One could say that this was one of the first permanent mold aluminum castings poured in the United States. Subsequent examinations have shown no evidence of hydrogen porosity or oxide inclusions. Aluminum-bronze was originally chosen since it was readily available, but Frishmuth recommended that the designers use aluminum. This cap was to be used as a lightning rod, which was connected to copper rods running to the ground, 555 feet below.

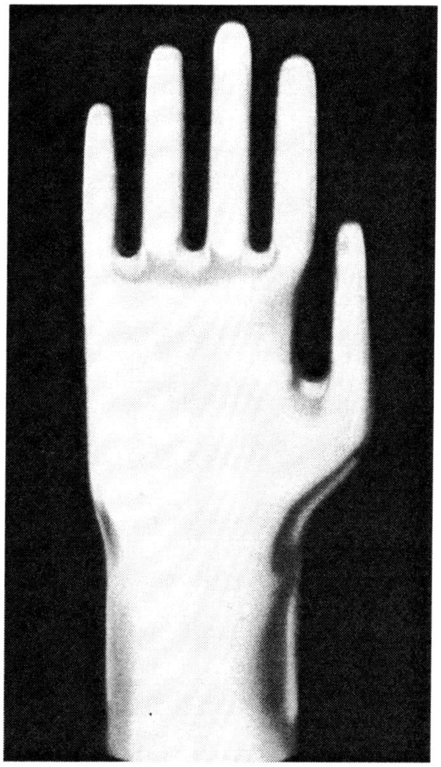

History aside, the following text will cover the most commonly used aluminum alloys in metalcasting.

In aluminum-base alloys, aluminum is the principal element. The primary alloying elements are silicon, copper, magnesium, nickel and tin. Iron is often present in small quantities, and, except in diecasting alloys, is considered an impurity. Manganese and chromium are also generally considered impurities. Titanium, boron, sodium, strontium, phosphorus and beryllium are used in very small amounts—alone or sometimes in combination—to control the quantity, size and shape of microfeatures such as grain size, and for silicon modification.

Fig. 9-2. Cast aluminum surgical glove mold shows extraordinary surface smoothness.

Fig. 9-3. This aluminum alloy casting is a 4-in. internal valve.

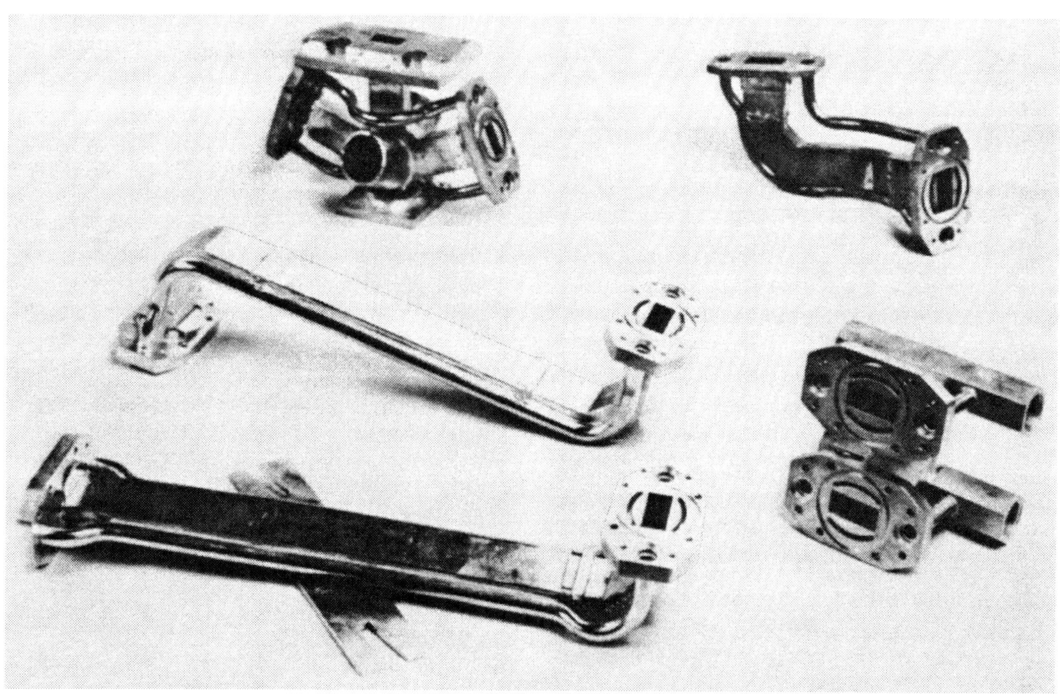

Fig. 9-4. Examples of high-precision investment-cast parts made of beryllium copper.

Aluminum castings (die, sand and permanent mold) are widely used in many applications due to aluminum's exceptional characteristics and its design flexibility.

For proper identification of alloys, the Aluminum Association established a standard casting alloy designation system for aluminum casting alloys as follows:

Casting Alloys Designations

Designation Number	*Primary Alloying Elements*
1XX.X	Aluminum, 99% minimum and greater
2XX.X	Copper
3XX.X	Silicon with added copper and/or magnesium
4XX.X	Silicon
5XX.X	Magnesium
6XX.X	Unused series
7XX.X	Zinc
8XX.X	Tin
9XX.X	Unused series

An example of this numbering system **(Table 9-1)** shows three different chemical compositions for Aluminum 356 alloy. The "0" (zero) following the decimal point indicates the chemical limits applied to a 356 alloy product—in this case, a casting. A "1" (one) following the decimal point indicates the chemical limits for the raw material, the ingot, used to make the product or, in this example, the 356.0 aluminum casting. A "2" (two) following the decimal point also indicates the raw material, the ingot, having somewhat different chemical limits (typically tighter limits) but still within the limits for 356.1 ingot.

Another item to consider is that 356.1 ingot can be supplied as a secondary product (remelted scrap, etc.). On the other hand the 356.2 ingot is made from primary aluminum (reduction cell)—in other words, it is made from aluminum that has never been used before. The former ingot is called "secondary ingot" and the latter "primary ingot".

Some alloy names or numbers include a letter. Such letters preceding an alloy number distinguish between alloys that differ only slightly in percentage of impurities or relatively minor alloying elements (for example, 356.0, A356.0, B356.0 and F356.0).

Space limitations prevent the printing of the complete Aluminum Association listing of aluminum alloy chemical compositions. However, **Table 9-2** is an abbreviated listing of the principal aluminum casting alloys, minus the footnotes found on the complete list. The AFS Publication, *Aluminum Casting Technology,* contains a complete "Registration Record of Aluminum Association Alloy Designations and Chemical Composition Limits for Aluminum Alloys in the Form of Casting and Ingot." Contact AFS or the Aluminum Association to obtain a copy of this material.

Following is a closer look at the aluminum alloys listed in **Table 9-2** with regard to physical and mechanical properties, castability and/or end uses:

Alloy 100.0—This alloy was designed especially for casting larger electric motor rotors. It is also used in other applications where high electrical conductivity is of prime importance.

Alloy 170.0—is designed primarily for casting small-diameter electric motor rotors. It is also used in connector castings for joining transmission cables and in other electrical applications.

Alloys 201.0 & A201.0—These alloys, which contain silver, provide exceptionally high mechanical strength after heat treatment. The silver is added primarily to improve resistance to stress corrosion cracking and also to increase the response of these alloys to the aging heat treatment. These alloys are used where the highest tensile and yield strengths and moderate elongation are required. Elevated temperature and impact strengths are very good. Applications for these alloys include, structural casting members, truck and trailer castings, cylinder heads, pistons, gear and pump housings, aerospace housings, and aircraft landing gear components. Resistance to hot cracking and solidification shrinkage is rated fair for these alloys and pressure tightness is good.

Alloy A206.0—This high-strength aluminum alloy is often used where damage-tolerant castings are required. Structural castings in the automotive and aerospace industries, where high tensile and yield strength and moderate elongation are needed, are poured from this alloy. Yield strength and fracture toughness are twice that of 356- and 357-type alloys. Castings where high fracture toughness is needed (rear end housings and truck spring hangers) make use of this alloy. High-temperature applications include cylinder heads and turbine and superchargers. As for castability, this aluminum alloy requires gating techniques that promote steep thermal gradients.

Alloys 242.0 and A242.0—These alloys are used extensively for applications where excellent strength and hardness at high temperatures are required. Heavy-duty pistons; motorcycle, diesel and aircraft pistons; aircraft generator housings; and air-cooled cylinder heads are typical applications. These alloys have good fluidity, are fair for pressure tightness, and have fair resistance to hot cracking and solidification shrinkage.

Alloys 295.0 & 296.0—High tensile properties and good machinability are characteristics of this alloy. It is used for gun control parts, gear housings, aircraft fittings, compressor connecting rods and railway car seat frames. Alloys 355.0 and 356.0 have largely replaced this alloy. When used, this aluminum alloy has good fluidity, pressure tightness and low solidification shrinkage. Resistance to hot cracking is fair.

Alloy 308.0—Typical applications of this alloy are ornamental grills, reflectors, and other permanent mold castings requiring good castability, weldability, pressure tightness and moderate strength.

Alloys 319.0, A319.0, B319.0 and 320.0—Alloys 319.0 and A319.0 have very good castability, weldability and pressure tightness and moderate strength. They are very stable alloys; that is, their good casting and mechanical properties are not affected seriously by fluctuations in their impurity content. Alloys B319.0 and 320.0 show higher strength and hardness than 319.0 and A319.0 and are generally used in the permanent mold process. Characteristics other than strength and hardness are similar to those of 319.0 and A319.0. Typical sand castings of these alloys are gasoline and diesel engine crankcases, gasoline and oil tanks, and oil pans. Permanent mold castings include water-cooled cylinder heads, typewriter frames, rear axle housings and engine parts. Pressure tightness, fluidity, resistance to hot cracking and solidification shrinkage tendencies are all classed as very good for this family of aluminum alloys.

Alloy 332.0—This aluminum alloy, as with 336.0, is used in applications where properties such as good strength at high temperatures, low coefficient of thermal expansion and good resistance to wear are required. This alloy has many characteristics similar to 336.0 and, in addition, offers improved castability and machinability. Typical applications include automotive and diesel pistons, pulleys and sheaves. Excellent fluidity and resistance to hot cracking are castability characteristics of this alloy, along with very good pressure tightness and a high resistance to solidification shrinkage.

Alloys 333.0 and A333.0—The combination of a low coefficient of expansion and good mechanical properties makes these alloys very suitable for casting applications such as gas meter and regulator parts, gear blanks, pistons and general automotive castings, and electric iron sole plates. Fluidity, pressure tightness and resistance to hot cracking are excellent along with a good rating for solidification shrinkage tendencies.

Table 9-1.
Three Different Chemical Compositions for Al 356 Alloy

Alloy	Form	Si	Fe	Cu	Mn	Mg	Zn	Ti	Each	Total
356.0	Sand, PM Castings	6.5 7.5	0.6 max	0.25 max	0.35 max	0.20 0.45	0.35 max	0.25 max	0.05 max	0.15 max
356.1	Ingot	6.5 7.5	0.5 max	0.25 max	0.35 max	0.25 0.45	0.35 max	0.25 max	0.05 max	0.15 max
356.2	Ingot	6.5 7.5	0.13 0.25	0.1 max	0.05 max	0.30 0.45	0.05 max	0.20 max	0.05 max	0.15 max

Table 9-2.
Aluminum Alloy Compositions

Alloy	Si	Fe	Cu	Mn	Mg	Cr	Ni	Zn	Ti	Sn	Al	Other Elements Each	Other Elements Total
100.0	0.15	0.6–0.8	0.10	—	—	—	—	0.05	—	—	99.0	0.03	0.10
170.0	—	—	—	—	—	—	—	0.05	—	—	99.70	0.03	0.10
201.0	0.10	0.15	4.0–5.2	0.20–0.50	0.15–0.55	—	—	—	0.15–0.35	—	bal.	0.05*	0.10
A201.0	0.05	0.10	4.0–5.0	0.20–0.40	0.15–0.35	—	—	—	0.15–0.35	—	bal.	0.03*	0.10
A206.0	0.05	0.10	4.2–5.0	0.20–0.50	0.15–0.35	—	0.05	0.10	0.15–0.30	0.05	bal.	0.05	0.15
242.0	0.7	1.0	3.5–4.5	0.35	1.2–1.8	0.25	1.7–2.3	0.35	0.25	—	bal.	0.05	0.15
A242.0	0.6	0.8	3.7–4.5	0.10	1.2–1.7	0.15–0.25	1.8–2.3	0.10	0.07–0.20	—	bal.	0.05	0.15
295.0	0.7–1.5	1.0	4.0–5.0	0.35	0.03	—	—	0.35	0.25	—	bal.	0.05	0.15
296.0	2.0–3.0	1.2	4.0–5.0	0.35	0.05	—	0.35	0.50	0.25	—	bal.	—	0.35
308.0	5.0–6.0	1.0	4.0–5.0	0.50	0.10	—	—	1.0	0.25	—	bal.	—	0.50
319.0	5.5–6.5	1.0	3.0–4.0	0.50	0.10	—	0.35	1.0	0.25	—	bal.	—	0.50
A319.0	5.5–6.5	1.0	3.0–4.0	0.50	0.10	—	0.35	3.0	0.25	—	bal.	—	0.50
B319.0	5.5–6.5	1.2	3.0–4.0	0.8	0.10–0.50	—	0.50	1.0	0.25	—	bal.	—	0.50
320.0	5.0–8.0	1.2	2.0–4.0	0.8	0.05–0.60	—	0.35	3.0	0.25	—	bal.	—	0.50
332.0	8.5–10.5	1.2	2.0–4.0	0.50	0.5–1.5	—	0.50	1.0	0.25	—	bal.	—	0.50
333.0	8.0–10.0	1.0	3.0–4.0	0.50	0.05–0.50	—	0.50	1.0	0.25	—	bal.	—	0.50
A333.0	8.0–10.0	1.0	3.0–4.0	0.50	0.05–0.50	—	0.50	3.0	0.25	—	bal.	—	0.50
336.0	11.0–13.0	1.2	0.50–1.50	0.35	0.7–1.3	—	2.0–3.0	0.35	0.25	—	bal.	0.05	—
354.0	8.6–9.4	0.20	1.6–2.0	0.10	0.40–0.60	—	—	0.10	0.20	—	bal.	0.05	0.15
355.0	4.5–5.5	0.6	1.0–1.5	0.5	0.4–0.6	0.25	—	0.35	0.25	—	bal.	0.05	0.15
C355.0	4.5–5.5	0.20	1.0–1.5	1.10	0.40–0.60	—	—	0.10	0.20	—	bal.	0.05	0.15
356.0	6.5–7.5	0.60	0.25	0.35	0.20–0.45	—	—	0.35	0.25	—	bal.	0.05	0.15
A356.0	6.5–7.5	0.20	0.20	0.10	0.25–0.45	—	—	0.10	0.20	—	bal.	0.05	0.15
357.0	6.5–7.5	0.15	0.05	0.03	0.45–0.6	—	—	0.05	0.20	—	bal.	0.05	0.15
A357.0	6.5–7.5	0.20	0.20	0.10	0.40–0.70	—	—	0.10	0.04–0.20	—	bal.	0.05**	0.15
359.0	8.5–9.5	0.20	0.20	0.10	0.50–0.70	—	—	0.10	0.20	—	bal.	0.05	0.15
360.0	9.0–10.0	2.0	0.6	0.35	0.40–0.60	—	0.50	0.50	—	0.15	bal.	—	0.25
A360.0	9.0–10.0	1.3	0.6	0.35	0.40–0.60	—	0.50	0.50	—	0.15	bal.	—	0.25
A380	7.5–9.5	1.3	3.0–4.0	0.50	0.10	—	0.50	3.0	—	0.35	bal.	—	0.50
B380	7.5–9.5	1.3	3.0–4.0	0.50	0.10	—	0.50	1.0	—	0.35	bal.	—	0.50
383.0	9.5–11.5	1.3	2.0–3.0	0.50	0.10	—	0.30	3.0	—	0.15	bal.	—	0.50
384.0	10.5–12.0	1.3	3.0–4.5	0.50	0.10	—	0.50	3.0	—	0.35	bal.	—	0.50
A390	16.0–18.0	0.50	4.0–5.0	0.10	0.45–0.65	—	—	0.10	0.20	—	bal.	0.10	0.20
B390	16.0–18.0	1.3	4.0–5.0	0.50	0.45–0.65	—	0.10	1.5	0.20	—	bal.	0.10	0.20
413.0	11.0–13.0	2.0	1.0	0.35	0.10	—	0.50	0.50	—	0.15	bal.	—	0.25
A413.0	11.0–13.0	1.3	1.0	0.35	0.10	—	0.50	0.50	—	0.15	bal.	—	0.25
443.0	4.5–6.0	0.8	0.6	0.50	0.05	0.25	—	0.50	0.25	—	bal.	—	0.35
A443.0	4.5–6.0	0.8	0.30	0.50	0.05	0.25	—	0.50	0.25	—	bal.	—	0.35
C443.0	4.5–6.0	2.0	0.6	0.35	0.10	—	0.50	0.50	—	0.15	bal.	—	0.25
A444.0	6.5–7.5	0.20	0.10	0.10	0.05	—	—	0.10	0.20	—	bal.	0.05	0.15
511.0	0.30–0.7	0.50	0.15	0.35	3.5–4.5	—	—	0.15	0.25	—	bal.	0.05	0.15
512.0	1.4–2.2	0.6	0.35	0.8	3.5–4.5	0.25	—	0.35	0.25	—	bal.	0.05	0.15
513.0	0.30	0.40	0.10	0.30	3.5–4.5	—	—	1.4–2.2	0.20	—	bal.	0.05	0.15
514.0	0.35	0.50	0.15	0.35	3.5–4.5	—	—	0.15	0.25	—	bal.	0.05	0.15
518.0	0.35	1.8	0.25	0.35	7.5–8.5	—	0.15	0.15	—	0.15	bal.	—	0.25
520.0	0.25	0.30	0.25	0.15	9.5–10.6	—	—	0.15	0.25	—	bal.	0.05	0.15
535.0	0.15	0.15	0.05	0.10–0.25	6.2–7.5	—	—	—	0.10–0.25	—	bal.	0.05	0.15
710.0	0.15	0.50	0.35–0.6	0.05	0.6–0.8	—	—	6.0–7.0	0.25	—	bal.	0.05	0.15
711.0	0.30	0.7–1.4	0.35–0.6	0.05	0.25–0.45	—	—	6.0–7.	0.20	—	bal.	0.05	0.15
712.0	0.30	0.50	0.25	0.10	0.50–0.65	0.40–0.60	—	5.0–6.5	0.15–0.25	—	bal.	0.05	0.20
713.0	0.25	1.1	0.40–1.0	0.6	0.20–0.50	0.35	0.15	7.0–8.0	0.25	—	bal.	0.10	0.25
771.0	0.15	0.15	0.10	0.10	0.8–1.0	0.06–0.20	—	6.5–7.5	0.10–0.20	—	bal.	0.05	0.15
772.0	0.15	0.15	0.10	0.10	0.6–0.8	0.06–0.20	—	6.0–7.0	0.10–0.20	—	bal.	0.05	0.15
850.0	0.7	0.7	0.7–1.3	0.10	0.10	—	0.7–1.3	—	0.20	5.5–7.0	bal.	—	0.30
851.0	2.0–3.0	0.7	0.7–1.3	0.10	0.10	—	0.30–0.7	—	0.20	5.5–7.0	bal.	—	0.30
852.0	0.40	0.70	1.7–2.3	0.10	0.6–0.9	—	0.9–1.5	—	0.20	5.5–7.0	bal.	—	0.30
853.0	5.5–6.5	0.7	3.0–4.0	0.5	—	—	—	—	0.20	5.5–7.0	bal.	—	0.30

**Silver: 0.40–1.0*
***Beryllium: 0.04–0.07*
Figures taken from Aluminum Association Specifications Jan. 1998.

Alloy 336.0—This is an aluminum alloy that offers good wear resistance, a low coefficient of thermal expansion and strength at elevated temperatures. Typical uses include automotive and heavy-duty diesel pistons, pulleys, sheaves and similar applications. Alloy 336.0 shows excellent fluidity and resistance to hot cracking. Pressure tightness is also very good.

Alloy 354.0—Alloy 354.0 is basically similar to C355.0, but its increased silicon and copper content give it greatly improved mechanical properties. Its tensile and elongation properties show almost no change up to temperatures of at least 300F (149C). The outstanding strength characteristics of this alloy make it especially useful in the aerospace industry, in applications where premium-strength castings are required. Alloy 354.0 has excellent castability.

Alloy 355.0—This aluminum alloy is a general-purpose, high-strength material with many desirable characteristics that make it popular for a wide variety of castings. Sand castings include air compressor pistons, printing press bed plates, water jackets, crankcases, electric motor fans and blowers, and accessory and gear housings. Some Alloy 355.0 permanent mold applications include impellers, aircraft fittings, timing gears, aircraft supercharger housings, passenger car wheels, jet engine compressor cases and engine blocks. Marine usage includes those applications where high strength is needed along with pressure tightness, corrosion resistance and high strength at elevated temperatures. Alloy 355.0 has excellent castability.

Alloy C355.0—This alloy is a modification of 355.0; however, its characteristics are much the same. Its elongation and strength are higher. Strengths at elevated temperatures are relatively higher and the castings are usually heat treated. Casting applications for this alloy are very similar to those of Alloy 355.0. Fluidity, pressure tightness, resistance to hot cracking and solidification shrinkage tendencies are all rated excellent.

Alloy 356.0—Applications for 356.0 are similar to those of 355.0. Alloy 356.0 is used extensively because of its excellent castability and has largely replaced Alloy 259.0. Some of the permanent mold castings made using Alloy 356.0 include machine tool parts, aircraft wheels and handwheels, bridge marine hardware, and castings such as aileron control sectors, rudder control supports, fuselage fittings and fuel tank elbows for airplanes and missiles. Automotive castings include spring brackets, cylinder heads, engine blocks and passenger car wheels. Castings made using the sand-casting processes include flywheel housings, automotive transmission cases, oil pans, rear axle housings, water-cooled cylinder blocks, etc. After heat treatment, this alloy has good machinability. The castability of Alloy 356.0 is excellent, which helps to make it one of the most widely used aluminum-silicon alloys.

Alloy A356.0—This alloy has greater elongation, higher strength and considerably higher ductility than Alloy 356.0. It has these better properties because its impurities are lower than 356.0. Typical applications of A356.0 include airframe castings, machine parts, truck chassis parts, aircraft and missile components, and structural parts requiring high strength. This alloy is considered to have good machinability. All casting characteristics are rated excellent.

Alloy 357.0—The characteristics and uses of this alloy are similar to those of 356.0. However, 357.0 exhibits greater ultimate and yield strengths. This alloy also shows high strength and good ductility after heat treatment. Applications of Alloy 357.0 include high-stressed aircraft and missile structures and high-velocity blowers and impellers. This alloy has excellent pressure tightness, fluidity, resistance to hot cracking and a slight solidification shrinkage tendency.

Alloy A357.0—provides very high mechanical properties and has excellent castability. This alloy is used for aircraft, missile components and other applications that require high tensile, yield and ultimate strengths. Its strengths at elevated temperatures are also very good. All of this alloy's castability characteristics are rated excellent.

Alloy 359.0—This alloy exhibits outstanding strength characteristics of particular usefulness to the aerospace industries. The alloy has excellent castability and exhibits little evidence of corrosion cracking when stressed at high levels. Alloy 359.0 has excellent castability and its fluidity is said to be better in permanent molds than Alloy 356.0.

Alloys 360.0 and A360.0—When excellent castability, mechanical properties and corrosion resistance are required, or when miscellaneous thin-walled and intricate parts are specified, these two alloys can be used. Strength at elevated temperatures is excellent. Typical castings include outboard motor parts, instrument cases, cover plates, a wide variety of general-purpose castings, and aircraft and marine items of many types. The excellent castability of these two alloys makes them favorite diecasting alloys for high-strength, thin-walled, complex castings.

Alloys A380.0 and B380.0—These two aluminum alloys are widely used for making general-purpose diecastings. They have good mechanical properties and are used to cast housings for lawnmowers and radio transmitters, air brake castings, gear houses and air-cooled cylinder heads. Their castability characteristics of fluidity, pressure tightness and resistance to hot cracking is said to be very good.

Alloys 383.0 and 384.0—These are high-strength alloys used for pistons and other components designed for severe service conditions. They are excellent alloys to use for castings with thin walls and large surface areas. Other uses include wringer frames, pulleys, automatic transmissions, air-cooled cylinder heads, and body and end shields for electric motors and toys. The fluidity of these alloys is excellent, pressure tightness is very good and the resistance to hot cracking is good.

Alloys A390.0 and B390.0—These companion alloys are hypereutectic aluminum-silicon alloys. (The term "hypereutectic" will be discussed in Chapter 14, Microstructure of Nonferrous Alloys.) The low coefficient of thermal expansion, high hardness and good wear resistance of these alloys make them suitable for internal combustion engine pistons and blocks, and cylinder bodies for compressors, pumps and brakes. When these alloys are used for diecastings or permanent molding, the castability is rated very good. When used for sand casting, the castability is good.

Alloys 413.0 and A413.0—These general-purpose alloys have good mechanical properties and also good characteristics for casting large, intricate parts with thin sections. These alloys are recommended for architectural ornamental, marine, and food and dairy equipment applications. Other uses include pressure-tight parts, large instrument cases, street lamp housings, typewriter frames, dental equipment and outboard motor pistons. Alloys 413.0 and A413.0 have excellent fluidity, resistance to hot cracking and pressure tightness.

Alloys 443.0, A443.0 and C443.0—The 443-type alloys are used widely for diecastings, where above average ductility is required. They are also popular for sand and permanent mold castings where moderate strength, excellent pressure tightness and castability are required. Diecastings include cookware, pipe fittings and marine fittings. Permanent mold castings include tire molds, cooking utensils and carburetor bodies. Pipe fittings, food-handling equipment, architectural and ornamental parts, and marine fittings are some of the castings made using sand molds. For diecasting, Alloy C443.0 has good fluidity and also has good pressure tightness and very good resistance to hot cracking. For sand casting, Alloys 443.0 and A443.0 have excellent pressure tightness and resistance to hot cracking, whereas the solidification shrinkage tendencies are very low. Finally, when Alloy 443.0 is used in permanent molds, it has excellent resistance to hot cracking and a high degree of pressure tightness with very good resistance to shrinkage.

Alloy A444.0—This alloy was developed for use in highway bridge guard castings, which require the unusual ability to absorb energy without rupture and have high-level impact properties. This alloy has medium strength with very high elongation and is suited for applications such as outboard motor propellers, cast furniture and bumper guards. Alloy A444.0 exhibits excellent resistance to hot cracking and excellent fluidity and shrinkage characteristics.

Alloy 511.0—Because this alloy has a more uniform appearance after anodizing (its silicon content is about one-third that of Alloy 512.0), this modification of 514.0 is used in ornamental hardware and architectural castings. This alloy is rated poor to fair in fluidity, pressure tightness, hot cracking resistance and solidification shrinkage tendencies.

Alloy 512.0—The increased silicon content gives this alloy improved foundry characteristics over 514.0, although the mechanical properties are reduced. This alloy has good tarnish and corrosion resistance, which makes it popular to use for cooking utensils, marine applications and pipe fittings, as well as general use castings. Alloy 512.0 has poor fluidity and resistance to hot cracking but fair pressure tightness and solidification shrinkage.

Alloy 513.0—The addition of zinc to this alloy makes it popular for permanent mold casting of ornamental hardware and architectural fittings. Alloy 513.0 has poor to fair fluidity and pressure tightness, fair resistance to hot cracking and a poor solidification shrinkage tendency.

Alloy 514.0—This alloy is used where excellent resistance to corrosion and tarnish is required. Alloy 514.0 has moderate strength and good ductility, and it anodizes readily. Typical uses for this alloy include fittings for chemical and sewage use, dairy and food handling equipment, tire molds, cooking utensils, hardware, ornamental housings and architectural applications. This alloy is rated poor in pressure tightness and resistance to hot cracking, and fluidity is rated poor to fair.

Alloy 518.0—Alloy 518.0 has excellent resistance to tarnish and corrosion, and high strength and ductility. However, it does not have the castability of Alloy 413.0 in small intricate castings, nor does it have the casting qualities of Alloys 360.0 and A380.0. This alloy does respond well to anodizing and mechanical finishing. Typical uses include architectural and ornamental castings, food and dairy equipment, escalator parts, aircraft and marine hardware and similar items requiring high strength, ductility and resistance to corrosive environments. Resistance to hot cracking and pressure tightness is poor. Alloy 518.0 has poor fluidity and very high-drossing tendencies.

Alloy 520.0—This alloy requires special foundry practice and is used primarily for sand casting, where excellent machinability, resistance to corrosion and high strength and elongation are required. In strength and corrosion resistance, Alloy 520.0 rates as one of the highest of all aluminum sand-casting alloys. Castings that must be shock resistant as well as strong are often made from this alloy. This alloy, however, is not recommended for use at temperatures exceeding about 250F (121C). Typical applications include aircraft fittings, railway passenger car frames, truck and bus frame sections, levers, brackets, and similar parts that will be subjected to severe conditions. Foundries must use the best foundry practices to yield completely sound castings. Alloy 520.0 has poor to fair fluidity and resistance to hot cracking. Pressure tightness and solidification shrinkage tendencies are poor.

Alloy 535.0—This aluminum-magnesium-type alloy possesses a high and stable combination of strength, shock resistance and ductility. It is ideally suited for parts in instruments and computing devices where dimensional stability is of major importance. This alloy does not have to be heat treated to attain the mechanical properties desired. In addition to the high ductility and tensile strength of Alloy 535.0, the Charpy impact is from 10 to 12 pounds, which makes it suitable for shock-resistant applications. Brackets, C-clamps and machined parts that need strength, as well as impellers, optical equipment and similar applications requiring a high polish or anodized finish are typical uses. In many cases, this alloy has replaced gray iron and malleable iron castings because it reduces weight without sacrificing strength. The machinability of this alloy is considered to be excellent. Alloy 535.0 has fair casting characteristics and attains its high physical and mechanical properties immediately upon solidification. The alloy shows fair fluidity with little serious tendency toward hot tearing.

Alloy 710.0—This self-aging alloy is used regularly for miscellaneous general-purpose castings that require subsequent brazing. Alloy 710.0 has high strength and ductility without heat treatment, although ductility is reduced during room temperature aging. This alloy has excellent machinability, fair castability with greater shrinkage tendencies than shown by the aluminum-copper alloys. Pressure tightness is good, but resistance to hot cracking is poor.

Alloy 711.0—This alloy has good strength and ductility without heat treatment. It is widely used for casting torque converter impeller blades and furniture parts requiring subsequent brazing, and has excellent machinability. The castability of this alloy is considered to be fair with pressure tightness and fluidity considered to be fair and the tendency toward solidification shrinkage and resistance to hot cracking being poor.

Alloy 712.0—This alloy is used when a combination of good mechanical properties without heat treatment is needed. It also has good shock and corrosion resistance and good machinability, along with dimensional stability. The machining characteristics of this alloy are excellent. Alloy 712.0 is widely used for marine castings, farm machinery, machine tool parts, and other applications where the part must have good strength and impact resistance. This alloy has fair to good castability. Although its pressure tightness and resistance to hot cracking are only fair, the alloy's fluidity and solidification shrinkage tendency are rated as good.

Alloy 713.0—This high-strength aluminum-zinc-magnesium alloy ages at room temperature and is ideal for castings too intricate to be heat treated. It has high yield strength, ductility, impact resistance and other final mechanical properties, which are achieved by self-aging for 21 days at room temperature. Alloy 713.0 achieves tensile and yield strengths comparable to non-self-aging alloys, and is dimensionally stable and will not expand or grow during aging as is common with some other aluminum alloys. Automotive parts, pumps, trailer parts and mining equipment are regularly cast in this alloy. Machinability characteristics of the alloy are excellent. The castability of this alloy is rated fair to good and is similar to other aluminum-zinc-magnesium type alloys. However, due to its narrow freezing range, large gates and risers are required to compensate for its shrinkage rate.

Alloys 771.0 and 772.0—These two aluminum alloys are high-strength, shock-resistant sand casting alloys that develop a high combination of physical and mechanical properties in the as-cast and room-temperature-aged conditions. Without heat treatment, mechanical properties for both alloys are superior to many conventional aluminum casting alloys after solution heat treatment. These alloys are completely dimensionally stable to 0.00001 in./in. after stress relieving. The machinability of these two alloys is classed as excellent. The castability of Alloys 771.0 and 772.0 is considered good. Fluidity and resistance to solidification shrinkage are above average for the 7XX system.

Alloys 850.0, 851.0 and 852.0—These alloys are most commonly used to produce bearings. They all have excellent bearing qualities and good compressive yield strengths.

Alloy 850.0—Major applications for Alloy 850.0 include bushings and journal bearings for railroads. Alloy 850.0's machinability is rated excellent; however, its castability is rated poor. The alloy has fair fluidity and solidification shrinkage tendencies, but its pressure tightness and resistance to hot cracking are poor.

Alloy 851.0—This alloy has less nickel and a great deal more silicon, has better castability and less hot shortness than does Alloy 850.0. However, it still shows the excellent bearing qualities, good compressive yield strength and machinability of Alloy 850.0. Rolling mill bearings are a typical use of this alloy.

Alloy 852.0—The castability and fabricating characteristics of Alloy 852.0 are similar to those of 850.0. Alloy 852.0 has excellent machinability and fair to good castability. Fluidity and solidification shrinkage tendencies are rated fair, but pressure tightness and resistance to hot cracking are poor.

Alloy 853.0—This experimental alloy has been used to a limited extent for heavy-duty bushing applications in diesel engines. Although there is not sufficient production test data to provide typical or guaranteed minimum property values, the results of laboratory tests show approximate levels of properties for test bars of 853.0 cast in sand and permanent molds.

Concentration has centered on the most commonly used aluminum alloys in aluminum foundries. There certainly are more aluminum alloys than those discussed and information concerning their properties, etc., can be found in handbooks on aluminum alloys. The melting, treatment and solidification of aluminum alloys will be discussed in Chapter 14, Microstructure of Nonferrous Alloys. This material will also include aluminum alloy microstructures and heat treatment.

COPPER-BASE ALLOYS

Of all the alloys cast, both ferrous and nonferrous, copper-base alloys are the oldest alloys known to be cast by man. There is evidence that copper was melted and cast into shapes as early as 3700 B.C. The oldest casting in existence is a copper frog believed to be cast in 3200 B.C. in Mesopotamia. The Bible reports that King Solomon had large bronze castings poured for the first temple in Jerusalem. Copper and tin were first combined to form bronze about 3000 years ago.

In the past, cast copper alloys were classified by a variety of systems including the ASTM letter-number designation, based on nominal composition, by trade names, and by descriptive terms such as "ounce metal," "Navy M," "gun metal" and so forth. However, with technological developments, new alloys were produced and existing alloys were modified, making the old designation systems inadequate and misleading. A new system was designed, based on a precise description of the composition range for each alloy, which is now the accepted alloy designation system used in North America, Australia and Brazil for cast copper and copper alloy products. Originally developed as a three digit system by the copper and brass industry, the designations now have been expanded to five digits that follow a prefix letter "C," and have been made part of the Unified Numbering System for Metals and Alloys. For example, copper alloy 836 in the original three-digit system is now C83600 in the UNS system. Today, the Copper Development Association administers this new designation system.

With this system, cast alloys are numbered from C80000 through C99999. The compositions are grouped into the following families of copper and copper alloys:

- *Coppers*—Metals that have a designated minimum copper content of 99.3% or higher.
- *High-copper alloys*—These cast high-copper alloys have a designated copper content in excess of 94%, to which silver may be added for special properties.
- *Brasses*—These alloys contain zinc as the primary alloying element, with or without designated alloying elements present (such as iron, aluminum, nickel and silicon). The brass alloys can be further divided into four major families:
  - *Copper-tin-zinc alloys,* also known as red, semi-red and yellow brasses.
  - *Manganese bronze alloys,* also called the high-strength yellow brasses.
  - *Leaded manganese bronze alloys,* also known as leaded, high-strength yellow brasses.
  - *Copper-zinc-silicon alloys,* also called silicon brasses and bronzes.
- *Bronzes*—Broadly speaking, bronzes are copper alloys in which the primary alloying element is not zinc or nickel. Originally, "bronze" applied to alloys with tin as the only or primary alloying element. Today the term is generally not used by itself but with a modifying adjective. The cast alloys include four main families of bronzes:
  - *Copper-tin alloys,* which are called tin bronzes.
  - *Copper-tin-lead alloys,* also known as leaded and high-leaded tin bronzes.
  - *Copper-tin-nickel alloys,* known as nickel silvers.
  - *Copper-aluminum alloys,* also called aluminum bronzes.
  - The family of alloys known as the *manganese bronzes,* in which zinc is the major alloying element, is included in the brass category.
- *Copper-nickels*—These are alloys with nickel as the primary alloying element, with or without other designated alloying elements present.
- *Copper-nickel-zinc alloys*—Commonly known as "nickel silvers," these are alloys that contain zinc and nickel as the primary and secondary alloying elements, with or without the presence of other designated elements.
- *Leaded coppers*—Leaded coppers comprise a series of cast alloys of copper with 20% or more lead, usually with a small amount of silver present, but without tin or zinc.
- *Special alloys*—Alloys whose chemical compositions do not fall into any of the above categories are categorized as "special alloys."

Table 9-3 lists some of the copper-base alloys, their chemical composition, mechanical properties and a few physical properties. This table by no means lists all of the casting alloys. For a complete and comprehensive publication of the copper-base casting alloys, refer to the "Standards Handbook," Part 7 Alloy Data (Cast Products), Fourth Edition, 1996. This book is available from the Copper Development Association (CDA), and lists chemical composition, applicable specifications, fabrication practices, machinability rating, typical uses, casting characteristics (castability), heat treatment, physical and mechanical properties. This wealth of information is too large to be included in this textbook.

It should be noted that the mechanical properties listed in **Table 9-3** are typical properties for these alloys; however, they can vary, based on the casting process used to make the casting. **Table 9-4** makes this comparison for three copper alloys, showing how the casting process can affect mechanical properties.

A new copper-base alloy, SeBiLOY, has been developed. The U.S. Federal Regulations now limit the amount of lead permitted in public drinking water supplies. Since most of the plumbing fittings, valves and faucets are made from leaded red brasses, this new regulation established an immediate need for a reduced lead or lead-free plumbing alloy. Since these castings undergo considerable machining, one of the plusses for the leaded red brasses, the new alloy would have to contain elements that would provide the same beneficial machining benefits as lead. Through research, it was found that a combination of bismuth (Bi) and selenium (Se) provided the same beneficial effect on machinability as does lead.

The chemical composition and mechanical properties of SeBiLOY I and II are presented in **Tables 9-5 and 9-6,** respectively. The UNS Number designations for these new copper alloys are C89510 (SeBiLOY I) and C89520 (SeBiLOY II).

Note in **Table 9-5** that a small amount of lead (0.25% maximum) is indicated. Although no lead is intentionally added, this allows for any trace content of lead from recycled materials.

Table 9-3.
Chemical Composition and Typical Properties of Copper-Base Alloys

Note: Alloys are repeated in each section.

Nominal Chemical Analysis:

UNS Number	Alloy Name	Cu	Sn	Pb	Zn	Fe	Al	Others
C81100	Copper	99.7	—	—	—	—	—	0.30
C83600	Leaded red brass	85	5	5	5	—	—	—
C84400	Leaded semi-red brass	81	3	7	—	—	—	—
C85200	Leaded yellow brass	72	1	3	24	—	—	—
C86300	Manganese bronze	63	—	—	25	3	6	4 Mn
C87500	Silicon bronze & brass	81	—	—	14	—	—	3.2 Si
C90500	Tin bronze	88	10	—	2	—	—	—
C92200	Leaded tin bronze	88	6	1.5	4.5	—	—	—
C93700	High-leaded tin bronze	80	10	10	—	—	—	—
C95400	Aluminum bronze	83.2	—	—	—	4	11	—
C96400	Cupronickel 30%	68	—	—	—	1	—	30 Ni, 1 Nb
C97600	Nickel silver	65	4	4	6	—	—	20 Ni

Properties:

UNS Number	Alloy Name	TS ksi/MPa	YS ksi/MPa	Elong. %	E Msi/GPa
C81100	Copper	25/172	9/62	40	17/117
C83600	Leaded red brass	37/255	17/117	30	13/93
C84400	Leaded semi-red brass	34/234	15/103	26	13/90
C85200	Leaded yellow brass	38/262	13/90	35	11/76
C86300	Manganese bronze	119/821	67/462	18	14/97
C87500	Silicon bronze & brass	67/462	30/207	21	15/106
C90500	Tin bronze	45/310	22/152	25	15/103
C92200	Leaded tin bronze	40/276	20/138	30	14/97
C93700	High-leaded tin bronze	35/241	18/124	20	11/76
C95400	Aluminum bronze	85/586	35/241	18	15/107
C96400	Cupronickel 30%	68/469	37/255	28	21/145
C97600	Nickel silver	45/310	24/165	20	19/131

UNS Number	Alloy Name	500 kg HB	Izod Impact ft.lb	Fatigue* Str. ksi/MPa	Spec. Gravity	Thermal Cond.% of Cu**	Elect. Cond. %IACS***
C81100	Copper	44	—	9/62	8.9	200/346.1	92/0.538
C83600	Leaded red brass	60	10	11/76	8.3/8.83	41.6/72	15/0.087
C84400	Leaded semi-red brass	55	8	—	8.7	41.8/72.4	16/0.095
C85200	Leaded yellow brass	45	—	—	8.5	48.5/83.9	18/0.104
C86300	Manganese bronze	225	15	25/172	7.8	20.5/35.5	8/0.046
C87500	Silicon bronze & brass	115	—	22/152	8.3	17.0/27.7	7/0.039
C90500	Tin bronze	75	10	13/90	8.7	43.2/74.8	11/0.064
C92200	Leaded tin bronze	65	19	11/76	8.6	40.2/69.6	14/0.083
C93700	High-leaded tin bronze	60	—	13/90	8.9	27.1/46.9	10/0.059
C95400	Aluminum bronze	170	16	28/193	7.5	34/58.7	13/0.075.0
C96400	Cupronickel 30%	140	78	18/124	8.9	16.4/28.5	5/0.028
C97600	Nickel silver	80	11	16/107	8.9	13/22.6	5/0.029

* *Endurance limit (100 million cycles)*
Btu·ft/hr·ft²·°F) at 68F / W/m·°K at 20C;* *%IACS at 68jF / Siemens/cm at 20C*
From Standards Handbook, Fourth Edition, 1996; for latest information, contact the Copper Development Association.

Table 9-4.
Comparison of Three Copper Alloys

Alloy Number	Tensile Strength	0.5 Yield Strength	Elongation % in 2 inches
C83600:			
Sand or Centrifugal Cast	30	14	20
Continuous Cast	36	19	15
C92200:			
Sand or Centrifugal Cast	34	16	24
Continuous Cast	38	19	18
C93700:			
Sand or Centrifugal Cast	30	12	15
Continuous Cast	35	20	06

Table 9-5.
Chemical Composition of SeBiLOY I and II

	SeBiLOY I	*SeBiLOY II*
Copper	86 to 88%	85 to 87%
Tin	4–6%	5–6%
Zinc	4–6%	4–6%
Bismuth	0.5–1.5%	1.6–2.2%
Selenium	0.35–0.75%	0.8–1.1%
Nickel	<1%	<1%
Antimony	<0.25%	<0.25%
Iron	<0.2%	<0.2%
Lead	<0.25%	0.25%
Phosphorus	<0.05%	<0.5%

Table 9-6.
Mechanical Properties of SeBiLOY I and II

	SeBiLOY I		*SeBiLOY II*	
Property	*Minimum MPa (ksi)*	*Average MPa (ksi)*	*Minimum MPa (ksi)*	*Average MPa (ksi)*
UTS	185(26.8)	209(30.3)	176(25.5)	210(30.4)
YS at 0.2% offset	119(17.3)	128(18.5)	121(17.6)	139(20.5)
Stress at 0.5% elongation	126(18.3)	136(19.7)	127(18.4)	146(20.9)
Elongation	8%	12%	6%	10%

Coppers (C80100–C812000)

The coppers are soft and ductile and are used almost exclusively for their excellent electrical and thermal conductivities. Thus, they are used for electrical terminals, connectors and water-cooled hot metal handling equipment such as blast furnace and cupola tuyeres. The coppers also have very high corrosion resistance, but this is usually a secondary consideration.

High Copper Alloys (C81400–C82800)

These copper alloys are used primarily for their unique combination of good conductivity and high strength. Their corrosion resistance can be better than that of copper itself. They are used in electrical contacts, clamps, welding gear and similar electromechanical hardware. The beryllium coppers have the highest strength of all of the copper alloys, and are used in heavy-duty mechanical and electromechanical equipment requiring ultrahigh strength and good electrical and/or thermal conductivity.

Brasses (C83300–C87900)

These alloys have zinc as the primary alloying element. This family of copper alloys finds a variety of uses as castings. Some the castings are plumbing fittings and valves, valve stems, water pump parts, boat hardware, low-pressure valve bodies, statuary and decorative pieces.

Red and Semi-Red Brasses, Unleaded and Leaded (C83300–C84800)

The most important brass poured, based on tonnage, are the leaded red brass C83600 (85-5-5-5). This family of copper alloys finds use in water valves, pumps, pipe fittings and plumbing hardware.

Yellow Brasses (C85200–C85800)

Some of the yellow brasses, such as C85400, C85800, are relatively low in cost and have excellent castability, high machinability and favorable finishing characteristics. Typical tensile strengths range from 34 to 55 ksi (234 to 379 MPa). Some of these alloys can be diecast due to their relatively narrow freezing range and good high-temperature ductility. Some of the castings produced using these copper alloys are gears and machine components in which relatively high strength and moderate corrosion resistance must be combined with superior machinability. They can also be used for architectural trim and decorative hardware.

High Strength and Leaded High Strength Yellow Brasses (C86100–C86800)

These alloys, also called manganese bronzes, are the strongest, as cast, of all the copper alloys, and are weldable but should be given a post-weld stress relief. These alloys are used mainly for heavy-duty mechanical products requiring moderately good corrosion resistance at a reasonable cost. The manganese bronzes have been supplanted, to some extent, by aluminum bronzes that offer comparable properties but have better corrosion resistance and weldability.

Silicon Brasses/Bronzes (C87200–C87900)

These are moderate strength alloys with good corrosion resistance and useful casting characteristics. Their solidification mode makes them very usable in diecasting, permanent mold and investment casting methods. Their applications range from bearings and gears to plumbing goods and intricately shaped pump and valve components.

Bronzes (C90200–C98840)

The term "bronze" originally referred to alloys in which tin was the major alloying element. Today, the UNS numbering system lists five subfamilies of bronzes: 1) tin bronzes; 2) aluminum bronzes; 3) copper-nickel alloys; 4) nickel silvers; and 5) leaded coppers.

Tin Bronzes (C90200–94900)

This category contains copper-tin-lead alloys, which are further broken down into Copper-Tin Alloys (C90200–C91700), Leaded Tin Bronzes (C92200–C92900), High-Leaded Tin Bronzes (C93100–C94500), and Nickel Tin Bronze (C94900). Tin-bronzes offer excellent corrosion resistance, reasonably high strength and good wear resistance. The tin-bronzes find their niche in bearings. However, the nickel-tin bronzes (C94700–C94900) are more versatile and are frequently used as valve and pump components, gears, shifter forks and breaker parts.

Aluminum Bronzes (C95200–C95800)

Aluminum strengthens copper and imparts oxidation resistance by forming a tenacious alumina-rich film. Iron, silicon, nickel and manganese are added to aluminum bronzes, singly or in combination, for higher strength and/or corrosion resistance in specific media. These alloys are very well known for their high corrosion resistance combined with exceptionally good mechanical properties. The aluminum bronzes are readily fabricated and welded, and have been used to produce some of the largest nonferrous cast structures in existence. Aluminum bronze bearings are used in heavily loaded applications. Resistance to seawater corrosion is exceptionally high in nickel-aluminum bronzes. Because of its superior resistance to erosion-corrosion and cavitation, nickel-aluminum bronze (C95500) is now widely used for ship propellers and other marine hardware.

Copper-Nickel Alloys (C96200–C96900)

These alloys are sometimes referred to as copper-nickels or cupronickels. Corrosion resistance and strength increase with increasing nickel content. These alloys offer excellent resistance to seawater corrosion. High strength and good fabrication ability have found them in a wide variety of uses in marine equipment. Typical products include pump components, impellers, valves, tailshaft sleeves, centrifugally cast pipe, fittings and marine products such as centrifugally cast valve bodies.

Nickel Silvers (C97300–C97800)

These copper alloys offer excellent corrosion resistance, high castability and very good machinability. They have moderate strength. Among their useful attributes is their pleasing silver luster. Valves, fittings and hardware cast in nickel-silver alloys are found in food and beverage handling equipment and as seals and labyrinth rings in steam turbines.

Leaded Coppers (C98200–C98840)

Leaded coppers offer the high corrosion resistance of copper and high-copper alloys, along with the low friction characteristics of the high-leaded bronzes.

Contact the Copper Development Association for additional information on copper-base alloys.

Table 9-7.
Standard Three-Part ASTM System of Alloy Designations for Magnesium Alloys

First Part:

- Indicates the two principal alloying elements.
- Consists of two code letters representing the two main alloying elements arranged in order of decreasing percentage (or alphabetically, if percentages are equal).

A = Aluminum	E = Rare Earth
H = Thorium	K = Zirconium
M = Manganese	Q = Silver
W = Yttrium	Z = Zinc.

Second Part:

- Indicates the amounts of the two principal elements.
- Consists of two numbers corresponding to rounded-off percentages of the two main alloying elements and arranged in same order as alloy designations in first part.
- Whole numbers.

Third Part:

- Distinguishes between different alloys with the same percentages of the two principal alloying elements.
- Consists of a letter of the alphabet assigned in order as compositions become standard.

A = First compositions, registered ASTM.
B = Second composition, registered ASTM.
C = Third composition, registered ASTM.
D = High-purity, registered ASTM.
E = High corrosion-resistant, registered ASTM.
X1 = Not registered with ASTM.

Table 9-8.
Nominal Compositions of Magnesium Casting Alloys for Sand, Investment and PM Castings

	Composition, %						
Alloy	Al	Zn	Mn	RE	Th	Y	Zr
AM100A	10.0	—	0.1 min	—	—	—	—
AZ63A	6.0	3.0	0.15	—	—	—	—
AZ81A	8.0	0.7	0.13	—	—	—	—
AZ91C	9.0	0.7	0.13	—	—	—	—
AZ91E	9.0	2.0	0.10	—	—	—	—
AZ92A	9.0	2.0	0.10	—	—	—	—
EZ33A	—	2.7	—	3.3	—	—	0.60
HK31A	—	—	—	—	3.3	—	0.70
HZ32A	—	2.1	—	—	3.3	—	0.70
QE22A(a)	—	—	—	2.0	—	—	0.60
EQ21A(a,b)	—	—	—	2.0	—	—	0.60
ZE41A	—	4.2	—	1.2	—	—	0.70
ZE63A	—	5.7	—	2.5	—	—	0.70
AH62A	—	5.7	—	—	1.8	—	0.70
ZK51A	—	4.6	—	—	—	—	0.70
ZK61A	—	6.0	—	—	—	—	0.70
WE54A	—	—	—	3.50(c)	—	5.25	0.50

Legend:

(a) These alloys also contain silver; that is, 2.5% in QE22A and 1.5% in EQ21A.

(b) EQ21A also contains 0.10% Cu.

(c) Comprising 1.75% other heavy rare earths in addition to the 1.75% Nd present.

MAGNESIUM-BASE ALLOYS

Nearly every conventional casting process—sand, permanent molding, shell, investment and diecasting—can produce magnesium castings. A large number of magnesium-base alloys are available for the production of castings. Sand and investment castings can be made in all of the magnesium alloys. However, not all alloys are suitable for production in all of the casting processes. Magnesium alloys cast by the permanent mold process are somewhat limited in number, and those used in diecasting are even more restricted.

The method of coding used in North America to designate magnesium casting alloys is taken from the ASTM Standard Practice B 275. The five principal alloying systems for magnesium and magnesium alloys include:

1. Magnesium having *aluminum and manganese* as the principal alloying elements is designated as type AM
2. Magnesium having *aluminum and zinc* as the principal alloying elements is designated as type AZ
3. Magnesium alloys containing *rare earths and zirconium* are designated EK, EZ and ZE
4. Magnesium having *zinc and zirconium* as the primary alloying elements is designated type ZK
5. Magnesium having *thorium and zirconium* as the primary alloying elements include types HK, HZ and ZH

There are other magnesium alloy systems that can be cast; however, the above five systems have become more widely used. **Table 9-7** describes the coding method. This coding method gives an approximate idea of the chemical composition of an alloy, with letters representing the main constituents and figures representing their percentage in the alloy.

Several examples are, namely: AZ91A, AZ91B and AZ91C. In these designations:

"A" represents aluminum, the alloying element specified in the greatest amount;

"Z" represents zinc, the alloying element specified in the second greatest amount;

"9" indicates that the rounded mean aluminum percentage lies between 8.6 and 9.4;

"1" signifies that the rounded mean of the zinc lies between 0.6 and 1.4;

"A" as the final letter in AZ91A indicates that this is the first alloy containing a composition qualified assignment of the designation AZ91;

"B" and "C" in the other two examples signify alloys subsequently developed whose compositions differ slightly from the first and from one another but do not differ sufficiently to effect a change in the basic designation.

The nominal compositions of the magnesium alloys used for sand, investment and permanent mold castings are shown in **Table 9-8. Table 9-9** gives the compositions of those magnesium alloys used in diecasting. The physical properties, such as specific gravity, density, thermal conductivity, coefficient of thermal expansion and electrical conductivity, are listed in **Table 9-10**. Note that these numbers are for sand casting alloys only. **Table 9-11** lists the mechanical properties found in these same sand casting alloys.

Magnesium alloy sand castings are used in aerospace applications because they offer a clear weight advantage over aluminum and other materials. These alloys are easy to cast but have other limitations. They exhibit microshrinkage when sand cast and are not suitable for applications in which temperatures of over 200F (95C) are experienced. The newer magnesium-rare earth-zirconium alloys were developed to overcome these limitations. Sand castings in the EZ33A alloy do show excellent pressure tightness with the elimination of the microshrinkage.

The magnesium-zinc-zirconium alloys, ZK51A and ZK61A, exhibit high mechanical properties, but suffer from hot-shortness and cracking, and are not weldable. For normal, fairly moderate temperature applications, 320F (160C), the two alloys ZE41A and EZ33A are finding the greatest use. They are very castable and can be used to make very satisfactory castings of considerable complexity, and only require a precipitation heat treatment.

With the aerospace industry demanding magnesium alloy sand castings be used in engine applications to retain higher mechanical properties at higher temperatures, (up to 400F (205C)), thorium was substituted for the rare earth metal content in alloys ZE and EZ, giving rise to the new ZH62A and HZ32 alloys. The alloys gave improved mechanical properties at elevated temperatures, as well as retained good castability and welding characteristics.

The AZ alloys have good castability and are widely used where good ductility and moderately high yield strength are required up to 350F (176C). Their yield strengths make them suitable alloys to use for aircraft landing wheels, levers, linkages, housings etc.

The magnesium alloys from which diecastings are made are mainly the AZ alloys (magnesium-aluminum-zinc). The two most commonly used alloys for many years have been AZ91A and AZ91B. The only difference between these two alloys is the higher allowable copper impurity in the AZ91B alloy.

The most important feature of magnesium alloys is their light weight, when compared to other casting alloys. However, magnesium alloys offer several other advantages over other alloys, and these include:

- Magnesium is an abundantly available metal
- Magnesium is easier to machine than aluminum
- Magnesium can be machined faster than aluminum, preferably dry.

Since World War II, the use of magnesium alloy castings has increased in the aircraft and aerospace industries. Also, as a result of the weight reduction program in the automobile industry, magnesium alloy diecastings have enjoyed increased usage.

Table 9-9.
Nominal Compositions of Mg Casting Alloys for Diecastings

	Alloying Element				
Alloy	**Mg**	**Al**	**Mn**	**Si**	**Zn**
AM60A	rem	6.0	0.13 min	—	—
AS41A	rem	4.25	0.35	1.0	—
AZ91A	rem	9.0	0.13 min	—	0.7
AZ91B	rem	9.0	0.13 min	—	0.7
AAZ91D (HP)(b)	rem	9.0	(a)	—	0.7

Legend:
(a) Manganese content to be dependent upon iron content.
(b) Proposed alloy to have very low limits for iron, nickel and copper.
HP = high purity.

Table 9-10.
Physical Properties of Sand Cast Mg Alloys

ASTM Alloy and Temper Designation	Specific Gravity g/cc 75 F (24 C)	Density lb/cu in.	Thermal Conductivity at 68 F (20 C) cgs units	Coefficient of Thermal Expansion per F 68-212 F (20-100 C)	Electrical Conductivity per cent I.A.C.S.
AM100A-F	1.81	0.065	0.12	14.5×10^{-6}	12
-T4			0.10		10
-T6			0.14		14
-T61			0.15		15
AZ63A-F	1.82	0.066	0.14		14
-T4			0.12		12
-T6			0.15		15
AZ81A-T4	1.80	0.065	0.12		12
AZ91C-F	1.81	0.065	0.13		13
-T4			0.11		11
-T6			0.13		13
AZ92A-F	1.83	0.066	0.12		12
-T4			0.11		10
-T6			0.14		14
EZ33A-T5	1.83	0.066	0.24		25
HK31A-T6	1.79	0.065	0.22		22
HZ32A-T5	1.83	0.066	0.26		27
K1A-F	1.75	0.063	0.30		31
QE22A-T6	1.82	0.066	0.25		25
ZE41A-T5	1.84	0.066	0.30		31
ZH62A-T5	1.86	0.067	0.26		27
ZK51A-T5	1.81	0.065	0.26		27
ZK61A-T6	1.83	0.066	0.26		27

Table 9-11.
Mechanical Properties of Sand Cast Mg Alloys

ASTM Alloy and Temper Designation	Specified Minimum Properties (B80-63)			Typical Properties					
	0.2 per cent Yield Strength psi	Ultimate Tensile Strength psi	Elong. per cent in 2 in.	0.2 per cent Yield Strength psi	Ultimate Tensile Strength psi	Elong. per cent in 2 in.	Brinell Hardness 500/10/30	Shear Strength psi	Endurance Limit psi 10^8 cycles R. R. Moore
AM100A-T6	17,000	35,000	—	22,000	40,000	1	70	22,000	10,000
AZ63A-F	11,000	26,000	4	14,000	29,000	6	50	18,000	11,000
-T4	11,000	34,000	7	13,000	40,000	12	55	17,000	13,000
-T5	12,000	26,000	2	14,000	30,000	4	55	17,000	11,000
-T6	16,000	34,000	3	19,000	40,000	5	73	20,000	12,000
AZ81A-T4	11,000	34,000	7	12,000	40,000	15	55	17,000	—
AZ91C-F	11,000	23,000	—	14,000	24,000	2	52	18,000	—
-T4	11,000	34,000	7	12,000	40,000	14	53	17,000	12,000
-T5	12,000	23,000	2	17,000	26,000	3	—	—	—
-T6	16,000	34,000	3	19,000	40,000	5	66	20,000	12,000
AZ92A-F	11,000	23,000	—	14,000	24,000	2	65	18,000	12,000
-T4	11,000	34,000	6	14,000	40,000	9	63	20,000	14,000
-T5	12,000	23,000	—	16,000	26,000	2	69	19,000	11,000
-T6	18,000	34,000	1	21,000	40,000	2	80	22,000	12,000
EZ33A-T5	14,000	20,000	2	15,000	23,000	3	50	—	10,000
HK31A-T6	13,000	27,000	4	15,000	32,000	8	55	—	12,500
HZ32A-T5	13,000	27,000	4	14,000	30,000	7	57	—	10,000
K1A-F	6,000	24,000	14	7,000	25,000	19	—	8,100	—
QE22A-T6	25,000	35,000	2	30,000	40,000	4	—	23,200	16,000
ZE41A-T5	19,000	28,000	2.5	20,000	30,000	3.5	—	—	—
ZH62A-T5	22,000	35,000	5	25,000	40,000	6	70	—	—
ZK51A-T5	20,000	34,000	5	24,000	40,000	8	65	22,000	9,000
ZK61A-T6	26,000	39,000	5	30,000	45,000	10	70	26,000	14,000

TITANIUM-BASE ALLOYS

Since the introduction of titanium and its alloys in the early 1950s, these materials have, in a relatively short period of time, become the backbone material for the aerospace, chemical and energy industries. These materials offer a high strength-to-weight ratio, excellent mechanical properties and corrosion resistance; thus, they find use in many critical applications. Titanium and its alloys are cast in rammed graphite molds and also in the investment casting process. Probably the investment casting process has given the use of these materials their greatest boost. In the 1970s, a derivation of the rammed graphite molding was developed using components of the more traditional sand foundries, along with inorganic binders (nobake molding).

Table 9-12 shows a comparison of the chemical composition of titanium alloys used for castings. **Table 9-13** shows the typical room-temperature tensile properties of titanium alloy castings. The titanium castings industry is relatively young when compared to other materials poured in foundries. The earliest application of titanium and its alloys was in the 1960s with the casting of corrosion-resistant pump and valve components. The aerospace industry's use of these materials began in the early 1970s with the use of rammed graphite molds. Aircraft brake torque tubes, missile wings and hot-gas nozzles were being added to the application of titanium alloys. With the advent of more precise investment casting technology and the commercial use of HIP (Hot Isostatic Pressing) in the mid-1970s, titanium alloy castings expanded into the critical airframe and gas turbine engine components. There is also continued growth in the casting of medical prostheses, such as hip and knee implants, using the investment casting process.

COBALT-BASE ALLOYS

Cobalt-base alloys were developed in the late 1930s for use as castings in aircraft turbochargers. These alloys offer better wear- and heat-resistant properties than do the iron-base alloys. While far surpassing nickel-base alloys in wear properties, under certain temperatures and/or duration conditions they also exhibit competitive or superior heat resisting properties. In addition all cobalt-base alloys offer excellent corrosion resistance in a variety of media.

Containing 50% or more cobalt and a relatively high chromium and refractory metal content, they differ in end application usage primarily because of carbon content. The heat-resisting cobalt-base alloys contain appreciable amounts of carbon, while the wear-resisting grades contain even more carbon. Cobalt-base alloy designations and typical applications are found in **Table 9-14.**

Although not possessing the strength of modern, vacuum-melted nickel-base casting alloys, their ease of weldability and good corrosion resistance at high temperatures allows cobalt-base alloys to compete with nickel-base alloys at temperatures above 1800F (982C), especially for long-term applications. Typically, these alloys will have nickel added to stabilize the high-temperature, face-centered-cubic crystal structure of cobalt.

The wear-resisting grades, while having significantly higher carbon than the heat-resisting grades, vary in refractory metal content. This has a direct bearing on the volumetric percentages of chromium and tungsten carbide present, which, in turn, relates to different wear and toughness properties.

Table 9-12.
Comparison of Cast Titanium Alloys

Alloy	Estimated Relative Usage of Castings	NominalComposition, Wt% O	N	H	Al	Fe	V	Cr	Sn	Mo	Zr	Special Properties*
Ti-6Al-4V	90%	0.18	0.015	0.006	6	0.13	4	—	—	—	—	General purpose
Ti-6Al-4V ELI	2%	0.11	0.010	0.006	6	0.10	4	—	—	—	—	Cryogenic toughness
Commercially pure Ti	5%	0.25	0.015	0.006	—	0.15	—	—	—	—	—	Corrosion resistance
Ti-6Al-2Sn-4Zr-2Mo	2%	0.10	0.010	0.006	6	0.15	—	—	2	2	4	Elevated-temperature creep
Ti-6Al-2Sn-4Zr-6Mo	<1%	0.10	0.010	0.006	6	0.15	—	—	2	6	4	Elevated-temperature strength
Ti-5Al-2.5Sn	<1%	0.16	0.015	0.006	5	0.2	—	—	2.5	—	—	Cryogenic toughness
Ti-3Al-8V-6Cr-4Zr-4Mo	<1%	0.10	0.015	0.006	3.5	0.2	8.5	6	—	4	4	Strength
Ti-15V-3Al-3Cr-3Sn	<1%	0.11	0.015	0.006	3	0.2	15	3	3	—	—	Strength
Total	100%											

** Superior, relative to Ti-6Al-4V*

Table 9-13.
Typical Room-Temperature Tensile Properties of Ti Alloy Castings (Bars Machined From Castings)
Specification Minimums are Less Than These Typical Properties

Alloy(s)	Yield Strength MPa / ksi	Ultimate Strength MPa / ksi	Elongation %	Reduction of area, %
Commercially pure (Grade 2)	448 / 65	552 / 80	18	32
Ti-6Al-4V, annealed	855 / 124	930 / 135	12	20
Ti-6Al-4V-ELI	758 / 110	827 / 120	13	22
Ti-6Al-2Sn-4Zr-2Mo, annealed	910 / 132	1006 / 146	10	21
Ti-6Al-2Sn-4Zr-6Mo, STA	1269 / 184	1345 / 195	01	01
Ti-3Al-8V-6Cr-4Zr-4Mo, STA	1241 / 180	1330 / 193	07	12
Ti-15V-3Al-3Cr-3Sn, STA	1200 / 174	1275 / 185	06	12

Solution-treated and aged (STA) heat treatments may be varied to produce alternate properties.

In addition to good resistance to wear, galling, abrasion, erosion and/or cavitation, the wear-resisting grades also offer good resistance to heat and oxidation; high hot hardness and toughness; good high-temperature dimensional stability; and good corrosion resistance in a variety of media.

A unique application of cast cobalt-base alloys is the use of ASTM F-75 for medical implant (prosthetic) devices. Essentially, a more highly purified (restricted) form of cobalt Alloy No. 21, it has the unique combination of high strength, corrosion and wear resistance necessitated by the human body's environment.

Typical mechanical properties of the cobalt-base casting alloys are given in **Table 9-15,** along with suggested heat-treat conditions.

In general, cobalt-base casting alloys are "foundry favorable." They have good castability such as good fluidity, low melting points, freedom from dissolved gasses and low alloys losses due to oxidation. On the down side, there are high cost, contamination of subsequent heats, crack prone wear resisting grades and poor machining qualities.

Table 9-14.
Nominal Compositions and Some Applications for Cobalt-Base Alloys

Alloy	C	Mn	Si	Cr	Ni	W	Fe	Others	Applications
				Composition, %*					
Wear-Resistant Alloys									
Co Alloy No. 3	2.0–2.7	1.00	1.00	29.0–33.0	3.00	11.0–14.0	3.00	—	Cutting tools, wear strips, valve seats
Co Alloy No. 6	0.9–1.4	1.00	1.50	27.0–31.0	3.00	3.5–5.5	3.00	1.5 Mo	Valve steats, punches, wear plates
Co Alloy No. 12	1.1–1.7	1.00	1.00	28.0–32.0	3.00	7.0–9.5	3.00	—	Bushings, saw teeth
Co Alloy No. 19	1.5–2.1	1.00	1.00	29.5–32.5	3.00	9.5–11.5	3.00	—	utting tools, bearings, rollers
Co Star-J	2.20	1.00	1.00	31.0–34.0	2.50	16.0–19.0	3.00	—	Cutting tools, wear parts
Co Alloy 98M2	1.7–2.2	1.00	1.00	28.0–32.0	2.0–5.0	17.0–20.0	2.50	0.8 Mo, 1.1 B, 4.2 V	Cutting tools, wear parts
Heat Resistant Alloys									
Co Alloy No. 21	0.2–0.3	1.00	1.00	25.0–29.0	1.75–3.75	—	3.00	5.5 Mo, 0.007 B	Turbine blades, combustion chambers to 815C (1500F)
Co Alloy No. 25	0.05–0.15	1.0–2.0	1.00	19.0–21.0	9.0–11.0	14.0–16.0	3.00	—	Gas turbine rotors and buckets
Co Alloy No. 31	0.45–0.55	1.00	1.00	24.5–26.5	9.5–11.5	7.0–8.0	2.00	0.5 Mo	Turbine blades
Co Alloy X-40	0.45–0.55	1.00	1.00	24.5–26.5	9.5–11.5	7.0–8.0	2.00	0.1 B	Gas turbine parts, nozzle vanes
Co Alloy X-45	0.20–0.30	0.4–1.0	0.75–1.0	24.5–26.5	9.5–11.5	7.0–8.0	2.00	0.01 B	Nozzle vanes
Co Alloy FSX-414	0.20–0.30	0.4–1.0	0.5–1.0	28.5–30.5	9.5–11.5	6.5–7.5	2.00	0.01 B	Gas turbine vanes
Co MAR WI-52	0.40–0.50	0.50	0.50	20.0–22.0	1.00	10.0–12.0	2.00	2.0 Nb	Gas turbine parts, nozzle valves
Co MAR M 302	0.78–0.93	0.20	0.40	20.0–23.0	—	9.0–11.0	1.50	0.01 B, 0.20 Zr, 9.0 Ta	Turbine vanes (815–1095C or 1500–2000F)
Co MAR M 509	0.55–0.65	0.10	0.40	21.0–24.0	9.0–11.0	6.5–7.5	1.50	0.20 Ti, 0.10 B, 0.50 Zr, 3.5 Ta	Gas turbine parts
Biomedical Alloy									
Co ASTM F75	0.35	1.00	0.40	27.0–30.0	1.00	—	1.50	6.0 Mo	Orthopedic implants

** In all compositions, cobalt makes up the balance.*

Table 9-15.
Mechanical Properties and Heat Treatment Conditions of Cobalt-Base Alloys

Alloy	UTS	0.2% YS	% El.	Heat Treatment Condition
Co Alloy #3	64,000	Near UTS	Nil	1650F/4 Hr/FC
Co Alloy #6	115,000	96,000	3	1650F/4 Hr FC
Co Alloy #12	107,000	Near UTS	Nil	1650F/4 Hr FC
Co Alloy #19	105,000	Near UTS	1	1650F/4 Hr FC
Co Alloy Star J	60,000	Near UTS	Nil	As cast
Co Alloy 98M2	80,000	Near UTS	Nil	As cast
Co Alloy #21	103,000	82,000	8	1500F/5-50 Hr/AC
Co Alloy #25	90,000	65,000	15	2200F/1 Hr/AC
Co Alloy #31	113,000	80,000	8	As cast
Co Alloy X-40	108,000	76,000	9	2150F/1 Hr/WQ + 1500F/4 Hr/AC + 1800F/2-4 Hr/OQ + 1350F/16 Hr/AC
Co Alloy X-45	108,000	76,000	9	2150F/1 Hr/WQ + 1500F/4 Hr/AC + 1800F/2-4 Hr/OQ + 1350F/16 Hr/AC
Co Alloy WI-52	109,000	85,000	7	1850F/2 Hr/OQ + 1350F/20 Hr/AC + 1200F/20 Hr/AC
Co Alloy FSX-414	107,000	64,000	2	As cast
Co Alloy MAR M 302	140,000	100,000	2	As cast
Co Alloy MAR M 509	113,000	85,000	3	2000F/2 Hr/WQ + 1450F/2 Hr/AC + 1325F/24 Hr/AC
Co Alloy F-75	110,000	84,000	9	2250F/3 Hr/AC

AC = Air Cooled; FC = Furnace Cooled; WQ = Water Quenched; OQ = Oil Quenched

NICKEL-BASE ALLOYS

Nickel-base alloys have great resistance to corrosion in the presence of most mineral acids, most organic acids and all alkalis. They are not resistant to corrosion by nitric acid, or by oxidizing salts such as ferric sulfates or copper sulfates. They have good mechanical strength, ductility and resistance to wear, although they cannot be used for bearings, except under light loads and at slow speeds.

An alloy of 70% nickel and 30% copper is known as Monel. When silicon is added to it, it can be age-hardened, and increased wear resistance results. An alloy of 80% nickel and 20% chromium, known as Nichrome, is used for electrical resistance heaters. An alloy of 80% nickel, 14% chromium and 6% iron, known as Inconel, is used where oxidation resistance with high strength at elevated temperatures is needed.

Alloys with a high percentage of nickel are used for chemical equipment, such as implements used in the dyeing of textiles and the manufacture of caustics, as well as for manufacturing water-softening equipment, valves and pump parts, and food handling equipment.

ZINC-BASE ALLOYS

Zinc-base alloys are primarily used in diecasting. The alloying elements are principally copper, aluminum and magnesium. The amount of each used is usually between 4 and 8% and depends on the properties desired. High-purity zinc is used as the base metal. Copper additions increase the strength, but reduce ductility. Addition of aluminum improves the strength of the alloy and delays the rate of attack of the alloy on steel dies, and thus improves the life of the die. Additions of magnesium improve the dimensional stability of a diecasting.

An alloy consisting of 20% zinc, 20% manganese, 1% aluminum, and the balance copper produces a white-bronze casting that is strong, ductile and corrosion resistant. It can be polished to a high silvery luster that makes it useful in architectural and marine hardware, plumbing fixtures, ornamental castings, hospital equipment and swimming pool equipment.

Zinc-aluminum alloys are a family of high-strength alloys designed for sand and permanent mold casting, both metal and graphite. They can be diecast as well as cast in molds of plaster, rubber and ceramics. **Table 9-16** shows the properties of three different zinc-aluminum alloys. The compositions for these alloys are as follows:

ZA8: 8.4% Al, 1% Cu, 0.02% Mg, balance Zinc

ZA12: 11% Al, 0.75% Cu, 0.o2% Mg, balance Zinc

ZA 27: 27% Al, 2.2% Cu, 0.15% Mg, balance Zinc

The zinc-aluminum alloys can be cast using various foundry methods without disadvantage. Sand casting is the most commonly used foundry process. The castings can be produced using existing patterns and matchplates designed for aluminum, gray iron and bronze. With a high degree of fluidity, thin castings are readily produced. Using gravity-poured permanent metal molds, a very smooth casting surface with close dimensional tolerances is produced. Zinc-aluminum alloys are readily machinable without galling, are corrosion resistant, non-sparking, and can be surface polished.

ALUMINUM COMPOSITES

A composite material is a heterogeneous solid consisting of two or more components that are mechanically or metallurgically bonded together. Each of the components retains its identity in the composite and has its characteristic structure and properties. By combining the components, the resultant characteristics are imparted, and are properties of the composite.

In a composite, discrete particles of one material are surrounded by a matrix of another material. A popular variety of composite material is the fiber-reinforced component, where thin fibers of one material are imbedded in a matrix of metal. This is commonly called metal matrix composite (MMC). The matrix supports and transmits loads to the fibers, which, in turn, supply strength and other mechanical or metallurgical properties to the structure.

The most promising applications of composites are in the automotive, aerospace/defense and industrial equipment. These applications involve moving components, and weight saving from composite materials will result in fuel efficiency. Composites with greater strength and stiffness can be made lighter. This allows other components of the same system to be lighter, as they now would be subject to lower inertial or vibrational loads. This improves the trend toward weight saving and fuel saving economies.

Castings produced from composite materials are characterized with excellent mechanical properties. These properties, combined with low density and lower cost, will permit composite castings to compete effectively with aluminum forgings as well as castings made from cast iron, steel, magnesium and titanium. Some typical castings are disk brake rotors, belt pulleys, aircraft structural and auxiliary components, and an aircraft camera gimbal, which is investment cast.

This concludes the discussion on nonferrous alloys, their properties and uses. There may be other less commonly used nonferrous alloys that can be cast; however, due to space limitations, they are not covered here. Contact the American Foundry Society's technical library for additional information on nonferrous alloys.

Table 9-16.
Mechanical Properties of Three Diecasting ZA Alloys

	ZA-8	ZA-12	ZA-27
UTS, MPa/ksi	371/54	400/58	421/61
YS (2% offset), MPa/ksi	290/42	317/46	365/53
% El. in 2 in.	6–10	4–7	1–3
Brinell Hardness (HB)	95–110	95–115	105–125
Impact Str., J/ft-lb (Notched Charpy)	42	28	—

BIBLIOGRAPHY

Alloy Data Sheet, (SeBiLOY) Low-Lead Red Brass Casting Alloys C89510 and C89520, Copper Development Association, Inc., New York, NY (1995).

Aluminum Casting Technology, 2nd Edition, The American Foundrymen's Society, Inc., Des Plaines, IL (1993).

Casting Copper Alloys, Copper Development Association, New York, NY (1994).

Casting Copper-Base Alloys, The American Foundrymen's Society, Inc., Des Plaines, IL (1984).

Comparative Properties of Casting Alloys, (ZA Casting Alloys), Eastern Alloys, Inc., Maybrook, NY (1983).

Design and Procurement of High-Strength Structural Aluminum Castings, The American Foundrymen's Society, Inc., Des Plaines, IL.

Metals Handbook, Ninth Edition, Volume 15, Casting, ASM International, Metals Park, OH.

Recommended Practices for Sand Casting Aluminum and Magnesium Alloys, The American Foundrymen's Society, Inc., Des Plaines, IL (1965).

Sadayappan, M., F.A. Fasoyinu, D. Cosineau, R. Zavadil, M. Sahoo and D. T. Peters, Casting Characteristics and Mechanical Properties of Bi/Se-Modified Yellow Brass (SeBiLOY III) in Permanent Molds, *AFS Transactions* (97-110), The American Foundrymen's Society, Inc., Des Plaines, IL.

Standards Handbook, Cast Copper and Copper Alloy Products, Part 7 Alloy Data (Cast Products), Fourth Edition, Copper Development Association, New York, NY (1996).

Whiting, L.V., Sahoo, M., Newcomb, P.D., Zavadil, R., Peters, D.T.; Detailed Analysis of Mechanical Properties of SeBiLOYs I and II, *AFS Transactions,* Vol 107, p 343 (1999).

Melt Furnaces 10

The melting process changes the metal from a solid state to a liquid state with the application of heat. The metalcasting process depends on the ability of a metal alloy to flow into and fill the mold cavity (or cavities) while in the liquid state. This change in state is accomplished by using specially designed furnaces.

Furnaces used to melt metal alloys in the foundry can be heated three different ways:

1) solid fuel as the heat source;
2) electrical energy used to produce the heat; and
3) gas (usually natural) as the heat energy.

Within the last two groups, there are several different types of furnaces, based on how the source of energy (and thus the heat) is applied to the metal. The different types of furnaces used in foundries will be covered later in this chapter.

FURNACE SELECTION

Furnace selection is dependent on several considerations, which include:

- Alloy Type
- Metal Quality
- Production Demand
- Economics

Alloy Type—One of the essential considerations is the type of metal alloy to be melted. Alloys have different melting temperatures and the different types of furnaces have, in some cases, maximum temperature limitations. For instance, a furnace required to melt ferrous alloys will have to achieve higher temperatures than will a furnace selected to melt aluminum alloys.

The furnace will sometimes be required to raise the charge materials (the materials to be melted in the furnace) *above* the melting temperature. This additional temperature is called "superheat." The superheat is needed so that the molten metal can be transferred from the furnace to the pouring area and still have enough heat (which affects its fluidity) to flow into the mold cavity. Also, certain metal alloys, while in the liquid state, will require different treatments before pouring and thus need this additional superheat. **Table 10-1** lists the pouring temperatures of some commonly poured foundry alloys.

Metal Quality—Different alloys also react differently to the conditions within the furnace during the melting cycle. Some alloys will oxidize more readily than others; thus, furnaces that reduce the exposure of the alloy to oxidizing conditions are preferred. In addition, some alloys, such as certain copper-base alloys, will see greater metal losses in some furnaces and not others. Metal loss means that some of the metal in the alloy will be removed from the alloy, usually by oxidation or vaporization, thus changing the chemical composition.

Deliberate changes in the chemical composition of the metal, and how often these changes will occur during the melting operation, should be considered. Some furnaces are more adept at accommodating these compositional changes than other furnaces.

Production Demand—Another important factor to be considered is the production demand to be placed on the furnace(s). Again, the different types of furnaces available have different speeds with which they can convert the solid charge materials to a liquid. If production demands include several or many chemical composition changes, then the appropriate furnaces should be chosen for the application. However, the more important of these two requirements is how fast the change from solid to liquid state can take place.

The furnace's melt capacity must also be considered when it comes to production demands. Melt capacity means how much volume of liquid metal it can produce, at the proper temperature and chemical analysis, per unit of time. Melt capacity is usually measured in tons or pounds per hour.

Most foundry melting is carried out on a batch basis from a cold charge. The exceptions to this are the continuously melting cupola used for cast iron production, and the melting of carbon steel in the direct arc furnace on a continuous basis.

Batch melting means that the furnace is completely emptied when the proper temperature has been reached. Once emptied, a new charge is placed in the furnace and very little if any more charge material is added during the melting cycle. In some cases, there is what is called a "tap-and-charge" operation. In this case, a "heel" or a planned proportion of molten metal is left in the furnace after the desired amount of molten metal has been removed. For instance, if three tons of molten metal is tapped out of the furnace, then, normally, three tons of new solid charge material will be added. The next charge is then placed into the furnace and thus also into the heel of molten metal. This can also be referred to as a "semi-continuous"

Table 10-1.
Average Pouring Temperature Ranges for the More Commonly Poured Foundry Alloys

Alloy	°F	°C
Zinc Alloys	650–850	345–455
Aluminum Alloys	1150–1350	620–735
Magnesium Alloys	1150–1350	620–735
Copper-Base Alloys	1650–2150	908–1180
Cast Irons	2450–2700	1340–1480
High-Mn Steel	2550–2650	1400–1455
Monel (70N,30Cu)	2500–2800	1370–1540
Ni-Based Superalloys	2600–2800	1430–1540
High-Alloy Steels	2700–2900	1480–1600
High-Alloy Irons	2800–3000	1540–1650
Carbon & Low Alloy Steels	2850–3100	1565–1700
Titanium Alloys	3100–3300	1700–1820
Zirconium Alloys	3350–3450	1845–1900

melting operation. Some furnaces lend themselves very well to the tap-and-charge operation.

Continuous-melting means that, once the furnace has been placed into operation, it is not shut off for the charging operation. In other words, molten metal is continuously tapped out of the furnace and solid charge materials are continuously added.

Economics—A melting furnace must be designed for effective heat transfer to the charge. Thermal efficiency is not the principal criterion in the choice of equipment and practice, however. The overall economics of melting operations depend upon many factors, including capital depreciation and degree of utilization. The operating costs themselves include maintenance and labor as well as fuel and power.

METALLURGICAL FACTORS

The aim in melting is to achieve close control of metal composition with low melting losses and avoidance of gas contamination and nonmetallic inclusions. The choice of melting is governed partially by the composition and quality requirement of the alloy and partially by the nature of the materials forming the charge. In many cases, the charge is virtually identical in composition to that required of the liquid metal to be poured into the mold(s). Usually, with rapid melting, there is a need for minor corrections to compensate for unavoidable melt losses. The practice of melting large melts often justifies chemical corrections to be made to the liquid metal before pouring.

The furnace charge may consist of pre-alloyed pig or ingot metal and hardener alloys, scrap from outside sources or scrap from internal cuttings of gates and risers. Externally purchased pig and ingot are normally supplied with a certified analysis. "Pig" normally refers to pig iron used in ferrous melting and "ingot" refers to the nonferrous alloys.

External and internal scrap materials should be sorted according to chemical composition. It is also good practice to chemically analyze purchased external scrap, so that the customer specifications can be met. The foundry should always be familiar with their scrap suppliers. Often foundries have a chemical analysis performed on the scrap before acceptance of the load. As few changes as possible should be made to the chemical composition of the molten metal before it is poured into the mold(s). In order to do this, the chemistry of what is being charged into the furnace should be known. **Figure 10-1** shows a charge material storage area for a high-production foundry that uses cupolas for melting cast iron.

Large, chunky, pieces of pig and ingot or heavy scrap have a small surface area and are less susceptible to melting losses and contamination. Though finely divided materials, such as turnings, are readily absorbed when fed directly into the liquid metal in the furnace, they are less desirable because they can cause dross, oxides and gas contamination. This is partially due to the tremendous amount of surface area of the turnings and any adhering cutting fluids. This becomes uneconomical as a foundry melting practice. Turnings should be sent to a refiner for production into briquettes or ingots.

Melt conditions can be varied with respect to the time-temperature sequence, the use of slags and fluxes and the atmosphere in contact with the charge. The furnace atmosphere can contain water vapor, CO, CO_2 and SO_2 as products of fuel combustion. The most common reaction is between metal constituents and oxygen present in the atmosphere and the refractory linings. The reactions are governed by the oxygen affinities of the elements present. It affects compositional control as losses of elements occur at different rates; therefore, the economic results must be studied. It also provides a means of refining by selective oxidation of impurities.

Vacuum melting, like inert gas melting, offers clean, neutral conditions for melting a charge with a minimum of chemical change. For instance, titanium is one such metal that is melted in a vacuum.

Oxidation loss is a function of the surface area-to-volume ratio of the charge and the time of exposure to its surrounding atmosphere. The highest losses of all are associated with the proportion of scrap and the cleanliness of that scrap. In copper-base alloys, melting losses can range from 1 to 4%. Greater loss is encountered with alloys containing zinc and aluminum. Losses are the highest in aluminum-magnesium alloys.

The final stage of melting often includes treatment of the molten alloy to influence the metallographic structure of the casting. In some cases, reagents are added to the molten metal while it is in the furnace or in a ladle. This treatment may be for deoxidation, degassing, desulfurization, grain refinement and/or modification.

Control of the chemistry of the molten metal must be maintained in order to produce the desired physical and mechanical properties of the cast metal. With the use of a spectrograph, the composition and quality of the metal analysis (elements in the alloy) can be quickly determined. Required additions can be introduced to the melt, bringing it up to the desired specification, and then the melt can be poured.

It cannot be stressed enough that the quality of the molten metal poured into the mold begins with the materials charged into the furnace. This includes measuring the amount of these materials accurately, which requires the proper maintenance of the weighing equipment. Faulty weighing equipment can lead to increased costs later in the production of the molten metal. Also, the chemistry of the materials in the charge must be known and the materials should be clean.

Fig. 10-1. Charge material yard in high-production foundry serving two cupolas. The preferred charge makeup method is from out of cars and only the trim is taken from metal storage bins. The cranes are on a common runway.

TYPES OF FURNACES

The types of furnaces used in the melting of nonferrous and ferrous alloys include the following:

- Cupola furnaces
- Crucible furnaces
- Electric/direct-arc furnaces
- Induction furnaces
- Electric resistance furnaces
- Reverberatory furnaces
- Dual-energy resistance furnaces
- Regenerative and Recuperative Burner Systems

The cupola differs from all the other furnaces in that the "charge" consists of coke (the fuel), fluxing materials (i.e., limestone) and the metallics. In all the other furnaces, the metallics are heated by an external source.

Cupola Furnaces

The cupola is one of a family of vertical-shaft furnaces, such as the blast furnace and some types of coal gasifiers, which operates on the counterflow principle, i.e., the solids move downward and the hot gases rise. A simple cupola is a refractory-lined stack with openings near the bottom for the exit of molten metal and slag. At a higher level in the stack, tuyeres introduce the blast air to the furnace. Above the middle of the stack, another opening permits the entry of the solid charges of coke, metallic charge materials and flux. A schematic diagram of a cupola is shown in **Fig. 10-2.** Actually, this is a refractory brick-lined cupola, which is not very common today. Cupolas used today to melt cast iron are discussed in this section.

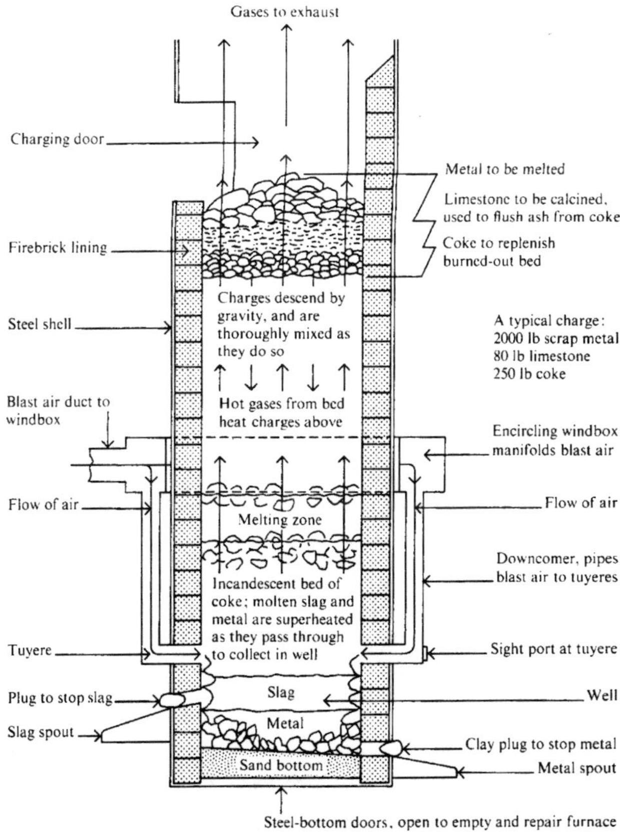

Fig. 10-2. Schematic diagram of a typical cupola.

Cupolas are not as common now as they were years ago. Most cupolas in use today are found in the larger, high-production foundries such as those producing pipe and automotive castings. A point of interest is that, many years ago, cupolas were used to melt some copper-base alloys. Today, it is unlikely that any cupolas are used to melt copper-base alloys.

Cupolas are sized by the inside diameter of the cupola shell. If the cupola has a lining above the tuyeres, then the lined inside diameter will have to be considered when setting the operating parameters. Diameters of commercial cupolas vary from 18 to 160 in. (46 to 406 cm). Cupolas are used to produce commercial grades of cast iron with carbon contents ranging from 2.40% to 3.80% carbon levels. Melting rates from cupolas can range from 1 to 100 tons per hour.

Cupolas can be operated on an intermittent tapping operation or on a continuous tapping operation. In the *intermittent* operation, the taphole is plugged shut (botted) and, at the proper time, it is unplugged to allow the molten cast iron and slag to run out of the taphole. This type of cupola operation is also referred to as a "bott-and-tap" operation. On the other hand, in the *continuous* operation, once the melting operation is under way, the taphole remains open, allowing molten cast iron and slag to leave the cupola. In the latter operation, the time from when the first molten cast iron leaves the cupola until the cupola is shut down is called a "campaign." A campaign can be any length of time from one day to many months.

All cupolas use foundry coke as the principal fuel. Other materials are used as supplements when carbon is absorbed during the melting process due to a deficiency in the regular coke or a reduction in the pig iron content of the charge. Such materials are pitch coke, petroleum coke, calcined pitch coke, carbon electrodes and graphite blocks. Over the years, attempts have been made to supplement the coke by injecting powdered coal, natural gas and fuel oil through the tuyeres; however, today, coke is the fuel of choice without supplements.

The cupola is a vertical cylindrical steel shaft that can be divided into several sections. The body section, or the lower part of the cupola, rests on a very firm foundation. **Figure 10-3** shows the bottom section of a cupola. Note that this section rests on legs and

Fig. 10-3. Bottom section of a conventional cupola.

has an opening in the bottom. Hinged doors close this opening. There is a charging door in the upper section, for introducing the raw materials. The windbox wraps around the body and can be detached from the body or attached as shown in **Fig. 10-3**. From the windbox, the blast air is directed down into the tuyeres and into the cupola.

Below the tuyeres is a section called the well, at the bottom of that well is the taphole located in front of the cupola. The taphole allows molten metal and slag, if it is a front-slagging cupola, to be withdrawn. The inner or working surface of the well is normally lined with either carbon block or brick. In order to minimize heat loss, a backing of firebrick is often placed between the carbon block and the steel shell. In the case of a rear-slagging cupola, the slag is removed through a slag hole, which is located just below the tuyeres and at the rear of the cupola. In front-slagging, the metal and slag discharge together through a single taphole, and pour into a small well in the runner, where the slag is skimmed off. When discussing ferrous alloys, the term slag is used. Slag is the impurities that result from the melting operation; it is considered a silicate material that must be disposed of and is considered as a contaminant in the metal casting.

The cupola shown in **Fig. 10-2** is a rear slagger, as the slag is removed through the slag hole in the back of the cupola. As stated earlier, this type of cupola is not as widely used as it was years ago.

When looking further at the cupola in **Fig. 10-2,** a few areas should be noted. In the bottom of the well and on the bottom of the cupola, there appear to be pieces of material that are solid. The figure shows that, just below the tuyeres and the melt zone, there is a bed of incandescent coke. This bed of incandescent coke extends down to the bottom of the cupola and is seen as rock-like material. The entire well of the cupola is filled with the incandescent coke and it is this bed that supports all of charge materials, or "burden," in the cupola. The burning-in of this bed of coke, at the start-up of the cupola, and maintaining its proper height are very critical parts of proper cupola operation.

Among the charge materials placed in the cupola is "flux." The flux can be either limestone or dolomite, and it lowers the fusion temperature and viscosity of the slag. The flux also fluidizes the natural slag-forming constituents, i.e., the coke ash with dirt and foundry sand introduced into the cupola with the metallic charge materials. In doing this, the flux improves combustion, as well as carbon pickup by the metal, and decreases the tendency for bridging and hang-ups in the cupola during the melting operation.

Two essential things are to be remembered when adding the flux. The first is that it should be evenly distributed and not concentrated into any small area of the cupola. Secondly, the flux should be of the correct size. The flux must be large enough not to be blown out of the cupola, yet small enough that it can begin its fluxing action as high in the cupola as possible. The cleaner the charge materials are (metallics and coke), the easier the fluxing. Cupola operators usually refer to the flux as "stone."

The next chapter will discuss "charging" the cupola, that is, placing the charge materials into the cupola.

Cupola Innovations—There have been many technical and design innovations made to cupolas over the years, innovations that have improved their metallurgical control, production rates and economical operations. The following technology has advanced the cupola melting to new levels of sophistication and efficiency. Some of the material is covered in more detail in Chapter 11, Ferrous Melting Practice.

Blast Air—Over the years different ideas were discussed regarding improving the air that is blown into the cupola (blast air). Initially, ambient air was blown into the cupola from a blower, through the windbox and finally the tuyeres. This meant the air temperature could change, as well as the humidity. One of the first attempts to control blast air characteristics was to preheat the air. The preheating of the blast air accelerates the combustion reactions and moves the combustion zone closer to the tuyeres. To the extent that heat energy is introduced with the blast air, fuel energy can be reduced such that the result is an increased melting rate with decreased coke requirements.

The blast air can be preheated in one of two ways: 1) externally fired or 2) recuperative hot blast systems. Externally fired systems employ an auxiliary fuel, such as gas or oil, which is burned in a combustion chamber in a heat exchanger through which the blast air passes. On the other hand, the recuperative hot blast system makes use of the heat contained in the cupola's exhaust gasses. These gasses are passed through a heat exchanger in which the blast air is preheated.

The temperature of the hot blast air is affected by the amount of refractory within the cupola. A conventional cupola, completely lined, is limited to blast air temperatures in the range of 700–800F (371–427C). On the other hand, liningless water-cooled or a combination lined and water-cooled cupola, both using water-cooled tuyeres, can have blast air temperatures in the range of 1000–1200F (538–649C). Hot blast air temperatures of 2000F (1093C) have been used.

Oxygen enrichment can be used to reduce cupola raw material costs and to increase melt rates. The oxygen can be added in the windbox, tuyeres or directly into the well. The amount of oxygen normally added is from 1 to 4% of the blast air volume. Another use of oxygen enrichment is at cupola start-up or immediately after a shutdown. Typically, 2–4% oxygen is added to the blast air when it is turned on. This can increase the tapping temperature very rapidly and can reduce the amount of cold iron pigging and casting defects due to cold metal. Oxygen enrichment of the blast air has been done since the early 1930s.

Humidity reduction in the blast air offers real advantages to the cupola operator and metallurgist, both in coke savings and in the many intangible savings such as those made possible by more consistent metallurgical control in the cupola. Several methods of dehumidifying the blast air are by refrigeration (similar to the home dehumidifier), adsorption method and absorption method.

There have been other attempts at improving cupola operation by adding other materials and gasses to the blast air. Space doesn't permit including these details herein; however, the reader is referred to the AFS Publication, *Cupola Handbook.*

Water-Cooled Tuyeres—Prior to water-cooled tuyeres, it was difficult to get the blast air to penetrate into the burden in the cupola. Extending properly sized clay tiles into the cupola was one of the attempts to alleviate this problem. To try to prevent them from breaking off due to the downward movement of the burden, the cupola operator would use refractory material around the protruding tuyere. That usually wasn't very successful. In time, these tuyeres would be broken off mechanically by the movement of the burden, or chemically attacked and "burned" back to the inside wall of the cupola.

Water-cooled tuyeres occurred in conjunction with the development of water-cooled shells, which shall be discussed next. The majority of water-cooled tuyeres are cast from high conductivity copper, (+99% pure copper), and are a hollow shell through which water passes. Water-cooled tuyeres are projected beyond the shell into the cupola to protect the bare shell from overheating and rapid deterioration.

Figure 10-4 is a sectional view of a cupola using water-cooled tuyeres. By projecting the tuyere into the cupola, there is better blast penetration into the center of the cupola. To further help blast air penetration, the tuyeres are installed at a downward angle of usually 10–15 degrees.

The water-cooled tuyere, along with the water-cooled shell, makes extended campaigns of one or more weeks possible without dropping bottom. The protruding tuyere minimizes the sidewall gas travel, thus helping to retard heat losses and excessive metal or slag buildup on the inside wall of the cupola. **Figure 10-5** illustrates the theory of tuyere velocity and the downward angle.

Another type of water-cooled tuyere is made of structural copper or copper coil wound around a steel cylinder. Titanium tubing has also been used for protruding tuyeres. Any type of protruding tuyere has to be made of a highly heat-conductive material.

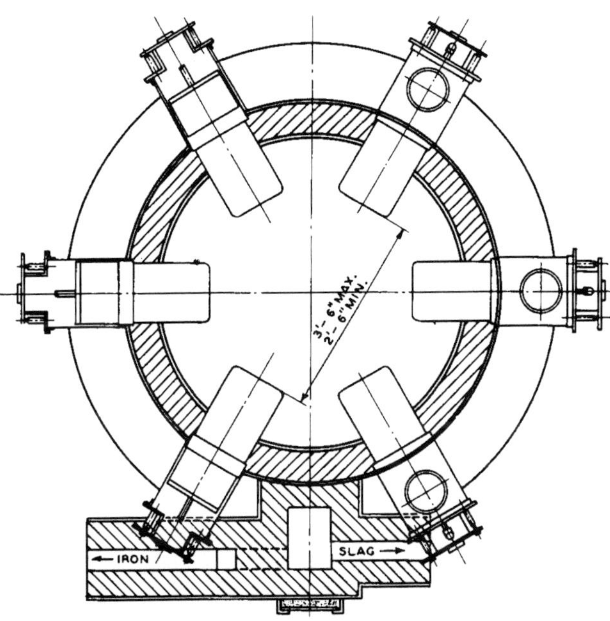

Fig. 10-4. Cutaway view of a cupola with water-cooled tuyeres.

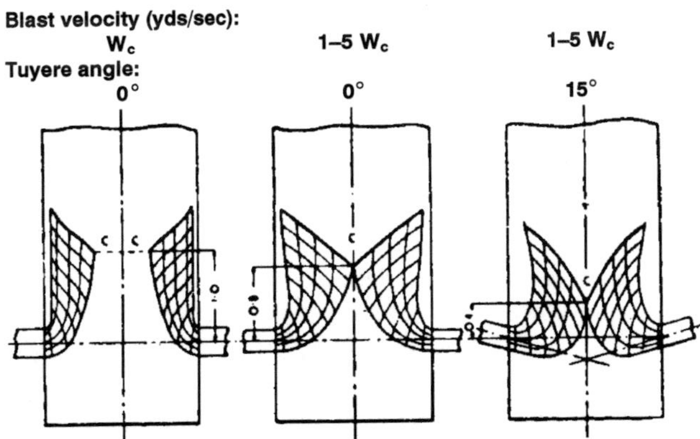

Fig. 10-5. Theory of blast velocity in tuyeres on iron temperatures in a liningless cupola.

Water Cooled Shell—In order to reduce the amount refractory used to line the inside wall of the cupola, foundry personnel sought a way around this economic burden. In some cases, the cupola had to be emptied of its burden by "dropping bottom." This was performed by opening the bottom doors. Once the cupola had cooled sufficiently, the cupola operator had to go into the cupola to repair the refractory lining. This practice was sometimes done on a daily basis. Thus, the water-cooled cupola was constructed.

Figure 10-6 is a cutaway view of a water-cooled cupola. In the early days of water-cooled cupolas, a water jacket, similar to a steam radiator, was built into the cupola shell, and water passed through it. One of the major problems with this method was that, if a leak occurred in the water jacket, some water would get into the inside of the cupola and affect its operation. Unfortunately, many of these leaks occurred inside the cupola.

Today, the majority of water-cooled cupolas have water cascading down the outside of the shell. The water-cooled part of the shell

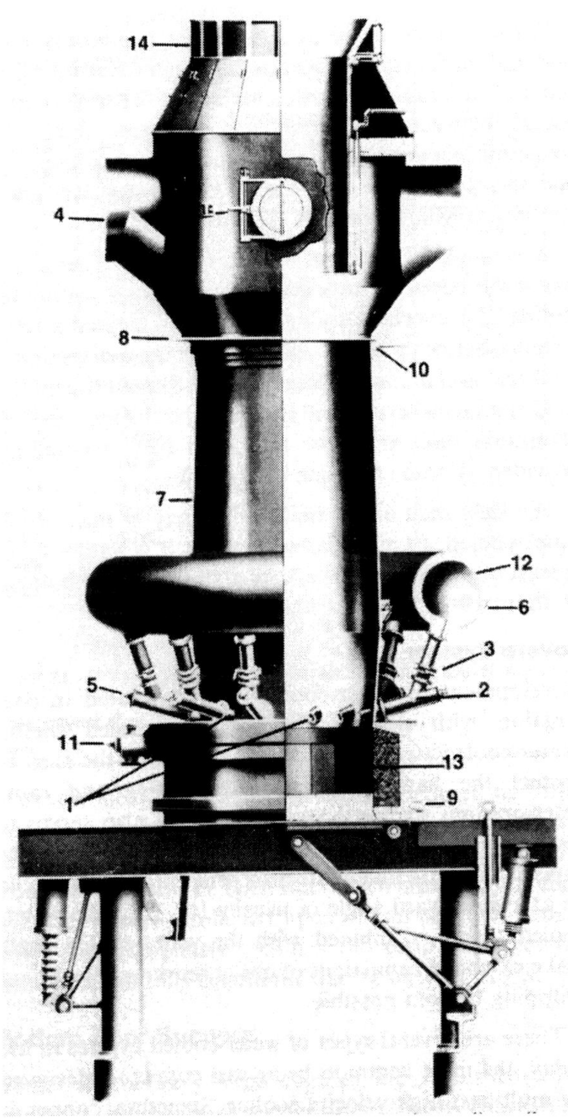

Fig. 10-6. Cutaway of a water-cooled liningless cupola.
Legend: 1. water-cooled tuyeres; 2. tuyere elbows; 3. control dampers; 4. gas take-off; 5. sight port; 6. detached windbox; 7. welded body; 8. dry expansion joint; 9. selective well-cooling; 10. upper spray-ring; 11. over-flow spout; 12. windbox lining; 13. carbon lining; 14. upper stack, independently supported.

runs from below the charge door down to below the tuyeres. Note that the shell in **Fig. 10-6** is tapered from below the charge door and gets larger until just below the tuyeres. This taper helps the cascading water to adhere to the outside surface of the shell. The water may be supplied to the surface of the shell by spraying or cascading at the top of the water-cooled section. Also, note in **Fig. 10-6** that the windbox is completely detached from the shell.

With the cascade-type cooling, the efficiency is related not only to water volume but also to the uniformity of its coverage. The water can be applied to the steel shell by means of an overflow moat surrounding the shell at the point of application. Regulation is possible by the use of valves, and adjustments of the flow pattern are made by weir height and tension of a neoprene wiper in the spreader-system assembly. This spray ring, as it is called, can be seen in **Fig. 10-7**. Some water-cooled cupolas will make use of another "redistribution ring" to distribute cooling water to the lower sections of the cupola to be cooled. Some advantages of the water-cooled cupola are:

1. Consistent internal diameter, which promotes operational uniformity.
2. Cupola can be run continuously or banked overnight.
3. Slag practice may be varied between acid to basic, if desired.
4. Since there is no refractory in the water-cooled zone of the cupola, refractory repair is kept to a minimum.

The major disadvantage of the liningless water-cooled cupola is the increased heat loss through the shell. Using a preheated blast usually compensates for this increased heat loss.

Prior to the 1970s, almost all cupolas operating in North America fell into one of two categories: completely refractory-lined, or the liningless water-cooled cupola with refractory below the tuyere level and above the charge door. The fully lined cupolas were typically used in small- to medium-tonnage melting operations and had campaign lives from several hours to 16 hours per day. On the other hand, the liningless water-cooled cupolas were favored in larger-tonnage melting operations and had campaigns of one week or more.

Prompted by the Arab oil boycott in 1973, energy conservation became the motto. Further impetus to reduce heat loss (energy) in cupola operations came from increases in foundry coke prices. As mentioned earlier, an increase in heat loss through the shell of water-cooled the cupola was a major disadvantage. Analysis of the energy distribution in the cupola melting process showed that large amounts of the total energy generated were consumed by thermal losses such as:

- sidewall loss to shell cooling water;
- heat loss to tuyere cooling water in the case of water-cooled tuyeres;
- top-gas sensible heat loss from gases exiting the cupola;
- CO gas formation inside the cupola during operation;
- slag and other miscellaneous heat losses in the cupola.

The sum of these losses can typically vary between 35% and 75% of the total energy used in the cupola's operation.

In the early 1970s, small- to medium-sized cupola operations with conventionally lined cupolas, that is refractory lining the entire inside of the shell, began modifications such as water-cooled protruding tuyeres and water-cooled shells. Campaigns increased from a few hours to 2–3 shifts per day. Some campaigns ran for a week before refractory repairs were needed. These successes were credited to the water-cooled protruding tuyeres and water-cooled

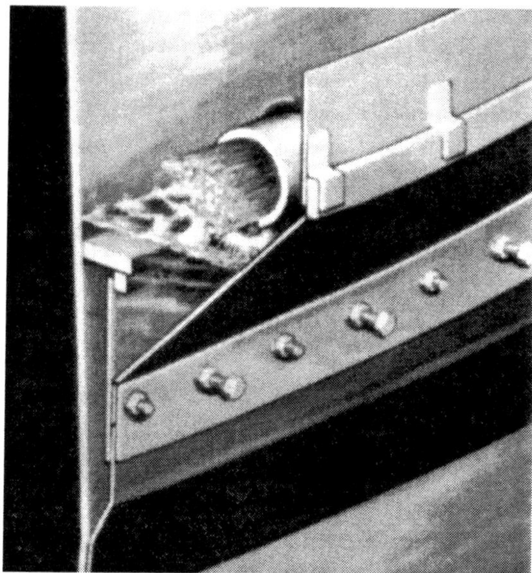

Fig. 10-7. Spray rings with adjustable neoprene wipers distribute cold water evenly to the cupola shell. They are placed to introduce coldest water to the hottest points.

shells, which protected the refractory lining from excessive heat so that, in some cases, campaigns of up to four weeks were reported.

In the later 1970s, the use of combination refractory-lined water-cooled sidewall construction, proven to work in smaller-tonnage cupola operations, was applied to the larger-tonnage hot-blast cupola operations. Substantial improvements in thermal efficiency over the bare shell design were obtained. Except for very large cupolas or operations requiring special slag practice, the combination refractory-lined water-cooled design has become an attractive choice for cupola melting operations.

Divided-Blast Cupola—The divided-blast cupola is equipped with two windboxes and two sets of tuyeres, as seen in **Fig. 10-8.** Each set of tuyeres is supplied with a measured and controlled quantity of blast air. The optimum spacing between the two rows of tuyeres was determined to be in the range of 30 to 42 in. (76 to 106 cm). The best results were obtained at approximately 36 in. (91 cm).

Results of the investigation of the divided-blast cupola further indicated that the blast should be equally divided between the two rows of tuyeres. Tapping temperature was increased or, alternatively, coke consumption decreased, and melt rate increased with constant melt temperatures. Consideration should also be given to the number and size of tuyeres at each level to maintain uniform introduction of blast air around the periphery of the cupola, and to achieve sufficient blast velocity.

Dry-Bottom Cupola—The dry-bottom cupola is a European development and is shown in **Fig. 10-9.** This design is characterized by a hearth, which is inclined toward the taphole, permitting all iron and slag that are melted to be discharged immediately through the taphole into one or two cylindrical slag separators. Refractory-lined vessels are attached to the outside of the cupola. The iron and slag leave the separator via two separate spouts.

With the dry hearth design, iron and slag are not collected in the well, but drop through the coke bed onto the hearth and are discharged and collected in the separator. The carbon pickup is determined solely by the difference in elevation between the tuyeres and the hearth, which is predetermined and accurately maintained.

The dry-bottom cupola hearth and separators are lined with an acid refractory. In the dry-bottom cupola refractory, wear area is moved to the separators and average campaign lengths of 4–6 weeks for the hearth are common. The separators are relined on a weekly or bi-weekly basis, and dropping bottom can be entirely eliminated. Using two separators allows one separator to be relined while the other is operational.

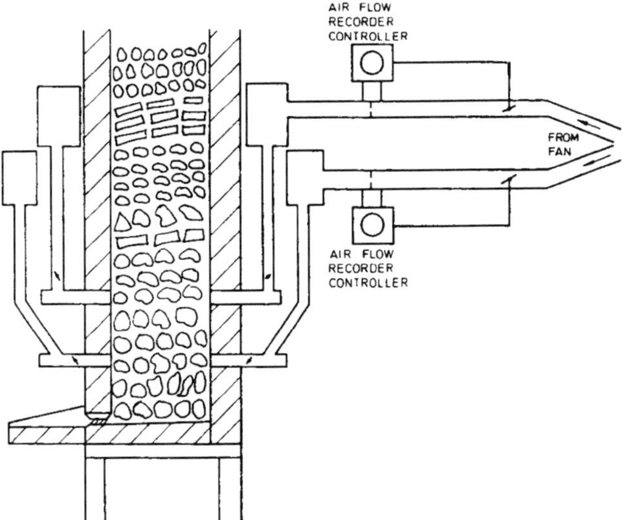

Fig. 10-8. Schematic of a divided blast cupola.

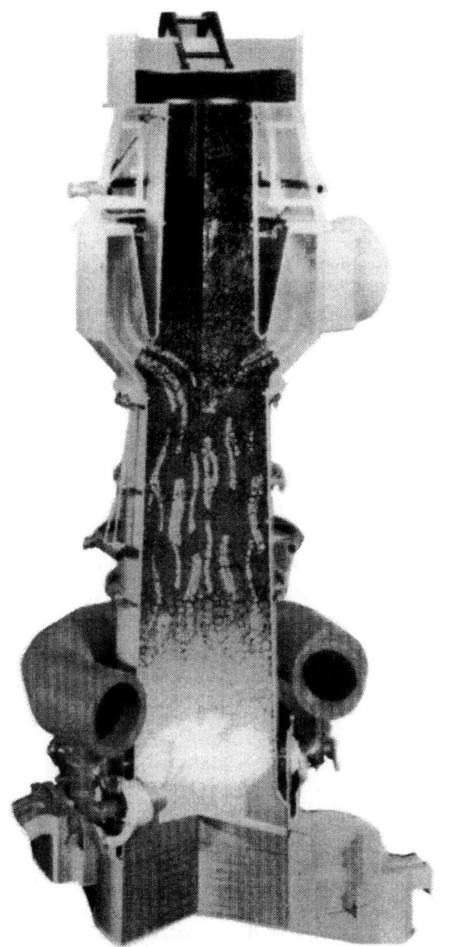

Fig. 10-9. Dry-bottom cupola.

Plasma-Fired Cupola—In the late 1980s, research was carried out using plasma in a cupola to melt cast iron. Plasma produced by exposing gas to a high-intensity electric arc can achieve temperatures in excess of 10,000F (5380C), which is much higher than the 2800F (1537C) practical limit for fossil fuel combustion. This high-temperature capability, along with such other characteristics as rapid heat transfer and excellent controllability, suggested that plasma might be an efficient and economical tool to affect physical and chemical change in cast iron production.

Figure 10-10 is a rough sketch of a plasma-fired cupola. The important component of the plasma-fired cupola is the torch that replaces the conventional tuyeres. This particular test unit had only one torch; however, larger units could have multiple torches.

Another unique feature of this test unit is the gas-recycling loop. This loop removes variable amounts of cupola top gas just below the charge door. Large particulate matter is removed from cupola top gas, and the gas is then introduced back into the system at the cupola-torch junction. This gas consists almost entirely of carbon monoxide and nitrogen. This mixture provides a nonoxidizing atmosphere that dramatically reduces coke consumption and precludes excessive metal loss due to oxidation when charge materials have high ratios of surface area to volume. This allows the plasma-fired cupola to melt loose borings and turnings without briquetting prior to charging.

Even though plasma-fired cupolas might not be operating in production foundries today, this is a novel idea that is worth mentioning here.

Environmental Control Equipment—The type and use of environmental control equipment used on cupolas will be discussed in a later chapter. However, at this point, it is safe to say that the need for environmental controls on the cupola has been an essential factor in many cast iron foundries switching to other methods of melting.

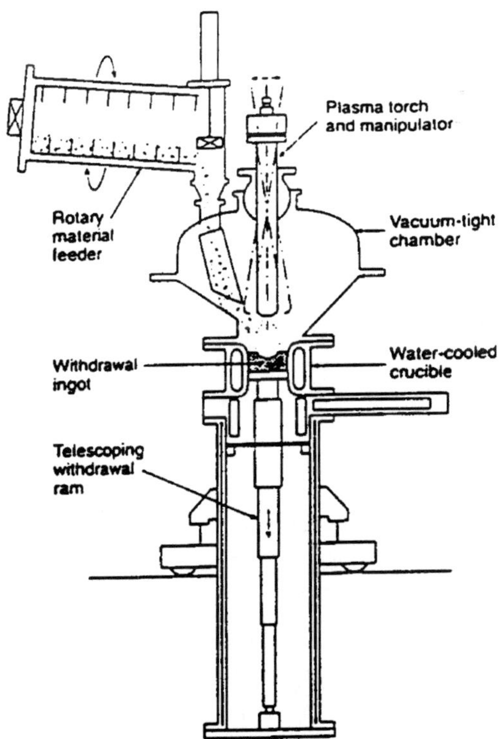

Fig. 10-10. Schematic of plasma-fired cupola.

Crucible Furnaces

After the cupola, the crucible furnace is the oldest type of furnace used to melt metals. It is one of the simplest furnaces to use for the melting of metals. These furnaces are used primarily to melt aluminum, copper-base and magnesium alloys.

There are three types of fuel-fired crucible furnaces used in foundries today: 1) stationary lift-out; 2) tilting and 3) stationary (bowl-type). The crucible-type *induction* furnaces will be discussed under "Lift-Coil and Push-Out Induction Furnaces."

- *Stationary Lift-Out Crucible*—The first is the stationary lift-out crucible type as seen in **Fig. 10-11.** In this type of crucible furnace, the crucible is removed from the furnace and the molten metal is poured directly from the crucible into the mold or another furnace or a molten metal holding device.
- *Tilting Crucible*—The second type of crucible furnace is the tilting crucible furnace seen in **Fig. 10-12**. The crucible is cemented into the body or shell of the furnace. To remove the molten metal, the furnace is tilted and the molten metal is poured into a pouring or transfer ladle. Since there is normally no cover on this furnace, the molten metal could also be hand-dipped or automatically dipped out, or even pumped out with an electromagnetic transfer pump.
- *Stationary (Bowl-Type) Crucible*—The third type of crucible furnace is the stationary crucible (bowl-type) furnace shown in **Fig. 10-13**. This type of crucible furnace is not tilted, so the molten metal must be removed from the crucible by hand or automatic dipping or by use of an electromagnetic molten metal pump.

A crucible furnace usually consists of a refractory-lined steel shell. The refractory can be either firebrick or a monolithic cast refractory shell that fits into the steel shell. In the case of the lift-out crucible furnace, there is a refractory-lined swing cover over the furnace. The two other types of crucible furnaces do not necessarily have covers. The inside bottom of the steel shell is also covered with a refractory lining.

A crucible made of either clay graphite or silicon carbide is used in the furnace. Silicon carbide is usually preferred for its better thermal conductivity, which means better heat transfer than clay graphite. In some cases, such as melting magnesium alloys, cast iron and steel pots or crucibles can be used. The capacity of the furnace is limited by the size of the crucible. Each furnace is designed for a given size crucible. It is important to remember this because it can affect crucible life and the proper distance between the crucible and the refractory lining.

Lift-out crucibles make possible the greatest amount of flexibility with respect to the range of alloys that can be melted. One can control the quality of the metal by using different crucibles for melting each different alloy; this prevents the contamination of one alloy by another.

Note in **Fig. 10-11** that the crucible is sitting on a base block or "stool." The base block plays a very important part in good crucible melting operations. The base block is especially dimensioned for the furnace and crucible size. It holds the crucible at the proper height in regard to the burner. Using the wrong size base blocks can lead to poor quality melts and poor crucible life. Base blocks can be made of clay graphite or silicon carbide.

Most crucible furnaces are fired by either gas or oil fuels that are readily available; the most commonly used fuel is natural gas. Adjusting the amount of fuel and combustion air can control the

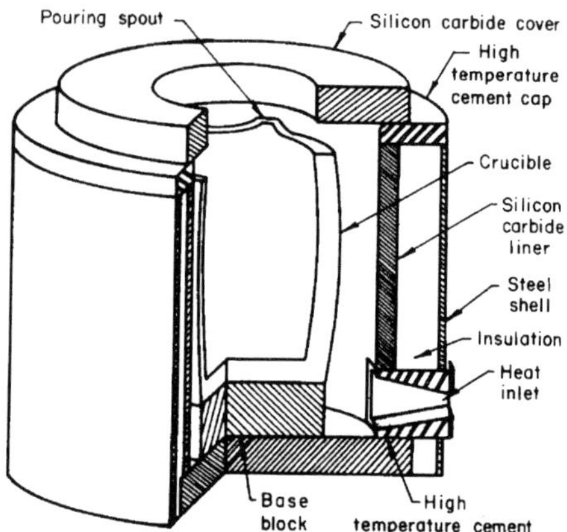

Fig. 10-11. Schematic of lift-out crucible furnace showing various components. Furnace may be either floor-mounted or installed in a pit, and may be fired with either gas or fuel oil.

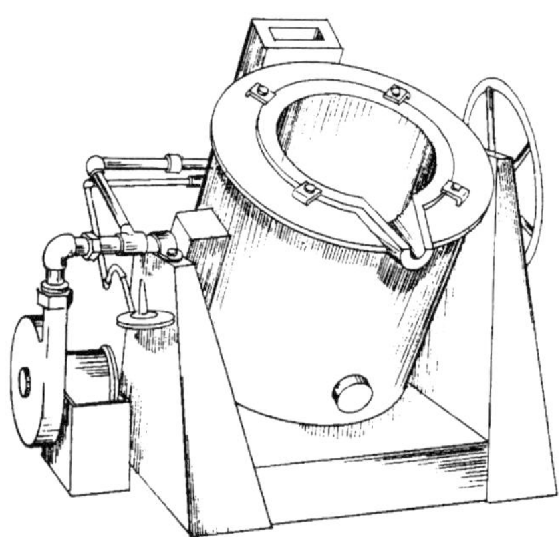

Fig. 10-12. Tilting crucible furnace.

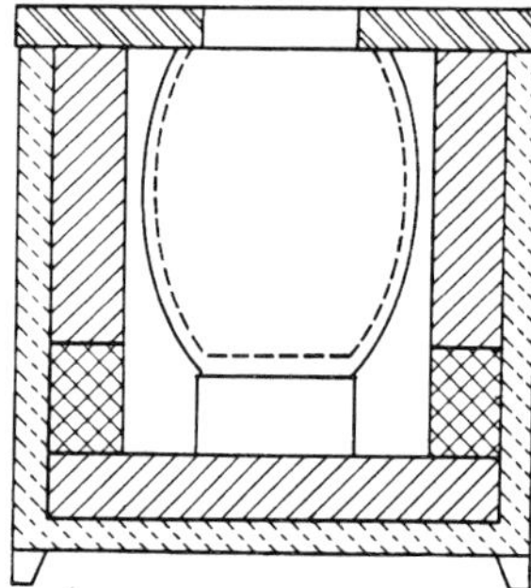

Fig. 10-13. Stationary crucible (bowl-type) furnace.

flame. Proper maintenance of the burner and its related equipment is paramount to high-quality molten metal. It should also be pointed out that the flame should not impinge directly onto the crucible. The flame should enter the furnace tangentially so that it swirls around the crucible and rises to the top. The flame should first touch the top of the base block and the bottom of the crucible. The exhaust gases or products of combustion exit the furnace through an exhaust port that is located in the center of the cover of the lift-out crucible furnace. The other furnaces have exhaust "chimneys" coming off the side of the furnace.

Crucible furnaces melt by conducting the heat from the flame through the crucible walls into the charge materials. The maintenance of the furnace wall plays an important part in the melting operation. The wall should be kept clean of any protrusions that can affect the movement of the flame up the sides of the crucible.

Electric/Direct-Arc Furnaces

Figure 10-14 shows a schematic of an electric/direct-arc furnace. History shows that, in 1800, Sir Humphrey Davy discovered the carbon arc. Pichon invented the electric furnace in 1853 and was granted a patent. However, the practical application of this type of furnace to the melting of metal is credited to Sir William Siemens. It began in 1878 when Siemens constructed, operated and patented furnaces operating on both direct and indirect electric-arc principles.

Siemens' first electric-arc furnace consisted of a crucible above which two horizontal electrodes were placed (indirect arc). One end of each of the electrodes was brought close to the other until an electric arc was struck between them. As far as is known, this type of electric arc furnace is no longer used in the U.S. foundry industry.

The next step in the development of electric-arc furnaces involved direct-arc furnaces. These furnaces were characterized by vertical electrodes with current travelling from the electrode to the charge materials and bath. Girod used a conducting bottom that contained studs to complete the electric circuit. Single, two- and three-phase units were used, where each electrode operated independently. These furnaces met with only limited success because of the inefficiency in current carrying capacity of the bottom materials.

Between 1888 and 1892, Heroult built a direct-arc furnace, which is the forerunner of today's electric/direct arc furnaces. This design was a three-electrode, three-phase unit, and was originally developed for the manufacture of ferroalloys and calcium carbide. The advantage of this furnace for the melting of alloy steels was soon recognized and used successfully.

The use of direct arc furnaces for the melting of cast irons occurred around 1930. The first applications of these furnaces were in foundries producing higher-quality iron castings for the automotive industry. These furnaces can be used in a duplexing operation with a cupola.

Two advantages of using the electric/direct arc furnace are the ability to use lower grade scrap and the ruggedness of the furnace. This furnace is also known to be a very good primary melting tool because of the speed at which the solid charge materials can be converted to molten metal and tapped. As will be learned later, the design of this furnace also lends itself to the treatment of carbon steel.

A electric/direct arc furnace consists basically of a shell, a moveable roof, and supporting equipment and controls. In any case, the interiors are lined with refractory. The shell has an access door located at the back of the furnace to allow additions of molten metal to the heat, rabbling (mixing) the heat, observing the progress of the heat, obtaining samples and removing slag from the heat. Large direct-arc furnaces may have an additional access door on the side of the shell for the same purposes. A pouring or tapping spout is located at the front of the furnace. The furnace has a tilting mechanism that allows it to be tilted forward to tap the furnace, and backward to slag off the melt.

The roof, on the other hand, has three ports through which solid cylindrical carbon or graphite electrodes pass and also a fourth port for a take-off for pollution control equipment. The roof can be swung aside to allow the top charging of the furnace.

The three electrodes, made of graphite, are connected to the electric power supply through bus bars and electric cables. Graphite

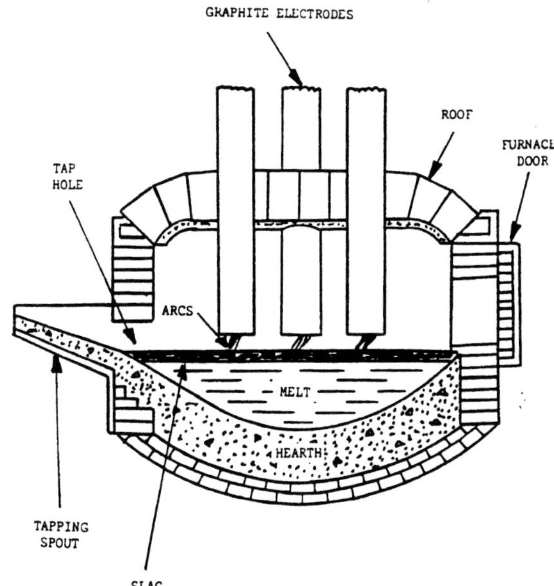

Fig. 10-14. Schematic of arc furnace.

electrodes are used because they are strong and have increased current-carrying capacity over other materials that were used for making electrodes. The electrodes are raised or lowered using electric motors or hydraulic rams. The gap where the electrodes pass through the roof is often sealed with a water-cooled gland to prevent a "chimney" effect, which increases oxidation loss of the electrodes.

The source of heat, which does the melting of the charge materials, is the arc that is struck between the electrodes and these materials. This arc is an extremely potent source of heat and reaches temperatures in excess of 7232F (4000C). The arc is a column of radiant energy that is essentially a hot column of gases and vapors. This arc is sometimes compared to a lightning bolt. The main source of heat is radiant energy that comes from hot spots occurring at the tip of the electrodes and the point where the arc hits the metal charge. The arc column also radiates heat onto the metal and into the adjacent scrap. The size and other characteristics of the arc are dependent on the electrical parameters that are fixed by the power supply and the electrode regulator mechanism, and by the nature of the materials involved. Most direct arc furnaces operate on a 3-phase alternating current.

The refractory selection for a direct arc furnace is a very important part of its proper operation. The demands placed on the refractories are severe, and, of course, the refractory materials must be economical. The refractory must be thermally stable and possess sufficient strength at temperature for the given application. The refractory must also be inert to the metal and slags encountered, and the material must allow ease of installation and maintenance.

One basis for classification of refractory materials relates to the chemical acidity or basicity of the material. For these discussions, materials high in SiO_2 (silica dioxide) are considered acid. On the other hand, those materials high in CaO (calcium oxide) or MgO (magnesium oxide) are basic. Other oxides, such as Al_2O_3 (alumina oxide) and Cr_2O_3 (chrome oxide), are considered neutral. In general, acid refractories are used in contact with acid slags, and basic refractories in contact with basic slags. **Figure 10-15** is a sketch of how these two types of refractories are applied in a direct electric-arc furnace. The use of basic and acid refractories will be discussed in Chapter 11, Ferrous Melting Practice.

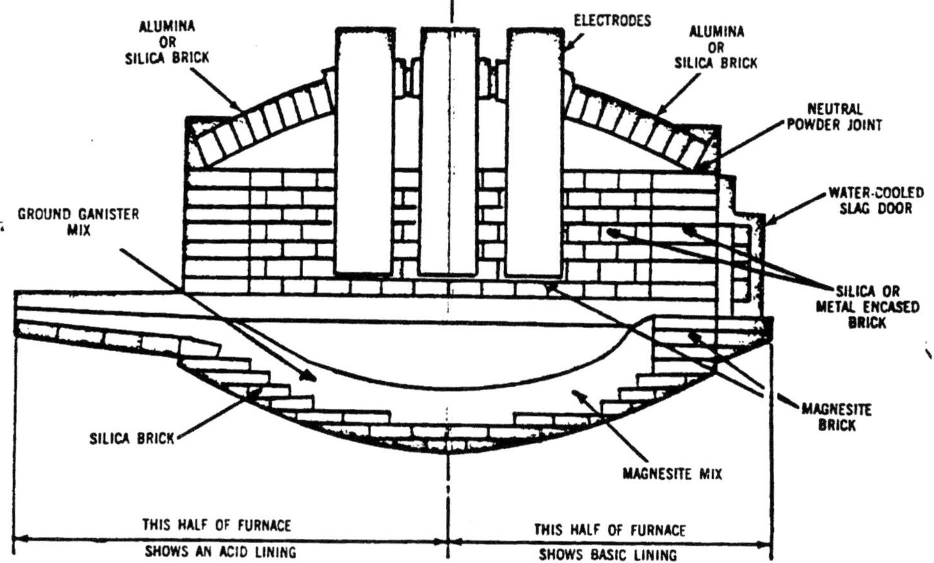

Fig. 10-15. General construction of an arc furnace lining showing acid and basic.

A unique use of arc melting principle can be found in the melting of titanium and its alloys. First, the furnace is situated in a vacuum chamber. The electrode, instead of being made of graphite, is made of pieces of titanium alloy welded together. The size and weight of the "electrode" is based on the amount of molten metal that will fill the mold. This is determined by the weight of the casting and the gating and risering system. A single electrode is used.

The arc is struck between the electrode and some pieces of titanium alloy on the bottom of the furnace. Melting is done under a vacuum. Since there is no air present in the vacuum chamber, the melt is gas- and slag-free. The mold is placed in a vacuum chamber, which is attached to the furnace's vacuum chamber, and the molten metal is tapped out of the furnace into the mold. Thus, all of the melting and pouring is done under a vacuum.

Modern arc furnaces are equipped with a variety of features to increase production rates, reduce heat times (melting time) and lower operating costs. Among these features are:

- Ultra High Power (UHP) transformers able to produce power levels of 600–900 kVa/ton are being installed.
- Water-cooled sidewalls and roofs to reduce refractory costs.
- Oxygen fuel burners to supplement heat input and improve melting efficiency.
- Foamy slags to shield sidewalls and roof from heat radiation from the arcs. This practice permits the use of maximum available secondary voltage using longer arcs with high power factors.
- Computer controls to optimize electrical power programming, automatic tap changing based on furnace condition and power demand. Systems that are more complex provide control of metallurgical parameters such as, tap temperature, timing process events, least cost charging calculations and data logging.
- Eccentric bottom tapping to reduce tap times, reduce temperature losses and avoid slag contamination in the ladle.
- Scrap preheating to recover energy from furnace waste gases.
- Single-electrode furnaces to reduce electrode consumption and noise.
- Continuous charging using preheated charge materials.
- Post combustion oxygen injection.

Induction Furnaces

The concept of induction melting has been around just about as long as that of electric-arc melting. At about the same time that the arc furnace was developed, the induction furnace was being developed by Ferranti in Europe and Colby in the United States. Theirs were low-frequency furnaces based on the idea of a ring furnace or a transformer core passed through a ring of molten metal. However, the foundry industry was very slow in accepting this new method of melting metal and very little work was done with this method. In 1911, the Kjellin Company in Sweden installed such a furnace.

In 1916, Wyatt demonstrated the first truly practical induction furnace. This induction furnace was of channel design and self-cooling. It was at this time that the brass industry began to grow, and by 1930, 800 of these furnaces were in operation. Since that time, induction melting has seen, and continues to see, many new and inventive ways of using these furnaces to melt and produce most of the metal alloys poured in the foundry industry.

Induction melting is the process of passing an electrical current through a wire or coil that causes a magnetic field to be set up around them. In the melting process, this field passes through the charge materials or molten metal causing current to flow in this material. The current flow in the charge material or molten metal is opposite that of the magnetic field. Heat is generated by this current flow due to the resistance of the charge materials or molten metal.

To make this come about, there has to be a wire—or coil, as it is called in induction furnaces. As a current of electricity is passed through this coil, a magnetic field is set up around the coil. **Figure 10-16** shows this magnetic field around a coil. Note, however, that the current in the coil can flow two ways. In addition, as the current direction changes, the magnetic field is reversed. **Figure 10-17** is a schematic of alternating current. Again, note that the electrical current alternately changes direction.

The number of times this change in electrical current reverses itself within one second is measured in cycles per second (CPS), or as it is called today Hertz (Hz). For instance, if the direction of electrical current flow changes 60 times per second we have a frequency of 60 Hz. Induction furnaces are sized by the amount of molten metal (pounds or tons) they will hold, and their frequency.

Electrically speaking, there are three ranges of frequencies discussed: low, medium and high.

Low Frequency, 60–100 Hz
Medium Frequency, 150–700 Hz
High Frequency, 701> Hz

The advantages and disadvantages of these frequencies will be discussed. First, some basic electrical terms will be bantered about when discussing induction melting.

Most people are familiar with the simple equation known as Ohm's Law: E = I x R; where E = voltage, I = current and R = resistance.

Thus, if any of these two factors are known, the third can be calculated. For instance, if the current were 10 and resistance was 11, then voltage would be 110. Simply put, voltage is the pressure in the electrical system that causes the current to flow. The current is the amount of electricity flowing in the system and is measured in amperes, usually referred to as "amps." The resistance, as the word says, is the resistance to the electrical current flow and is measured in ohms.

Power is the energy expended or work done per second by an unvarying current of one amp under pressure of one volt and is calculated as: P = I x E.

Power is measured in watts, and is equal to the product of voltage and current (W = I x E). It is normally referred to as kilowatts (kW). One kW is equal to 1000 watts or 3413 Btu. Electrical energy is measured in watt-hours, or kilowatt-hours, where one kWh equals 100 Wh. One watt-hour is the energy expended when one watt is used for one hour. Energy consumption in an induction furnace is measured by a watt-hour meter. An experienced furnace operator can predict the status of the melt by observing the watt-hour meter. Experience has shown that in a 60-Hz coreless furnace, it takes 19 kWh to raise the bath temperature of one ton of iron 100°F (56°C).

One more important thing to remember is that not all of the power in an electric circuit is used. This is due to the reactive power in the electric circuit. [Picture a glass of beer. The total contents of the glass would be the *apparent power;* the foam at the top would be the *reactive power;* and the beer would be the *real power.]*

In addition, when operating an induction furnace, one should try to get as close to a power factor of one as possible. Power Factor or PF is found by dividing kW by kVA. kW is the power actually used and kVA is the power actually entering the circuit. The power factor can indicate when one is getting the most out of the electrical energy being paid for.

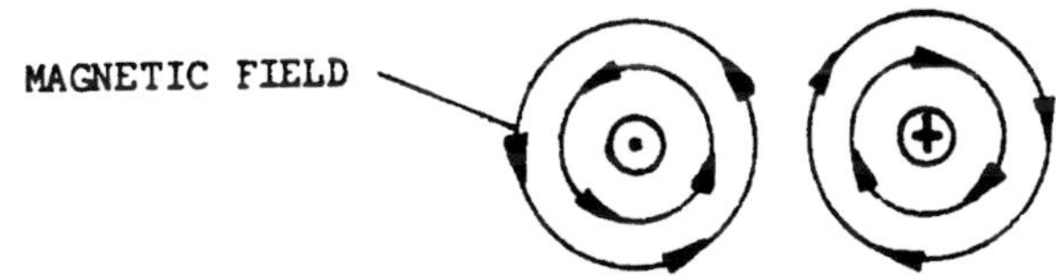

Fig. 10-16. Section of magnetic field.
Dot in middle of left circle = current flowing toward you;
+ in middle of right circle = current flowing away from you.

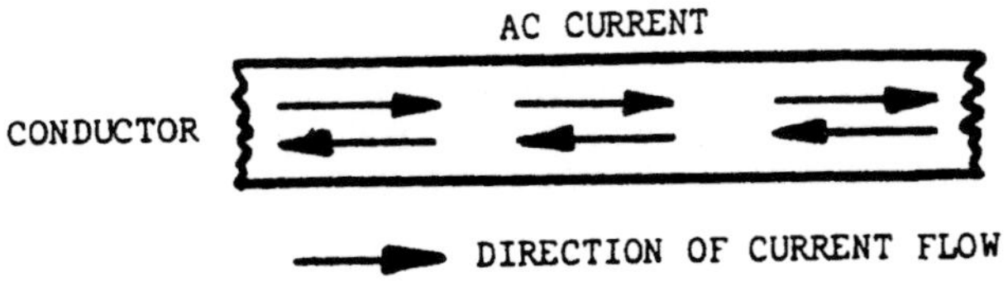

Fig. 10-17. Schematic of alternating current.

The density of the alloy to be melted and the size of the furnace determine the frequency of the induction furnace. Dense alloys, such as copper-base alloys, require a low- to medium-frequency furnace. On the other hand, light alloys, such as aluminum alloys, require a medium- to high-frequency furnace. Today, many foundries, regardless of alloys melted, use a medium-frequency furnace.

There are two types of induction furnaces used for melting metal alloys: channel (or core) and coreless.

Channel Induction Furnace

The channel-type induction furnace will be discussed first. **Figure 10-18** shows a vertical bath (or shell) channel furnace and **Fig. 10-19** shows a cutaway view of the same style channel furnace.

The channel furnace can be described as a large step-down transformer with a shorted secondary winding. The furnace coil is the primary winding to which a voltage is applied. The channel itself, the path through which molten metal flows, is the secondary

Fig. 10-18. Vertical bath (or shell) channel furnace.

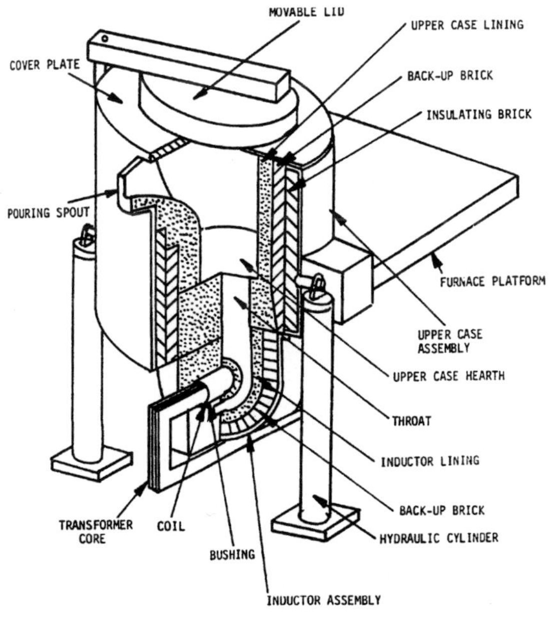

Fig. 10-19. Cutaway view of the vertical bath furnace shown in Fig. 10-18.

winding and has one turn. On the channel furnace, this transformer is called an inductor. There are two ways this inductor can be designed.

The first design is that of a single-loop inductor with a single coil and loop wrapped around a single iron core. This is illustrated in **Fig. 10-20**. This type of inductor is the least bulky of the two types and the easiest to line with refractories. The loop must be kept larger, however, to keep current densities down.

The second type is the twin-loop inductor, as seen in **Fig. 10-21.** This type of inductor uses two loops of molten metal wrapped around two coils, using either one or two cores. Twin-loop inductors generally have the greatest bulk. The twin-loop inductor, since it is two single loop inductors, can have greater power input.

There are two types of twin loop inductors, the "D" inductor and the "T" inductor. The D inductor uses two coils that are generally wired in series wrapped around one core, and both coils operate from one control. The "T" inductor uses two coils and two cores. Each coil is operated independently and may be considered as two completely independent inductors.

Regardless of the type of inductor used, it must be cooled. Cooling can be done by either water or air. In the case of air cooling, blowers are used to accomplish the right amount of airflow to keep the inductor cool. Most water-cooling systems are of the closed-loop recirculating type. Included in the water-cooling system is a heat exchanger for cooling the recirculating water.

The vertical channel furnace is a large bull ladle or crucible with an inductor attached to the bottom. To pour molten metal, the furnace is tilted by hydraulic cylinders and pivots on the same axis as the spout or lip of the furnace. This allows the spout to remain at the same elevation during the entire tapping operation. This type of channel furnace also uses a swinging roof, which is swung aside for charging and slagging operations.

The vertical bath (or shell) channel furnace is also well suited for cold charging or the pouring of molten metal into the furnace. Cold charging means solid charge materials. Slagging is also easily done. This furnace is often used as a holding or duplexing furnace in conjunction with other types of furnaces.

The other type of channel furnace is the drum type as seen in **Fig. 10-22.** This furnace utilizes a shell that lies horizontal to the floor. One or more inductors can be attached to the drum. In some cases, one inductor is attached to the bottom of the drum and the second is attached toward the top of the drum. Only one inductor is used at a time. Should the operating inductor fail, the drum is rotated so the inductor at the top of the drum is now at the bottom and operating. This allows the removal of the failed inductor without having to empty or shut down the furnace. **Figure 10-23** shows a cutaway view of a drum-type channel furnace.

Drum-type channel furnaces are used for holding or duplexing operations. That is, molten metal is poured or laundered into the furnace through a spout. Cold charging can be done through a charging door in the shell. Drum-type channel furnaces have no real size limitation since multiple inductors can be used at the same time. These types of channel furnaces have been installed with capacities up to 250 tons (254,000 kg) using three single-loop inductors and a power input of 3000 kW.

Fig. 10-20. Single-loop inductor.

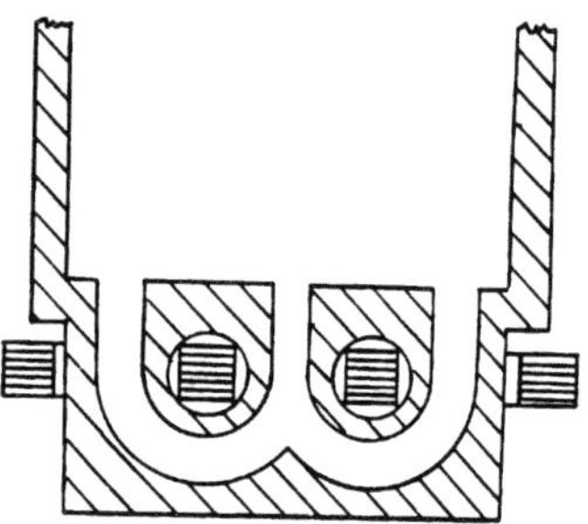

Fig. 10-21. Twin-loop inductor.

Fig. 10-22. Drum-type channel furnace.

Regardless of the type of channel furnace used, the inside of the shell is refractory lined. Vertical channel furnaces normally use a monolithic lining in both the shell and inductor; whereas the drum type of channel furnace uses a refractory brick lining, along with a monolithic lining. Refractory practice for induction furnaces, in general, will be covered later in the section on Induction Furnace Refractory Practice.

One big advantage of the channel induction furnace is that, of all the induction furnaces, it is electrically the most efficient. In other words, it offers excellent service for the amount of power used.

Depending on how the induction furnace is to be used, the channel furnace does have several drawbacks. First, as long as the furnace is being operated, there has to be molten metal in the inductors and, to some level, in the bath. If the molten metal loop in the channel is broken, the best thing to do is empty the furnace. In addition, to start up an empty channel furnace, it requires a source of molten metal to fill the inductor channel and part of the bath. The channel(s) have to be kept open and not allowed to plug up with slag or dross (impurities from the metal being melted).

Coreless Induction Furnace

Figure 10-24 shows a coreless induction furnace and **Fig. 10-25** shows a cutaway view of a coreless induction furnace. The coreless induction furnace can be described as a refractory crucible surrounded by an induction coil. An electric current flows through the coil setting up a magnetic field that passes through the refractory wall of the crucible and couples with the charge materials in the crucible.

When designing this type of induction furnace, the designer tries to achieve an optimum balance between factors such as furnace capacity, heat loss and electrical coupling between the furnace charge and the coil. If the thickness of the refractory lining is increased to prevent heat loss, then the furnace capacity and electrical efficiency are reduced. On the other hand, a reduction of refractory lining thickness can increase furnace capacity and increase electrical efficiency. This can be done at the expense of thermal efficiency, refractory life and safety margin.

The copper coil must be continuously cooled with flowing water passing through it, to prevent thermal failure of the copper of which the coil is made. The number of turns in the coil can vary from as few as five for a high-frequency furnace to 20 or more for a line-frequency furnace.

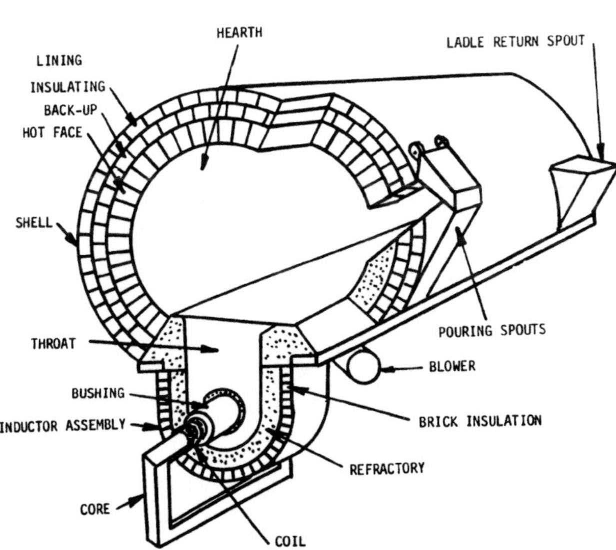

Fig. 10-23. **Cutaway view of drum-type channel furnace.**

A big advantage of the coreless induction furnace is that, once the charge materials are all liquid, a "stirring" action takes place in the bath. This stirring action is depicted in **Fig. 10-26** by the four separate steps of arrows in the bath. This stirring action gives excellent uniformity of chemistry and temperature to the bath of liquid metal. The stirring action also causes a meniscus or convex surface to appear on the top surface of the bath. For a given metal alloy and furnace design, this stirring action is directly proportional to the power induced.

Fig. 10-24. Coreless induction furnace.

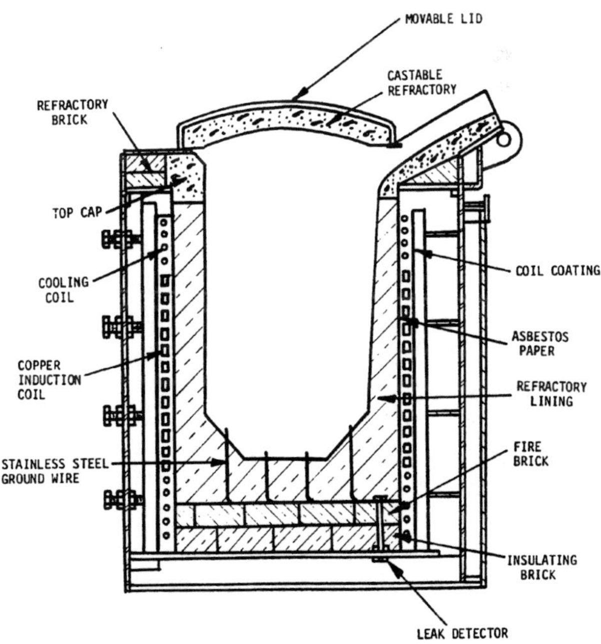

Fig. 10-25. Cutaway view of a coreless induction furnace.

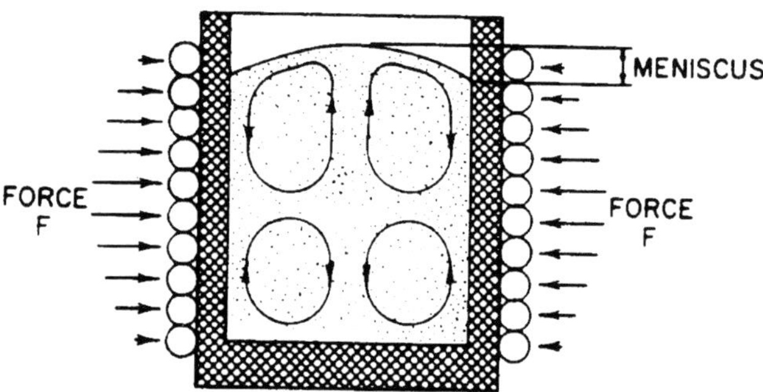

Fig. 10-26. Stirring action in an induction furnace.

Most coreless induction furnaces are equipped with hydraulic cylinders to allow the furnace to be tilted around the axis of the pouring spout. The tilting of the furnace is an important aspect of the induction melting operation. It must be smooth and easily controlled to prevent excessive shock loading of the refractory. In some cases, the furnace can be tilted backward to facilitate the slagging-off operation. In addition, most coreless induction furnaces have moveable lids that are used to reduce radiant heat loss from the bath and swing aside to allow the charging of charge materials. Hydraulic cylinders are also used for this operation.

The hydraulic systems must use nonflammable hydraulic fluid because of the close proximity of molten metal. A phosphate-ester type is normally used. Another precaution should be that all gaskets, seals and packing are compatible with the hydraulic fluid used.

A water-cooling system is necessary on a coreless induction furnace, due to the high current density in the coil and the heat loss through the crucible walls. Therefore, the cooling water is a very important part of the system.

Most cooling systems are of the closed-loop recirculating type. This is especially important with the emphasis these days on water pollution. Generally, two types of heat exchangers are used to cool the water. They are the water-to-water and/or an evaporative heat exchanger. For small coreless induction furnaces, the water-to-water type is most commonly used. For larger coreless induction furnaces, the evaporative heat exchanger is usually more economical. The application of recirculating systems offers the following advantages:

- It prevents buildup in the induction coil and the water piping.
- It allows operating with a cooling water inlet temperature above the wet bulb temperature, which assures the prevention of condensation on electrical components.
- The total water cost is considerably reduced when using an evaporative-type heat exchanger.

Except for installations where the electrical switchgear is totally enclosed and water-cooled, a ventilation system must be provided to cool the electrical equipment. Filtered air is blown into the switchgear room at a slight positive pressure, to eliminate dust contamination. The air must be thermostatically controlled to prevent freezing during the winter.

Another big advantage of the coreless induction furnace is that it can be started with a cold charge. That means it can be started with only solid charge materials in the crucible. (Remember that the *channel* induction furnace required molten metal to be started after a shut down.) This leads to another advantage: it is much easier to change the chemistry of the metal alloy being melted by simply emptying the furnace and starting it up with all new charge materials. In addition, the stirring action makes this operation even easier, without emptying the furnace.

There are also two terms normally used when using the coreless induction furnace. These terms are "batch melting" and "tap-and-charge" operations. *Batch melting* is where the furnace operator melts a batch of metal and, at the proper time, empties the entire contents of the furnace. This operation is again followed for the next batch. In the *tap-and-charge* operation, the operator taps out only a given amount, pounds or tons, of molten metal and then charges an equal amount of solid charge materials or molten metal from another source into the molten metal bath left in the furnace. Tap-and-charge operations can be carried out also with the channel furnace; however, batch melting cannot—at least to any extent.

Batch melting operation, for years, was limited to small coreless induction furnaces on a limited volume basis. Today, however, high-volume induction batch melting has swept through the foundry industry. Since there is no need to maintain a molten heel of metal in the furnace, smaller furnaces can be used, electrical energy is used more effectively, alloy changes are easier and charging safety is enhanced. Quite often, small coreless induction furnace batch melting has been replaced by high-volume batch melting operations. High-volume batch melting is a result of new technology becoming available to the foundry industry and of a deeper understanding of induction melting itself.

Lift-Coil and Push-Out Coreless Induction Furnaces

There are two other types of induction melting furnaces used for melting small batches of both ferrous and nonferrous alloys. These two furnaces are the "lift-coil" and the "push-out" coreless induction furnaces. Here, the coil is located around a separate crucible that rests on a pedestal.

In the lift-coil furnace, the crucible is filled with the charge material and the coil is dropped down, surrounding the crucible. When the charge has been melted and is at the proper temperature, the coil is lifted up off the crucible. The crucible, which is on a cart, is moved away from under the coil box. The crucible is then removed from the cart and the molten metal is poured into molds or transferred to another furnace. **Figure 10-27** shows a lift-coil induction furnace.

In the push-out furnace, the crucible is seated on a hydraulic ram. The crucible is placed on the ram and lowered into the coil. When the charge material has been melted, the ram pushes the crucible up and out of the coil. The crucible is then removed from the top of the ram, and the molten metal is poured into molds or to another type of furnace. The push-out furnace can be a single unit or a double unit. The double unit is shown in **Fig. 10-28.**

The crucibles used in these furnaces can be made of a clay-graphite mixture or silicon carbide. The open-top crucible is the type most commonly used in these two types of induction furnaces. The one big advantage of either of these two furnaces is that quantities of metal alloys, from a few pounds to several hundred pounds, depending on the density of the metal alloy, can be melted at one time. In addition, if many different metal alloys are poured, a separate crucible(s) for each alloy can be used. This use of different crucibles for different alloys can help prevent contamination of one alloy to another.

Induction Furnace Refractory Practice

Refractory practice (or lining the furnace) is one of the most important phases of using the channel and coreless induction furnaces for melting metal; therefore, considerable time will be spent discussing refractory practice.

The operation of an induction furnace places severe demands on the refractory; therefore, the refractory materials must be economical. Looking back at **Figs. 10-19, 10-23 and 10-25**, one can see the use of refractories in the channel and coreless induction furnaces. The close proximity of water-cooled coils in the coreless furnace and water-cooled inductors places even greater performance characteristics on the refractory. The refractory must also be thermally stable and possess sufficient strength for the given application. Another characteristic of the refractory is that it must be inert to the metals, slags and dross it encounters. The refractory should also have good installation and maintenance characteristics.

Fig. 10-27. Lift-coil induction furnace.

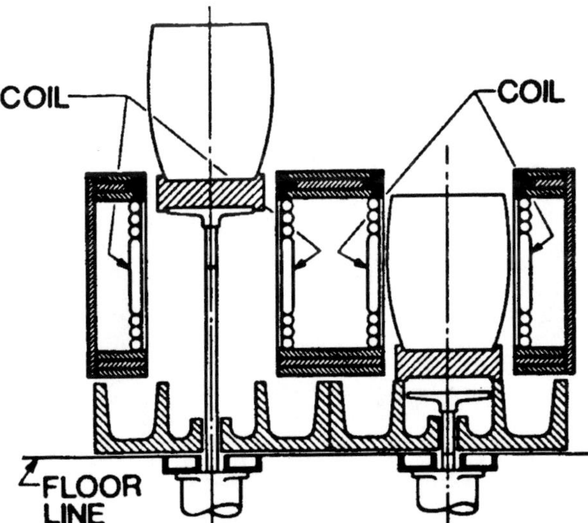

Fig. 10-28. Schematic of double pushout-type, coreless induction furnace (crucible in melting position in left-hand view).

Classifying refractories relates to the basicity or acidity of the material. Materials high in SiO_2 are considered acid, while those high in MgO or CaO are basic. Other oxides such as Al_2O_3 and Cr_2O_3 are considered neutral. In general, acid refractories are used where acid slags are encountered and basic refractories where basic slags or dross may be encountered. On the other hand, neutral refractories are generally resistant to both basic and acid slag and dross.

Another method of classifying refractory materials is in their physical form. The common types are:

- Ramming mixes
- Refractory plastics
- Patching/maintenance materials
- Refractory sealer washes
- Prefired shapes
- Gunning mixes

Refractory ramming mixes are of low moisture content (normally below 6%), are designed for installation using retaining forms, have excellent grain sizing of aggregate to allow maximum installed density, and exhibit reduced workability in hand forming operations.

Plastic refractories can be manufactured by the extrusion method because of their high clay and moisture contents. They offer a high degree of plasticity and workability. Plastic refractories normally do not need retaining forms due to their high green strength and their bonding mechanisms.

Patching and maintenance refractories are normally used to repair refractory areas that have been eroded, spalled, mechanically abused, etc. These materials can range from common fireclay or ganister to high-purity magnesia mixes. The selection of proper patching material relates to the material being patched, the slag or oxides present, and the metal alloy and temperature conditions.

Sealer washes are fine-grained refractories used to fill voids, fissures, shrinkage cracks, thermal shock cracks, etc. A sealer wash should have good bonding properties to insure adherence to the refractory.

Prefired shapes are refractory materials, such as bricks or special shapes, that are formed by pressing them into a mold or die, and then they are fired at a high temperature in a kiln to produce a ceramic bond or to activate the chemical bonding agent.

Gunning mixes are special monolithic refractories that are blown into place using a pneumatic gun, and are used for patching furnace walls and arches.

The bonding mechanism—how well the refractory aggregates stick together under melting operations—is of great importance. Chemically bonded materials normally do not require high burning temperatures to develop maximum strength. Chemically bonded materials utilize phosphate additives as the main constituent in the bonding process. They will generally display excellent strength in the intermediate temperature ranges, as compared to clay bonded materials.

Clay bonded refractories are quite prevalent in the foundry industry due to the availability of various types and grades of clay and their adaptability to bonding different types of refractory aggregates.

Channel Induction Furnace Linings

With channel induction furnaces, there is concern about both the inductor section and the furnace body section. The design of the furnace, especially the inductor, makes repairs difficult, and heavy-duty, long-life refractories are required. The most common refractories used in channel furnaces are the alumina-based materials. Some furnace operators prefer to use the 85% alumina ramming mix in the inductor section and the 65–70% alumina material in the furnace body lining. Due to the high thermal conductivity of the high-alumina refractories, it is normal practice to use insulating brick as a backup layer.

The conditions encountered in the inductor loop are unique in that the refractory is not exposed to the slag in the melt (under normal operating conditions), and it is continuously operating, making thermal shock conditions negligible. Remember that the channel induction furnace always has to have molten metal in it, unless it is shut down for repairs or other reasons. For this reason, the use of magnesia castable refractories is recommended.

An important point to consider is that, in most cases, these refractory materials exhibit less erosion in the inductor channel or loop. With alumina inductors, the erosion takes place progressively until the loop has enlarged to the point where the inductor has to be replaced.

The disadvantage of magnesia is that it is more susceptible to thermal shock than is alumina. Should the inductor loop freeze, expansion cracks will form, which will probably not close before the loop is remelted. This can lead to a runout in the inductor, which can cause major problems when the molten metal reaches the coils.

The throat area, where the molten metal leaves the channel, is often lined with magnesia castable or alumina brick or ramming mix. This is the point where the velocity in the molten metal bath is the highest, thus, a point where severe erosion can take place. In any case, it should be remembered that the refractory material must be compatible with the slag or dross and any conditioning fluxes that are used. The lining of a typical cylindrical drum-type of channel furnace is shown in **Fig. 10-29.**

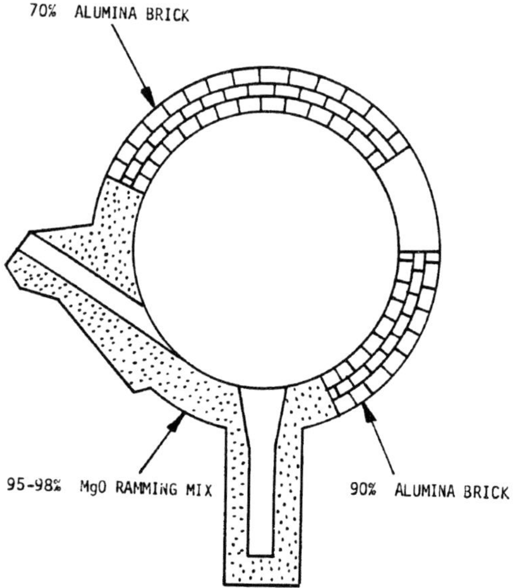

Fig. 10-29. Channel induction furnace refractory construction.

In duplexing operations, the area in contact with the slag materials is especially susceptible to erosion, and a zoned lining construction may be employed to balance overall lining life. Silica-free high-alumina refractories have shown considerable promise for this application. A large double-inductor cylindrical drum-type channel induction furnace lining is illustrated in **Fig 10-30.**

Coreless Induction Furnace Linings

The most widely used ramming mix for coreless induction furnaces is a Swedish quartzite that is over 99% pure SiO_2. Domestic silica has been used successfully but to a lesser extent. The refractory characteristics of the Swedish silica make it an ideal material to use for iron melting in the coreless induction furnace.

Silica exhibits unique thermal expansion characteristics compared to other common refractories, as illustrated in **Fig. 10-31.** This was discussed in the chapter on Molding Sands. As can be seen in the figure, silica undergoes a rapid expansion between room temperature and approximately 800F (427C) and little or no expansion above 1200F (649C). The other refractory materials, such as magnesia, chromite, alumina and zircon, exhibit expansion and contraction throughout the entire temperature range. The expansion characteristics indicate that the silica receives little or no thermal shock when operating in the normal temperature range of 2000F (1093C). The curve indicates that the temperature must be increased slowly, up to about 1200F (649C), to minimize expansion stresses. On the other hand, the silica lining should not be allowed to cool below 1500F (815C) if maximum service life is to be maintained.

The strength of the silica refractory aggregate is developed by a sintering and fritting (heat bonding) action caused by the boric acid added to react with the silica fines. The amount of boric acid used is dependent on the characteristics of the silica and the operating temperature.

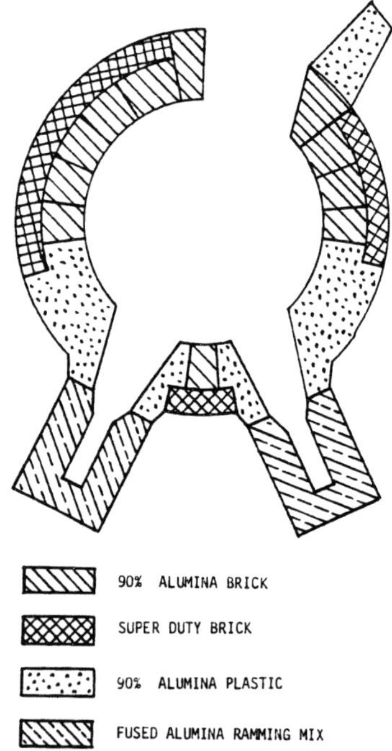

Fig. 10-30. Large channel induction furnace.

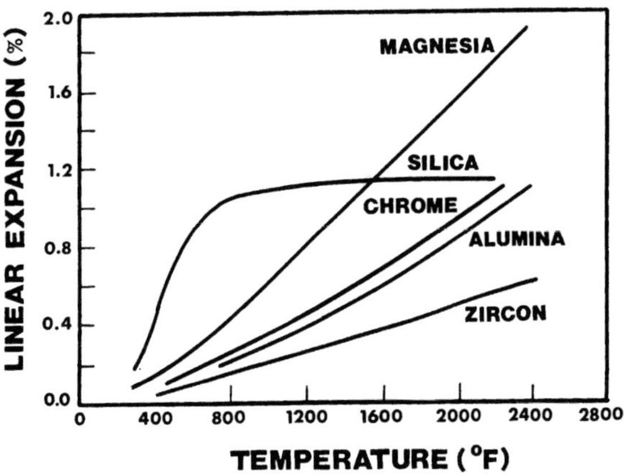

Fig. 10-31. Linear expansion of various refractories.

The silica lining will form three distinct zones upon sintering, due to the variation in temperature from the hot to the cold faces of the lining. The three zones are illustrated in **Fig. 10-32.**

1. *Sintered Zone.* This zone is subjected to the highest temperature, and maximum transformation occurs. This zone has the highest strength and lowest porosity due to the substantial amount of glass phase formed by reaction of the silica fines with the boric acid.
2. *Fritted Zone.* This zone is exposed to lower temperatures and is not in direct contact with the molten metal. Less transformation occurs and the zone exhibits more porosity. When wear of the sintered zone occurs, the fritted zone is subjected to more heat and is transformed into the sintered zone.
3. *Quartzite Zone.* This untransformed zone is a relatively loose granular mass that allows for vertical expansion without stressing the transformed mass.

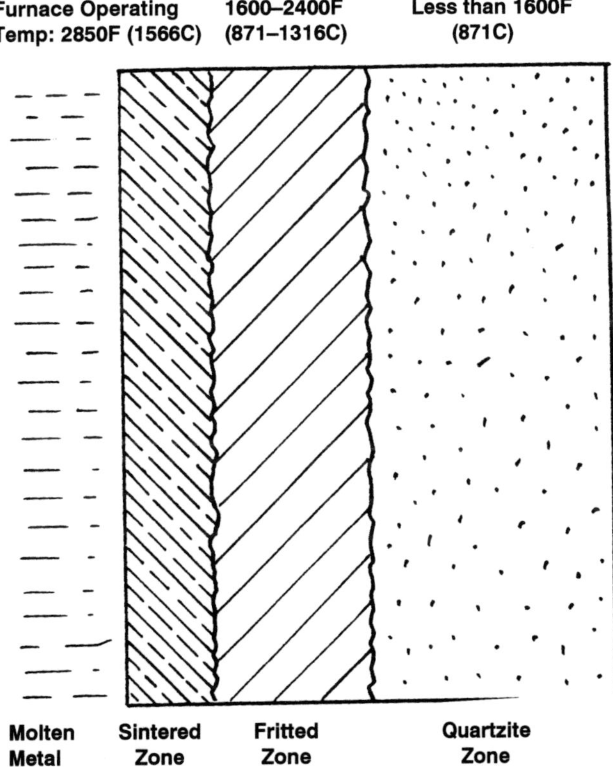

Fig. 10-32. Three-zone lining.

The operating temperatures, slag conditions and other variables will greatly affect the wear characteristics of the refractory lining. A desirable wear pattern exhibits uniform wear of the crucible sides and bottoms. Nonuniform wear characteristics indicate improper operation of the furnace.

The installation of a lining requires careful attention to insure the life of the lining. In other words, poor installation practice will lead to poor lining life, and the opposite is true for good installation practice. Although the boric acid can be mixed at the time of installing the lining, today, premixed silica blends are becoming popular. The ramming or compaction of the refractory material can be done by mechanical, electrical or pneumatic vibration techniques.

The sintering of the freshly installed lining can be done through induction heating or a combination of fuel firing and molten metal additions. The new refractory material is held in place using a metal form. This form is left in the furnace after the ramming of the lining is completed. The power is turned on in the furnace and the form heats up and finally melts. This action sinters the lining.

Regardless of the method of sintering used, the refractory should be slowly heated up to 1500F (816C).

The operating cycle of the furnace should be such that the lining operates at the lowest average temperature. In addition, the lining should not be exposed to long cooling periods below 1500F (816C). Inspection and maintenance procedures should be scheduled on a regular basis.

Other refractory systems are used in coreless induction furnaces. For instance, high alumina castable refractories are used for maintenance and patching materials, particularly with the rammed silica linings. Both castables and plastics in the 90% plus alumina range are used frequently in the upper cap and pouring spout areas of the coreless furnaces. These refractories provide resistance to iron oxide splash and thermal shock, and are easy to maintain. Induction furnace linings of 95% alumina have proven to work well in furnace lid linings as they provide resistance to mechanical abuse, thermal shock and oxide splash.

When large, high-powered coreless induction furnaces are used, the rammed silica linings may not be adequate. Such factors as greater metal throughput, increased stirring action and washing action of metal and slag, and more frequent tapping and charging can all lead to lining deterioration. When these conditions exist, heavy-duty brick linings are used.

This type of lining is popular among large furnace operations melting the high-alloy and specialty steels, particularly with vacuum melting. However, this particular lining practice can also be related to the cast irons and low-alloy steels.

The preferred lining design comprises double-brick construction, with all joints staggered between the working and safety or back-up courses. High-quality alumina bricks are generally used, along with 85% alumina mortars.

Electric Resistance Furnaces

Another type of furnace that uses electrical energy as the source of heat to melt or hold molten metal is the electric resistance furnace (these furnaces work like a household electric toaster). They can be used for melting; however, their main use is for holding molten metal, and used mainly with nonferrous alloys. Standard sizes range from 1000 to 4000 lb (453 to 1812 kg). **Figure 10-33** is a sketch of an electric resistance furnace.

The lining of this type of furnace is made of lightweight insulating brick, which is placed in front of insulation slabs. Rugged nickel-chrome elements are mounted on the sidewalls of the furnace lining. The heat created by the elements is transferred by radiation and conducted through a crucible into the molten metal. The molten metal can be tilt-poured from the furnace or ladled out of a crucible.

The furnace shown in **Fig. 10-34** is a pressurized electric resistance holding furnace used in the low-pressure permanent mold casting process discussed in Chapter 2. The mold is placed over the sealed furnace cover, and molten metal is forced through the tube, into the mold, by air pressure on the surface of the bath.

One of the beauties of this type of furnace is that there is very little, if any, movement in the molten metal bath. This is very advantageous for melting those nonferrous alloys that are prone to gas pickup and dross formation.

Electric Globar Furnaces

This type of furnace also uses electrical energy to produce heat. In this furnace, a silicon-carbide rod is held in place over the top of the molten metal bath. There is a resistance to the electric current as it is passed through the rod. This generates heat, which is radiated from the surface of the rod, heats the exposed refractory surfaces in the furnace and is radiated onto the surface of the molten metal. The heat radiated from the refractories also heats the molten metal.

These types of furnaces are normally used for holding molten metal. They have been used in large automotive foundries pouring gray iron, where the furnace is placed alongside the pouring lines. Molten metal is transferred from the primary melters or other holding furnaces to the Globar furnaces and then tapped into pouring ladles.

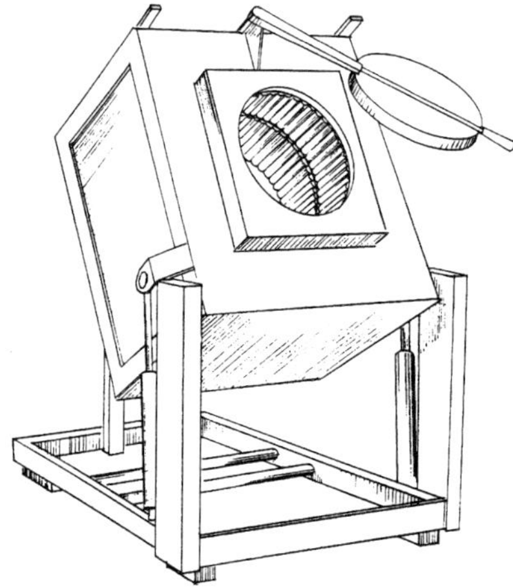

Fig. 10-33. Electrical resistance furnace.

Reverberatory Furnaces

Reverberatory furnaces are sometimes referred to as open-hearth furnaces. Using the latter name can lead one to believe that they are similar to the open-hearth furnaces that were used for melting the carbon steels; this is not so. Reverberatory furnaces find their greatest use in the aluminum foundries, especially foundries requiring large quantities of molten aluminum. This includes foundries such as permanent mold and diecasting plants.

Reverberatory furnaces may be of many different designs and have capacities ranging in size from 2,000 to over 200,000 lb (900–90,000 kg) capacity. These direct-fired furnaces are usually stationary, but there are some models that tilt. They may have one or two chambers, use gas or electricity for the source of heat, and be charged in several different ways.

Today's aluminum reverberatory furnace has the potential of many years of service life, provided proper operating practices are followed. For lining the bath area, refractory manufacturers have developed a brick that is high in alumina and specially treated for high resistance to wetting by and absorption of aluminum. Any refractory will last much longer if not exposed to thermal shock and severe temperature changes.

Experience has shown that optimum furnace lining life is obtained from furnaces that:

- always contain a substantial heel (bath of molten aluminum) close to capacity loading;
- are maintained at a temperature close to the operating point;
- are protected from rapid chilling caused by frequent opening of large charge doors;
- are not subjected to extreme firing rates.

Direct-Fired Reverberatory Furnace

With the increased demand for aluminum castings, this furnace has gained in popularity. **Figure 10-35** shows a front-charging, one-chamber direct-fired reverb (short for reverberatory) furnace. The heat energy comes from a direct-fired gas flame(s). Melting in this furnace is accomplished by the flame(s) making direct contact with the solid and molten metal.

Because conditions in this furnace can cause large amounts of dross and a high rate of gas absorption into the molten aluminum,

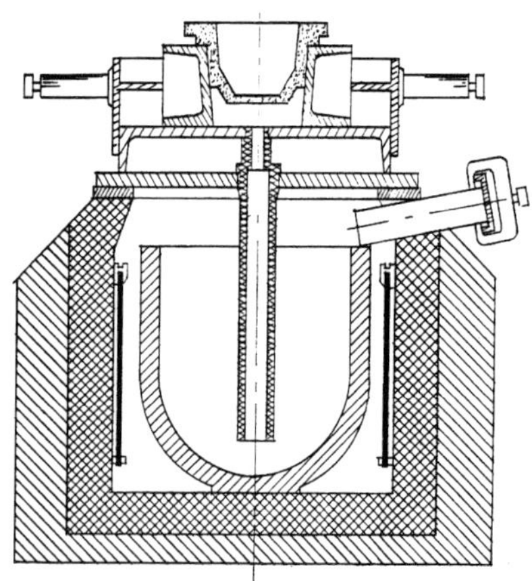

Fig. 10-34. Electrical resistance furnace for low-pressure processes.

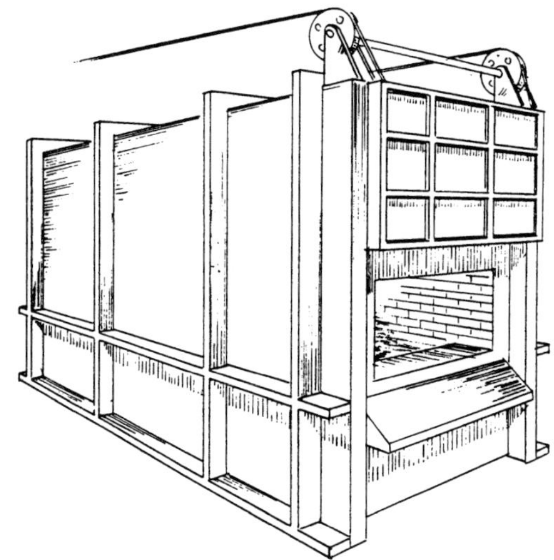

Fig. 10-35. Front-charging, single-chamber reverberatory furnace.

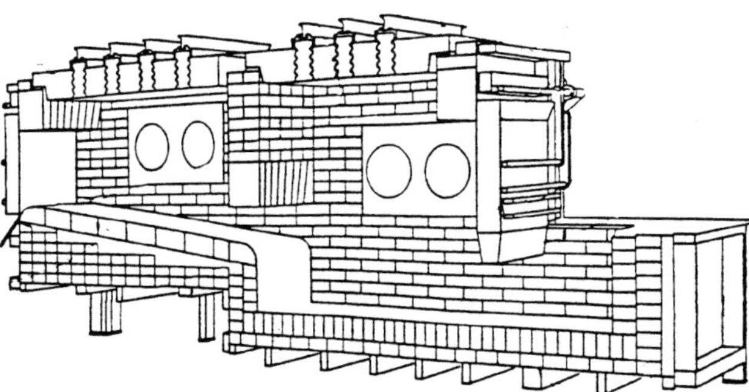

Fig. 10-36. Sloping dry-hearth reverberatory furnace.
114-mm thick, 85% high-alumina firebrick throughout molten metal contact area. 64-mm thick insulating firebrick that withstands 1095C used on holding well floor; 114-mm thick used on vertical walls.

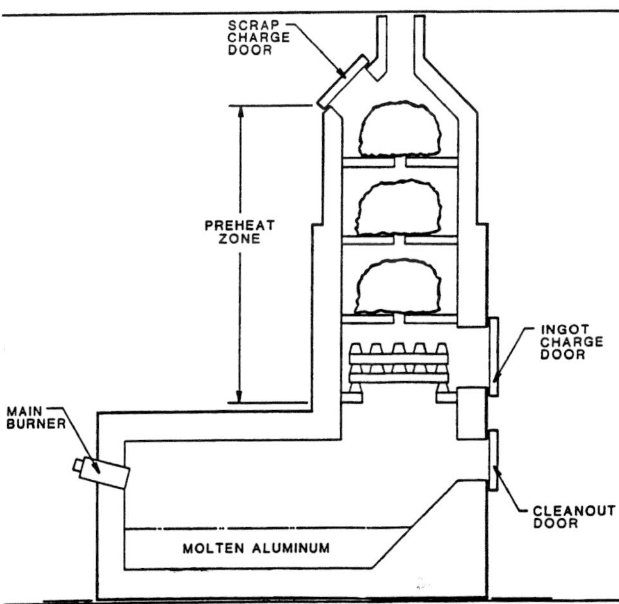

Fig. 10-37. Side-charging stack melting furnace. [Drawing courtesy of K. Broeckenhauer, Seco/Warwick Corp.]

proper operating procedures should be followed. Direct contact of the gas flame on the molten metal, or even close to it, can cause turbulence that would promote gas absorption and the formation of dross. The controlling of charging procedures and combustion furnace atmospheres can protect against these hazards.

Sloping Dry-Hearth Reverberatory Furnace

The sloping dry-hearth reverb furnace is seen in the cutaway view shown in **Fig. 10-36.** The scrap and ingot is placed on the sloping hearth and, as it melts, it runs down the hearth into the bath in the secondary chamber. As seen in **Fig. 10-36**, there is a section of the front of the furnace where the molten metal is exposed to the atmosphere outside of the furnace. This section of the furnace is referred to as the well, and it is here that the molten metal is removed by ladling.

The melt loss in this reverb, compared to other reverbs, is one of the highest. Melt loss is calculated by weighing the amount of metal charged into the furnace versus the molten metal actually removed from the furnace. The loss can range from 2% to 12% for every pound of aluminum charged. This type of reverb also has a very poor fuel efficiency rating.

One more thing to keep in mind is that, when natural gas is burned, one of the by-products of combustion is water. These products of combustion will pass over the charge materials as they work their way down the sloping hearth. Aluminum has an affinity to absorb hydrogen from the moisture in the combustion products, which can contact the charge materials. Proper maintenance of the burners and the gas-to-air ratio burned is important.

To help reduce melt loss, flame impingement should be minimized, and the designated melting rate of the furnace should not be exceeded. The best way to reduce melt loss in this type of furnace is to minimize the amount of light bulky scrap.

Stack-Melting Reverberatory Furnace

The stack or tower-type of reverberatory furnace is a high-volume, high-efficiency melter. These furnaces have melting capacities of 1000–4000 lb/hr (450–810 kg/hr). An improved efficiency of 40–60% is achieved by using the flue gases to preheat the metallic charge materials before they enter the molten bath. Stack discharge temperatures are reduced, and melt losses usually average about 2–3% of the charge.

In stack melters, **Figs. 10-37 and 10-38,** the charge is loaded into the top of the stack. In the stack, the charge is heated to a mushy state as it moves down the stack before reaching the dry hearth. At this point, final melting occurs, and the molten metal flows into the holding bath.

Figure 10-37 shows that the burner is located at the end of the furnace, farthest from the stack. More modern designs of this type of furnace are called "jet" and have the burner located at the base of the stack, as seen in **Fig. 10-38**. In one design, a second burner is located on the side of the furnace over the bath of metal.

The molten metal is tapped from the furnace through a spout located on the side of the furnace, or from the end of the furnace opposite the burner. These furnaces are found in large high-production aluminum foundries.

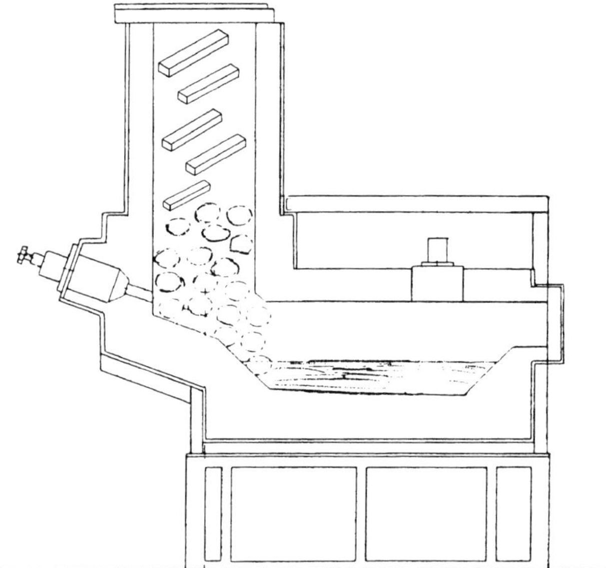

Fig. 10-38. Tower-type "jet melter" furnace.

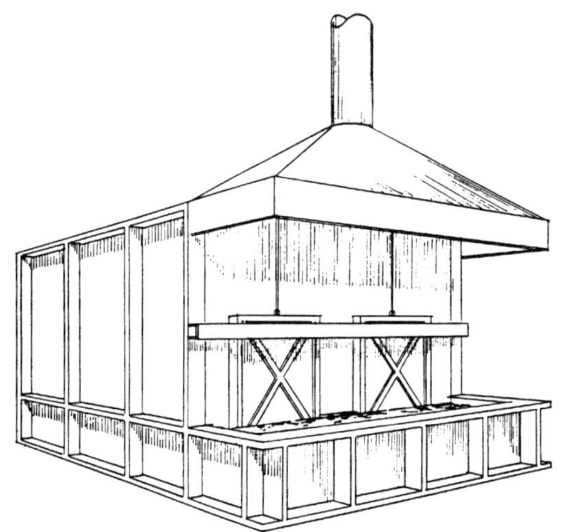

Fig. 10-39. Direct-fired, well-charged reverberatory furnace.

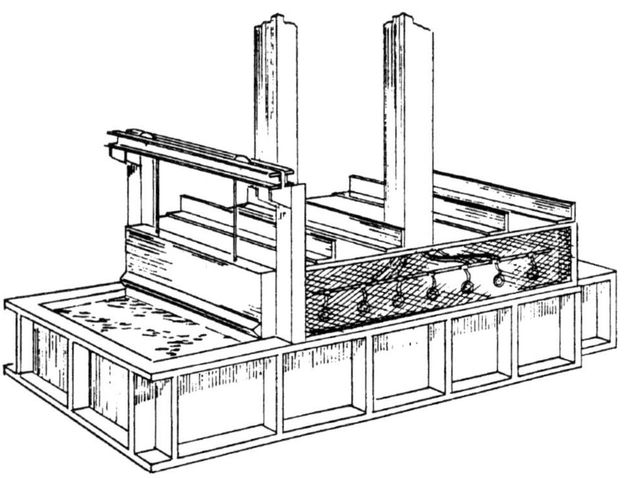

Fig. 10-40. Single-chamber electric reverberatory furnace with well.

Wet-Bath Reverberatory Furnace

The wet-bath reverberatory furnace **(Fig. 10-39)** has become one of the most-used systems for melting aluminum. It essentially consists of an exterior charging well that is separated from the main bath and heat chamber by a submerged refractory arch. It may have a ladling well on the opposite end of the charging well or a bottom spout used to tap molten aluminum from the furnace. Recent designs have greatly reduced the molten metal-to-furnace roof distance, and the furnaces now use roof mounted gas-radiant burners. These innovations have increased fuel efficiency while appreciably reducing oxide and metal loss problems. An essential safety precaution when operating this type of furnace is that all charge materials must be dry before being placed in the charging well.

The wet-bath reverberatory furnace is not normally used for batch charging. This furnace is best run on a prescribed continuous charging program with the correct volume of charge per hour. This furnace has relatively low melt loss rate when compared to other gas-fired furnaces. Based on cold charging, the melt loss per pound of aluminum charged is from 3–5% in normal operation.

Electric Radiant Reverberatory Furnace

The electric radiant reverb furnace is similar to the oil- or gas-fired reverb furnaces (See Electric Globar Furnaces, mentioned earlier). It has a remote charging well, separated by a submerged refractory arch (or door) from an isolated heat chamber incorporating silicon carbide heating elements over the molten bath. **Figure 10-40** shows this type of reverberatory furnace.

Based on a cold charging operation, this furnace has a melting efficiency of about 70–75%. The furnace can melt from a cold start, which means that it does not need a heel of molten metal in it when starting up after draining the furnace. The melt loss is approximately 1% per 1 lb (0.45 kg) of aluminum melted. Because there is no combustion process, no flues are needed, and the thermal head runs much lower than it does in a fossil fuel-fired furnace.

In the heat chamber, the metal is tranquil with no agitation to cause dross formation or encourage gas pickup. The short distance between the bath and radiant rods provides a greater amount of heat transfer of Btu into the bath. There is less furnace cleaning required and there is no noise or excess heat load.

Front-Charging Reverberatory Furnace

Smaller reverb furnaces with one or more dipping wells (wells from which the molten aluminum is ladled) are frequently used in preparing foundry melts. Reverberatory furnaces that have a dipping well(s) have a charging door on the opposite end of the furnace. At this time, the charging, fluxing, degassing and skimming of dross are done before ladling and pouring the melt directly into molds. Alloyed aluminum ingots and scrap are charged into the furnace more or less continuously.

The single-well, dip-out type of furnace is one of the most used for aluminum melting. This type of furnace is seen in **Fig. 10-41.** Many large aluminum foundries use several of these furnaces as holding furnaces for die and permanent mold casting operations. The molten aluminum for these furnaces is usually supplied to them from larger reverb furnaces. When these units are used as holding furnaces, the burner equipment need only be sufficient to handle the normal heat loss of the furnace itself.

Dual-Energy Reverberatory Furnace

Unpredictable energy conditions led to the development of a reverb furnace that could be the ultimate in melting flexibility. **Figure 10-42** shows a sketch of a dual-energy reverb furnace. This type of furnace can melt with fossil fuels and electricity in the same chassis.

When operating in the electric mode, the furnace consists of an upper chassis with electric radiant elements, plugged flue and plugged burner ports. Should the electric power be interrupted, the furnace can then be switched over to using the fossil fuel burner(s). Depending on the size of the furnace, this changeover can take from one to four hours, during which the molten aluminum stays liquid. This is a rather new concept and appears to have promising possibilities.

Regenerative and Recuperative Burner Systems

Heat reclamation systems improve the efficiency of reverberatory furnaces by preheating combustion air, using heat from the flue gases. There are two basic types of heat exchanger systems for waste heat recovery: recuperators and regenerators.

Recuperators are heat exchangers that separate the flue gas from the combustion air by a ceramic or metallic wall through which the heat transfer occurs. The combustion air is preheated as it passes through the recuperator on its way to the burner(s).

Regenerators have two completely separate chambers that contain a media (material) that can store heat as the flue gases pass through it. The stored heat is then used to heat combustion air as it is allowed through during alternate combustion/exhaust cycles. When two burner/recuperator chambers are fired alternately, a constant heat source is obtained.

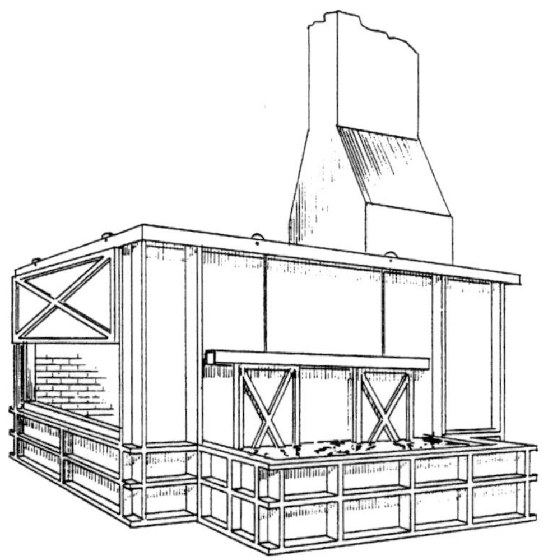

Fig. 10-41. Single-chamber gas-fired reverberatory furnace with well.

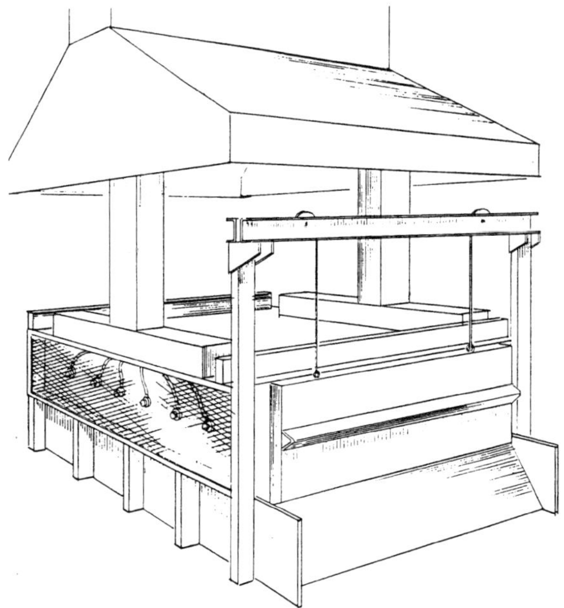

Fig. 10-42. Dual-energy reverberatory furnace.

The next two chapters will discuss the actual melting practices for both ferrous and nonferrous alloys using the different types of melting equipment discussed in this chapter.

BIBLIOGRAPHY

Aluminum Casting Technology, 2nd Edition, 1993, American Foundrymen's Society, Des Plaines, IL.

Casting Copper-Base Alloys, 1984, American Foundrymen's' Society, Des Plaines, IL.

Cupola Handbook, Fifth Edition, 1984, American Foundrymen's Society, Des Plaines, IL.

Electric Melting Technology, Vol 1, Induction Melting, 1973, Cast Metals Institute, Des Plaines, IL.

Electric Melting Technology, Vol II, Arc Melting of Gray & Ductile Iron, 1973, Cast Metals Institute, Des Plaines, IL.

Electric Melting Technology, Vol III, The Arc Melting of Cast Steel, 1973, Cast Metals Institute, Des Plaines, IL.

Heine, H.J., "Plasma-Fired Cupola Promises Melting Economics, Foundry Management and Technology," March 1986, Cleveland, OH.

Heine, R.W., Loper, C.R., Rosenthal, P.C., *Principles of Metal Casting,* 2nd Ed, McGraw-Hill, New York, 1968.

Mortimer, J.H., Batch Induction Melting: A Success Story, *Foundry Management and Technology,* August 1994, Cleveland, OH.

Schleg, F.P., Class notes from CMI Courses, 1965-92.

Steel Casting Metallurgy, 1984, Steel Founders' Society of America, Des Plaines, IL.

Ferrous Melting Practice

11

The previous chapter discussed the various types of furnaces used by foundries to melt the alloys. This chapter will discuss how some of these furnaces are used to melt the ferrous alloys used to produce castings.

The ferrous alloys to be discussed include the family of cast irons, carbon steels and the high-alloy or specialty steels. The two types of furnaces discussed will be the cupola and the electric furnace. At the end of this chapter, the text will cover the testing and analysis instruments and methods used for chemistry control, chemical analysis and temperature measurement.

CUPOLA MELTING OF FERROUS ALLOYS

The cupola is the oldest type of furnace. This furnace uses foundry coke as part of the charge materials and it remains in use today to melt cast iron. It is the blast air reacting with the carbon in the coke that generates the heat necessary to melt the metallic charge materials.

In cupola melting, the basic charge consists of three parts: coke, metallics and flux. The flux is added to the charge to remove sand, ash and dirt in the form of slag. (Flux was explained in Chapter 10 and slag removal will be discussed later in this chapter.)

Today, the majority of cupolas used are found in large, high-production cast iron foundries. They are usually used in the continuous-tap mode and are being continuously charged with the charge materials. The following text takes a look at how the cupola accomplishes this daily duty.

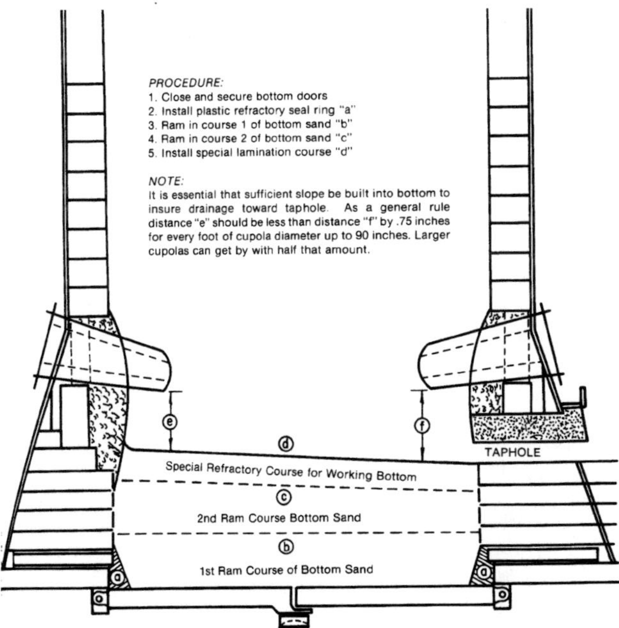

Fig. 11-1. A typical approach to cupola bottom design.

Cupola Bottom Installation

Assuming that the cupola has been emptied and the inside of the shell has been inspected, it is now ready to return to operation. The first step is to inspect and close the bottom doors and install a new cupola bottom. **Figure 11-1** shows a cupola bottom and its installation. The materials used for the cupola bottom will vary, depending on the needs of a particular cupola. **Table 11-1** lists the materials that can be used with acid practice. Many foundries install a simple bottom of 100% sand; others will build-in refractories along with the sand.

For reasons of convenience, it has been a long-standing practice in some foundries to use sand from the molding line as the cupola bottom sand. This practice will generally give satisfactory results for short campaigns. Long-campaign cupolas may install a bottom completely of high-quality refractory.

Table 11-1. Materials Used in Cupola Bottom Installation With Acid Slag Practice

Silica sand—Silicon dioxide (SiO_2) **Comments:** The traditional choice for use in cupola bottoms, where limited demands prevail.
Fireclay—Aluminum silicates **Comments:** For bonding sand grains together. A high heat-duty grade (pyrometric cone equivalent 31-32) should be used to insure good quality bottom sand.
Pitch, ground—Carbon (C) **Comments:** By-product of coke-making used to improve hot strength of molding sand.
Kyanite—Silicate of $Al_2O_3 \cdot 2SiO_2$ **Comments:** Tends to enhance the quality of bottom sand essentially due to resulting higher alumina level.
Water—H_2O **Comments:** Necessary to activate fireclay. Should be clean and uncontaminated.
High-alumina refractory mixes—Mixtures of $Al_2O_3 \cdot SiO_2$ in ranges of 45 to 90% alumina **Comments:** Lower high-alumina levels are used for low-demand situations. The alumina level can be increased to fit demands. Generally used to laminate over silica sand, but may be used for total bottom if heats of several weeks duration without dropping bottom prevail.
Cheesecloth—Cotton fabric **Comments:** Can be used directly on bottom doors to serve as a filter, to prevent loss of sand grains where excess weeping exists and causes grains of sand to be lost with possible undermining of sand bottom.
Cupola gun-patch—Clay-gannister materials **Comments:** Acceptable for cupola heats of short duration only. Except for the formulated products, these are naturally occurring deposits wherein the clay may not be of sufficiently high pyrometric cone equivalent (PCE) to insure good performance.

For cupolas that will be run for longer periods, molding sand may be totally inadequate. Quite often, the binders used in the molding sand make it unsuitable for cupola bottom use. It is highly recommended that special bottom sand be prepared for this type of cupola operation. The use of these new materials should guarantee a cupola bottom of consistent properties and ramming characteristics. Given below is a suggested formulation that has performed well for one-week campaigns.

Suggested Bottom Sand Mix

- 1600 lb (720 kg) silica sand (AFS 55–60 gfn)
- 400 lb (180 kg) kyanite (35 mesh)
- 400 lb (180 kg) fireclay
- 40 lb (18 kg) pitch
- 9 gal water

A common practice of installing several inches of a selected refractory on top of the sand bottom (to ensure against erosion of the bottom-working surface) is called *laminating*. The material most widely used for laminating is silica-alumina in ramming-mix form.

The actual bottom installation should not begin until reconditioning of the cupola interior has been completed and all of the tools and materials for installing the bottom have been assembled. Should it be necessary to mix the bottom sand in advance, it should be placed in a container that can be covered, to retain moisture. Contamination of the bottom sand is not to be tolerated.

Following are the main considerations in cupola bottom installation.

1. Bottom doors should be completely closed and support members should be installed so the doors will remain tight and rigid.
2. A plastic refractory seal should be installed in the area designated "a" (small triangular area on either side of area "b") in **Fig. 11-1**. This is especially important if the doors fit poorly.
3. One-half of the total bottom sand should be placed into the cupola and rammed into place. This is important because it is difficult to achieve proper ramming throughout a thick mass.
4. The balance of the bottom sand should then be placed into the cupola and rammed into place. If a lamination practice is used, such as shown in **Fig. 11-1** (area "d"), the second course must be finished to an elevation some 3–3.5 in. (7.5–8.75 cm) lower than the final bottom elevation, to allow room for the lamination or working course.
5. The desired material for the final course of the cupola bottom should next be deposited into the cupola and rammed into place. If the bottom is too low, add more bottom sand and ram it into place. If the bottom is too high, remove the excess bottom sand and ram again. There is no compromise in this matter. This final course must match up with the bottom of the tap hole on the front side.

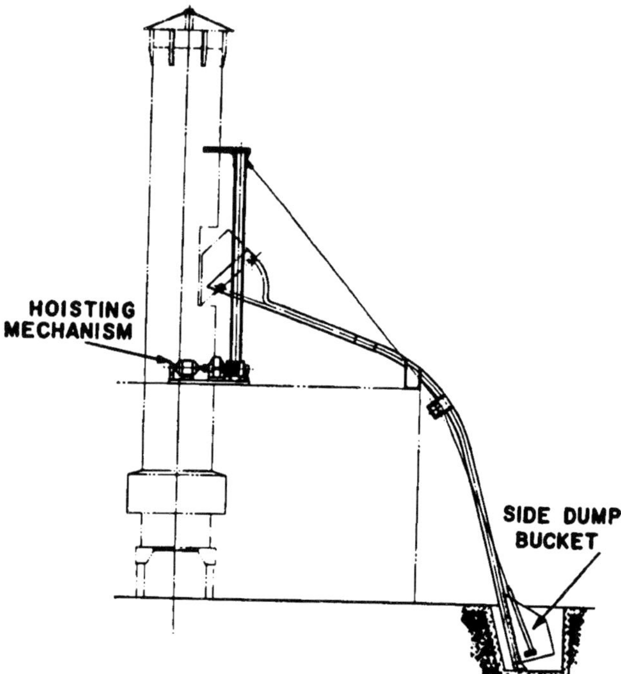

Fig. 11-2. Vertical skip-hoist with side-dump skip charge bucket.

After the bottom has been installed, bed coke can usually be dropped onto an unprotected sand bottom. However, if slab wood is used to achieve coke ignition, extreme care must be used to position the slabs, flat on the sand bottom. If this isn't done, there is a chance that a piece of wood may be driven into the sand bottom by the falling coke.

The proper installation of a cupola bottom is a demanding operation. A plan of approach is critical to any installation where performance is important to a successful operation. Cupola bottom problems can be averted if excellent engineering design, high-quality materials and superior workmanship go into every cupola bottom.

Charging and Feeder Charging Mechanisms

Once the cupola bottom has been prepared, the charge materials are placed into the cupola, the first material will be coke. In the previous chapter, it was learned that the charge door in the cupola is located toward the top of the stack. There are various ways that the charge buckets, containing the charge materials, are raised to the charge door and placed into the cupola.

Charge Buckets

Figure 11-2 shows a drawing of one method of bringing the charge materials up to the charge door. This is called a *vertical skip hoist* and, in this particular case, a *side-dump skip charge bucket* is being used. Another method is to have a straight vertical lifting mechanism, similar to a crane, that lifts the charge bucket up to the charge door where it is inserted into the cupola and emptied.

There are different types of charge buckets, and these same charge buckets can be used for charging induction and direct arc furnaces. The charge bucket's primary job is to distribute the charge materials evenly across the top of the charge materials already in the cupola.

Figure 11-3 is a sketch of *side-dump skip charge bucket*. The advantages of this style of charge bucket are:

1) it can do a fair job of distributing the charge in small cupolas;
2) larger-sized scrap can be charged due to the open top;
3) the vertical or inclined skip using the side-dump bucket is less expensive than other types of chargers.

Disadvantages of the side-dump skip bucket are:

1) coke and charge materials will be thrown against the far side of the cupola, causing the charge to pile higher on one side. This results in blast channeling and impact of the charge against the lining (if the lining is up to that level in the cupola), which tends to be destructive;
2) small-sized materials, like limestone, segregatc and tend to drop off the discharge end of the bucket and concentrate on the near side of the cupola. This segregation creates excessive fluxing of the lining and increases the tendency of bridging at the tuyeres on the far side.

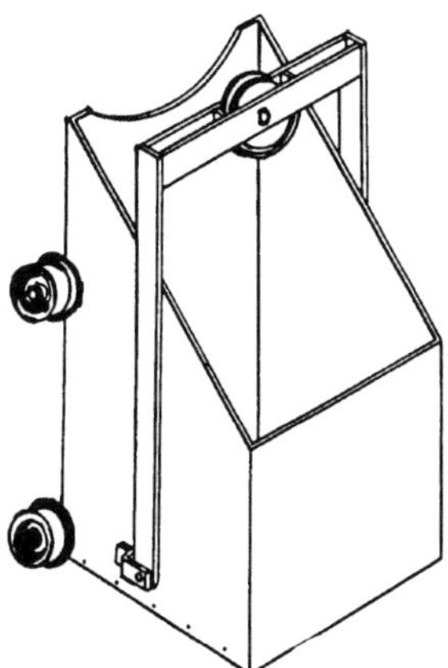

Fig. 11-3. Side-dump skip charge bucket.

The *double-leaf hinged-bottom buckets* (**Fig. 11-4**) are widely used for charging large-sized metal scrap, which makes it necessary for the bucket to have a full open bottom.

Advantages of the double-leaf hinged-bottom bucket are:

1) there is full volume capacity for any given size bucket;
2) manual labor is not required to hook or unhook, as is required with another type of bucket;
3) there is more uniform distribution of charge materials than with the side-dump bucket.

Disadvantages of the double-leaf hinged-bottom bucket are:

1) possibility of the latch being tripped, accidentally causing the full load to be dropped prematurely;
2) the entire load may suddenly be dropped from the charging door level with considerable impact, and damage may be done to the bed coke;
3) the bucket has a tendency to pack the charge materials tightly at the center of the stack and loosely around the circumference, thus permitting channeling around the cupola lining or wall.

The *cone-bottom bucket* is generally preferred over other methods of handling charge materials when the size of the material has to be controlled. **Figure 11-5** shows a cone-bottom bucket. What can't be seen is that the bottom of the bucket is cone-shaped so that when the bottom is opened the cone shape causes the charge materials to be discharged around the periphery of the bucket.

Fig. 11-4. Double-leaf hinged-bottom charging bucket in discharging position in the cupola. The bucket is suspended from the bucket carriage of the skip hoist.

Fig. 11-5. Cone bottom bucket suspended from hook on monorail trolley. Note the flange on the bucket for supporting the bucket-shell on the wishbone.

Advantages of the cone-bottom bucket are:

1) due to excellent distribution of the charge materials in the cupola, the charge is more open in the center. This provides for good blast air distribution;
2) the charging crane operator can control the stock discharging from the bucket;
3) the wishbone method of supporting the bucket shell may be used on a crane or a monorail.

Disadvantages of the cone-bottom bucket are:

1) the stem in the center of the bucket (attached to the top of the cone) may interfere with certain materials. The stem, to which the cone bottom is attached, pushes the cone bottom down and out of the bucket, allowing the charge materials to leave the bucket;
2) material must be properly sized and must often be of premium quality;
3) the diameter of the bucket must be sufficiently less than the inside diameter of the cupola at the charging door, to allow space for discharging the materials;
4) the use of this type of charge bucket requires a very large charge opening;
5) an additional charge crew employee will be required.

The small *orange-peel cone bucket* shown in **Fig. 11-6** has four leaves that pivot from the square top and are latched at the bottom.

Advantages of the orange-peel cone bucket are:

1) it can be used with skip hoist;
2) it provides good distribution.

Disadvantages of the orange-peel cone bucket are:

1) the cross bar may interfere with filling the bucket;
2) maintenance of the leaf-latching mechanism is required.

Feeder Charging

Feeder charging was the offshoot of pollution control regulations requiring small charge door openings that do not allow the use of the full cone or double-leaf charge buckets. The charge feeder receives the charge materials from a charge bucket and then, with the use of a vibratory feeder, these materials are fed into the cupola. **Figure 11-7** shows a feeder charge car receiving the charge materials from a charge bucket. **Figure 11-8** shows the feeder charge car placing the charge materials into the cupola. The feeder charge car can be pivoted so that it can be used to feed two cupolas operating side by side.

Fig. 11-6. Small orange-peel cone bucket. Note that the four leafs (normally closed) are open to show cone and latch mechanism.)

Coke Bed Fundamentals

The most important factor in successful cupola operation is the preparation of the coke bed. Any effort to economize on a phase of coke bed preparation can lead to numerous melting problems. Before actually getting into the procedure for installing a coke bed, foundry coke should be fully understood.

Foundry Coke

Many times, coke is thought of as a raw, carbon-bearing residue remaining after a batch of coal has been heated long enough to drive off most of the volatiles. Actually, coke is a highly complex, versatile finished product made to close specifications. **Figure 11-9** shows a battery of ovens producing foundry coke. Good foundry coke begins with the selection of as many as six different grades of coal. A high percentage of low-volatile, low-sulfur coals must be used to make good foundry coke.

Coke plays a dual role in the cupola: it not only acts as a melt fuel, but it also provides a source of carbon for the cast iron. Because coke serves in a chemical as well as thermal capacity, its use requires proper care on the part of the cupola operator. Coke's dual role, unfortunately, is the basic reason for most of the misunderstanding relative to its performance, and why it is frequently blamed for poor cupola performance.

The following are characteristics of foundry coke that should be looked for:

1. *Proper chemistry:* less than 0.8% sulfur
less than 8.0% ash
less than 1.0% volatiles
over 90% total carbon
2. *Uniformity of Size.* Uniformity affects blast permeability of the coke bed, gas distribution and combustion efficiency. Coke is sized to two dimensions. Coke 4x6 in. (10x15 cm) can be shorter or longer than 6 in. (15 cm). There is no practical method for sizing coke to three dimensions without excessive cost to foundries.
3. *Stability and Strength.* Sufficient resistance to abrasion and size degradation. Adequate strength to support the cupola charge weight, even in the incandescent state. Ash is necessary to coke's strength.
4. *Proper Density.* Weight-to-volume ratio affects combustibility or "reaction rate."
5. *Color.* Consistent dark gray-to-black metallic, except grayish (cauliflower) where exposure to oven heat is greatest at the wall. Discoloration does not mean inferior quality coke. Brown spots indicate the presence of iron oxide, and not inferior quality.

Companies producing foundry coke do not ship coke fines. When the coke comes off the sizing screens for shipment, it is clean. Fines in coke occur later, from abrasion during transportation and rehandling, rough handling, mechanical handling systems, power scoops, etc. Coke should be handled as little as possible. The more handling, the more chances there are to shatter the coke.

Coke Bed Preparation

No phase of cupola operation is more important than a properly prepared and burned-in coke bed. A poor coke bed produces off-spec irons, which may require several hours for the coke bed to be adjusted and the metallurgy corrected. In very severe cases, it may mean having to drop bottom and start the preparation procedure all over again. (Dropping bottom will be covered later in the chapter.)

Fig. 11-7. Feeder car receiving charge from side-discharge bucket. Car is at center position and is capable of pivoting to left or right to charge one of two cupolas.

Fig. 11-8. Feeder car charging right cupola. Alternate cupola is shown in background.

Fig. 11-9. Battery of ovens producing foundry coke.

The initial height of the bed above the tuyeres, and the degree to which the bed is burned through, are critical factors governing the metal temperature and melting rate obtained in the early stages of the melt. It is usually impossible to remedy any defects in the bed preparation in less than an hour after the start of the melt.

The coke to be used to make the bed may be hand-picked to obtain uniformly large pieces of coke. This provides greater void area in the bed and improves the penetration of air blast, and the volume of molten metal in the well. Generally, coke lump size should be between 1/10 to 1/12 of the inside diameter of the cupola. Many cupola operators feel that uniformity of size may have a greater effect than size alone on efficient cupola operation. The cupola can be operated with coke that is less than optimum size, as long as the coke is uniform in size.

The charging bucket is loaded with the proper amount of hand-selected coke. The coke is dumped through the charging door into the cupola and onto the cupola bottom. The next two steps in coke bed preparation are ignition and burning-in. Some methods of igniting the coke bed are 1) burners or torches, 2) hot blast, 3) externally ignited coke and 4) kindling and oily material.

Ignition and Burning-In—In conformance with the Clean Air Act of the 1970s, certain methods of igniting the coke bed are not recommended. The use of kindling, paper or oily materials may be restricted by this act. Torches or burners can be used with either natural gas or oil. Some torches are also designed to use oxygen along with the fuel, increasing fuel efficiency.

The torches must have enough pressure to ensure penetration into the center of the coke bed. These torches can be placed into the cupola through special holes called "ignition ports" in the sides of the cupola shell. These ignition ports are several inches above the cupola bottom. The torches can be inserted through the tuyeres themselves. If a single burner or torch is used, it should be moved around the shell so that there is even ignition of the coke bed. Tuyere covers can be opened to ensure adequate burning of the additional coke that is added, after the coke bed is burning, to achieve the proper coke bed height.

If the cupola is equipped with a hot blast system capable of blast air temperatures of 800–1000F (475–540C), the use of torches or burners is unnecessary. The preheated blast air entering the cupola at normal blast rates is sufficient to ignite the bed. (Blast Air Blowers and Controls will be covered shortly.) When burning-in with hot blast, the tap hole may be left open to ensure ignition of the coke bed below the tuyere level. With the hot blast temperatures of 800–1000F (427–538C), ignition and burning-in times are much shorter than with other conventional methods.

Some foundries use a procedure where the first portion of the bed coke is ignited outside of the cupola in a special grate-bottomed bucket. When the coke in the bucket is fully ignited, it is dumped into the cupola through the charge door. This method does relieve some of the first-ash buildup on the bottom, and certainly can be observed for complete ignition.

After the first portion of the coke bed is thoroughly and uniformly ignited, coke is added until the top of the bed is about 4–6 in. (10–15 cm) lower than the required height. The tuyere covers are then closed and ignition of the coke bed is completed by gently blowing air through the tuyeres for a few minutes. The coke bed can be visually checked from the top of the cupola, and the burning coke bed can be checked by looking through the tuyere sight glasses and checking the color of the burning coke.

Coke Bed Height—When the fan has been turned off, the tuyere covers are opened and the coke bed can be poked through the tuyeres to close and consolidate any possible voids. This operation may often reduce the coke bed height by 4–8 in. (10–20 cm). The bed height is then measured with the bed-measuring gauge seen in **Fig. 11-10**. Additional coke is then added to bring the coke bed to its predetermined height.

During the heat, there are three main items to be considered in maintaining a balance in the cupola: coke, metallurgy (for charge makeup) and air. It is generally accepted that the cupola should be operated at a constant air volume or weight. With all three items in balance, there should be good control of temperature, melt rate and

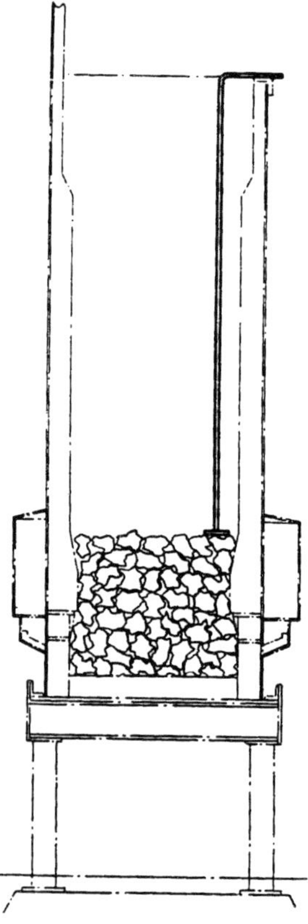

Fig. 11-10. Gauge for measuring height of cupola bed.

metal chemistry. Unless there is an obvious cause, such as a cupola shutdown, cold iron is nearly always the result of the coke bed height and incoming air becoming unbalanced.

There are several ways to correct an imbalance between the air and coke bed height. These procedures are fully covered in the AFS publication, *Cupola Handbook*.

Coke Bed Performance

In order to maintain proper cupola operation, the coke bed must be maintained and checked for possible problems. Constant observation during actual cupola operation is essential. Checking coke bed performance during operations is not an easy task. However, several fundamental symptoms may occur during the operation that will alert the operator to possible improper coke bed performance.

1. The most obvious symptom is a sudden drop in metal temperature for no apparent reason. In most cases, this sudden change is due to a low coke bed. This drop will continue until the blast air and coke bed height are in balance again.
2. The color of the slag coming out of the cupola is another way to check the coke bed. Basic slags will have a gray to light-gray appearance under normal conditions. Normal acid slags are grayish-green in color when the bed height and air input are in balance. If the slag color gradually changes from the normal color to a dull, black color this indicates that the coke bed height has fallen below its normal operating height. Steps must be taken to re-establish the proper air-input-to-coke bed height ratio.
3. If the cupola operator has a means of checking the carbon equivalent (CE) of the iron at the spout, a dropping CE would indicate a dropping coke bed height. (Carbon Equivalent will be covered later in this chapter under Electric Arc Melting (Cast Iron)). Also, changes in the chemistry of the iron (such as a drop in silicon and manganese) or any sudden changes in melt rate will indicate poor coke bed performance.

A good cupola operator should always be alert to any changes, even gradual changes, taking place in the operation of the cupola and be ready to take corrective action. The operator should also keep accurate records of all phases of the cupola's operation: coke quality, coke bed performance, burning-in, measurement of coke bed height and adequate data to monitor coke bed performance.

Fig. 11-11. Cut-away view of positive displacement blower.

Cupola Blast Air System

The function of the cupola blast air system is to supply sufficient oxygen units to combine with coke carbon, in order to produce the heat required for the melting process. More than one ton of air is required to provide sufficient oxygen to melt one ton of iron, and eight tons of air for each ton of coke charged. This is why some consideration should be given to the cupola blast air system. Blast air for the cupola is very seldom considered as part of the charge materials going into the cupola, when in reality it is one of the most important.

Today's cupolas are normally run on a continuous basis; thus, the properly run cupola requires a steady flow of blast air in known and controllable amounts. Therefore, one has to understand the cupola melting and combustion process and properly select the required blast system components. Once this system is selected, proper maintenance of the system is necessary.

The basic components of the blast air system are:

- blast air blower
- blast air controls
- blast air duct system
- cupola wind drum and downcomers
- cupola tuyere
- blast air conditioning equipment

Blast Air Blowers and Controls

Once the amount of blast air required per ton of iron to be melted is determined, the equipment required to supply and control this amount can be selected. There is a general rule-of-thumb calculation for this: consider that cupolas will melt somewhere between 2 and 2.5 tons of iron per hour for each 1000 scfm of blast air used. Melt rates in the 3-ton per hour range for each 1000 scfm of blast air used can be achieved with high-efficiency hot blast cupolas.

Blowers used for providing blast air to the cupola are one of two types: positive displacement or centrifugal. The use of positive displacement blowers has diminished because of their size, initial and operating costs, and horsepower requirements. The positive displacement blower seen in **Fig. 11-11** has two lobes mounted on parallel shafts. These lobes rotate in opposite directions within a suitable housing. The impellers do not touch one another or the housing.

Centrifugal blowers can be categorized into three general types: single-stage turbo, multistage turbo and centrifugal compressor. The single-stage turbo is essentially a high-performance fan consisting of an impeller rotating inside a close-fitting housing. This type of blower will usually meet the requirements for a cold blast low stock (charge materials)-height cupola system. **Figure 11-12** shows a single-stage turbo blower.

Multi-stage blowers consist of two or more impellers mounted on the same shaft. The air enters at the center of the first impeller and is thrown by centrifugal force against stationary vanes to the inlet of the second impeller. The pressure is then further boosted as it passes from one impeller to the next until it is finally discharged from the blower. **Figure 11-13** shows a cutaway view of a multi-stage turbo blower. There are three types of multi-stage turbo blowers and they differ in how the impeller shaft is supported.

Since the early 1960s, the centrifugal compressors have been frequently used as blast air blowers in large-production foundries. Their typical volume would range from 14,000 scfm to 35,000

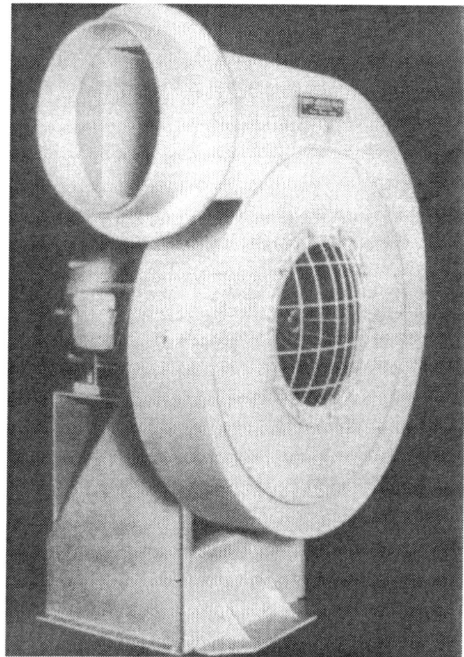

Fig. 11-12. Single-stage blower of the type used on small cupolas. Blower is suitable for 16–32 oz discharge pressure.

scfm, with pressures of 80 to 120 oz. **Figure 11-14** shows a single-stage scroll design centrifugal compressor. Multi-stage centrifugal blowers are also available.

Several improvements have been made in helping to maintain the blower in proper operating condition. Solid-state vibration monitoring devices have been designed to protect rotating equipment against serious damage. Malfunctions may be detected when an increase in vibration occurs. There are also temperature measuring devices to track the bearing and electric motor stator phase winding temperatures. Variable-speed drives are also being used on more blast blowers. These devices allow a greater degree of flexibility in pressure and volume range of the blast blower.

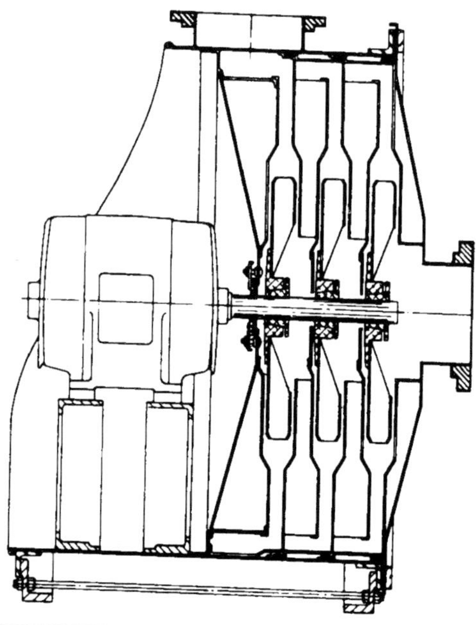

Fig. 11-13. Standard multi-stage blower, with impeller mounted directly on blower motor.

Blast Air Measurement

Until early in the twentieth century, blast air was manually controlled with the use of pressure gauges. Windbox pressures were usually taken to indicate the quantity of air being supplied to the cupola. The next step in measuring blast air was to place a differential volume meter in the system. Cupola operation was greatly improved by the use of volume measurement. To improve this method even further, features were added that allowed the instrument to be set for any quantity of air to be delivered into the cupola.

Today's equipment for controlling the cupola blast air provides a predetermined constant quantity of oxygen in blast air. Current equipment for controlling blast air takes into account temperature and atmospheric pressure that will affect the weight of oxygen in the air. **Figure 11-15** shows a typical automatic air weight controller.

Blast Air Duct Systems

Blast air for the cupola is usually drawn from outside the foundry. The inlet should be located so that there is little chance of drawing in airborne dusts and gas discharge from adjacent stacks. Meanwhile, the blast air blower should be in an area where it is convenient to maintain, and some distance from the cupola to avoid possible damage from normal cupola procedures. Carbon steel construction is normal for inlet ducts. As is true for all elements of the blast system, the layout should minimize the number of elbows and transitions to minimize pressure drop.

The wind drum's purpose is to distribute the blast air as evenly as possible to the individual tuyere downcomers. The wind drum is the part of the blast system that is wrapped around the outside circumference of the cupola shell. Wind drums are normally rectangular in cross section, and the blast air entrance is toward the top of the wind drum.

Tuyere downcomers, when used for regular blast cupolas, are box-shaped. For hot blast cupolas, the downcomers are usually circular in cross-sectional area.

Fig. 11-14. Single-stage scroll design centrifugal compressor. Impeller is supported by its own shaft, independent of the compressor drive.

Fig. 11-15. Automatic air weight controller.

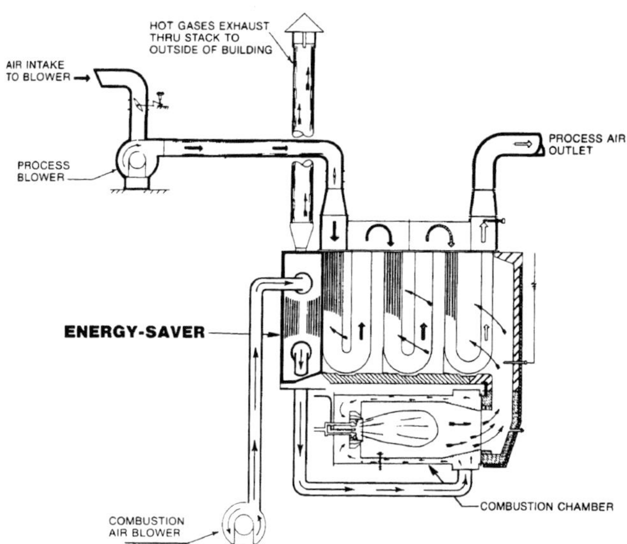

Fig. 11-16. Energy saver in tube-type hot-blast heater using heat in exhaust air to preheat air for combustion.

Blast Air Conditioning

In the early days of cupola melting, the air blown into the cupola was the ambient air found inside or outside the foundry. As time went on, the metalcaster learned that the condition of the air being blown into the cupola could play a very essential role in improving cupola operation. The three common ways used today to condition the blast air are:

1) preheating the air, called *hot blast;*
2) adding oxygen to the blast air, called *oxygen enrichment;*
3) removing humidity from the blast air, called *dehumidification.*

Hot Blast—When a cold blast cupola is converted to hot blast, the net coke savings are greater than the Btu value of the hot blast input. The overall thermal efficiency of the system is improved. Normally, the less efficient the cupola, the greater the coke savings will be after such a conversion. The recoverable energy will also be reduced substantially.

The use of hot blast involves preheating the cupola blast air from above ambient to temperatures usually between 570F and 1110F (298C and 598C). Temperatures as high as 1830F (998C) have been used. Preheating the blast air provides a faster reaction rate between the oxygen in the hot blast and the cupola coke (carbon) in the charge, as well as a reduction in the chilling effect of the blast air in the melting process.

Some of the general benefits of hot blast are:

- a reduction in coke consumption per ton of iron melted;
- an increase in the iron's temperature at the spout of the cupola;
- an increased tolerance for lower quality coke;
- lower metallic charge costs;
- a reduction in oxidation losses;
- improvement of the cast iron chemistry;
- increased carbon pickup and carbon content of the cast iron at the spout;
- increased melt rate;
- lower man-hour operating costs per ton of cast iron poured;
- reduced refractory consumption per ton of cast iron poured for refractory-lined cupolas.

There are three methods used to preheat the blast air: external, indirect recuperative and direct recuperative.

Figure 11-16 shows a schematic of an *externally fired* hot blast system. In this type of a preheater, the exhaust gases from the combustion chamber preheat the blast air. The source of heat comes from the burning of either oil or natural gas. The hot combustion gases circulate around the outside of the heat exchanger tubes, and the blast air flows in a counter-current direction from the combustion gases. Modulating the burner fuel and air input regulates the blast air temperature.

The *indirect recuperative* method takes the cupola exhaust gases from just below the charge door. These gases are then quenched, and all sensible heat is removed through water evaporation. The cooled, saturated gases are cleaned and used in a conventional, externally fired hot-blast system. The rest of the cleaned gas is then available as a low-Btu gas for use elsewhere in the foundry. The indirect, recuperative hot-blast systems have been used with high-production, water-cooled cupolas in excess of 100 in. (254 cm) in diameter.

Figure 11-17 shows the general arrangement of a *direct recuperative* hot-blast system. These units take the spent exhaust gases from the cupola below the charge door. When this type of a system is used, some means have to be provided to clean the surfaces of the heat exchanger's external surfaces that are in contact with the spent gases. There are several different types of this system that clean the hot gases; it is best to check the manufacturer's guidelines.

Oxygen Enrichment—The use of oxygen enriches the cupola blast air and has become a common means to reduce cupola raw material costs and to increase melt rates. Cupola oxygen enrichment has been used since the early 1930s. Ever since that time, there has been an improvement in air separation techniques, and bulk quantities of oxygen have been readily available for use in cupolas.

Oxygen enrichment is used as a cost-saving measure in cupola operation by permitting charge coke reductions, substituting pig iron with less expensive charge metallics, and permitting ferroalloy reductions. Increases in melt rates can be obtained at a low cost per additional ton with the use of oxygen enrichment. The option to

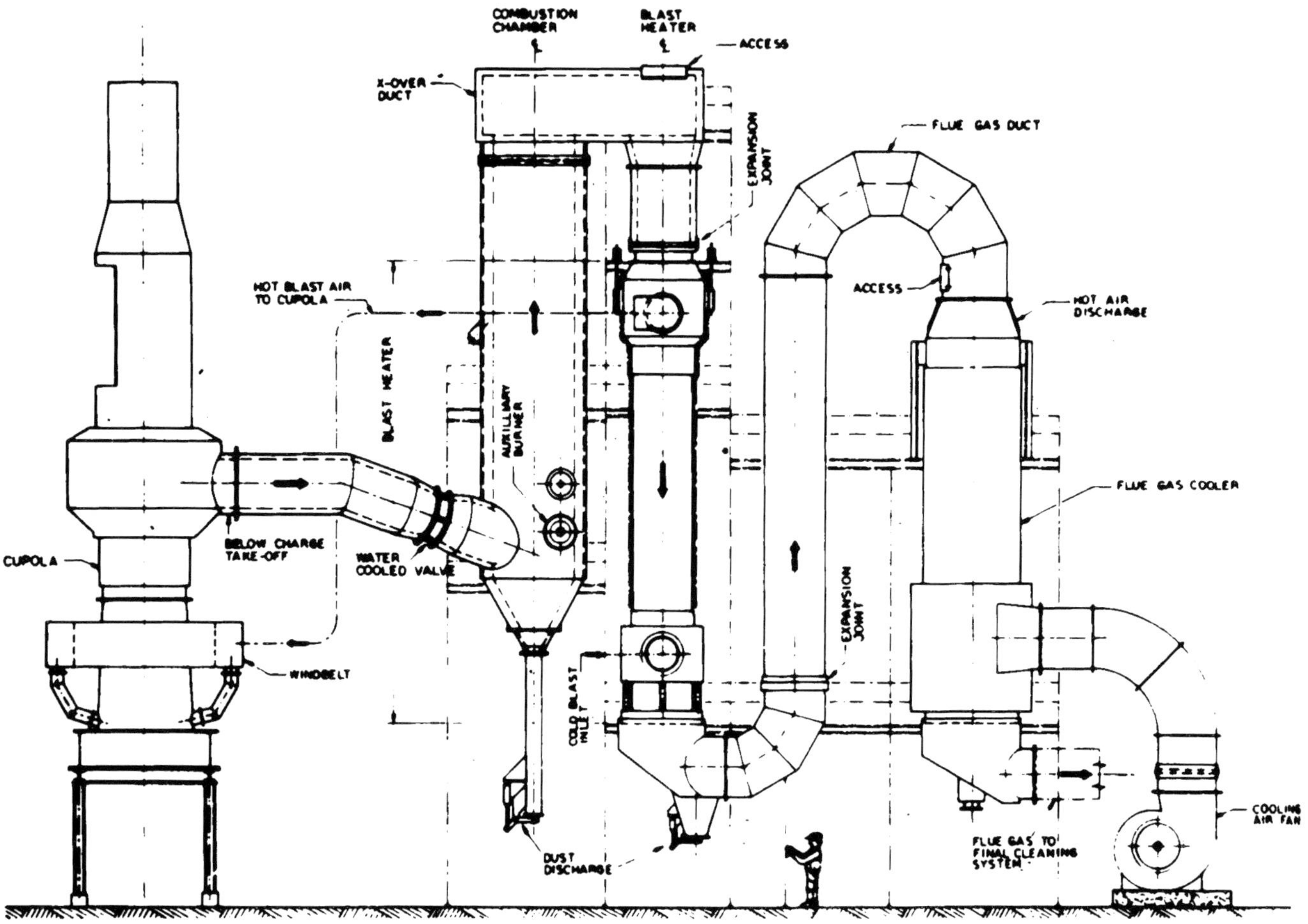

Fig. 11-17. General arrangement of recuperative hot blast and flue gas system.

instantaneously vary oxygen enrichment levels allows the cupola operator an added degree of control over constantly changing cupola operating conditions such as coke and metallic charge quality, blast humidity and temperature. Oxygen enrichment gives the cupola operator the power to almost instantly increase the melt rate and temperature of the cast iron at the spout of the cupola. Why does all of this happen?

Atmospheric air, when introduced into the cupola, contains 21% oxygen and 79% nitrogen. Oxygen enrichment allows the increase of the oxygen content of the blast air by 1–4%, to levels of 22–25%. This means that, in the combustion zone of the cupola, there is an increased flame temperature. This fact is shown in **Fig. 11-18.** Within the cupola, heat is being transferred by all three methods of heat transfer: conduction, radiation and convection. Heat transfer efficiency is a function of flame temperature. Higher flame temperatures increase the efficiency of heat transfer to the metallic charge.

In the case of cupola operations, an increase in flame temperature means that more heat energy is absorbed by the iron below the melt zone and that less heat energy is lost in the stack gas as a result of radiation increasing the heat transfer. In addition, the percentage of oxygen and nitrogen introduced into the combustion zone of the cupola directly affects combustion performance. With oxygen enrichment, the cupola operator has control over this matter.

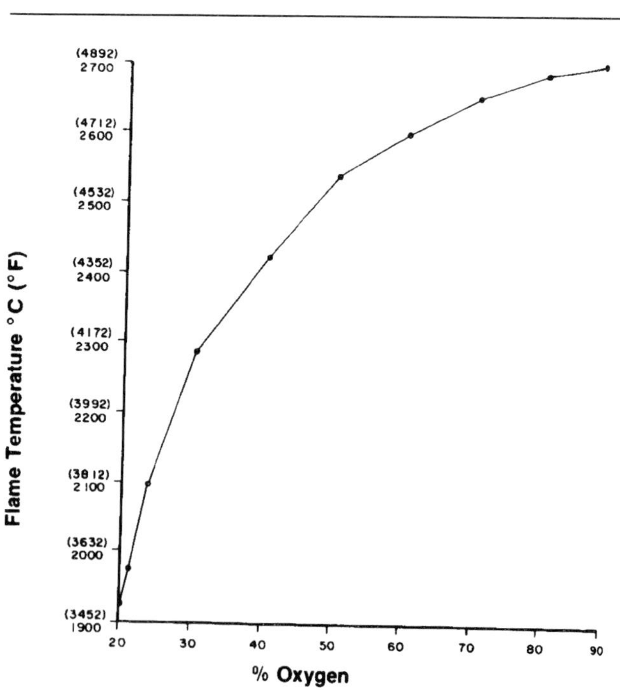

Fig. 11-18. Theoretical flame temperature vs. percent oxygen.

How is this oxygen injected into the blast air system? There are three methods used today for introducing oxygen into the combustion zone: diffuser enrichment of the cupola blast air, tuyere injection and well injection (**Fig. 11-19**).

The greater majority of cupolas using oxygen enrichment use *diffuser enrichment* for getting extra oxygen into the blast air. This method is the simplest and least expensive way to provide oxygen-enrichment to a cupola. Gaseous oxygen is fed into the wind main (or wind box) through a diffuser and mixes with the blast air before it goes into any tuyeres. This method results in a uniform oxygen distribution to all tuyeres and requires the least maintenance.

Tuyere injection involves a flow of gaseous oxygen into each individual tuyere (or every other tuyere) by means of stainless steel or Monel injector tubes. Each injector tube should have a flow control valve so that oxygen levels can be balanced around the cupola's combustion zone as conditions warrant. Tests have indicated that when this method of oxygen injection is designed and operated properly it can be more effective than diffuser enrichment.

With the *well injection* method, oxygen is injected into the coke bed beneath the tuyeres, using water-cooled injectors. Virtually 100% of the oxygen is directed into the combustion zone, which makes this method of oxygen enrichment highly effective in providing optimum heat transfer to the cast iron. Frequent monitoring and maintenance of the injector tubes is required, since they are located in an area where they are vulnerable to molten metal and slag attack. These associated problems seriously hamper this enrichment method.

Dehumidification—It was reported back in 1929 and 1931 that, on humid days, controlling chill depth and combined carbon content of the cast iron became very difficult. Increasing the coke per charge or holding the molten cast iron in the cupola well for longer periods could reduce this problem. This problem was eliminated when the blast air was sent through a refrigeration system that reduced the moisture in the blast air.

From a chemical standpoint, water vapor (humidity) reacts with the incandescent coke according to the water-gas reaction:

$$H_2O + C + heat = CO + H_2$$

Cupola operation and metal quality are affected by the following considerations:

1. The reaction consumes coke to form carbon monoxide.
2. The reaction is endothermic; thus, it absorbs heat that would normally be available to melt cast iron.
3. The water vapor contacts the molten cast iron at a high temperature and can affect the properties of the cast iron.

Today, there are three methods used to remove the humidity from the blast air: refrigeration, adsorption and absorption.

The *refrigeration* system works the same as a residential dehumidifier. In the foundry, the air is cooled to a temperature low enough to condense excess moisture from the air. This method involves passing the air through a chilled water spray or a cold, finned coil. A refrigerating medium is required to chill the water or the coil, directly. The cost of the equipment and operating it to dehumidify the air is expensive. Also, this cool air, when blown into the cupola, has to be heated by adding more coke to the charge.

The *adsorption* method removes moisture from the air by chemical means. Adsorption materials are usually made of solids

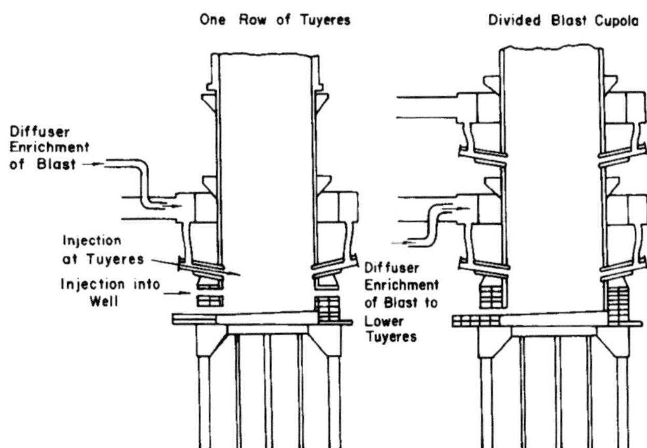

Fig. 11-19. Methods of introducing oxygen into the cupola.

that do not change chemically as they adsorb the moisture. These systems function by automatically switching the flow of air from one bed of adsorbent to another, so that one bed is removing moisture while the other is being reconditioned by heating. This system is disadvantageous because it provides no accurate means of humidification. If a humidity level higher than the ambient air humidity is required, there is no way this system can bring the moisture level up to where it should be.

Absorption involves the removal of moisture from the ambient air by a chemical change of the absorbing material. Absorbents are usually hydroscopic salts used in a water solution. In order to maintain a desired blast air moisture level, an absorption process system is usually required. When used properly, this system can keep the moisture level of the blast air constant.

For more detailed information on blast air conditioning, the reader is referred to Chapter 14 in the AFS publication, *Cupola Handbook.*

Charge Materials and Makeup

Now that the proper coke bed is burned-in, it is time to begin charging the cupola. It must be mentioned here that what is said about the charge makeup area and charge materials holds true for all of the methods of melting metal alloys, whether they are ferrous or nonferrous. It can be accurately said that high-quality molten metal begins in the charge makeup area. A clean, well-organized charge makeup area, with well-maintained equipment, is the first step in producing high-quality metal alloy castings.

This text will concentrate on the charge materials and makeup area for *ferrous* alloys. Ideally, the charge materials storage area should be under cover (**Fig. 11-20**). The beauty of such a storage area is that the charge materials can remain dry, and during the winter it can prevent water from freezing on the charge materials. When ferrous alloys are stored outside, they rust, and this oxide will have to be dealt with in the melting operation.

The charge materials should be segregated by alloy chemistry, purchased scrap or foundry returns such as gates and risers, and foundry rejects. The coke and flux (limestone, called "stone") that are used in cupola melting are also stored in this area. It is doubly critical that the coke be stored indoors, because it is porous and has fissures in which water can be trapped. Also, should this water freeze, it will expand and can cause the coke to split into smaller sizes than desired.

Fig. 11-20. Charge material yard in a high-production foundry serving two cupolas. The preferred charge makeup method is from out of cars and only the trim is taken from metal storage bins. The cranes are on a common runway.

The outside charge materials storage area shown in **Fig. 11-21** is usually found in larger production foundries. However, it's a good idea for foundries to have a small storage area inside, where charge materials can be brought in from the outside storage area a few days prior to use. Note the bins that are used to segregate the various charge materials. With this type of storage area, the charge materials are exposed to the elements. Again, it is good practice to bring the necessary charge materials indoors several days before they are to be used in the melting operation.

The charge materials are moved around with larger magnets (if they are ferrous alloys) or some other means, such as clamshell buckets or an orange-peel-type of bucket. A mobile crane or an overhead crane can move these carrying devices about. The metallic charge materials are then taken to the loading platform where they are placed into a metal weighing hopper or, in some cases, directly into the charging bucket. There is some type of weighing device used for the materials that have been placed in the container. The crane operator has been told how much of each material is to be placed in each charge. If the operator has placed too much material in the container, the excess is removed and placed on the trim platform. Alternatively, in case there is not enough material, the necessary material can be taken from the trim platform.

Fig. 11-21. Charge material yard with concrete metal-storage bins.

The coke and stone (limestone) are usually placed into some type of overhead storage bin. A weigh hopper or some type of carrying container is placed under the storage bin and the coke and stone are deposited directly into the container. From here, the container is taken to the charge makeup area. Another method of moving the coke and stone would be to use some other type of conveying device, such as a belt container, and having the conveyor place the coke and stone directly into a charge bucket. It is good practice to handle the coke as little as possible to prevent it from shattering into smaller pieces.

In all cases, scales are used to weigh the charge materials. These scales can be digital readout or dial types of scales. No matter which type is used, they must be properly maintained. The scale's load cells take a tremendous amount of shock abuse and can easily become out-of-calibration. Some type of a maintenance and calibration practice schedule should be set up. The schedule can be by the shift or daily. If the scales are out-of-calibration, the charge makeup will not meet the specifications required to produce quality molten metal. **Figure 11-22** shows a dial scale and **Fig. 11-23** shows a digital read-out scale being used in conjunction with two different types of weigh hoppers.

Fig. 11-22. Tilting-type metal weigh hopper. Note alloy material in skid-type hoppers and small spring-type scale for accuracy in weighing alloys.

Quite often, ferrous melting practices require the addition of a special alloying material to the charge materials. When this is required, it should be done on separate scales that can read to the smallest division of weight required such as ounces or increments of one pound (0.45 kg) or five pound (2.25 kg). Alloying materials cost money and the addition of the wrong amount can cause out-of-specification molten metal. These scales should have an even more restrictive maintenance schedule than the other scales used in the charge makeup area.

Now that the charge materials are properly weighed, it is time to place them in the charging bucket or, in the case of some cupola operations, into the skip hoist-charging bucket. If melting takes place in an induction furnace or a direct arc furnace, the metallic charge materials are placed into the charge bucket and made ready for the charging operation. However, in the case of the cupola, the charging of the charge materials can be done in several different ways. The coke may be placed in the charge bucket first, followed by the stone, and then the metallics; thus, creating a layered effect. In some cases, such as with the vibratory type of weigh hopper, the cupola charge materials may become mixed when they are placed into the charge bucket.

Cupola Combustion

In reality, the cupola melts the charge materials by using gases that are generated in the cupola itself. The economic advantage of cupola melting over other melting forms is due to the efficient energy conversion in the cupola, which is the generation of these gases. This conversion involves the combustion of the coke, with combustion gases transferring heat to the charge materials.

The overall chemical reactions involved in the burning of the coke are well known; however, the specific individual reactions and reaction rates are still the subject of much discussion and research. Knowing and understanding these reactions can lead to improved cupola operation.

The primary combustion or oxidation zone **(Fig. 11-24)** is where the blast air enters the cupola through the tuyeres. When this air contacts the burning coke, the following two gases are formed: carbon dioxide (CO_2) and carbon monoxide (CO). Which of these two gases is formed first is still a matter of discussion and research.

Fig. 11-23. Independently mounted feeder weigh hopper. Note digital read-out.

Water vapor in the blast air will be reduced to hydrogen by the hot coke and carbon monoxide. Any moisture in the charge materials will be driven off in the preheat zone of the cupola.

Combustion reactions in the cupola can become quite involved; the discussion here will be limited to the two main combustion reactions. These two major exothermic combustion reactions involve the carbon in the coke and the oxygen in the blast air. These two substances react to form CO_2 and CO, and the major heat energy is formed during these reactions. *Exothermic* means that these reactions will generate heat. However, there are two endothermic reactions that take place, and these are the reduction of CO_2 and H_2O. *Endothermic* reactions are the opposite of exothermic reactions, in that they will absorb heat. Thus, these two endothermic reactions should be avoided.

The carbon-oxygen reactions of major importance are:

$$C + {}^1/_2O_2 = CO \quad (1)$$

$$C + O_2 = CO_2 \quad (2)$$

Again, there is disagreement between researchers as to which of these reactions occurs first. Two other reactions occur and are considered in the carbon-oxygen system:

$$CO + {}^1/_2O_2 = CO_2 \quad (3)$$

$$CO_2 + C = 2CO \quad (4)$$

The first three reactions (Equations 1–3) are *exothermic* and generate heat energy, whereas the fourth reaction (Equation 4) is *endothermic* and absorbs heat energy. Following are the heat energy results of these reactions:

Reaction	*Btu Generated*	*Reaction*	*Btu Generated*
1	4,350	3	10,200
2	14,550	4	5,850

Btu stands for British thermal units, or the amount of heat required to raise the temperature of one pound of water one degree Fahrenheit or a unit of heat measurement equal to 0.2520-kilogram calories.

Remember the opening statement in this section: *the cupola melts the charge materials by using gases that are generated in the cupola itself.* Reactions 1, 2 and 3 generate the gases that, in turn, supply the heat energy (Btu) necessary to melt the metallic charge materials placed into the cupola. Further information can be obtained in Chapter 12 in the AFS publication, *Cupola Handbook.*

Cupola Zones

Referring to **Fig. 11-24,** the various zones within the cupola will be discussed. Before looking at the various zones, it should be explained that the word *burden* is used to describe the charge materials placed in the cupola.

Well Zone

In the cupola well, liquid or molten metal surrounds the coke and the slag separates by floating to the top of the molten cast iron. The well zone extends from just below the tuyeres to the cupola bottom. The main reaction occurring in this zone is the dissolution of carbon from the coke by the molten cast iron. In fact, a large amount of the carbon in the cast iron is picked up in this zone. The reactions in this zone are more important for the chemistry of the cast iron than for combustion control.

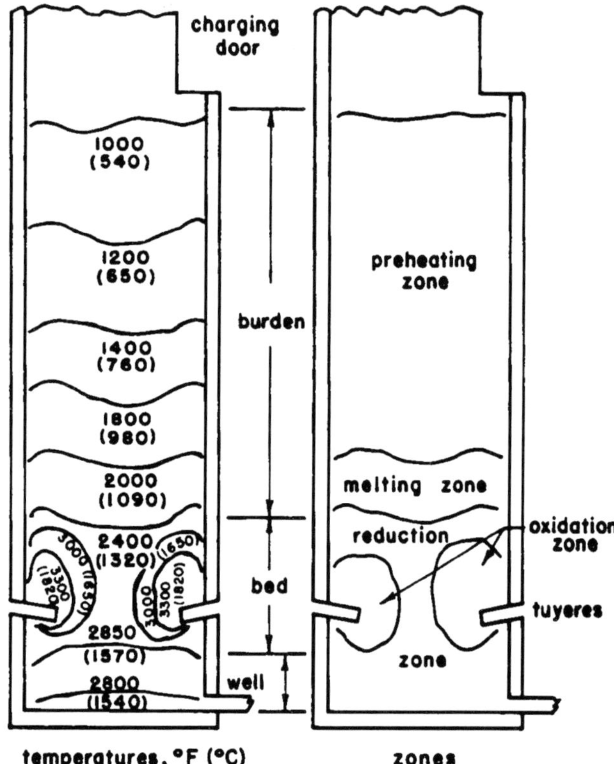

Fig. 11-24. Location of cupola zones.

Oxidation (Combustion) Zones

The oxidation or combustion zones are the regions where the blast air enters the cupola from the tuyeres. It is in these zones that the CO and CO_2 are formed by the blast air reacting with the hot coke. The volume of these zones is governed by various factors such as blast rate, blast volume, blast temperature, coke size and reactivity, cupola diameter, method of tuyere cooling, wall lining type, divided blast and oxygen enrichment.

The gas temperatures in this zone can change very rapidly. Various computer simulations estimate that the gas temperatures can range from 3092F to 3812F (1700C to 2100C). At these temperatures, oxygen reacts rapidly with the coke and carbon monoxide until it is reduced to about a 1–2% level that remains unreacted as it passes up the stack (cupola).

The primary purpose of this zone is to supply the additional heat energy for additional superheating (super temperature) of the molten cast iron. It also provides the hot gases to preheat, melt and superheat the charge materials in the zones above the tuyeres. This is the zone where the heat energy is generated and is the most critical zone for successful cupola operation. Two major operational parameters—tap temperature and melt rate—are determined by the events that take place in oxidation zone.

Reduction Zone

The reduction zone ranges from the top of the well zone to the bottom of the melt zone and surrounds the oxidation zones. Superheating the cast iron is the main function of this zone. However, in this zone, there is also the possibility of oxide reduction of the iron, manganese, silicon and other metals. The reduction of oxides is desirable in the region between the oxidation zones and the well zone. The concentration of CO is very high in the lower part of the reduction zone, as the CO_2 present reacts with the coke to form additional CO. The amount of carbon dissolved in the iron and the molten metal temperature increases as the length of this zone increases.

Remember the four reactions that take place in the cupola? The fourth reaction (Equation 4)—$CO_2 + C = 2CO$—was an *endothermic* reaction, which reduces the amount of sensible heat that will enter the melt and preheat zones. In addition, this reaction consumes carbon, which could be used in the oxidation zones, and for carbon pickup in the well zone. Control of this reaction has to do with controlling the coke. A coke with a low reactivity tends to reduce the amount of CO formed. Also, large pieces of coke have a small surface area for a given mass and the amount of CO_2 reduced is directly related to the coke surface area.

Melting Zone

The melting zone goes from the top of the coke bed to the point where melting first occurs in the metallic charge. This zone is bounded by the reduction zone on the bottom and the preheating zone on the top. The heat of fusion, or the heat necessary to melt or liquefy the metal, is supplied by the hot gases coming from the upper reduction zone. Slag also begins to liquefy in this zone.

If low-carbon steel is used in the metallic charge, it can pick up some carbon in this zone from the stack gases and coke and, thus, lower its melting temperature. The size and shape of the metallic charge affect what happens in this zone because of their effect on heat transfer conditions. Large pieces of bulky steel scrap may not melt in this zone and, if these materials do not melt in this zone, the final cast iron composition will be lower in carbon and the temperatures will be lower than normal.

Preheating Zone

The preheating zone is the portion that extends from the top of the melt zone to the top of the burden. The primary functions of this zone are to preheat the charge materials from room temperature to melting temperature, to evaporate moisture from the charge materials, and to decompose the limestone or calcium carbonate into carbon dioxide and lime. Some carbon dioxide may react with the coke to form carbon monoxide.

Heat transfer in this zone is very critical, as the primary function of this zone is to preheat the metallic charge materials to their melting temperature. The driving off of moisture from the charge materials is an *endothermic* reaction and should be avoided by charging dry material, as was discussed earlier. Also, the charging of wet materials will drive the melting zone closer to the oxidation zone and reduce the size of the reduction zone. This will mean colder cast iron at the spout and lower carbon pickup.

Since the CO concentrations in this zone are high, the reduction of iron oxides or hydroxides in the form of rust can occur. These reactions are *endothermic* and will adversely affect the heat balance; thus, oxidized (rusty) metallic materials will require higher coke rates.

When the charge materials reach the melting zone, melting begins. The droplets of molten metal pass down through the oxidation zones and the reduction zone. As the droplets of molten metal pass through these zones, they contact the coke and pick up some carbon. The droplets then collect in the well, and it is here that the greatest amount of carbon is picked up.

Slagging and Tapping

Slagging

The major objective of cupola melting is to produce a high-quality cast iron with properties that meet or exceed specifications. If good slag control is not practiced, an inferior cast iron will result.

Slag is the fused product formed in the melting operation by both the oxidation products of the metallic charge and the nonmetallic charge materials, such as coke ash and refractories, and flux materials such as limestone. The primary purpose of the slag is to provide a means of removing nonmetallic materials from the cupola. The secondary purpose is that it can be used to help control the chemistry of the cast iron. There are normally two slag practices used in cupola melting: acid and basic.

Acid Slag Practice—Cupola iron melting using the acid slag practice is relatively simple due to lower slag volumes and less complex slags than used in basic slag practice. Acid slag practice has less silicon loss and less carbon derived from the coke; however, they do not remove phosphorus or sulfur from the molten metal. The latter point means that charge materials must be of the highest quality.

Basic Slag Practice—A basic slag is obtained when the basicity ratio is one or greater. The basicity of the slag is determined by the following calculation:

$$\text{Basicity} = CaO + MgO / (SiO_2 + Al_2O_3)$$

where CaO is lime, MgO is magnesia, SiO_2 is silica and Al_2O_3 is alumina.

A basic slag practice can be used in the melting of cast iron that is formulated into ductile and compacted graphite irons. Ductile iron requires a very low sulfur level and a basic slag can help to lower the sulfur level in the molten cast iron exiting the cupola. However, if a silica lining is used in the cupola, the basic slag will attack the lining and cause excessive lining loss, and less sulfur will be removed. Basic slags are not as viscous as acid slags.

A thick slag could cause bridging, which would result in poor stack permeability. This could result in the desired reactions not taking place in the cupola and help to produce a cast iron that is out-of-specification. Lack of good slag control may also result in a cast iron with insufficient silicon, a very essential element in the production of quality cast iron. Adding limestone as a flux and controlling the slag increases the carbon and silicon levels in the molten metal to acceptable levels.

Tapping

The shape of the tap hole can be round, square or rectangular in cross section. The total area must be only large enough to allow all of the molten materials to pass through.

A round tap hole allows the concentration of slag attack at the 12 o'clock position. On the other hand, the other two cross-sectional shapes spread the slag attack over a larger surface area, thus reducing erosion of the tap hole. The slag hole can be formed by using fireclay blocks with the proper sized tap hole in them or can be formed when the breastplate is formed using monolithic lining practice. Special refractory tap hole blocks are available for use in basic cupolas. For more information on refractories used and the construction of tap holes and slag holes and notches, check Chapter 10 of the AFS *Cupola Handbook,* 6th Edition.

Slag holes and tap holes are located at various places on the cupola for the purpose of drawing off the slag and tapping the molten metal. As learned in Chapter 10, there are two basic ways of removing (tapping) the molten cast iron from the cupola: intermittent and continuous.

Intermittent Tapping, Rear Slagging Cupola—The intermittent tap cupola is best suited for a jobbing foundry in which a constant molten cast iron supply is not required. The intermittent tap cupola is operated with one tap hole for the molten cast iron and another for the slag. The bottom tap hole is for the molten cast iron to flow out of the cupola and is considered to be in the front of the cupola. The slag tap hole is in the rear of the cupola and a distance above the cupola bottom, but not high enough to allow slag to enter the tuyeres. The height of the slag tap hole will determine the volume of molten cast iron per tap. **Figure 11-25** shows schematically an intermittent tapping, rear slagging cupola.

Since this type of cupola is a relatively small tonnage producer, the expense of air pollution control equipment required for its operation has forced the retirement of most intermittent tap cupolas in favor of electric melting. Thus, a true intermittent tapping, rear slagging cupola may never be seen.

Continuous Tapping, Front Slagging Cupola—The continuous tapping cupola is used when a constant supply of molten cast iron is required. In the continuous tapping, front slagging cupola, the cast iron and slag flow continuously through a common tap hole, as illustrated in **Fig. 11-26**.

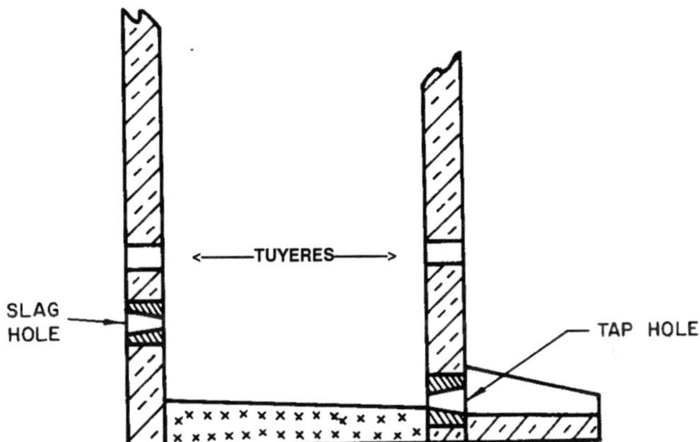

Fig. 11-25. Intermittent tapping, rear slagging cupola.

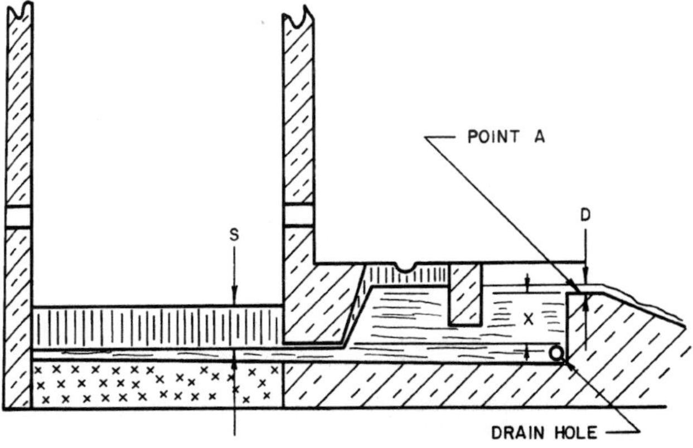

Fig. 11-26. Continuous tapping, front slagging cupola. Letters in figure are used in equations to determine slag height in various parts of the cupola: Point A = top of dam; S = slag; D = iron depth; X = vertical distance from taphole top to A.

Melting begins with the tap hole botted (closed). Deciding when to unbott or open the tap hole is a very critical point in the cupola's operation. If the tap hole is opened too soon, it may blow...that is; the blast air will blow any accumulated slag and molten cast iron out of the tap hole. If the tap hole is opened too late, there is a good chance that slag may enter the tuyeres. Some operators set a time, while others may watch through tuyeres for the first droplets of cast iron to appear and then allow a certain number of minutes to pass before opening the tap hole. It is good practice to keep a record of tapout.

As the slag and cast iron flow through the tap hole, they enter a reservoir that contains the "slag skimmer." The slag skimmer holds both the slag and molten cast iron back in the reservoir to allow the slag to float to the surface and flow out the slag notch. The molten cast iron, being heavier, flows under the skimmer into the launder.

The launder is a refractory lined trough that directs the molten cast iron into either a forehearth or a holding furnace. The forehearth is a refractory lined unheated vessel in which the molten cast iron is stored and, if needed, its chemistry is adjusted to specifications. From the forehearth, the molten cast iron can go directly to the pouring floor or to another holding furnace.

The operation of the continuous tapping, front slagging apparatus is quite complex and involves the densities of the molten cast iron and slag, and also the pressures inside the cupola and windbox, slag and iron pressures, and the iron dam height. For more information, it is suggested that the reader study Chapter 18 in the AFS publication, *Cupola Handbook,* 6th Edition.

Continuous Tapping, Rear Slagging Cupola—In a continuous tapping, rear slagging cupola operation, the slag and molten cast iron flow out through separate holes in the cupola. **Figure 11-27** shows a schematic of this type of operation. Note that, in this case, the slag and molten cast iron are separated inside the cupola, with the slag going out the slag hole in the rear of the cupola and the molten cast iron out the tap hole in the front of the cupola. (The front of the cupola is always considered to be where the molten cast iron tap hole is located). Again, the operation of this system is quite complex and dependent on the same conditions as with the continuous tapping, front slagging cupola.

One thing to note in **Figs. 11-26 and 11-27** is that both of these systems have a drain hole. This hole is unbotted when the cupola is being shut down and allows the cupola to be drained of any molten cast iron before dropping bottom.

Dry-Bottom Tapping and Slagging Cupola—The dry-bottom cupola is used more often in Europe than in the United States. In this case, the cupola well and bottom are lined with carbon block and brick. This type of tapping and slagging operation can be seen in **Fig. 11-28**. The molten cast iron and slag leave the cupola through a common passageway to a refractory lined separator. With this type of operation the slag can be removed through a separate hole or can go through a common tap hole with the molten cast iron and be separated as with the continuous tapping, front slagging operation.

One advantage of the dry-bottom cupola is that it is well suited for long melting campaigns. Some cupolas are equipped with two separators that are used alternately. This allows the relining of one separator while the other is being used. In addition, since there is no slag layer in the cupola, carbon and silicon recovery will be improved along with less variation. One disadvantage is the poor accessibility of the tap hole.

Slag Collection and Disposal

In recent decades, there has been a near revolution in cupola design, operation, etc. However, no aspect of cupola operation has benefited less from these advances than slag collection, drop removal and disposal. This is regrettable because these operations are, in part, among the most inherently hazardous of all cupola operations. When these hazards are understood and respected, these cupola operations can be carried out safely, responsibly and economically. One should be aware that some of the practices described here might not always be desirable or permissible.

Good slag practices are fundamental to successful cupola operation. In cupola melting, as with other metallurgical melting operations, the quality of the final product is closely affected by the physical properties and chemical composition of the slags. Although it is a bothersome waste material, cupola slag beneficially reacts with the molten cast iron to remove many impurities from the cast iron—principally oxides, sulfides and phosphides.

The amount of slag generated during the cupola melting operation can vary depending on many factors. Although discussed earlier, some of these factors are worthy of review. Chief among these factors, if using a lined cupola, is the attrition of the refractory lining. Other factors include coke ash and nonmetallics charged into the cupola such as oxides attached to the metallic charge (rust),

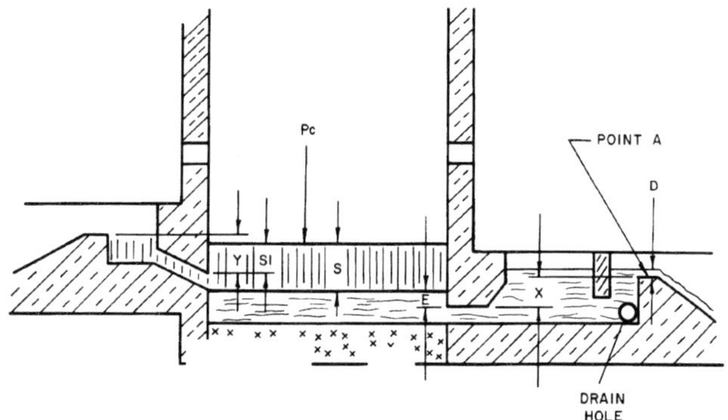

Fig. 11-27. Continuous tapping, rear slagging cupola. Letters in figure are used in equations to determine slag height in various parts of the cupola: Point A = top of dam; S = slag; D = iron depth; X = vertical distance from taphole top to A; Pc = pressure inside cupola; Y = slag dam height; S1 = slag depth above top of slag taphole; E = iron depth above top of iron taphole.

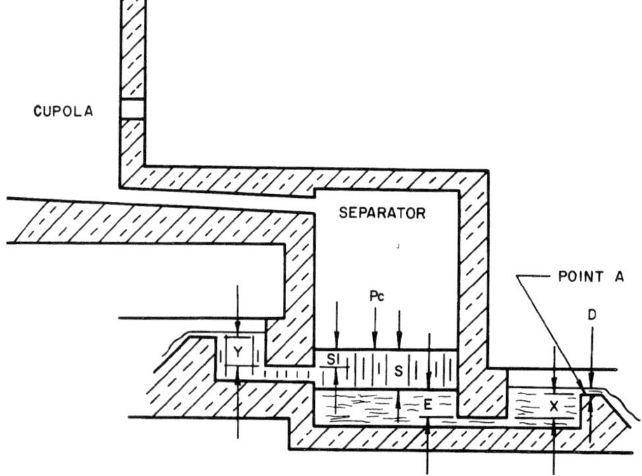

Fig. 11-28. Dry-bottom tapping and slagging cupola. (See Fig. 11-27 for legend.)

foundry sand on charged foundry returns, gates and risers, fluxes and just plain dirt picked up in the metallic and coke storage areas. Normally, it can be said that the cleaner the charge material, the less slag will be generated. In addition, by using a clean charge, the pollution control equipment is not over burdened.

As the slag exits via the slag notch in front slagging, or over the slag dam in rear slagging, some means of handling and disposal has to be available. Smaller cupolas have relied mostly on the "dry" methods and larger cupolas have used "wet" methods. In recent years, due to environmental considerations, this distinction as begun to be blurred. Some larger cupolas have been converted to the dry methods because of the increasing costs of wastewater treatment and disposal.

Dry Slag Collection—In its simplest form, the slag is allowed to collect on the floor, directly below the slag spout, especially with a small- to medium-sized cupola run for short campaigns. After the slag has cooled, it is hauled out to a disposal site. In this instance, some cast iron may also be inadvertently disposed of.

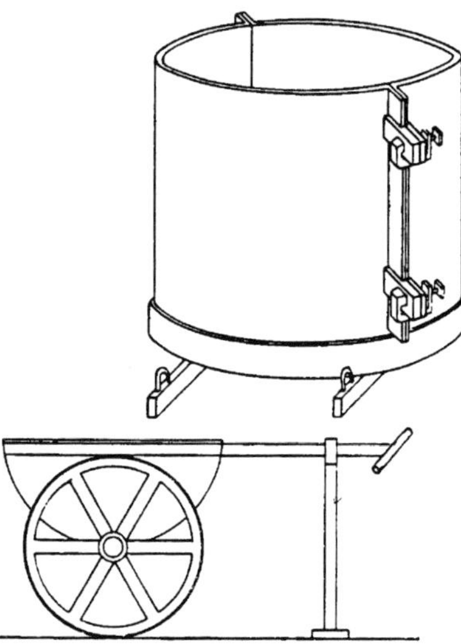

Fig. 11-29. Two types of slag containers that usually catch slag directly from the cupola. The containers are lined with fire clay.

Figure 11-29 shows examples of round-bottomed and flat-bottomed slag containers. These containers can have a dry sand bottom and a thin coating of clay wash on the interior, to allow for easy dumping. If a container with a round bottom is used **(Fig. 11-30)**, entrapped iron will tend to settle where it can be recovered more easily. **Figure 11-31** shows a solidified slag cake being dumped.

Fig. 11-30. Cast iron slag pot mounted on a truck. The pot is tilted to the side by a gear bracket.

Cooling masses of slag must be treated with respect. What is dark and solid on the outside may be very hot and liquid on the inside. Some drawbacks to dry collection of slag include diminished working conditions around the cupola and the continuing and significant expenses in maintaining the slag collection containers.

Wet Slag Collection—Larger cupolas, especially those running a basic slag, can generate large amounts of slag over short periods. Thus, handling these amounts of slag using the dry methods can cause a "traffic jam" around the cupola. Faced with these problems, operators of large cupolas have chosen to use water to quench, granulate and transport the slag generated in their cupolas. Smaller cupolas can use this method of slag collection also; however, the economics may not justify using this method.

Some of the claimed advantages of the wet method are:

- minimal material handling costs;
- improved, safer working conditions;
- easier trapped metal recovery;
- much improved disposal conditions, in some cases producing a salable product, depending on local conditions.

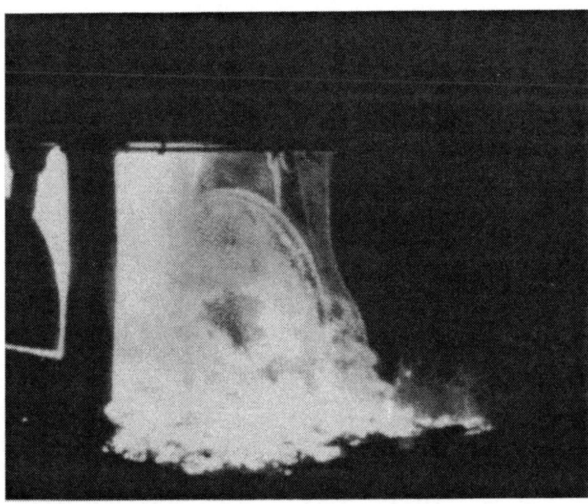

Fig. 11-31. Cupola drop of solidified slag.

Wet slag collection equipment usually consists of a sluice-head or other means of combining slag with water. The water has to be of sufficient volume and velocity to quench and granulate the slag into small particles. **Figure 11-32** shows a sluice-head impinging water onto a stream of slag. The slurry of slag and water is flushed through the trough system to collect in a holding tank that is often some distance from the cupola. It is in this tank that the slag settles out of the water. This is commonly called dewatering the slag. The water is then recirculated to the sluice-head using dedicated pumps and piping.

The slag in the water is very abrasive and can rapidly abrade the refractory lining of the trough. Replaceable liners are often applied at critical points of wear. This action is especially severe at critical points of wear. This action is also severe in the area where the water impinges on the slag stream due, in large part, to the considerable amount of heat in this area.

Another approach to wet collection is the use of a cylindrical-shaped vessel with a conical bottom. Water is introduced into the vessel, tangentially, near the top, and centrifugal force forms a water wall several inches thick in the vessel. The slag stream falls into this wall of water and is spun downward into the conical section in a vortex, exiting into a discharge pipe or trough. When the slag stream hits the water wall it is granulated and quenched. Upon leaving the vessel, the slurry of slag and water is sluiced away to a settling tank or pit.

Slag Disposal—Slag disposal is a substantial problem to the foundry, usually second in volume to that of spent foundry sand. Now, cupola experts are not aware of any place where cupola slag is considered a hazardous waste material. This does not mean there are no such instances or that there will never be.

An ideal solution would be to find constructive reuses for the slag. This is more likely for slag than sand because it is a reasonably stable and uniform material. Among the possibilities being explored, so far, is the use of slag as ordinary fill, as railroad ballast and as an aggregate in concrete. Some counties and townships are using a "slag seal coating" to help preserve their asphalt roads.

Dropping Bottom

There are a few cupolas in operation that are designed so that there is no need to drop bottom. However, the greater number of cupola operators are faced, at intervals, with having to open the bottom doors and spilling the cupola's contents on the ground or floor. This operation is called "dropping bottom." This operation is necessary to allow the repair and restoration of refractories inside the cupola and other repairs that may have to be made to the cupola.

Here, the cupola bottom has to meet two contradictory requirements. First, it has to be strong and completely contain the molten cast iron, slag and corrosive products of melting iron. In other words, the cupola bottom cannot allow any leaking to take place during the entire heat or campaign. Then, when the heat or campaign is ended, it must not impede the dropout of the cupola contents.

Fig. 11-32. Overhead view of tank receiving sluiced slag. Pumps (shown at left) convey the slag to a dewatering and disposal area.

The over-riding emphasis in dropping bottom must always be *safety*. The entire bottom-dropping operation requires careful planning and a well-trained crew. Practices should be developed to react calmly to any delays or other difficulties encountered. A major consideration is the floor or ground onto which the cupola's contents will fall. The area under the cupola where these materials will fall must be dry! Remember that, depending on the cupola bottom method used, the contents of the cupola will be hot and contain some molten metal and slag. When the dropped materials hit the floor, they will turn any water on the floor or in the ground into steam. As learned in science classes, when water turns to steam its volume increases about 1600 times. Should this reaction occur under the cupola's dropped contents, a serious explosion can occur.

Cold Bottom Dropping

This method of bottom dropping was developed to greatly reduce many of the dangers inherent in conventional hot bottom dropping practices. In this method, a large volume of water is introduced through the tuyeres, at the proper time, to thoroughly quench the coke bed before opening the bottom doors.

The procedure requires that all of the metallics in the cupola be melted. After this has occurred, all of the various drain holes in the cupola are opened to allow the molten metal and slag in the well and spout to drain out. Blast air can be used to help remove the slag. After no more slag can be made to flow, the blast air is turned off, the tuyeres are opened and the water is introduced into the cupola through the tuyeres. When water begins to drip from around the bottom doors and through the vent holes, the water is turned off. Schedules permitting, a cooling down period of one hour or more takes place. After this, the bottom door's support is pulled out and the bottom doors open and the cupola's contents drop out.

Hot Bottom Dropping

With this method of dropping bottom, it is very critical that the ground or floor under the cupola be dry. Should there be moisture present in this area, there is a very good chance that a "drop explosion" can take place. If there is any doubt about moisture being present in this area, it is best to play it safe and remove the sand under the cupola and replace it with dry sand, to ensure a dry area.

A hot bottom drop should include planning and preparation for the drop—sometimes hours before the drop is to take place. By doing this, the operator can ensure a safe and quick systematic approach to the dropping procedures. Once the cupola blast is shut off, the materials inside the cupola, including the bottom material, begin to cool down. Any lengthy delay at this stage of the bottom dropping operation can cause difficulties in completing the drop.

The blast is shut off after the last charge is melted out and molten cast iron stops flowing through the tap hole. At this point, all drain holes are opened to allow the exit of any molten cast iron from the well. Blast air may be used at this point to help with the draining operation. This step is important in that it can prevent any molten cast iron or slag from splashing during the drop and causing a hazard.

After the well has been drained, the bottom door supports are pulled out and the bottom doors open, allowing the bottom and the coke bed to drop out of the cupola. If some type of a container is not used, the drop will fall directly onto the ground or floor under the cupola. It is essential at this point for the drop to be quenched by spraying water onto it. **Figure 11-33** shows a commonly used water quenching system. This quenching should be done quickly to save any usable coke and minimize heat damage to the cupola structure.

Some cupola operators will build a three-walled steel or concrete enclosure with a doorway closing off the fourth side, if desired, under the cupola. This effectively contains the initial flare, expedites quenching and serves to contain the drop for safety reasons and easy removal.

When it is desirable to catch the drop in a container, car or skid, special bottom door locking devices can be used. These locks are usually not designed to support the bottom doors of a fully charged cupola, but are relied upon only after the burden in the cupola has been substantially melted out. The bottom door supports are removed and the container is moved quickly into place. The locking device is then disengaged and the bottom doors open.

After the reusable coke has been collected, the drop is then disposed of. As with slag, this drop material is not considered a hazardous waste material, at least to date. Thus, it can be disposed of at any landfill. However, most slag and drops are sent to *regulated* landfills.

Computer-Assisted Operation

Because cupola performance is very sensitive to a large number of variables, a cupola is much more difficult to operate than are other melting processes. There are many different reactions taking place inside the cupola, which all have a bearing on the final cast iron coming out of the tap hole. For this reason, it is necessary to spend more time discussing cupola operation.

In many respects, the cupola has rightfully gained the reputation of requiring skilled operators. It is inexact in its output, and no two cupolas, although identical, operate the same way. These characteristics are not consistent with today's demands for precisely controlled cast iron from the cupola.

Fig. 11-33. Drop quenching system.

To help meet these demands, the industry's technical people have embarked on using computer models to help eliminate the undesirable characteristics of cupola performance. To do this, cupola process models have been developed that integrate the effects of the numerous variables in cupola operations. These computer models should guide cupola operations toward optimal performance. The ultimate goal of these projects, not yet achieved, is complete computer control of the cupola operations.

This limited space cannot discuss all of the ramifications of operating a cupola. Those wanting to delve further into this subject should obtain a copy of the AFS publication, *Cupola Handbook,* 6th Edition. In particular, Chapter 27 discusses computer-assisted cupola operation.

ELECTRIC FURNACE MELTING OF FERROUS ALLOYS

Chapter 10 of this book discussed the various electric furnaces used in the foundry industry. Among these furnaces were the induction furnaces (coreless and channel) and the electric/direct arc furnace. These electric furnaces can be used to melt any of the ferrous alloys discussed in an earlier chapter. The induction furnaces are used more frequently in cast iron foundries, while the direct arc furnace finds its biggest use in melting carbon steels. The discussion will begin with electric induction furnaces.

Induction Furnace Melting

One of the main advantages of an induction furnace is the small amount of alloying material that is lost during melting. For all practical purposes: what goes in, goes out. This allows for good composition control through careful weighing of alloy additions. If the composition of the charge materials is accurately known, precise metallurgical control is rather easily realized. This is very important when expensive alloying elements are used.

On the other hand, refining, such as that used in melting steel, is rather limited. Lower grade scrap cannot be used because it cannot be refined in the conventional manner. No distinct slag layer is developed due to the stirring action in an induction furnace, and the slag:metal volume ratio is not high enough to permit normal refining.

The gas content of induction-melted heats is generally somewhat lower than heats melted in other types of furnaces. The constant movement of the bath, especially in the coreless furnace, tends to allow bubble nucleation and removal as the molten metal passes through the bath surface. In addition, the absence of the products of combustion, such as CO, CO_2, H_2 and H_2O, eliminates a potential source of gas.

Coreless Induction Melting (Cast Irons)

Furnace Lining—When melting cast iron in a coreless induction furnace, the first thing to check is the furnace lining. If the furnace has been completely relined, it should be checked to verify that the lining has been completely sintered-in (bonded by the application of heat). If the lining was shaved (removing only the surface of the lining to a determined depth), the new portion of the lining should have been sintered. Any patching of the lining should be thoroughly checked. To put it simply, proper refractory installation and maintenance is very critical to successful induction melting.

Charge Materials—The next area to consider is the charge material to be used. Basically, the charge consists of foundry returns, scrap and alloying elements. For producing quality cast iron, the charge material selection will involve a compromise between quality and cost. Remember, induction furnaces are not refiners. Quality charge materials will produce better quality heats. Somewhere along the line, money has to be spent to produce quality molten gray iron. This determination is up to the metalcaster. Most foundries operate between the two extremes, attempting to balance quality with cost.

If possible, high-quality scrap should be used. Such materials generally have the following attributes:

1. Predictable, uniform and desirable physical size.
2. Known, uniform and desirable chemical composition.
3. A minimum of undesirable constituents.

Nonmetallics, such as wood, oil, water and sand, are undesirable, but are often found in low-grade scrap. Trace elements, such as chromium, molybdenum and tin, are undesirable in some applications, and should be controlled.

The use of iron and steel scrap is very common in the production of cast iron in coreless induction furnaces because of cost factors. The required carbon is added using carbon-raising materials and/or picked up using ferroalloys. It is very important that the scrap be clean, free of off-grade material and free of residual alloys (or contain a known amount of alloy) for ease of control.

A major portion of the charge is usually made up of foundry returns such as gates and risers. It is important that these be segregated by alloy content or type if several different grades of cast iron are produced in the foundry. This scrap should be clean of any adhering foundry sand or other foreign materials. It is said that in induction melting, each 1% of foundry sand adhering to charge materials decreases the melt rate by 1.5% and increases power cost per ton by 1.5%.

The scrap, as for melting in any type of furnace, should be stored so that there is a minimum of rust formation. Ice and snow is also a problem with scrap stored outside, and, in certain melting operations, can present a severe hazard.

Induction melting has allowed an increase in the use of steel scrap (specifically, carbon steel) in the melting operation. Although these materials have a low carbon content, adding carbon can raise the carbon in the melt. Carbon has a lower specific gravity than iron and will tend to float toward the top of the melt. If this is allowed to occur, the carbon can become oxidized; and if there is no movement of the iron within close proximity of the carbon particles, the iron closest to the carbon particles will become saturated with carbon. However, with the stirring action in induction furnaces, the carbon particles are kept in motion, preventing a carbon-saturated layer from forming in the melt. The latter is particularly true with the coreless induction furnaces. A key factor to remember is that a finite time is required for carbon dissolution, which depends on bath chemistry, temperature and percent of carbon required.

Following are some typical charge makeups for gray, ductile and malleable iron:

Gray Iron—A typical charge for gray iron base metal consists of steel bushelings, foundry returns, lump ferrosilicon, lump ferromanganese and charge chrome. This is melted to yield a base iron composition as follows:

Total Carbon, 3.2–3.4%
Silicon, 2.2–2.4%
Manganese, 0.50–0.55%
Sulfur, 0.03–0.04%
Chromium, 0.26–0.30%
An inoculant can be added to the ladle for chill control.

Ductile Iron—A typical charge for ductile iron may consist of steel bushelings, foundry returns and a low-sulfur carbon material containing 0.25% maximum sulfur. After treatment and inoculation, the composition analysis was:

Total Carbon, 4.00–4.10%
Silicon, 0.95–1.05%
Manganese, 0.35–0.40%
Sulfur, 0.020–0.025%

Malleable Iron—A typical charge for malleable iron consists of steel bushelings, foundry returns, lump ferrosilicon, ferroboron and ground petroleum coke (with 1.00–1.25% sulfur). This charge makeup resulted in a tap-out composition of:

Total Carbon, 2.50–2.70%
Silicon, 1.30–1.50%
Manganese, 0.35–0.40%
Sulfur, 0.03–0.04%
Boron, 0.0030–0.0035%

Chapter 13 will cover the quality control checks made on the molten metal before it is poured. This will also include an explanation of inoculation and treatment.

Preheating—At this point, it is important to determine if the coreless induction furnace is batch or tap-and-charge operated. Why is this important? Remember, from Chapter 10, when using the tap-and-charge method, a measured amount, by weight, of molten cast iron is tapped out of the furnace and replaced by a similar amount of solid charge. Alternatively, in the case of channel induction furnaces, there always has to be a given amount of molten metal in the furnace. Both of these cases will present a hazard should there be any moisture in the charge materials being placed in the furnace. To overcome this potential hazard, the charge can be preheated.

Preheating the charge materials for induction furnace melting operations may offer the operator many advantages, the main one being improved safety conditions, especially when a heel of molten iron is left in the furnace. Other advantages may be:

- economic;
- ability to increase production rates;
- reduced air pollution within and outside the foundry;
- reduced nonferrous contamination for iron melting operations.

Because of the safety and economic advantages, many foundries using induction melting have gone to preheating or are considering it.

The first equipment used were simply "dryers" that would evaporate any moisture and burn off some oil that may be present in the charge materials. There were some disadvantages to this method, among which was that the cycle in the drier was longer than the melting cycle in the furnace. With increased interest in the use of induction melting processes, new concepts of "preheating" were introduced. These new methods provided high rates of heat input that allowed the preheating cycle to keep up with the melting cycle. The higher preheating temperatures also reduced the amount of oxidation and scale formation on the surfaces of the charge materials.

Either electricity or gas may be used in a preheater to provide the heat energy to preheat the charge materials. The charge materials are usually placed on a conveyor that moves the charge materials through the preheater. Another method of preheating is to use a charge bucket to which gas burners can be attached at the bottom, and then disconnected. After preheating, the bucket can then be used to place the preheated charge materials into the furnace. Any flammable materials in the charge to be preheated are burned in a preheater. With the excess dilution air and high temperatures, most of these materials burn completely, and the small amount of smoke is well diluted. The amounts of emissions discharged, and their collection requirements are beyond the scope of this textbook. For "Current Emission Factors," see the Environmental section of the AFS web site.

The high temperatures in the preheater also tend to reduce any contamination from nonferrous materials in the charge. Such materials as lead, tin, zinc and aluminum tend to burn off or melt in the preheater. However, removal of these materials is not complete, so the results may only be a minor benefit.

Thin scrap along with high burner temperatures can lead to "bridging" in the preheater. This is caused by the premature melting of the thin scrap. In addition, the sizing of the material to be preheated must be such that there is sufficient permeability to allow the hot gases to flow through the charge material without restrictions.

Cold-Start Batch-Type Melting—For cold charging in a batch-type operation, there is no need to preheat for safety reasons, because the furnace is empty (although the lining may be hot). If the furnace has been shut down for repairs and the lining has cooled down, charge materials of large enough size and/or starter blocks should be used. Large enough means that the charge materials should be bulky and heavy. Starter blocks are produced in the foundry by pouring cast iron into cylindrical blocks that will fit into the furnace. These blocks should not fit very tightly because they will expand before melting and cause damage to the furnace lining. The lid of the furnace should be swung shut, the power should be applied to the furnace and the melting cycle will begin.

The melting cycle begins with the magnetic field penetrating the charge materials, as well as the starter blocks, if used. Remember, the magnetic field enters the charge materials and meets resistance to its passage. This resistance causes heat to be generated, which, in turn, starts the melting process. The magnetic field does not pass completely through the charge materials but only to a given depth. The depth of this penetration is determined by:

1) change in direction of current,
2) frequency of change—high frequency-less penetration, low frequency-greater depth penetration.

A check of the watt-hour meter will indicate the status of the melt, or the furnace lid can be swung open to get a visual check of the melt. While the lid is open, one can see the meniscus (curved surface of a liquid in a vessel) that is formed as the power is turned on. This meniscus is formed by the stirring action within the melt. The greater the power applied to the furnace, the larger the meniscus will be.

If needed, additional charge materials may be added to the furnace. To get the maximum usage of the power applied to the furnace, the melt should be at least as high as the top of the coil. If more charge material is to be added to the bath of molten metal, it is highly recommended that this material be preheated for safety reasons. After the charge material has been added to the bath of molten metal, the bath temperature will drop some. The lid should be swung closed and the power applied to the furnace to continue the melt cycle.

When the melt cycle is ended (again, this can be checked by looking at the watt-hour meter), the lid is swung open. The next operation is the slagging-off operation. Some furnaces are equipped so that they can be tilted backward a given amount to aid in the removal of slag from the surface of the bath or melt. This operation can be aided by placing a slag coagulant on the surface of the slag. This material causes the slag to become "crusty" (like a scab), which helps makes removal easier. If the furnace cannot be tilted backward, the bath is slagged by "scooping" the slag off the surface of the bath. This can also be aided by using a slag coagulant.

After the slagging operation is completed, the temperature of the bath is taken. If the temperature is within specification, the furnace is tapped. Since this is a batch-type operation, the bath is tapped into a ladle or ladles and taken to the pouring area or to a holding furnace. In some cases, the molten metal is tapped into a launder, which directs it into a holding furnace.

After the furnace has been completely emptied, the lining can be visually inspected and necessary repairs can be made. These repairs would only be patching type repairs to very small areas of the lining. For repairs greater than this, the furnace would have to be taken out of service.

Tap and Charge Operation—In this operation, the coreless induction furnace is never emptied until the end of a campaign. In other words, there is always some amount of molten cast iron in the furnace until it is turned off. The operation of the furnace, in general, is the same as just discussed...the big difference being in the charging. In this case, the proper preheating of the charge materials is highly recommended, in order to have a safe melting operation.

The starting up of a cold furnace that is to be used for tap and charging is the same as that for a batch-melting furnace. Once the charge materials have been melted, the power is turned off and the bath slagged off. A given amount, by weight, of molten metal is tapped from the furnace. Next, a given amount, by weight, of charge material is placed into the bath. The bath temperature will drop; however, when the power is turned back on, the temperature will begin to rise and the solid charge materials will begin to melt and dissolve into the bath. The stirring action in the bath will homogenize both the composition and temperature of the bath.

Again, it is to be stressed that the bath level be maintained higher than the coil in the furnace to get the best use of the electrical power being placed into the furnace. Of course, as the time approaches to shut the furnace down, the bath level is allowed to drop and the furnace emptied out completely.

Coreless Induction Melting (Carbon Steels)
Two major developments opened the door of opportunity to use coreless induction furnaces for melting carbon steels. First, the technology was developed to confine the melting operation to a furnace and the refining operation to a vessel. Second, the furnace manufacturers developed sound, strong and reliable solid-state power devices. In the melting of the cast irons, the molten metal does not have to be refined before it can be poured. In the case of the carbon steels, there is the need to do some refining of the molten metal before it can be poured into molds.

The conventional carbon steel melting/refining process, called the *carbon boil,* is not normally done in the coreless induction furnace. The carbon boil can be done in the direct arc furnace, the other type of furnace commonly used to melt the carbon steels. The carbon boil, as will be learned later, is the process by which the carbon level is lowered for producing carbon steels. It is, however, possible to do limited refining in the coreless induction furnace. If a carbon boil is desired, it may be initiated by introducing iron ore, nickel oxide or gaseous oxygen to the bath. Taconite pellets can also be used. Temperature must be very carefully controlled in order to control the rate of the carbon boil.

Coreless induction furnaces can be used to melt all of the common grades and alloys of steel poured in the foundry industry. The size is dictated by the economics of the specific application, and the refractories can be a limiting factor, in some cases. The requirement for air pollution abatement equipment is also a governing factor. Most induction-melted steel is melted in medium- to high-frequency coreless furnaces. This process is essentially a "dead-melting" operation, in that the oxidation level of the bath is intentionally kept as low as possible and no carbon boil is induced. The melting process consists of charging the furnace, melting the charge, making the required alloy additions to adjust the chemistry, adjusting the temperature and tapping the heat.

One of the main advantages of using the coreless induction furnace is the small amount of alloying elements that is lost in the melting operation. As stated before, "what goes in, goes out." This allows for good composition control of the carbon steel through careful weighing of alloy additions. If the composition of the charge materials is accurately known, precise metallurgical control can be realized. This is especially important if expensive alloying additions are used.

Again, the stirring action gives a uniform composition and temperature of the bath. This is especially advantageous when making alloying additions.

The selection of scrap material to be melted is another story. Since the refining practices normally used in direct-arc melting practice cannot be used in the coreless induction furnace, lower grade scrap should not be used. The reason for this is that no distinct slag layer is developed due to the stirring action and the slag volume:metal volume ratio is not high enough to permit normal refining. There are few melting/refining operations performed, therefore, they will be discussed later in direct arc melting of steels.

Steel melting with line frequency coreless induction furnaces is not as common as using the higher frequency furnaces. The reason is that, traditionally, silica linings are used in the line frequency furnaces. The silica lining is not as refractory as some of the other lining materials used and the erosion of the lining takes place faster. In addition, active aluminum, calcium, magnesium, carbon and titanium attack the silica lining if added early or in large quantities in the melting cycle.

The replacement of the silica lining with other refractories allows oxidation of silicon and manganese at lower temperatures and carbon at higher temperatures. These refining reactions are the same as with other steelmaking processes. However, adequate slag handling provisions have to be available. The ease of controlling the temperature and the stirring action favor this process.

Coreless Induction Melting (High-Alloy Steels)
Heat-resistant and corrosion-resistant high-alloy steels, stainless steels, etc., can be melted in induction furnaces. Investment casting foundries almost universally use induction furnaces for melting purposes.

Since there are no electrodes (direct arc furnace) or fuels (cupola or gas fired furnaces) used, the molten metal in the furnace contacts only the refractory and the atmosphere. Melting in a vacuum or under an inert gas cover can eliminate the latter. **Figure 11-34** shows a sketch of an argon cover being used to melt with an inert gas. In general, with good refractory practice, there is no contamination during melting, except that involved in the interaction with the atmosphere. Therefore, the alloy can be melted with almost no change in composition.

Again, the stirring action in the coreless induction furnace is beneficial in temperature control and homogeneity of the chemistry of the molten metal. With only limited refining capability, compared to other melting processes, induction melting does offer the best method of holding tight specifications on a variety of alloys.

Refractory practice and the types of refractory used in induction furnaces, when melting the high-alloy steels, can be very involved. This information is beyond the scope of this text and best left to the refractory producers and suppliers.

Metal-to-slag interactions typically, although not exclusively, may involve the following:

1. Interaction with the atmosphere.
2. Intentional slag covers.
3. Intentional slag pH adjustments.
4. Slag viscosity adjustment additions.

Slag on the surface of the bath may be intentional due to the interaction with the atmosphere, originating from "dirty" charge materials and/or resulting from metal-to-refractory reaction. The slag can be acidic, neutral or basic and viscous or non-viscous in nature. The typical slag that forms naturally during melting in the atmosphere is highly oxidized and acidic. Because of this, there is an attraction of readily oxidizable elements to the slag. This also

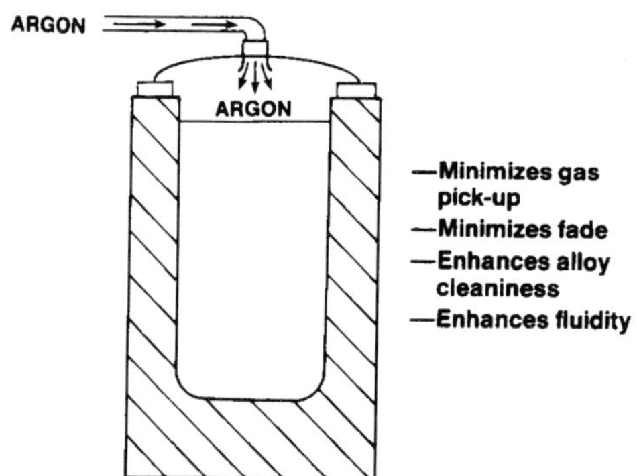

Fig. 11-34. Induction furnace argon cover gas arrangement.

causes the fade of certain elements and subsequent enrichment of others in the bath until equilibrium is reached in the slag between the bath and the atmosphere.

Intentional additions of lime (CaO) or calcium carbide (Ca_2C) to the bath surface can improve sulfur removal (desulfurization) efficiencies. On the other hand, intentionally adding silica (SiO_2) to the bath creates a highly acidic slag, which facilitates the removal of boron or the optimal addition of elements that will improve machinability.

Ideally, the refractories used in induction furnaces are considered inert, with their service life limited by erosion and mechanical (charging) wear, abrasive wear and thermal fatigue situations. Chemical interactions between the melt and the lining are neither desired nor normally considered. However, lining life/wear (reactivity) in contact with the melt is directly related to the state of oxidation, contamination, superheat temperature and time at temperature.

Even without the occurrence of severe lining breakdown and/or erosion, there are many nonmetallic particles present in the bath. These are due to refractory particles, oxides from the charge materials, additions, and oxides formed during meltdown. The stirring action in the bath serves to force the molten metal in the bath away from the refractory wall, while attracting nonmetallic materials toward the wall. This buildup can proceed to where melting efficiency is reduced, pouring is affected and charging interference is encountered. An example of this phenomenon is shown in **Fig. 11-35**.

Fig. 11-35. Nonmetallic particle buildup on an 88% Al_2O_3 crucible.

The sequence of the various grades of high alloy steel to be melted should be taken into account. Some of the elements in one grade may be bad for the next grade to be melted in the same furnace. The elements can be found in the skull left in the furnace from the previous heat, or they could have leached into the refractory wall. While it would be ideal to change the refractory lining or run a wash heat with each grade or alloy change, it is nearly always impractical from an economic and productivity standpoint. A wash heat would be when an alloy is melted that will leach out the elements in the refractory wall and absorb the skull. This metal would then be tapped out of the furnace. The foundry metallurgist and furnace operator should schedule the sequencing of alloys or grades to be melted.

Temperature control of the bath is very critical. Too high a temperature can cause excessive lining wear, gas pickup in the bath, certain casting defects and additional energy costs. On the other hand, tapping to cold can cause certain casting defects and excessive furnace skulling.

The fade or loss of elements during melting can be related to 1) reactivity to the atmosphere and 2) partial pressure relationships. Of these two, the atmosphere is the prime cause of fade in air-melted alloys. When the bath contacts the atmosphere, various elemental constituents in the alloy will react with the oxygen and nitrogen in the air. The reaction of the molten metal with oxygen is greater than with the nitrogen.

Elemental fade results are dependent on:

- alloy type
- furnace size
- melt frequency (stirring action)
- melt time
- melt temperature
- melt stock quality

The primary gases in metal—hydrogen, nitrogen and oxygen—can originate from several sources. These sources include the atmosphere, furnace refractories, charge materials, rust and oil additions and turbulence in the bath. In actual practice, degassing additions can cause more harm than good because they can increase nonmetallic inclusions or put elements, such as silicon, above the specification.

In some instances, degassing additions are made to alloys that don't need degassing due to their tolerance for dissolved gases. Also, some degassing additions that work effectively on some alloys will work marginally or create problems in other alloys. **Table 11-2** lists the alloy systems and their relative accommodation of dissolved gas. Those listed as "low" most often will require degassing. Those in the "moderate" group may require degassing upon successive remelting (such as gates and risers). Those listed as "high" most often can be melted and poured without degassing additions.

Channel Induction Melting

As learned in Chapter 10, the channel induction furnace has to have a bath of molten metal in it at all times during operation. This means that if the furnace has been shut down for repairs and is ready to be placed back into service, it has to have a source from which *molten* metal can be obtained. Once the furnace is back on line, subsequent charging operations can be made into a molten metal bath.

Table 11-2.
Relative Rating of Alloy System to Accommodate Dissolved Gas

Low	Medium	High
Carbon Steels	Tool Steels	Mg Steels
Low-Alloy Steels	400 Series SS	300 Series SS
Ni-Cu Allopys	17-4	Ni Alloy B
Si Irons	15-5	Ni Alloy C
Sealing alloys		Ni Alloy X
		General Ni-Base
		Co-Base
		N-155

The channel induction melting furnace lends itself well to melting, holding and duplexing and, as mentioned earlier, it is the most electrically efficient of the two types of induction furnaces used in foundries. Holding or duplexing means that the molten metal is held at the same temperature as it was when it came out of the primary melter, or raised, if necessary, until it is needed for pouring into the molds. Some people say that, during duplexing, final chemical adjustments can be made to meet specifications.

This furnace can be used in conjunction with a cupola, coreless induction furnace and direct arc furnace in the melting operations; these latter furnaces being the primary melting furnaces with the channel furnace being either a holding or duplexing furnace.

The channel induction furnace, by its very nature, is a tap-and-charge-operated furnace. Molten metal is "charged" into the bath when molten metal has been tapped out. Very seldom, if ever, is the furnace completely emptied, except for major repairs or planned maintenance.

However, the channel furnace *can* be used as a primary melter. When it is used in this manner, it is very important to pay attention to the charge materials, for safety reasons. It is strongly recommended that the charge materials be properly preheated. Remember, when charging into a liquid bath, wet charge materials can cause an extreme safety hazard.

The same types of charge materials can be used as in the coreless induction furnace. Because the melt cycle will be started by pouring molten metal into the furnace, there will be no need for starter blocks. The same quality checks should be made of all solid charge materials being placed in the channel induction furnace. The same statement can be made when using this furnace as was made for the coreless induction furnace: "garbage in, garbage out."

The amount of stirring action is dependent on the height of the bath above the inductor channel throats, the density of the molten material and the power applied to the furnace. The higher the bath in the furnace and the denser the material, the higher the ferrostatic pressure. Thus, greater power will have to be used to increase the movement of the molten metal through the inductor channels upward into the bath. In addition, by gradually reducing the cross-sectional area of the channel(s) from the inlet to the outlet, velocity at the outlet of the channel(s) can be increased. Also aiding in the stirring action are the convection currents in the bath. The hotter molten metal leaving the channel inductor will rise to the surface of the bath and the colder molten metal will sink toward the bottom.

Although not so much of a problem in ferrous melting, precautions should be taken to prevent the channels from plugging. Slag or dross normally causes the plugging of the channels. Clean-out plugs are built into the inductor so that the channels can be periodically cleaned, if necessary.

Minor repairs to the refractories in the bath area of the furnace can be made by lowering the bath level or, if necessary, draining the furnace. Again, molten metal will have to be available to restart the furnace if it has been drained. Inductor lining repair of a vertical bath furnace will require draining the furnace, removing the inductor and replacing it with another inductor. However, in the case of a drum-type channel furnace with two separate inductors, the drum can be rotated and the bad inductor can be removed without draining the furnace.

Electric/Direct-Arc Melting (General)

The electric/direct-arc furnace was described in Chapter 10. The intent here will be to discuss the actual operation of the arc furnace for producing steel and cast iron. As learned in Chapter 10, the name of this metal melting method describes how this melting is made to take place. An electric arc is struck between an electrode and the metallic charge materials. This arc can reach temperatures up to 7000F (4000C).

The function of the electrodes in an arc furnace, or single electrode in a direct arc furnace, is to bring electrical energy into the furnace. Through the formation of an arc, the electrical energy is converted to heat.

Graphite electrodes are manufactured in two grades: regular and premium. The big difference is the pitch impregnation process used with the premium grade. Graphite electrodes can be seen in **Figure 11-36.** The raw materials are a mixture of selected and sized petroleum cokes and a coal tar pitch binder. These raw materials are then crushed, milled and sized. The grade and size of the electrode to be made dictate the amount of these materials used. The cooled mixture is then extruded into rounds under closely controlled temperature and pressure. These interim products are called "green" electrodes.

The green electrodes are next baked in a gas-fired furnace. The baking cycle takes up to four weeks, and reaches peak temperature at 1700F (926C), which carburizes the pitch, getting rid of decomposition products and bonding the coke particles. Premium grade electrodes are then impregnated with pitch, under pressure in an autoclave, to obtain greater strength and density.

The baked electrodes are next packed into an electrical resistance furnace for graphitizing. At temperatures of over 5000F

Fig. 11-36. Electrodes in use in a large arc furnace. [Photo courtesy of SGL Carbon]

(2760C), the hard carbon converts to graphite. This cycle takes up to four weeks. The graphite electrodes are then machined to provide accurate faces and threaded sockets.

Electrode diameters range from 3 in. (7.62 cm) for small furnaces to 28 in. (71.12 cm) for a 400-ton (406,400-kg) furnaces. Although the transformer rating and the diameter and capacity of the furnace are the primary factors governing electrode diameter, there are other factors to be considered. **Table 11-3** shows the current carrying capacities of various diameter electrodes.

In general, the smallest electrode diameter should be used that will permit the maximum rate of production without excessive electrode consumption, due either to overheating (oxidation) or breakage. Such a rule should be based on comparative trials in the foundry or experiences under similar conditions at another foundry.

Electrodes are manufactured in lengths of 5–8 ft (152–244 cm) with a threaded socket in each end. Separate nipples are supplied to screw into the sockets to join electrode sections into an electrode column as shown in **Fig. 11-37**. An electrode column is usually made of two or three sections of electrodes joined together. The electrode columns are held in place with electrode clamps. When the upper section becomes to short to be held by the clamp, a new section is added to the upper end of the column.

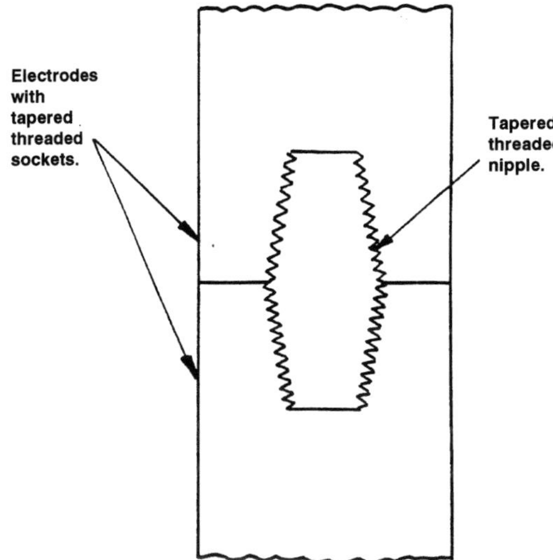

Fig. 11-37a. Sketch of typical electrode joint.

Fig. 11-37b. Nipple of one electrode section. [Photo courtesy of SGL Carbon]

Table 11-3.
Typical Properties and Current Carrying Capacities of Graphite Electrodes

Electrode Diameter (In.)	Cross Sectional Area (In.^{2})	—Current Carrying— —Capacity (Amps)*— Regular	Premium
3	7.1	—	1,200–2,500
4	12.6	—	1,600–3,500
5-1/8	20.6	—	2,500–4,800
6	28.3	—	3,500–6,000
7	38.5	4,600–7,800	6,000–9,000
8	50.3	5,400–9,100	7,000–11,500
9	63.6	7,000–11,500	9,000–13,000
10	78.5	8,000–13,000	11,000–18,000
12	113.0	11,500–17,400	15,000–25,000
14	154.0	15,000–24,000	20,000–35,000
16	201.0	20,000–30,000	25,000–40,000
17	227.0	24,000–33,000	28,000–45,000
18	255.0	26,000–36,000	30,000–48,000
20	314.0	30,000–42,000	35,000–55,000
24	453.0	40,000–55,000	48,000–85,000

**These capacities may vary from the listed limits, depending on operating conditions.*

The rates of electrode losses due to oxidation are mainly a function of furnace temperature, the oxidizing nature of the gases in the furnace and the volume flow rate of gases through the furnace. The taller the furnace and the larger the electrode, the greater the surface area exposed to oxidation.

The rate of tip wear of the electrode is greatest during meltdown when the highest inputs and arc currents are used. During this period, the furnace temperature is low—1650–2730F (899–1499C)—making oxidation losses on the electrode surface low. After meltdown, the power inputs and arc currents are greatly reduced, and tip wear is minimized. However, the temperature in the furnace is now about 2730–3450F (1499–1899C), and oxidation loss at the surface increases. Thus, it is obvious that the judicious control of power input can result in reduced electrode consumption.

Graphite electrodes, by nature, are brittle and easily damaged. By far, the most frequent cause for electrodes breaking is due to the falling and caving in of scrap during meltdown. In order to minimize this problem, large, heavy scrap should be placed toward the bottom of the charge. Large plates and long rails should be carefully placed. Nonconducting material in the charge can also lead to electrode breakage. For instance, sand-coated foundry returns, ladle skulls, etc., beneath an electrode can cause the arc to go out and cause the automatic electrode control device to drive the electrode into the charge, causing it to break.

Finally, proper maintenance of the electrode masts, clamps and regulators is required. In addition, the ports through the furnace roof must allow easy movement of the electrodes. Electrode joints must be tightened to the recommended torque to prevent loose joints that can lead to electrode breakage.

Electric Arc Melting (Carbon Steel)

There are two types of arc melting practice used for melting steel: acid melting practice or basic melting practice. These two practices differ in the pH of the furnace slag and refractories.

Acid Melting Practice

The term acid practice implies that the bath of molten steel is contained by acid refractories (silica) and that the slag is acid in chemical behavior. Sulfur and phosphorus, which can lead to steel casting defects, cannot be removed to any significant degree in the acid practice. This results in the need to carefully control the level of these two elements in the charge and slagmaking materials.

Charge Materials—The acid electric arc furnace charge materials normally consist of foundry returns such as gates and risers, scrapped castings and purchased scrap. The purchased scrap can be forgings, springs, rails, stampings and other such materials. These purchased materials should contain low amounts of sulfur and phosphorus. The incoming scrap should be sampled for these chemical elements. Basic electric steels are excellent charge materials for the acid melting practice.

The amount of foundry returns used in the charge is usually 30–40%. However, up to 100% foundry returns can be used if the chemical analysis is carefully controlled. In addition, the foundry must have a method of segregating the various steel alloy grades poured in the foundry. Bad heats of molten steel have been made because the wrong foundry returns were used.

The scrap should be below 0.05% for both sulfur and phosphorus, and oxidation losses tend to concentrate the sulfur and phosphorus to a slight degree. Most steel specifications set maximum limits of 0.05% on these two chemical elements. Other residual chemical elements that can cause problems in heat treating and welding must also be controlled. These residual elements, in addition to sulfur and phosphorus, include copper, tin, chromium, nickel, molybdenum, tungsten and cobalt.

The size and physical condition of the scrap should be carefully controlled. The following observations have been made concerning visual scrap inspection:

1. Oversize scrap can cause furnace damage and delays.
2. Excessive rust can cause low metallic yield and a low meltdown carbon.
3. Excessive oil is a source of hydrogen and air pollution.
4. Sealed containers are a potential explosive hazard.
5. Suspicious scrap should quickly be set aside.
6. Siliceous materials will increase slag volume.
7. Some foreign scrap has entered the U.S. that was radioactive, and may continue if not caught at the ports of entry. (This is also true for scrap used to produce cast iron and high-alloy steels.)

The placement of the charge materials in the charging bucket can also effect the furnace operations. A recommendation for loading the bucket is illustrated in **Fig. 11-38**. The bottom of the charge bucket is made up of borings, turnings and light scrap. These materials will act as a cushion as the charge is placed in the furnace, and protect the refractories on the bottom of the furnace. Next, heavy scrap is placed in the electrode triangle that prevents the heavy scrap from falling against the electrodes and possibly breaking them. Above the heavy scrap is placed the bundles and medium scrap, which absorb heat and protect the furnace sidewalls. The top of the charge consists of the same materials placed in the bottom of the charge bucket. When the furnace power is turned on, these top materials will allow fast penetration of the electrodes, which protects the roof and sidewalls of the furnace from arc glare. A typical charge is made up of 35% heavy scrap, 40% medium scrap and 25% light scrap.

Usually, the entire charge is placed in the furnace at one time. However, if the available scrap is not dense enough, it may be necessary to add the second bucket (backcharge) after 60–70% of the first charge has been melted. The main charge is placed in the furnace after the roof of the furnace has been swung open. Once the charge is in place, the roof is swung back over the furnace and melting can begin. If an additional charge is necessary, the procedure is repeated and melting can be started. Additional small charges or alloy additions can be made through the charging door in the back of the furnace.

Melting and Oxidation—The number one transformer tap is usually used for meltdown; however, the number two tap may be used to begin meltdown to help start the electrodes into the charge. By getting the electrodes buried into the charge quickly, excessive wear on the roof and sidewalls is reduced. After the electrodes are buried deep enough in the charge, the operator may switch back to the number one tap and continue the meltdown cycle. A pool of molten steel will form in the bottom of the furnace. The electrodes keep burrowing through the charge until they reach this pool of molten steel. Since there is thin, light scrap at the bottom of the furnace, this pool is quickly built up. When the electrodes reach this pool, they stop burrowing, back off slightly and maintain the arc. This is done automatically by the automatic controls on the furnace. If this pool were not formed, the electrodes would continue to burrow and could burn through the bottom of the furnace.

From this point on, the remaining charge materials will be melted by the heat radiation coming off the pool of molten steel, and arc splash off the electrodes. Some solid charge materials may fall into the pool of molten steel and actually be dissolved. As the pool of molten steel increases in volume and rises, the electrodes continue to rise, still maintaining the arc.

After the charge has been melted, the voltage to the electrodes is reduced and the refining period begins. This is done because the arc is now being struck between the electrodes and the slag covering the bath. By reducing the power into the electrodes, slag splashing from under the arc and the arc splash itself against the sidewalls and roof of the furnace is reduced, thus reducing refractory wear.

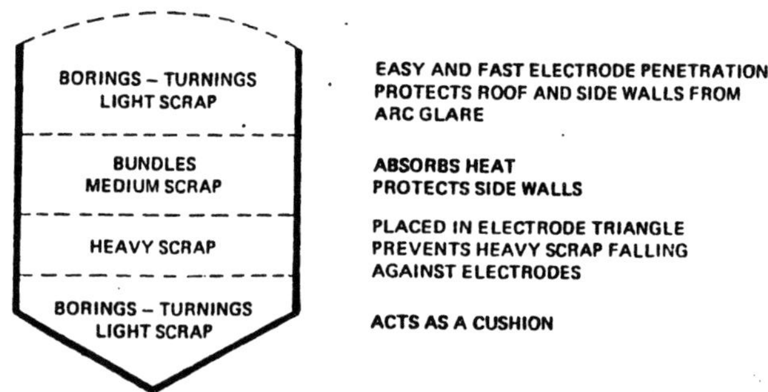

Fig. 11-38. Ideal charge makeup for loading the charge bucket.

It should be pointed out that the arc acts as lightning when it passes through the air during a storm. Lightning changes the nitrogen in the air from diatomic nitrogen (N_2) to nascent nitrogen (2N). The nascent nitrogen is absorbed by the rain and carried into the ground. Similarly, the arc between the electrode and the bath does the same thing. However, in this case, the nascent nitrogen is picked up by the heat of molten steel. This nitrogen has to be removed from the heat in order to prevent nitrogen-type defects in the steel castings.

Slag Control—Slag materials in the acid melting practice are silicon dioxide (SiO_2), iron oxide and manganese oxide, plus oxides of calcium and aluminum. The SiO_2, or silica, comes from the erosion of the furnace bottom and sidewall refractories, silica sand on foundry returns, sand shoveled into the furnace and a small amount from the reaction of iron oxide and silicon to form SiO_2. The calcium oxide comes from the addition of limestone or lime, and manganese oxide comes from the deoxidation product of the reaction of manganese and iron oxide. The iron oxide comes from the rust on the scrap charged into the furnace and its oxidation while the steel is being melted. Aluminum and magnesium oxides are usually present in very small percentages and of little concern when producing acid electric steel.

The effects of the slag constituents on slag fluidity can be stated as follows: the higher the total of basic materials, the higher the fluidity will be; and the higher the percentage of SiO_2, the more viscous the slag will be.

It is very important that the furnace operator know the approximate analysis of the slag when using the acid melting practice. If the heat melts with a high silica (SiO_2) content, the melt becomes very sluggish. This can cause an actual pickup of silicon into the melt, due to a lack of oxygen in the melt. In heats of this type, it is very hard to get an active carbon boil. On the other hand, if the heat melts with an extremely high iron oxide (FeO) content in the slag, it is an indication that over-oxidation has occurred in the meltdown.

Pouring a slag test cake can check the amount of iron oxide in the slag. The color of the slag surface and the fracture surface gives an indication of the iron oxide content. The problem is that slag colors change considerably with the manganese and calcium oxide contents in the slag. The density of the slag cake fracture is considered a better way of estimating the iron oxide content of the slag.

Slag sample extractions and the slagging operation are performed through the charge door. The slagging operation consists of "raking" the slag off the heat and out the charge door into a container or pit.

Refining—At the end of meltdown, the carbon levels are usually too high. To reduce the carbon levels, a "carbon boil" is used. The carbon boil can be done by introducing mill scale or iron ore to the bath or by introducing gaseous oxygen to the bath. The generally accepted practice is to use gaseous oxygen. The gaseous oxygen is introduced into the bath through a lance that is placed into the furnace through the charging door and submerged into the bath. The lance is moved around in the bath so that the gaseous oxygen is well distributed throughout the bath. The principal reaction during the carbon boil is the oxidation of carbon to carbon monoxide (CO).

The CO bubbles rising through the bath produce a vigorous bubbling action. As these bubbles rise through the bath, hydrogen and nitrogen in the bath diffuse into the CO bubbles and are removed. This, by the way, is the only practical way of reducing dissolved hydrogen and nitrogen, other than using a vacuum melting practice. The vigorous bubbling also causes a stirring action that produces a more even chemical composition and temperature distribution in the bath.

When the lab tests come back to the operator, adjustments will be made to the carbon level by recarburization (increase) or more carbon boil (decrease) to the required carbon level. A final check is made on the carbon level and, if it is satisfactory, the heat is "blocked" by adding ferrosilicon to the heat. Blocking stops the carbon boil and oxidation in the heat.

Final alloy additions are made to the heat to bring it up to final chemistry. After these additions are made, the heat temperature is raised for a few minutes to ensure that the additions have melted and that the desired tapping temperature has been reached.

Deoxidizing and Tapping—After the addition of deoxidizers for the blocking action and other additions are made, the heat is ready for tapping. Final slag and temperature adjustments are made at this time. After the deoxidizers are dissolved, the heat should be tapped as soon as possible, or else the heat will pick up hydrogen and nitrogen. The elapsed time between blocking and tapping should be a maximum of eight minutes, and preferably shorter if possible.

With the preheated ladle in place, the furnace is tilted rapidly, which allows the molten steel to be poured into the ladle. This rapid tilting allows the slag to be held back and the molten steel flows into the ladle first. The tapping of an electric furnace is seen in **Fig. 11-39**.

Alloys and final deoxidizers are added at the base of the molten metal stream as it fills the ladle. Deoxidation is carried out by adding deoxidizers to the molten steel. The most commonly used deoxidizer is aluminum, since it is the most potent deoxidizer. Calcium-silicon alloys can also be used. Good tapping practice ensures a compact molten metal stream, uniform tapping times and slag-metal separation. This results in minimum reoxidation of the molten metal, minimum nitrogen pickup, more uniform and higher alloy recoveries, simplified alloy addition practice, and a stirring action in the ladle.

Fig. 11-39. Tapping of an electric arc furnace.

In order to obtain uniform alloy recovery, temperature control and chemical control, the following is recommended when tapping the furnace:

1. Add ferroalloys to the stream or at the base of the stream in the ladle.
2. The proper sequence of alloy addition should be first carbon, then manganese and silicon followed by calcium.
3. All additions should be made by the time the ladle is 75% full.
4. Ladle additions should be minimized.
5. Additions should be carefully weighed.
6. Proper alloy size should be based on heat size.

Basic Melting Practice

Many aspects of the basic melting practice are similar to that of the acid melting practice. The basic melting practice will be discussed in this section in terms of operating practices and controls as they differ from the acid melting practice.

The term "basic practice" means that the heat of molten steel is held in a basic refractory lined furnace and that the slag is generally basic in chemistry. The main reason for using the basic practice is that sulfur and phosphorus may be removed from the molten steel in the furnace. This is required for the production of steel castings that need improved impact properties. In addition, in some cases, the basic practice can be more economical for plain carbon grades, where lower quality scrap may be upgraded. The basic melting practice is almost universally used for high-alloy and stainless steels as well as manganese steel.

Charge Materials—Charging practices in the acid melting operation also pertain to the basic melting operation. The typical charge may consist of 30–50% foundry return scrap, 30–50% purchased scrap, 5–15% basic pig iron and 3.5% limestone. The charge is generally limited to 0.10% phosphorus and 0.08% sulfur, and will frequently run at lower levels.

The foundry returns, such as gates, risers and rejected castings, should be free of excess foundry sand, since this material will lower the basicity of the slag and reduce the slag's dephosphorizing ability. All materials in the charge, and all later additions, should be free of moisture to prevent hydrogen pickup in the heat. As discussed earlier in this chapter, some operators will preheat the charge and the additional materials before placing them in the furnace.

Melting and Oxidation—Meltdown is accomplished in the same manner as with the acid melting practice. The voltage to the electrodes is systematically reduced as the meltdown nears completion. The limestone or lime is added to give a calcium oxide (CaO) content equal to about 2.5% of the charge. Should the meltdown carbon be low, carbon should be added to the heat to ensure a minimum carbon level of 0.30% above the desired finish carbon level, or even 0.40% for high grade alloy steels.

The carbon boil is done in the same manner as with the acid melting practice. Plain carbon steels are usually "blown" (carbon boil) in one increment, while the alloy steels are blown in two increments. The reason for the latter is to allow for an interim chemical analysis. Sometimes fluorspar additions are made during the carbon boil to control the foaming of the slag and to improve slag fluidity. The addition of burnt lime is usually found necessary toward the end of the oxidizing period to maintain a highly basic slag.

The function of a basic slag is listed as follows:

1. Removes phosphorus and sulfur from the melt.
2. Prevents surface oxidation of the molten metal.
3. Protects the basic refractories.
4. Removes oxidizable elements.
5. Assists in maintaining a stable arc.

The oxidizing slag, which develops during the carbon boil, appears to be black and brittle when cold but will be reasonably fluid. A basic slag, which appears to be excessively fluid usually, contains excessive amounts of silicates and insufficient calcium oxide. Burnt lime should be added to help control slag viscosity.

Refining and Slagging—There are two common methods used in the refining operation: one is the *single-slag practice* and the other the *double-slag practice*. The single-slag practice allows only *minor* sulfur removal, while the double-slag practice allows *substantial* sulfur removal from the heat. The single-slag practice does produce steel with better fluidity and takes less time.

In the *single-slag* practice, the oxidizing slag is allowed to remain on the heat in the furnace, and calcined lime is added to adjust the lime-silica ratio (V-ratio) in the slag. V-ratio is another name for lime:silica ratio. The amount of CaO to add is best determined through practice with the aimed-for lime-silica ratio of 3:1. Phosphorus is oxidized and is transferred to the oxidizing slag.

The carbon boil follows the same practice as discussed in the acid melting practice. It is good practice to bring the heat to finishing temperature, gradually, rather than leaving it to fall at the end of the refining period. This allows a slight carbon boil to continue throughout the heat, which helps prevent gas pickup at the end of the heat.

After the heat has reached the desired carbon level, it is ready for a partial block. Most heats are blocked with ferrosilicon or silicomanganese. Chrome and nickel additions are made shortly after the block.

Ferromanganese is added and the heat is tapped within 1.5–2 minutes. Final additions are made to the ladle and the heat is deoxidized with aluminum, calcium-silicon alloys or other deoxidizers. Frequently, lime is added to the slag in the ladle to chill the slag, and inhibit phosphorus reversion.

The *double-slag* practice is so named because one slag is used to remove phosphorus and another slag is used to remove sulfur. The first oxidizing slag is completely removed after dephosphorization, to prevent the phosphorus from reverting into the heat during the reducing conditions present during desulfurization.

The carbon boil proceeds in the same manner as in the single-slag practice. When the carbon level has reached the desired level, and the carbon boil has begun to slow, the oxidizing slag is completely removed from the bath. The electrodes are raised, and the slag is removed by using rakes.

After slagging, the oxygen content of the bath should be substantially reduced before applying the reducing slag. Again, adding lump ferrosilicon, ferromanganese, or both, to the bath accomplishes this initial blocking. Usual practice is to add about half of the total ferrosilicon and ferromanganese required for the heat at this time.

The material used to produce the reducing slag is premixed lime, fluorspar (spar) and powdered carbon. An amount of this mixture equal to about 3% of the bath weight is added to the furnace. Should these materials not be mixed, the lime and spar are added first, and the resulting slag dusted with the powdered carbon. The slag should be stirred (rabbled) to ensure good slag-metal contact.

The voltage to the electrodes should be reduced, at this time, to prevent overheating the bath and undue fluxing of the roof and sidewall refractories. The slag, when cooled to room temperature, will be black during the first part of the refining period and go through some color changes as the refining period progresses. When the slag reaches a white or gray color, it indicates the formation of excess stable calcium carbide. The continual addition of powdered carbon to the slag allows the reactions to proceed and refining of the sulfur to continue.

Deoxidizing and Tapping—When the slag reaches the white or gray stage, final deoxidizers and alloy additions should be made, and the heat can be tapped. The final additions of manganese, silicon and chromium can be made through the white or gray slag. If recarburization is necessary, it is best done by adding low phosphorus, low-sulfur pig iron, or by carbon injection with an inert carrier gas.

About ten minutes after the alloys have been added, the heat is ready for tapping. Bath temperature is adjusted for tapping, usually in the range of 2950–3050F (1621–1676C). Tapping and final deoxidation are carried out the same as for the single-slag practice.

Electric Arc Melting (Alloy Steel)

The melting practice for low-alloy steels is essentially the same as for plain carbon steels. Alloys containing nickel, copper and molybdenum are generally added with the charge, as they are not readily oxidized. On the other hand, alloys such as chromium and manganese are not added until a carbide slag (reducing slag) is obtained, since they oxidize very readily.

In cases of large alloy additions, care should be taken not to chill the bath too abruptly, and time should be allowed to ensure complete melting. Thorough rabbling of the bath will ensure complete mixing.

Stainless steel is seldom melted using a carbide slag. Instead, powdered ferrosilicon is added to the slag to produce reducing conditions, or an alumina slag is used. If stainless steel scrap is used, the single-slag practice is generally used.

High-manganese steel (12–14%) is always melted with the basic slag melting practice. The furnace bottom is made of sintered magnesia. The single-slag practice is used with lime being added during the melt period. A good basic slag is maintained during melting. After meltdown, the slag is deoxidized with pulverized coke and spar is added to adjust slag viscosity. The heat is blocked with ferrosilicon.

Slag practice plays a large part in the melting of the plain carbon and alloyed steels. These materials also undergo a refining practice that brings about the final desired chemistry.

AOD Refining

In 1967, Joslyn Steel produced the first commercial heat of steel using the Linde AOD process. By the end of 1977, approximately 70–80% of U.S. stainless steel production was made using this process. Most of this production was wrought material, but the AOD process has been found to be a steel processing method that can be used in the foundry as well.

AOD stands for argon-oxygen decarburization, which describes the process. It is the decarburization of molten metal using oxygen with argon or some other inert gas in a vessel similar to the Bessemer converter. The vessel is filled with molten steel from a furnace and the oxygen-inert gas mixture is blown into the bath through tuyeres located in the lower sides of the vessel. **Figure 11-40** shows an AOD vessel. If necessary, up to 30% cold charge can be added to the vessel without significantly affecting the refining efficiency.

The procedure for using the AOD process is as follows: The heat is first melted down in an electric furnace, either induction or electric arc, and the chemistry is somewhat adjusted. The heat is tapped into a ladle, sampled, slagged and transferred to the AOD vessel. A slag is built up in the vessel and the molten steel's temperature is taken.

Oxygen, along with argon or some other inert gas, is injected into the bath to get the temperature to about 3100F (1704C). A sample and temperature are taken again at the end of this cycle. Oxygen and argon are then injected at a lower oxygen-to-argon ratio for a determined length of time, based on the carbon of the sample taken at the end of the first cycle. A still lower oxygen-to-argon ratio blowing cycle is often used at the end, to reach the final carbon level. This last cycle is sometimes skipped if high pouring temperatures are required, because the blowing action reduces the melt temperature.

The temperature of the molten steel is again taken and argon is blown into the heat if the temperature has to be brought down, or oxidizing silicon is used to raise the temperature, if necessary. When the final chemistry adjustments have been made and the temperature is correct, the heat is tapped and ready for pouring. By controlling the AOD process, it is often not necessary to wait for a final analysis before tapping the heat.

Casting quality is often improved by virtue of some of the inherent properties produced by the AOD process. One of the benefits is the way sulfur is reduced without any special steps. The AOD process can be a very useful tool in the steel foundry industry, as its advantages far outweigh the disadvantages.

Fig. 11-40. Five-ton AOD vessel.

Desulfurization

Acid-melted steels can be desulfurized in the ladle. However, for this to be done in the steel foundry, several limitations related to foundry melting have to be overcome.

First, because of the relatively small ladle sizes in the foundry, temperatures losses in the molten steel, during ladle desulfurization, must be overcome. This is done by adding extra superheat in the furnace and having properly preheated ladles to assure that, after the desulfurization cycle is completed, there is still enough temperature in the molten metal to pour the casting(s).

Second, for acid melting, an efficient means of preventing furnace slag from entering the desulfurization ladle has to be developed. If this is not accomplished, efficient desulfurization will not occur. Aluminum losses and altered recovery rates of other oxidizable elements must be established as a function of the desulfurization variables employed for foundry-size heats.

Ladle Slag Control—Chemical reaction-wise, this process works the same way as those in the furnace acid slag practice. The deoxidized molten steel is poured into a ladle in which a high basicity slag has been added. Dry slag mixtures normally used in this process consist of lime, spar, silica, alumina and deoxidizers such as aluminum and silicon. Slag fluidizers, such as sodium and potassium, are also part of this mixture. The slag mixture is added as the molten steel is being tapped into the ladle. The stream of molten steel acts as the mixing force. Extremely low sulfur levels can be reached by this technique if the steel and slag are mixed by external forces such as gas bubbling or magnetic stirring.

When the ladle slag desulfurization practice is used, care must be taken that the acid furnace slag does not enter the ladle. Acid- or basic-lined ladles can be used; however, lining wear will be greater for acid-lined ladles. In addition, the efficiency of sulfur removal will be less for acid-lined ladles.

Because ladle slag desulfurization can be used for refining acid melted steels, they are of great interest to the steelworker.

Powder Injection/Wire Injection—Injection of reactive metals, such as calcium, manganese or rare earths, into the molten steel has come upon the scene. Ultra low levels of sulfur can be obtained by this technique.

The basis of reactive metal injection desulfurization is the chemical combination of the reactive metal with sulfur dissolved in the molten steel. Because the reactive metals are strong deoxidizers, it is necessary to have a low level of oxygen in the molten metal before starting the injection process. Aluminum is used as the deoxidizer. If there was oxygen in the melt, the reactive metal would react with the oxygen, not the dissolved sulfur. It is also necessary to have a slag present that will hold the sulfur removed by the reactive metal.

Electric Arc Melting (Cast Iron)

The electric arc furnace is used as a primary melting tool for cast iron, not as a refining tool, as was the case in cast steel melting. To produce quality molten cast iron in the electric arc furnace, careful attention should be paid to the selection of the materials to be melted and the operating procedure for the furnace.

Charge Materials

The selection of the ferrous charge materials for melting cast iron in an electric arc furnace usually involves a compromise between quality and cost. High-quality charge materials usually imply high cost, whereas lower-quality charge materials imply lower cost. Most foundries operate between these two extremes, attempting to balance quality and cost.

High-quality scrap materials generally exhibit the following attributes:

1. Predictable, uniform and desirable physical size.
2. Known, uniform and desirable chemical composition.
3. A minimum of undesirable constituents.

Low cost, low-grade scrap often contains nonmetallics, such as wood, oil, water, dirt and sand, which are undesirable. Trace elements, such as chromium, molybdenum and tin, are also undesirable and should be controlled.

The ferrous charge materials most commonly used in the electric arc furnace to produce cast iron are pig iron, iron and steel scrap, ingot iron and metallized (pre-reduced) ores. The pig iron can be obtained in many varying chemical compositions. In addition, it is available as primary, secondary and imported pig iron.

Due to cost factors, the use of iron and steel scrap is very common. Carbon is added through carbon-raising materials and/or it can be picked up from ferroalloys. While many sizes and grades are available, the most commonly used are:

- low-phosphorus plate and punchings
- cut structural plate
- electric furnace bundles
- foundry steel, 2 ft (61 cm) and under
- alloy-free turnings

Other types of purchased scrap, such as fragmentized scrap, ingot iron and metallized ore, are not commonly used.

A major portion of the charge is made up of foundry returns. This will include gates, risers and rejected castings. As in the other melting practices, it is important to segregate these materials by alloy content, if several grades are produced. In addition, the materials should be free of any adhering sand or other undesirable foreign material.

All scrap and other charge materials should be stored so that excessive rust and other oxidation is minimized. The scrap should be stored indoors, if possible, to keep moisture (such as snow and ice) from collecting on or in the scrap. If this can't be done, then some of the scrap to be melted should be brought inside a day or two before it is to be melted to give it a chance to dry out.

A typical charge usually consists of steel scrap, foundry returns, lump ferrosilicon, lump ferromanganese and pig iron. This is melted to produce a base cast iron composition as follows:

	Gray Iron	*Ductile Iron*
Total Carbon	3.15–3.30%	3.70–3.90%
Silicon	2.00–2.20%	1.40–2.00%
Manganese	0.60–0.70%	0.40–0.60%
Sulfur	0.05–0.06%	0.035% max.
Phosphorus	0.5–0.06%	0.080% max.

As with the charging of the electric arc furnace for steel melting, the arrangement of the materials in the charging bucket and in the furnace is important. A typical arrangement of the charge materials is illustrated in **Fig. 11-41**. This arrangement in the furnace will help reduce electrode breakage and increase refractory life and carbon recovery. Normally the electric arc furnace is top-charged with an orange peel or clamshell-type of drop-bottom charge bucket. **Figure 11-42** shows an electric arc furnace being charged.

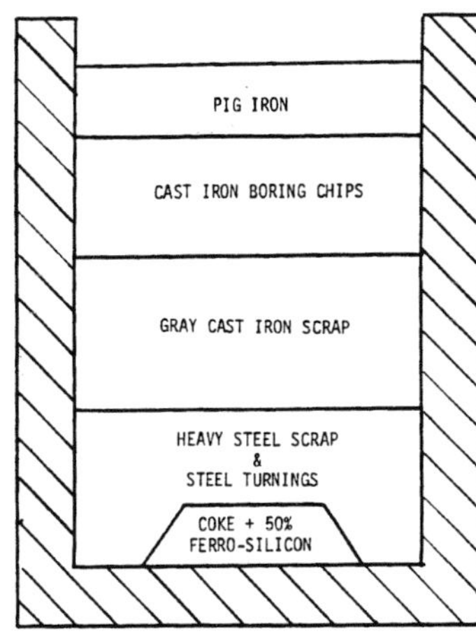

Fig. 11-41. Schematic showing position of charge materials in furnace.

Melting Procedure

Melting of the charge materials begins after the furnace roof has been returned to its operating position. Melting usually begins with the highest voltage transformer tap to give a rapid meltdown time. However, in some cases, meltdown may begin on a lower tap to protect the roof and sidewall refractories from excessive radiation from the arc. Once the electrodes have bored holes into the charge, the switch to high tap is made.

When melting is complete, the temperature is brought up to tapping temperature, and carbon equivalent and silicon analyses are made. At this time, any slag on the surface of the bath may be skimmed off through the charging door. Since it is usually not economical to use the slag for refining, as is done in steel melting, the principles of slag chemistry will not be discussed.

When using an electric arc furnace it is said that the meltdown should be done rapidly, and that when the molten metal reaches the proper tapping temperature the furnace should be tapped. The reason for this is that, as with cast steel melting, the arc is very hot and disassociates the nitrogen (N_2) in the air and can cause nitrogen pickup by the molten metal. In addition, the high temperature of the arc can cause the loss of some of the critical chemical elements in the heat.

Fig. 11-42. Electric furnace being charged.

CE and Silicon Analysis

When the carbon equivalent (CE) and silicon (Si) analysis are determined, the carbon content of the heat is calculated using the following formula:

$$\%C = CE - 1/3\% \text{ Si}$$

The carbon content is then adjusted by either the injection or the addition method. In the *injection* method, carbonaceous material is blown into the heat using a carrier gas and a lance. The operator stirs the bath by moving the lance during the injection period. The carrier gas may be dry nitrogen or dry, filtered air. A fine-grained carbonaceous material, such as powdered graphite with low sulfur content, is used for the injection method. The upgrading of the carbon content using the injection method gives consistent carbon recoveries and good chemical and metallurgical control.

The *addition* method involves placing an appropriate carbonaceous material in the charge or in the ladle. The recoveries in this method are dependent on the type of carbonaceous material used and the heat temperature.

After the chemistry has been adjusted, the furnace is rapidly tilted to fill the receiving (tapping) ladle. A tight, compact stream is desirable to minimize splashing and oxidation of the molten cast iron.

Desulfurization

The desulfurization of the cast iron is done in a ladle or other desulfurization vessel. Sulfur exerts a marked effect on the behavior of gray iron. It affects the undercooling and chilling tendencies and affects the type of graphite flakes produced.

On the other hand, sulfur content is very critical to the economical production of ductile iron. Excessive sulfur in the base iron will combine very readily with the high-cost magnesium treatment materials used in producing ductile iron. A basic melting practice in the furnace can produce a base iron with sulfur content as low as 0.004%. However, the cost of this method of melting the base iron usually outweighs its use. With the careful selection of the charge materials, acid-lined furnaces are capable of melting base irons with 0.030–0.040% sulfur.

The desulfurization processes used in the production of ductile iron will be discussed in Chapter 13, Microstructure of Ferrous Alloys.

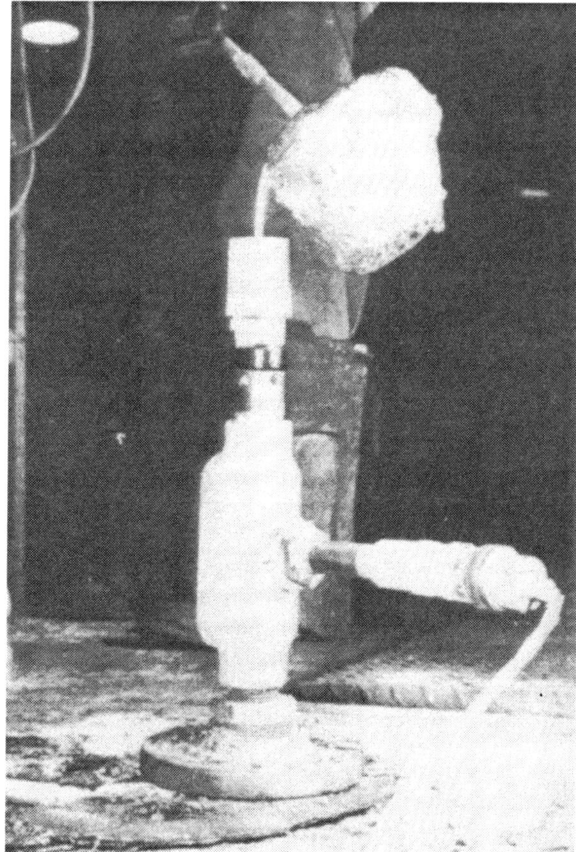

Fig. 11-43. Pouring mold (cup) for thermal arrest to determine carbon equivalent.

Fig. 11-44. Pouring cup to determine CE carbon and silicon values shown in digital read-out (inset).

TESTING AND ANALYSIS

This portion of the chapter will cover the more commonly used instruments and methods of controlling the properties of the alloys being produced, whether they be in the cast iron family, carbon steel family or high-alloy steel family. Details of how various instruments work or the specifics of each test can be found in the manufacturer's manuals.

Metal Chemistry Control

As has been emphasized throughout this chapter, metal chemistry control begins in the charge material selection, storage and weighing practice. Some of the instruments used to check the chemistry of the metal alloy, before it is poured into the mold, can also be used to check the chemistry of the charge materials.

Eutectometer

This instrument is mainly used for checking the carbon and silicon levels in cast irons. A small sample of molten cast iron can be obtained from the furnace or ladle and poured into a bonded-sand cylindrical, cup-shaped mold, 2 in. (5 cm) high and 1.25 in. (3.75 cm) in diameter. This cup contains a chromel-alumel thermocouple, which is connected to an indicating strip-chart recorder with a range of 2000–2500F (1093–1371C). As the liquid cools, the recorder traces the cooling curve of the sample, as well as the momentary thermal arrests when the freezing temperatures are reached. The instrument then makes a comparison of these arrests with a conversion chart that gives the operator the %CE within 2–3 minutes. **Figure 11-43** shows a sample being poured into the cup and **Fig. 11-44** shows the digital readout equipment of the CE and silicon values.

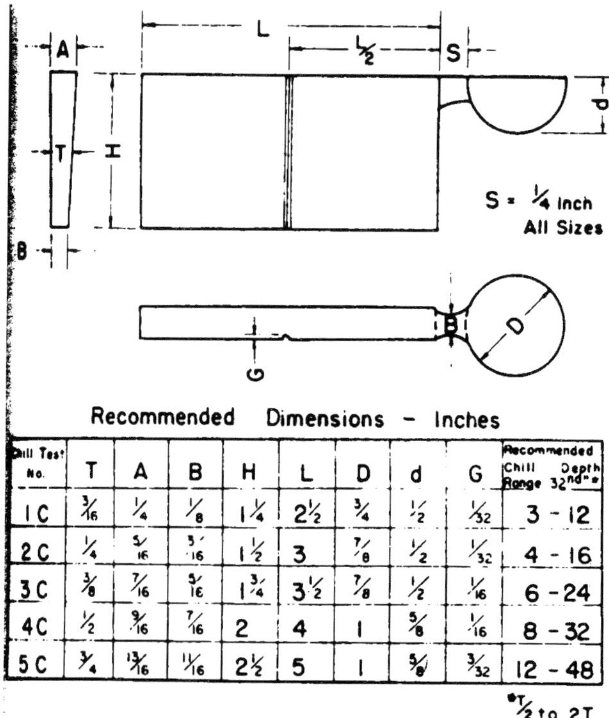

Recommended Dimensions – Inches

Chill Test No.	T	A	B	H	L	D	d	G	Recommended Chill Depth Range 32nd"s*
1C	3/16	1/4	1/8	1 1/4	2 1/2	3/4	1/2	1/32	3 - 12
2C	1/4	5/16	3/16	1 1/2	3	7/8	1/2	1/32	4 - 16
3C	3/8	7/16	5/16	1 3/4	3 1/2	7/8	1/2	1/16	6 - 24
4C	1/2	9/16	7/16	2	4	1	5/8	1/16	8 - 32
5C	3/4	13/16	11/16	2 1/2	5	1	5/8	3/32	12 - 48

*T/2 to 2T

Fig. 11-45. Recommended dimensions for chill test specimens ASTM designation A367-58. Casting to be made in dry sand core. Allow 3/4 in. of sand on all sides.

Chill Test

Chill tests have been used for many years in cast iron foundries. However, useful interpretation of the test results does take some experience. These tests are inexpensive and easy to perform.

Figures 11-45 through Fig. 11-47 show the dimensions for three different chill tests.

A chill test specimen is made by pouring a molten metal sample into a bonded sand mold in the shape of a wedge. In some cases, the mold is placed on a metal plate so that the molten metal is exposed to the plate, which serves as a chill. The casting freezes first in the thin section and progresses slowly toward the thick section. In the case of gray iron, the thinnest section freezes as white iron and the thickest section as gray cast iron.

Various types of cast irons have widely different chilling tendencies, so it is important to select a chill wedge design that shows a reasonable chill level for the cast iron being produced.

Figure 11-48 shows the fractures of chill wedges and their chill depth. The white material is the chilled cast iron or white iron.

Fluidity Spiral

Although not commonly used, the fluidity spiral checks the ability of the molten metal to flow readily. **Figure 11-49** shows a scale drawing of a fluidity spiral. The spiral mold is usually made of dry sand but can be made of any molding sand. The fluidity is measured by the length that the spiral fills with the molten metal until it freezes off.

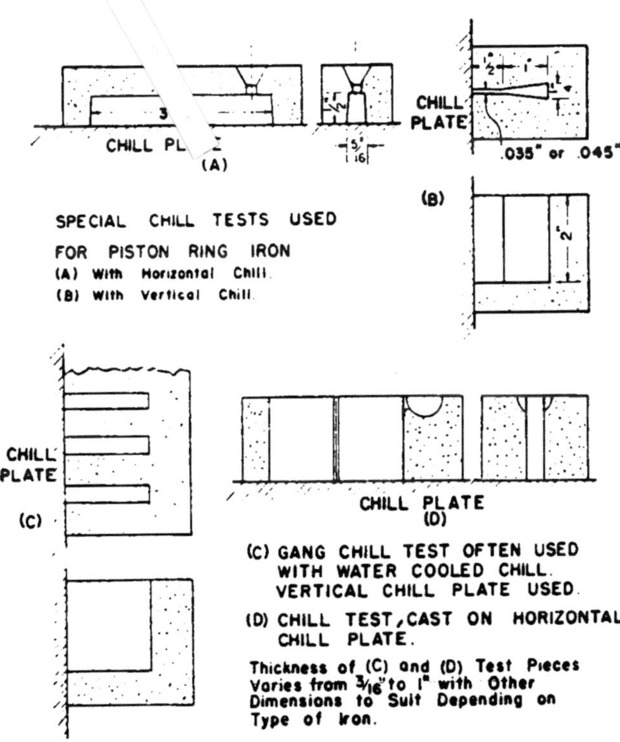

Fig. 11-46. Representative types of chill tests used by gray iron foundries.

Chemical Analysis

Due to the complexity and working environment required, the instruments used to determine the chemistry of the molten metal are normally found in a laboratory. This environmentally controlled laboratory is usually found somewhere near the melting area. These chemical tests can be made on instruments, such as carbon and sulfur determinators, in just a few minutes. Silicon, manganese, chromium and other metallic elements can be checked using an atomic absorption instrument. Wet chemistry can also be used, but is more time consuming.

The air or vacuum spectrometer finds much use in foundries. A chilled sample is prepared on the melt deck and sent to the laboratory to be read on the spectrometer. The sample can be conveyed to the laboratory by a pneumatic transfer system. The spectrometer checks the sample and records its chemistry. This chemistry can then be digitally read in the laboratory as well as on the melt deck. If necessary, adjustments can be made to the molten metal's chemistry or the molten metal can be tapped from the furnace as is.

Temperature Measurement

The three common instruments used to measure molten metal temperature are optical pyrometer, radiation pyrometer and immersion thermocouples. The accuracy of these instruments is largely based on the quality of the instrument maintenance program and the thoroughness of the operator-training program.

Optical Pyrometer

This type of temperature measuring instrument has been used in the foundry industry for many years. Optical pyrometers are brightness measuring instruments. There are two types of these instruments and either one requires that the operator receive proper instruction on its adjustment and use. The temperature readings of these instruments are based on the thermal emissivity of the molten metal being checked. (Emissivity is the ratio of the radiant energy emitted by a surface to that emitted by a blackbody at the same temperature.) It is not recommended that the temperatures of a bath of molten

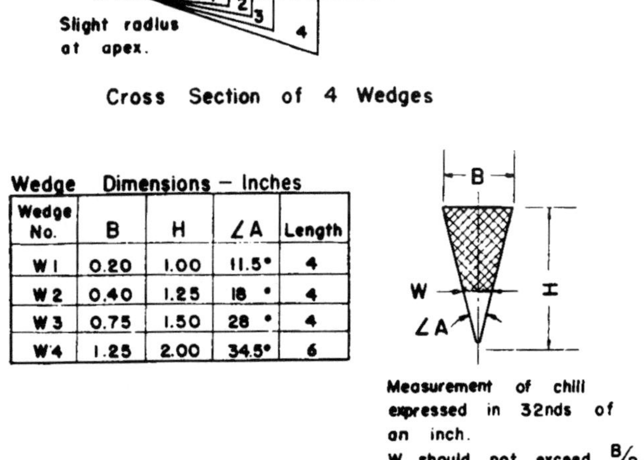

Wedge Dimensions – Inches

Wedge No.	B	H	∠A	Length
W1	0.20	1.00	11.5°	4
W2	0.40	1.25	18°	4
W3	0.75	1.50	28°	4
W4	1.25	2.00	34.5°	6

Fig. 11-47. Recommended dimensions for test wedges ASTM designation A367-58.

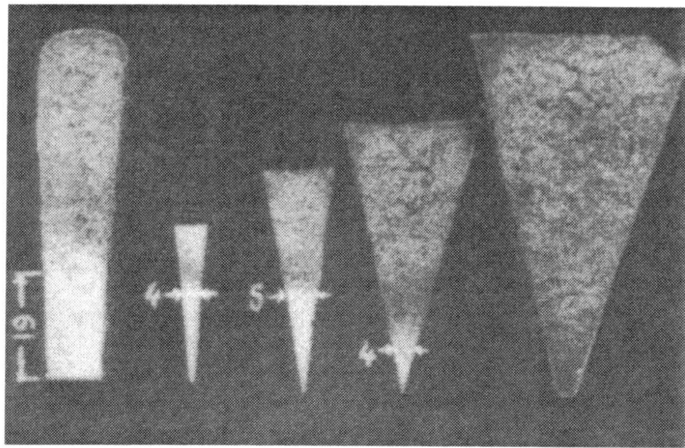

Fig. 11-48. Fractures of test wedges showing depth and width of chill. Whole numbers represent 1/32 inch.

metal in an electric arc furnace or channel induction furnace be taken with the optical pyrometer. However, it can be used for checking temperatures in a *coreless* induction furnace. It is important that the operator understands that the emissivity increases as turbulence increases, such as in a falling stream of molten metal during furnace tapping and during the pouring of castings.

Radiation Pyrometer

This pyrometer is relatively inexpensive and produces an electrical signal readily measured by a conventional recorder. For these instruments to give a proper reading, the field of view must always be full of the molten metal. The radiation pyrometer is not suitable to be used for sighting into an electric arc furnace or a channel furnace. It can be used with *coreless induction* furnaces, but only when the furnace is full and the furnace cover is swung out of the way. The radiation pyrometer can be used to measure the temperature of the stream of molten metal as it is being tapped from the furnace or poured into a mold. A hand-held, direct-reading pyrometer using photoconductive cells has been used for measuring temperatures in the foundry and would be subject to the same limitations.

Immersion Thermocouple (Pyrometer)

A properly constructed and maintained immersion thermocouple, also called a pyrometer, is the only reliable way of getting accurate temperature measurements in all types of electric furnaces and tapping, transfer or pouring ladles. Some furnace operators claim to be able to accurately tell the temperature of the molten metal by just looking at its color. They may come close but are not accurate. The immersion thermocouple, when properly maintained, is the most accurate way to measure the true temperature of the molten metal.

There are two types of immersion pyrometers: the *repeated use* type and the *expendable* type. The *repeated use* type, also called the "dip stick," has been used to measure molten metal temperature in ladles for many years. With this type of pyrometer, the thermocouple in the end of the instrument is used repeatedly. The thermocouple end of the instrument is dipped into the molten metal and the temperature is read on the instrument itself.

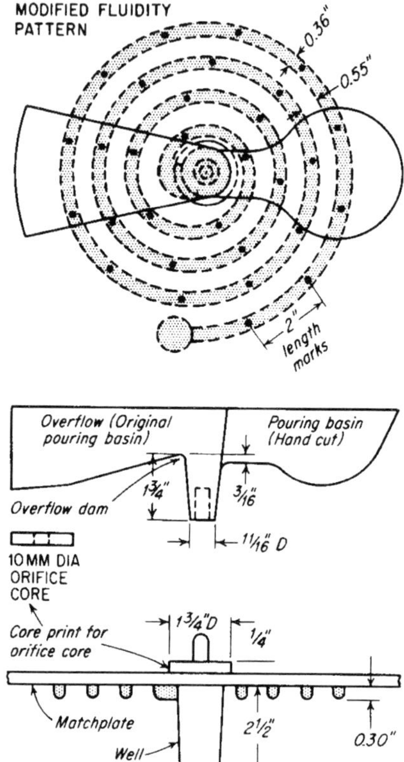

Fig. 11-49. Fluidity spiral.

The *expendable* type uses a thermocouple that is encased in a small-bore, thin-walled protection tube. The thermocouple is not permanently attached to the end of the instrument but has to be attached prior to use, and can be removed after each use or for maintenance. The temperature is not read on the instrument itself, but on a recording device that is visible to the person taking the reading. This instrument reaches the molten metal temperature faster than the repeated use instrument. The expendable type of thermocouple does not have the working life of the repeated use type. In some cases, if not used properly, the expendable type of thermocouple may have a one time use life.

More information, details on the melting, and control of ferrous alloys can be found in the references listed in this chapter's bibliography. The material discussed in this chapter has been gleaned from the references listed.

BIBLIOGRAPHY

Brown, P.A., Medium Frequency Induction Melting, Pillar Corp., Menomonee Falls, WI.

Cervellero, P.B., *Control of Carbon (and Silicon) in the Production of Synthetic Iron,* Inductotherm Corp., Rancocas, NJ.

Cupola Handbook, 6th Edition, American Foundry Society, Des Plaines, IL (1999).

Electric Arc Furnace Steelmaking...The Energy Efficient Way to Melt Steel, CMP Tech Commentary, Center for Metals Production, Pittsburgh, PA (1985).

Electricity in Steelmaking, CMP Tech Commentary, Center for Metals Production, Pittsburgh, PA (1988).

Harris, G., Guidelines for Proper Coreless Furnace Maintenance, Inductotherm Corp., *Modern Casting* (Apr 1999).

Heine, R.W., Loper, Jr., C.R., Rosenthal, P.C., *Principles of Metal Casting,* 2nd. Ed. McGraw-Hill, New York (1968).

Induction Melters Examine Furnace and Pouring Technology, Refractories, AFS Induction Melting, Holding and Pouring Conference, Jan 1999, *Modern Casting* (July 1999).

Installation, Sintering, and Patching Procedures for Damp Ramming Mixes in Coreless Induction Furnaces, Harbison-Walker Refractories Co., Pittsburgh, PA.

Klemp III, T., Kiely, J.P., *Remelting Practice for the Precision Caster,* Cannon-Muskegon Corporation, Muskegon, MI.

Mortimer, J.H., "Batch Induction Melting: A Success Story," Inductotherm Corp., *Foundry Management and Technology* (Aug 1994).

Spear, W.N., "Induction Melting of Cast Stainless Steels," Nickel Development Institute, *1987 Electric Furnace Proceedings.*

Steel Casting Metallurgy, Steel Founders' Society of America, Des Plaines, IL (1984).

Svoboda, J.M., *Electric Melting Technology, Vol. 3: Arc Melting of Cast Steel,* Cast Metals Institute, Des Plaines, IL (1973).

Svoboda, J.M., *Electric Melting Technology, Vol. 2: Arc Melting of Gray and Ductile Iron,* Cast Metals Institute, Des Plaines, IL (1973).

Svoboda, J.M., *Electric Melting Technology, Vol. 1: Induction Melting,* Cast Metals Institute, Des Plaines, IL (1973).

Swaney, R.U., "Induction Furnace and Developments and Industry Trends," Inducto Steel, *1987 Electric Furnace Proceedings.*

Sylvia, J.G., *Cast Metals Technology,* American Foundrymen's Society/Cast Metals Institute, Des Plaines, IL (1990).

Understanding Electric Arc Furnace Operations for Steel Production, CMP Tech Commentary, Center For Metals Production (1987).

Vukovich, N., *Arc Furnace Melting Practices,* (Foseco, Inc.), Cast Metals Institute, Des Plaines, IL (1987).

Nonferrous Melting Practice

12

The melting of nonferrous alloys, as with ferrous alloys, must be regarded as more than changing solid metals into a liquid state. **Table 12-1** indicates the melting points of various nonferrous metals. The melting operation is an important part in producing sound and strong metal castings. Well-maintained melting equipment plus the use of proper melting techniques are the primary requisites for achieving a nonferrous metal casting that, within practical limits, will retain the desired metallurgical properties of the original alloy from which it was cast.

This chapter will not spend much time discussing the various furnaces used to melt nonferrous alloys, as much of this information has been covered in previous chapters. The discussion will be centered on the use of gas-fired, induction and resistance rod furnaces as nonferrous alloys melting tools. The sections are divided into the alloy families, as follows:

- Aluminum Alloys
- Magnesium Alloys
- Copper Alloys
- Titanium Alloys

ALUMINUM ALLOYS

For pure aluminum, the melting point is 1221F (660C), the heat of fusion is 169 Btu/lb and the specific heat at 212F (100C) is 0.226 cal/g. Note that these values are for *pure* aluminum. Because foundries pour aluminum *alloys,* these values will change, depending on the other metals and elements in the aluminum alloy. Like other metals, aluminum comes to its melting temperature quite readily. Then it must take on additional heat—the heat of fusion—to change to the liquid state. **Figure 12-1** is a graph that shows the heat content of pure aluminum at various temperatures and illustrates this important characteristic.

Aluminum alloys do not behave well in all aspects of the melting process; there are certain characteristics that cause reactions to take place that must be controlled to produce a satisfactory melt and product. For instance, at elevated temperatures, the aluminum combines very rapidly with oxygen to form oxides. During the melting process, the quantity or volume of these oxides increases as temperature increases. The problem with this is that the oxides and the aluminum alloy have specific gravities that are very close together. Thus, if there is any agitation in the molten metal, the oxides and the liquid aluminum alloy will be easily mixed together. If not removed from the molten aluminum alloy, these oxides will become unwanted inclusions in the metal casting.

Another problem to be considered and controlled is that of gas pickup by the molten metal. The most troublesome gas is hydrogen. This hydrogen can be picked up from the surrounding atmosphere and, during melting with gas, from the products of combustion or exhaust gases. Dirty charge materials can also introduce hydrogen into the melt.

There are other problems confronting the aluminum foundry when melting or holding aluminum. These problems will be addressed when discussing the furnaces in which they are most pronounced.

Table 12-1.
Melting Points of Various Nonferrous Alloys

Metal	Melting Point
Aluminum	1215F (652C)
Cobalt	2714F (1490C)
Copper	1981F (1083C)
Magnesium	1204F (651C)
Nickel	2645F (1452C)
Titanium	3272F (1800C)
Zinc	787F (419C)

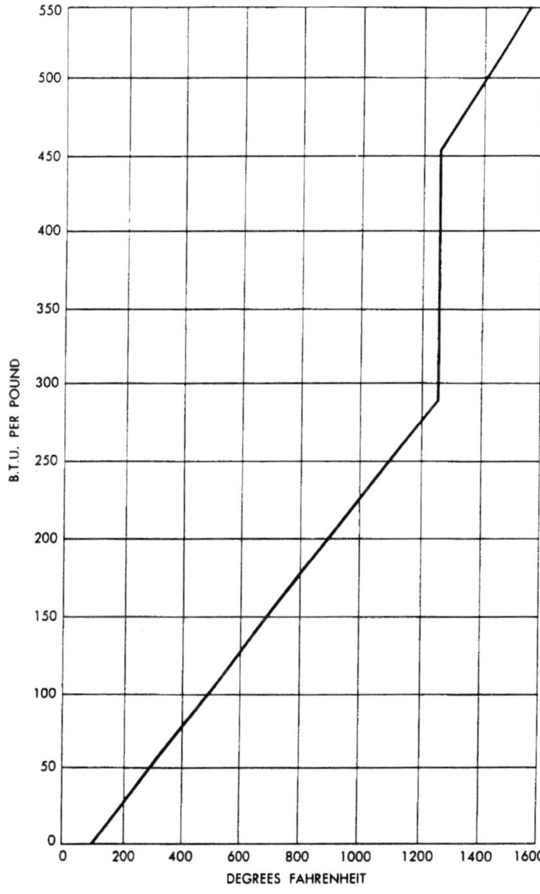

Fig. 12-1. Thermal capacity of pure aluminum. Note that aluminum doesn't just get hot enough to melt. After reaching the melting temperature, it requires more than 160 additional Btu/lb before the aluminum actually takes the liquid form.

Charge Materials

When melting aluminum alloys, the charge usually consists of aluminum alloy ingots and sows, foundry returns and some purchased aluminum alloy scrap.

Aluminum Alloy Ingots and Sows—Ingots can be purchased or produced in the foundry. Purchased ingots can be of two types: primary or secondary.

The *primary* ingot comes directly from the smelter and has never been used to make aluminum alloy castings or other aluminum alloy products. The chemical analysis of this ingot is held to closer specifications than the secondary ingot. Aluminum foundries pouring premium aluminum alloy castings will usually use only primary ingots as the charge material.

Secondary ingots are made from aluminum alloys that have been used for castings or other aluminum alloy products. The chemical analysis specifications are not as stringent for secondary ingot as they are for primary ingot. **Figures 12-2 and 12-3** show two different ingot forms. **Figure 12-2** is continuous-cast aluminum foundry ingot, while **Fig. 12-3** shows individually cast aluminum foundry ingots.

Large, single ingots, called *sows* are also used in large furnace applications **(Fig. 12-3).**

Fig. 12-2a. Typical compact skid of continuous-cast aluminum foundry ingot. Storage areas usually can be reduced by as much as 50%.

Fig. 12-2b. Stacked ingots protected by sturdy plastic are securely bundled and ready for shipment. Alloy and heat treatment numbers are stamped on each ingot.

Fig. 12-3. Sows, usually used in large furnace applications.

Some large-volume aluminum foundries produce aluminum alloy ingot from foundry returns such as gates, risers and defective castings. They place these materials in large reverberatory furnaces for melting, and then adjust the melt's chemical analysis, if required. The molten metal is then poured into ingot molds. This ingot is then used in their melting operations.

Foundry Returns—When using foundry returns as charge material, it is very important that they be clean and dry. Also very important is that all of the foundry scrap used in a given charge is of the same chemical analysis as that specified in the end casting. This means that the foundry will have to set up a system by which gates, risers and defective castings can be identified and segregated.

Purchased Scrap—Purchased scrap is normally used in the high-volume aluminum foundries. This scrap is purchased according to the foundry's chemical specifications and may consist of turnings, chips and other aluminum products. Again, it is important that the scrap be clean and dry and of known chemical analysis. In the foundry, it should be kept in a dry atmosphere and at rather stable temperatures. If the scrap, and ingot for that matter, gets cold and then enters a warm space, it will begin to "sweat." This moisture, if it gets into the melt, can cause a safety hazard and can introduce hydrogen into the melt.

Melting Safety

Any form of moisture can be very dangerous in molten aluminum. The moisture turns to steam, and the expansion reaction causes an explosion. Although it is very rare, a secondary reaction can occur that has tremendous explosive power.

Tools and Their Maintenance

Regardless of the type of furnace used to melt the aluminum alloy, the melter will use a variety of tools to prepare the melt. These tools include stirrers, plungers, skimmers, hand ladles and degassing tubes.

The tools are usually made of steel. Stirrers are used to carefully stir additions into melt, whereas plungers are used to plunge volatile materials (degassers) to the bottom of the melt. Skimmers, as their name implies, are used to skim the dross (oxides) off the surface of the melt. Hand ladles are used to transfer molten metal from the furnace to the molds, whether they be sand molds, permanent molds, or to a diecasting machine. Degassing tubes are used to introduce a degassing agent and flux into the melt.

The steel tools should be coated with a refractory material. Molten aluminum is an aggressive "solvent" that will attack and deteriorate and uncoated steel tool. Molten aluminum will pick up some of the iron from the steel, which is detrimental to aluminum castings if the level gets too high. After coating, the tools should be preheated to a near-red glow before being inserted into the melt. This preheating will ensure that the refractory coating is moisture-free and will not become a safety hazard.

Graphite tools are sometimes used, especially graphite fluxing tubes for degassing. These tools are more fragile than steel; however, they are not sources of metallic contamination. These tools should also be thoroughly preheated before being inserted in the melt.

Temperature Control and Measurement

Temperature control of the melt is very critical when it comes to melting quality aluminum alloy heats. In fact, gating systems (which will be discussed in a later chapter) should be designed to permit pouring the casting at the lowest temperature possible. Overheating and extended holding times at these elevated temperatures can result in coarse-grained aluminum alloy castings with low mechanical properties. Also, extended holding times can lead to the absorption of more gas and more dross formation.

Residual heat in the furnace can cause the melt's temperature to continue to rise, even after the furnace is turned off. Automatic temperature control devices that minimize this problem are available.

Temperatures of molten aluminum alloys are normally measured using chromel-alumel thermocouples (also referred to as pyrometers). **Figure 12-4** shows two types of thermocouples. The ends of these thermocouples can be either open or closed. The use of the radiation-reading instruments, as used in ferrous alloys, will not give proper melt temperatures. The reason is that aluminum has very poor emissivity and doesn't radiate heat or light. A person can easily be burned on aluminum alloy castings that have just been poured. One can literally touch the casting before feeling the heat of the casting.

In order to read the molten aluminum alloy's temperature, the thermocouple or pyrometer must be dipped into the melt. These temperatures can first be taken in the furnace and then in the pouring ladles or other pouring devices. The use of a portable pyrometer is shown in **Fig. 12-5**.

As with all quality control instruments used in the foundry, thermocouples must been well maintained. This means that they should be checked for accuracy, and recalibrated if needed. Since the end of the thermocouple will be dipped into the melt, it must be dry to prevent a possible reaction.

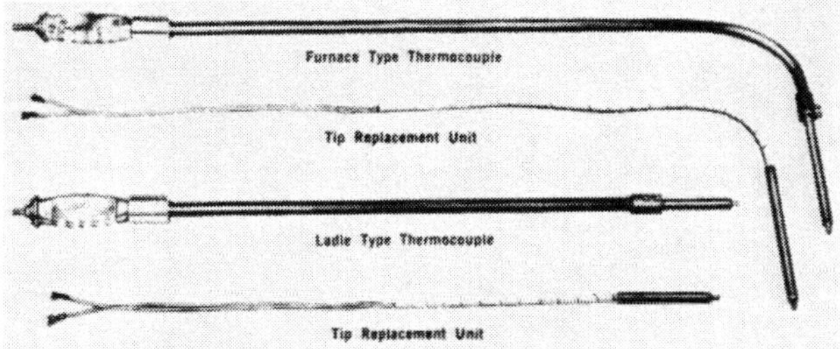

Fig. 12-4. Quick tip changes are possible with tip replacement units for closed-tip thermocouples.

Fig. 12-5. Checking melt temperature with a portable thermocouple. Melt temperature is read on a compact, lightweight pyrometer.

Automatic Ladling Devices

Reliability of process control in aluminum alloy casting production is improved through the proper application of automatic ladling and pouring devices. This equipment helps assure quality, and minimizes wasted production time and casting defects such as misruns (incomplete casting). In high-volume casting production foundries, the money saved by reducing cycle times can be significant. This equipment is also available for use in all types of high-production casting.

Figure 12-6 shows an automatic ladling device being used with the expendable pattern or lost foam casting process. The arm holding the ladle is operated by a robot. The ladle is lowered into the melt, filled with the molten aluminum alloy, and then raised out of the melt. The ladle is swung to a position over the mold, and the molten aluminum alloy is poured into the mold.

Another type of automatic pouring device is shown in **Fig. 12-7**. Here, two pouring chambers are filled pneumatically, using a vacuum. In this case, the green sand molds are moved into position on a conveyor. When the chambers have been filled with the right amount of molten aluminum alloy and the molds are in position, the vacuum is released and pressure is applied to the chamber. The molten aluminum alloy runs down the launders or troughs, which direct it to the pouring basins in the molds.

Fig. 12-6. This automatic ladling device provides a consistent pouring rate when pouring expendable pattern castings.

Another automatic ladling and pouring device uses electromagnetic pumps to move the molten aluminum alloy at exact rates, and to release it in precise doses. The pumps used for aluminum alloys are rectangular channels and made of ceramic materials. This method of ladling and pouring has the advantage of no moving parts.

Fig. 12-7. Green sand molding line with pressure-pour auto ladling. [Photo courtesy of CMI-Wabash Cast.]

Furnaces Used for Melting Aluminum Alloys

Crucible Furnaces and Crucibles

The crucible melting furnace has been used for many years as the workhorse in many aluminum foundries. This is especially true for smaller foundries doing jobbing work that requires the melting of various aluminum alloys in small volumes. These types of crucible furnaces were discussed in Chapter 10, Alloy Melting Furnaces.

Crucible Selection and Care—Since all of these furnaces require a crucible to hold the melt, some time should be spent discussing the selection and care of crucibles. The composition of the crucible can be either clay-bonded graphite or carbon-bonded silicon carbide for gas-fired furnaces. In the case of induction melting, ceramic-bonded oxide refractories can be used when ferrous and other high melting temperature alloys are to be melted. Some of these ceramic-bonded crucibles are rated for temperatures up to 5000F (2760C).

The carbon-bonded silicon carbide crucibles are preferred because of improved conductivity of heat. These crucibles are strong, quickly reach temperature and are available in large sizes. These crucibles, when used in coreless induction furnaces, can develop "hot spots," which can cause spalling (cracking/flaking).

Clay-bonded crucibles are somewhat pliable when hot, and retain heat longer when removed from the furnace. Because of these properties, clay-bonded crucibles are used as transfer ladles. **Figure 12-8** shows typical crucible shapes and sizes.

As soon as new crucibles are received, they should be inspected for damage during shipping. To check for cracks, the "ring test" is recommended: Tap the crucible with a wooden hammer or mallet or a coin. If the crucible is cracked it will give a dull or flat sound, whereas a sound crucible will give a ringing sound. The new crucibles should be stored in the vertical position with the opening at the top. New and used crucibles should also be stored in a warm dry place off the floor.

The proper positioning of the crucible in the furnace is also critical. The crucible should be centered in the furnace (**Fig. 12-9**) so that there is equal distance from it to the furnace walls. In addition, it should be placed on the proper sized pedestal block for the furnace, so that its bottom is on the same level as the centerline of the burner port. If the crucible is to be lifted out of the furnace after melting has been completed, a piece of cardboard should be placed between the crucible bottom and the top of the pedestal block. This will prevent the crucible from sticking to the pedestal block. If the inside of the furnace or the pedestal block is hot, a piece of wet cardboard should be used.

When a new clay-bonded graphite crucible is used for the first time, it should be brought up to temperature slowly (particularly true for large sizes) to insure proper annealing and the safe removal of any moisture in the crucible. On the other hand, carbon-bonded crucibles should be brought up to melting temperature very rapidly to insure the proper functioning of the protective glaze. To further protect either type of crucible, a slightly oxidizing atmosphere in the furnace, during the melting operation, is recommended.

With the furnace roof swung open, and before starting up the crucible, the inside surface of the furnace lining should be checked for any protrusions. If any protrusions are found, they should be removed so that they will not interfere with the smooth swirling action of the flame as it swirls upward around the crucible. After this has been done, the pedestal block or stool should be positioned properly and a piece of cardboard placed on its top surface.

Charging—Next, the crucible is charged with the smaller charge materials first. Precautions should be taken so that the charge materials are not packed tightly in the crucible. As the charge

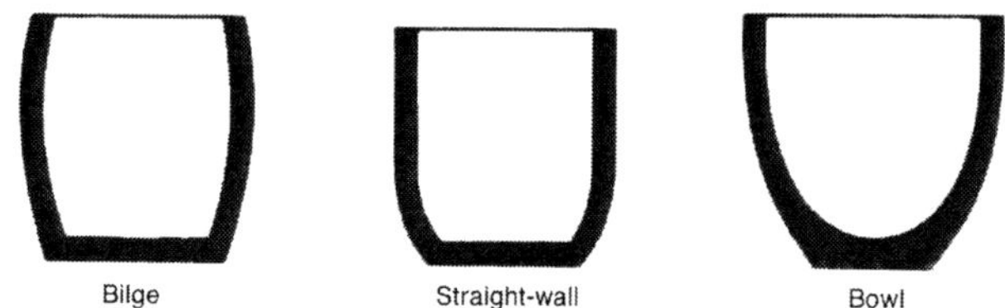

Fig. 12-8. Three basic crucible shapes: bilge, straight-wall and bowl.

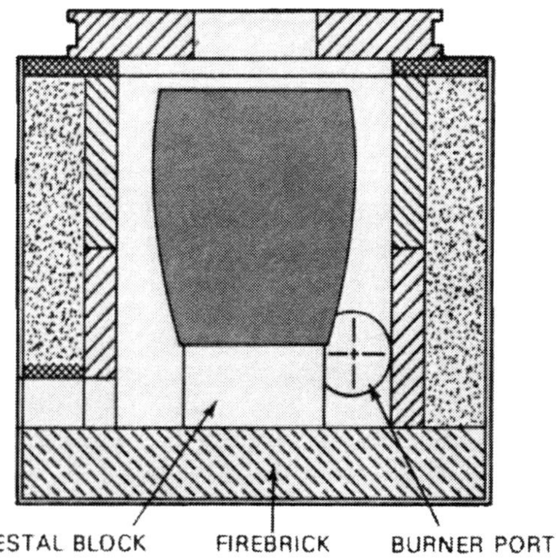

Fig. 12-9. Proper positioning of crucible in furnace.

materials heat up, they will expand. If there is no room to accommodate this expansion, the crucible could be cracked or ruptured. Ingot is charged later, usually after the first charge materials have melted.

The air blower is turned on for a short period to rid the furnace interior of any gas that may have settled to the bottom of the furnace. Then, with the air blower still on, the gas valve is opened and the automatic ignition ignites the gas and air mixture. With the burner flame on, the roof is then swung shut.

The flame is then adjusted to the proper operating condition, which is done by manipulating the amount of gas and air. The flame used to melt aluminum alloys should be slightly oxidizing. This can be examined visually by observing the flame and exhaust gases coming out of the exhaust port in the furnace roof. The color of the flame should be slightly green. The oxidizing flame will also improve the life of the crucible.

Melting Operation—With the furnace operating, the next charge materials to be used can be placed on the roof the furnace. Care should be taken that these materials don't cover any portion of the exhaust port. Placing these materials on the roof or cover of the furnace will preheat them and drive off any moisture. (Remember that these next charge materials will be placed into a molten bath of metal.)

Never place any charge materials in the exhaust port, as this will disrupt the proper exodus of the flame and exhaust gases from the furnace. If this done, the surface of the melt will be disturbed. Remember that water vapor is present in the exhaust gases and, if it is not allowed to escape from the furnace, can find its way into the melt, creating hydrogen porosity.

The surface of the melt will have an oxide skin on it. This surface, if not disturbed, will greatly help in reducing the amount of gas absorption. In addition, if the surface is broken open, more oxide will form that will later have to be removed.

When the first charge materials are molten, the preheated charge materials can be placed into the molten metal. Tongs are used to pick up and place these materials into the bath. This operation is done normally with the flame turned off. The roof can be swung open or the materials lowered into the bath through the exhaust port. With the new charge materials in the crucible, the flame is reignited, the roof closed and, if necessary, adjustments made to air/gas mixture to get a slightly oxidizing flame. With large-capacity crucibles, several charging operations may have to be made.

After a predetermined amount of time, the flame is shut off and the temperature of the molten metal is measured with a pyrometer. The temperature should be high enough to allow treatment of the molten metal and still have enough temperature to be poured into the mold(s). Remember that gas absorption is time- and temperature-related. The higher the temperature of the molten metal and the longer it is held at those temperatures, the more gas will be picked up.

Hydrogen Degassing—The gas to be removed from the molten aluminum alloy is hydrogen. If the temperature is high enough, the crucible full of molten metal is removed from the furnace. If the crucible is small, it can be lifted out of the furnace manually, with crucible tongs, as seen in **Fig. 12-10**. The molten metal mat be tapped into a transfer or pouring ladle for degassing.

The degassing operation can also be performed while the crucible is in the furnace. In addition, the degassing should be done just prior to pouring when the melt temperature begins to drop. If the molten metal temperature is still climbing, wait before degassing.

The five methods of determining the hydrogen content in the molten metal are as follows:

1. Reduced Pressure Test (Straube-Pfeiffer Test or "vacuum" test)
2. Quantitative Reduced Pressure Test (Severn Science)
3. Density Measurement
4. Initial Bubble Test (Hycon Test)
5. Recirculating Gas Test (Telgas, Alscan)

The most commonly used method in the foundry is the *Reduced Pressure Test* (RPT) and will be briefly discussed here. This test is also the most effective if the sampling is done in a repeatable sequence that eliminates or reduces variables. Care should be taken to keep the testing conditions the same from test to test, to avoid inconsistent results.

This test for assessing the gas content of the molten aluminum alloy consists of solidifying a sample of the melt under reduced pressure (usually 1–100 mm Hg or 28–29 in. Hg). While the sample is in the molten state, the gas is in solution. However, as the sample begins to solidify, the gas is rejected by the solidifying aluminum alloy and it forms holes. Because of the reduced pressure atmosphere, these holes get larger than if the samples were allowed to solidify in atmospheric pressure.

There are several different manufacturers of the testing apparatus used to perform this test and they are similar to the one shown in **Fig. 12-11**. The sample cup is preheated to a cherry-red color and dipped into the melt to obtain the sample to be tested. The bell jar with a vacuum seal is placed over the sample and the vacuum pump started. The vacuum pump then pulls the necessary reduced pressure or vacuum in the bell jar. Some bell jars have a clear top, which

Fig. 12-10. Small crucible being manually lifted out of furnace.

Fig. 12-11. Reduced pressure tester for detecting gas in molten aluminum alloys.

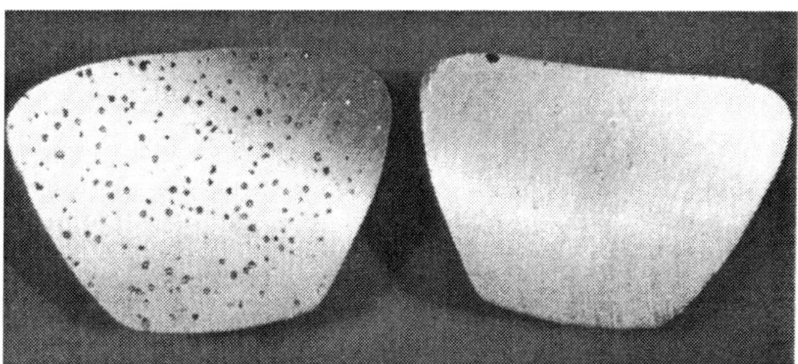

Fig. 12-12. Metal samples solidified at reduced pressure using an instrument similar to that shown in Fig. 12-11. At right, the cross section of a degassed specimen reveals sound metal with no porosity. Nondegassed metal (left) shows porosity caused by entrapped gases that were not removed from the melt.

allows the operator to see how the sample is reacting in the bell jar under vacuum. If the sample has a mushroom type of dome, it contains large quantities of gas; if the sample top is relatively flat, it has some gas; and if the top is sunken, the sample is normally free of gas.

To further check the amount of gas content, the sample is cut in half, vertically. Some operators will use a fine grit abrasive paper to smooth the cut surface and to remove any metal that may cover some holes caused by the sawing of the sample. **Figure 12-12** shows two different samples that have been tested and cut in half. One permanent mold foundry has devised a chart of cut samples for comparison purposes and for porosity quality control. This chart is seen in **Fig. 12-13**.

If, after the test, it is determined that the molten aluminum needs to be degassed, the actual degassing operation begins. Degassing is typically carried out by introducing sparging gas into the molten aluminum alloy. Sparging means to agitate the melt by use of compressed air or gas entering through a lance or pipe. As the gas bubbles move up through the bath, hydrogen diffuses into the bubbles. The method of using sparging gas differs in two ways: the type of sparging gas used and the method used to introduce the gas into the melt. Another method that can be used is to plunge chlorine-base tablets to the bottom of the melt.

Begin by using a *sparging gas.* The secret to the successful use of a sparging gas is not how big the bubbles are, but how small and well distributed they are. These gases can be introduced into the melt through a lance, porous refractory (plug or tile), rotary impeller or a porous plug.

The lance can be made of steel or graphite. If a steel lance is used, it must be coated inside and out with a refractory wash that is thoroughly dry. A graphite lance, on the other hand, does not require a refractory coating. The use of a graphite lance is seen in **Fig. 12-14** for degassing a melt in a stationary crucible furnace. Lances generally produce large gas bubbles that rise quickly through the melt and create turbulence at the surface, as seen in the water model in **Fig. 12-15.** Note that the bubbles tend to rise close to the lance. For this reason, if a large melt is to be degassed, several lances must be used or a single one moved constantly. This movement has to be made carefully so as not to introduce more hydrogen into the melt by unduly disturbing the surface of the melt.

By attaching a plug into the end of either the steel or graphite lance, a finer dispersion of the sparging gas takes place. However, again, the bubbles tend to rise only in the area of the plug. This again limits the amount of molten metal that contacts the sparging gas and may require the lance to be moved through the melt. Some operators will plug the end of the tube so that no sparging gas can escape there and then drill many small holes along the length of the lance to try to obtain more small bubbles and better dispersion.

The most effective way of degassing is by the use of the rotary impeller. This method uses a special impeller head that rotates very rapidly, producing many fine bubbles and dispersing them throughout the melt. These small bubbles rise slowly through the melt, which gives the hydrogen a longer time to react with them. This method of degassing does not generally create the turbulence associated with the lance method. **Figure 12-16** shows a water model of the rotary impeller method of degassing.

The use of a porous plug to introduce the sparging gas into the well of reverberatory furnaces has been accomplished in one manufacturer's furnace. The porous plug is placed in the center of the bottom of the well, and the sparging gas, under pressure, is forced through the porous plug into the melt. The gas is dispersed through the melt in the form of many small bubbles that rise slowly to the surface. Since these furnaces are run on a continual basis, the degassing can take place on a continual basis or intermittent, if desired.

The sparging gases may be one of two types or a combination of the two types. *Inert gases*, such as dry nitrogen or argon, can be used to degas aluminum alloys. They are called inert because the gas does not react with the molten aluminum alloy. On the other hand, *reactive gases,* such as chlorine, Freon and fluorine, can be used separately or in combination with other inert gases. They are called reactive gases because they will undergo a reaction with liquid aluminum or the alloying elements or additives in the melt. (The use of pure chlorine has been largely discontinued because of the toxicity of the fumes generated and the corrosive effect it has on metal surfaces.) A reactive gas should not be used if the melt has been modified using strontium or sodium. Degassing should be done before modification if a reactive gas is to be used.

For many years, chlorine and dry nitrogen have been the most common degassers for molten aluminum alloys. As these gases are bubbled through the melt, they remove hydrogen and help to float oxide particles to the surface of the melt. For this reason, chlorine and dry nitrogen can be considered *cleaning fluxes.* With the OSHA

Fig. 12-13. Comparison standards of cast A356.0 poured under 100 mm pressure, which can be used for porosity quality control. For example, No. 1 may be specified as the standard for high-quality castings, and No. 3 for general commercial quality. The center number for each indicates percent surface area porosity, the bottom number the density. [Courtesy of Stahl Specialty Co.]

and EPA Clean Air Acts, straight chlorine has been replaced by combinations of 10% chlorine-90% nitrogen or 10% chlorine-90% argon. In addition, fluxing with 100% nitrogen or argon has become more popular.

Tablets or cakes made up of degassing salts that contain some chlorine can also be used to degas the smaller sized melt. Probably the most popular of these degassing salts is hexachloroethane. The tablets are plunged to the bottom of the melt and held there with the plunger. The plunger is held in this position until the reaction in the melt has ended, indicating that the entire tablet has dissolved.

Hexachloroethane tablets can also act as a flux to help remove nonmetallics from the melt at the same time that it is degassing the melt. Sometimes, more than one treatment with these tablets will be required to achieve the contained gas specifications for the heat. Most of these compounds liberate more noxious, though less toxic, fumes than chlorine and chlorine-nitrogen mixtures. After the reaction appears to be ended, it is best to let the melt sit a few minutes, in order to be sure that the reaction is complete.

Fluxing the Molten Metal—Aluminum foundries are relying more on fluxing to assure the best physical and mechanical properties in aluminum alloy castings. There are many commercially available and acceptable fluxes, providing many opportunities to improve casting quality and to cut casting cost by using the appropriate fluxes.

Common fluxes fall into four broad categories:

1. Covering fluxes, which are used to prevent gas pickup, reduce dross formation by oxidation and thereby reduce metal loss.
2. Cleaning fluxes, which are used to remove solid nonmetallic inclusions.

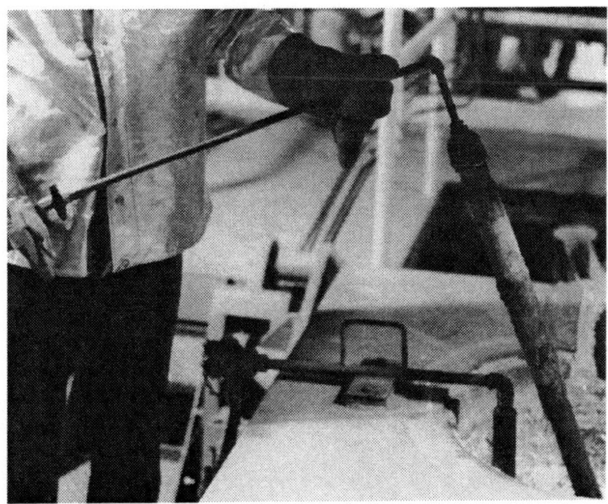

Fig. 12-14. A lance is inserted into molten aluminum for degassing.

3. Degassing fluxes, which are added to the melt to remove the entrapped gases (especially hydrogen) that can contribute to casting porosity and blowholes.
4. Drossing-off fluxes, which are used to recover metal from drosses. **Figure 12-17** shows the results of using a drossing-off flux.

There is no such thing as a universal flux. There are many variables to be considered when selecting a flux. The reason the flux is being used and the composition of the aluminum alloy will determine the ingredients in the flux. A flux that is suitable for one particular aluminum alloy may not be good for another because of differences in oxidation rates. Melting and pouring temperatures can affect flux compositions. Finally, the casting method must also be considered. The treatment of an aluminum alloy for permanent molding is usually different than for high-pressure diecasting or sandcasting.

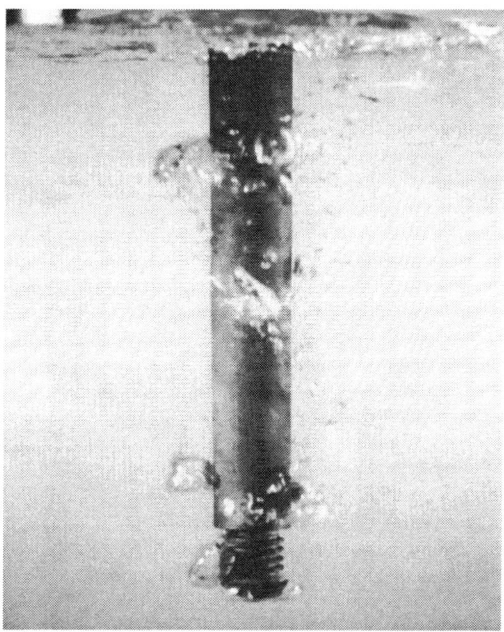

Fig. 12-15. Large bubbles are obtained with a lance in a water model study. Note the turbulence at the surface.

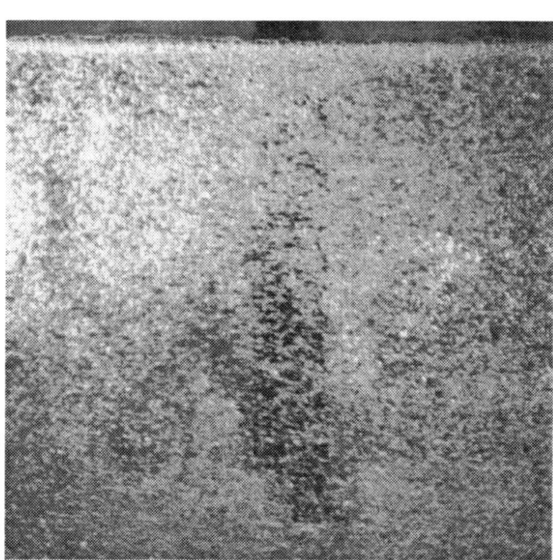

Fig. 12-16. Small, well-dispersed bubbles are obtained in a water model study when using a rotary impeller. Compare the relatively undisturbed surface with the agitated surface in Fig. 12-15.

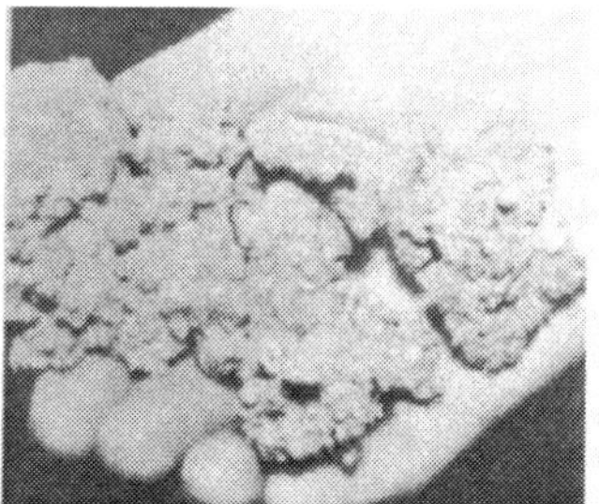

Fig. 12-17. (Left) Lumpy, metallic dross resulting from an unfluxed melt. (Right) Powdery, metal-free dross results when an effective drossing-off flux is used.

Two more treatments can be done to the melt before it is poured into the mold. These two treatments are grain refining and modification, and will be discussed in **Chapter 14, Microstructure of Nonferrous Alloys.**

Drossing the Molten Metal—The next step in preparing the molten aluminum alloy for pouring is to skim off the dross on the surface of the molten metal. If the molten metal is to be poured manually from a crucible, it is set in a pouring shank after being taken out of the furnace. **Figure 12-18a** shows a pouring shank. After the dross has been removed, a final temperature check can be made and then the pouring opera-tion can take place. **Figure 12-18b** shows the use of a pouring shank to pour molten aluminum alloy.

In larger crucibles or reverb bath-type furnaces, the molten aluminum is often poured from a refractory-coated hand ladle that is dipped into the furnace **(Fig. 12-19).**

After the pour has been made the smaller crucible can be placed back in the furnace and charged for the next melt of the same aluminum alloy. If the next alloy to be melted is different, the crucible designated for that alloy should be used. The just-used crucible is emptied of any remaining molten metal and the inside cleaned. To clean the inside, the crucible is set on its side in a special rack, if available, and scraped out.

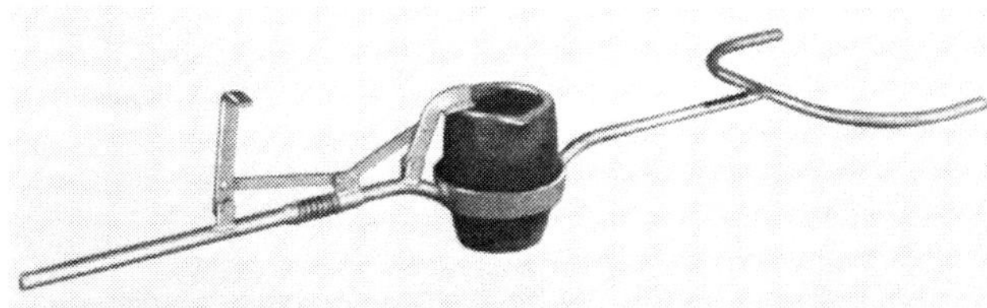

Fig. 12-18a. Crucible shank has a single handle and a safety lock to prevent accidental dropping of the crucible.

Fig. 12-18b. Crucibles with double handles are frequently used.

Fig. 12-19. Hand-ladling.

Induction Furnaces

Both the *coreless* and *channel* induction furnaces can be used to melt aluminum alloys. Other than the standard induction furnaces, discussed in ferrous melting practices, there are other types of induction furnaces that can be used. These furnaces are coreless furnaces in which a crucible is used to hold the molten metal. These latter furnaces can be divided into the push-up, swing-away, lift-coil and the tilting type. In the *push-up type,* the crucibles are pushed up out of the coil with a hydraulically actuated pedestal. With the *swing-away furnace,* the coil is raised and swung away from over the crucible so that it can be removed from the base of the furnace. In the *lift-coil,* the coil is raised high enough so that the crucible can be removed. The *tilting-type* induction furnace uses a stationary crucible surrounded by the coil and is tilted to tap the melt. With the exception of the tilt furnace, the coil can be placed around one crucible while an adjacent crucible is being poured or charged. **Figure 12-20** shows these four types of induction furnaces.

Clay graphite crucibles are usually used in the higher frequency range over 540 cycles because of the electrical characteristics. For frequencies in the 180–540 cycle range, either the clay graphite or the carbon-bonded crucibles can be used. Whether using the standard coreless or the channel furnace, high frequencies are normally used. Although the stirring action in the coreless furnace is good, it should not be of such magnitude that it causes severe turbulence in the melt. Should this severe turbulence take place, it can cause extreme gas pickup and oxide formation.

When using any of these four furnaces, the location of the crucible within the coil is critical. **Figure 12-21** shows the correct and incorrect placement of a crucible in the coil. The induced power and heating of the charge materials will be influenced by the type and mass of the charge. The charge materials should be carefully distributed to eliminate any sizable void areas and avoid developing uneven heating of the crucible.

(12-20a) push-up

(12-20c) lift-coil

(12-20b) swing-away

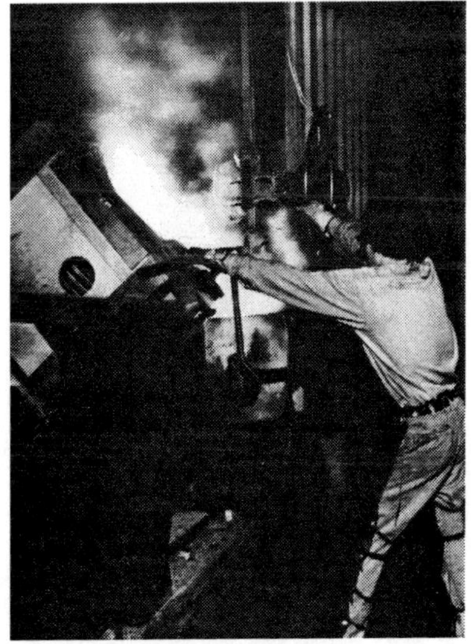

(12-20d) tilting

Fig. 12-20. Four types of induction furnaces: push-up, swing-away, lift coil and tilting.

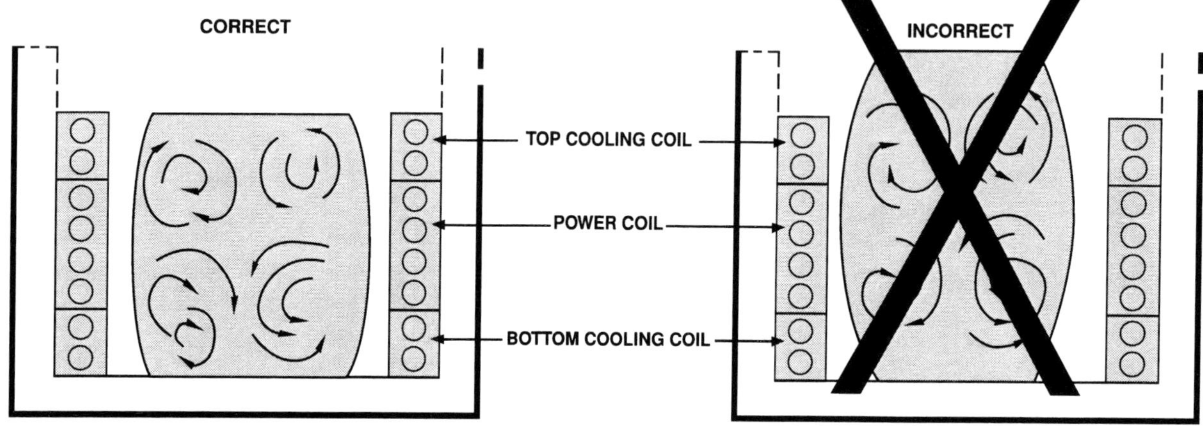

Fig. 12-21. Correct and incorrect placement. Crucible should be equidistant between the furnace walls and positioned vertically so that the center of the bilge is opposite the center of the power coil.

When molten metal is being held, it is suggested that the furnace be turned off rather than idled. It is also recommended that, when the tilting furnace is being tilted, the power be turned off to prevent uneven heating of the crucible, which can crack the crucible.

After the melt has been completed and is at the desired temperature, it is subject to the same treatments—degassing and oxide removal—as when using the gas-fired crucible furnaces. However, if done correctly, induction-melted aluminum alloys should contain less gas than those melted using gas as the heat source.

Channel induction furnaces, when used to melt aluminum alloys, are prone to the buildup of nonmetallics (oxides) in the channels, which will require periodical removal. These furnaces are normally used in foundries to prepare melts of both ingot and scrap. Their greatest use is as holding furnaces for molten aluminum alloys melted in other furnaces.

Coreless induction furnaces have been used for many years at a high frequency (above 60 Hz). However, in recent years, coreless induction furnaces operating on 60 Hz have been developed for the melting of aluminum alloys. The other advantages of the coreless furnace (over the channel furnace) are the same as those discussed in Chapter 10, Alloy Melting Furnaces.

Two-chamber coreless induction furnaces for continuous operation are frequently used in aluminum foundries. Melting is accomplished in one chamber and pouring from the other. Good temperature control can be achieved on the pouring side if care is taken when charging the metal.

Once the melt is at the proper temperature and before it is poured, it is subjected to the same degassing and oxide removal practices discussed with gas-fired crucible melting. If the induction melting has been done properly, there should be less gas in the melt when compared to gas fired melting. In addition, melting time is normally shorter.

Electric Resistance Furnaces

These furnaces, as described in Chapter 10, are used mainly as holding furnaces. They may be used to melt aluminum alloys in some situations. If they are used to melt solid charge materials, and already have a liquid bath, it is imperative that these materials be thoroughly dry. Charging should be done with care to prevent the oxide skin from being dragged down into the melt. In addition, whether with molten metal or solid charge materials, charging should be done with care so as not to cause undo gas entrainment by the melt.

Degassing of these melts can be done on a continual basis, especially if the molten aluminum alloy is hand or automatically dipped out of the furnace. If the melt is tilt-poured out of the furnace into the ladle, the same degassing techniques mentioned earlier can be employed.

Reverberatory Furnaces

As mentioned in Chapter 10, these furnaces are used where high volumes of aluminum alloys are to be melted and held. Depending on the type of reverberatory furnace used, charging can be done on a hearth, into a stack or into a bath of molten metal.

It should be realized by now that, when charging solid materials into a bath of molten aluminum alloy, it is imperative that all materials are dry. This is the advantage of the dry hearth and stack method of charging the furnace. These two furnaces will preheat and drive off moisture and possible oil contaminants before the charge materials reach the molten metal. In fact, in most cases, these materials will begin to melt on the hearth. Where gas is used as the fuel, care must be taken to properly maintain the flame and furnace atmosphere so that the charge materials are not oxidized as the exhaust gases pass through and over the pieces of charge material. Remember these exhaust gases have water vapor in them.

In the electric reverberatory furnace, there are no exhaust gases to be an issue. In addition, the metal bath is kept hot and liquid by quiet, nonturbulent radiant heat. Thus, the only hydrogen to be picked up by the molten metal and the only oxide formation is from contact with the atmosphere in the furnace. Normally, melts from the electric reverberatory furnaces will have less gas and oxide inclusions in them. Charging of this furnace is done through a charging well opposite the dipping well. This furnace is also much cooler and less noisy while in operation.

Hydrogen Degassing—With the exception of the stack-type reverberatory furnaces, the degassing operation is continuous in the well where the molten metal is removed from the furnace. As with all degassing techniques, it is not the size of the degassing bubbles but how many and how diffused they are in the melt. When using a lance, or more than one lance, they must be moved around to make sure that the degassing bubbles are diffused throughout the melt. With the use of rotary impeller equipment or a porous plug, this diffusion can be obtained with fewer problems.

With the stack reverberatory furnaces, degassing can take place in the transfer ladle if the molten metal is to bc poured into molds. If this isn't the case, the molten metal is transferred to a holding furnace and the degassing can take place there.

Again, as with all of the furnaces used to melt molten aluminum alloys, the effectiveness of the degassing operation can be checked by using one of the hydrogen concentration measuring methods.

Handling Molten Aluminum

Regardless of how carefully aluminum alloys are melted, it all can be ruined by mishandling the molten metal after it leaves the furnace.

One of the problem characteristics of aluminum is that it and its oxide have nearly the same specific gravity. For aluminum, the specific gravity is 2.7 and that of aluminum oxide is 3.5. This means that aluminum oxide will slowly settle in molten aluminum. This can be seen by the amount of aluminum oxide sludge that is removed from the floor of a reverberatory furnace when it is shut down for cleaning. The drosses that *do* float do so because they have become spongy with air and gas.

The speed with which aluminum oxide will form can be seen when skimming a ladle of molten aluminum. Before one can reach to the other side of even a small ladle, the surface becomes dull, indicating the oxide skin formation. Thus, when handling molten aluminum, the oxide surface or skin should be protected from being broken and forming more oxides. This means that operations, such as tapping from furnaces, dipping, fluxing, degassing, skimming and pouring into molds, should be done with a minimum amount of turbulence.

Adherence to fine handling methods will pay many dividends. The procedures necessary to produce aircraft-quality structural castings would not be economical for just any type of casting; however, adherence to the recommendations for properly handling molten aluminum alloys will improve the quality and appearance of any casting without undue cost.

MAGNESIUM ALLOYS

Magnesium and its alloys behave differently than do aluminum alloys when being melted. One of the major differences is the type of furnaces and crucibles used. The majority of the magnesium alloys are melted in gas-fired crucible furnaces. In some cases, electric resistance crucible furnaces are used. In addition, the crucibles used are made of steel and not of the same materials as used for crucibles used to melt the aluminum alloys. The different chemical and physical properties of magnesium alloys dictate the use steel crucibles, and different refractories and designs of the crucible furnace. **Figure 12-22** shows a cross section of a fuel-fired crucible furnace used to melt magnesium alloys.

When magnesium becomes molten, it tends to oxidize and burn. Therefore, care must be taken to protect the molten metal surface from oxidation. These alloys form a loose, permeable oxide coating on the surface of the molten metal. This permeable oxide allows oxygen to pass through and support burning below the oxide at the surface. To overcome this problem, a flux or a protective gas cover must be used.

The design of the furnace also comes into play to overcome this problem. Note in **Fig. 12-22** that the exhaust gases are directed out of the furnace through an exhaust stack. The crucible itself is used to seal off any chance of allowing the exhaust gases from coming in contact with the melt in the crucible. Should the exhaust gas contact the surface of the melt, it would expose the molten magnesium to oxygen. This furnace design is also known as a *bale-out furnace* because the crucible is stationary in the furnace and the molten magnesium alloy is dipped (baled) out for pouring into molds or diecasting machines.

The furnace refractory lining is very important, as magnesium reacts very violently with some refractories. High-alumina refractories and high-density "super-duty" firebrick of a typical 57% Si and 43% Al have been found to work well.

When a flux cover is used, there is normally a formation of sludge that will settle in the bottom of the crucible. This sludge has very poor thermal conductivity, which will lead to a loss of thermal efficiency for melting in the furnace, and overheating in areas of the crucible. This can lead to premature failure of the crucible. Thus, a program should be set up where these crucibles are cleaned out over a given period. This can be done by filling the crucible with water and allowing it to soak, which will remove all buildup.

Melting Procedures for Magnesium Alloys

There are two main procedures that can be used when melting magnesium alloys: *fluxing* and *fluxless*. Either of these procedures can produce good molten magnesium and its alloys for metal castings. Many of the precautions taken apply equally to both procedures.

Fluxing Procedure—The basic requirement when melting magnesium is to keep oxygen away from the molten metal. There are separate proprietary fluxes for each type of magnesium alloy: magnesium-aluminum types and magnesium-zirconium types. Supplier instructions must be followed and the right flux must be used for the type of alloy.

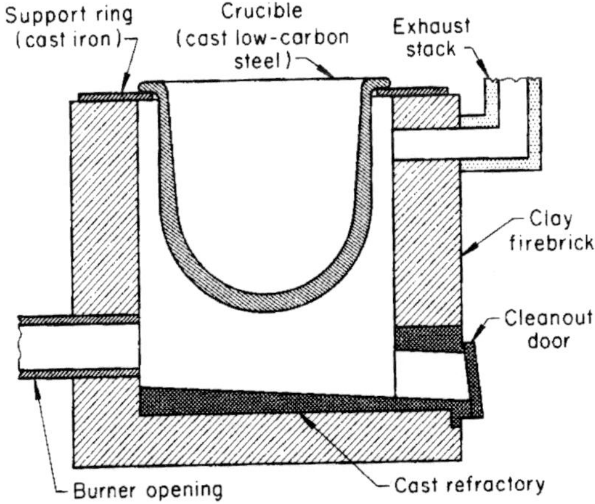

Fig. 12-22. Cross section of a stationary fuel-fired furnace for open-crucible melting of magnesium.

A typical fluxing procedure is to place the proper amount of flux in the crucible and heat the crucible to a dull red color. Next, the charge materials are placed in the crucible. These materials must be clean, dry and free from oil, oxide, sand and corrosion when placed in the crucible. Also, there should be no foreign metals in the charge. Any charge materials that contain dirt or oxide should be first melted separately and poured into ingots before being placed in a production heat.

Bridging (local freezing across a mold) of the metal charge during melting should be avoided. The remaining charge materials should be progressively placed in the crucible to maintain an advancing level of molten magnesium. During this procedure, additional flux is lightly sprinkled onto the melt surface.

Fluxless Procedure—With the fluxing procedure, particularly in diecasting, the presence of flux presents some operational problems. Flux inclusions in the diecastings are not uncommon. This problem can be overcome by using the fluxless procedure. In this procedure, gas mixtures are used to flux the molten magnesium. The gases used are air/sulfur-hexafluoride or air/carbon-dioxide/sulfur-hexafluoride gas mixes. The type of casting process also plays a role in the amount of each component found in the gas mixture. For instance, for magnesium castings made in sand molds, the gas mixture is higher in sulfur-hexafluoride, and sometimes, when the magnesium-zirconium alloys are to be poured, the parent gas is carbon dioxide.

In this procedure, it is also possible to have less melting losses. With the fluxless procedure, the amount of sludge found at the bottom of crucibles is less than with the fluxing procedure.

Degassing Procedure

Gas is not soluble in molten pure magnesium. Where the problem comes in is when magnesium-aluminum alloys are melted. The aluminum brings with it the ability to absorb hydrogen. This hydrogen is removed when the alloy is grain refined with carbon inoculation and hexachloroethane. On the other hand, when melting the magnesium-zirconium alloys, the zirconium additions do the degassing and grain refining.

Handling Molten Magnesium

When using the fluxing procedure, the flux is carefully skimmed off the surface of the molten metal. A wire brush is used to brush off any flux that may be on the rim and pouring lip of the crucible. Until pouring, the surface of the molten metal is protected by a cover gas or dusting of coarse sulfur and fine boric acid. During pouring, this skin is pushed back to keep it from going into the mold. A dusting with sulfur can protect the falling stream of molten metal.

With the fluxing procedure, there will be some residual flux-bearing material at the bottom of the crucible. This material should not be allowed to get into the mold. With the zirconium alloys, this is even more important, and at least 15% of the charged weight should be left in the crucible after pouring.

For the fluxless procedure, the cover gas is temporarily turned off and the oxide skin is removed from the surface of the molten metal. After this is done, the gas cover is again turned on over the surface of the molten metal, and kept on until pouring is completed. This same gas cover can also be used to protect the falling stream of molten metal.

COPPER ALLOYS

When melting copper-base alloys, the term "melt control" means controlling the furnace and the atmospheric conditions under which the metal is being melted. There are several other processing variables that can affect the quality of the copper-base castings. These variables include the following:

1. *Furnace Selection*—The type of furnace used to melt copper-base alloys influences the melt quality control. For all practical purposes, the same furnaces used to melt the aluminum alloys can be used to melt copper-base alloys. The one exception to this is the reverberatory furnaces. Although some reverberatory furnaces are used in larger copper-base melting operations, such as ingot production, very few are found in the average copper-base foundry. Depending on the type of furnace used, different melting procedures may have to be followed.
2. *Fluidity/Pouring Temperature*—The complexity of the castings and their section thickness will normally dictate the pouring temperature. Higher pouring temperatures will make chemistry and gas control of the melt more difficult.
3. *Mold Materials/Gas Generagion*—All mold and core materials can generate gas to varying degrees. These gases, along with those obtained from the melting operations, can lead to gas defects in the castings.
4. *Gating/Gas Pickup*—A poorly designed gating system can result in gas pickup by the molten metal as it flows through the system and enters the mold cavity. The gas derived from poor melting practice and that from a poorly designed gating system can lead to defective castings.
5. *Solidification/Shrinkage Porosity*—Since both gas porosity and shrinkage may occur in the portions of the casting that are last to solidify, they are often not easy to tell apart. Thus, a good job of gas control in the melting process will make the job of defect analysis easier.
6. *Mechanical Properties*—Good chemistry plays an important role in achieving the proper mechanical properties for the copper-base alloy being poured. Certain chemical elements can be reduced through improper melting procedures.

Charge Materials

Charge materials used to produce molten copper-base alloys can be broken down into five mains groups. These groups are:

1. *Commercially Pure Raw Materials*—In this case, the foundry would be using these materials to produce the copper alloy chemistry desired, by melting the proper amount of each alloy. Melting the alloy in this manner is seldom cost effective, except for new copper alloy development. Most copper-base foundries do not compound their own alloys, except when short-run specialty chemistries are required.
2. *Purchased Scrap*—The opposite extreme of using commercially pure raw materials is the use of an all-purchased scrap charge. The suggested way to use this material is to run a chemical analysis on it, melt it, and then pour ingots for future use. The economic advantages of scrap metal charges are best used by large copper-base foundries that have easy access to chemical analysis equipment and the ability to suitably modify the chemistry during the melting operation.
3. *Secondary Ingot*—Most copper-base foundries rely on secondary ingot with a known chemistry for their melting

operations. A 100% charge of secondary ingot can be used; however, most copper-base foundries will use secondary ingot in conjunction with scrap and foundry returns. Secondary ingot producers will provide metal chemistries within the specified chemistry limits and, if necessary, provide modifications in chemistry or tolerance limits to meet special customer needs.

4. *Returns*—Returns consist of gating systems, risers and occasional defective castings. In addition, if the foundry has its own machining operation, the turnings and borings generated can also be melted. When these materials are used, they must be clean, dry and free from any cutting fluids. These materials should also be segregated by alloy so that the proper returns are used as specified. Care should also be taken that deoxidizers or impurities are not building up in these materials because they can cause casting defects.
5. *Late Additions*—Small quantities of special materials, such as deoxidizers, grain refiners, degasifying agents and alloy additions to make up for melt losses, can be added to the melt. Although these additions may be small, their effect on casting integrity may be large.

Furnaces Used for Melting Copper Alloys

Gas-Fired Crucibles

The same types of gas-fired crucible furnaces and crucibles can be used for copper-base alloys as were used for aluminum alloys. In addition, the same precautions for charging the crucibles and preheating charge materials should be followed. However, in the case of the copper-nickel alloys, it is recommended that induction furnaces be used; in the case of "pure" copper, gas-fired crucibles should be used. **Figure 12-23** shows a battery of gas-fired lift-out crucible furnaces.

Most copper-base alloys melted in a gas-fired furnace are melted under an oxidizing atmosphere. This is done to ensure a higher dissolved oxygen content in the melt, and thus lower hydrogen content. Hydrogen is very difficult to remove from the melt, whereas oxygen is easily removed by standard deoxidation techniques.

Induction Furnaces

Both the *coreless* and *channel* induction furnaces can be used to melt copper-base alloys. Most of these alloys are melted in 1000 or 3000 Hz units, depending on the size of the furnace. Very large melts are usually handled with 1000 Hz or less frequencies.

When melting the copper alloys that are prone to dross formation, clogging of the channels in the channel induction furnace can become a problem. In addition, the stirring action, if it becomes too great, can cause dross formation and gas pickup in either of these furnaces.

The push-up crucible, lift-coil, swing-away and tilting coreless induction furnaces **(see Fig. 12-20)** are found in many smaller copper-base foundries that pour many different types of alloys. Crucibles can be used for specific copper alloys only. This practice helps avoid the use of wash heats in larger induction furnaces and holding chemistries specified for the alloy to be melted.

Melting Considerations

Metal Loss Control

Regardless of the type of charge materials used for melting copper alloys, the melt chemistry can change due to "metal losses." The most common cause of metal losses are the formation of dross due to the reaction of the melt with the atmosphere in and around the furnace, ladle material, and loss due to the low-boiling-point elements vaporizing.

Dross Formation—The greatest cause of dross formation in melting copper alloys comes from the atmosphere contacting the molten metal. This dross can be formed in gas-fired furnaces when the products of combustion, which contain water vapor, contact the melt. When induction furnaces are used, the extreme stirring action caused by the magnetic field will cause dross formation, especially on days with high humidity.

When melting copper alloys, it is recommended to use lower temperatures to reduce the amount of chemical reaction that can cause dross formation. In addition, shorter holding times will help reduce dross formation.

The use of crucibles or refractories that do not react with the melt will also help reduce the formation of dross. The presence of some oxides in a clay graphite crucible will allow some dross to be formed. Ideally, a silicon carbide crucible is best; however, cost may be a deterrent.

Melt covers or fluxes can be used to help combat dross formation, but with mixed results. Melt covers are primarily used to try to prevent the atmosphere from getting to the melt surface. Charcoal is one example of a melt cover. However, if nickel-bearing copper alloys are being melted, the use of a charcoal cover is prohibited due to the solubility of carbon in the melt. The selection of fluxes must also be made with care, as they may react with the refractories, crucible and mold materials.

Vapor Losses—Using the same techniques to reduce dross formation will also help to reduce vapor losses. The most notable element loss in copper-base alloys is zinc. If the melt temperature is kept low, the zinc loss will be low. However, some "zinc flare" can be advantageous to gas removal. It should be pointed out that some vapor losses, such as lead and beryllium, can cause a health hazard and should be monitored.

Fig. 12-23. Battery of gas-fired, lift-out crucible furnaces; note ingots preheating on furnace covers.

Gases and Degassing

It must be pointed out at the onset that metallic copper begins to absorb hydrogen at 1500F (816C), which is several hundred degrees below its melting point. The amount of hydrogen absorbed increases as the melt temperature increases and reaches the maximum at the melt point. It should also be pointed out that the presence of gases in the melt is not always detrimental.

Types of Gas—Two important examples of the positive role of gases in copper alloys are 1) their use to disperse shrinkage and 2) their use as degasifiers. The use of gas for the dispersal of shrinkage has been used successfully in producing discontinuous porosity in copper alloy castings, which creates better pressure-tightness and surface quality.

Gases that can be considered detrimental to copper alloy casting quality and integrity can be categorized as either *simple* or *complex.*

Simple Gases—Hydrogen is an example of a simple gas. Higher temperatures of the liquid melt will increase gas solubility as seen in **Fig. 12-24.** All of the gas contents are given at standard temperature and pressure (20C and 760 mm Hg). Some important concepts pointed out in **Fig. 12-24** appear to be limited since, in actual casting conditions where a copper alloy melt is exposed to pure hydrogen at 1 atm, pressure is unrealistic. However, the data shown are still useful because of the concept of chemical equilibrium.

Complex Gases—In many copper-base alloys, the most important gas causing porosity is water vapor. Water itself is not dissolved in the melt, but both hydrogen and oxygen are. During solidification of the copper-base alloy, these two gases combine to create gas porosity in the casting. Both of these gases are not found in the liquid metal at the same time in large quantities. If the hydrogen is high, the oxygen level will be low and vice versa. However, as the casting begins to solidify, the gases move into the remaining liquid metal, thus increasing its gas content.

In the case of nickel-bearing copper alloys, carbon pickup occurs and the important complex gas is *not* hydrogen-oxygen but carbon-oxygen. The gas evolved now becomes carbon monoxide (CO). However, again, an increase in one gas provides a decrease in the other gas.

Testing for Gas—There is one way to test the gas content of a wide-freezing range copper-base alloy such as 85-5-5-5 (C83600). The test consists of pouring a sample of the alloy and watching it solidify. All else being equal, the greater degree of "sink," the lower the gas content. In 1949, AFS research produced the fracture test. This test involved pouring a special piece of alloy against a chill. The piece was then fractured, and the fractured surface was evaluated for color and texture as a measure of the gas content.

A quantitative approach to measurement of the gas content was needed. In 1972, the International Copper Research Association and the AFS jointly sponsored research to develop such a test. This research developed a type of *reduced pressure test,* similar to the one used to check aluminum alloys for gas content. A rough drawing of the equipment used for the copper-base alloy reduced pressure test is shown in **Fig. 12-25.**

The key to this reduced pressure test is to establish a solid surface skin and directional solidification to produce a wave-front freezing pattern in the sample. This freezing wave front forces the gas into the remaining liquid. When exposed to the reduced pressure, the gas in the remaining liquid and the atmosphere above the sample attempt to come to equilibrium. The results of this action are depicted in **Fig. 12-26.**

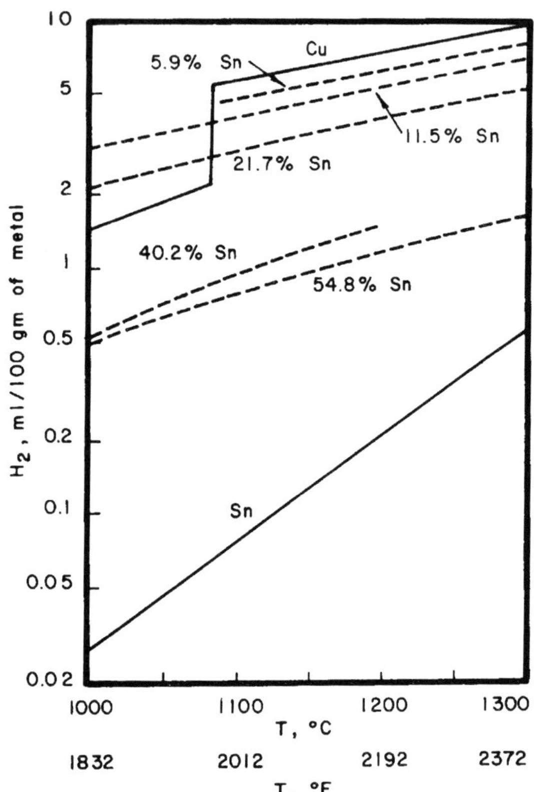

Fig. 12-24. Solubility of hydrogen at atmospheric pressure in pure copper, pure tin and copper-tin alloys.

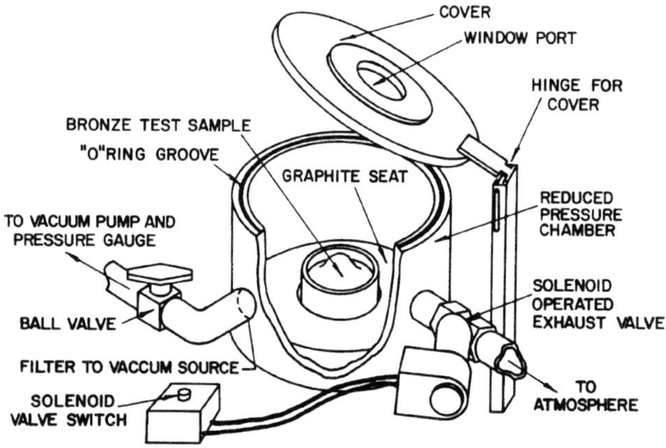

Fig. 12-25. Test chamber as used in the reduced pressure test for copper alloys.

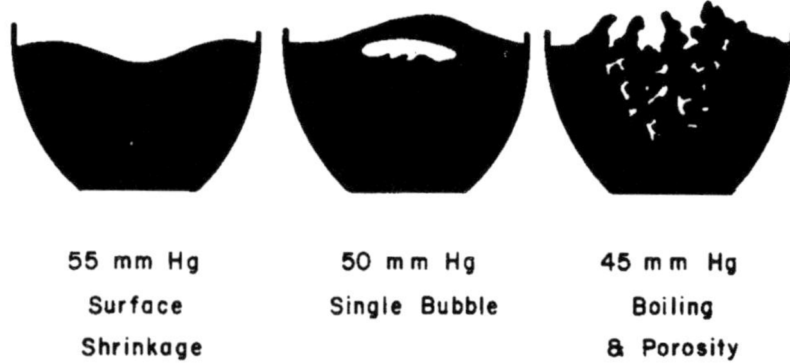

Fig. 12-26. Cut-away samples display the different responses of same amount of gas dissolved in the liquid metal when exposed to different pressures in the test chamber of the reduced pressure test.

However, since copper-base alloys have different freezing ranges, varying cooling rates are required to produce the proper wave front freezing. This is done by a selection of different crucible materials and graphite seats or heat sinks on which the crucible will sit to ensure the necessary thermal gradients. **Table 12-2** lists the three combinations of crucibles and seats that have proven successful for this test.

Quantitative analysis can be used to help determine the correct degassing technique. However, these tests are expensive and their interpretation is difficult. The average copper-base foundry finds the reduced pressure test to be more useful in respect to cost and speed of results.

Degassing—Removing gases from the molten metal can be done either by mechanical or chemical methods. Mechanical methods depend on the principle that the melt attempts to come to equilibrium with the surrounding atmosphere. Chemical methods use an agent, which is added to the melt, and which has a greater affinity for the gas than does the molten metal.

Mechanical Degassing—A melt's pickup of gas in a gas-fired furnace occurs because the surrounding atmosphere is rich in combustion by-products such as H_2O, CO_2, CO and excess oxygen. On the other hand, the magnetic stirring action in a coreless induction furnace can result in degassing the melt. This is true only if the surrounding atmosphere is low in detrimental gases such as water vapor. The removal of gases will be greater when the surrounding atmosphere is dry rather than wet. Thus, on high-humidity days, the amount of gas removed using this technique will be less.

The most common method of mechanical removal of gases from the melt is by purging with a second gas. If the purging gas is inert, such as dry nitrogen, and the dissolved gas is hydrogen, the hydrogen will migrate to the nitrogen gas bubble and will rise and be swept away. It is important that the purge gas be introduced deep in the bath and that the bubbles be small and numerous. This will maximize the contact time and the surface-to-volume ratio of the purge gas.

Another mechanical technique is to plunge the melt with solid additions that will form gas bubbles at the liquid-metal temperature. One such commercial product consists of briquettes of polytetrafluoroethylene (Teflon) which produces a purging gas when plunged beneath the surface of the melt.

Chemical Degassing—As mentioned earlier, chemical methods add an agent to the melt that has a greater affinity for the gas than does the molten metal. In red brass, it is a common practice to add phos-copper shot to the molten metal just prior to pouring. The P-Cu shot chemically ties up the oxygen in the melt. The oxygen present in the molten metal reacts with the phosphorus to form an insoluble phosphorus-oxygen compound. The use of P-Cu shot does not remove hydrogen from the melt directly.

In the case of aluminum bronzes, research has shown that considerable amounts of oxygen and hydrogen may be present in the melt at the same time. Chemical additives are commercially available in the form of proprietary "degassers." These degassers combine with both of these gases and remove them from the melt. These degassing agents usually contain highly reactive metal such as titanium and zirconium. Rapid pouring of the molten metal after the addition has been made should be avoided, since a considerable amount of time is required to properly rid the melt of the dissolved gases.

Table 12-2.
Crucible and Seat Combinations for the Copper Alloy Reduced Pressure Test

Alloy Type	Crucible	(Graphite)	Time Before Exposure to Reduced Pressure
Pure copper (i.e., C81100)	Preheated porcelain	Flat with small depression	15–30 sec
Short freezing range (i.e. C95400)	Preheated porcelain	Loose fitting to top of crucible lip	30–45 sec
Wide freezing range (i.e., C83600)	Steel	Tight fitting to top of crucible lip	5–10 sec

The one big disadvantage of adding solid agents to the melt is their residual level in gates, risers and sprues. These items normally become part of the subsequent charge materials used in future melting operations. High-residual phosphorus has been found to cause problems, such as: 1) the surface finish of the casting can become rougher than normal; and 2) it becomes difficult to suppress hydrogen absorption by the melt. In the first case, the phosphorus increases the fluidity of molten metal due to the reaction between the phosphorus oxides and the sand-bonding agents. In the second case, the oxygen in the melt is tied up by the phosphorus; thus, the amount of dissolved oxygen in the melt is lessened, making it easier for hydrogen to be absorbed.

On the other hand, insufficient levels of residual phosphorus can also be troublesome, since the deoxidized molten metal is subject to reoxidation during pouring. This problem can occur when the pouring takes place in high humidity and when the molten metal is poured into green sand molds. Thus, it is important that the foundry determine the correct level of residual phosphorus that can be tolerated for its particular operations.

Table 12-3 is a summary of gases found in copper-base alloys after melting and degassing.

Basics of Good Copper Alloy Melting

The following are major points to keep in mind for proper melting of copper-base alloys.

- *Charging Practice*—Only clean, dry, preheated charge materials should be used; the use of purchased scrap, returns, turnings, borings, etc. should be minimized.
- *Furnaces*—Fuel-fired furnaces contain high water vapor levels in the products of combustion that contact the melt, causing the strong opportunity to pick up gas. The atmosphere around induction furnaces is generally low in water vapor content and high in oxygen; therefore, there is low potential for gas pickup. Reaction and diffusion are aided by the magnetic stirring.
- *Melt Time*—Melting time, especially at high superheat, should be minimized, especially in fuel-fired furnaces. Power to induction furnaces should be controlled to avoid violent stirring, which can contribute to oxidation and metal losses.
- *Melt Atmosphere*—The melt atmosphere should be mildly to strongly oxidizing, depending upon how readily the alloy will dissolve oxygen as compared to how readily it will form dross. Where oxidation losses are of particular concern

Table 12-3.
Gas Content of Copper Alloys After Melting and After Degassing

Alloy Family	Gases Present	Remarks
Pure copper	H_2O, H_2	Approximate H_2O/H_2 ratio of 1:0. Higher purity increases the H_2O level and lowers H_2.
Cu-Sn-Pb-Zn alloys	H_2O, H_2	Pb does not affect the gases present. Increased Sn lowers the total gas. Increased Zn increases the H_2, with a loss in H_2O.
Aluminum-bronze	H_2O, H_2, CO	The presence of 5 wt% Ni in C95800 causes CO to occur rather than H_2O. Lower Al leads to higher total gas.
Silicon brass and bronze	H_2O, H_2	Approximate H_2O/H_2 ratio of 0:5. Increased Zn decreases the H_2 and increases H_2O. (This is opposite to the effect of when Sn is present as inthe Cu-Sn-Pb-Zn alloys.)
Cupronickels	H_2O, H_2, CO	All three species present up to 4 wt% Ni, beyond which only CO and H_2 are present. H_2 increases with Ni to 10 wt% Ni. H_2 is decreased at 30 wt% Ni.

(aluminum bronzes, high-strength yellow brasses, etc.), a proper melt cover material should be used.

- *Melt Temperature*—Superheat temperatures should be minimized. Increased temperatures dramatically increase gas pickup, dross and metal losses.
- *Degassing*—Alloys that *do not* dissolve oxygen (i.e., aluminum bronzes) should be purged with inert gases to remove dissolved hydrogen. Alloys that *do* dissolve oxygen readily (copper, red brasses, cupronickel) or moderately (tin bronzes) should be melted under oxidizing conditions to suppress hydrogen pickup and should be deoxidized with just enough deoxidant to prevent reoxidation during transfer and pouring. High-zinc alloys are self-degassing because of the high vapor pressure of zinc.
- *Melt Covers*—Oxidizing melt covers (copper oxide, silicate-borate mixtures) can be used to remove hydrogen or maintain it at low levels, and to consolidate drosses and oxides for easy removal. Neutral melt covers (glass and dry silica) form a mechanical barrier between the melt and the furnace atmosphere. This can reduce hydrogen pickup and, at the same time, reduce oxygen pickup. These materials are not very reliable for gas control but do help with dross removal and the reduction of vapor losses. Reducing melt covers (charcoal, graphite) prevent excessive oxidation losses but may be a source of hydrogen pickup if they are moist or contain hydrocarbon additives. If used in excess, they may prohibit oxygen pickup, thus allowing hydrogen pickup. Reducing covers are useful in retaining low oxygen levels in the molten metal, after deoxidation and prior to pouring.

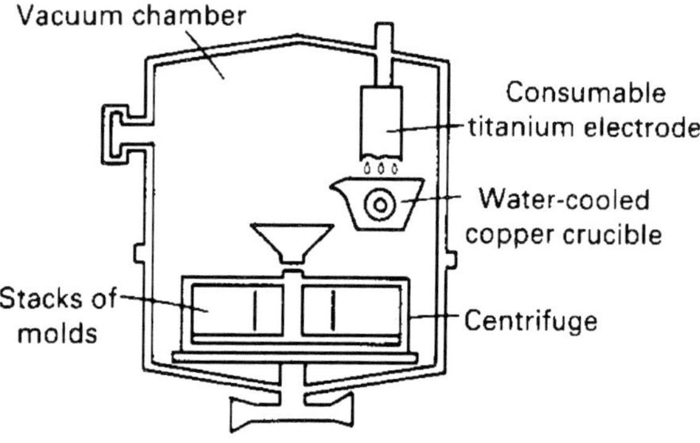

Fig. 12-27. Drawing of a centrifugal vacuum furnace for melting titanium alloys.

TITANIUM ALLOYS

The dominant method of melting titanium is by consumable electrodes. In this method, a titanium electrode is lowered into a water-cooled copper crucible, which is inside a vacuum chamber. This is known as *skull* melting. This technique prevents the highly reactive liquid titanium from dissolving the crucible because it is contained within a "skull" of solid titanium, which has been formed between the liquid metal and the crucible. When an adequate quantity of molten titanium has been melted, the consumable electrode is quickly retracted, and molten metal is poured into molds. The skull remains in the crucible for a subsequent pour, or it can be removed.

The consumable electrode practice does not afford much opportunity for superheating the molten metal. This is primarily due to the cooling effect of the water-cooled crucible. It is common practice to pour castings centrifugally or by using preheated molds. After the molten metal has been poured into a mold, the cooling process takes place in a vacuum or an inert gas atmosphere until the molds can be safely removed to air without causing oxidation of the titanium.

The consumable electrodes are made of ingot metallurgy forged billets, consolidated revert wrought material, selected foundry returns, or a combination of all of these. **Figure 12-27** is a drawing of a centrifugal vacuum furnace for melting titanium alloys.

BIBLIOGRAPHY

American Foundry Society, Inc., Des Plaines, IL, *Aluminum Casting Technology,* 2nd. Edition, 1993.

American Foundry Society, Inc., Des Plaines, IL, *Casting Copper-Base Alloys,* 1984.

American Foundry Society, Inc., Des Plaines, IL, *Recommended Practices for Sand Casting Aluminum and Magnesium Alloys,* 1965.

American Society for Materials, Metals Park, OH, "Casting Magnesium," *Metals Handbook,* 9th Edition.

Barone, R. V., Godfrey, T. F., Rosenberg, R. A.; "Bronze Alloys Electric Melting and Quality Control," American Foundry Society, Inc., *AFS Transactions,* 1973.

Clark, K., Technical Marketing Manager, Magnesium Elktron, Creston, IA, technical information, June 2000.

Crucible Institute, Inc., Jersey City, NJ, *Use and Care of Crucibles.*

Crucible Institute, Inc., Madison, CT, Notes on Crucible Melting Copper-Base Alloys.

Hampton, D., "Rotary Degassing Units Improve Quality, Lower Costs," *Foundry Management and Technology,* February 1994.

Klemp, T.; "Casting Cobalt Alloys," *Modern Casting,* December 1988.

Modern Equipment Co., Port Washington, WI, Jet Melter Brochure.

Schleg, F.P., CMI course notes from nonferrous courses, 1965-1992.

Microstructure of Ferrous Alloys

13

FERROUS AND NONFERROUS MICROCTRUCTURE

The study of the casting properties of alloys is closely related to the study of the structure and properties of liquid metals, as well as the study of crystallization theories. Practical data and theories of physics and chemistry agree that liquid metals, and alloys near the melting point, have structures that resemble those of crystals of the solid metal.

The material that will be covered in the first part of this chapter will apply to ferrous and nonferrous alloys alike.

The structure of a casting is a function of alloy composition and casting geometry, which plays an essential role in the speed at which a casting or section of a casting will solidify. "Freeze" is another term used to describe how a metal casting solidifies. Preliminary treatment of the liquid metal alloy, with additions of various chemical elements before pouring or during pouring, can also affect the metallic structure of the casting. Variations of cooling rates (heat transfer) also affect the casting structure as it solidifies in the mold.

Atomic Arrangement

The mechanical properties of a casting are largely affected by the grain size developed during the solidification (or freezing) of the casting. The manipulation of grain shape, size and crystallographic orientation will also affect the cast structure and its properties. More about this will be discussed here and in the following chapter.

In a given body of molten metal, the coolest parts begin to solidify first. Solidification normally begins at the metal-mold interface where there is a steep thermal gradient. A thermal gradient is the temperature difference over a given unit of time or distance. The steeper the thermal gradient, the faster solidification can occur. Since the steepest thermal gradients occur at the metal-mold interface, microscopic crystallites of metal, called nuclei, form and grow in a tree-like fashion. This tree-like formation is also called a dendritic structure **(Fig. 13-1)**. The arms or branches continue to grow on one another until the mass of liquid metal is completely solid. The basic metal structure consists of orderly arrangements of atoms. When molten metal solidifies, the atoms arrange themselves into various configurations called crystals. The smallest group of atoms in a lattice structure is known as a unit cell.

This growth pattern is an ordered structure of many unit cells packed upon and around one another. The atomic arrangement of these cells plays an important part in determining the microstructure and properties of a metal casting. One arrangement of these cells might give the metal casting good ductility while another might yield high strength. In crystalline materials, such as metals, atoms are arranged in a geometric pattern called a unit cell. **Figure 13-2** shows three types of unit cells. These unit cells may be repeated in space to form a crystal similar to a stack of building blocks.

The simple, cubic unit cell has an atom located at each corner of the cell. This arrangement of atoms can be repeated in three dimensions to generate a crystal of larger size **(Fig. 13-3)**. The cubic cell is not often encountered in metallic systems. In metallic systems, the unit cells are found to have arrangements that are more complex. The body-centered-cubic (BCC) cell is typically found in iron at room temperature (ferrite) and can be found in chromium and tungsten. In this unit cell, there is an atom in the center of the

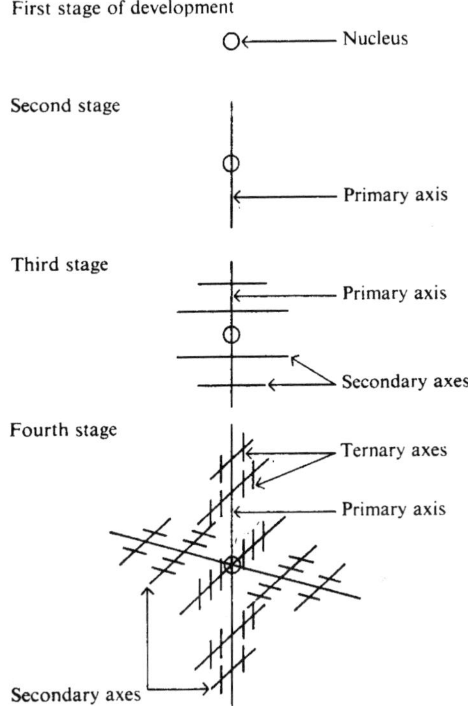

Fig. 13-1. Sketch of dendrite growth. Branches grow gradually, filling in to complete a solid mass of metal.

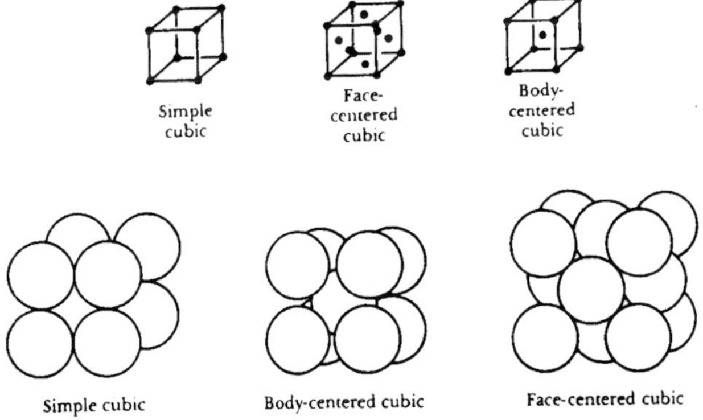

Fig. 13-2. Simple cubic, body centered cubic (BCC) and face centered cubic (FCC) unit cells.

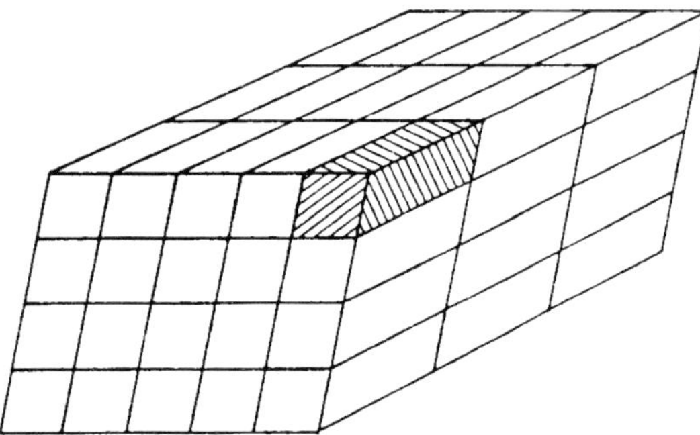

Fig. 13-3. Repitition of unit cell (cross hatched) in three dimensions generates a crystal.

unit cell along with the atoms located at the corners. The BCC unit cell reflects somewhat denser packing of atoms than is present in the simple cubic cell.

The more common arrangement of atoms found in metallic systems is the face-centered-cubic (FCC) cell. The FCC has atoms at each corner of the cell, as well as one atom in the center of each face of the unit cell. The packing of these atoms is denser than the BCC pattern and can be found in iron at elevated temperatures (austenite), copper, aluminum, lead, silver and nickel.

These cellular arrangements of atoms develop, for example, during solidification. Atoms in the liquid state are randomly distributed, or nearly so, and have no real structure. Consider a pure metal, cooled to its freezing temperature or slightly below that temperature. At this point, crystallization has begun at a number of centers, or nuclei, as seen in **Fig. 13-4(a).** The unit cells form space lattices that grow with the aggregation of more unit cells that have been formed from the liquid or, in this case, molten metal.

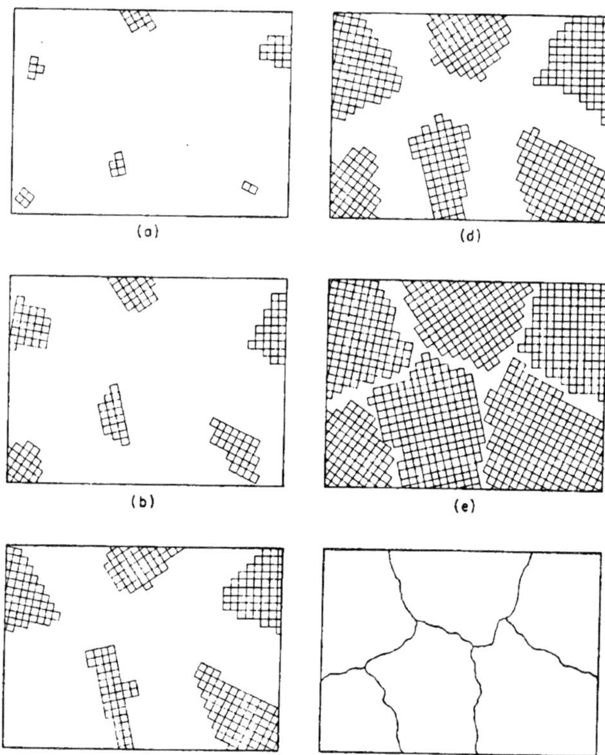

Fig. 13-4. Stages in the process of solidification of a metal.

The lattice structures continue to grow in the directions of the axes of the lattice unit until they meet an adjacent growing lattice, or until they contact the vessel or mold wall. The resulting structure is made up of grains, the size and shape of which are determined by the solidification rate (how fast or slow solidification takes place). These grains are individual, single crystals, separated from each other by grain boundaries, which is where the orientation of the adjacent lattices mismatch (**Fig. 13-4(f)**). The interference of these grain boundaries causes them to be revealed when a piece of metal is polished and etched. The polishing, etching and subsequent examination is accomplished by the metallurgical practice called *metallography.* **Figure 13-5** shows a sample of the grain structure of iron.

Phase Diagrams

Matter is usually found in one of three phases: solid, liquid or gas. In the case of metalcasting, the solid and liquid phases apply. A phase is defined as having the following characteristics:

1. A phase has the same structure or atomic arrangement throughout.
2. A phase has essentially the same composition and properties throughout.
3. A definite interface exists between the phase and its surroundings, or an adjoining phase.

Some examples of these phases can be seen in **Fig. 13-6**. The melting of an ice cube, made of pure water in a vacuum chamber with reduced pressure and temperature, is an example where all three phases can be found, solid ice, water and water vapor. In this case, the three phases would coexist as solid water, liquid water and gaseous water. Each of these phases would have its own unique atomic arrangement, have different properties, and a well-defined boundary would exist between them. All of these phases would also have the same composition.

A phase does not have to be a pure material, and several phases may combine to form a single phase. An example of this can be seen in **Fig. 13-6(b)**. Regardless of the amount of water and/or alcohol introduced into the container, one phase would result. Also, these two components are soluble in each other, and they exhibit unlimited solubility.

Copper and nickel exhibit unlimited solubility in the *liquid state,* which results in a single liquid phase, regardless of the amounts of copper and/or nickel. When this alloy solidifies, only one solid phase forms, because there is no interface existing

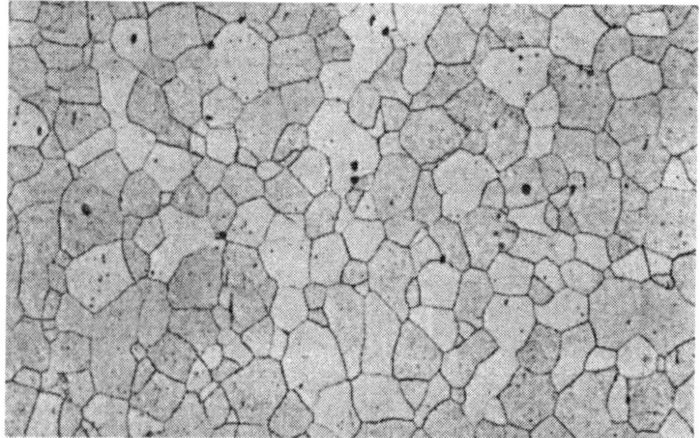

Fig. 13-5. Grain structure of iron (100X).

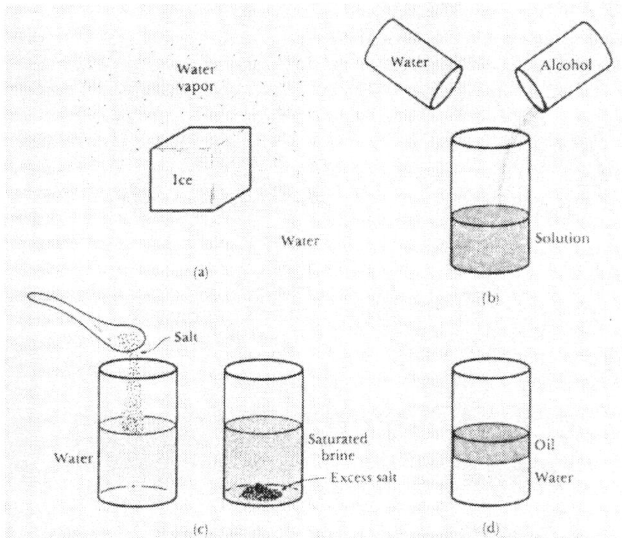

Fig. 13-6. Illustration of phases and solubility. (a) the three forms of water (solid, liquid and gas); (b) unlimited solubility of water and alcohol; (c) limited solubility of water and salt; (d) insolubility of oil and water.

between the copper and nickel atoms. Copper and nickel exhibit unlimited solid solubility (also referred to as a solid solution).

On the other hand, copper and zinc (brass) have unlimited solubility when in the *molten state.* However, only a limited amount of zinc can be dissolved in the crystal structure of copper. The excess zinc is found in the microstructure of the alloy as a copper-zinc compound, and the two solid phases exist together.

In some cases, virtually no solubility exists. Such is the case with water and oil in the *liquid state as well as solid state.* The two phases remain distinctly separate from each other. This same condition exists if one were to try to mix copper and lead. These two elements are immiscible (nonsoluble).

The metallurgist uses a "road map" called a phase diagram to determine how a specific metal alloy or pure metal will solidify. This diagram indicates the temperature at which solidification will begin and end. The phase diagram also will tell what is occurring during the various stages of solidification, i.e., how much of the metal alloy is solid or liquid; and based on the temperature and alloy content, what stage of solidification is taking place. Phase diagrams are unique to each individual alloy. Other names for phase diagrams are equilibrium diagrams and constitution diagrams.

Cooling Curves

Phase diagrams are a compilation of separate cooling curves for a given metal or alloy system. Solidification can be depicted graphically by cooling curves, as shown in **Fig. 13-7.** The solidification process is a function of time and temperature. When a pure metal changes from a liquid to a solid, the physical change takes place at a constant temperature level. An alloy in which the metals are in solution solidifies with a constant downward change in temperature.

Figure 13-7 illustrates cooling curves of an antimony-bismuth alloy system. The cooling curve on the left is for pure antimony and the one on the far right is for pure bismuth. In between are cooling curves for various antimony-bismuth alloy conditions. Below each cooling curve is an illustration of what the alloy would look like, when completely solid and properly prepared for viewing under a microscope.

As can be seen in the cooling curve of **Fig. 13-7(a),** pure antimony shows an arrest, or stop, in the drop in temperature at point "x" and an end at point "y." If one were to study the metal in the crucible, one would find that solidification began at "x" and ended at "y." Solidification, in this case, occurred at a constant temperature (1167F/632C) because of the liberation of the heat of fusion.

An alloy of 25% bismuth and 75% antimony would have a cooling curve as shown in **Fig. 13-7(b).** In this case, the cooling curve continues downward in a constant manner until it reaches point "x" or 1095F (587C), at which point a break in the curve occurs and solidification begins. Another break in the cooling curve occurs at point "y" or 800F (426C), at which point solidification ends. An examination of the structure formed in this alloy would show that a solid solution of antimony and bismuth was formed during solidification.

Other alloys of different ratios in the antimony-bismuth system are evaluated, and their cooling curves are developed and plotted on a phase diagram. The phase diagram for these alloy systems can be seen in **Fig. 13-8**. Another name for this type of phase diagram is a "binary" diagram. In other words, this phase diagram is produced for an alloy with two constituents of which at least one is a metal. A "ternary " diagram encompasses three constituents producing an alloy. Phase diagrams can be developed for four elements in a single alloy system.

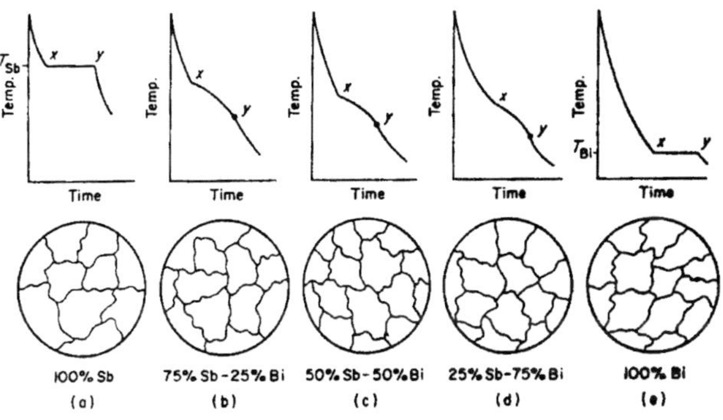

Fig. 13-7. Cooling curves for the antimony-bismuth system.

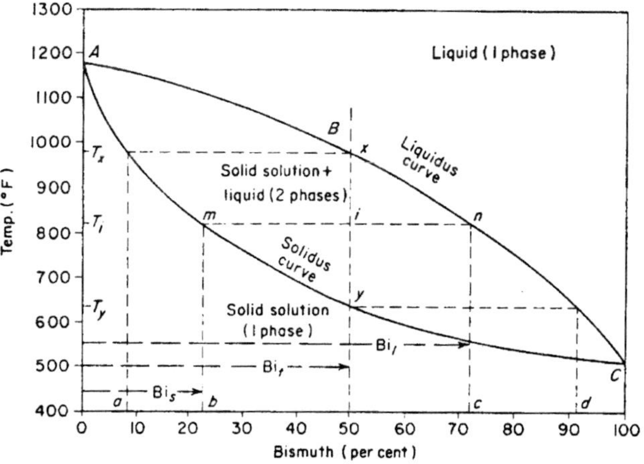

Fig. 13-8. Equilibrium diagram for the antimony-bismuth system.

On the phase diagram in **Fig. 13-8,** all of the points "x" from the cooling curves are connected to form the liquidus curve: the line at which solidification begins. All of the points "y" from the cooling curves are connected to form the solidus curve: the line at which solidification has ended. Everything above the liquidus curve is liquid, and everything below the solidus line is solid.

During the course of solidification, the composition of the liquid and solid phases changes. Within the liquidus and solidus lines, both solid and liquid phases are present. A very unscientific term for this mixture of the two phases is "mush." At a temperature of T_x, a 50% bismuth alloy is completely liquid or in the liquid phase. This can be determined by finding "50%" on the horizontal axis and following the dashed line from this point up to the liquidus curve at point "x." When this alloy is cooled to below T_x, it enters the two-phase (mushy) region and forms a solid composition given by point "a" on the horizontal axis. As the temperature continues to drop, the composition of the solid phases changes along the solidus curve from point "a" to point "m," to point "y." At the same time, the composition of the liquid changes along the liquidus curve from point "x" to point "n" and, finally, to a liquid of the composition of point "c." This change in composition during the course of solidification is typical for most alloys.

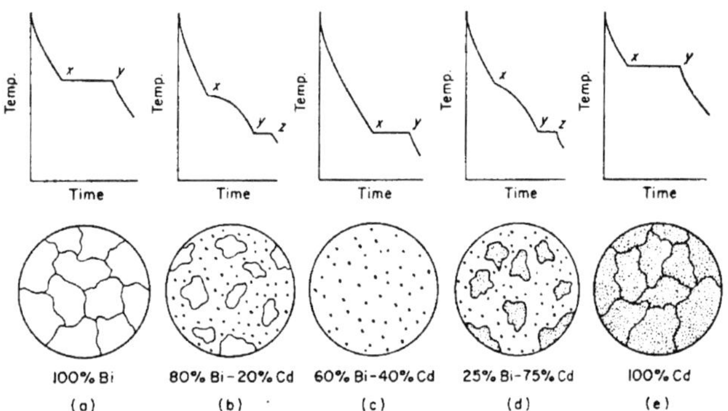

Fig. 13-9. Cooling curves for the bismuth-cadmium system.

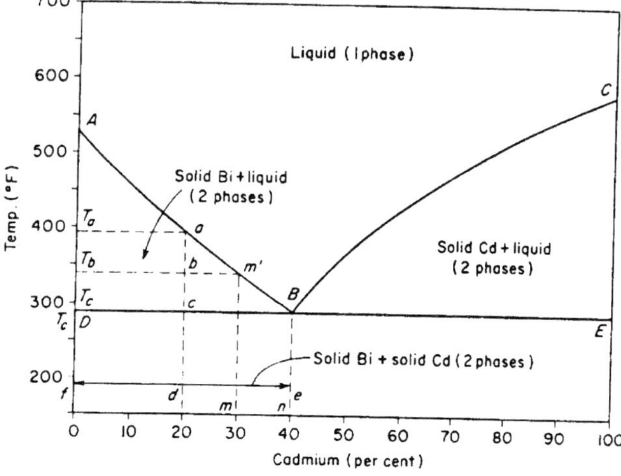

Fig. 13-10. Equilibrium diagram for the bismuth-cadmium system.

Lever Arm Principle

The phase or equilibrium diagrams also can be used to help determine the chemical or alloy composition of the phases in equilibrium and the amount of the phases at any given temperature. To make this determination, the lever arm principle, (or lever rule) can be applied anywhere in the phase diagram where two phases coexist.

Again, look at the phase diagram for a 50% bismuth alloy in **Fig. 13-8**. The previous study of this phase diagram indicated that this alloy is composed of a solid composition "m" and a liquid composition "n" at temperature "T_i." Considering the concentration of bismuth present in the solid phase, in the liquid phase and in the alloy as a whole, it can be shown that the amount of solid solution and the amount of liquid will be inversely proportional to the distances from the point representing the composition of the alloy to the point representing the composition of the phase in question.

Mathematically, this is how it works: Again, using temperature "T_i," it is evident that the total mass of *liquid phase* is given by the relation or fraction "m_i/m_n." On the other hand, the total mass of the *solid phase* is given by the relation or fraction "i_n/m_n." Multiplying each fraction by 100 will determine the percentages of each phase present, which is the lever rule.

Again, it should be noted that, below the solidus temperature, all alloys in this antimony-bismuth alloy system are single phase. This homogeneous solid solution type of system is also referred to as an isomorphous system.

Component Solubility

Alloy Components Completely Soluble in the Liquid Phase but Insoluble in the Solid Phase—These have a completely different set of cooling curves and phase diagrams than for the antimony-bismuth alloy. Bismuth-cadmium is such an alloy. **Figure 13-9** illustrates a set of cooling curves developed by the freezing of this alloy. The cooling curves for pure bismuth **(Fig. 13-9(a))** and pure cadmium **(Fig. 13-9(e))** are the same as seen in **Fig. 13-7** for the pure metals in that alloy. However, as these two metals are mixed, a different shaped cooling curve is developed.

When an alloy of 20% cadmium is cooled, a break in the cooling curve is seen at point "x." It is at this point that solidification of this alloy begins. The cooling curve then continues downward, but at a slower rate, until it reaches point "y." At point "y," there is what is called an "arrest;" in other words, the temperature remains constant for a period. This arrest continues until point "z" is reached, where solidification has been completed.

The microstructure of this 80%Bi and 20%Cd alloy would be similar to that in **Fig. 13-9(b)**. The large grains of bismuth were formed during the cooling from point "x" to point "y." On the other hand, the intimate mixture of bismuth and cadmium was formed during the arrest between points "y" and "z."

Figure 13-9 shows that when there is 40% cadmium in the alloy, the cooling curve has a single arrest from point "x" to point "y." The microstructure of this alloy is composed of an intimate mixture of bismuth and cadmium, referred to as a *eutectic.* Since this eutectic is formed in this particular alloy, 60% Bi and 40% Cd is called a eutectic alloy.

The phase or equilibrium diagram for the bismuth-cadmium alloy is seen in **Fig. 13-10.** Remember, this phase diagram was created from the data points in the cooling curves in **Fig. 13-9.** It should be pointed out that, normally, there are more cooling curves generated than are shown in the illustrations. These cooling curves are plotted to form the phase diagram for an alloy system.

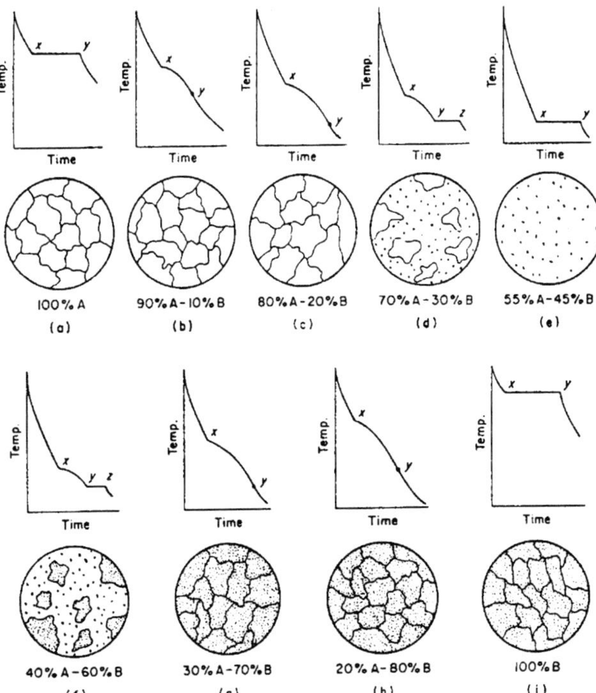

Fig. 13-11. Cooling curves for a hypothetical A-B system.

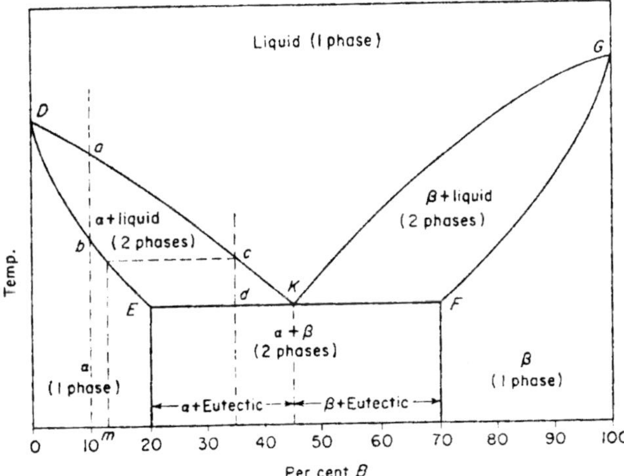

Fig. 13-12. Equilibrium diagram for the A-B system.

The phase regions of this system are as follows:

Above curve ABC - homogeneous liquid solution (one phase).
Area ABD - solid bismuth plus liquid (two phases).
Area BCE - solid cadmium plus liquid (two phases)
Area DBE - solid bismuth plus solid cadmium (two phases).

At point B, the structure of the alloy is completely eutectic, and is known as the "eutectic point" of this alloy. At this eutectic point, the alloy solidifies in a manner similar to a pure metal. Other terms that should be explained are: hypoeutectic or hypereutectic alloys. The eutectic point of the alloy is the dividing point for these two terms. In this case, if the alloy has greater amounts of bismuth (to the left of the eutectic composition) it is a hypoeutectic alloy. If the alloy has greater amounts of cadmium (to the right of the eutectic composition) it is a hypereutectic alloy.

It should also be pointed out that the lever arm principle can be applied to this phase diagram also. This principle, as discussed earlier, can help determine the composition of the phases that will be in equilibrium and the amount of the phases at any given temperature.

Components Completely Soluble in the Liquid Phase and Partially Soluble in the Solid Phase—There are a few alloy systems that are completely insoluble in the solid state, as previously discussed. However, the most common type of alloy system is partially soluble in the solid state. Consider a hypothetical alloy made up of components A-B. **Figure 13-11** shows the cooling curves for this hypothetical alloy. A number of alloys have been selected covering this hypothetical system, and it can be noted that some of the alloys solidify in a manner similar to the solid solution alloys discussed in **Fig. 13-7**. In addition, several of the alloys solidify in a manner identical to the eutectic system discussed in **Fig. 13-10.**

Looking at the cooling curves shown in **Fig. 13-11,** it can be seen that points x, y and z establish the phase diagram for our hypothetical alloy made up of components A and B shown in **Fig. 13-12.** The single-phase solid on the left of the phase diagram is referred to as the (α) or alpha phase. This phase is a solid solution of component B in component A, with a maximum solubility of 20% component B. The other single phase shown in this phase diagram is referred to as (β) or beta phase and is found on the right side of the diagram. This solid solution is made up of component A in component B with a maximum solubility of 30% component A. The phase areas of this diagram are as follows:

Above DKG - homogeneous liquid solution (one phase)
Region DEK - α solution plus liquid (two phases)
Region GKF - β solution plus liquid (two phases)
Below DE - α solid solution (single phase)
Below GF - β solution (single phase)
Below EKF - α solid solution plus β solid solution (two phases)

More typically, alloy systems like the hypothetical one just discussed are somewhat more complicated and are beyond the intended scope of this book.

As mentioned at the beginning of this chapter, the discussion, so far, applies to ferrous as well as nonferrous alloys.

FERROUS ALLOY MICROTRUCTURE

The following text will look specifically at the structures of ferrous alloys. The discussions will begin with the family of cast irons, then steels.

Gray Cast Iron Microstructure

It is important to know the chemistry of the metallic charge materials used in the melting furnace. **Table 13-1** lists some elements that can be found in the metallic charge materials and that can affect the structure of the gray iron. Note that some of these elements will promote graphitization while others will promote the formation and stabilization of carbides. Thus, as mentioned in Chapter 11, Ferrous Melting Practice, the production of the desired structure in gray iron (and the entire family of cast irons) begins in the charge makeup area.

In order to understand how and why certain structures are found in gray cast iron, one first must look at the iron-carbon equilibrium diagram in **Fig. 13-13.** This impressive phase diagram tells the story, from a metallurgical standpoint, of both the family of cast

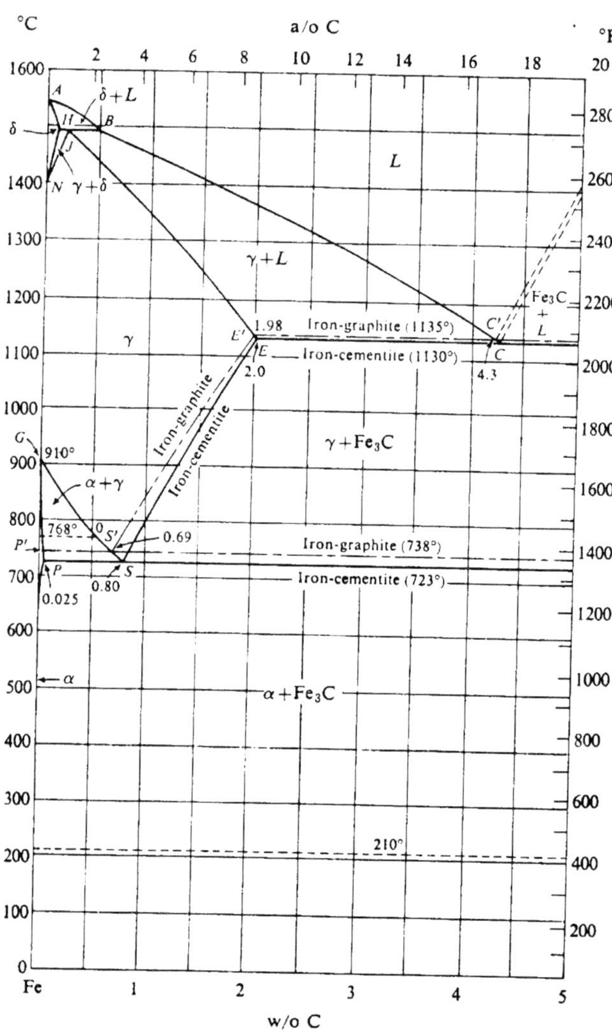

Fig. 13-13. Iron-carbon phase diagram. The metastable iron-iron carbide system is represented by solid lines. The stable iron-graphite system is represented by broken lines.

Table 13-1.
Structural Effects of Some Elements in Cast Irons

Element	Effect During Solidification	Effect During Eutectoid Reaction
Aluminum	Strong graphitizer	Promotes ferrite and graphite formation
Antimony	Little effect in amounts used	Strong pearlite stabilizer
Bismuth	Carbide promoter, but not carbide former	Very mild pearlite stabilizer
Boron up to 0.15%	Strong graphitizer	Promotes graphite formation
Boron greater than 0.15%	Carbide stabilizer	Strong pearlite retainer
Chromium	Strong carbide former Forms complex carbides, which are very stable	Strong pearlite former
Copper	Mild graphitizer	Promotes pearlite formation
Manganese	Mild carbide former	Pearlite former
Molybdenum	Mild carbide former	Strong pearlite former
Nickel	Graphitizer	Mild pearlite promoter
Silicon	Strong graphitizer	Promotes ferrite and graphite formation
Tellurium	Very strong carbide promoter, but not stabilizer	Very mild pearlite stabilizer
Tin	Little effect with amount used	Strong pearlite retainer
Titanium under 0.25%	Graphitizer	Promotes graphite formation
Vanadium	Strong carbide former	Strong pearlite former

irons and carbon steels. On the vertical axis is the temperature and on the horizontal axis are the iron and carbon levels. The carbon level reads toward the right from 0 to 5% C. Some phase diagrams have carbon levels above 5%. There can be a division made on this phase diagram between the cast irons and carbon steels. This division is made at or very close to 2% carbon. The carbon steels are found below 2% C and cast irons above this point. When studying the structures in cast iron, the concentration will be located to the right of 2% carbon area of the phase diagram.

The complexities of the development and reading of this phase diagram will be left up to the metallurgists and those studying the subject of metallurgy. As discussed earlier, these diagrams were developed by producing a series of cooling curves for various iron and carbon contents and plotting them together to form one large diagram. The lever arm principle discussed previously can also be used with this diagram.

Another simplified version of the iron-carbon equilibrium diagram can be seen in **Fig. 13-14.** The dividing line between carbon steel and cast iron is shown at 2% carbon. Another version of this diagram shows the stages of solidification taking place in cast iron **(Fig. 13-15)**. Again, the carbon level is shown on the horizontal axis and the temperature on the vertical axis. Also, note that this diagram begins at 2% carbon.

To begin with, some of the words and terms found in **Fig. 13-15** should be defined. *Melt* means that all of the elements in this alloy system are in the liquid state as a soluble liquid phase above the super heat zone.

Austenite is a solution of carbon and the high-temperature face-centered cubic (FCC) crystalline form of iron, which occurs during solidification. Later, during slow cooling, it can change to pearlite or ferrite or a mixture of the two. To the left of the eutectic point, austenite is the first constituent to turn solid. To the right of the eutectic point, carbon is the first element to turn solid. The eutectic point or eutectic composition of cast iron has a composition factor of 4.3%. The composition factor is determined by the formula: CF = %TC + 1/3 %Si. (The composition factor (CF) equals the percentage of total carbon (TC) in the alloy plus 1/3 the percentage of silicon (Si) in the alloy. If the CF is below 4.3%, it is a hypoeutectic cast iron; if the CF is above 4.3%, it is a hypereutectic cast iron.) CF is also referred to as carbon equivalent or CE.

By using the diagram in **Fig. 13-15**, the progress of solidification of gray cast iron having a CF indicated by line "A" can be followed. At point 1, solidification begins with the formation of solid austenite dendrites, for hypoeutectic alloys. A schematic of a dendrite was seen in **Fig. 13-1**. This dendrite formation takes place in the primary freezing range. If it were a eutectic alloy, as shown by the letter "B," this step would be eliminated. During eutectic freezing, the solids that form can either be a mixture of austenite and carbides or of austenite and graphite. If the former occurs, the cast iron is freezing as white iron, whereas, if austenite and graphite occur, the alloy is freezing as gray or ductile iron.

Three things can affect the results that occur during eutectic freezing. First, if the silicon level is high and there is slow cooling, graphite will form. Second, if the silicon level is low and faster cooling takes place, carbides will form. Graphite is one of the soft, crystal forms of carbon. Third, carbide or cementite (FeC_3) is a solid solution of iron and carbon, which is very hard and brittle. Further cooling takes place between points 3 and 4, and during this time the precipitation of graphite from the austenite takes place. The excess carbon in the austenite comes out as carbide in white irons and as graphite in gray and ductile iron.

The graphite can take several forms during solidification. In gray iron, the graphite comes out as flakes; in malleable iron, temper carbon or graphite aggregates are developed through the heat treatment of white iron. Spheroidal or nodular graphite is developed when cast irons are treated with special elements, which will be discussed shortly. The amount, size, shape and distribution of graphite in cast irons has a great influence on its properties.

The last change in cast iron structure takes place between points 4 and 5 **(Fig. 13-15)** as the austenite transforms over this temperature range. This change takes place in the solid state during cooling and is very complex. The material that forms at this stage of solidification is called the matrix. This matrix helps impart the mechanical properties to the family of cast irons.

Gray Iron Graphitization

When cast iron alloys are heat treated, there is a change in the matrix. However, once the graphite has been formed, the only way to change its form is by giving it the ultimate heat treatment: remelting. With the most favorable graphitizing (forming of graphite) conditions, only ferrite is formed in gray and ductile irons. Ferrite is a solution of iron and small amounts of carbon. Ferrite is relatively soft, ductile, and has moderate strength. In cast irons, ferrite contains the silicon present in the alloy that hardens and strengthens the ferrite. **Figure 13-16** shows graphite flakes surrounded by a matrix of ferrite.

When the graphitizing conditions are less severe, ferrite and pearlite, or only pearlite, are formed. Pearlite is a mixture of ferrite and cementite (iron carbide) arranged in alternate layers. **Figure 13-17** is photomicrograph of the polished surface of a slice of gray iron. The black "lines" are flake graphite, and the material around the flakes is pearlite. The alternate layers of cementite and ferrite are difficult to see under this magnification. Etching of the surface of this metallographic sample would also help make the pearlite stand out more clearly. Pearlite in cast irons is strong, moderately hard and has some ductility. (In ductile iron, mixed structures of ferrite and pearlite form as "bull's-eyes" of ferrite around the graphite spheroid.) Cooling below point 5 **(Fig. 13-15)** to room temperature produces little change in the iron.

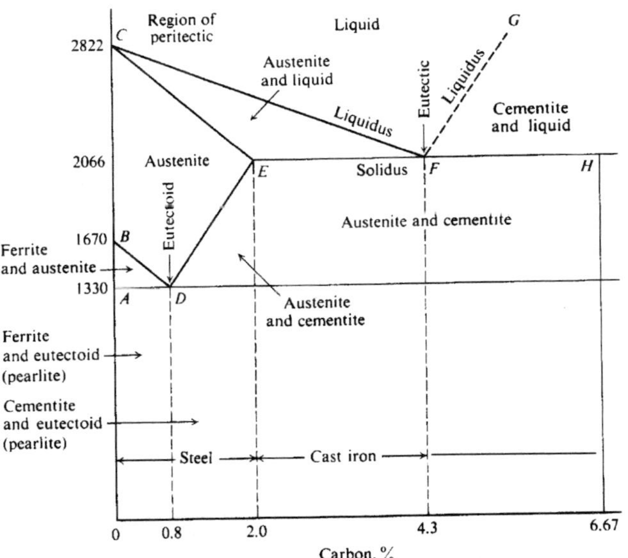

Fig. 13-14. Schematic representation of iron-carbon diagram. (The reader can disregard the letters A through H.)

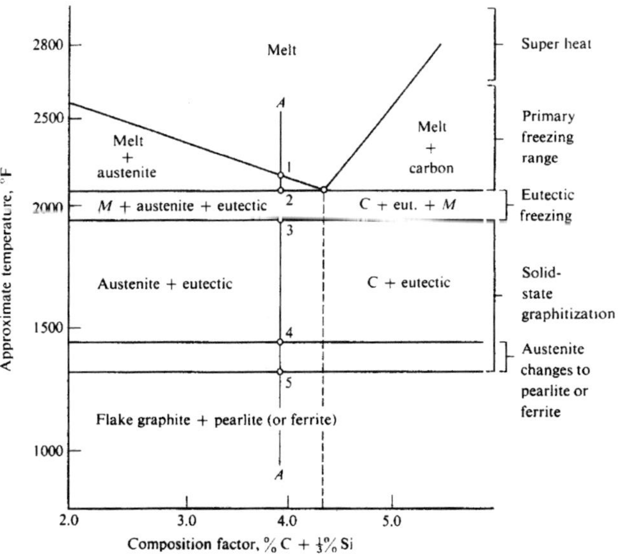

Fig. 13-15. Diagram showing approximate range of temperature of solidification in cast irons.

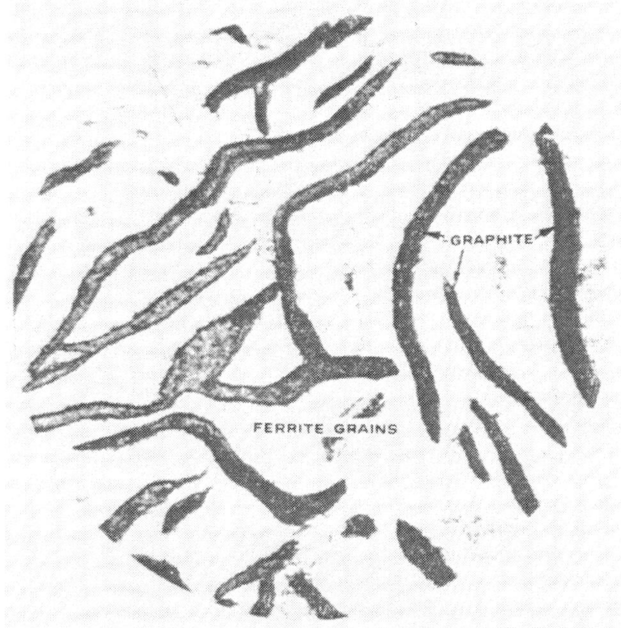

Fig. 13-16. Ferritic structure of gray cast iron (etched, 200X approximately) [Courtesy A.P. Gagnebin–Internaltional Nickel Co.]

Referring back to **Table 13-1,** there are chemical elements that, when added to the cast iron alloy, can help stabilize the pearlite. Stabilizing means that the pearlite, once formed, will not break down or it will have a difficult time breaking down. These elements include manganese (Mn), antimony (Sb), copper (Cu), tin (Sn) and niobium (Nb) also known as columbium (Cb). The latter element, niobium, is not found in **Table 13-1** because it is not normally added to the melt as a pearlite stabilizer. This element is used in the making of high-strength, low-alloy steels, also known as HSLA steel. These alloys are finding their way into the ferrous metals scrap stream and thus could unknowingly be charged into the melting furnace.

Another solid mixture that can occur in cast irons, especially in gray iron when phosphorus is present, is steadite. Steadite is a eutectic of iron and iron phosphide, and of low melting temperatures (about 1750–1800F/954–982C). Phosphorus segregates into areas that will solidify late in the freezing process, and steadite areas often reveal a cellular pattern. Iron phosphide is very hard and can raise the hardness and brittleness of gray iron. The formation of steadite can be controlled by the amount of the phosphorus content in the gray iron. **Figure 13-18** shows the microstructure of steadite.

The most potent promoter of graphitization in gray iron is silicon. This graphitization, as just discussed, takes place at three important stages:

1. Graphitization during solidification (liquid to solid)
2. Graphitization by carbon precipitation from austenite (solid state)
3. Graphitization during the eutectoid transformation (solid state).

Silicon is present in gray iron from about 1.0 to 3.5%, by weight. Silicon can be added to the gray iron along with the charge materials, and just prior to pouring the molten gray iron into the mold.

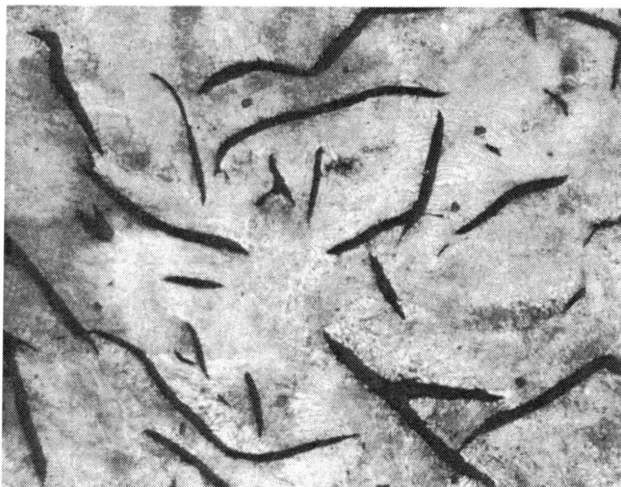

Fig. 13-17. Most of the pearlite matrix is resolved, 500X.

Increasing the amount of silicon in the gray iron will shift the eutectic point of the iron-carbon diagram to the left. This shift in the eutectic point is described by the following relationship:

Eutectic carbon % = 4.30 – 1/3 x % silicon in the iron

Note that this formula is different than the one used to determine carbon equivalent. The latter term, carbon equivalent (CE), is a useful expression because many of gray iron's properties have been found to be related to it. If the gray iron is hypereutectic, the graphite forms first during solidification and the melt is said to form kish. This kish floats to the surface of the melt and pops out into the air and can be seen as sparkly graphite flakes floating on the surface of the iron or in the air above the iron. The silicon occurs in a dissolved state in the ferrite of gray iron, which in turn hardens and strengthens the ferrite.

Silicon promotes graphitization. However, low percentages of silicon are not sufficient to cause graphitization during solidifica-

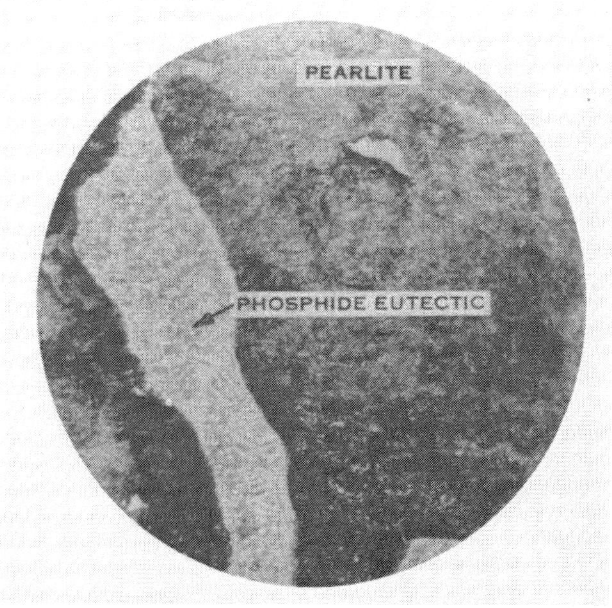

Fig. 13-18. Microstructure of steadite, the iron-iron phosphide eutectic, 1000X. [From AFS Cast Metals Handbook, 1957]

tion. As will be learned further on, silicon does cause nucleation and graphitization in the solid state at high temperatures. Nucleation means that the silicon will act as "sites" for the graphite to cling to and grow into flakes of graphite. Silicon also helps to cause the decomposition of iron carbide (Fe_3C). Certain silicon percentages will cause limited graphitization to take place during solidification, the result of which is a mottled iron, partly white and partly gray.

Sulfur and manganese are two other elements found in gray iron. Sulfur may be present in the gray iron up to about 0.25% and is one of the important modifying elements present in gray iron. A low-sulfur iron-silicon-carbon alloy (under 0.010% S), will graphitize most completely, that is, the greater percentage of carbon in the alloy will form graphite flakes. On the other hand, high sulfur content favors the retention of a completely pearlitic microstructure in gray iron; thus causing sulfur to be known as a carbide stabilizer. Above about 0.25%, sulfur in gray iron is considered to contribute to retardation of graphitization.

The sulfur in the iron reacts with the manganese that is also present in the iron. Sulfur, when alone in the iron, will form iron sulfide (FeS). This compound will segregate into the grain boundaries during the latter stages of solidification. However, when manganese is present, MnS or complex manganese-iron sulfides are found, depending on the manganese content. Manganese sulfides precipitate early and continue to do so during the entire solidification process. They are usually found randomly distributed in the microstructure.

Figure 13-19 shows a photomicrograph of manganese sulfide inclusions. When MnS is formed, the effect of sulfur is diminished and the formation of a pearlitic microstructure is diminished. Manganese alone tends to retard the graphitization process; thus, manganese above that needed to react with sulfur will assist in the retention of a pearlitic microstructure. The following rules have been put forth regarding sulfur and manganese in gray iron:

1. %S x 1.7 = %Mn; chemical equivalent of S and Mn percent needed to form MnS.
2. 1.7 x %S + 0.15 = %Mn; the Mn percent that will promote a maximum of ferrite and a minimum of pearlite.
3. 3 x %S + 0.35 = %Mn; the Mn percentage that will develop a pearlitic microstructure.

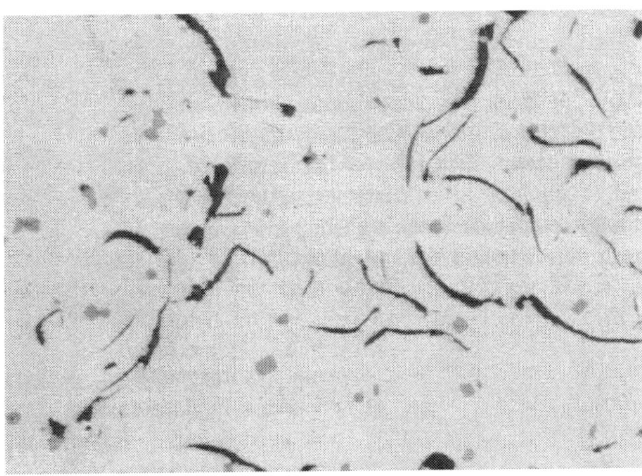

Fig. 13-19. Manganese sulfide inclusions (light gray) in gray iron, unetched, 250X. [Courtesy of L.F. Porter]

For commercial gray irons, which require a pearlitic microstructure, rule 3 offers a favorable combination of manganese and sulfur percentage.

Phosphorus, as mentioned earlier, will help to form steadite in gray iron. The phosphide of iron is hard and brittle, just like carbide. Increasing the phosphorus content in the gray iron will cause a proportional increase in the hardness and brittleness of the gray iron, especially above about 0.30% P.

Thin-section gray iron castings, such as bathtubs, fry pans and sinks, will contain higher percentages of phosphorus than will other gray iron castings. To a limited degree, improved fluidity of the molten gray iron can be obtained, and the molten metal can better completely fill the thin sections of the casting. It must be mentioned, however, that the addition of phosphorus for this purpose must be made carefully.

The actual changes in graphite size, number of flakes and their distribution are all related to fundamental metallurgical nucleation and growth principles. **Figure 13-20** shows the AFS and ASTM graphite-flake size rating charts. **Figure 13-21** shows the types, indicated by letters A through E. **Table 13-2** lists the types of graphite and factors that go into a particular type of graphite formation. Large flakes, randomly distributed, are formed when the nucleation rate is low and there is ample time for diffusion to take place. Under these conditions, graphitization (formation of graphite flakes) occurs readily, and the formation of Type A and B graphite flakes will occur. Type A flake graphite is what is found in most normal gray iron.

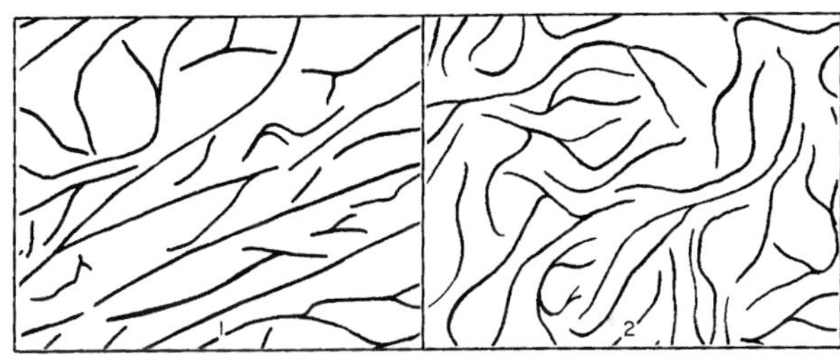

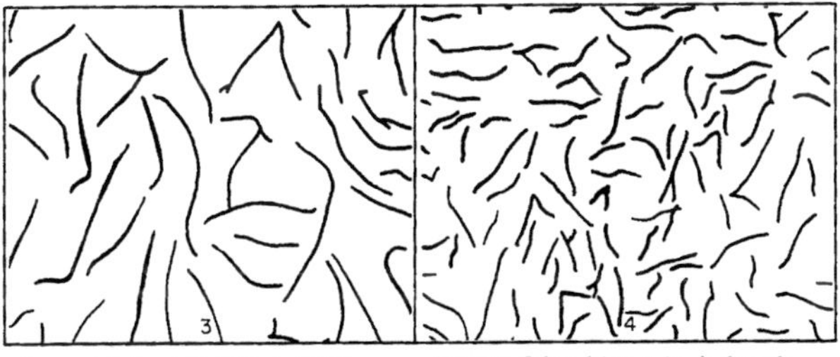

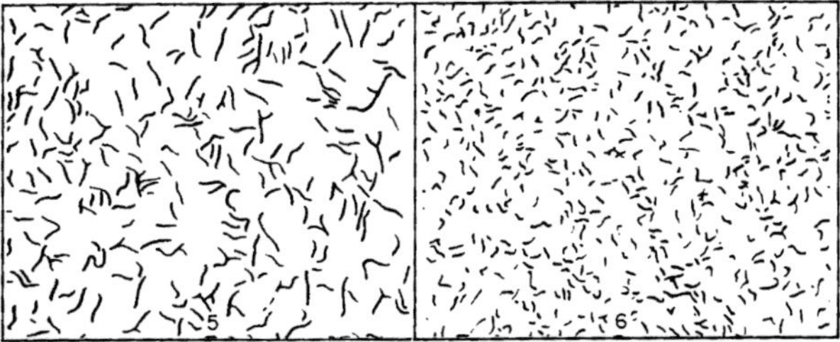

Fig. 13-20. AFS and ASTM graphite-flake size rating charts, 100X. [From AFS]

Table 13-2.
Graphite Formation

Temp. of Eutectic Solidification	Type of Structure	Factors
2100-2040F (1149–1116C)	Type A graphite, type B at lower temperatures.	High carbon percentages, near eutectic carbon content, and ladle inoculation favors type A graphite.
2040–1970F (1116–1077C)	Increasing types E and D graphite as temperature decreases.	Increased cooling rate, carbide stability favors type E. Super-heating also favors types E and D graphite.
1900–1950F (1038–1066C)	Mottled and white iron, type D and/or E graphite present.	Increased cooling rate (chilling) favors under-cooling and white iron.

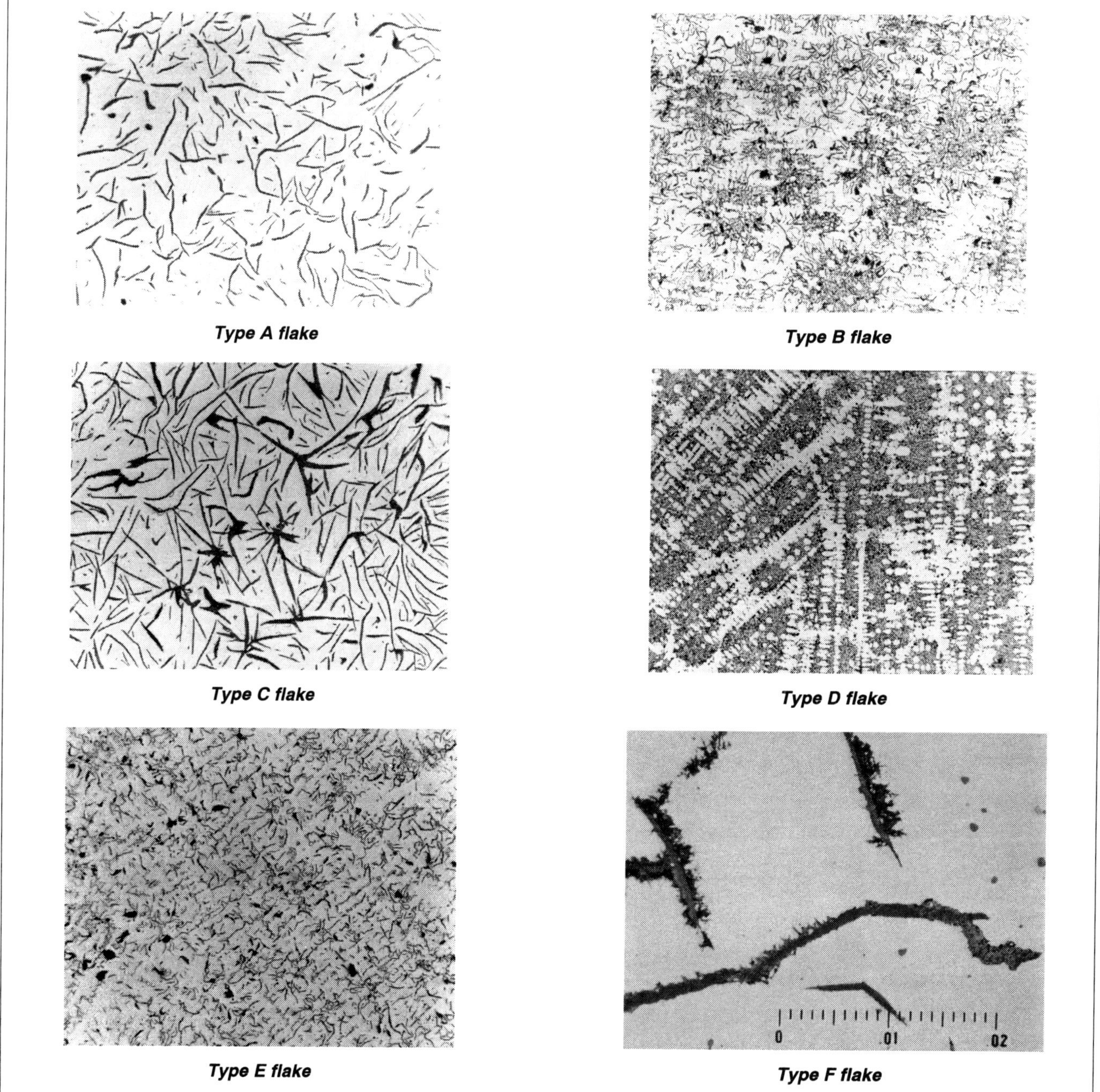

Fig. 13-21. The six types of flake graphite (A–F).

Figure 13-22 shows an example of large, coarse flakes of graphite called kish. Kish is normally found on the surface or very near the surface of heavy-section, hypereutectic gray iron. In this photomicrograph, the matrix material has been etched away to show the actual flakes of graphite.

On the other hand, small graphite flakes are encouraged to form when there is rapid nucleation due to the moderate undercooling, but still time for diffusion and graphitization. Undercooling occurs as the alloy temperature is dropping. When the alloy temperature reaches the liquidus line, it can slide down below that line for a very short period of time and then return to the line.

Severe undercooling inhibits or prevents nucleation of the graphite and results in chilled or white iron. When this type of cooling and solidification occur, types D and E graphite flakes will be produced. Type D and E graphite flakes will produce an abnormal gray iron.

Figure 13-22 actually shows two different types and sizes of graphite flakes in a hypereutectic gray iron. The large flakes are kish graphite and the smaller flakes are eutectic graphite.

Superheating the gray iron will also affect the type and size of graphite flakes produced. Superheating means that, in the melting operation, the molten metal's temperature is increased to above about 2750F (1510C). If the gray iron is superheated, undercooling during solidification is likely to occur, and the graphite flake size is reduced. In this case, types D and E would occur. Thus, superheating of the gray iron should be done with caution since it plays a role in the type of graphite flakes that will be produced. Chilled and mottled gray irons in thin sections are also more likely to occur in superheated gray irons unless they are properly inoculated.

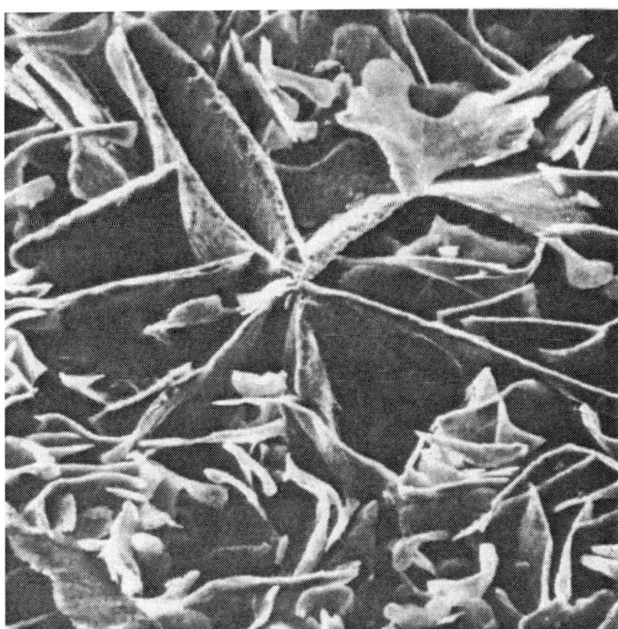

Fig. 13-22. Lower magnification view showing both primary kish graphite and eutectic graphite in hypereutectic gray iron, originally 175X.

Effect of Gray Iron Inoculation

Inoculation is the act of adding an inoculant to the molten gray iron, which produces effects far out of proportion to any resulting change in chemical analysis. Inoculating gray iron will cause a marked change in the resultant type of graphite flake, compared to the same gray iron that isn't inoculated. When ferrosilicon or some other graphitizing agent is added in small amounts, 0.05–0.25%, type A graphite flake formation is favored. It appears that by inoculating the gray iron, undercooling is prevented from taking place. This effect is most pronounced when a superheated gray iron is inoculated. By combining good melting and inoculation practice, type A graphite flakes of the kind desired is most consistently produced in commercial gray irons.

The quality of the inoculant added to molten gray iron is also very important. First of all, the inoculant must be of the proper mesh size. The size of the inoculant is determined by mesh size. The size is determined by passing the inoculant through a series of screens or sieves. The openings in these screens determine whether the inoculant particle will pass through it or not. This is very similar to sizing refractory particles used in making sand molds and cores. The size of the inoculant can range from lumps to a fine granular material.

Calcium silicide was one of the first materials to become widespread in use as an inoculant. Today, a number of other substances are known to act in a way similar to calcium silicide, either singly or in combination and in varying degrees of effectiveness. Among these substances are silicon, aluminum, zirconium and carbon. Some inoculating materials are undesirable because they also produce unwanted side effects. Others fade too quickly; e.g., their effect becomes lost and the molten gray iron reverts to its condition prior to inoculation.

Inoculation plays a big role in reducing the section sensitivity in gray iron to a minimum. Section sensitivity means that the thickness of the casting section helps determine its mechanical properties. For instance, very thin sections may be chilled, brittle and unmachinable, while the same gray iron in thicker sections may be opened-grained, soft and have very low strength. A properly inoculated gray iron of the same composition will not show these extremes.

The most common inoculant used today is a ferrosilicon alloy also containing calcium and aluminum. The calcium and aluminum are the two ingredients in the inoculant that provide the major share of nuclei for the graphite in the molten gray iron to grab onto and grow into flakes. This action is very similar to the seeding of clouds to promote rainfall. Clouds consist of water vapor that does not fall because it is not heavy enough. Flying above the clouds, an inoculant such as small particles of silver iodide are dropped into the cloud. The water molecules in the cloud grab onto these particles and build up into water droplets, which fall to the earth. Dust in the air also acts as an inoculant, helping water droplets to form and cause the rain to fall.

In the case of gray iron, graphite grows from eutectic cells as the iron solidifies, and each of these cells grows from a nucleus. Inoculation treatment increases the number of nuclei in the molten iron and, therefore, causes the production of a larger number of eutectic cells with their clusters of graphite flakes. In any given volume of gray iron, the larger the number of eutectic cells, the smaller their size. Inoculation treatment is also a means of controlling grain size and density. As the eutectic cell count increases and their size diminishes, the flake size of the type A graphite becomes smaller. When these changes occur, there will be an improvement in mechanical properties.

As mentioned earlier, effects of inoculation can begin to fade over time, and the molten iron will revert to the condition it was before the inoculant was added. Therefore, it is critical that the inoculant be added to the molten iron at the proper time. It is best to inoculate the molten iron as close as possible to the time that it will be poured into the mold. The inoculant can be added to the transfer ladle or pouring ladle; injected into the metal stream as it fills the ladle or mold; or by wire injection into the stream as it fills the ladle or mold; and also placed into the gating system where it will be dissolved as the molten iron flows through the system. The latter method of inoculating the molten iron is also called the in-mold inoculation process. **Figure 13-23** shows different shapes and sizes of inoculant inserts placed in the gating system. Other methods of inoculation would be wire injection and direct feed of the inoculant into an autopour stream.

Because of the many factors involved, the microstructure of the gray iron in a commercial casting may vary over the entire range discussed thus far. The essential points involved for each microconstituent are summarized as follows:

1. *Graphite*—Type and size of graphite are established during solidification of the iron. Chemical composition, undercooling, superheating, inoculation and cooling rate are important factors controlling graphite size and shape.
2. *Ferrite*—Ferrite in the microstructure is promoted by strong graphitizing conditions. Slow cooling rate and a chemical composition causing effective graphitization during freezing and in the solid state are important factors.
3. *Pearlite*—Restriction of graphitization during the eutectoidal transformation, cooling rate and the proper balance of manganese and sulfur in the iron are important factors.

Effect of Heat Treatment on Gray Iron

Gray irons may be heat treated to 1) improve machinability, 2) improve wear resistance, 3) improve strength and 4) provide dimensional stability and stress relief. In heat treating gray irons, the temperature of the gray iron is raised into an austenitic temperature range. At these temperatures, gray irons are amenable to many of the heat treatments applied to the carbon steels. Remember that heat treating gray iron does not change the graphite flake type or size but does change the matrix.

To improve machinability, gray irons are annealed or normalized, which will soften them and improve their machinability. To improve wear resistance, the gray iron is hardened and tempered. In this case, quenching in oil or water will produce a martensitic structure, giving the hardness characteristics of hardened steels. The quenched iron may then be tempered at various temperatures to reduce hardness to that desired.

Effect of Alloying on Gray Iron

An alloyed gray iron is one that has been combined with elements, such as chromium, copper, nickel or molybdenum, to obtain some beneficial effect. Alloyed gray irons are not produced to the same amount or tonnage as regular gray iron. However, their special properties make them more desirable for certain applications. The effects of alloying gray iron fall into two categories: 1) effect on the microstructure, the metal matrix and the graphitization process; and 2) effect on properties.

Fig. 13-23. Mold inoculant inserts.

The effect on the microstructure depends on the alloying element's tendency to form carbides, dissolve in ferrite, alter pearlite or influence the graphite size and distribution. The effect on properties depends on the alloying element's tendency to improve strength, resist corrosion, etc. Some alloying elements are capable of having multiple effects, others are added for one purpose. The following text will explain what the various alloying elements can do for gray cast iron.

Chromium—Chromium, in small amounts, is used to produce a fully pearlitic microstructure. A pearlitic microstructure of higher than 0.60% combined carbon content develops higher hardness and strength. Chromium may also be used to improve the gray iron's resistance to oxidation. High alloyed irons may contain as much as 30% chromium. These irons are known as white irons and are used for very special conditions involving scaling resistance and corrosion resistance.

Molybdenum, Molybdenum-Nickel—Molybdenum is very effective in strengthening and hardening irons. This element causes the austenite to transform to fine pearlite or bainite, a hard material. A combination of molybdenum and nickel are effective in causing this transformation to take place. Alloyed irons using molybdenum or a combination of molybdenum-nickel may have to be heat treated to obtain their optimum properties.

Nickel—Nickel mildly promotes graphitization in gray iron. It also dissolves in and hardens ferrite. When used alone, nickel may reduce the amount of silicon that must be present to produce gray iron. Ni-resist™ is an austenitic gray iron used for applications requiring corrosion resistance. **Figure 13-24** shows an austenitic gray iron. For exceptional wear resistance, gray irons may be alloyed with nickel, molybdenum and chromium to form a martensitic microstructure in the as-cast condition.

Silicon—Silicon can be used as a special alloying element in cast irons. When 6–8% silicon is added to cast iron, there is improved resistance to scaling. When higher amounts of silicon (16–18%) are added, an alloy having corrosion resistance to sulfuric and other acids and corrosion media is obtained.

Copper—When added to gray iron, copper acts as a mildly graphitizing element and dissolves in the ferrite phase. When used in amounts up to 3%, it will increase wear resistance in sliding friction such as is found in brake drums and cylinder liners. It can also help improve corrosion resistance in mildly acid and atmospheric conditions. Most gray irons do contain some copper, as it is found as a residual element in some raw materials.

Aluminum—Aluminum is not used very much as an alloying element in gray irons. However, it is present in many of the ferroalloy additions made to gray iron. In small amounts, less than 0.25%, it is known to act as a powerful graphitizing element both during and after solidification. On the other hand, an alloy containing over 8% aluminum and 3.0% carbon will freeze as white iron and does not graphitize, even if heat treated. Thus, aluminum, depending on the amount present, can act as a graphitizer or carbide stabilizer.

Titanium—Titanium behaves similarly to aluminum when added to gray iron. Small percentages of titanium will promote graphitization, reduce chilling tendencies and refine graphite flake size. When present in amounts greater than 0.25%, titanium will form titanium carbide (TiC), which is a stable carbide that cannot be decomposed by heat treatments.

In summation, the effects of alloying elements on gray irons is complex. In general, their use may be simply listed as follows:

- Increased mechanical strength
- Increased resistance to wear
- Increased resistance to corrosion
- Increased resistance to scaling or oxidation
- Increased resistance to abrasion.

Malleable Iron Microstructure

Typical chemical compositions of white irons, which can be heat treated to form malleable iron, are shown in **Table 13-3**. By using the carbon equivalent (CE) formula studied earlier in this chapter, the CE of these malleable irons can be determined. As an example, averaging the %C and %Si in these three malleable irons will result in an average CE for each of these three malleable irons: the average CE of ASTM No. 32510 is 2.86; ASTM No. 35018 has an average CE of 2.59 and the cupola-malleable has an average CE of

Table 13-3.
Typical Chemical Composition of White Irons Heat-Treatable to Malleable Iron

Element %	ASTM 32510	ASTM 35018	Cupola-malleable
Carbon	2.30–2.65	2.00–2.45	2.80–3.30
Silicon	0.9–1.40	0.90–1.30	0.60–1.10
Manganese	0.25–0.55	0.21–0.55	Less than 0.65
Phosphorus	0.18	Less than 0.18	Less than 0.20
Sulfur	0.05–0.18	0.05–0.18	Less than 0.25

3.33. Next, by looking back at the iron-carbon diagram found in **Fig. 13-14,** one can determine where these CEs would be positioned on the diagram. (To the left of the eutectic point would be correct.) Note that the eutectic point has a CE of 4.3; therefore, malleable irons are hypoeutectic alloys.

Malleable iron, when it freezes, is white iron. In other words, it has the carbon tied up in the form of iron carbide and pearlite. This is because malleable irons tend to have lower silicon levels than do gray irons. In addition, the percent carbon in malleable irons tend to be less than in gray irons. White iron, as learned earlier, is a very hard, brittle material with very little or no ductility.

Figure 13-25 shows white iron microstructures at two different magnifications. To get this hard, brittle material, with no ductility to speak of, to have more ductility than gray iron, it has to be heat treated.

Most malleable iron castings are lightweight, generally under 50 lb (22.5 kg). (Heavier malleable iron castings can be poured.) Section thicknesses are generally thin and under 2 in. (5 cm). The pouring of these thin sections also helps in the formation of white iron. After the castings have solidified, they are shaken out of the molds and allowed to cool. At shakeout, core sand (if cores were used) is drained out of the cored passageways. The gating system and the risers are then removed and the castings are ready for heat treating.

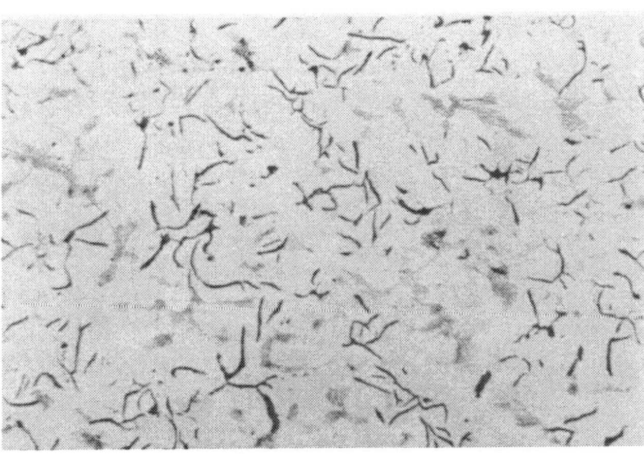

Fig. 13-24. Microstructure of austenitic gray iron. Chemical composition: 2.70% C, 1.85% Si, 15.0% Ni and 2.0% Cr. Mechanical properties: 130 Bhn, 30,000 psi ultimate tensile strength. Etched with 5% nital, 500X. [Courtesy of International Nickel Co.]

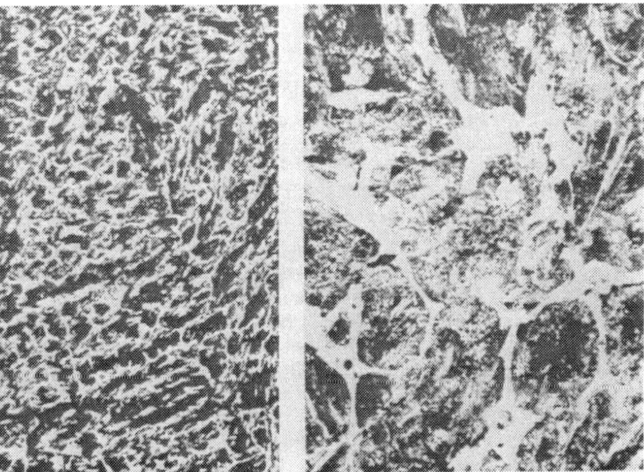

Fig. 13-25. Etched white iron at low 100X magnification (left) and high 500X magnification (right). The white areas are spines of carbon in the combined form, called cementite or Fe_3C, a very hard and brittle constituent. The matrix is substantially fine pearlite. [Courtesy Malleable Iron Foundry Society]

Effect of Heat Treatment on Malleable Iron (Annealing or Malleablization)

Remember, the initial castings are hard, brittle and have very poor ductility. However, the word *malleable* means that the castings are workable. In other words, they can be mechanically deformed by hammering, bending, etc. In order to change these castings into malleable castings, they have to be annealed or heat treated. This operation, called malleabilizing, will change their microstructure to one that will allow the castings to be "worked."

The annealing method can be either batch or continuous. In the batch method, the castings are carefully packed into metal baskets and placed in a heat-treat furnace, where the annealing cycle begins. Since the furnace and casting mass are large, heating and cooling are slow, and the cycle can be a long one. In the continuous method, the castings are placed on a metal belt and conveyed through a tunnel-type of heat-treat furnace. In this type of operation, the casting is annealed either in its own atmosphere or in a specially generated atmosphere. **Table 13-4** gives a summation of these two types of annealing processes.

Figure 13-26 illustrates a typical annealing cycle. Note that the castings are raised to a temperature of 1600–1750F (871–954C) rather quickly. This first step involves the nucleation of graphite, which occurs mainly during this rise in temperature and somewhat during the very early part of the holding period. The second step involves holding the castings at a temperature of 1600–1750F (871–954C), during which the first-stage graphitization (FSG) takes place. The objective of the second step is to eliminate massive carbides from the iron structure.

The third step is slow cooling through the allotropic transformation range of iron, where further graphitization takes place. This

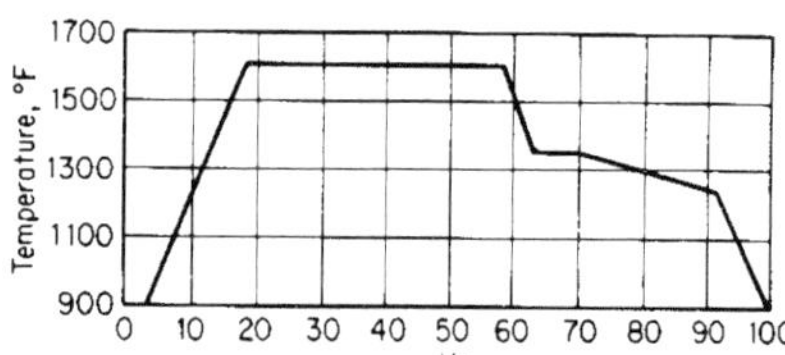

Fig. 13-26. Cycle of temperature and time for malleableizing white iron. Actual duration of cycle may be much less or more than indicated. [Courtesy of Malleable Founders' Society]

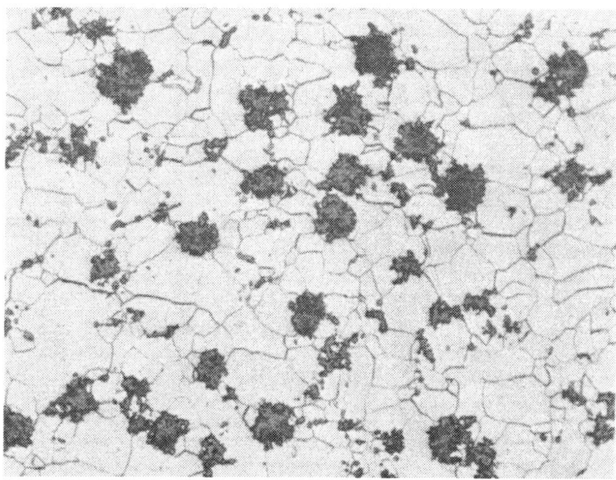

Fig. 13-27. Microstructure of standard malleable iron: ferrite and graphite, 125 Bhn, nital etched, 150X.

Table 13-4. Annealing Cycles

Type	Heating	Holding at first stage temperature	Cooling and second stage of graphitization	Total
Periodic oven, pot-annealing, packed	To 1650F (899C) in 40 hr	40 hr at 1650F (899C)	1650–1100F (899–593C) in 75 hr	155 hr
	To 1600F (871C) in 45 hr	45 hr at 1600F (871C)	1600F–1100F (871–593C) in 60 hr	150 hr
Continuous furnace	To 1700–1750F (927–954C) in 3–5 hr	5–13 hr at 1700–1750F (927–954C)	Cooling to 1300F (704C) in 8–36 hr	14–60 hr

third step is also known as second-stage graphitization (SSG). The objective of the third step is the formation of a completely ferritic matrix, free of pearlite and carbides. **Figure 13-27** shows a microstructure of standard ferritic malleable iron; i.e., graphite surrounded by a ferritic matrix. **Figure 13-28** is a photomicrograph of the graphite structure found in malleable iron. Note that the graphite is neither round nor flake in shape, but it still can be called nodular. However, this shape of graphite does make the casting malleable.

Because the number of nodules developed and their distribution in the microstructure is so important, a means of measuring the number and distribution of the nodules has been developed. This is done by counting the number of nodules observed under a microscope at a given magnification. This number may be used as such or converted to nodules per square millimeter (N/mm^2). The size and distribution, shape and number of the graphite nodules may vary.

Another type of malleable iron is *pearlitic malleable iron*. This cast iron has increased strength and wear resistance combined with reasonable toughness over that of ferritic malleable iron. A microstructure of pearlitic malleable iron is seen in **Fig. 13-29**. Note the pearlitic matrix and the ferrite surrounding the graphite nodules. Pearlitic malleable is made in the following ways:

1. By preventing complete SSG by adding alloying elements such as manganese, molybdenum, or chromium.
2. By arresting the annealing cycle during SSG.
3. By heat-treatment of standard malleable iron.

There are three types of pearlitic malleable iron. Type 1 often contains manganese between 0.5 and 0.90% to retain pearlite

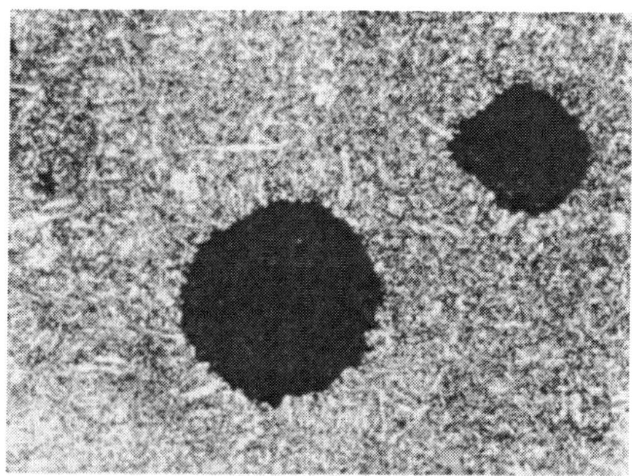

Fig. 13-28. Typical microstructure of pearlitic malleable iron. (Originally 500X)

during the regular annealing cycle. Type 2 depends on rapid cooling through the transformation-temperature range to produce pearlite rather than ferrite. Type 3 is made by reheating ferritic malleable iron to a temperature just above the critical range and then quenching it in air or liquid medium. The quenched malleable iron is then tempered to produce the desired Bhn and tensile properties.

Ductile Iron Microstructure

Ductile iron consists of graphite spheroids dispersed in a matrix, similar to that of carbon steel. The only significant difference between gray iron and ductile iron is in the shape of the graphite; however, the matrices may be similar. The graphite in ductile cast iron is spheroidal, or round, in shape, which gives this alloy improved ductility over malleable or gray cast iron. The introduction of ductile iron has significantly reduced the size of the malleable iron foundry industry and has had its effects on the carbon steel industry. Ductile iron is also known as spheroidal graphite iron and nodular iron.

A typical range of chemistries for ductile iron can be as follows:

Total Carbon, 3.5–4.0%
Silicon, 2.0–2.80%
Manganese, 0.2–0.6%
Phosphorus, 0.02–0.08%
Sulfur, 0.005–0.04%
Residual Mg, 0.02–0.07%

Again, taking the average of the carbon and silicon, and using the CE formula shows that ductile iron would have a CE of approximately 4.55. Referring back to the iron-carbon diagram found in **Fig. 13-14,** this time, the CE of ductile iron would fall to the right of the eutectic point, which means that ductile iron is a hypereutectic alloy. This coincides with the need for a high carbon content to produce satisfactory ductile iron.

Figure 13-30 shows a microstructure of ductile iron. Note that the graphite is spheroidal in shape although not perfectly spheroidal. In this particular microstructure, the spheroids of graphite are surrounded first by ferrite and then pearlite. This type of ductile iron is sometimes called "bull's-eye" ductile iron. **Figure 13-31** shows a nodule or spheroid of graphite found in ductile iron. (This

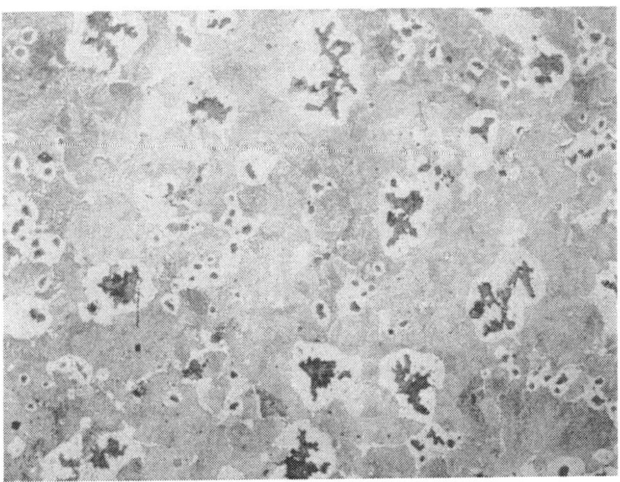

Fig. 13-29. Microstructure of one type of pearlitic malleable iron. Nital etched, 100X.

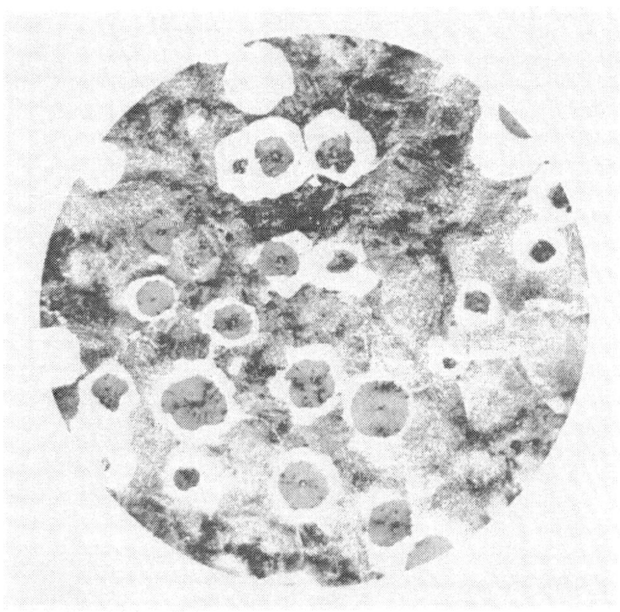

Fig. 13-30. Typical as-cast microstructure of ductile iron showing graphite spheroids surrounded by ferrite. The darker, lamellar structure in the matrix is pearlite. Etched, 250X. [Courtesy of International Nickel Co.]

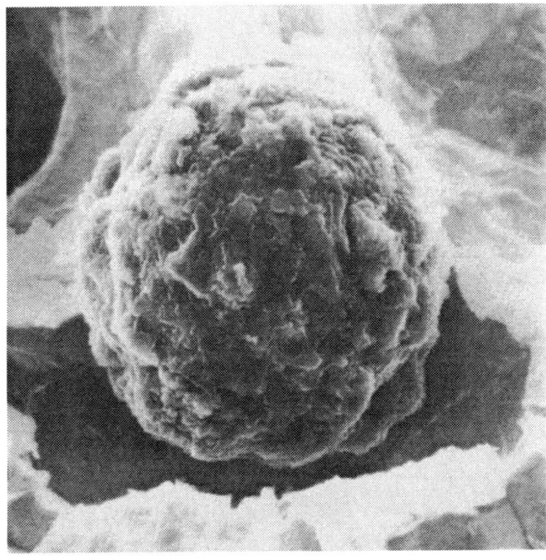

Fig. 13-31. Appearance of typical nodule in as-cast material, after etching away the matrix, 950X.

scanning electron microscope photo was taken after some of the matrix around the nodule was etched away.) The problem now is how to get the graphite to form spheroids.

The key to producing good ductile iron begins in the charge makeup and melting areas. Ductile iron is gray iron that has been treated with an alloy that causes the spheroids to form. This gray iron is called the base iron and is the start of producing ductile iron. A poor base iron will normally end up producing a poor grade of ductile iron. A poor base iron can be upgraded by the use of special alloy additions and treatments, but this can become quite expensive. For this reason, ductile iron producers have found it necessary to improve their normal melting practices and to have better control of these practices than those used for gray iron.

Examining the chemical composition of ductile iron shows that the sulfur level is very low. The reason for this will be discussed shortly. These sulfur levels can be kept low by selecting charge materials that are low in sulfur content and by controlling melting practices. The base iron can be melted in any of the furnaces mentioned in the chapter on furnaces or ferrous melting practices. In the case of cupola melting, the amount of sulfur found in the coke should be monitored carefully. A basic refractory lining can also be used to reduce sulfur levels in the melt. The use of a basic slag in the melting practice is also a means of reducing sulfur content.

Effect of Desulfurization on Ductile Iron

If the sulfur levels are above that required, desulfurization of the molten base iron will be required. Should the molten base iron have a higher-than-specified sulfur level at the time of treatment, an appreciable amount of the treatment alloy will be consumed before the expensive treatment alloy can cause graphite spheroidization to occur. Desulfurization can be done using several different desulfurization agents and methods.

Some of the more practical desulfurizing agents include:

1. Caustic soda (NaOH)
2. Soda ash (Na_2CO_3)
3. Burnt lime (CaO)
4. Limestone ($CaCO_3$)
5. Calcium carbide (CaC_2)
6. Calcium cyanamide ($CaCN_2$)

Of these desulfurizing agents caustic soda is normally ruled out because of health hazards; limestone is first reduced to CaO before it desulfurizes; and calcium cyanamide is not recommended because it will increase the nitrogen of the base iron. Limestone is most commonly used in *basic* melting practice (rather than *acid* practice) to help desulfurization to take place in the furnace. Soda ash may require two treatments to lower the sulfur to specified percentages in the melt.

Calcium carbide is probably the most commonly used desulfurizing agent today. Calcium carbide, when added to the base iron, combines with the sulfur to form calcium sulfide (CaS), which floats to the surface of the melt as dross.

Calcium carbide can be added to the base iron using several different methods. One method, as illustrated in **Fig. 13-32**, injects fine calcium carbide into a desulfurization ladle. The ladle is equipped with a porous plug in the bottom, through which an inert gas, such as nitrogen, is injected. This gas causes a stirring action that mixes the calcium carbide with the melt, and desulfurization takes place.

Another method is the "shaking ladle." In this method, calcium carbide is added to the melt and the shaking ladle gyrates, causing the molten base iron to mix with the calcium carbide, causing desulfurization to take place. **Figure 13-33** shows the shaking ladle process. It has been reported that sulfur levels as low as 0.02% have been achieved at 70–75% efficiency.

Sometimes a can, similar to a coffee can, is filled with the required amount of calcium carbide and is added to an empty ladle that has been preheated. When the base iron is poured into the ladle, the can melts, exposing the calcium carbide to the molten base iron. The filling of the ladle causes a stirring action that distributes the calcium carbide in the molten base iron. Some foundries will also have a porous plug in the bottom of the ladle through which an inert gas is passed into the ladle after the ladle has been filled, causing this stirring action to continue.

Other Elemental Effects

Some of the elements that should be controlled in the base iron (i.e., silicon and phosphorus) are as follows:

Silicon—The normal range of silicon in ductile iron is 1.80–2.80%. Remember that silicon affects the carbon equivalent value and affects the number of spheroids and the occurrence of flotation. Silicon also increases the amount of ferrite formed and helps to strengthen the ferrite. Silicon is best used as an inoculant for spheroidal-graphite control.

Phosphorus—As with gray iron, phosphorus forms a very brittle structure in ductile iron, called steadite. Since phosphorus adversely affects toughness and ductility, a maximum of 0.05% is usually specified.

Other elements may be used in ductile iron for various reasons. Time and space limitations prohibit their discussion here. A metallurgist can be consulted to learn more about these elements and their effect on ductile iron.

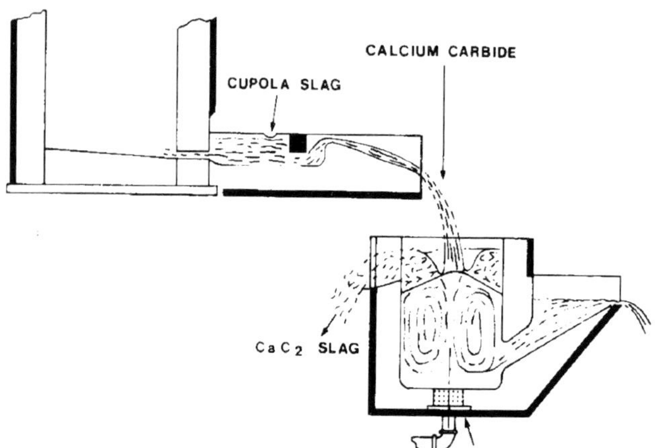

Fig. 13-32. Porous plug desulfurization at the cupola spout.

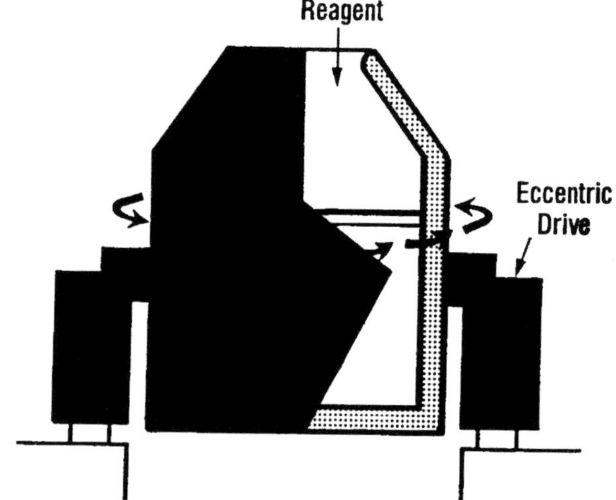

Fig. 13-33. The shaking ladle desulfurization process.

Spheroidal Graphite Formation in Ductile Iron

How and why the spheroidal graphite is formed in ductile iron is still debated and, for all practical purposes, unknown. However, the use of certain elements to help form these spheroids *is* well known.

Solidification of spheroidal graphite occurs at temperatures higher than that at which eutectic flake graphite forms for similar carbon equivalents. Therefore, the number of graphite spheroids is determined in the early stages of solidification. As the temperature of the ductile iron continues to fall, graphite will precipitate onto the existing spheroids at temperatures down in the eutectoid range.

As with gray iron, the cooling rate through this range and/or alloy treatment determines the matrix structure. Bull's-eye patterns like those seen in **Fig. 13-30** are typical of the ferritic-plus-pearlitic matrix of as-cast ductile iron. As with gray iron, subsequent heat treatment, after the ductile iron casting has completely solidified, can change the matrix.

The importance of the number of graphite spheroids necessary to obtain a fully spheroidal graphite structure must be stressed. Should an insufficient number of spheroids form, there will be an inadequate number of sites to which the carbon in the liquid can attach itself. If this should occur, and if the composition and processing variables are favorable, flake graphite and iron carbide will form from the liquid as cooling continues. Should both of these alternatives occur, the results would produce inferior properties when compared to a fully spheroidal graphite structure.

Magnesium Treatment—The most commonly accepted method of promoting spheroidal graphite is by adding magnesium to the base iron of either hypo- or hypereutectic alloys. Other elements, such as cerium, calcium and yttrium, have been tried but have been proven inadequate.

Magnesium does three things. First, it acts as a deoxidizer and desulfurizer in the base iron. Should the oxygen and/or sulfur content of the base iron be too high, a substantial amount of the magnesium will be used to form magnesium oxides and sulfides. Second, magnesium promotes the formation of graphite spheroids. As mentioned earlier, why this happens has not been determined. Third, magnesium prevents the nucleation of graphite flakes during solidification, and thus promotes the formation of graphite spheroids.

The name used to define the addition of magnesium to the base iron is *Mg-treatment.* In other words, the base iron is treated with magnesium. The amount of magnesium added is dependent largely on the amount of oxygen and sulfur content in the base iron.

The base iron temperature at treatment time is usually 2800–2850F (1537–1565C). This temperature range is considerably higher than the boiling point of magnesium; thus, when it meets the base iron, it vaporizes and the reaction may be quite violent if not carried out properly. This reaction is similar to the old-time photographer using the vaporization of magnesium to cause a "flash" when taking photos.

The recovery, or the effectiveness of the magnesium treatment to cause the formation of graphite spheroids, is determined by the depth of the base iron through which the magnesium vapor rises before reaching the surface. In addition, the time to cover the treatment alloy and the depth to which it is covered are important factors in magnesium recovery. Higher base iron temperatures also lower the magnesium recovery rate.

Mg-Treatment Methods—There are several ways in which the Mg-treatment of the base iron can be carried out.

One way is simply to place the treatment alloy on the bottom of an empty ladle and pour the base iron into the ladle. When using this method, the way in which the ladle is filled is very critical. The stream of molten metal must be directed away from the treatment alloy and the ladle filled very rapidly. If this is not done, there is the chance of having the treatment alloy float to the surface and burn.

A very popular variation of this type of treatment technique is the "sandwich method." **Figure 13-34** shows a cross section of the ladle and its recommended dimensions. The treatment alloy is placed in a "pocket" in the bottom of the ladle and covered with small pieces of sheet steel, steel punchings or chips. The steel cover is thought to keep the base iron from contacting the treatment alloy for a short period of time, thus allowing the ladle to be filled to a point at which the reaction can be better contained in the ladle. This also causes the magnesium vapor to travel upward through a larger volume of base iron, thus giving better recovery rates. **Figure 13-35** shows the "sandwich method" being used to treat the base iron.

Another method of treating the base iron is by using the "covered ladle" method. **Figure 13-36** shows a cross section of design of the covered ladle. Using this method limits the amount of oxygen available (for burning of magnesium) to that contained in the air in the ladle after the lid is in place. There is no seal between the cover and the top of the ladle. The filling of the ladle with the base iron displaces the air in the ladle, which passes out between the cover and the top of the ladle. As seen in **Fig. 13-36**, the tundish limits the

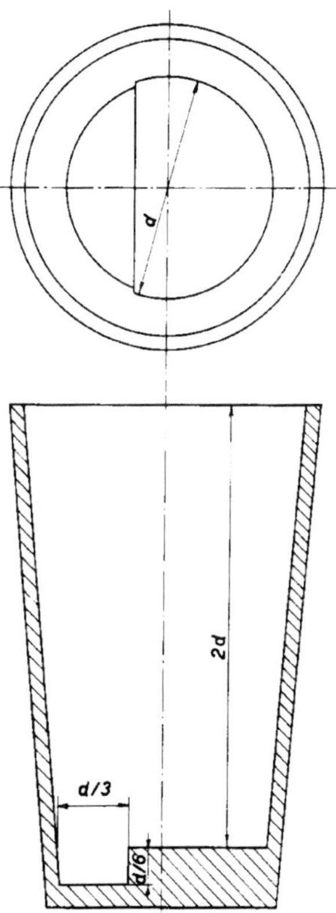

Fig. 13-34. Diameter/height relationship of a ladle suitable for open-ladle transfer treatment method. Well on bottom facilitates the "sandwich" treatment.

rate of filling of the ladle. However, magnesium recovery improves substantially, and the treatment reaction and MgO dust is contained in the ladle.

The use of a porous plug in the ladle, which causes a stirring action, can be used to obtain better recovery rates than the previously mentioned methods of treatment. The stirring action tends to keep the treatment alloy in the molten base iron most of the time. This advantage should be weighed against the cost and maintenance of the stirring equipment.

The "plunging" method uses a vented graphite or refractory bell that is connected to a refractory-covered plunging rod. The treatment alloy is placed in the bell and plunged into the ladle that is filled with the base iron. The treatment alloy can be placed in a steel foil package, a can, or it can be impregnated into coke or magnesium-iron aggregate that is briquetted. **Figure 13-37** shows the equipment used in the plunging treatment method. This method of treatment offers better control over the residual magnesium level and a higher magnesium recovery rate than the other open ladle treatment methods.

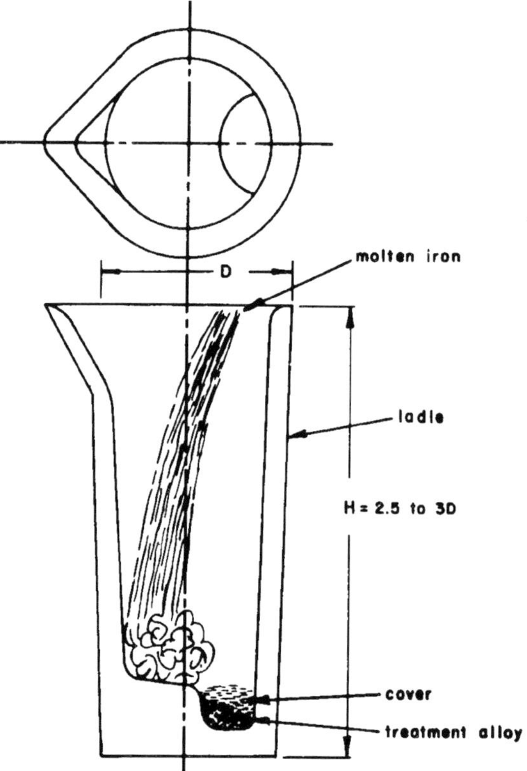

Fig. 13-35. Schematic of the sandwich method for treating ductile iron.

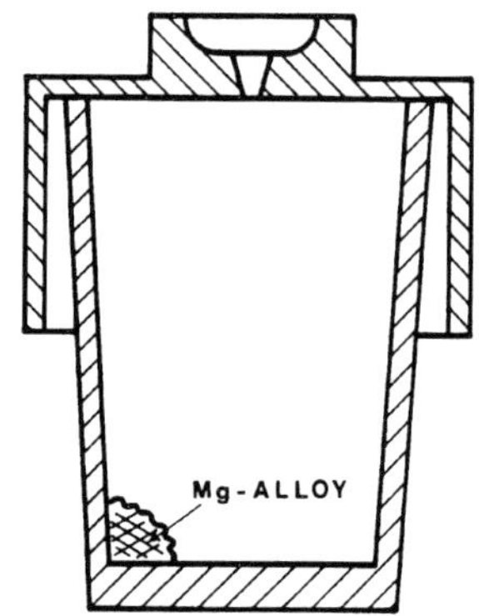

Fig. 13-36. Covered ladle for magnesium treatment.

Another treatment method uses a "mechanical feeder," which injects treatment alloy that is continually being added to the stream of base iron as it leaves the furnace or transfer ladle. This method of treatment offers continuous and uniform treatment of the base iron, and generally results in a higher magnesium recovery rate.

There are also two other methods of treating the base iron: Flowtret and In-mold (flow-through and in-the-mold). These two methods do not require a ladle for the treatment process.

The first of these methods is called "Flowtret," and it is a patented method of treating the base iron with the treatment alloy. A schematic of the vessel used to treat the base iron is seen in **Fig. 13-38**. In this method, the treatment alloy is placed into a chamber in the Flowtret device, and the base iron is treated as it passes through the device. The base iron can be tapped into the device directly from a furnace or a transfer ladle. In either case, the treated base iron is deposited into the pouring ladle for pouring directly into the mold(s), or deposited into a transfer ladle and poured into an automatic pouring device. This is a batch-type operation in that new treatment alloy is placed into the device after each batch. Reported magnesium recoveries are exceptionally good and the treatment is practically environmentally acceptable.

The second process is called the "In-mold" process, a patented process that does not require a treatment ladle. This process works

Fig. 13-37. Equipment for magnesium treatment with the plunging method.

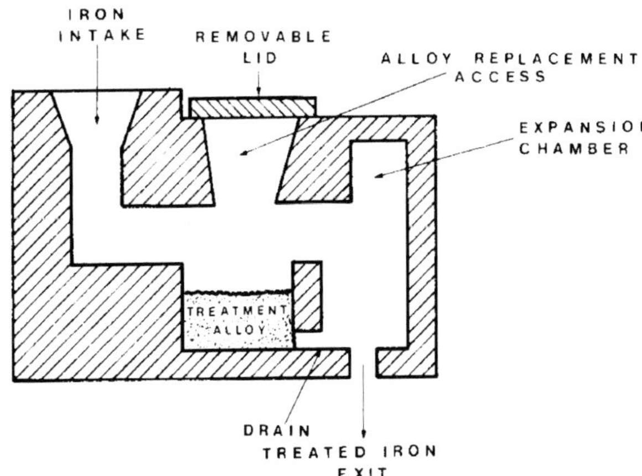

Fig. 13-38. Schematic illustration of the flow-through method.

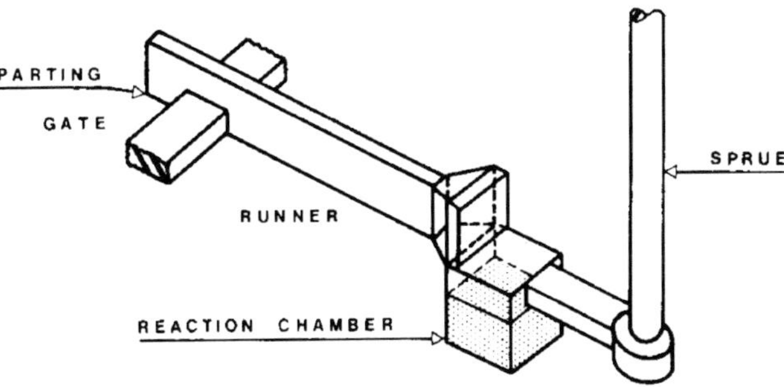

Fig. 13-39. In-mold treatment.

similarly to the in-the-mold inoculation process discussed earlier. In this process, the treatment takes place in the mold as the base iron flows through the gating system. **Figure 13-39** shows a schematic of this method of treating the base iron. In this case, the reaction or treatment chamber is designed to be part of the gating system. Again, this is a batch-type operation in that the only base iron treated is that which flows through the chamber. The big advantage of this method of treatment is that only the base iron flowing through the system is treated and not the remaining base iron in the ladle or pouring device. Therefore, if there is an equipment breakdown, the remaining base iron in the pouring ladle or automatic pouring device will not have been treated with the treatment alloy.

As will be discussed shortly, there is a window of time in which the treated base iron, if not poured, will begin to lose its nodularity. Thus, this treated base iron will have to be pigged or poured back into the furnace, and a new batch treated. Other advantages of this method are reported to be excellent recovery of magnesium, and it is environmentally friendly.

There are some other methods for treating the base iron; however, the ones previously discussed are the more common methods used in the United States and Canada. Discussion of other less commonly used methods can be found in the literature on ductile iron production.

Effect of Inoculation on Ductile Iron

The reason for inoculation, as discussed under Gray Cast Iron, is to provide "seeds" or nuclei onto which the graphite can attach itself and grow into nodules. In other words, it is the magnesium's task to have the graphite form nodules and the inoculant's task to allow more nodules to be formed. The more nodules that are formed, the more graphitization will take place. This, in turn, will decrease the tendency for chill or iron carbides to form during solidification. The formation of these carbides will result in greater hardness and generally inferior mechanical properties of the ductile iron; therefore, inoculation is practiced to offset those results. Postinoculation also promotes the amount of ferrite produced during the solidification of the ductile iron.

Since ferrosilicon is considered the most effective inoculant to use with ductile iron, it has become the most commonly used inoculant. Ferrosilicon comes in several different grades, with 85% silicon being the most widely used grade. Postinoculation has been found more effective than increasing the silicon content of the base iron.

It is best to treat the base iron first and then inoculate. It has been shown in Canadian research that the postinoculation is more effective if it follows the treatment of the base iron. One of the reasons given is that the temperature of the base iron is lowered during the treatment cycle, thus making the inoculant more productive in forming nuclei. As with the inoculation practice for gray iron, the effectiveness of inoculation also fades with time in ductile iron. Should the treated and inoculated ductile iron be transferred into different ladles before pouring, it is suggested that a small amount of ferrosilicon be added to the ductile iron at each transfer site.

If the treated and inoculated ductile iron is not poured within a prescribed amount of time, the nodules will begin reverting back to flake graphite...and not a good flake graphite, at that. Therefore, it is apparent that the scheduling for producing quality ductile iron is very critical. Each foundry producing ductile iron castings will have to determine this time frame for their own operations. Remember that simply increasing the superheat of the base iron will not help the situation; in fact, it will help diminish the effectiveness of the postinoculation.

A recent report concerning the inoculation of ductile iron discusses the use of a new inoculant in which cerium (Ce), calcium (Ca), sulfur (S) and oxygen (O) are present. This inoculant is used for ductile iron and is said to make the ductile iron more machinable with a more homogeneous microstructure along with less shrinkage tendencies. The amount of Ca and Ce in the inoculant is controlled to minimize chill (white iron) formation and neutralize bad trace elements that prevent the formation of nodular graphite. The controlled amounts of S and O are available to react with the Ca and Ce when the inoculant is added to the molten iron.

The introduction of this particular inoculant into the base iron is to promote controlled concentrations of nonmetallic elements, such as S and O, with the metallic elements. From a balanced and controlled metallurgical standpoint, this inoculant deliberately produces a greater density of nucleation sites. The increased number of nucleation sites is the result of the reaction between the metallic elements (Ca and Ce) and the nonmetallic elements (S and O) in the inoculant. The S benefits graphite formation and the O plays an important role in the inoculation process. The additional nucleation sites will then perform in parallel with the traditional sites formed during reactions between the nodulizer, inoculant and the base metal.

Process Control of Ductile Iron

Although the final CE (carbon equivalent) content of the ductile iron is usually hypereutectic, the base iron's analysis before treatment and inoculation is hypoeutectic. Good chemical control of the base iron is an important factor in producing quality ductile iron. Chemical analysis of the ductile iron is done in the same manner as for gray iron, with the exception that special attention is given to the amount of residual magnesium.

Graphite shape and size are also process control checks for ductile iron. Two charts have been developed to help determine the graphite shape and nodule size. **Figure 13-40** shows a chart used to determine graphite shape and **Fig. 13-41** shows the six different nodule sizes. These charts are from ASTM (American Society for Testing and Materials) and are expected to be updated during the year 2002. They are the A247 comparison charts.

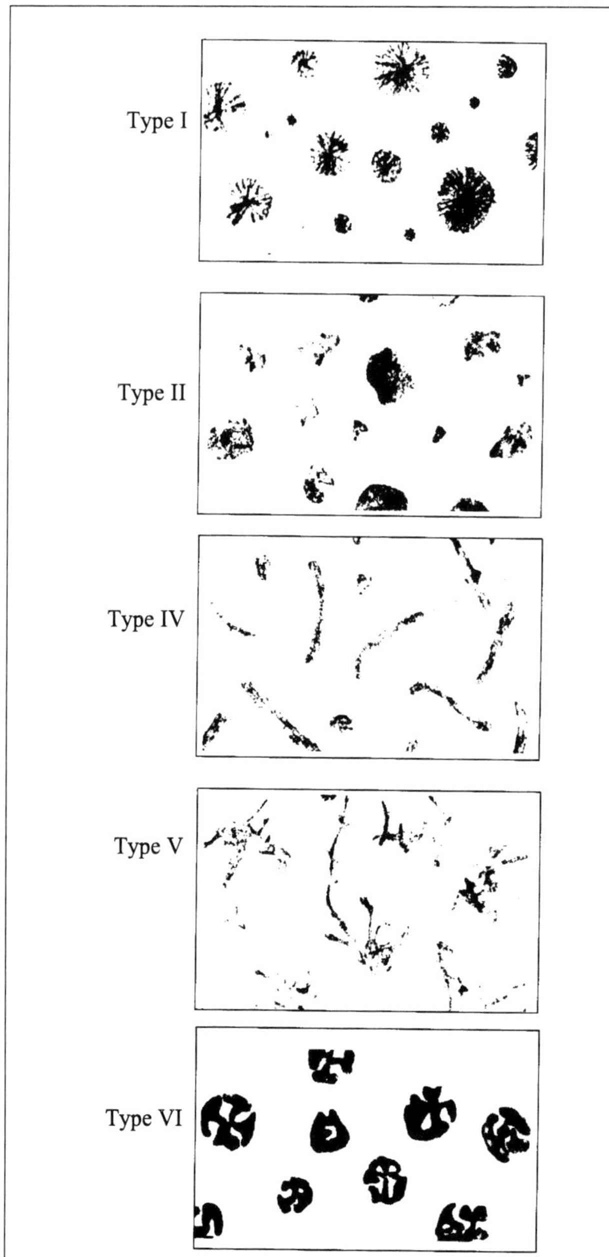

Fig. 13-40. Current classification of graphite shapes in ductile cast iron, 100X. Size of images have been changed for publication purposes.

The following factors help to determine the shape of the nodules formed:

1. Pouring temperature
2. Casting section size
3. Amount of effective magnesium
4. Postinoculation practice
5. Analysis of the base iron

All of these factors play a part in producing quality ductile iron. For instance, low pouring temperatures, heavy section sizes, not enough magnesium treatment, lack of inoculation, and a low carbon content base iron will result in poor graphite shapes.

When discussing ductile iron casting defects, the term "dross" will often be used. Remember that one of the first functions of magnesium was to act as a desulfurizer and the second function was as a deoxidizer in the base iron. When these functions are performed, magnesium sulfide and magnesium oxide are formed. These compounds are called dross and can appear on the cope surface of the casting. This defect is compounded by high magnesium additions, high pouring temperatures and turbulence in the gating system and mold cavity as it fills.

Effect of Heat Treatment on Ductile Iron

Ductile iron is a very heat treatable alloy, and thus its mechanical properties can be changed through heat treatment, if desired. Another term often heard is "as-cast ductile iron." This means that the ductile iron casting can be used "as is" without any heat treatment to change its mechanical properties. Two other terms also commonly used are "ferritic ductile iron" or "pearlitic ductile iron." Of course, these terms refer to the matrix that surrounds the nodules. When ductile iron is heat treated, it is the matrix that is affected and

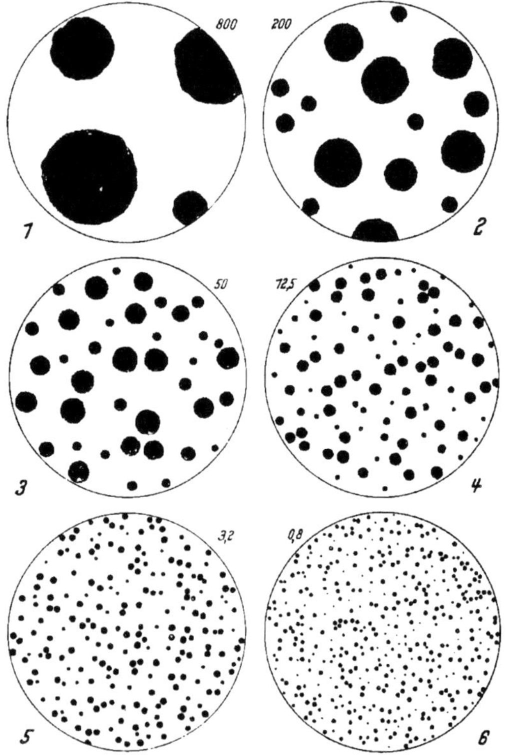

Fig. 13-41. Current classification of spheroidal graphite sizes based on size of spheroid at 100X. [From C.K. Donoho]

not the nodules. As with gray iron, the only way to change the shape of the graphite formed during solidification is by giving the casting the "ultimate heat treatment"..... in other words, remelt the casting and begin again.

In changing the matrix, the carbon content can range from almost zero to over 0.80%. This can be accomplished by metal analysis, alloying elements, foundry process control and/or heat treatment. This can change the matrix from an all-ferritic matrix to a mixture of ferrite and pearlite or all pearlite, martensite, tempered martensite or bainite. The latter three materials are very hard. **Figure 13-42** shows a ferritic grade of ductile iron.

Returning to **Fig. 13-30** ductile iron is shown with a mixture of ferrite and pearlite as a matrix. On the other hand, **Fig. 13-43** shows a quenched ductile iron resulting in a matrix of martensite and bainite. Bainite is formed in the heat treating cycle during the transition from pearlite to martensite.

It is common to heat treat ductile iron castings for relieving stress or obtaining desired properties. Heat treating can also be used for annealing, normalizing and tempering, quenching and tempering, austempering and martempering. The actual heat treating sequences will not be discussed in this book, but the reader is referred to the available literature on the heat treating of ductile iron in the Bibliography.

Compacted Graphite Iron (CGI) Microstructure

With the inception of ductile iron back in the late 1940s, researchers and metalcasters have found that, when ductile iron was treated with not enough magnesium, the cast iron produced was high in sulfur content and had "quasi-flake" graphite. This quasi-flake graphite was neither spheroidal in shape nor sharp-edged as found in gray cast iron. These quasi-flakes of graphite were pudgy and had rounded edges as seen in **Fig. 13-44**. For comparison, this figure includes electron scanning microscope photos of gray, compacted graphite and ductile irons.

The production of quality CG iron begins with the same process controls required for ductile iron. For years, the making of this alloy was dependent on the undertreatment (not enough magnesium), so that the graphite did not come out as spheroids. Another thing that hampered its consistency, and therefore the ability to be considered compacted graphite, was the fact that graphite shape is partly dependent on cooling and solidification rates.

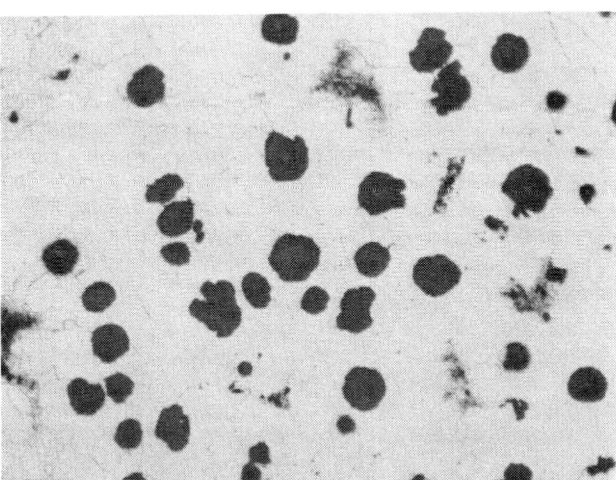

Fig. 13-42. Ferritic grade in final condition, etched in nital. Nodules of graphite in matrix consisting of equaxed grains of ferrite, 100X.

In the process of further studying CG iron, it was found that the addition of titanium to ductile iron helped in the formation of these quasi-flakes of graphite. It may have been said that compacted graphite iron is ductile iron that has been "poisoned" with the addition of titanium. Now, a proprietary treatment alloy has been developed, which is used to treat the base iron. This alloy contains the titanium that causes the compacted graphite flakes to form as seen in **Fig. 13-44**. There may be some spheroids of graphite found in compacted graphite iron and the percent of nodularity is kept as low as possible usually between 0 and 10%, especially in performance-critical sections.

The matrix material in this cast iron alloy can be either ferritic or pearlitic or a combination of the two through alloying and control of the cooling cycle. If required, compacted graphite can also be heat treated using any conventional cast iron heat treating process.

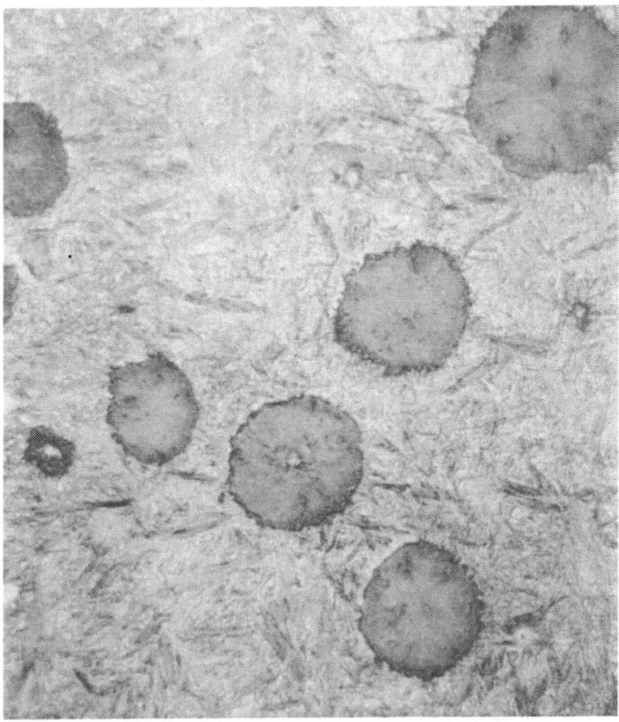

Fig. 13-43. Ductile iron quenched from 1650F (899C) into an oil bath resulting in this acicular mixture of martensite and bainite. Etched, 500X. [Courtesy of International Nickel Co.]

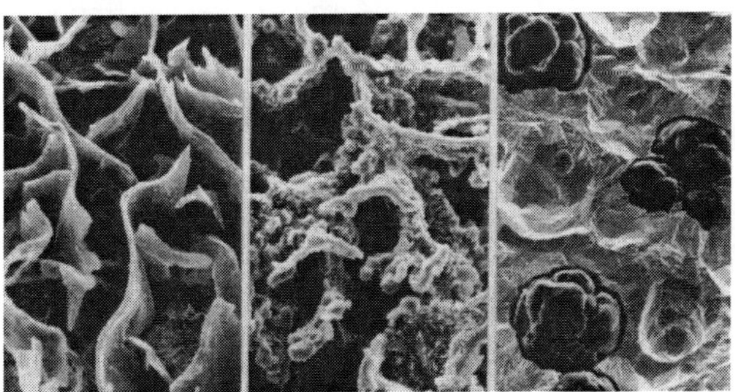

Fig. 13-44. Deep-etched SEMs show the 3D shape of graphite for (left to right) gray iron, compacted graphite iron and ductile iron.

Austempered Ductile Iron (ADI) Microstructure

Austempered ductile iron was conceived approximately 25 years ago. Again, the same process controls used to produce high quality ductile iron are used for ADI. The ductile iron should have:

1. Uniform nodule distribution with a minimum nodule count of 100 N/mm^2.
2. Nodularity in excess of 80%.
3. Maximum level of 0.5% carbides plus nonmetallic inclusions.
4. Maximum allowable volume of 1% porosity and/or microshrinkage.

The chemical composition is also controlled to retard the formation of pearlite during cooling. The three elements used to accomplish this are copper, molybdenum and nickel. Since the ductile iron will be heat treated, it is also good practice to establish chemical composition specifications with the heat treater.

Some people will say that ADI's microstructure is made up of nodules of graphite surrounded by bainite. This is incorrect! The matrix in austempered ductile iron is made of acicular ferrite in a high-carbon austenite called "ausferrite." **Figure 13-45** shows a photomicrograph of austempered ductile iron after heat treating. If the heat treatment procedure is not performed properly, bainite can result.

The heat treatment procedure, called austempering, consists of three steps:

1. Austenitize in the temperature range of 1500–1750F (840–950C) for a time sufficient to produce a fully austenitic matrix that is saturated with carbon.
2. Rapidly cool the entire casting to an austempering temperature in the range of 450–750F (230–400C) without forming pearlite or allowing the formation of ausferrite to begin.
3. Isothermally treat at austempering temperature to produce ausferrite with an austenite carbon content in the range of 1.8–2.2%.

Fig. 13-45. This 500X photomicrograph of Grade 1 ADI (etched with 5% nital) illustrates the ausferrite matrix.

Carbon Steel Microstructure

Carbon steel is iron containing very little carbon. Therefore, a photomicrograph of a carbon steel microstructure would not show any "free" graphite in the form of flakes or nodules. There are three classes of carbon steel:

1. Low carbon (less than 0.20% C)
2. Medium carbon (between 0.20 and 0.50% C)
3. High carbon (above 0.50% C).

Figure 13-46 shows the portion of the iron-iron carbide diagram in which the carbon steels would fall. Note that the area between the liquidus and solidus line is labeled "austenite." Austenite is a solid solution in which initially all of the alloying elements are dissolved. The transformation of the austenite plays a very important role in the final microstructure of carbon steel castings.

Just as cast irons had a eutectic point, so do the carbon steels. However, in their case, it is given a different name. The eutectoid point of carbon steel is very close to 0.80% carbon. Thus, carbon steels found to the left of the eutectoid point are called hypoeutectoid and those to the right are called hypereutectoid.

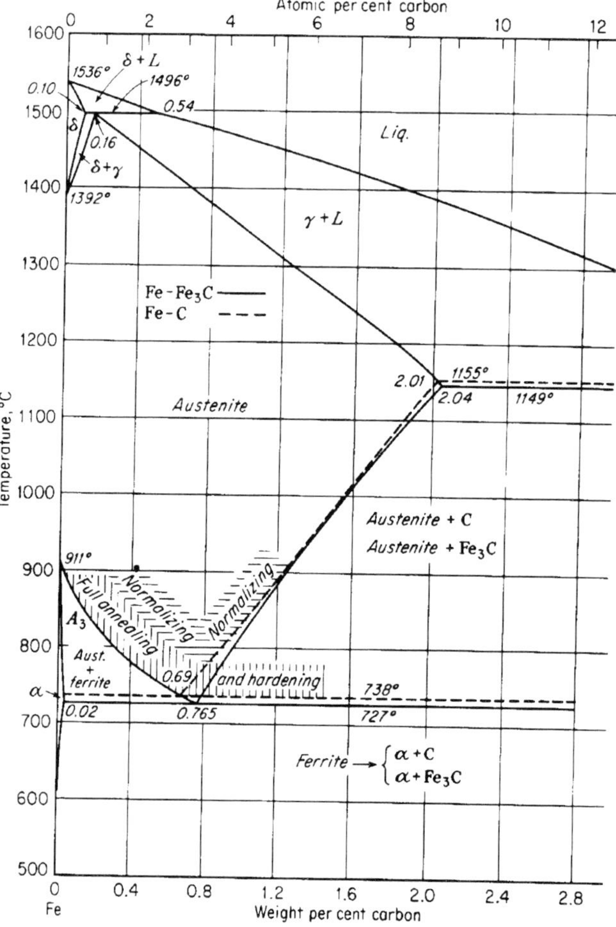

Fig. 13-46. A portion of the Fe-Fe$_3$C and Fe-C diagrams. [After Kirkaldy and Purdy, courtesy of American Institute of Mining, Metallurgical and Petroleum Engineers]

Austenite Transformation of Carbon Steel

The subject of austenite transformation can become quite involved. The following text will just cover the basics. Should the readers wish to learn more about this particular subject, it is strongly suggested that they contact a carbon steel foundry metallurgist or delve into the literature listed in this chapter's Bibliography.

Consider a eutectoid steel (0.80% C) that is at a temperature above 1333F (723C) in the region of single-phase austenite. This area was pointed out a few paragraphs ago. On cooling at this temperature, the entire alloy changes (transforms) from austenite, at 0.80%, to α-ferrite, which has a carbon content of 0.025%, plus iron carbide (cementite) with a carbon content of 6.67%. The resulting structure of this transformation is called pearlite. Remember that pearlite was discussed under cast irons. **Figure 13-47** shows a photomicrograph of this structure. These lamellar layers of ferrite and cementite form in patches or nodules.

On the other hand, consider a hypoeutectoid steel having a carbon content of 0.30%, which has been heated to a temperature where the structure will be fully austenite in the form of grains. As the temperature continues to drop from the austenite region, primary ferrite begins to form at about 1472F (800C). This ferrite is found in the grain boundaries between the austenite grains. As the temperature continues to drop to just above the eutectoid temperature 1333F (723C), the amount of ferrite increases until it consists of ferrite at 0.025% carbon and austenite at 0.08% carbon. When the temperature goes below the eutectoid temperature, the austenite transforms to pearlite and the final structure contains 35% primary ferrite and 65% pearlite. A microstructure resulting from this transformation process can be seen in **Fig. 13-48**. Note that the white areas are primary ferrite and the shady areas are pearlite.

It should be mentioned here that other elements that can be in the carbon steel alloy can be found in the ferrite and/or pearlite upon completion of solidification.

The transformation of hypereutectoid carbon steel is similar to hypoeutectoid, except that the primary phase forming from the austenite is not ferrite but cementite. As this carbon steel continues to cool from a fully austenitic structure, cementite starts to form at, and within, the grain boundaries. This transformation continues until just above the eutectoid temperature of 1333F (723C). At this temperature, the austenite composition is 0.80% carbon and the cementite has 6.67% carbon.

Upon cooling to a temperature below the eutectoid temperature, the austenite transforms to pearlite. The final structure of this alloy would have 25% primary cementite and 75% pearlite, and under the microscope would look like the sample in **Fig. 13-49**. Note the cementite in the grain boundaries. At times, this brittle cementite found in the grain boundaries is undesirable in hypereutectic alloys. This network structure of cementite can be modified through heat treating measures.

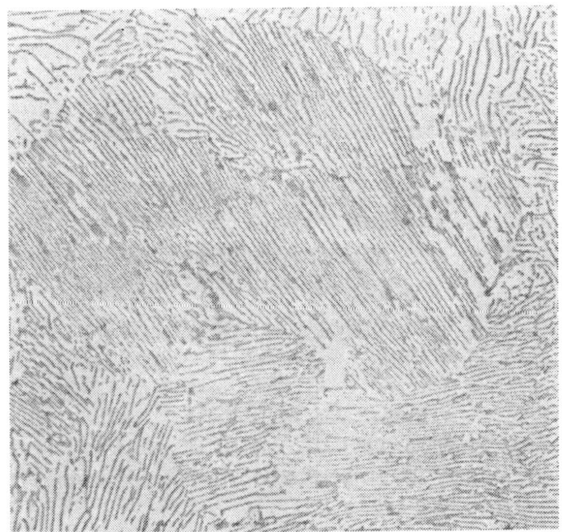

Fig. 13-47. Structure of eutectoid steel, pearlite, 1000X.

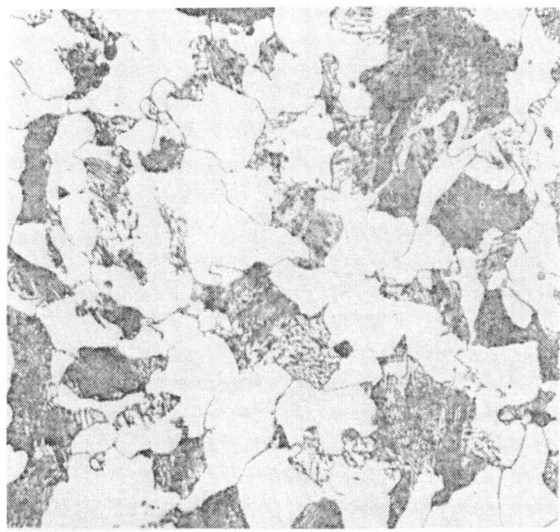

Fig. 13-48. Structure of hypoeutectoid steel, ferrite plus pearlite, 1000X.

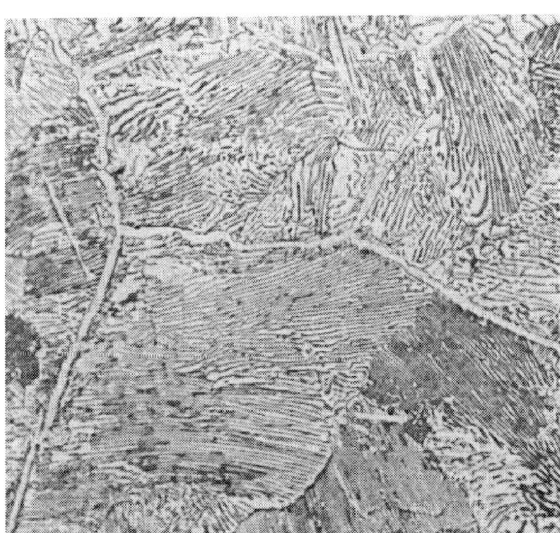

Fig. 13-49. Structure of hypereutectoid steel, cementite plus pearlite, 1000X.

Segregation in Carbon Steel

One of the greatest problems that can occur in the microstructure of steel castings is the segregation of alloying elements. This segregation occurs during the solidification of the casting. Normally, steel castings solidify with a dendritic structure. The region between these dendrites is where the greatest amount of alloying elements will normally be found. The effects this segregation can be partially reduced by heat treating. **Figure 13-50** shows the dendritic structure in a steel casting.

The degree or amount of this segregation is affected by the rate at which the steel casting solidifies. In the case of steel castings produced in sand molds, slow cooling takes place and results in a very strong dendritic structure. Rapid cooling, which takes place in areas where chills are used or in thin sections, produces a very fine structure where the dendritic pattern may be very microscopic in nature.

Most of the alloying elements used in steel castings segregate toward the interdendritic regions, as mentioned earlier. This causes a very nonhomogeneous distribution of the alloying elements. Some of the problems that can result are: nonuniform microstructures, nonuniform response to heat treatment, and phases in the interdendritic regions due to the high concentration of alloying element there.

To overcome this problem it is suggested that increased cooling rates reduce the dendritic arm spacing and thus reduce the extent of alloy element segregation. Under these conditions, heat treatments can be devised that will produce more uniform distribution of the alloying elements throughout the casting.

Effect of Heat Treatment on Carbon Steel

Most steel castings will be heat treated in some manner. The heat treatments used to change the mechanical properties of steel castings are the same treatment as used for wrought steels. Since there are specific texts written for this particular subject, heat treatment will not be detailed herein. However, several additional features should be mentioned here.

As previously mentioned, segregation of the alloying elements in steel castings can lead to problems. When steel is heat treated for rather long times, and at elevated temperatures to achieve a better distribution of the carbon and alloying elements, it is called homogenization. The temperatures at which this heat treatment takes place can vary, based on section thicknesses. A heavy-section casting will require higher temperatures at longer periods than will thin-section castings. The data indicates that some benefit may be received from homogenization for some steel castings of a certain composition and for some steel castings where extremes in service requirements are encountered. However, for the most part, this heat treatment does not appear to be justified on the basis for improving mechanical properties and the costs involved therein.

Annealing of steel castings is done the same way as for wrought steels. The casting temperature is raised to the austenitic stage and then allowed to cool in the heat treat furnace. The purposes of annealing are:

Fig. 13-50. Dendrite structure in a steel casting.

1. To refine the austenitic grain structure
2. To soften the casting for machining
3. To relieve stress
4. To improve toughness

Normalizing and annealing are very similar processes, with the main differences being that higher temperatures are used in normalizing and that the castings are allowed to cool in ambient air. This heat treatment gives higher strength and hardness than does annealing and is used as the final heat treatment where strength requirements do not go higher than 100,000 psi.

Stress-relief annealing will relieve the stresses in a casting. Stress-relief can also be done using subcritical temperatures. For instance, holding at 750F (400C) will reduce stresses about 50% whereas holding at 1000F (537C) will reduce stresses by over 90%. Time at temperature is also a factor. However, if a liquid quench and temper are employed, the stress relief is accomplished by the tempering.

Liquid quench and temper of steel castings follows the same principles and techniques used for wrought steel products. If a steel casting is cooled fast enough to hinder the transformation of cast steel to a ferrite-pearlite product, it will produce a martensitic product. Martensite is the hardest product available for a given carbon content. By reheating martensite to a subcritical temperature (tempering), a dispersion of fine carbide in ferrite will result, which, in turn, will lead to a slow softening and toughening of the steel casting.

Quenching and tempering treatment will provide a way of controlling the properties of a given steel within broad limits and will give the steel casting the best combination of properties obtainable. A steel casting tempered to a given tensile strength will have the highest ductility, toughness, or yield strength, when compared to other heat treating methods achieving the same hardness level.

High- and Low-Alloy Steel Microstructure

Using sufficient alloying elements in steel castings to place them in the high-alloy steel family can inhibit the normal transformations in steel, to the point that the austenitic structure is retained. These types of steel are found in the stainless and heat-resistant ranges. The primary alloying element is chromium, and with 13% of this element, the structure is martensite. Adding more chromium alone to the steel will stabilize the ferrite in the structure. Nickel also has to be considered, as it is also an important constituent in high-alloy steels. Most of these alloys contain both Ni and Cr. An example of this is the well-known 18 (Cr) 8 (Ni) varieties that are predominantly austenitic in structure. When these alloys are in this condition, they are nonmagnetic.

Because their structures are mainly a solid solution, austenitic stainless steels are not particularly strong. To improve their strength, they are given a homogenizing heat treatment and cooled very rapidly. Their strength is increased by the amount of ferrite that is retained. In addition, because of their structure, the austenitic high-alloy steels are not susceptible to an increase in strength by hardening and tempering heat treatments.

Following are some microstructures of low- and high-alloy steels. **Figures 13-51** and **13-52** show the microstructures of a low-alloy steel at the same magnification but with different etchants. **Figure 13-53** is of a 13% Cr steel, as cast. Note that the microstructure is generally martensitic. Finally, **Fig. 13-54** is of the previously mentioned 18/8 Cr-Ni steel. This microstructure was taken of an as-cast sample.

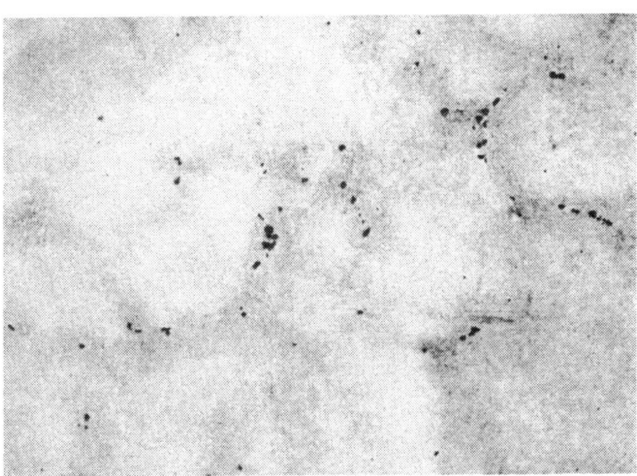

Fig. 13-51. General view of low-alloy steel after etching in picral (4% picric acid in alcohol); structure is not resolved at this magnification, but segregation and Type III inclusions are visible, 100X.

Fig. 13-53. Martensitic structure in as-cast 13%Cr steel, 100X.

Fig. 13-52. General view of low-alloy steel after etching in nital (2% nitric acid in alcohol) gives a structure of darker appearance but with little or no evidence of microsegregation, 100X.

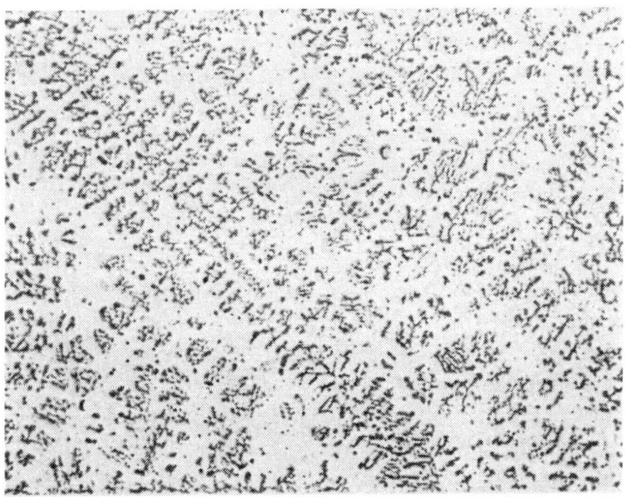

Fig. 13-54. Dendritic structure in as-cast 18/8 CrNi steel (electrolytically etched in oxalic acid) 20X.

BIBLIOGRAPHY

This chapter has taken a brief look at ferrous alloys, their structures and how they are obtained. This subject is much more involved than the space allotted. Many fine references can be obtained for a more in-depth understanding and working knowledge of this subject. Much of the material in this chapter came from these references.

Bailey, A.R., Samuels, L.E., *Foundry Metallography,* Metallurgical Services, Inc., Niagara Falls, NY.

Dawson, S., Schroeder, T., "Compacted Graphite Iron Offers a Viable Design Alternative," American Foundrymen's Society, Inc., *Engineered Casting Solutions* (Spring 2000).

Ductile Iron Data for Design Engineers, QIT America, A Division of High Purity Iron, Inc., Chicago, IL (1990).

Hayrynen, K.L., "ADI: Another Avenue for Ductile Iron," American Foundrymen's, Society, Inc., *Modern Casting* (Oct 1998).

Heine, R.W., Loper, C.R., Rosenthal, P.C., *Principles of Metal Casting,* Second edition, McGraw-Hill, New York (1967).

Inoculation of Gray Cast Iron: Part I, Foseco Foundry Practice, No. 43 (Mar 1967) Foseco, Inc., Cleveland, OH.

Inoculation of Gray Cast Iron: Part II, Foseco Foundry Practice, No. 44. (May 1967) Foseco, Inc., Cleveland, OH.

Iron Castings Handbook, Iron Castings Society, Inc., 1981.

Karsay, S., *Ductile Iron Production,* Quebec Iron and Titanium Corporation, Sorel, Quebec, Canada (1976).

Kovacs, B.V., "Austempered Ductile Iron—Fact or Fiction," American Foundrymen's Society, Inc., *Modern Casting,* pp 38-41 (Mar 1990).

Patterson, V.H., "Some Basic Considerations in Controlling the Mechanical Properties of Cast Iron—Part VII," *Foote Foundry Facts,* Foote Mineral Company, Exton, PA.

Schleg, F.P., Personal Class Notes from American Foundrymen's Society, Inc., Cast Metals Institute (CMI) Ferrous Metallurgy, Melting and Metallography Courses, 1965-1992.

Skaland, T., Elkem ASA, "Inoculation Material Improves Graphite Formation in Ductile Iron," AFS, *Modern Casting* (Dec 2001).

Steel Casting Metallurgy, Steel Founders' Society of America, Barrington, IL (1984).

Microstructure of Nonferrous Alloys

14

In Chapter 13, the first section contained information that applied to both ferrous and nonferrous alloys (atomic arrangement and phase diagrams). This chapter will examine the microstructure of nonferrous alloys, primarily in aluminum and copper alloys. As with the ferrous alloys, the microstructures of nonferrous alloys determine the mechanical properties of the casting. Also, as with ferrous alloys, the chemical composition of the nonferrous alloy, the speed at which solidification takes place in the mold, whether or not a form of inoculation is used, and subsequent heat treatment (if any) all play a part in determining the microstructure of nonferrous alloys.

NONFERROUS PHASE DIAGRAMS

As learned in the previous chapter, nonferrous alloys also have phase (equilibrium) diagrams to help foundry personnel understand what is occurring in the casting as it is solidifying in the mold. It also shows the changes that are taking place and what state in the solidification process the alloy is at when the temperature and chemistry are known. These phase diagrams are read in the same manner as the ones discussed in the previous chapter.

Figures 14-1 and 14-2 show two different aluminum alloy phase diagrams. **Figure 14-1** is a phase diagram for a simple aluminum-silicon alloy. Note that this phase diagram has curved liquidus lines and a straight solidus line, and that in between these lines is a mixture of solid and liquid materials called "mush." The eutectic point of this alloy can be seen at 11.6 wt.% Si.

Another phase diagram, that of aluminum-copper, is shown in **Fig. 14-2**. This phase diagram is much more complicated than the one seen in **Fig. 14-1** and is more commonly used by the aluminum metallurgist and researcher in their work.

Another type of phase diagram is a ternary phase diagram. As the name implies, this phase diagram is used for an alloy consisting of three chemical elements. An example of a ternary phase diagram is shown in **Fig. 14-3.** This particular ternary diagram is for a copper-tin-zinc alloy. The base of the triangle indicates the zinc content; the left side of the triangle indicates the tin content; and the right side, the copper content. Again, it shows the various phase compositions.

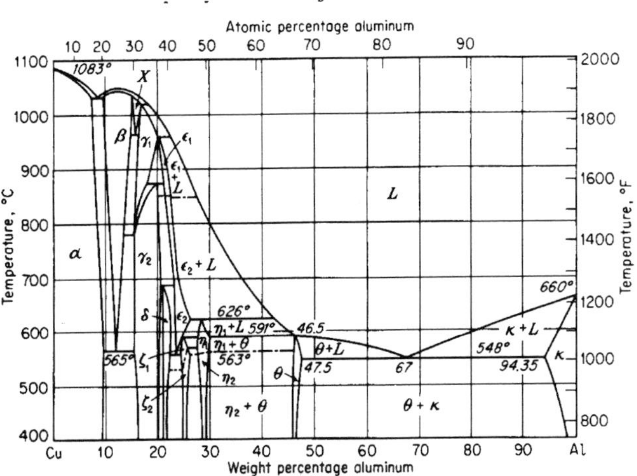

Fig. 14-2. Entire diagram of the Al-Cu equilibrium system. [From American Society for Metals]

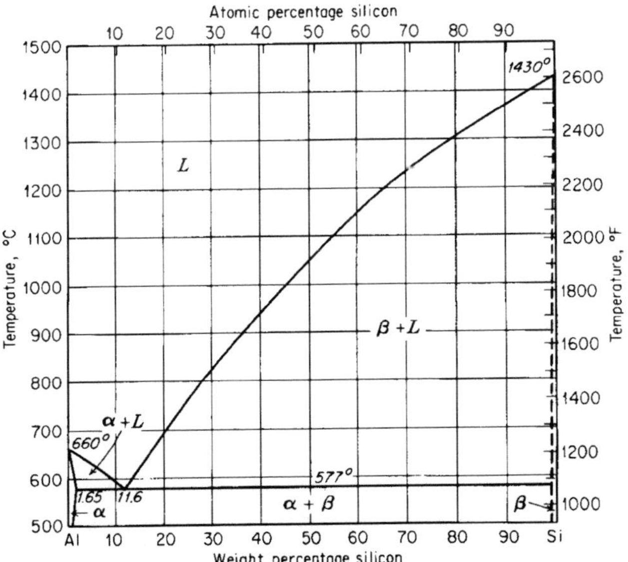

Fig. 14-1. The Al-Si equilibrium system. [From American Society for Metals]

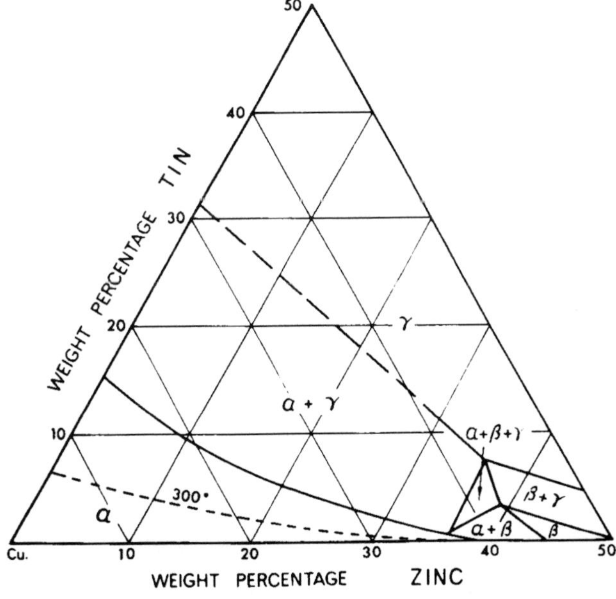

Fig. 14-3. The 500C (932F) isothermal section of the Cu-Sn-Zn ternary equilibrium diagram.

NONFERROUS ALLOY MICROSTRUCTURE

Aluminum-Silicon Alloy Microstructure

Aluminum casting alloys, as learned in Chapter 9, are divided into groups based on their primary alloying element. For instance, the 300 and 400 series contain silicon as the primary alloying element. The most commonly used of these two series is the 300 series, which will be reported first. The aluminum-silicon alloys can be divided into three broad groups: 1) hypoeutectic compositions having 4–6% Si; 2) eutectic alloys having 10–13% Si; and 3) hypereutectic alloys having approximately 18–24% Si.

Effect of Grain Refinement

When the Al-Si alloys solidify, they will normally form coarse, equiaxed and columnar grains. **Figure 14-4** illustrates these types of grains after solidification has been completed. As with other alloys, the degree of coarseness or length of the columnar crystals depends on the pouring temperature of the molten metal, how fast the molten metal solidifies in the mold and how many grain-forming nuclei are present in the molten metal. Grain size can be decreased by adding the most common alloying elements used with aluminum, which will help improve the mechanical properties.

A form of inoculation called grain refinement is used to help form a fine grain structure. This is very similar to the inoculation practice used in cast irons to promote graphitization. The grain refiner is added to the molten aluminum alloy just prior to pouring it into the mold(s). The grain refiners most commonly used are titanium, boron or zirconium, with a compound of titanium/boron being the most popular. The grain refiners provide additional nuclei, which produce a finer, equiaxed grain during solidification. **Figure 14-5** shows two different samples of a chill-cast 319.0 alloy: one that was not grain refined (unrefined) and one that was refined. Note the larger columnar grains of the unrefined sample versus the fine equiaxed grains of the refined sample.

A simple test has been developed to evaluate the grain size after grain refinement. This test consists of pouring the molten metal into iron rings that are 3 in. (76 mm) in diameter and 1 in. (25 mm) high. The ring is set on a fused silica block, which serves as an insulator. The molten metal is poured into the ring and allowed to solidify. After solidification has been completed and the sample has cooled, it is removed from the ring. The sample does not have to be polished or cut because the grain structure is sharp and clear. **Figures 14-6a and 14-6b** show this test and the results. Note that this test shows the grain structure for permanent mold, sand and die casting, based on which surface is being examined.

Grain refinement offers the following advantages for aluminum casting alloys:

- Less hot tearing during solidification, which coarse-grained aluminum alloys are prone to do.
- Less shrinkage in grain-refined aluminum alloys.
- Porosity is smaller and more evenly distributed in grain-refined aluminum alloys, offering increased pressure tightness in the casting.
- Improved mechanical properties in heavier sections of castings poured using grain-refined molten aluminum alloys.

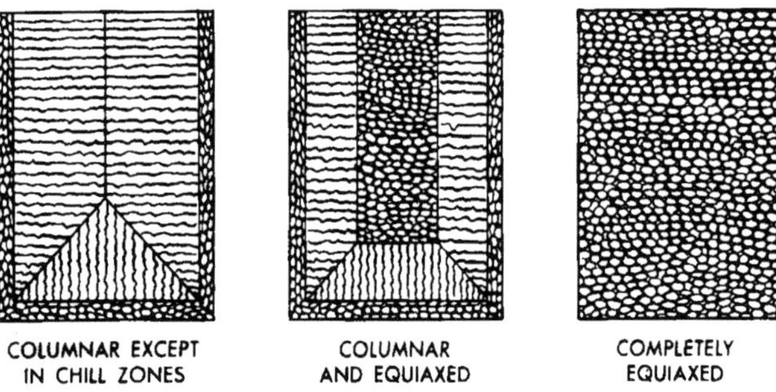

Fig. 14-4. Grain structures in aluminum castings.

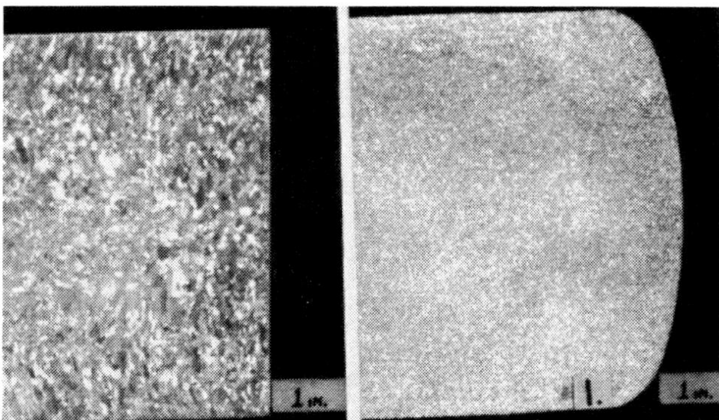

Fig. 14-5. Sections of chill-cast 319.0 alloy poured under identical conditions. These two polished casting sections were chemically etched to show their grain structure. The casting on the left, made without grain refining the melt, shows large, columnar crystals at the edges. Grain refining the melt produced the equiaxed grain structure shown on the right.

Fig. 14-6a. Test samples for evaluating aluminum grain refiner additions are formed by pouring molten metal into iron rings set on a block of fused silica.

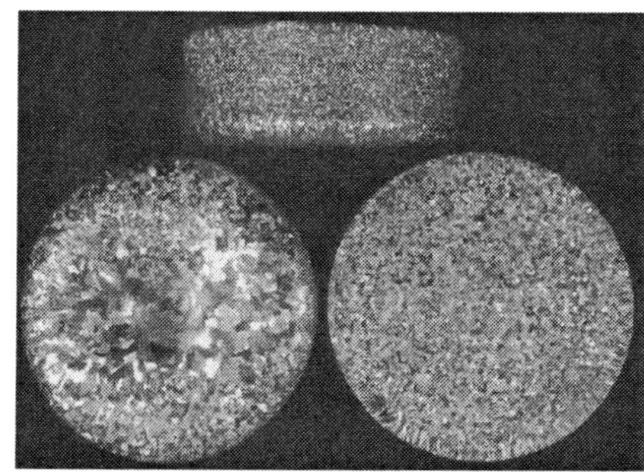

Fig. 14-6b. The three surfaces of each test sample show grain structures produced by the most widely used aluminum casting methods. The sample's bottom (lower right) resembles a die casting; the sample's top (lower left) a sand casting; and the sample's side (top) a permanent mold casting. [Photos courtesy of KB Alloys, Inc.]

Effect of Structural Modification

Interest has always been shown in the structural modification of hypoeutectic and eutectic aluminum-silicon alloys. These two types of Al-Si alloys fall in the 5–13% Si range.

Included in this range of silicon are A322.0, 319.0, 356.0 and 357.0, which are often used in the production of aluminum alloy castings. When these alloys solidify, the aluminum and silicon solidify as separate entities. The silicon that solidifies as pure silicon remains as pure forms of acicular (needle) shaped platelets. Some of the silicon may combine with the aluminum and form aluminum-silicide. The aluminum itself forms the matrix surrounding the silicon.

As with the graphite flakes in gray cast iron, these plates of silicon decrease the ductility of the Al-Si alloys. In order to change the size and shape of the silicon platelets in the Al-Si alloys, modifiers can be added to the alloy. Among the modifiers are sodium, calcium, strontium and antimony. Of these modifiers, strontium is the most popular in the United States and Canada, and antimony has found acceptance in Europe. Due to environmental and handling concerns, sodium has lost its popularity as a modifier of Al-Si alloys. Calcium is not as effective a modifier as sodium.

Modification of the Al-Si alloys is done just prior to pouring the molten alloy into the mold(s). Almost all producers of these aluminum alloys modify them prior to shipping to foundries. However, modification has been found to vary in effectiveness depending on how the foundry processes the alloy. Thus, foundries have found it best to fully control the modification process by adding the modifier just prior to pouring. **Figure 14-7** shows an A356.0 alloy that was not modified before pouring. Note the needle-like dark shapes, which are the silicon platelets.

Strontium is added to the molten Al-Si alloy as an aluminum masteralloy containing 10% Sr and 14% Si. This aluminum masteralloy has been found to be a convenient way to add the modifier and eliminate the uncertainty associated with sodium modification. Although at first a somewhat greater amount of strontium has to be added to modify the Al-Si alloy properly, as compared to sodium, much smaller amounts have since proved to be adequate. Another advantage of strontium over sodium is that it retains its modification longer. It has also been reported that small amounts of sodium and calcium in the melt interferes with strontium modification.

Small additions of antimony to Al-Si alloy can be used to structurally modify these alloys. When used, antimony remains as a permanent constituent of the alloy and, thus, performs as a permanent modifier. It is also reported that antimony is unaffected by holding time, remelting or degassing.

Strontium and sodium are not compatible with antimony and their modification abilities are adversely affected. In addition, antimony-modified alloys are not recommended for food contact applications. Antimony's use as a modifier in North American foundries has been limited because of concerns about potentially toxic antimony compounds; however, this concern has not yet been fully resolved.

The following figures look at some Al-Si microstructures with and without modification. **Figure 14-8** shows the effects of modification of a 356.0 aluminum alloy. **Figures 14-9 and 14-10** show high magnification scanning electron microscope photos of unmodified and modified silicon platelets. Note how much more compact the modified silicon platelets are than the unmodified.

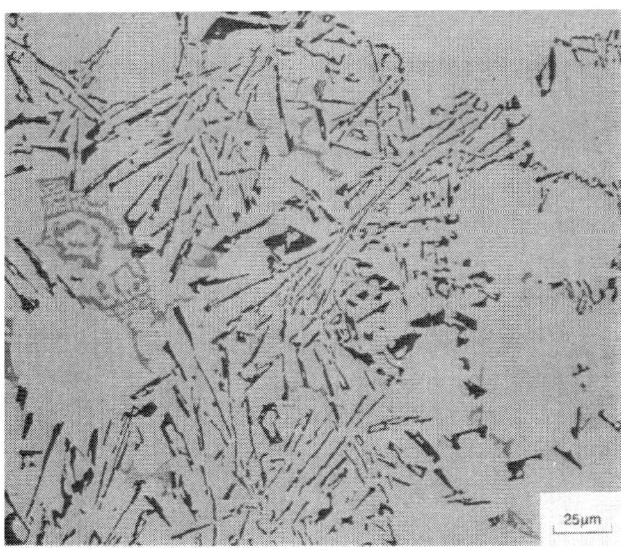

Fig. 14-7. As-cast structure of A356.0 cast without a modifier. Rectangular and needle-like dark shapes are plate-like particles of silicon. Lighter background area is aluminum. [Photos courtesy of KB Alloys, Inc.]

The purpose for modifying Al-Si alloys is "ductility." Looking more closely at the unmodified silicon, it looks similar to the graphite flakes in gray cast iron. The ductility of gray cast iron is reduced because of the flakes of graphite. The same is true for the Al-Si alloys. The silicon platelets reduce the ductility of Al-Si alloy castings. However, when these alloys are modified, their ductility is increased, as are other mechanical properties.

An instrument has been developed to help determine how well the Al-Si alloy has been grain refined and modified. This instrument works the same as the eutectometer used to measure the carbon equivalent and silicon level in cast iron. A sample of the alloy to be tested is poured into a specially prepared sand cup containing a thermocouple. The thermocouple is connected to an instrument that develops a cooling curve for the alloy. Based on the points where solidification begins, the thermal arrest begins and solidification has been completed, the instrument gives a reading of the quality of the grain refinement and modification treatments.

Figure 14-11 shows a thermal analysis instrument and its component parts. The instrument gives both a digital display and printed readout of its findings. A grain-size chart, as seen in **Fig. 14-12,** can be used to give a graphic display of the grain size number digitally displayed on the instrument. As already indicated, the finer grain structure yields enhanced mechanical properties of the aluminum alloy casting.

Another means of determining the mechanical properties of an aluminum alloy casting has been developed and should be mentioned here. It is a well-known fact that, as aluminum alloys solidify, dendrites form. The closer the dendrites are to one another, the more improved the mechanical properties (i.e., tensile strength) will be. A means of measuring the distance between these dendrites has been developed and can be used for determining the mechanical properties of the aluminum alloy upon solidification. The use of dendrite arm spacing (DAS) as a tool to help determine mechanical properties in an aluminum alloy casting was developed in Sweden. There is literature available on this subject in the Bibliography of this chapter.

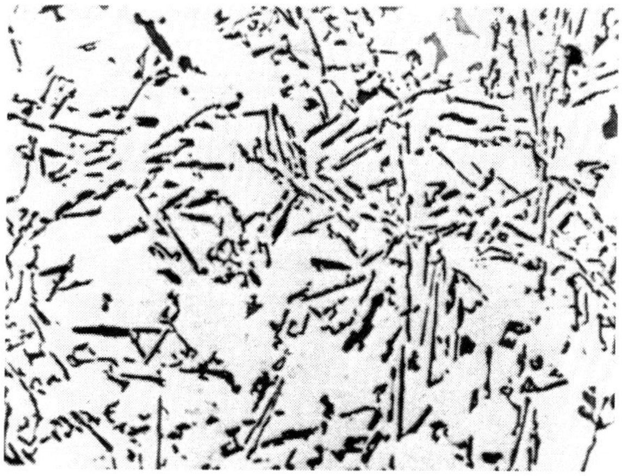

(14-8a)

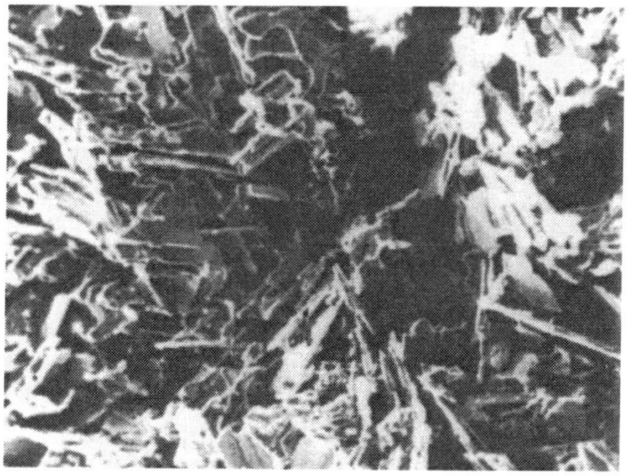

(14-8b)

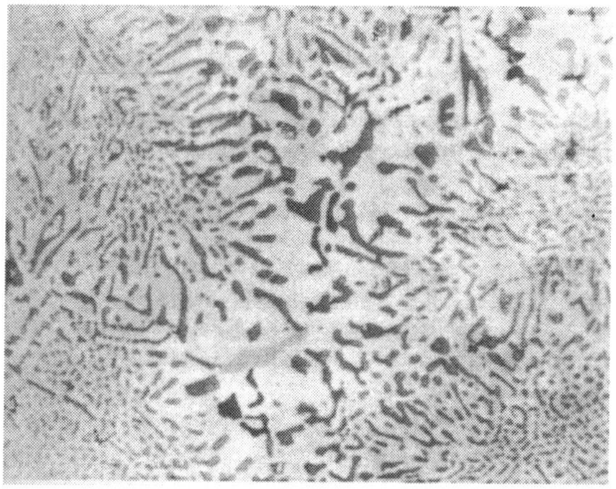

(14-8c)

(14-8d)

Fig. 14-8. Effect of modification on the morphology of the eutectic phase in 356.0. The SEM samples (b and d) were deep-etched to bring out a 3D perspective of the eutectic phases. (a) Unmodified, 0.5% HF, 300X. (b) Unmodified, deep-etched, 300X. (c) Modified, 0.5% HG, 800X. (d) Modified, deep-etched, 2000X. [Photos courtesy of G.K. Sigworth]

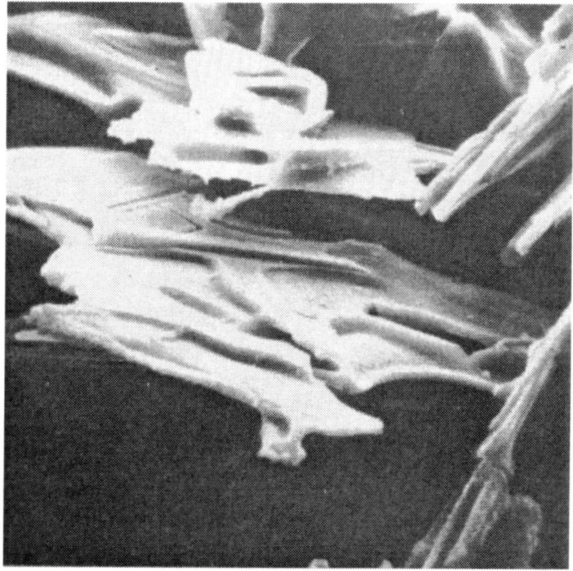

Fig. 14-9. SEM photographs of silicon crystals. Unmodified, 875X. (Enlarged by 25% for publication.)

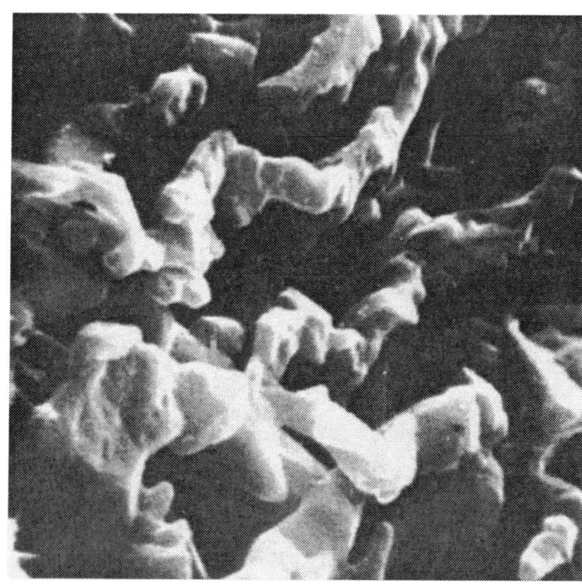

Fig. 14-10. SEM photographs of silicon crystals. Modified, 1000X. (Enlarged by 25% for publication.)

Fig. 14-11. A thermal analysis instrument used in assessing the grain refinement and modification levels in Al-Si foundry alloys.

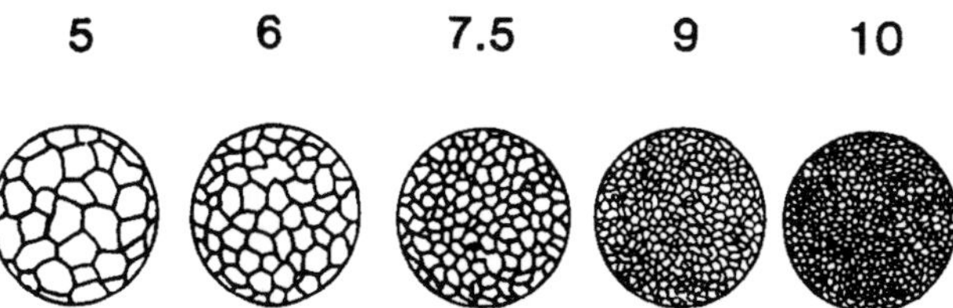

Fig. 14-12. Schematic representation of grain sizes in a casting as evaluated and displayed by a thermal analysis instrument.

Effect of Heat Treatment

All aluminum alloys can be heat treated, since they can all be annealed. However, not all aluminum alloys can be heat treated to change mechanical properties. Aluminum castings are cast from both heat-treatable and non-heat-treatable alloys. Non-heat-treatable alloys are alloys whose strengths are not appreciably affected by solution treatment, quenching and artificial aging. Aluminum alloy castings made from a non-heat-treatable alloy are most often identified as being in the "F" (as-cast) temper. Both heat-treatable and non-heat-treatable aluminum alloys can be given specific thermal treatments for purposes other than improving their strengths. The heat-treatable aluminum alloys are those having copper, magnesium or zinc, and combinations of magnesium and silicon.

Diecasting alloys are not normally solution heat treated, although some diecasting alloy compositions in the 300 series would realize substantial increases in their strength. The nature of conventional die castings is such that they tend to blister when subjected to solution heat treating temperatures. This is due to highly compressed gases entrapped in the diecastings during the casting process. Improvements in the diecasting process have resulted in better internal quality and compatibility with solution treatment practices, so some of these diecasting alloys may eventually be included in the heat-treatable category.

Heat treatment does improve the properties; however, many aluminum alloy castings meet specifications without heat treatment and can be used as cast. The letter "F" identifies those castings that can be used in the as-cast state. However, in the as-cast state, high residual stresses may be present in the casting.

Depending on the aluminum alloy and the casting process used, as-cast aluminum alloys are susceptible to dimensional growth and low strength, and a decrease in toughness- and corrosion-resistance. To overcome these problems, and to enhance strength and ductility, aluminum alloy castings can be heat treated by one or more heating and cooling cycles. This heat treatment includes three basic operations: solution treatment, quenching and aging. Combinations of these heat treatments include stabilization and stress relief; annealing and age hardening are called tempers. These tempers are given a letter and number designation. These designations and their explanations are as follows:

T2: Casting annealed
T4: Solution heat treat, quenched
T5: Artificial aging treatment only, low temperature
T6: Solution heat treat, quenched, aged
T7: Solution heat treat, quenched, over aged

The details regarding each of these heat-treating processes can be found in some of the literature listed at the end of the chapter in the Bibliography.

Aluminum castings are sometimes heat-treated for one or more of the following reasons:

- Homogenization;
- Stress relief;
- Improved dimensional stability and machinability;
- Optimized strength, ductility, toughness and corrosion resistance.

Microstructures of aluminum alloy castings in the as-cast state (F) are not homogeneous. The reason for this is that the alloying and impurity elements tend to segregate into networks of eutectic constituents. Homogenization will distribute these elements evenly throughout the matrix so that the casting properties will be uniform.

Residual stresses are created in the casting as it cools from the elevated casting temperature and solution temperatures. These residual stresses can result from the cooling and contraction of the casting, poor casting design, or molds and cores that restrict the contraction of the casting. Heating the casting to an intermediate temperature can relieve residual stresses. To maintain tight dimensional tolerances during and after machining, castings should be heat treated to form stable precipitate phases and to deplete the matrix of excess solutes.

The greatest reason for heat treating aluminum alloy castings is to improve their mechanical and corrosion-resistance properties. This can be accomplished by spheroidizing constituent phase particles and by precipitation hardening. Corrosion resistance is improved by homogenizing the distribution of alloying elements and by dissolving easily oxidized phases such as magnesium silicide (Mg_2Si). Most often, heat treatment is a compromise maximizing the important properties at the expense of others. With this in mind, age-hardening principles are exploited to tailor the heat treatments to each particular case.

The subject of heat treating aluminum alloy castings is quite complex and beyond the intended scope of this book. There are several excellent references listed in this chapter's bibliography: "Aluminum Casting Technology" and "Principles of Metalcasting."

Other Aluminum Alloy Microstructure

Silicon is unquestionably the most important alloying element in the vast majority of aluminum casting alloys. It not only improves the castability of the aluminum alloys but also plays a role in improving the mechanical properties. There are two other elements commonly used to improve aluminum alloys for specific reasons: copper and magnesium.

Aluminum-Copper Alloys

The 200 series of aluminum alloys contains copper as the primary alloying element. Copper has the single greatest impact of all the alloying elements used for increasing strength and hardness. However, if the copper content exceeds about 12%, it can deteriorate mechanical properties. Under equilibrium conditions, about 6.5% copper is soluble in aluminum at solidification temperatures (1018F/548C). At room temperature, the solubility drops to less than 0.5% copper. This decrease in solubility accounts for the susceptibility of Al-Cu alloys to heat treatment. It is the copper-aluminide, an intermetallic compound, that allows this alloy to be heat treated. The highest strength aluminum casting alloys available are those in the 200 series alloys that also contain some magnesium. **Figure 14-13** is a photomicrograph of an Al-Cu alloy with about 4% copper.

Aluminum-Magnesium Alloys

The aluminum-magnesium alloys, with magnesium being the primary alloying element are found in the 500 series alloys. The most commonly used alloy in this series is 535.0, formerly known as AlMg 35. Aluminum-magnesium alloys have excellent mechanical properties, corrosion resistance and machinability. Magnesium up to about 15% is soluble in aluminum at the solidification temperature of the alloy, which is 844F (441C), although the solubility decreases to less than 2.5% at room temperature. An Al-Mg alloy with more than 6% Mg can be heat treated because magnesium combines with the silicon in the alloy to form magnesium silicide, which is an intermetallic compound. All aluminum alloys contain some amount of silicon, and this intermetallic compound is formed with the addition of magnesium. **Figure 14-14** is a photomicrograph of a 535 Al-Mg alloy.

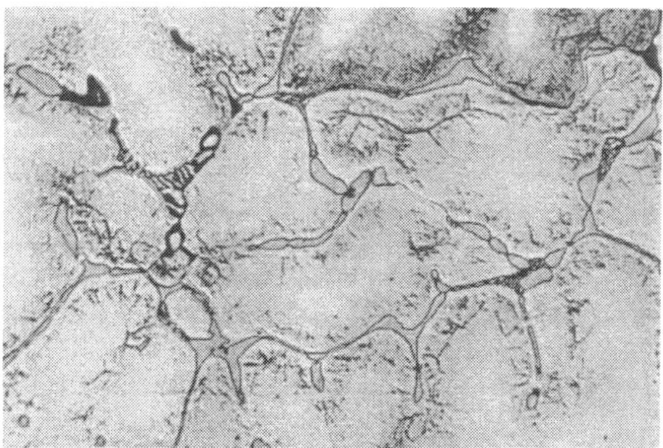

Fig. 14-13. Microstructure of commercial aluminum casting alloy with about 4% Cu as the principal alloying element. Sand-cast alloy 295F (formerly 195F). Almost continuous interdendritic network of (theta) Al-Cu and (alpha)-Al-Fe-Si surrounding cored-aluminum solid-solution dendrites. Coring is indicated by the precipitated (theta) Al-Cu in the shaded areas. Keller's etch, X250.

Aluminum Composites

Aluminum composites, as learned near the end of Chapter 9, consist of two or more dissimilar components that are mechanically or metallurgically bonded together. Each of the components retains their identity in the solidified composite and has its characteristic structure and properties.

Aluminum composites are normally made up of a 300 series aluminum alloy along with silicon carbide, copper graphite or powdered ceramic. When cast, about 20% of the liquid aluminum composite is occupied by the composite material. When all of the prescribed melting, pouring and cooling operations are followed, the particles of the composite material will be uniformly distributed throughout the aluminum matrix of the casting. **Figure 14-15** shows a photomicrograph of an Al-Si carbide composite.

Aluminum composite castings can be heat-treated to improve their mechanical properties using the normal heat-treating practice for aluminum alloys, which was covered earlier in this chapter.

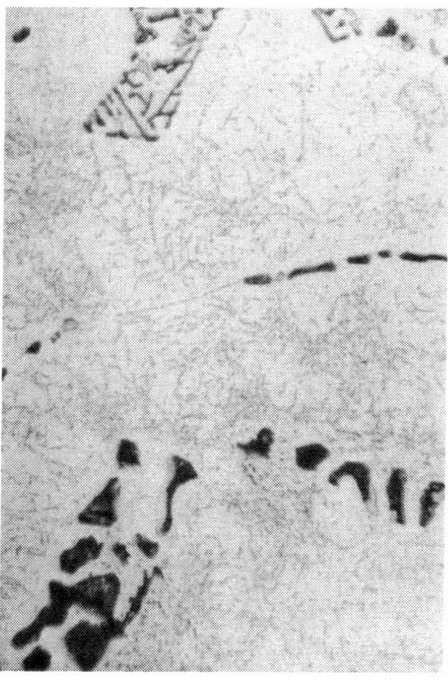

Fig. 14-14. Photomicrograph of a 500-series alloy.

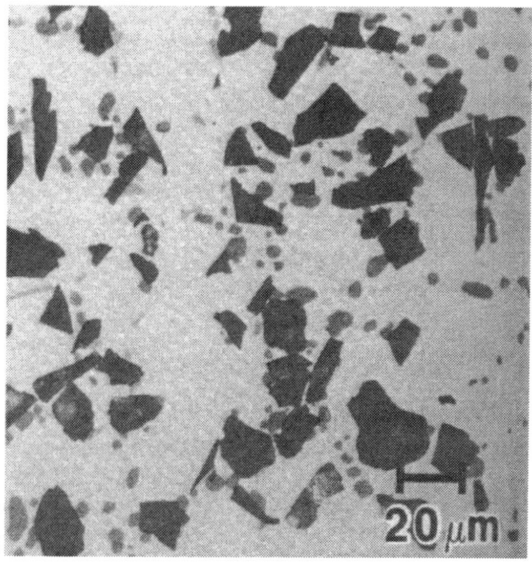

Fig. 14-15. Microstructure of a 359/SiC/20p tensile bar.

Magnesium Alloy Microstructure

In Chapter 9 we learned that the magnesium alloys mainly used for castings can be divided into five systems, as follows:

1. Magnesium having *aluminum and manganese* as the principal alloying elements and designated type AM.
2. Magnesium having *aluminum and zinc* as the principal alloying elements and designated type AZ.
3. Magnesium having *rare earths and zirconium* present which included type EK, EZ, and ZE.
4. Magnesium having *zinc and zirconium* as the principal alloying elements and designated type ZK.
5. Magnesium having *thorium and zirconium* as the principal alloying elements that includes types HK, HZ and ZH.

Broadly, magnesium alloys can be divided into two groups: those containing aluminum and those grain refined with zirconium.

Degassing is often required for those magnesium alloys containing aluminum. However, if zirconium is present in the alloy, it reacts with and precipitates hydrogen as zirconium hydride from the melt along with other impurities. Thus, if a sufficient amount of zirconium is present in the final alloy, it will guarantee that hydrogen has been removed along with the excess zirconium that is lost.

Pure magnesium, when it solidifies, has a structure consisting of coarse columnar grains. Having limited deformation systems, pure magnesium would have low strength and poor ductility. When alloying magnesium with aluminum or zirconium, there is considerable grain refinement, with zirconium being the better grain refiner of the two. Unlike aluminum, where grain refining mainly improves casting soundness, a fine grain structure with magnesium is essential to obtain suitable mechanical properties. Tensile strength and, especially, ductility are greatly increased, and impact resistance is improved. These alloys can be heat-treated to improve their mechanical properties.

Figure 14-16 shows a magnesium AZ alloy, as cast. The etched structure shows a white network of magnesium-aluminum compound in a solid solution matrix. This compound can be found not only at the grain boundaries of the solid solution, but also at cell boundaries within the grains. **Figure 14-17** shows this same structure after deep etching and at a higher magnification. The photomicrograph shows the distinction between matrix grains. **Figure 14-18** is the same magnesium alloy, only this time the alloy has been solution heat-treated. This heat treatment has dissolved practically all of the white magnesium-aluminum compound and produces a set of equiaxial grains of essentially uniform composition.

Figure 14-19 shows a photomicrograph of a magnesium ZK alloy, as cast. The dark areas are most likely compounds of zirconium with impurity elements. The lighter particles are probably a magnesium-zinc compound, while the grain boundaries are zinc-rich, and the white particles in the grain boundaries are a magnesium-zinc compound. Normal heat treatment processes applied to this alloy will not significantly affect the microstructure.

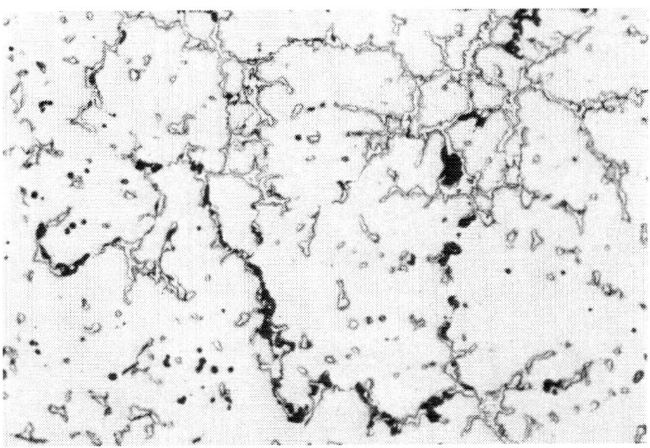

Fig. 14-16. General view of Mg alloy (Type AZ) alloy after etching in 2% nitric acid in alcohol, 100X.

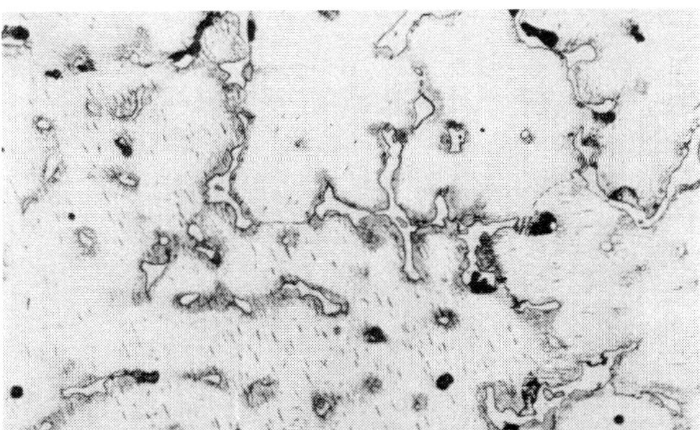

Fig. 14-17. The same structure as shown in Fig. 14-16 shows heavy chemical attack during polishing followed by etching in 2% nitric acid in alcohol and gives distinction between matrix grains, 500X.

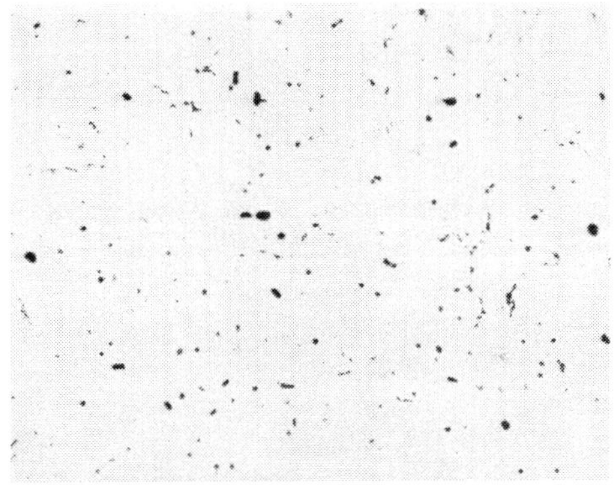

Fig. 14-18. General view of solution-treated alloy (shown in Figs. 14-16 and 14-17) after etching in 2% nitric acid in alcohol.

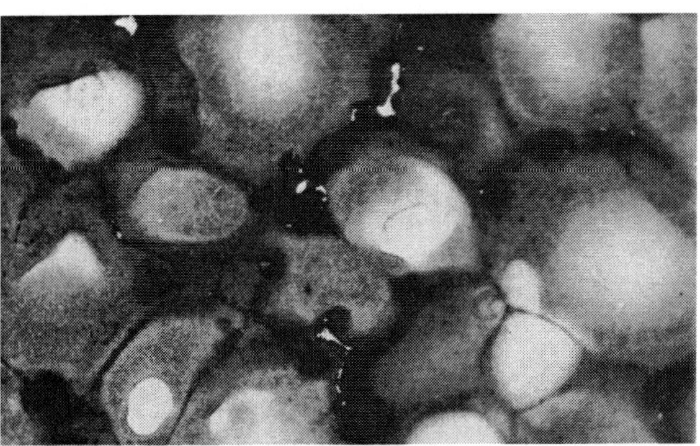

Fig. 14-19. Magnesium-zinc-zirconium alloy after etching in 2% acetic acid: grains of solid solution showing concentric and shell-like coring with white particles of Mg-Zn compound at the boundaries, 500X.

Copper Alloy Microstructure

This section will examine some selected microstructures of copper-base alloys. The actual study of microstructures and their interpretations can become very complex. Individuals interested in pursuing a career as a copper-base metallographer or metallurgist might get an improved understanding of these fields by studying photomicrographs.

Figure 14-20 shows a sample of a sand-cast leaded red brass. The term "coring" is defined as the variable composition in solid-solution dendrites, with the center of the dendrite being richer in one element.

Figure 14-21 is a photomicrograph of a manganese bronze (C86300 alloy) microstructure. The small, black inclusions in the microstructure are particles of a complex, iron-rich intermetallic. Iron is essentially insoluble in these alloys and is added primarily for grain refinement.

The metallurgy of the aluminum bronzes is complex because these alloys often contain four or more major alloying elements. In addition, the microstructures obtained are dependent on the cooling rate from the pouring temperature. Some of the aluminum bronze alloys are heat treatable and others are not. Those alloys that can be heat treated can have their strengths increased from the as-cast strengths. There is probably more information published on the metallurgy of the aluminum bronze alloys than any other copper-base alloy system. The reader is referred to the Bibliography following this chapter for an introduction to this literature. **Figure 14-22** shows the grain structure of a C95700 high manganese aluminum bronze alloy.

Finally, **Fig. 14-23** shows a microstructure of a C96400, 70-30 copper-nickel alloy. This photomicrograph shows nickel-rich primary dendrites (white), vein-like nickel silicides and particles of chromium silicide that are precipitated in the interdendritic regions.

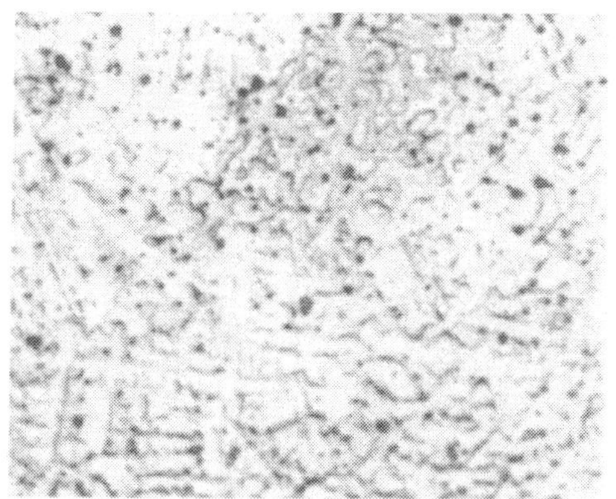

Fig. 14-20. "Coring" in a sand-cast leaded red brass. Ferric chloride etch, 50X.

Fig. 14-21. As-cast (sand) microstructure of alloy C86300 (63% Cu, 25% Zn, 3% Fe, 6% Al, 3% Mn). Black specs are iron-rich particles of intermetallic phase, 100X.

Solidification (Freezing Ranges)

Because the solidification range varies greatly in copper alloys, it is convenient to divide copper alloys into three groups to discuss solidification and the development of their structure: short, long and intermediate freezing ranges (also called narrow, wide and medium). Two of these groups represent extremes in the solidification, or freezing range, of copper alloys; thus, some of these alloys fall into an intermediate classification. There is more information in Chapter 16, Solidification of Metals and Alloys.

Group I (Short Freezing Range)—Copper-base alloys that solidify by skin formation and have a very short freezing range fall into this group. In the phase diagram, these alloys have a very small "mushy" area or distance between the liquidus and solidus lines. This group includes pure copper in its many forms, the near-eutectic alloys such as many of the aluminum bronzes, and certain yellow brasses.

Solidification of this group of alloys begins with the formation of numerous small crystallites at the mold wall. The crystallites that are most favorably oriented begin to grow rapidly, both sideways and inwards, linking up with their neighbors and forming a continuous skin or shell of solidified metal, called the solidification front. **Figure 14-24** depicts this mode of solidification. This front is the boundary between liquid and solid metal and continues to move toward the center of the casting section.

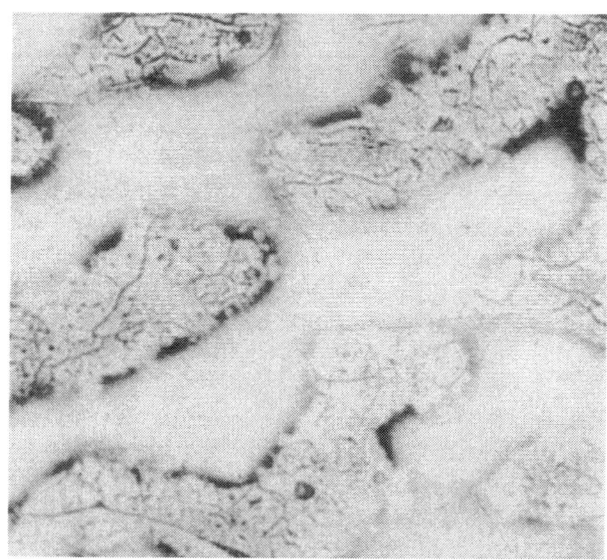

Fig. 14-22. Overall grain pattern in MnAl bronze C95700 alloy after etching in alcoholic ferric chloride, 20X.

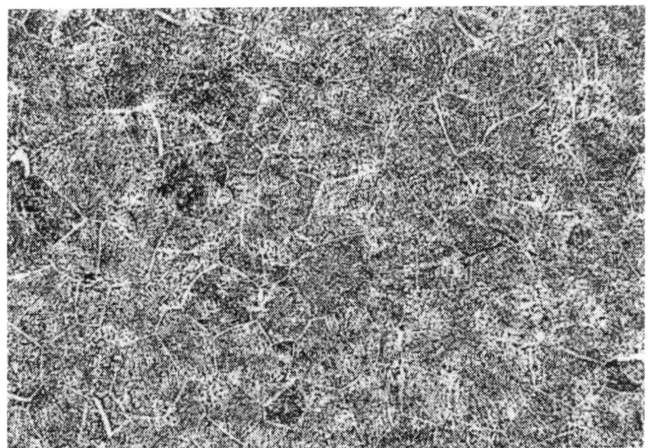

Fig. 14-23. Microstructure of 70/30 Cu-Ni (C96400) alloy containing 0.26% Si and 1.4% Cr. Shows Ni-rich primary dendrites (white). Vein-like Ni silicides and particles of chromium silicide are precipitated in the interdendritic regions. Etched in alcoholic ferric chloride, 500X.

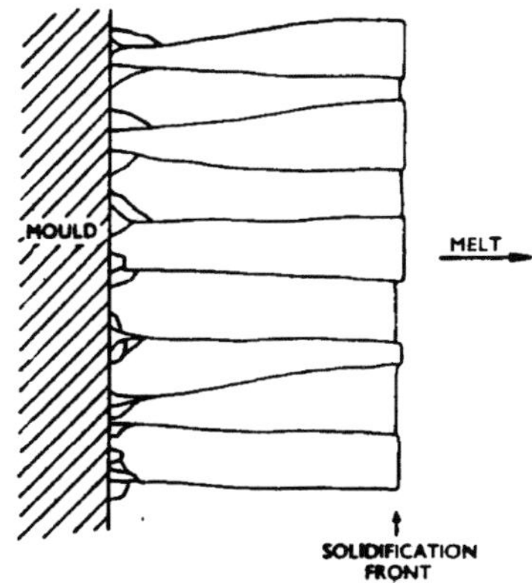

Fig. 14-24. Solidification front in copper of high purity (short freezing range).

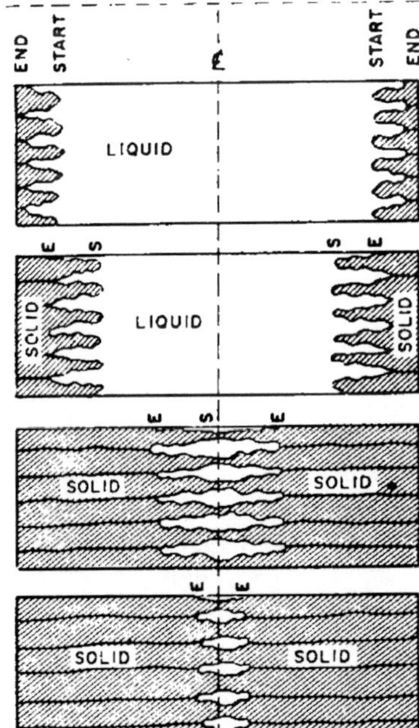

Fig. 14-25. Stages in the (short) freezing of copper alloys of low alloying content. Freezing by skin formation.

Typically, only the pure coppers and alloys of exact eutectic composition solidify in this manner. Copper-base alloys with very low alloying element content or impurities solidify in the same manner. However, instead of having a smooth front, the solidification front is somewhat roughened due to a tendency toward dendritic growth of the crystals. This mode of solidification is seen depicted in **Fig. 14-25**.

The main characteristic of all the copper-base materials in Group I is their very short freezing range (or none at all). The macrostructures of castings made from these materials usually exhibit columnar crystals. Macrostructure means that these structures can be seen with very little or no magnification.

Group II (Long Freezing Range)—The copper-base alloys that fall into this group solidify in a totally different manner. Again, looking at a phase diagram of these alloys shows that the alloys have a "wide" mushy area or longer distance between the liquidus and solidus lines. Thus, a long freezing range. Common copper-base alloys falling into this group are tin bronze, leaded bronze, red brass, semi-red brass and nickel silver.

Figure 14-26 depicts the solidification pattern of the Group II alloys. Solidification does begin at the mold wall with the formation of numerous crystallites on the mold walls. Next, the growth of these crystals is retarded and can be stopped temporarily. The first crystals formed are considerably poorer in alloying elements than the liquid from which they solidified and this, in turn, causes a delay in growth.

As the liquid temperature continues to fall, a second batch of crystallites nucleates just outside the alloy-enriched area. This second crop of crystallites is also restricted, and a third batch begins to grow just beyond the enriched region, which now surrounds the second batch. This process continues until small crystallites have been nucleated throughout the casting.

Figure 14-27 shows these stages of freezing in the Group II alloys. The crystals closest to the mold walls grow slightly faster than those near the center of casting sections. This process of freezing can be compared to the setting of concrete. First, the concrete is liquid, then mushy and finally solid. With this in mind, it is easy to understand why the macrostructures of these castings show equiaxial crystallization.

Group III (Intermediate Freezing Range)—As mentioned earlier, some copper-base alloys fall somewhere between these two groups. In this intermediate mode of solidification, crystallites are again formed at the mold surface. These crystallites then begin to form dendrites that struggle to grow toward the center of the casting sections. **Figure 14-28** depicts this solidification mode. Sometimes, these spindly dendrites will continue to grow until they reach the center of the casting sections. In this case, the macrostructure of the casting will be wholly columnar.

On the other hand, they will frequently diminish, and the central region of the casting sections will solidify in a mushy manner typical of the Group II alloys. When this occurs, the macrostructure in the center of the casting sections will be equiaxial grains and the outer regions will be columnar. Note in **Fig. 14-28** that, in these alloys, there may be a liquid pathway that extends to the mold wall, almost until the end of solidification.

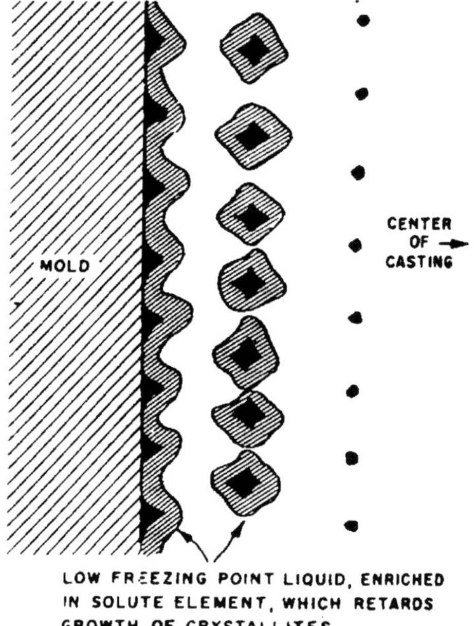

Fig. 14-26. Diagrammatic representation of restriction of crystal growth in the (long) freezing of copper alloys of high alloy content.

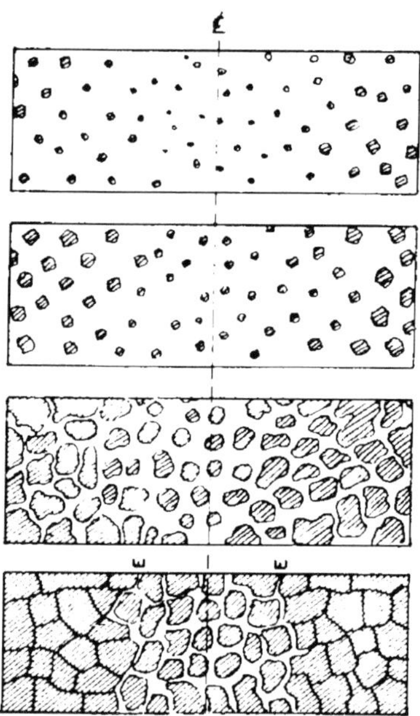

Fig. 14-27. Stages in the (long) freezing of copper alloys of high alloy content.

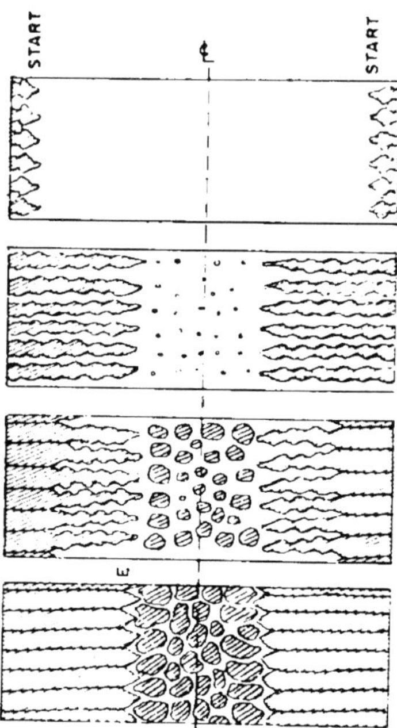

Fig. 14-28. Stages in the freezing of a copper alloy in the intermediate range.

Effect of Grain Refinement

The number of crystals nucleated in long freezing range copper-base castings is determined by the grain size. The number of nuclei present for crystallization in the melt (when it is poured and prior to pouring, to the onset of freezing) determines the grain size. Therefore, if it is possible to increase the number of nuclei present in the melt, the size of these grains can be decreased, and the casting's mechanical properties will be improved.

In the early days, the metalcaster would vibrate the mold shortly after pouring by hitting it with a sledge hammer. This would vibrate the mold and the new-forming casting inside. The theory was that this vibration would send shock waves through the molten metal in the casting and cause the dendrites that were forming to break into small pieces. Each one of these small pieces would then become a nuclei and form new crystals. Later, mechanical means were used to do the same job. These vibrational methods of "grain refinement" proved to be unreliable, and thus other means of grain refining were found.

Grain refinement of copper-base alloys has never been used extensively. However, methods of grain refining aluminum bronze, red brass and gunmetals have been developed using chemical grain refinement. Red brass can be grain refined with the addition of 0.05% zirconium to the melt, or about a 1% iron addition has also proven to produce some grain refinement. Aluminum bronzes and manganese bronzes can also be grain refined using iron additions of this order. However, iron is normally found in these alloys to start with, and these alloys will normally solidify with a fine grain structure.

The iron-free aluminum bronzes and brasses may be refined by the addition of 0.03% zirconium and 0.02% boron, or by the addition of boron alone. In the case of the iron-free aluminum bronzes and red brass castings, the grain refinement methods produce a useful improvement in strength. The effects of grain refinement on bronzes and gun metals are more complex.

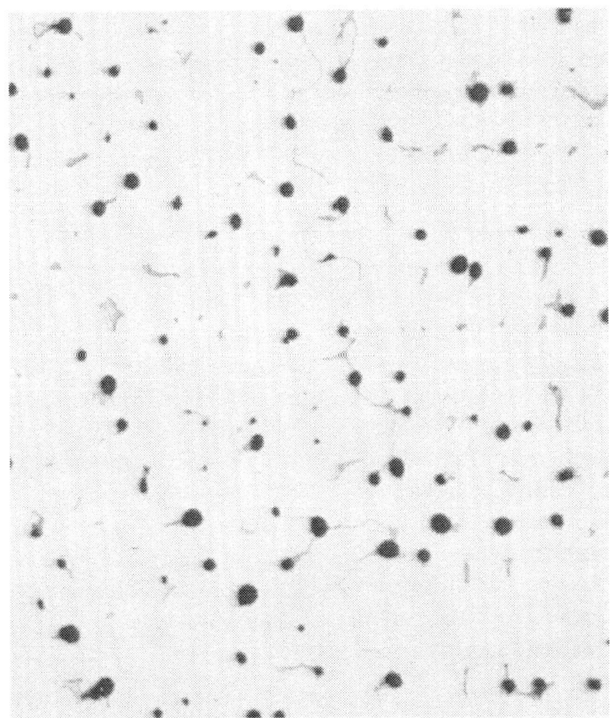

Fig. 14-29. Evenly distributed microshrinkage in C90500 alloy sample having an average density of 8.6 g/cc, 100X.

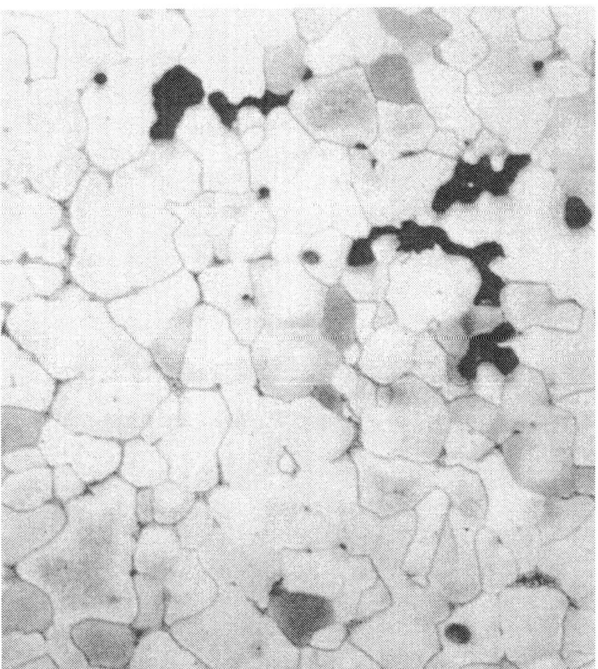

Fig. 14-30. Localized, interconnected microshrinkage in C90500 alloy sample having an average density of 8.75 g/cc. Despite its higher average density, this sample is much less pressure-tight than that shown in Fig. 14-29.

There is a downside to all of this. It is a well-known fact that the majority of bronze and red brass castings are not well fed during solidification, and the results of grain refining could prove disastrous. Grain refining, for reasons that are not fully understood, is liable to convert ordinary well-dispersed microshrinkage into layered microshrinkage or concentrated, interconnected microshrinkage. Should this occur, both the mechanical properties and pressure tightness will be greatly reduced. Therefore, the grain refinement of these copper-base alloys should be done very judiciously and with much forethought.

Figures 14-29 and 14-30 show the microstructure of a tin bronze (C90500), also known as gun metal. These two photomicrographs were chosen because they demonstrate what was discussed in the previous paragraph. **Figure 14-29** shows an evenly distributed microshrinkage that appears as black spots. This type of microshrinkage can be tolerated without adverse effects on mechanical properties or pressure tightness. However, **Fig. 14-30** shows localized and interconnected microshrinkage, the black areas, which can extend to the surface of a casting and allow leakage to occur. This photomicrograph also shows the grains and grain boundaries very well. It is in these grain boundaries that most impurities and "junk" in the alloy will end up when the casting has completely solidified. In many cases, it is this material that can hurt the mechanical properties of a casting. (This is true for most alloys, both ferrous and nonferrous.)

BIBLIOGRAPHY

There is a wealth of material available on the subject of nonferrous microstructures.The reader can contact the AFS library for more information on the subject.

Aluminum Casting Technology, 2nd Edition, American Foundrymen's Society, Inc., Des Plaines, IL (1993).

Aluminum Composites Now Feasible for Investment Castings; Doors Open to New Aerospace & Commercial Markets, *INCAST,* Investment Casting Institute (Dec 1990).

Bailey, A.R., Samuels, L.E., *Foundry Metallography,* Metallurgical Services Laboratories Ltd., Betchworth, Surrey, England, (1971).

Casting Copper-Base Alloys, AFS, Des Plaines, IL (1984).

Copper Casting Alloys, Copper Development Assn., New York (1994).

Heine, R.W., Loper, C.R., Rosenthal, P.C., *Principles of Metal Casting,* Second Edition, McGraw-Hill, New York (1967).

Jorstad, J.L., Reviewed and Written Material for this Book (1993).

Oswalt, K.J., Misra, M.S., "Dendrite Arm Spacing (DAS): A Nondestructive Test to Evaluate Tensile Properties of Premium Quality Aluminum Alloy (Al-Si-Mg) Castings," *AFS International Cast Metals Journal,* American Foundrymen's Society, Inc., Des Plaines, IL (March, 1981).

Schleg, F.P., Personal Class Notes from American Foundrymen's Society, Inc., Cast Metals Institute (CMI) Nonferrous Metallurgy Courses and Workshops (1965-1992).

Sylvia, J.G., *Cast Metals Technology,* AFS, Des Plaines, IL (1990).

Tuttle, B.L., Principles of Thermal Analysis for Molten Metal Control.

Gating Practice

15

Before delving into the subject of gating practice, the definition and history of gating will be detailed. Gating is the means by which the molten metal enters the mold cavity. Simply put, the gating system is a series of channels through which the molten metal will flow from the lip of the pouring device to enter the mold cavity(s). The proper design of the gating system is very critical to the success of producing a quality metalcasting.

Before the late 1940s, the design of a gating system was more or less up to the individual molder if hand molding, or up to the individual patternmaker if the gating system was to be mounted on the pattern plate. Many times, the success of the gating system came only after much trial and error and scrapped castings. At that time, gating was considered an "art."

During and right after World War II, the aluminum metalcasting industry became involved with producing castings for the aircraft industry. The aluminum foundries were having trouble meeting the stringent quality requirements set by the aircraft industry for these castings. The big problem was the entrainment of dross in the castings. These foundries came to the American Foundry Society (AFS) seeking their help in overcoming this dross inclusion problem. The AFS contracted with the Battelle Memorial Institute to study this problem and develop a gating system that would help to alleviate this problem. This was the first known effort to study how to design a gating system to help produce high-quality metalcastings.

Battelle started by studying how water flowed through various Plexiglas gating systems. To enhance this study, they took high-speed movies of the water flowing through the various designs. When they determined that they had a gating system designed to solve the problem, they made sand molds with those gating systems for testing. Both horizontal and vertical molds were made.

Molten aluminum 535 Al-Mg was poured. After the castings had been shaken out and cleaned, they were x-rayed. The same procedure was followed for those gating systems that showed poor flow characteristics. The x-rays backed up the same patterns of flow as was shown when water was poured in the Plexiglas molds. The castings poured using the good gating system passed radiographic examination and the others failed. This gating system became known as the Battelle System and/or the AFS System.

Of course, skeptics said, "that was water flowing in Plexiglas molds, or that was aluminum being poured and we pour iron or some other alloy." In the 1980s, the Canadian Bureau of Mines and Energy Resources, (now known as CANMET) in Ottawa followed up the Battelle work by using the same gating design and pouring carbon steel alloy into sodium silicate-CO_2 molds. They filmed the flowing carbon steel using real-time radiographic techniques. Again, they used both good and bad gating designs based on Battelle's work. They obtained the same results, as did Battelle. (Both Battelle's and the Canadian work are available on videotape from the AFS.)

Today, computerized programs to design a proper gating system can be used. Programs available will actually show the mold cavity(s) filling and predict the amount of turbulence in the mold cavity(s). These programs, as well as the work done earlier, are based on the basic principles of "fluid flow." These principles have been challenged by metalcasters, many times, in the design of their gating systems...and so many times they have lost.

FLUID FLOW OF MOLTEN METALS

A brief and basic review of the laws of fluid flow and how these laws affect the flow of molten metals is in order. The main thing to remember is that molten metal is a fluid and will follow the same laws as do water or air. One can study designs of heating, ventilating and air-conditioning systems for examples of fluid flow principles at work.

Bernoulli's Theorem

The basic law of hydraulics or fluid flow, known as Bernoulli's Theorem, gives relationships between the factors that influence the fluid behavior of molten metal. One can apply this theorem to gating systems with satisfactory results. The theorem states that, at any location in a system, the sum of potential head (head = height above a given reference plane), pressure or metallostatic head, and velocity or kinetic head equals a constant.

To derive Equation 1, Bernoulli's Theorem can be summed up in terms that state: *the sum of the potential energy, the velocity energy, the pressure energy, and the frictional energy of a flowing liquid are equal to a constant.* When energy losses occur due to turbulence and friction, this loss must also be considered.

The factors involved in Bernoulli's Theorem are shown schematically in **Fig. 15-1.** The potential energy is naturally a maximum at the highest point in the system, or at the top of the pouring basin. As the metal passes through the gating system, the potential energy is rapidly changed to kinetic or velocity energy and pressure energy. Once flow is established, the potential and frictional heads are virtually constant; the velocity is high when the pressure is low and vice versa. While metal is flowing, there is a constant loss of energy in the form of fluid friction between the metal and the walls of the gating system. Though Bernoulli's' theorem does not consider it, there is a heat loss, which reduces the solidification time and the fluid life of the molten metal. The theorem can be expressed as:

$$wZ + wPv + \frac{wV^2}{2g} + wF = K \tag{1}$$

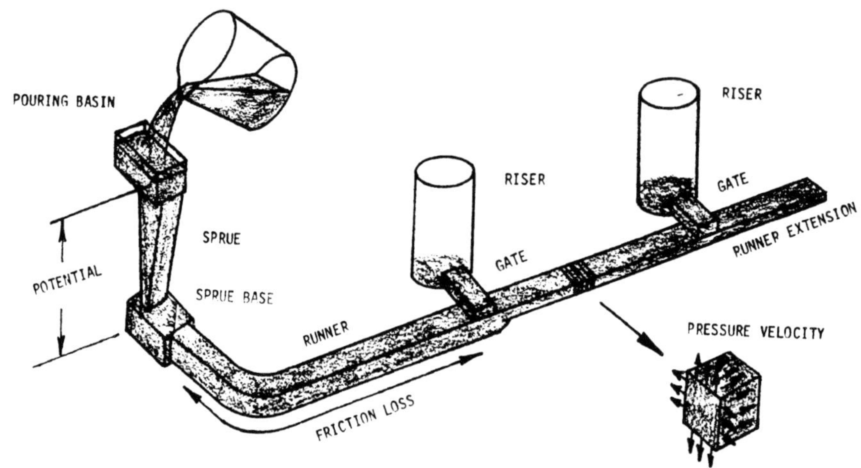

Fig. 15-1. Application of Bernoulli's Theorem to a gating system. [After Wallace & Evans]

where w = total weight of fluid flowing (lb or kg)
Z = height of liquid (ft or cm)
P = static pressure in liquid (lb/in.^{2} or kg/cm^2)
v = specific volume of liquid (ft^3/lb or cm^3/kg)
g = acceleration due to gravity (32 ft/sec^2 or 975.35 cm/sec^2)
V = velocity (in. or ft/sec or cm/sec)
F = frictional losses (ft or cm)
K = a constant

When the equation is divided by w, all the terms have the dimensions of length, and may be considered to represent:

Potential head, Z
Pressure head, Pv
Velocity head, $V^2/2g$
Frictional loss of head, F

It is most important to be able to establish the proper flow system rapidly. This means that when liquid metal enters the sprue, it should be flowing under conditions that resemble those present when a full flowing system has been established. In order for Bernoulli's Theorem to prove true, the entire system (in this case, the gating system) has to be completely filled with molten metal.

The application of Bernoulli's Theorem can be illustrated in the horizontal gating system shown in **Fig. 15-1**. This theorem aids in explaining the aspiration of gas into the flowing stream of molten metal in some locations in the gating system, and the variation in molten metal pressures and velocity of the flowing molten metal.

The potential energy is at maximum when the molten metal enters the pouring basin **(Fig.15-1)**. Potential energy quickly changes to kinetic energy and pressure energy as the flow of molten metal is established. When this flow is established, the potential and frictional heads are relatively constant. It is at this point that the kinetic and pressure energy become significant considerations. Therefore, when the velocity is high, the pressure is low and vice versa.

Behavior of the flowing molten metal stream in a horizontal runner can be explained with this theorem. If potential energy and friction losses are relatively constant, what happens when the kinetic and pressure energy are considered? If the runner in **Fig. 15-1** maintained the same cross-sectional area for its entire length, the velocity at the first gate would be high and the pressure would be low. Just the opposite would be true at the second gate. This situation would cause unequal volumes of molten metal to flow through the gates; that is, more molten metal would flow through the gate furthest from the sprue. However, when the runner cross-sectional area is reduced by an amount equal to the cross-sectional area of the first gate as it is passed, as shown in **Fig. 15-1**, the velocity and pressure energy are equalized. This condition produces the same or near equal volume of molten metal flow through both gates at nearly the same time.

When next attending a basketball game, look up into the rafters and study this theorem. Note how the heating and cooling engineer has designed the "air gating system." The main duct has been reduced in cross-sectional area as each outlet or vent is passed, in order to get equal volumes of air out each vent.

When these principles are used, turbulence can be reduced by avoiding sharp corners, e.g., sharp changes in cross-sectional area. However, as was emphasized earlier, all portions or parts of the gating system must be filled with flowing molten metal. Under pressure that is slightly higher than atmospheric pressure (or air), mold and core gases can be aspirated into the molten metal as it flows. These entrapped gases can be the source of casting defects. **Figure 15-2** illustrates the effects of pressure head and change in velocity based on how the molten metal is forced to change direction.

This discussion of Bernoulli's Theorem may seem arbitrary at this point; however, this is because its meaning requires an understanding of the second major principle or law of fluid flow, namely the Law of Continuity.

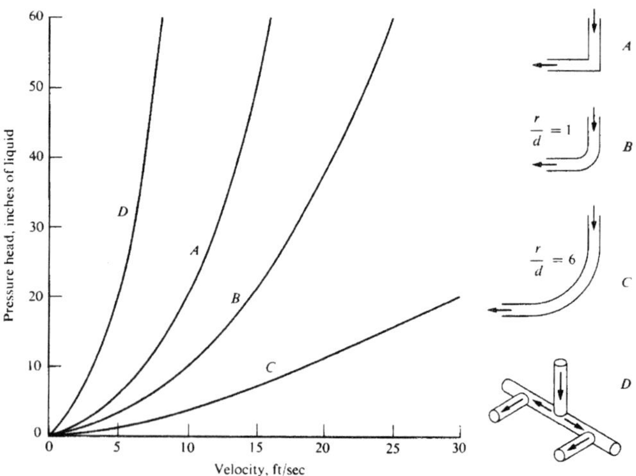

Fig. 15-2. Effect of pressure head and change in gate design on velocity of metal flow.

Law of Continuity

The second fundamental principle, basic to the flow of molten metal in a gating system, is a mass balance equation commonly called the "Law of Continuity." This principle can be stated as follows: *the volume of molten metal flowing in a full channel is the same at all points in the channel.* More simply stated, what goes in one end must come out the other. Mathematically, this is stated as:

$$V = a_1v_1 = a_2v_2 = a_3v_3 \text{ etc.} \quad (2)$$

where V = volumetric flow rate (in.^{3}/sec or cm^3/sec)
a_1 = cross-sectional area at point 1 (in.^{2} or cm^2)
v_1 = velocity at point 1 (in./sec or cm/sec)
a_2 = cross-sectional area at point 2 (in.^{2} or cm^2)
v_2 = velocity at point 2 (in./sec or cm/sec)
etc.

This means that if a given volume of molten metal flows past one point of the gating system at a given period of time, then the same volume or quantity must flow past another point downstream in the gating system in the same period. If the cross-sectional areas of the two points are the same, then the velocity of the metal at the two points will be equal. On the other hand, if the cross-sectional area of the runner is changed, or reduced, then the velocity of molten metal will increase.

Consider a runner in which the cross section is changed. **Figure 15-3** shows a runner that begins with a cross-sectional area of 2 in.^{2} (12.9 cm^2) and is reduced to 1 in.^{2} (6.45 cm^2). The volume or quantity of molten metal flowing in this system is 15 in.^{3}/sec (245.85 cm^3/sec.) or 3.75 lb (1.68 kg) of gray iron per second. To find the velocity at Point 1, divide the volume by the cross-sectional area and find that the velocity is 7.5 in./sec (19.05 cm/sec). At Point 2, the same volume of gray iron is flowing, but now the cross-sectional area is 1 in.^{2} (6.45 cm^2), and the velocity has increased to 15 in./sec (38.1 cm/sec).

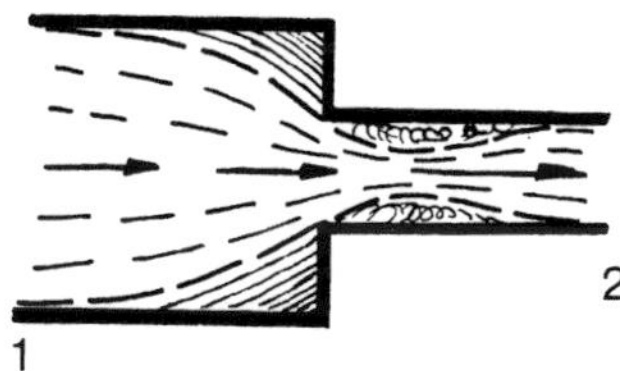

Fig. 15-3. Effect of change of area of channel section, where $V = a_1v_1 = a_2v_2$.

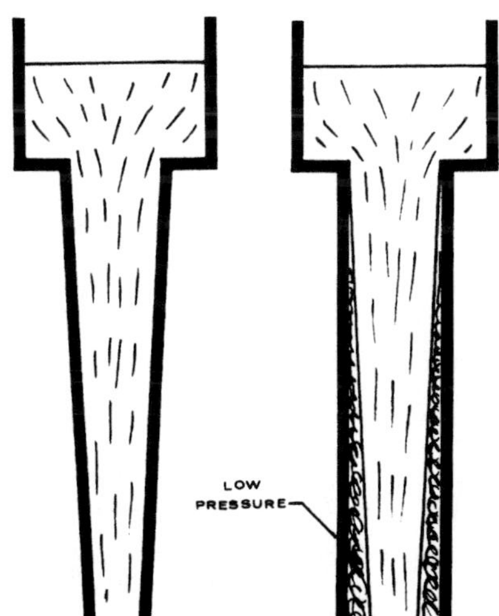

Fig. 15-4. Fluid flow in tapered (left) and straight (right) sprues.

As can be seen from the example, the velocity of the molten metal can be controlled in a gating system. To slow the velocity, increase the cross section of the runner; to speed up the velocity, decrease the cross-sectional area. Using Equation 2, if any two of the three parts are known, the third can be found. That is, if the cross-sectional area and the volume of molten metal are known, the velocity can be found, etc.

Without realizing it, most people have made use of the Law of Continuity. When playing in the yard with the garden hose, trying to get friends wet, the stream of water coming out of the end of the hose just wouldn't quite make it to where they stood. If there was no nozzle on the end of the hose, a thumb stuck over the end could *reduce the cross-sectional area* of the opening. If successful in reducing this opening enough, the stream of water would reach the target.

Based on the Law of Continuity, the *velocity and pressure in the water stream were increased* and carried the water to where it was aimed. This same principle is what allows firefighters to stand out of harm's way while battling a fire.

Velocity

So far, the term "velocity" has been used in studying Bernoulli's Theorem and the Law of Continuity. Velocity is the speed at which the molten metal flows past a given point in the gating system and is usually measured in in./sec (cm/sec). The initial velocity in a gating system is determined by the following formula:

$$v = \sqrt{2gh} \quad (3)$$

where v = velocity (in./sec or ft/sec or cm/sec)
2 = a constant
g = the acceleration due to gravity (32 ft/sec^2 or 386.4 in./sec^2 or 981 cm/sec^2)
h = the distance of the fall (in. or ft or cm or m)

It should be remembered that this equation is theoretical and that frictional losses are not considered. Even air can impart some frictional loss on a falling stream or object. This loss is rather insignificant, and its effect will be ignored in these calculations.

In this case, the height of the fall is from the lip of the pouring ladle or device to the point at which the fall is stopped and made to change directions. In the case of bottom-pour ladles, it would be from the top surface of the molten metal in the pouring ladle to where the straight fall is ended. Thus, if the person pouring the molten metal is aiming the stream directly into the sprue, he or she has control of this initial velocity by simply raising or lowering the lip of the pouring ladle. This initial velocity affects the velocities in the rest of the gating system.

Since the Law of Continuity states that the volumetric flow rate will remain constant, it follows that the cross-sectional area of a free-falling stream of molten metal will decrease as the velocity increases. Quite possibly, if this stream were allowed to fall far enough, it would break up into droplets of molten metal. This accounts for the tapered shape of a free-falling stream, as seen in the two sprues shown in **Fig. 15-4**. Actually, the falling stream shown in the right-hand sprue would not have straight sides as shown, but would be "flopping" from side to side. This fact was demonstrated by the work done at CANMET (mentioned earlier).

Momentum

There is one final law of physics that pertains to fluid flow, and it must be considered if gating systems are to be scientifically designed. Sir Isaac Newton's first law states: "Bodies in motion in a particular direction tend to stay in motion in that direction unless acted upon by opposing forces." In other words, a flowing stream of molten metal, which is flowing in a particular direction, does not want to change its direction easily. To make it stop or change direction, a force has to be applied. This tendency of a fluid to continue to move in a straight line and its opposition to change direction is a result of "momentum." Experienced drivers and highway engineers are well aware of momentum; thus, they design gradual curves in highways rather than sharp, 90-degree corners.

Consider Newton's first law in the design of the gating system. **Figure 15-5** shows three situations that can be found in a gating system. Situation "B" shows a 90-degree turn. Note that the molten metal, as it flows, runs into an opposing force at the outside corner. This area is a high-pressure area, depicted by the closely packed small dashes. On the other hand, a low-pressure area is set up at the inside corner, depicted by the swirling lines. These two areas were set up because of momentum. The low-pressure area, in particular, can cause problems such as the aspiration of air and mold and core gases into the molten metal stream as it flows through this area. Another problem is that the sharp corner, at the low-pressure area, can be eroded away and the molding sand can become entrapped in the molten metal stream.

In **Fig. 15-5 "A,"** there is a smaller channel connected to a larger channel. Since there is higher velocity in the smaller channel, the momentum carries the flowing molten metal into the larger channel until a point is reached where the velocity is slow enough to overcome the momentum effects. This, then, will allow the stream of molten metal to "open up" and fill the larger channel. Again, a low-pressure area is set up in the inside corners of the larger channel. The same problem will be found here as was found in the situation "B" low-pressure area.

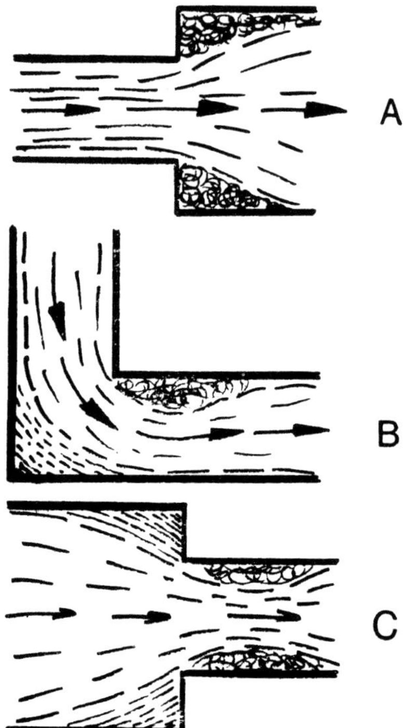

Fig. 15-5. Development of low pressure areas.

Figure 15-5 "C" shows both high and low pressure areas being set up, where the larger channel is connected to the smaller channel. Again, there will be aspiration problems in the low-pressure areas. The sharp corner of molding sand also stands a chance of being eroded away.

The problems illustrated in **Fig. 15-5 A, B and C** can be overcome by following the design principles used by the highway engineer: streamline the gating system. Use curved corners to help the molten metal make the turn without undo problems. Use gradual changes in cross-sectional areas in channels or runners. In other words, do everything to overcome the effects of momentum in the gating systems.

Frictional Forces

As mentioned earlier, frictional forces do play a role in the flow of molten metal through the gating system. These forces cause a reduction in the velocity of the flowing stream of molten metal. Some areas where the effects of frictional forces can reduce velocity are:

1. Velocity loss at the entrance of the sprue.
2. Velocity loss due to friction of the walls of the sprue.
3. Velocity loss due to the change of direction at the base of the sprue.
4. Velocity loss due to the friction of the runner walls.
5. Velocity loss due to filters, if used.
6. Velocity loss due to turns in runners and ingates.

These frictional losses may be considered as individual components; however, since the frictional losses are essentially constant, once flow is established, the overall system loss is usually sufficient. Calculating frictional losses is beyond the scope of this text; however, the literature contains numerous examples that can be studied and applied to frictional losses.

Reynolds Number

It is very important to deliver the molten metal to the mold cavity(s) as quickly as possible, with the least amount of turbulence. If the design of the gating system is not correct, there will be varying degrees of turbulence in the flow. If the design is correct, there will be a peaceful and tranquil (laminar) flow of molten metal in the gating system.

In laminar flow, the liquid particles follow well-defined, nonintersecting paths, without turbulence. In turbulent flow, the paths of the liquid particles cross and recross one another in an intricate pattern of interlacing lines and eddies. The degree of turbulence can vary considerably from slight to violent, depending on the conditions of flow.

Figure 15-6 illustrates laminar flow, turbulent flow and severely turbulent flow.

Under each illustration is a number known as the Reynolds number (usually, N_R). The Reynolds number is a dimensionless number that is used to characterize the flow of liquids. Since molten metal is a liquid, we can use this number to determine the condition of flow in our gating systems. This number is obtained by using the following formula:

$$N_R = v_d / \nu \quad (4)$$

where v = the velocity of the liquid, in./sec or cm/sec
d = hydraulic diameter of the channel, cm or in.
ν = kinematic viscosity of the liquid, cm^2/sec or $in.^2/sec$

In the case of channels that are not round, "d" is taken to equal four times the hydraulic radius or:

d = 4 x cross-sectional area / perimeter of cross section

The representative values of the kinematic viscosity of various metal and water are given in **Table 15-1.**

Experimental work has shown that for an N_r of 2,000 or less, flow is invariably smooth or laminar. **Figure 15-6-top** illustrates the laminar flow in a runner. When the N_r is greater than 2,000, but less than 20,000 (**Fig. 15-6-middle**,), there is turbulent flow. Finally, when the N_r exceeds 20,000 there will be severe turbulence, as illustrated in **Fig. 15-6-bottom**. The N_r for most practical gating systems is considerably greater than 2,000, therefore, there will some measure of turbulence.

On closer study, **Fig. 15-6-middle** shows that when the N_r is smaller than 20,000, there is an undisturbed boundary of flowing molten metal along the surfaces. The turbulence is confined within this boundary, which also will prevent the aspiration of air and mold gases into the turbulent flow where it could be entrained. Also, erosion of molding sand from the surface of the runner is reduced or eliminated.

However, when N_r exceeds 20,000, the surface layer of the molten metal stream is ruptured, and severe turbulence occurs as illustrated in **Fig. 15-6-bottom**. When these conditions exist, slag and dross formation can occur. Along with air, mold gases, slag and dross can be entrained in the severely turbulent flowing stream. In the case of sand molds, the chances of sand erosion is greatly increased.

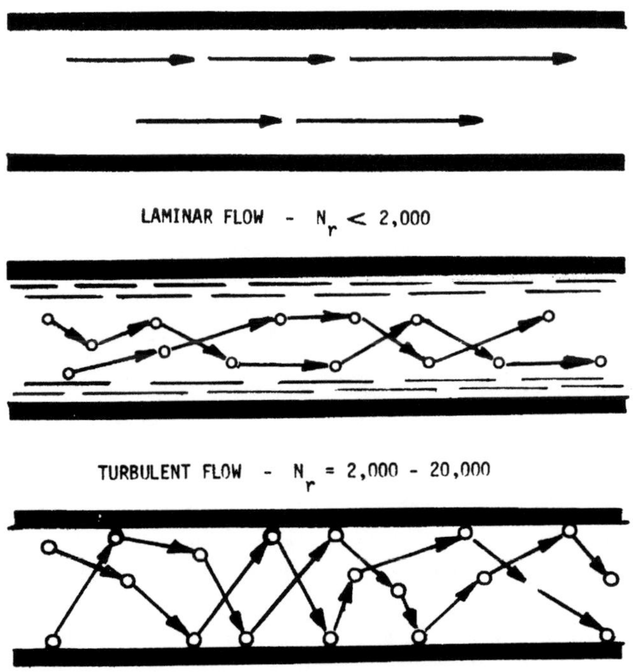

Fig. 15-6. Relationship between Reynolds Number and flow characteristics.

Table 15-1.
Values of Kinematic Viscosity

	Density lb/in.³	Kinematic Viscosity in.²/sec
Water	0.036	0.00155
Aluminum	0.097	0.00197
Magnesium	0.063	0.00124
Copper	0.282	0.00062
Gray Cast Iron	0.250	0.00070
Steel	0.280	0.00068

The N_r can be used to predict the amount of turbulence in the gating system, and with this prediction in mind, a gating system can be designed accordingly. Regardless of the alloy being poured, it is advisable to design a gating system that will minimize turbulence. This is especially true of aluminum alloys, magnesium alloys, some copper alloys and high-alloy steels. To achieve true laminar flow would, in most cases, require a gating system that would be too large and very expensive.

Although the N_r is not used directly in the design or calculation of gating systems, its ability to predict turbulence is a very useful concept and is the basis for many of the equations and other relationships in the gating design.

Fluid Life (Fluidity)

Fluid life, or fluidity, is another important variable relating to the flow of molten metal in the gating system and mold filling characteristics. To the metalcaster, this does not mean the reciprocal of viscosity. Fluid life tests can be poured either using the standard fluidity spiral or the serpentine-style mold. **Figure 15-7** shows a drawing and the dimensions of the standard fluid life spiral. The foundry may want to design its own fluid-life-testing device. One precaution for using these testing devices is that the mold should be made out of the same molding process used to pour the castings for which fluidity is being checked. In other words, if the molten metal being checked is to be poured in nobake molds, then the test mold should be made of nobake sand. These tests need not be run for every heat poured, but can be used as a standard.

Both the metal alloy and mold properties affect fluid life. Metal alloy properties are:

1. Superheat
2. Alloy composition
3. Alloy viscosity
4. Surface tension
5. Surface oxides
6. Absorbed gas films
7. Suspended inclusions
8. Inclusions precipitated during cooling
9. Metallostatic pressure

Superheat and alloy composition, are the two essential factors affecting fluid life. It is obvious that a given molten metal alloy heated to a higher superheat will have a longer period in the mold before it freezes; therefore, it will flow farther and fill the finer details of the mold than the same alloy poured at a lower temperature. The same thing is true of ordinary pancake syrup. It will flow faster when it is heated than it will at room temperature.

The manner in which alloy composition effects fluid life depends on the freezing characteristics of the alloy and the chemical elements that make up that alloy. The best fluid life is observed in pure metals or eutectic alloys. In addition, those alloys having narrow freezing ranges will have better fluid life because they have little or no mushy region during the solidification process.

An alloy having a long or wide freezing range will show poorer fluid life. This is characterized by interlacing dendrites and, in some cases, many islands of solid material, surrounded by liquid near the freezing temperature. This mushy condition during flow can be seen in the fluid-life test mold.

Certain chemical elements when added to different alloy families can help increase their fluid life. Silicon is such an element. When silicon is added to carbon steel and aluminum alloys, their fluid life is improved. On the other hand, phosphorus when added to gray cast irons and certain copper-base alloys will increase their fluid life. However, be sure to check with the foundry metallurgist before adding any chemical element to the metal alloy(s) to improve fluid life.

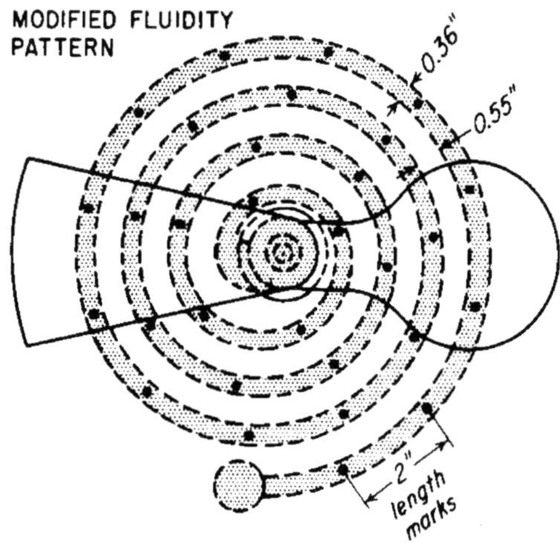

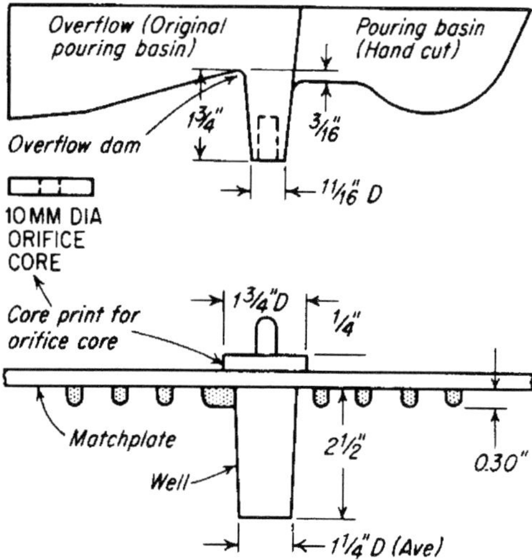

Fig. 15-7. Fluidity spiral pattern, a modification of the fluidity test of Saeger and Krynitsky. [From Porter & Rosenthal]

GATING SYSTEM REQUIREMENTS

Many factors must be controlled if a good metalcasting is to be obtained. For example, the soundness and cleanliness, externally and internally, of castings can be affected by the way the molten metal flows and enters the mold cavity(s) and solidifies. In order to determine the correct design of a gating system, one needs to know the flow and solidification characteristics of molten metal.

A gating system is a network of channels through which the molten metal flows, from the pouring lip of a ladle or pouring device to the innermost areas of the mold and, when necessary, into reservoirs of molten metal called risers. (Risers will be discussed in more detail in Chapter 17.) In other words, it is a system of delivery: an interrelated and complete assembly of channels, from pouring basin (or cup) to sprue to runner, ingates and riser, that ensures complete filling and compensatory feeding of the casting cavity with sound, clean metal to form a usable casting.

In this process, yield (weight of metal that remains a usable casting compared to the weight poured, including gating and risering systems) is of importance. The quality of the metalcasting should determine how good the gating and risering system need to be, and thus how much metal is needed in the gating and risering system.

An ideal gating system for the production of quality metalcastings must meet the following criteria:

1. The metal should flow through the gating system with a minimum of turbulence in order to avoid the oxidation of metal, the entrainment of air, the aspiration of mold gases, and the inclusion of undesired matter because of mold or core erosion, slag or dross entrapment.
2. The metal should enter the mold cavity so that it will produce proper thermal gradients with the surrounding mold surfaces, and also within the solidifying metalcasting, so that directional solidification takes place toward the source(s) of feed metal (risers).
3. The gating system must be large enough to accommodate the casting to be poured, and yet not be excessive in size, so that proper casting yield is obtained for the quality of metalcasting required.
4. The gating system should be designed to take the "art" out of pouring, especially if manual pouring is used. Pouring technique varies markedly between individuals, and may even vary considerably for a given individual. In short, the gating system should be designed so that the entire gating system can be filled very quickly and then kept full during the entire pour.

POURING PRACTICE

Before discussing gating system design, the means by which the molten metal gets to the mold and how it is poured into the gating system should be discussed. This is the most critical part of producing a quality metalcasting and the one operation in the foundry that receives minimal attention.

Manual Pouring

For all practical purposes, when manual pouring is being used the gating system begins with the individual doing the pouring. As learned earlier, there are fluid flow principles that will control the flow of molten metal into the mold cavity(s). These principles are governed by onc important factor: *the gating system has to be completely filled with flowing molten metal.* If the pourer cannot accomplish this essential task, it is either the fault of the individual, the gating system design or a poorly maintained pouring ladle. The pourer must be able to get the lip of the pouring ladle as close to the pouring basin as possible. **Figure 15-8** shows a manual-pouring ladle on which the pouring lip has been extended to help the pourer keep it close to the pouring basin.

Another type of pouring ladle is the bottom-pour ladle. **Figure 15-9** shows a bottom-pour ladle being used to pour a metalcasting. This type of pouring ladle is used mainly for pouring carbon steel and larger cast iron castings. In this case, the nozzle of the ladle is located over the pouring basin, and a stopper rod mechanism is actuated, which allows the molten metal to leave the ladle.

No matter which of these two types of pouring ladles are used, their proper maintenance is of prime importance. The lip of the ladle or the stopper rod and the nozzle must be well maintained. Also, the refractory linings in these ladles should be kept in good condition and properly preheated before use.

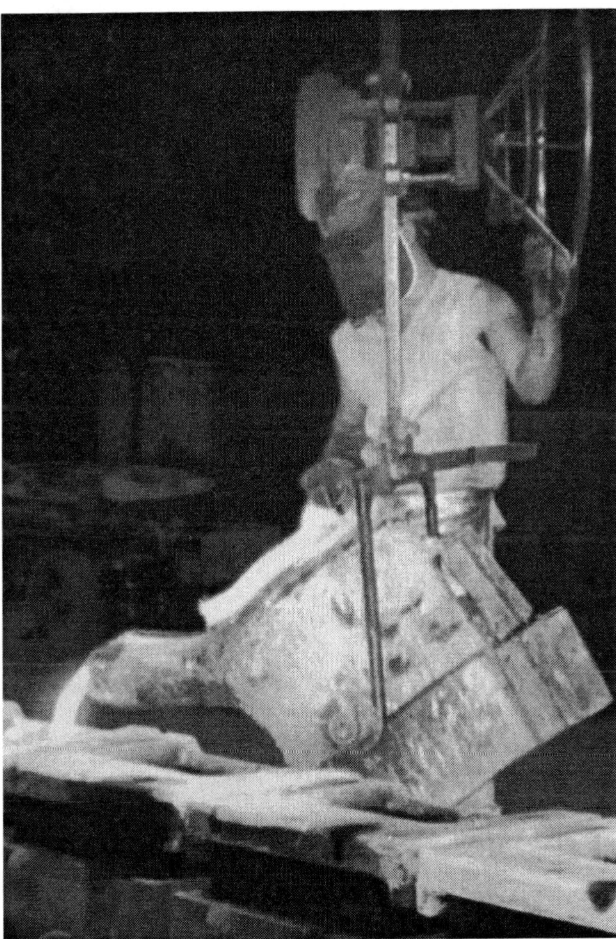

Fig. 15-8. Manual pouring ladle with extended lip. [Photo courtesy of Hunter Automated Equip. Co.]

The key to proper manual pouring is a well-trained pourer. If not properly trained, this individual can cause many castings to be defective (even if equipped with excellent molds and cores, molten metal of the highest quality, and a well-designed gating system). He or she should be given the proper safety equipment to wear, and should work in a well-lighted and ventilated area.

Automatic Pouring

Automatic pouring has replaced manual pouring in many high-production foundries. In the case of automatic pouring, the human element has been taken out of the operation. Some of the advantages of automatic pouring are as follows:

1. Close control of molten metal temperature is provided, usually by means of an electric induction holding furnace.
2. There is a means for dispensing a metered quantity (by weight, volume or sensing device) of molten metal within a time span corresponding to the required pouring cycle.
3. The molten metal is directed, without turbulence, into the sprue basin (incorporating some type of tracking technique if the mold is moving during pouring).
4. Sequential indexing of the pouring spout from one mold to the next is provided, with an indicator to signal the start of pouring.

Automatic pouring systems were seen as the practical answer to overcome the problems encountered with manual pouring, and during the past three to four decades, much research and development has gone into this method of pouring. There are dozens of patented systems for automatic pouring. Virtually all automatic pouring systems are based on the concept that the molds pass directly beneath the automatic pouring device, which is located adjacent to a holding furnace and the primary melter. In this process, mold size and sprue location are standardized.

Fig. 15-9. Bottom-pour ladle.

One of the more common types of automatic pouring is the pressure pouring device. **Figure 15-10** is a schematic of the basic design of such a device. The molten metal is transferred from either a holding furnace or the primary melter and poured into the automatic pouring device, where it is held, normally, in a channel induction furnace. Air pressure is then applied to the inside of the holding furnace and the molten metal rises in both the fill and pour tube. Since the pour tube exit is lower than that of the fill tube, molten metal can only leave the furnace through the pour tube. The weight or volume of molten metal dispensed from the automatic pouring device is set by the degree and duration of the applied pressure, the working level of the bath of molten metal, and the diameter of the pour tube.

Figure 15-11 is a schematic of a gravity-pouring device using a stopper rod and nozzle. As with the pressure-pour device, this pouring device also receives the molten metal either from a holding furnace or directly from the primary melter. It also has a channel induction furnace in which the molten metal is held until it is poured. In this case, the stopper rod is raised and the molten metal exits through the nozzle directly into the mold's gating system. A sensor (seen on the right in **Fig. 15-11**) stops the pour by lowering the stopper rod.

In both of these cases, the mold is brought to a brief stop for the actual pouring. In many situations, however, the molds keep moving on a conveyor and do not stop for the pouring process. Look again at **Fig.15-10,** which shows a moving pouring basin. At the

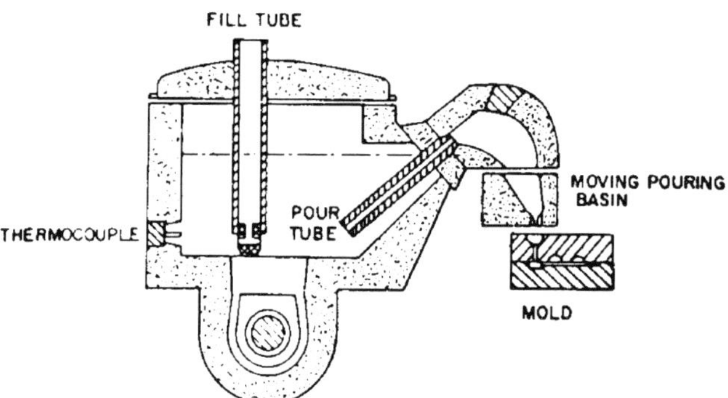

Fig. 15-10. Schematic view of channel-induction furnace with fill tube and pour tube, designed for automatic pressure-pouring.

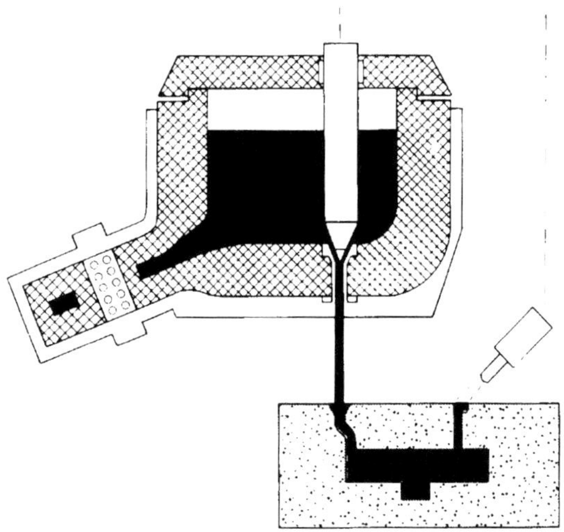

Fig. 15-11. One gravity pouring machine design.

beginning of the pour, this basin is indexed into position so that its exit orifice is directly over the sprue opening of the mold to be poured. The basin locks mechanically to the mold and travels along with the mold as the metal is being dispensed from the furnace. The length of the basin is such that its opening remains under the pour tube until enough molten metal has been received to fill the mold. The basin continues along with the mold until all of the molten metal in it has been dispensed into the mold. At this point, a limit switch disengages the basin and it is returned to its original index position.

As with all automatic mechanical devices, proper maintenance is an essential factor in doing their job correctly. Defective castings have been the result of poorly maintained automatic pouring devices. An important factor is maintaining the proper cross-sectional area of the nozzle and stopper rod or the pouring tube. These can have a buildup of slag and oxides, which will reduce their cross-sectional area.

GATING SYSTEM DESIGN

Two important factors in designing a gating system are how the gates are positioned and where the choke is located.

The choke is the smallest total cross-sectional area in the gating system and is normally the point of highest velocity. It could also be said that the choke is the valve in the gating system that controls the quantity or volume of molten metal flowing in the system. To increase the volume of molten metal flowing in the gating system, the cross-sectional area of the choke would have to be increased. It can be compared to wanting more water to come out of the faucet; it has to be opened wider. (The choke calculations will be discussed in more detail near the end of this chapter.)

The majority of gating systems used today are attached to matchplate patterns. Following this practice ensures that the gating system for a given matchplate is the same from mold to mold. The same holds true for permanent molds, diecasting dies and shell molding patterns. Very few gating systems are cut into the mold by hand. Besides slowing production, there would be as many different gating systems as there are molds!

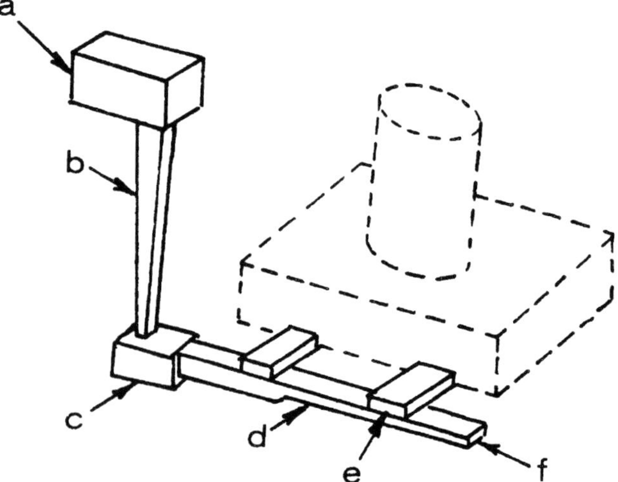

Fig. 15-12. Components of AFS horizontal gating system: a = pouring basin; b = sprue; c = sprue basin; d = drag runner; e = cope ingate; f = runner extension.

Horizontal Gating System

Starting with the design of gating systems as they are positioned in the mold, the first is the horizontal gating system as seen in **Fig. 15-12.** This gating system runs parallel to the parting line and is in the horizontal position.

Pouring Basin

The system should begin with the use of a pouring basin or cup to make it easier for the pourer to maintain a full system and provide the required flow of molten metal. A pouring basin is used to help separate the dross and slag from the molten metal before it flows through the gating system.

Figure 15-13 shows the design of a pouring basin and illustrates its use. It is recommended that the lip of the pouring ladle, when pouring, be in the position indicated in the sketch. If the ladle lip was on the end of the pouring basin opposite the sprue, the molten metal's momentum could carry some of the slag or dross over the dam and down the sprue, causing cold shot to form at the bottom of the sprue. (The sprue is the vertical passageway in the gating system and will be discussed further.) The dam, if designed properly, will keep the molten metal from immediately going down the sprue forming a pool of molten metal in the well of the pouring basin.

If necessary, the molten metal can be poured from the side of the pouring basin into the well. To help prevent slag or dross from getting close to the sprue opening, a skim core can be placed in the pouring basin. The bottom edge of the skim core is normally placed right above the dam. However, care must be taken to see that the cross-sectional area between the skim core and top of the dam does not become the "choke" in the gating system.

The well (or bottom) of the pouring basin, when filled with molten metal, serves as a "cushion" to absorb the kinetic energy of the falling stream of molten metal. This action can be compared to filling a glass with water. First, there is tremendous turbulence at the bottom of the glass. However, after a very short period, a heel of water is established in the glass and the filling action becomes more tranquil. The "heel" of water acts as the cushion.

Once the well is filled, the molten metal runs over the dam and begins to fill the sprue. The molten metal backs up into the basin once the sprue is filled. Depending on the skill of the pourer or automatic pouring device, this occurs very quickly. Once the basin is filled to the proper level, all that has to be done is to maintain the molten metal at that level.

Should the level of molten metal drop too low, a "bathtub" effect will occur. Fill a bathtub with water to a depth of about 4–5 in. (10–12 cm), pull the plug and watch the action of the water over the drain hole. When the water reaches a certain level, a whirlpool begins to form and continues until the bathtub is just about empty. The technical term for this whirlpool is vortex.

A vortex will form over the sprue should the molten metal level get too low in the pouring basin. Should this vortex form, it will suck air and slag or dross down the sprue and into the molten metal stream.

Basins can be molded into the cope half of the mold, or can be made separately out of green sand or chemically bonded sands. In the latter case, a few more inches or centimeters of head height is obtained. Separately molded pouring basins are normally used with large floor molds and pit molds.

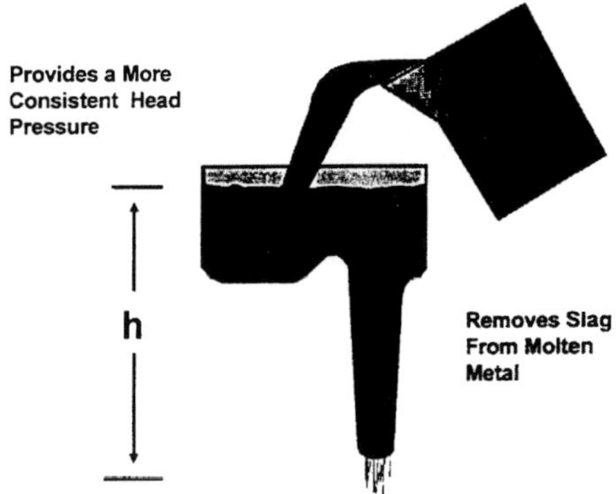

Fig. 15-13. Proper position for pouring ladle lip.

The junction of the sprue and the pouring basin should have a generous radius to avoid certain problems. One problem is that the molten metal stream could contract and become the choke in the gating system. Another problem is that gases can be entrained in the falling molten metal stream due to the low-pressure area set up at this location.

Pouring Cup or Sprue Cup

Many foundries will use pouring cups instead of basins. If designed and used properly, pouring cups can be used successfully. **Figure 15-14** is a sketch of a standard pouring cup. From the top, this type of pouring cup would look similar to a bull's-eye target. One incorrect pouring technique uses the phrase, "pour it down the hole." When this is done, the pourer (using a lip pour ladle) controls the initial velocity of the molten metal in the gating system by the distance that the lip is held above the pouring cup, causing the velocity in the gating system to vary, which is a poor practice. To correctly use the pouring cup, the stream of molten metal should be directed against the side of the pouring cup. However, now another problem comes into play: vortexing. Being circular in shape, the molten metal can very easily form a vortex in the pouring cup. Aiming the stream of molten metal is intended to reduce the distance the metal must fall and thus reduce the initial velocity; however, the round shape of the pouring cup causes a problem.

Figure 15-15 shows sketches of two recommended shapes for pouring cups. Pouring cup "A" is semi-round in shape. The flat part of the pouring cup will break up the tendency of the vortex to form when the pourer pours against the side of the pouring cup. Pouring cup "B" is rectangular. In this case, when the pourer pours against the side of the cup, a vortex will not begin to form. However, for

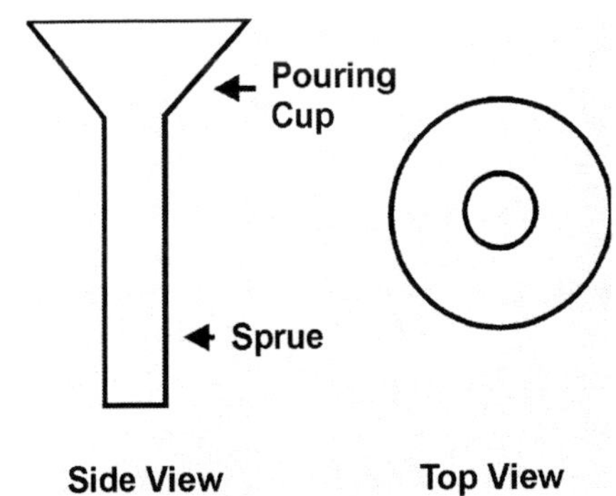

Fig. 15-14. Two views of the standard pouring cup.

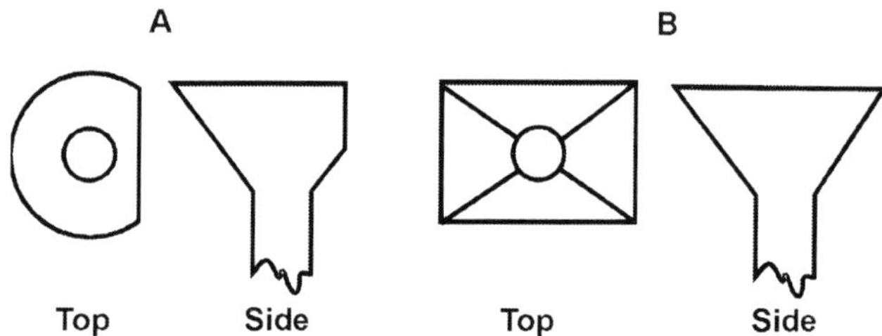

Fig. 15-15. Two recommended geometries for pouring cups.

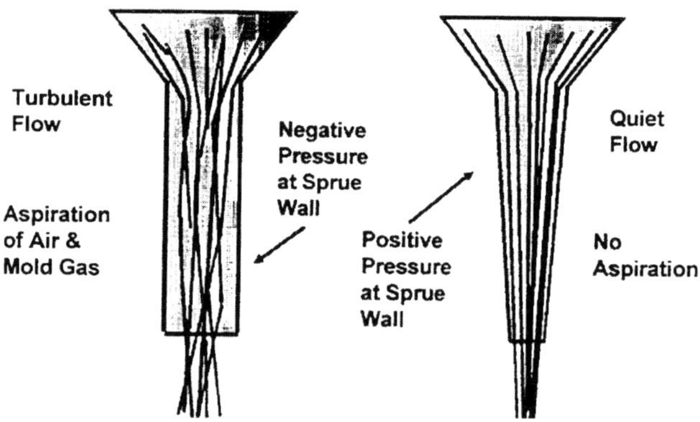

Fig. 15-16. Straight sprue vs. tapered sprue.

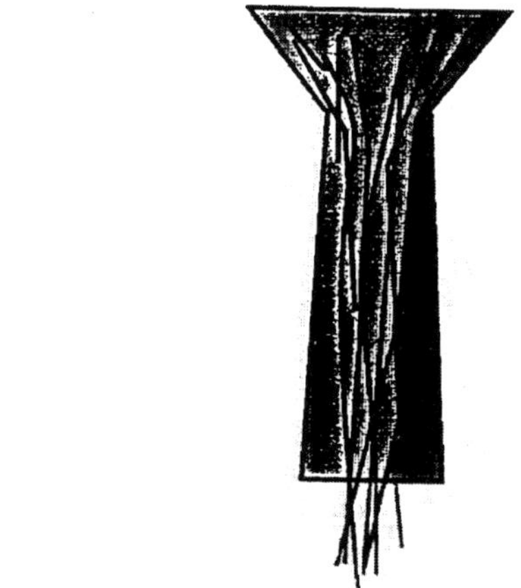

Fig. 15-17. Reverse tapered sprue.

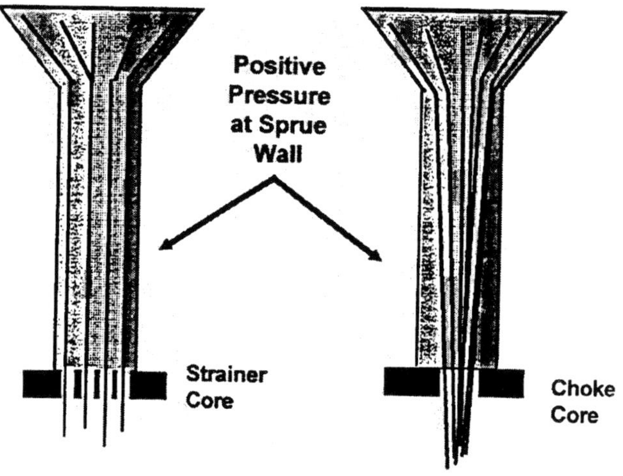

Fig. 15-18. Straight sprue with choke core or strainer core.

these two types of pouring cups to be used successfully, the molten metal should not be poured directly down the sprue. If asking "why not use a square pouring cup?" the answer is that in any geometric shape in which a circle can be inscribed and touch all of the sides of the pouring cup (such as a square), a vortex can form.

Sprue

A properly designed sprue is an extremely important feature in a good gating system. As the stream of molten metal moves down the sprue, it increases in speed and becomes smaller in cross section. When a sprue has the same cross section from top to bottom, a low-pressure area is set up between the molten metal and the sprue walls. Under these conditions, air and mold gases can be drawn into the stream of molten metal. This trapping of gases during pouring is known as aspiration and can be prevented by using a tapered sprue (with the smaller end located at the bottom of the sprue). **Figure 15-16** illustrates a falling stream of molten metal as it moves through either a straight sprue or a tapered sprue with the small end down.

The problem of aspiration in the sprue is accentuated when the sprue pin is attached to the cope side of the pattern plate. The sprue pin must have positive draft in order to allow the cope of the mold to be drawn off the pattern. In order to do this, the sprue should be wider at the bottom where it is attached to the cope pattern plate and narrower at the top of the sprue pin. This type of sprue in the mold is called a "reverse tapered" sprue and is illustrated in **Fig. 15-17.**

Strainer/Choke Core

Figure 15-18 illustrates two ways the aspiration problem can be eliminated. The reverse taper sprue and the straight sprue can be fooled into thinking they are a tapered sprue with the small end down. One way of doing this is to place a *strainer core* at the bottom of the sprue, as shown in the illustration. To clear up any misconception regarding the strainer core, it will only strain (out of the flowing molten metal stream) those particles that are larger than the holes in the strainer core. The strainer core acts as a choke core.

The strainer core also acts as the outlet of a funnel. No matter how much liquid is poured into a funnel, the amount to leave the funnel is dictated by the cross-sectional area of the funnel outlet. In this case, the excess liquid will fill and, if not controlled, overflow the funnel. The reason the pourers thought that the strainer core kept slag from entering the casting was that it "strained" the slag from the molten metal. What the strainer core actually did was to help fill the sprue very quickly, just like the funnel outlet did for the funnel, and this action kept the slag at the top of the sprue.

A negative characteristic of the strainer core is that it breaks up the falling stream of molten metal into as many smaller streams as there are holes in the strainer core. This increases the exposed surface area of the molten metal and increases the amount of oxidation. This oxidation will continue until the molten metal completely fills the area below the strainer core. In addition, when these individual streams come together again, turbulence can occur.

The solution to these problems is the use of a *true choke core,* as seen in **Fig. 15-18-right.** Again, the choke core acts similarly to that of the funnel outlet discussed earlier. The choke core will not let all of the molten metal coming down the sprue pass through its opening. Thus, the sprue fills very quickly and the molten metal maintains a positive pressure against the walls of the sprue. Slag and dross will then be kept on the top surface of the sprue. Also, the surface area of the molten metal stream leaving the choke core is less than a strainer core having the same cross-sectional area when all of the openings are added together. **Figure 15-19** illustrates this point.

If desired, the choke core can become the choke in the gating system. It will be the "valve" that controls the amount or volume of molten metal that flows through the gating system. In order to do this properly, it has to be the point of smallest total cross section and the point of highest velocity.

Gating research has shown that good results can be obtained with rectangular tapered sprues, small end down. This type of sprue was used in the research done at Battelle. The chances of forming a vortex in this sprue are nonexistent, because one cannot inscribe a complete circle in a rectangle and have it touch all four sides at the same time. There will be more sprue surface area in contact with the molten metal than there will be in a round sprue holding the same volume of molten metal. Some metalcasters claim that there will be more heat loss in the rectangular-shaped sprue because of the increase in surface area. This will prove to be negligible, because the sprue's surface becomes very hot, almost immediately.

Sprue Well (Base)

The next part of the gating system that the molten metal will enter is the sprue well, which is located at the base of the sprue. (This can also be called the sprue base.) It is in the sprue well that the falling stream of molten metal will reach its highest velocity. It is also here that the molten metal will change direction, from a vertical drop to a horizontal flow. As one can imagine, there will be a tremendous amount of turbulence in the well. **Figure 15-20** illustrates sprue well designs for two gating systems that will be discussed shortly. Note that, in both cases, the well extends below the depth of the runner. This "sump" serves the same purpose as did the well in the pouring basin discussed earlier. As the molten metal builds up in the well it acts as a cushion and absorbs the kinetic energy of the falling stream of molten metal. Also, the increased area in the well also slows the velocity of molten metal and helps make the transition from vertical to horizontal flow more tranquil.

The geometry of the well can be circular, rectangular or square. The dimensions of a round well are given in **Fig. 15-20** and the dimensions of a rectangular well are given in **Fig. 15-21.**

The important part of a properly designed well is its sump. First, the sump should be deep enough to form a pool of molten metal to absorb the kinetic energy of the falling stream of molten metal. To do this, it has to be deeper than the runner or the runners attached to it. Normally, it is designed to be twice the depth of the runner(s). In addition, the bottom surface of the sump should be flat or convex in shape. With the latter shape, the molten metal tends to roll off this surface with little turbulence. This is ideal; however, a flat surface will also do nicely.

The worst shape for the bottom of a well is concave, as seen in **Fig. 15-22.** The research work done at Battelle showed that the turbulence in this shaped sump never really quit until the water

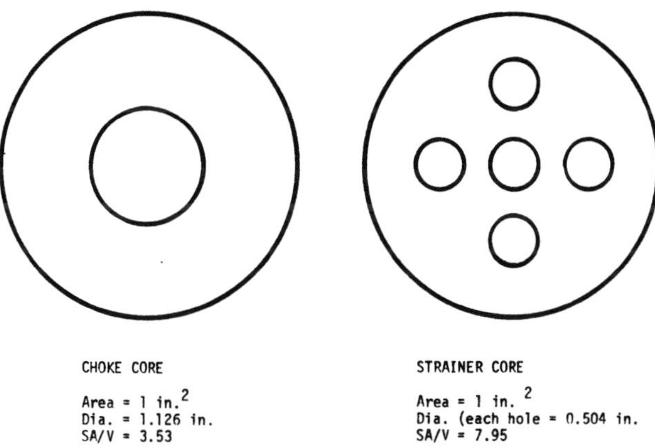

Fig. 15-19. Surface area exposed to air for choke and strainer cores.

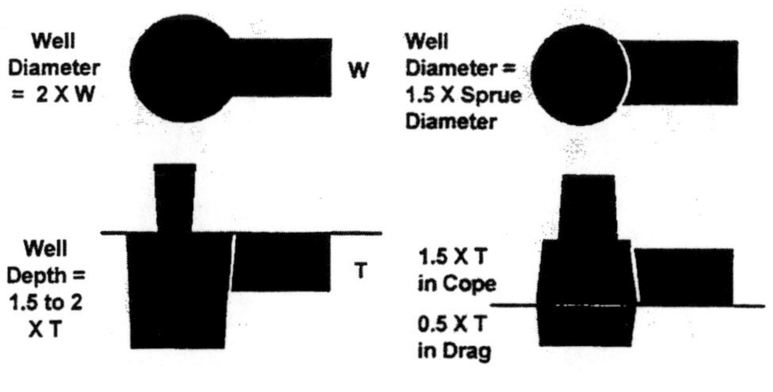

Fig. 15-20. Well designs for two gating systems.

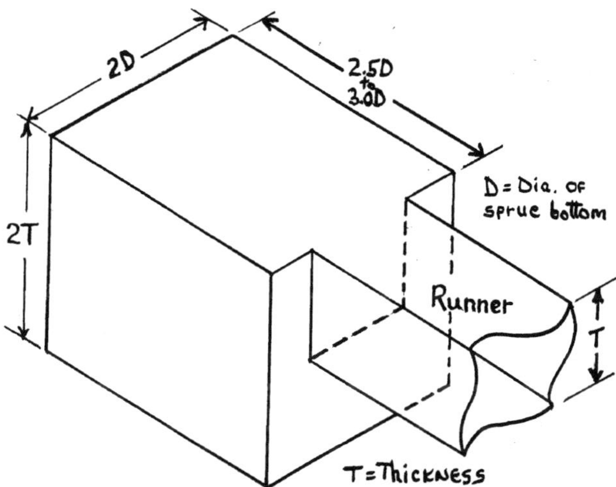

Fig. 15-21. Dimensions for a rectangular well.

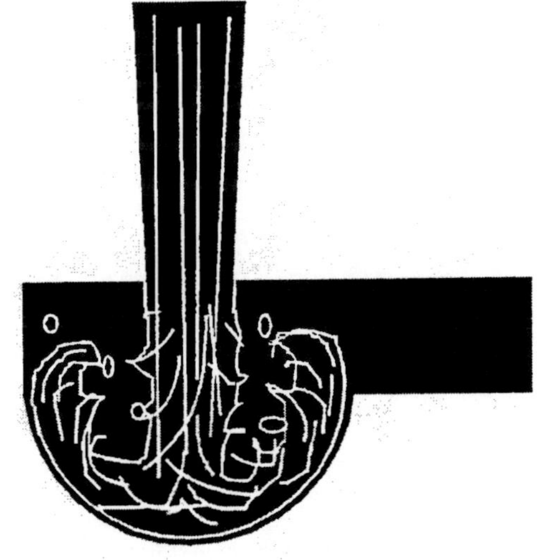

Fig. 15-22. Turbulence.

stopped flowing. This can be proved by holding a soup ladle under a faucet. This turbulence in the well can cause erosion and gas entrainment in the molten metal. Foundry workers will sometimes use this shaped bottom because they see it used on the bottom of risers. As will be learn learned later, when discussing risers in Chapter 17, the hemispherical bottom of a riser is used for an entirely different reason than molten metal flow control.

When pouring very large ferrous alloy castings in floor molds or pit molds, the bottom of the well can be made of chemically bonded sand or firebrick. This is done to help prevent or reduce the amount of erosion that can take place due to the large volumes of molten metal impinging and flowing over this surface.

Runners

The next part of the gating system is the runner(s). There may be one or more runners attached to the sprue well. It is the function of the runner to carry the molten metal to the gate(s) or ingate(s), which, in turn, will direct the molten metal into risers, or directly into the mold cavity. The runner should be streamlined to help prevent turbulence in the flowing stream of molten metal. This was discussed in the section on Fluid Flow.

Another essential part played by the runner is to distribute equal amounts of molten metal when more than one ingate is used. **Figures 15-23 and 15-24** illustrate the proper distribution of molten metal. **Figure 15-23** shows a straight runner: the runner is not reduced in size once it leaves the well. Here, the farthest gate from the sprue delivers most of the molten metal needed to fill the mold cavity, while the nearest two gates to the sprue actually pull molten metal out of the mold cavity. This action is due to the low-pressure area set up in these two gates as the molten metal passes by. This type of distribution does *not* meet the second requirement of a gating system: *The metal should enter the mold cavity so that it will produce proper thermal gradients with the surrounding mold surfaces, and also within the solidifying metalcasting, so that directional solidification takes place toward the source of feed metal (risers).*

On the other hand, the runner in **Fig. 15-24** is reduced in cross-sectional area as it passes the first two gates. This type of runner is called a step-down runner. Note that all of the ingates are feeding molten metal into the mold cavity at the same time, which leads to a more even distribution of molten metal into the mold cavity. Also, note that no molten metal is being pulled out of the mold cavity. Momentum will help carry the molten metal past the first two gates so there will not be a perfectly even distribution of molten metal into the mold cavity. The reduction in cross-sectional area in the step-down runner is equal to the cross-sectional area of the ingate just past it. Some foundry workers and patternmakers will simply taper the runner from the bottom surface, upward, to approximate the step-downs.

Runner Extensions—A critical part of any runner is the runner extension, which extends the runner beyond the last gate.

As the molten metal continues along the runner, it can pick up any loose particles of molding sand or other materials and carry them into the mold cavity. These inclusions are normally found in the leading edge of the molten metal stream. This leading edge is also the molten metal that took a severe thrashing in the sprue well before being pushed into the runner.

This "contaminated" molten metal will be trapped in the runner extension and not allowed to enter the mold cavity. The length of the runner extension should be such that it does not allow the leading edge of the flowing molten metal to "bounce" back into the last gate. In some cases, the bottom surface of the runner extension is tapered upward, or the top surface tapered downward, to help prevent "washback" into the last gate. A very important part of the runner extension is a vent, which allows the molten metal to force out any trapped air and completely fill the extension.

Many times, there may not be sufficient room to have a runner extension long enough to do its job because the flask size will not allow it. If this is the case, the runner extension can be made to turn the corner (with a generous radius) at the end of the runner. A simpler solution is to place a sump at the end of the runner and allow the leading edge of molten metal to fall into it. In either case, a vent must be used to allow the trapped air to escape from the mold.

Ideally speaking, the horizontal gating system should be located in the mold so that the molten metal fills the mold cavity from the bottom upward. This is not always possible and, in these cases, the gating system should be located so that molten metal entering the mold cavity does not have to fall an excessive distance. An excessive drop of molten metal can damage it in the mold cavity and cause defective castings.

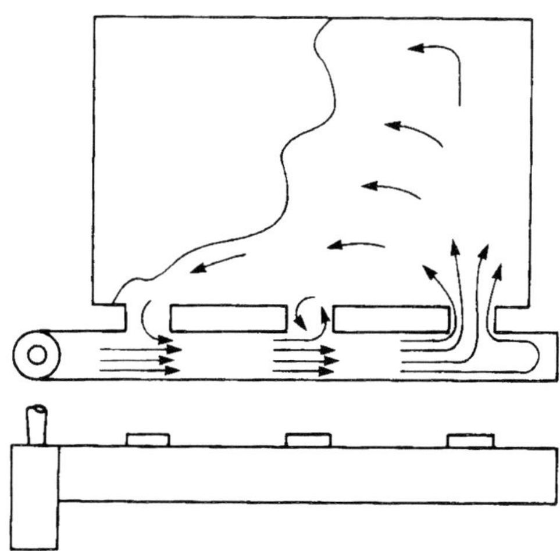

Fig. 15-23. Unbalanced gating system resulting from incorrect runner design.

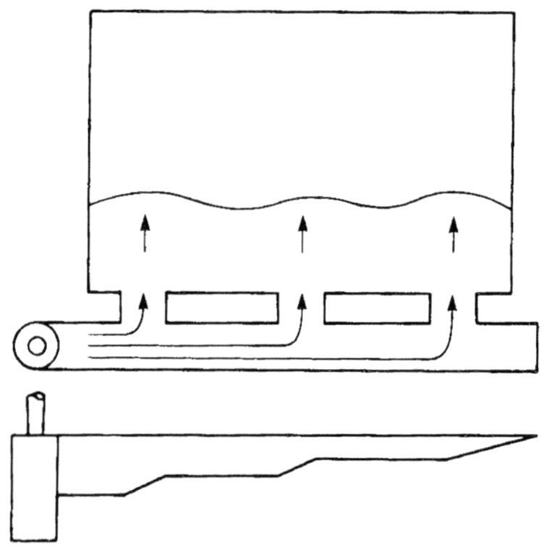

Fig. 15-24. A tolerably balanced gating system with step-down runner.

Vertical Gating System

The second classification of a gating system, based on position in the mold, is the vertical gating system. This type of gating system is found in vertically parted sand molds, permanent molds, investment casting molds, and in large molds such as pit molds. In this case, the gating system is in a vertical position within the mold.

A vertical gating system is normally characterized by a taller sprue, when compared to a sprue used in a horizontal gating system. As one would suspect, there would very high velocities at the base of the sprue. Both Battelle and CANMET performed gating research on vertical gating systems and both found the same results. In both cases, it was found that a choke at the sprue outlet, followed by opening the gating system up from that point worked very well.

Figure 15-25 illustrates the gating system designed at Battelle. Note that this vertical gating system has the same component parts as the horizontal gating system. They used a pouring cup, a rectangular tapered sprue (small end down), an enlargement or well, runner and runner extension with a vent. The design of these parts is similar to those of the horizontal gating system just discussed. However, in this case, the gate is fed vertically into a vertical side riser that was connected to the plate casting with a web. A web is the thin section that connects the riser to the casting. When using this vertical gating system, the aluminum plate casting was found to be radiographically free of inclusions and gas bubbles.

There are many times when the casting design does not lend itself to a continuous web gate as seen in **Fig. 15-25. Figure 15-26** shows a simplified example of what happens when the sprue is connected to the mold cavity with separate gates. Note that the bottom gate does the feeding and molten metal leaves the upper part of the mold cavity via the two upper gates. The water analogy studies done at Battelle showed that if the vertical gating system was designed as seen in **Fig. 15-27,** there would be sequential flow through the gates as the mold cavity and reverse sprue filled. This same gating system could be used with each gate feeding a casting. **Figure 15-28** shows one foundry's use of this style vertical gating system to pour ductile iron castings in shell molds. The larger part of section A-A is a riser.

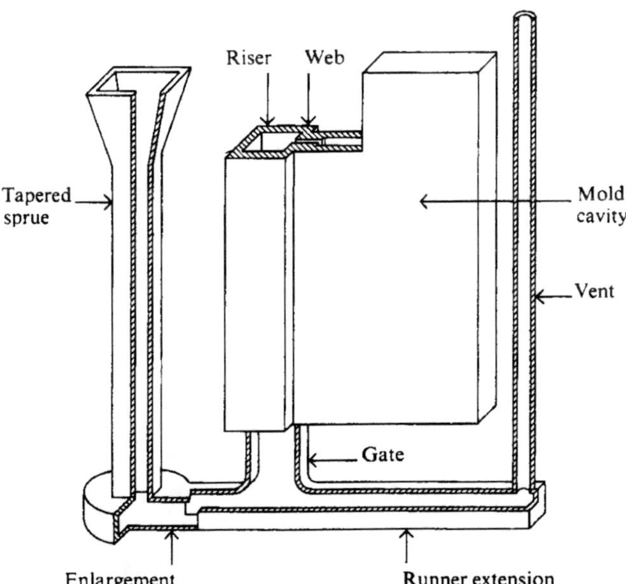

Fig. 15-25. Example of vertical gating system in a light metal alloy.

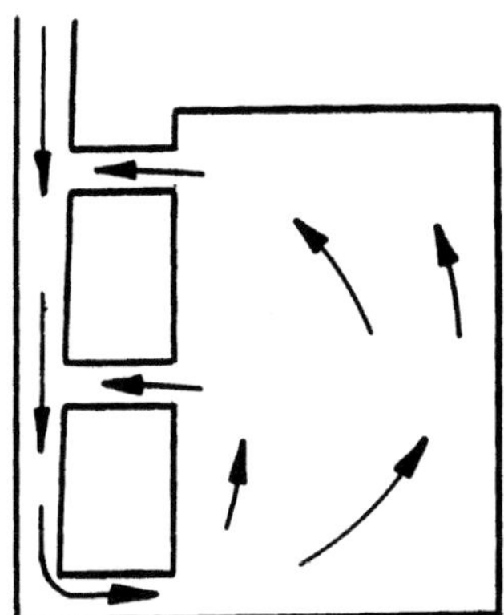

Fig. 15-26. Improperly designed step gate illustrating flow pattern.

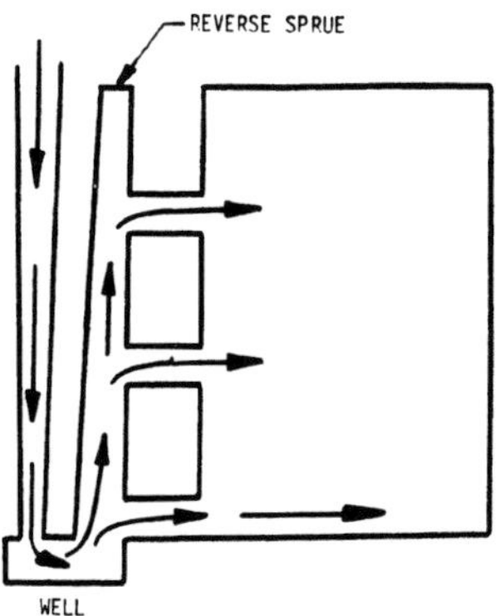

Fig. 15-27. Properly designed step gate illustrating flow pattern.

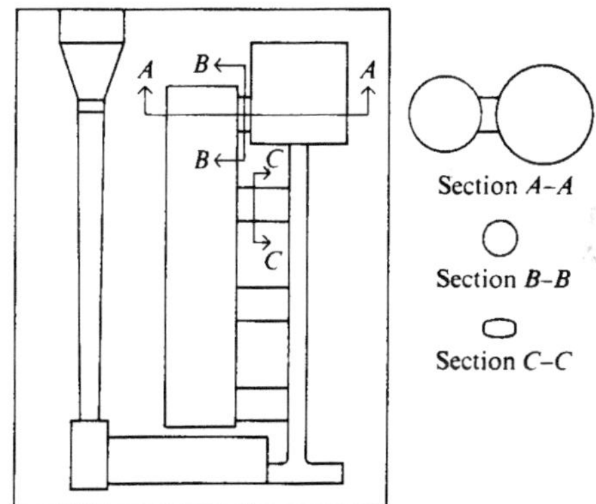

Fig. 15-28. Example of vertical gating system in a ductile iron shell mold.

Another way this style of vertical gating system can be used is by bottom gating into a casting or castings. **Figure 15-29** illustrates the use of bottom gating for two castings. This same vertical gating system could be used to produce a number of smaller castings by simply gating into risers, instead of the two castings, and then having gates feed directly from the riser into the individual castings.

All of the vertical gating systems shown thus far have the choke located at the base or outlet of the sprue. Next, vertical gating systems with the choke located at the gates will be covered.

Figure 15-30 is an illustration of a vertical gating system in which the choke is at the ingates. This system has a pouring basin onto which two top runners are attached. From these runners, the molten metal drops down into two down runners or risers. Individual gates then feed the molten metal into the mold cavities. The choke in this system can be located where the down runner or riser is attached to the top runners, or it can be located at the gates.

Figure 15-31 illustrates another vertical gating system having the choke located at the gates. In this case, the pouring basin or cup empties into a tapered sprue with a well at the bottom. The molten metal fills the sprue quickly and then travels to the runners. The gates are taken off the side-bottom of the runner. Adding up the cross-sectional area of each gate where it is attached to the runner would show that this total cross-sectional area would be the *choke* in this gating system.

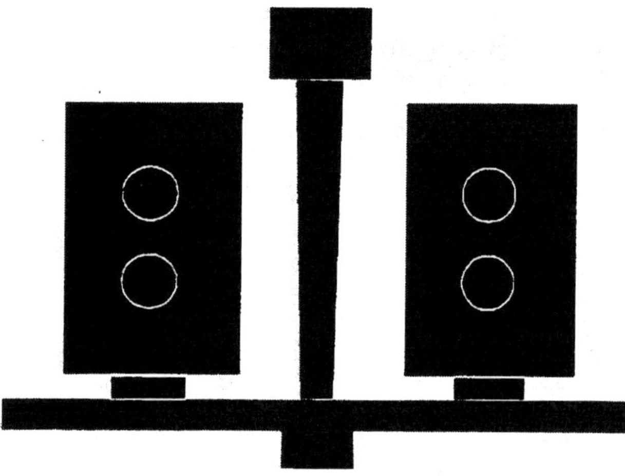

Fig. 15-29. Vertical gating design in nonpressurized system.

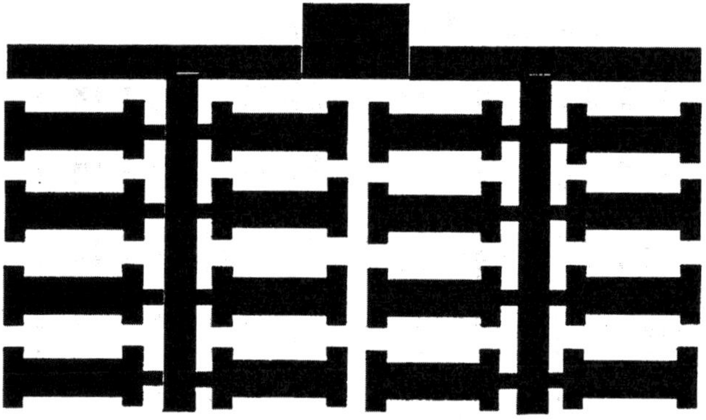

Fig. 15-30. Vertical gating design in pressurized system with basin attached to runners.

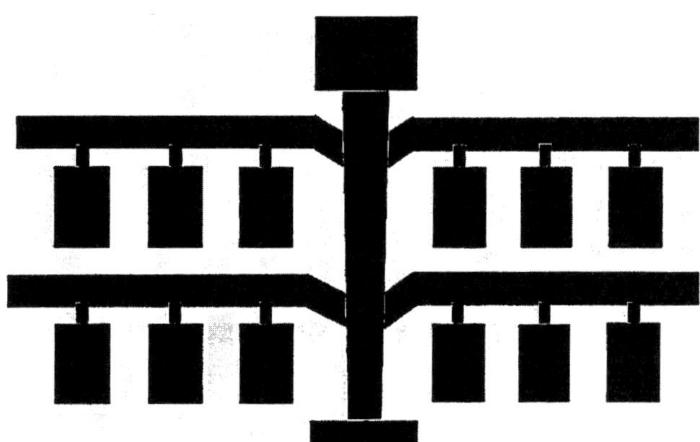

Fig. 15-31. Vertical gating design in pressurized system with basin attached to sprue.

Tile Gating System

When pouring very large ferrous alloy metalcastings, tile gating systems are used. As one would expect, tile gating is used to help protect the molding sand from the tremendous amount of heat generated by the flowing molten metal, which, in turn, can cause molding sand erosion problems. Tile gating can also be used for very large nonferrous alloy castings.

The tile is normally made of ceramic and is purchased according to the inside diameter and length of the individual tiles. Users of tile gating systems are limited to those inside diameters produced by the tile manufacturer. The tile sections are joined together and assembled into the desired gating system. After assembly, the sand mold is compacted around the tile gating system.

Figure 15-32 shows the various shapes and parts available to construct a tile gating system. The ceramic tile material is permeable and will allow the aspiration of air and mold gases into the molten metal stream; therefore, the same precautions should be taken to keep the gating system completely filled with molten metal, as discussed previously. The same principles of fluid flow apply to tile gating as is applied to gating systems made in sand molds.

Pressurized and Nonpressurized Gating Systems

Besides the gating system's position in the mold, there are two other means of distinguishing between gating systems, whether it is *pressurized* or *nonpressurized.* The essential difference between the pressurized and nonpressurized gating system is the location of the choke. In the pressurized gating system, the choke is located at the gates. In the nonpressurized gating system, the choke is located at or close to the sprue exit.

The term nonpressurized is misleading in that, according to Bernoulli's Theorem, there has to be pressure in the gating system as the molten metal flows. A better name for this gating system would be the expanding gating system. However, since this gating system was developed at Battelle back in the late 1940s and early 1950s, it has been called the nonpressurized gating system. It has also been known as the AFS gating system.

Gating Ratio—To help determine where the choke is located in the gating system, a gating ratio is used. This ratio compares the total *sprue* cross-sectional area to total *runner* cross-sectional area to

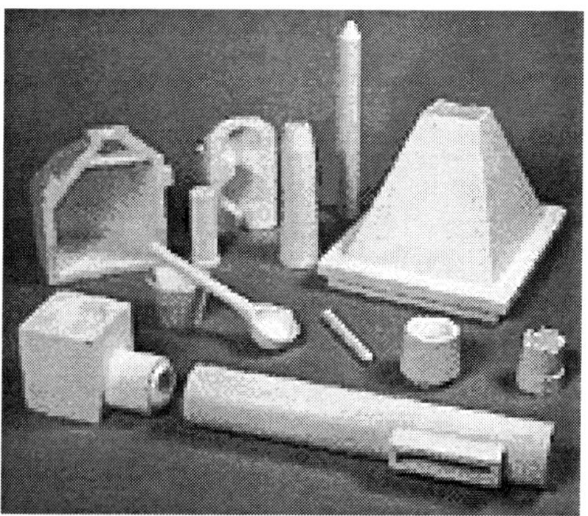

Fig. 15-32. Shapes and parts available for construction of tile gating system.

total *gate* cross-sectional area. An abbreviation of this ratio would be S:R:G (for sprue:runner:gate). It would work as follows:

1. Use the total cross-sectional area at the sprue exit. If there are two sprues in the gating system, then it is the total cross-sectional area of both sprues.
2. Next, take all of the runners, find out the cross-sectional areas where they are attached to the sprue well and add these together.
3. Finally, find the cross-sectional area of all of the gates attached to these runners and add them together. In this case, it is assumed that all of the gates are distributing the molten metal into the mold cavity(s) at the same time.

When these numbers are plugged into the gating ratio, the choke location will be determined because it will be at the smallest number in the ratio.

Turbulence Sensitivity—Some of the criteria used in selecting either a pressurized or a nonpressurized gating system are sensitivity to turbulence, choke location and gate velocity.

The first is what is called the sensitivity of the metal alloy being poured. The metal's sensitivity to turbulence, and its ability to form slag or dross due to turbulence, is one measure to be considered. Following are various alloys and their sensitivity to turbulence:

1. ***Group I***
   Alloys not greatly sensitive to turbulence
   - Gray Iron and Malleable Iron
   - Leaded Red Brasses, Leaded Semi-Red Brasses
   - Tin Bronzes and Leaded Tin Bronzes
   - High Leaded Tin Bronzes
2. ***Group II***
   Alloys somewhat sensitive to turbulence
   - Ductile Iron
   - Carbon, Low and High Alloy Steels
   - Silicon Bronzes and Brasses
   - Copper Nickel Alloys
3. ***Group III***
   Alloys that are very sensitive to turbulence
   - All Aluminum and Magnesium Alloys
   - Aluminum and Manganese Bronzes
   - Yellow Brasses
   - Chrome Copper
   - White Manganese Brass

What separates the pressurized gating system from the nonpressurized gating system is the location of the choke. By definition, the choke is the smallest total cross-sectional area in the gating system. Remembering the Law of Continuity, it is also known that the choke is the highest velocity point. High velocities cause turbulence, and turbulence, in turn, causes slag and dross to form.

In the case of the pressurized gating system, the choke must be located at the gate(s) and, thus, the highest velocity would be at the gate(s). This would cause big problems for those alloys that are very sensitive to turbulence (Group III) and could cause problems with the somewhat sensitive alloys (Group II). On the other hand, in the nonpressurized gating system, the highest velocity is at the exit of the sprue or very close to it, and from there on the velocity is decreased, especially at the gate(s). Thus, the nonpressurized gating system is highly recommended for the very sensitive alloys (Group III) and can be helpful with the somewhat sensitive alloys (Group II).

Also keep in mind that high gate velocities can cause erosion if the outlet of the gate (into the mold cavity) is close to the sand mold wall or core surface. If this occurs, these materials will be found in the casting. This can be compared to removing the nozzle from a garden hose and turning on the water. When holding the end of hose close to the ground, there will be very little, if any, soil erosion. However, by holding a thumb over the hose outlet, which reduces the opening, there will be soil erosion due to the higher velocity.

Pressurized Gating System

As mentioned earlier, the pressurized gating system has the choke located at the gate(s). Thus, the smallest total cross-sectional area would be the last number in the gating ratio. Three of the more common pressurized gating ratios used are 4:8:3, 12:11:10 or 14:12:10. Note that in all three of these ratios, the smallest number is the last or the gate(s). Other pressurized gating ratios can be used; however, in whatever ratio is used, the smallest number has to be the last number in the ratio.

The physical design of a pressurized gating system is found in **Fig. 15-33.** The ratio indicated here is 4:8:3; however, the design would be the same for any pressurized gating system. There is a sprue, well, runner and runner extension. The runner in the pressurized gating system is in the cope and the gate(s) come off the bottom side of the runner. The latter point is seen in the end view drawing.

At this point, it should be mentioned that the aim of any gating system should be to trap out any dross, slag or sand being carried along in the flowing molten metal. This can be done if the molten metal is not flowing at high velocities. Following are the densities of some common alloys and the density of the slag or dross formed in their melting and pouring.

Metallic Densities (g/cc)

Aluminum	2.41	Steel	7.81
· Al_2O_3	3.96	Cast Iron	6.97
Copper	8.00	· Fe_2SiO_4	4.34
· CuO	6.00	· SiO_2	2.85

The density of silica sand (SiO_2) is also given. Note that the densities of the slag, dross and silica sand are all less than the alloy, with the exception of aluminum.

The pressurized gating system works similar to the funnel discussed earlier. The molten metal enters the runner and slows while filling the runner. Since the gate(s) are the smallest total cross-sectional area in the gating system, they function like the outlet of the funnel. Because a greater volume of molten metal is flowing into the runner than the gate(s) will allow to leave the runner, it will fill very quickly. In the case of the copper and ferrous alloys, the dross, slag, molding sand or any other foreign material lighter than the molten metal will begin to float to the surface of the runner. This action is shown in **Fig. 15-34.**

With the velocity in the runner slowed, the lighter materials in the molten metal will float out and make contact with the cope surface of the runner. Once this material makes contact with the surface of the molding sand, it will stick there as if it were glued to the surface. This will then help to ensure that clean molten metal passes through the gate(s), since they come off the bottom side of the runner. Should the gate(s) be removed from the bottom side of the cope runner, there is a chance of some of this material being dropped, along with the molten metal, through the gate(s) before the runner is filled.

Although **Fig. 15-34** does not show it, the runner can be stepped down to promote an equal distribution of the molten metal into the mold cavity(s). Since the foreign material will be floating to the cope surface of the runner, it shouldn't be stepped down to bring it closer to the gate(s). In this case, the runner can be stepped down by bringing in the side or sides of the runner, if there are gates on both sides of the runner. The step down would be calculated the same way discussed earlier.

An advantage of the pressurized gating system is that it will have small ingates that help in the cleanup of the metalcasting:

1. In the case of gray and malleable iron, the gates will break off during shakeout. In the case of the copper alloys, the gates can be sheared off.
2. If designed correctly, the gates will trap slag, dross and other impurities found floating in the molten metal.
3. In some cases, the pressurized gating system will be smaller and require a lesser volume of molten metal than will the nonpressurized gating system.

Nonpressurized Gating System

The nonpressurized gating system was designed to be used with those alloys that are very sensitive to turbulence (Group III). However, it has found its place also with the other two groups of alloys, which are not as sensitive to turbulence. In fact, caution should be used when aluminum and/or magnesium are mentioned in the alloy's composition.

The physical design of the nonpressurized system is seen in **Fig. 15-35**. The choke in this gating system is found at the outlet of the sprue, or somewhere close to the sprue outlet. If a reversed taper sprue is used without a choke core, then the choke can be placed in the runner(s) where it connects with the well. This is seen in **Fig. 15-36**.

The following may explain why the nonpressurized gating system is designed the way it is. It would be preferable to use a tapered sprue, small end down, for the reasons discussed previously. However, due to pattern and molding constraints this is not always possible. If, for some reason, it is not possible to use a choke core at the outlet of the sprue, it will become necessary to use a runner choke as seen in **Fig. 15-36**. Of course, there will be a well at the base of the sprue.

Note that, in this gating system, the runner is located in the drag and the gates branch off the top of the runner. This is done so that the runner fills completely with molten metal before any of it can pass through a gate in to the mold cavity(s). As with the pressurized gating system, if it is possible get the slag, dross or other impurities in the molten metal to float and touch the cope surface of the runner, that is where they will stay. In order for this to happen, the runner has to be completely filled with molten metal.

Also, in order for the impurities in the molten metal to float, they have to be lighter than the molten metal, and the velocity of the flowing molten metal has to be such that it will allow them to "float out." The velocity can be slowed by opening up the cross-sectional area of the runner, where it is attached to the well. In the case of a runner choke, the opening up of the runner takes place where it leaves the runner choke. Ideally, the sprue well can help reduce the velocity if a tapered sprue (small end down) or a choke core is used.

It is the aim of the nonpressurized gating system to trap the impurities on the cope surface of the runner before it can reach the first gate. Slowing the velocity of the flowing molten metal will greatly assist this; however, it will take some distance to do this before reaching the first gate. The distance between the sprue and the first gate should be at least four times the depth of the runner. Many times, this becomes a problem, especially when trying to maximize the number of castings produced from each mold.

As with the importance of the gate location, in reference to the runner in the pressurized gating system, the same holds true for the nonpressurized gating system. As seen in the end view of the runner in **Fig. 15-35**, the gate is taken off the cope surface of the runner. Note also, that the gate does not completely overlap the cope surface of the runner. This method of connecting the gate to the cope surface of the runner gives a little more surface to trap the impurities, should

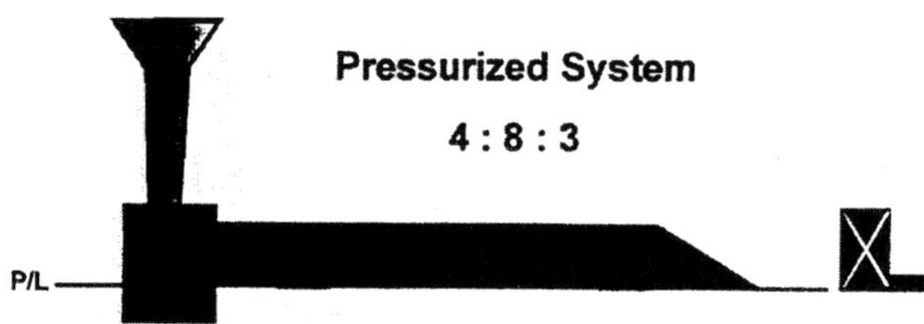

Fig. 15-33. Pressurized system—4:8:3.

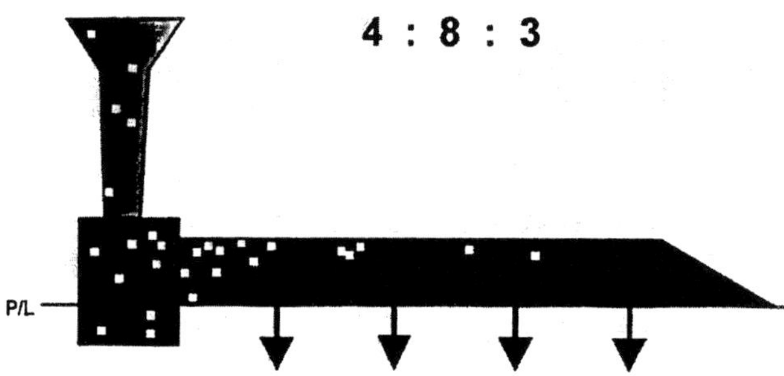

Fig. 15-34. Slag separation in a pressurized system—4:8:3.

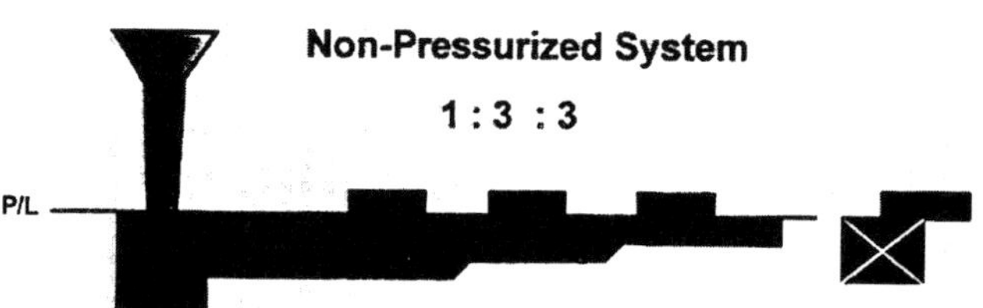

Fig. 15-35. Nonpressurized system—1:3:3

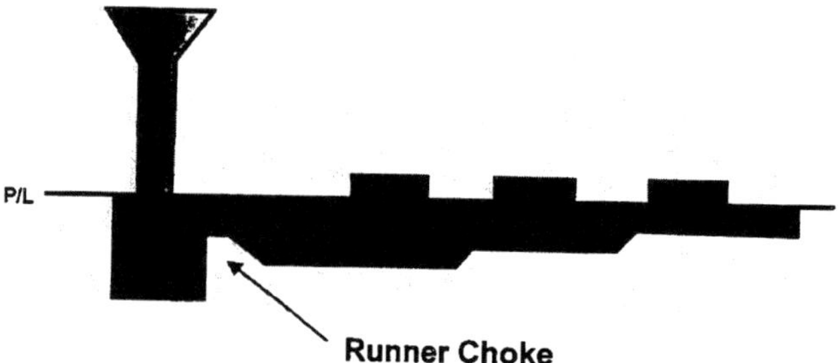

Fig. 15-36. Nonpressurized gating system—1:3:3, with choke in runner.

they get this far. However, should there be two gates at the same location to feed from both sides of the runner, a complete gate overlap is recommended.

Should the gates be connected to the runner any other place than that recommended, the runner will never fill completely in time to trap out the impurities on the cope surface of the runner. The reason for this is that the molten metal will flow through the gates before the runner can be completely filled. In this case, the runner will not be filled until the molten metal in the mold cavity reaches a level above the cope surface of the runner. Before this happens, the impurities can flow into the mold cavity(s) during the time it is taking to fill the runner.

The runner can be stepped down or tapered to obtain near equal distribution of the molten metal through all of the gates at roughly the same time. The step-downs are equal to the cross-sectional area of the gate, just past. This step down should not be placed directly below the gate but just beyond the gate and within about 3/4 in. (2 cm) of the gate. At this point, the following question may arise: If the cross-sectional area is reduced, won't the velocity be increased, according to the Law of Continuity? The answer is that, when the runner is stepped down, some of the flowing molten metal is forced through the gate, thus decreasing the quantity or volume of molten metal beyond that gate. Thus, we have changed two details in the Law of Continuity equation, namely cross-sectional area and quantity.

This raises another interesting point. What if a very thin section of the metalcasting is being fed by a gate, and it is felt that a greater amount of molten metal is needed to feed that section to prevent a defect, such as a misrun. (A misrun is when the casting is not fully formed when removed from the mold.) Most metalcasters will reply that they have to fill that thin section faster. (This doesn't mean at a higher velocity, but with a higher flow rate or quantity of molten metal.) This can be done by simply increasing the cross section of the gate and increasing the step down of the runner just beyond the gate. Many times, it is thought that all that can be done is to pour the molten metal "down the hole" and hope for the best. This is not true! The molten metal can be controlled as it flows through a properly designed pressurized or nonpressurized gating system.

As with the pressurized gating system, a runner extension or sump is placed at the end of the runner. The design of the runner extension or sump is the same as discussed earlier.

Earlier it was mentioned that the nonpressurized gating system should be used with the aluminum, magnesium and some of the copper alloys. Now comes the reason for wanting to do this. Remember, in the pressurized gating system, the choke is located at the gates; thus, this becomes the point of highest velocity and highest turbulence in the system. Also, remember that these alloys are very prone to forming oxides when this occurs. Should these oxides be formed at the ingates, the only place for them to end up is in the casting(s). With high metal velocities at the gates, magnesium oxide can be formed in ductile iron. The magnesium added to produce graphite nodules in ductile iron is easily oxidized due to turbulence. The magnesium oxide dross that is created can produce microinclusions in the casting that reduce mechanical properties of the part, or produce surface defects.

In the case of aluminum alloys, the breakup of the aluminum oxide skin formed around the molten metal should be prevented. This skin forms almost instantaneously, which can be seen when skimming the dross off the surface of a ladle or pot of aluminum alloy. This skin, when not broken, helps to prevent the formation of more aluminum oxide and helps prevent the pickup of hydrogen, so prevalent in aluminum alloys. Keeping the turbulence to a bare minimum helps maintain this oxide skin cover. In other words, it achieves as close to a laminar flow as is possible.

For carbon, low- and high-alloy steels, it is desirable for the molten metal to enter the mold cavity(s) in a more gentle manner than in gray iron. These alloys are poured at very high temperatures and, if forced against mold and core surfaces, can very well cause erosion problems. Carbon steels also have poor fluid life, and the nonpressurized gating system is of help in overcoming this problem. In addition, many of the low- and high-alloy steels contain elements that are very prone to forming oxides. The same can be said for titanium and cobalt, both because of oxidation and high pouring temperatures, which could damage molds and cores.

In summary, when designing the nonpressurized gating system the following guidelines are recommended. Gating ratios such as 1:4:4 and 1:3:3 would be suggested for the very sensitive alloys (Group III). On the other hand, ratios such as 1:2:2 or 1:1.5:2 could be used for the Group II or intermediate sensitive alloys. As with the pressurized gating system, this ratio can be altered until arriving at one that will meet the needs of producing a quality metalcasting. However, to be a true *nonpressurized* gating system, the smallest number in the ratio has to be at the outlet of the sprue or where the runner attaches to the well. It cannot be at the gate(s).

GATING SYSTEM AIDS

Gating systems aids help to produce quality castings by following proper melting, slagging or dross removal procedures, *before pouring.* Followed (with equal importance) by properly trained pourers and pouring practices, well maintained automatic pouring devices and, finally, properly designed gating systems (with mold allowances made for the gating aids). However, in many cases, production demands and "yield" become the chief determinators as to gating system design.

Over the years, metalcasters have attempted to find the ways and means to overcome problems caused by not following the aforementioned practices. Certain accessories, practices and calculations have been devised to help solve these problems.

Filters/Screens

There was a time when some aluminum alloy foundries used a wad of steel wool at the base of the sprue. Supposedly, this was done "to cushion the fall of and help clean up the molten metal." Next, mica screens were used. These mica screens consisted of pieces of mica that had many, tiny holes drilled into them. Then followed fiberglass materials, which were used to help clean up the molten metal. Remember, the aluminum foundries have the problem of dross being heavier than the molten aluminum. Thus, some means had to be devised to get the dross out of the flowing stream of molten aluminum before it went into the mold cavity(s).

Figure 15-37 shows an aluminum bronze casting in which screens were used to clean the molten metal. This is a vertical gating system and the screens are located at the base of the sprue and where the runner is attached to the base of each side riser. By the way, the casting itself consists of the three vertical cylindrical shapes and the rectangular shapes surrounding the cylinders. The rest of the aluminum bronze surrounding the casting are risers, which will be discussed in Chapter 17, Risering Practice.

To help aluminum metalcasters, ceramic foam filters were developed in the 1970s and have been used successfully in aluminum alloy gating systems ever since. Over time, several other ceramic filters have been developed for use with both ferrous and nonferrous alloy gating systems. The other two classifications of ceramic filters are the pressed and extruded filters. **Figure 15-38** shows the three classifications of ceramic filters and their sizing by the openings in the filter.

Cell characteristics, or holes and pores, are measured as follows:

1. *Pressed Filters*—Hole diameter and number of holes
2. *Foam Filters*—Pores per linear inch (centimeters)
3. *Extruded Filters*—Cells per square inch (centimeters)

It must also be remembered that not all of the filter surface area will be used to filter the molten metal. Some of the filter area will be isolated from the molten metal because they will be the "prints" which will locate and hold the filter in place in the mold. The usable filter area is the total filter area minus the print area.

Fig. 15-37. Aluminum bronze casting using the principles of vertical gating with screens and web gates.

The pour rate of a filter is the quantity of molten metal that can pass through the filter, measured in pounds or kilograms per second. The capacity of the filter is the volume or quantity of metal (in lb or kg) that will pass through the filter before blockage occurs.

The ceramic filters used today can also be classified by the method in which they clean the molten metal stream. These classifications are:

1. *Screening*—Inclusion particles in the molten metal stream are retained on the front surface of the filter because they are larger in cross-sectional area than the openings in the filter.
2. *Cake Filtration*—Inclusion particles in the molten metal stream are separated out by developing a layer or "cake" of previously deposited particles.
3. *Deep Bed Filtration*—Solid inclusion particles retained by the filter are generally much smaller than the pore openings of the filter. Surface forces cause inclusion particles to attach to the surface of the filter.

The first two methods of cleaning the molten metal stream are the basis on which the *extruded* and *pressed* ceramic filters work. The presence of inclusion particles that effectively block the entrance of the filter media is required for cake filtration.

The ceramic *foam* filter works primarily by deep-bed filtration. In this ceramic filter, inclusion particles can be retained throughout the filter thickness. The reticulated structure of the ceramic foam filter allows for a high internal surface area and enhances the deep-bed filtration mechanism. Filtration efficiency is improved by reducing the filter pore size, thus increasing the internal surface area of the filter.

The first molten metal to reach the filter makes contact and freezes or solidifies on the surface of the filter. This metal is then remelted by the molten metal that follows it. The metallostatic head pressure of the molten metal in the gating system overcomes the surface tension in the molten metal and causes it to flow through the filter. This action is called priming the filter.

Alloy characteristics can affect the molten metal flow through the filter. Alloys with good fluid life flow readily through the filter. Narrow freezing range alloys work the best due to their better fluid life. Alloys with poor fluid life and high surface tension make filter priming difficult.

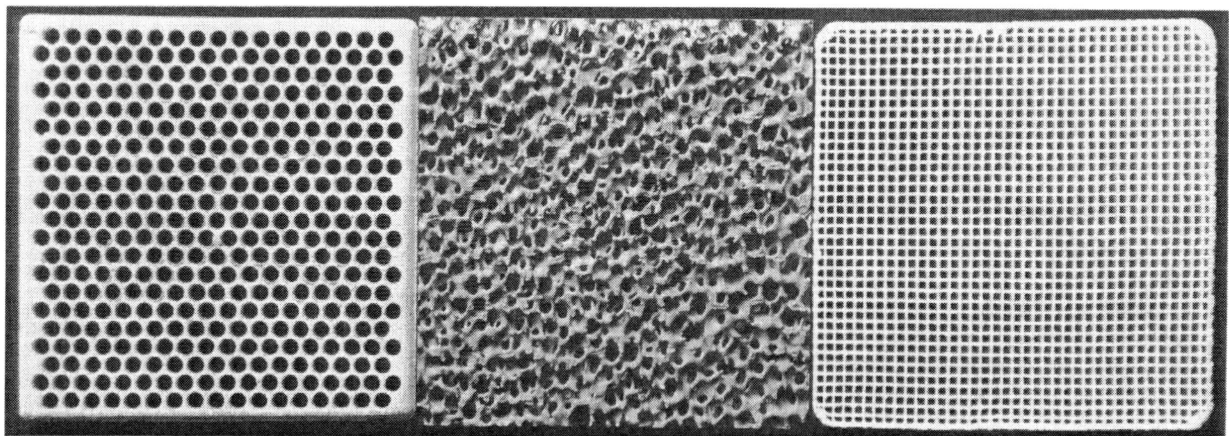

Fig. 15-38. Three types of ceramic filters (l-r): pressed, foamed and extruded. Each is a 55x55x13 mm (2.17x2.17x0.5 in.). The pressed filter contains 367 holes with hole sizes of 0.090 in. (2.28 mm). The foamed filter has 25 pores/in. (ppi). The extruded monolith filter has 300 cells/sq.in. [Photo courtesy of Selee Corp.]

Filter Size

As was said earlier, the choke is the flow control valve in the gating system. In other words, it determines the number of pounds or kilograms of molten metal that will flow through the gating system per unit of time (normally seconds or minutes), depending on the weight of the metalcasting. (See Gating System Choke Calculations near the end of this chapter.)

The choke has to remain totally open during the entire pour to maintain the proper flow rate. If filters do their job (which they will), the amount of area for the molten metal to pass through will begin to decrease due to the holes and pores beginning to plug up. If this should happen very quickly, and the plugged area becomes large enough, the filter could become the choke, and thus regulate the flow rate. If this flow rate is not the "designed" flow rate, casting defects will occur.

When determining how large the filter should be, the first thing that should be considered is the alloy being poured. For instance, for gray iron, the usable filter area must be at least three times as large as the choke area. On the other hand, when pouring ductile iron, the usable filter area must be at least five times as large as the choke area. (Time and space does not allow the listing of the ratios for other alloys; however, filter manufacturers can be contacted for this information.)

Filter Placement

It is recommended that filters be placed in the runner(s) of the gating system. It is *not* recommended to place filters in the top or bottom of a sprue. Ideally, filters should be placed at each gate, because this is the place the cleanest molten metal should enter the metalcasting. However, this can become cost prohibitive, except for those metalcastings requiring the highest quality specifications such as premium aerospace metalcastings. For the majority of metalcastings produced today, it is recommended that filters be placed in the runner(s).

Figure 15-39 illustrates the location of filters in two nonpressurized-gating systems used for aluminum alloys. An enlargement is designed to accommodate the filter, as shown. Where possible, the filter should be placed close to the first gate attached to the runner. The same design can be used for pressurized gating systems; however, in this case, the runner(s) will be placed in the cope and the gates will be attached to the runner, as discussed previously. It is also recommended that the mold cavity be filled from the bottom up, in order to prevent dropping the molten metal down into the mold cavity and over cores, if cores are used. If this is not done, there is the risk of damaging the molten metal, which has just been "cleaned."

Direct-Pour System Filtering

A unique use of ceramic foam filters is that in the DYPUR® (or direct pour) system shown in **Fig. 15-40**. In this system, a ceramic foam filter is placed in an insulating sleeve (which will be discussed in Chapter 17, Risering Practice). The molten metal is poured directly into the insulating sleeve passing through the filter directly into the cavity. This "gating" system was designed to be used with aluminum and copper alloy castings.

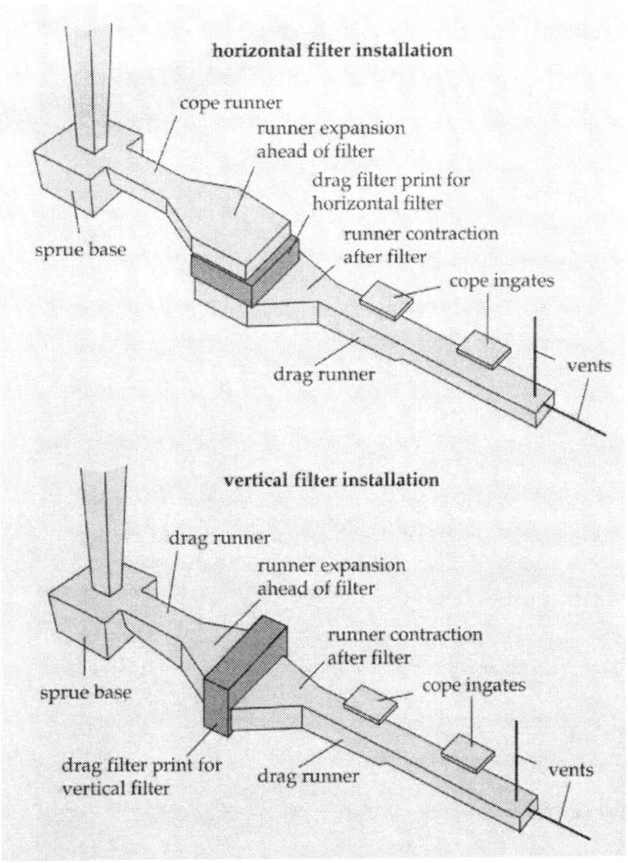

Fig. 15-39. The two ways that filters can be placed in the runner.

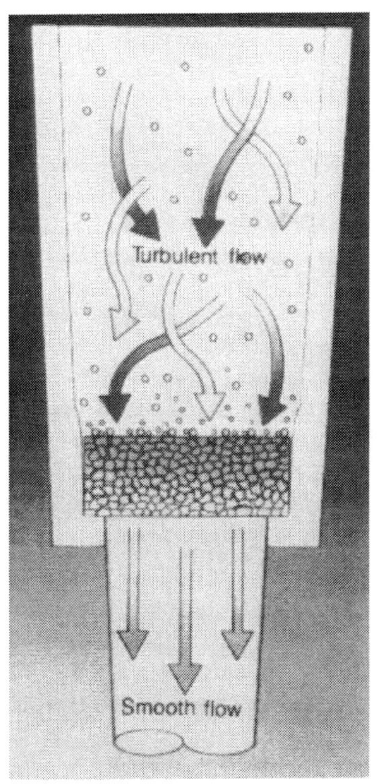

Fig. 15-40. The DYPUR® direct-pouring system. In this method, an insulating sleeve and a ceramic foam filter reduce turbulence and remove inclusions. They also serve as the gating system in many cases, allowing metal to be poured directly in the mold cavity through the top of the casting. [Drawing courtesy of Foseco, Inc.]

This method of pouring metalcastings promotes directional solidification and it improves casting yield because conventional gating systems can be eliminated. By using a foamed ceramic filter, inclusion particles are also filtered from the molten alloy. In this system, the molten metal left in the insulating sleeve acts as a riser, thus the gating system and riser are one.

Casting dimensions are critical when using the DYPUR system. When producing large, chunky metalcastings or metalcastings where the area immediately below the filter is open, the distance of the drop of molten metal should be less than 4 in. (10 cm), for optimum results. This distance can be stretched, in some instances, to 6-in. (15-cm). The open free-fall of the molten metal, when it leaves the filter, will cause the velocity and turbulence in the falling molten metal to increase due to the effects of gravity.

However, when pouring thin-walled metalcastings (0.5 in., 1.25 cm or less), the molten metal drop can be extended without increased turbulence during the fall. It has been found that, even in tight areas, a molten metal drop of greater than 12 in. (30 cm) should be avoided.

Advantages of Filtering

Some of the advantages of using filters in either the pressurized or the nonpressurized gating system are:

1. Removal of minute inclusion particles from the molten metal as it passes through the filter.
2. Removal of these minute inclusion particles can improve mechanical properties in the metalcasting.
3. Filters can reduce the amount of turbulence in the gating system by producing a more laminar flow.
4. In the case of the direct-pour system, the usual gating system is eliminated.
5. Improved molten metal fluid life.

GATING FOR ALUMINUM COMPOSITES

When the nonpressurized gating system was used for pouring an A356 matrix 15% by volume silicon carbide reinforced aluminum composite casting, poor casting quality resulted. It was shown that conventional gating techniques will not always work for aluminum composites. The main problem with the test metalcastings was the appearance of gas bubbles, internally and on the surface of the metalcastings. Additional research showed that if a bubble trap or open riser was placed where the molten metal entered the runner(s), the occurrence of these defects was reduced. Further research showed that if the choke was placed in the runner(s) where the molten metal left the bubble trap, the castings were radiographically free of gas bubbles.

Figure 15-41 illustrates the gating system used to produce high-quality aluminum composite castings. The bubble trap in this drawing is larger than required. Research has shown that the cross-sectional area of the bubble trap need only be 1–1.5 times the cross-sectional area of the runner(s). The bubble trap does not have to extend through the cope; however, it must be vented to the atmosphere.

It is very important that the choke be located within 0.5 in. (1.25 cm) after the bubble trap. It is also recommended that the sprue outlet be no larger than 10–12% of the combined cross-sectional areas of the runners beyond the chokes. For aluminum composites, the pouring or flow rates are generally best if kept lower than those used for conventional aluminum alloys. A commonly used flow rate is in the range of 1.3–1.5 lb/sec (0.58–0.67 kg/sec).

The gating ratio used for designing gating systems for aluminum composites is different than those discussed earlier. Instead of using three numbers in the ratio, four numbers are used. For instance, a gating ratio may be 1.55:1:10:10 for a given casting. The same as the ratios mentioned earlier, these four numbers represent total cross-sectional areas. The 1.55 represents the sprue outlet cross-sectional area, 1 is the cross-sectional area of the choke, and the two 10s represent the total cross-sectional areas of the runner(s) and gate(s).

It is recommended that the gates used in the aluminum composite gating system be located at the bottom of the casting. This will prevent the molten metal from dropping into the mold cavity(s), which can cause turbulence. The aluminum composite alloy must be handled in a nonturbulent manner. If this fundamental fact is respected, then many successful gating systems can be designed and used.

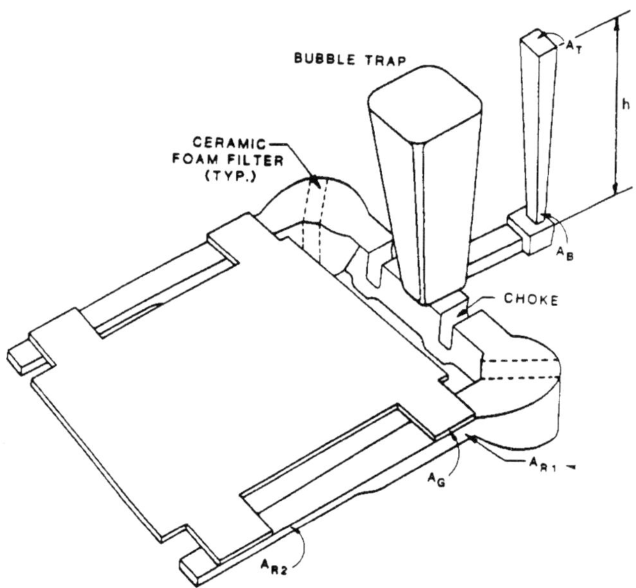

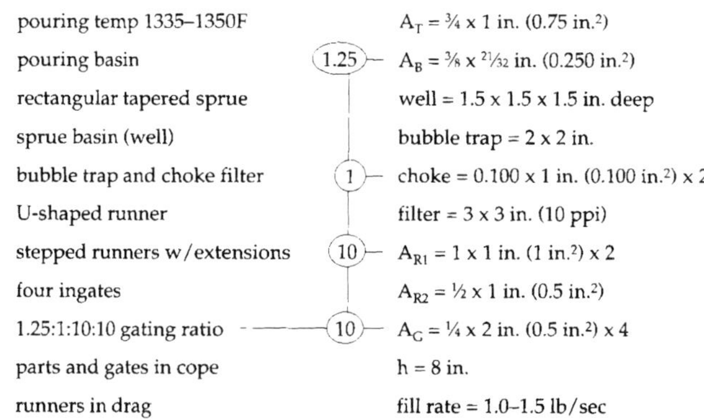

Specifications Used

pouring temp 1335–1350F		A_T = ¾ x 1 in. (0.75 in.²)
pouring basin	1.25	A_B = ⅜ x 21/32 in. (0.250 in.²)
rectangular tapered sprue		well = 1.5 x 1.5 x 1.5 in. deep
sprue basin (well)		bubble trap = 2 x 2 in.
bubble trap and choke filter	1	choke = 0.100 x 1 in. (0.100 in.²) x 2
U-shaped runner		filter = 3 x 3 in. (10 ppi)
stepped runners w/extensions	10	A_{R1} = 1 x 1 in. (1 in.²) x 2
four ingates		A_{R2} = ½ x 1 in. (0.5 in.²)
1.25:1:10:10 gating ratio	10	A_G = ¼ x 2 in. (0.5 in.²) x 4
parts and gates in cope		h = 8 in.
runners in drag		fill rate = 1.0–1.5 lb/sec

Fig. 15-41. Gating system incorporating filters used to pour the best aluminum composite alloy trial casting.

GATING SYSTEM CHOKE CALCULATIONS

Computer programs available today will help calculate a gating system and actually show how the gating system will fill the mold cavity(s). The size of the various gating system components can be calculated by hand, using proven formulas. Actually, most of the computer gating programs use these formulas.

Time and space does not allow for the calculation of an entire gating system; however, the following is a formula used to calculate the choke area recommended for a given metalcasting.

The formula used to determine the choke area is:

$$a = \frac{W}{d \cdot \sqrt{2 \cdot g \cdot H} \cdot c \cdot t} \qquad (5)$$

where a = total choke area ($in.^2$ or cm^2)
W = weight of casting, (lb or kg)
d = density of alloy being poured ($lb/in.^3$ or kg/cm^3)
g = acceleration due to gravity ($386.4 \ in./sec^2$ or $980 \ cm/sec^2$)
H = effective sprue height (in. or cm)
k = frictional coefficient for sprues
0.88 - round tapered sprue small end down
0.47 - round straight sprue
0.74 - square tapered sprue small end down
t = pouring time (sec)

(In Equation 6, there is an example choke area calculation using English units for a 50-lb casting.)

To determine the density of the alloy being poured, the following approximate densities for various alloys are:

Alloy	*lb/in.^{3}*	*Alloy*	*lb/in.^{3}*
Aluminum	0.10	Carbon steel	0.28
Magnesium	0.07	Stainless steel	0.27
Zn-Al alloys	0.22	Inconel	0.30
Copper	0.31	Monel	0.31
Cast irons	0.26		

Referring to Equation 5, because gray cast iron will be poured, $d = 0.26 \ lb/in.^3$. The effect of gravity will be $g = 386.4 \ in./sec^2$. The effective sprue height, H, is determined using one of the drawings (b) and formula found in **Fig. 15-42.**

Since the entire metalcasting will be contained in the cope, the formula $H = h - C/2$ will be used, where h = height of molten metal in the sprue and pouring basin, and C = the height of the casting in the cope. In this example, h = 10 in. and C = 6 inches. Placing these numbers in the formula, it is found that H = 7 inches. A round tapered sprue, small end down, will be used, so k = 0.88

The last number required for the formula is the pouring time, t. This is the number of seconds needed to take to pour the casting. In the case of high-production foundries using automatic molding machines, the pouring time is often limited by the cycle time of the machine. The pouring time chosen should be less than the cycle time of the molding machine. For simplicity's sake, use a pouring time of 5 seconds.

Now that all of the required numbers necessary to determine the choke area are known, they can be plugged into the original formula to determine the choke area:

$$a = \frac{50 \, lb}{0.26 \, lb/in.^3 \times \sqrt{2 \times 386.4 \times 7} \, in./sec \times 0.88 \times 5 \, sec} = 0.6 \, in.^2 \qquad (6)$$

Calculating this formula would determine that a total choke area of approximately 0.6 $in.^2$ is needed to pour this 50-lb gray iron casting in 5 seconds.

Once the total choke area has been calculated, this number can be plugged into the gating ratio desired, followed by calculation of the size of the remaining components in the gating system.

SUMMARY

This chapter has attempted to provide a brief description of what a gating system is and does. Remember that the gating system cannot overcome poor pouring practice or poorly maintained automatic pouring devices. Poor pouring practice, along with a good or poorly designed gating system, can ruin perfectly good molds, cores and molten metal and cause metalcasting defects to occur.

To learn more about gating system design, etc., the reader is encouraged to consult the information provided in the Bibliography or attend an AFS Cast Metals Institute course on Gating and Risering Practice and Design.

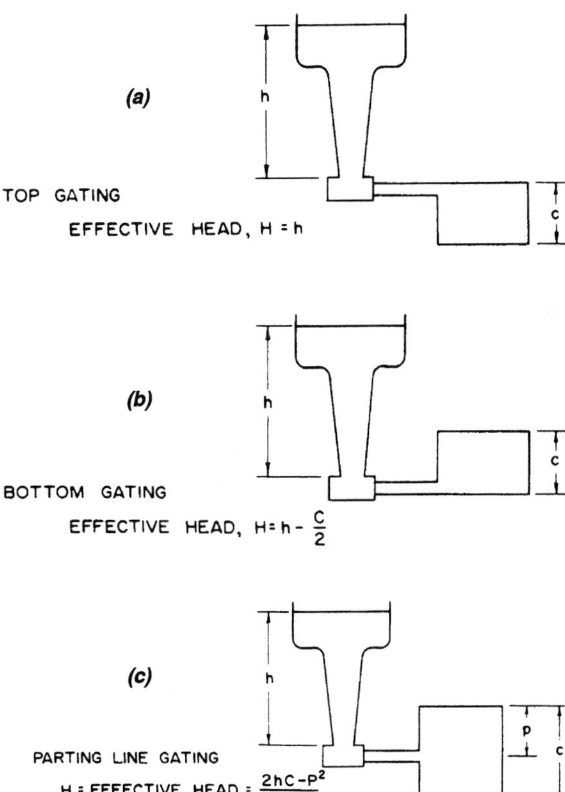

Fig. 15-42. Effective head of molten metal in different gating systems. [After Dietert]

BIBLIOGRAPHY

Aluminum Casting Technology, 2nd. Edition, American Foundrymen's Society, Inc., Des Plaines, IL 1993.

Beeley, P.R., *Foundry Technology,* Halsted Press, Div. John Wiley, New York, 1972.

Campbell, J., *Castings,* Butterworth-Heinemann Ltd., Oxford, England, 1991.

Casting Copper-Base Alloys, American Foundrymen's Society, Inc., Des Plaines, IL 1984.

Cupola Handbook, American Foundrymen's Society, Inc., Des Plaines, IL, 1999.

DYPUR , Technical Bulletin, Foseco, Inc., Cleveland, OH.

Eastwood, L.W., "Symposium on the Principles of Gating," *AFS Transactions,* vol 59, page 25, 1951.

Filter Application (Cifunsa Presentation), Hamilton Technical Ceramics, Paris, Ontario.

Gauckler, L.J., Waeber, M.M., Conti, C., Jacob-Duliere, M., "Ceramic Foam for Molten Metal Filtration," *Journal of Metals,* Sep, 1985.

Grube, K., Eastwood, L.W., "A Study of the Principles of Gating," *AFS Transactions,* vol 58, page 76, 1950.

Grube, K.R., Kura, J.G., "Principles Applicable to Vertical Gating," *AFS Transactions,* vol 63, page 35, 1956.

Grube, K.R., Lang, R.M., Kura, J.G., "Modifications in Vertical Gating Principles," *AFS Transactions,* vol 64, page 54, 1956.

Heine, R.W., Rosenthal, P.C. Loper, C.R, *Principles of Metalcasting,* McGraw-Hill, New York, 1955.

Introduction to Metalcasting, Handouts & Reference Materials, AFS.

Kannan, S., Duque, R.A., *Effective Molten Metal Filtration: A Partnership in Quality,* SELEE Corporation, Hendersonville, NC.

Karsay, S.I., *Ductile Iron Production I,* Rio Tinto Iron & Titanium Corp., Rosemont, IL, 1976.

Meader, R.F., "Gating of Gray Iron Castings," Whiting Machine Works, Whitinsville, MA, n.d.

Mollard, F.R., Davidson, N., *Ceramic Foam—A Unique Method of Filtering Molten Aluminum in the Foundry,* Consolidated Aluminum Corp., St. Louis, MO, Apr, 1978.

Murphy, F.E., Jackson, G.J., Rosenberg, "Bronze Valve Vertical Gating in Shell Molds," *Modern Casting,* page 81, July 1961.

Polich, R.F., Saunders, A. Jr., Flemings, M.C., "Gating Premium Quality Castings," *AFS Transactions,* vol 71, page 418, 1963.

Sandford, P., Pischel, R.P., *New Applications for Ceramic Foam in Gravity Permanent Mold Casting,* Foseco, Inc. Cleveland, OH.

Schleg, F.P., Personal class notes from AFS-Cast Metals Institute, Inc., Gating and Risering Courses and Nonferrous Workshops, 1965–1995.

Schmahl, J.R., Davidson, N.J., *The Application of Ceramic Foam Filter Technology in the Aluminum Foundry,* Consolidated Aluminum Corp., St. Louis, MO.

Svoboda, J.M., *Basic Principles of Gating & Risering,* AFS-Cast Metals Institute, Inc., Des Plaines, IL.

Sylvia, J.G., *Cast Metals Technology,* AFS-Cast Metals Institute, Inc., Des Plaines, IL, 1990.

Taylor, H.F., Flemings, M.C., Wulff, J., *Foundry Engineering,* John Wiley, New York, 1959.

Wallace, J.F., Evans, E.B., "Principles of Gating," *Foundry,* page 74, Oct 1959.

Solidification of Metals and Alloys

16

Any discussion of the solidification of metals should begin with a basic understanding of heat transfer as it applies to the metalcasting process. One could say that the metalcasting process is based on heat transfer. First, solid metal is heated until melted, poured into a mold, then the heat is removed so it can return to a solid.

Proper melting practice not withstanding, the most critical step is how the heat is removed from the casting. Properly handled, good mechanical properties will result; improperly handled, casting defects will occur. The well-known metalcasting researcher, William S. Pellini, Head/Metal Processing Branch, Metallurgical Division, Naval Research Laboratory, Washington, D.C. wrote the following:

"Heat transfer is the very basis of the foundry industry. In great part, the foundrymen's everyday work relates to transferring heat in and out of the metal in such a way as to end up with a saleable casting and a profit. The efficiency with which he can accomplish this is dependent on his knowledge of established practices and on the correctness of his intuitive deductions in cases such that the established know-how falls short of his requirements. The foundryman can never know enough regarding his most important working tool: practical heat transfer."

This quote is taken from the 1953 *AFS Transactions*, "Practical Heat Transfer: An Interpretive Report." Pellini's report is based on research he had conducted under the auspices of the AFS Heat Transfer Committee.

This chapter will focus on *basic* "practical heat transfer" as it applies to the production of quality metalcastings.

HEAT TRANSFER

Heat transfer can be a very complicated subject. As first stated, in the metalcasting process, solid metal is heated to a specific temperature and poured into a mold. The mold is cooled (heat is removed) returning the metal to solid in the form of a casting. Alloy melting was covered in Chapters 11 and 12; the latter stage of heat transfer will be discussed here.

Measurement of Heat

Heat is a form of energy. In the case of metal, the heat is contained in the binding forces between atoms and in the vibrational movement of these atoms. As heat is added to the solid metal during the melting process, these atoms begin to vibrate causing the binding force holding the atoms together to weaken. Finally, after sufficient heat has been added to this vibrating mass, it becomes liquid. Additional heat, called superheat, is added to the liquid mass so there is adequate time to prepare the molten metal for pouring into the mold.

The quantity of heat is measured in calories (cal) or British thermal units (Btu). By definition, a calorie is the amount of heat required to raise the temperature of one gram of water one degree centigrade, at standard barometric pressure and at a standard temperature. A Btu is defined as the amount of heat required to raise one pound of water one degree Fahrenheit, at or near its point of maximum density.

Temperature is a measure of the amount of heat in an object or mass. Temperature is determined by three properties of the object or mass, namely:

1. *Specific heat of solid*—the amount of heat required to raise one unit of material one degree of temperature.
2. *Latent heat of fusion*—the amount of heat required to melt one unit of material at constant temperature.
3. *Specific heat of liquid*—the amount of heat required to raise one unit of liquid material one degree in temperature.

The effects of these three properties can be seen on a simple cooling curve such as that seen in **Fig. 16-1.** Note that, as the temperature of the molten metal drops, it remains liquid until point "A." During this temperature drop, the liquid is surrendering its "specific heat of liquid." At point A, solidification begins and ends at point B. Note that between points A and B, the temperature of the metal remains the same. Between these two points, the metal is losing its "latent heat of fusion." Then from point B on down, the metal is solid and gives up its "specific heat of solid." This cooling curve represents a pure metal or eutectic alloy. Cooling curves for a solid solution or an insoluble solution alloy would be different.

As a comparison of these three properties of metals, **Table 16-1** shows a chart listing the necessary Btu's to raise aluminum, copper and iron from 70F (21C) to a superheat of 200F (111C).

It is interesting to note that it takes more Btu's to raise the aluminum to a superheat of 200F than it does to raise the iron to the same superheat. Look at the chart more carefully to see why this is true. Note that the *specific heat of solid* and the *latent heat of fusion* for aluminum are greater than that for iron. The major concerns in producing quality metal castings are the control and dissipation of superheat and the latent heat of fusion.

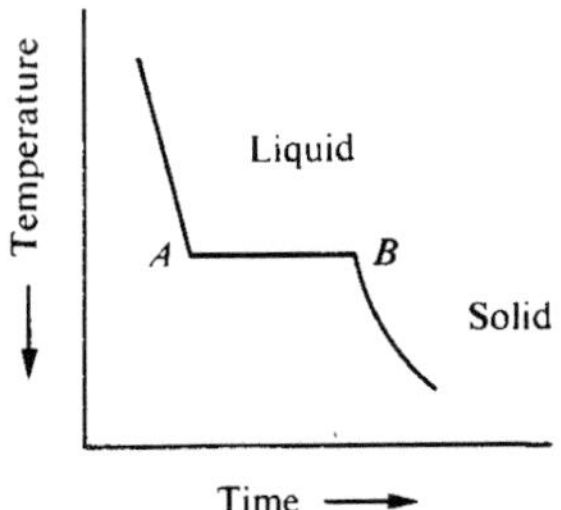

Fig. 16-1. Cooling curve indicating change that takes place during solidification of a pure metal.

Table 16-1.
Proportions of Heat Needed to Melt Metal

	Aluminum	Copper	Iron	Units
Melting temperature	1220	1981	2798	°F
Specific heat of solid	.215	.092	.110	Btu/lb °F
Latent heat of fusion	170	91.1	117.9	Btu/lb
Specific heat of liquid	.258	.118	.146	Btu/lb °F
Heat to raise 1 ton from 70°F to melting temp.	494,000 53%	352,000 61%	600,000 67%	Btu
Heat to melt 1 ton at melting temperature	340,000 36%	182,000 31%	235,000 26%	Btu
Heat to provide 200°F to 1 ton	103,000 11%	47,000 8%	57,000 7%	Btu

Measurement of Thermal Gradient

A piece of metal with a lower temperature than another has less heat contained within. Since, nature seeks balance it will try to equalize the temperature of the two pieces of metal. The closer in proximity these two pieces of metal are, the easier it is for nature to do its job. In doing so, nature will cause heat to flow from the hotter piece of metal to the colder piece. Simply put, heat flows from the hot object to the cold object, or in this case, from the hot molten metal to the surrounding air, ladles, mold and cores, if used.

In order for heat to flow, there must be a driving force to make it happen, and this force is the temperature difference. In order for the molten metal to solidify in the mold, heat transfer must occur. The speed at which this heat transfer occurs is called a temperature gradient or, more commonly, *thermal gradient.*

Thermal gradients are measured by the difference in temperature between two points, and are designated in degrees Fahrenheit or Celsius per unit of time or distance. In most cases, a thermal gradient is designated in units of distance. **Figure 16-2** shows an example of how a thermal gradient can be calculated. In this case, the molten metal is an aluminum alloy and the mold is formed from green sand.

The steeper (larger) the thermal gradient, the quicker heat transfer will occur, and the quicker the molten metal will solidify. As one would suspect, as the heat is transferred into the mold material, the mold material's temperature will rise and the molten metal temperature will decrease. As time progresses, the thermal gradient slows, and solidification in the molten metal begins to slow. This reaction can be compared to downhill snow skiing. A skier racing down a steep slalom course is traveling faster than a skier is on the gentle beginner slope. Initially, the thermal gradient between the molten metal and the mold is similar to the skier racing down the steep slalom course. As the thermal gradient decreases, it is similar to the skier slowly gliding on the beginner hill. The metalcaster can control thermal gradients through gating design, the use of various mold and core materials, and the use of chills.

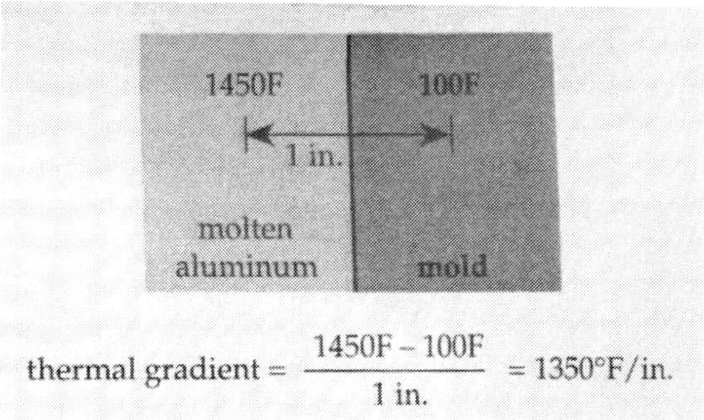

Fig. 16-2. How to measure a thermal gradient in the mold, based on distance.

Methods of Heat Transfer

There are three forms of heat transfer, which are classified by their method of moving heat: conduction, convection and radiation. In the case of producing a metalcasting, all three methods are employed.

Conduction—Scientifically speaking, conduction is a process by which heat flows from a region of higher temperature to a region of lower temperature within a solid, liquid or gas, or between different mediums, by direct physical contact. In conduction heat transfer, the heat energy is transmitted by direct atomic communication without appreciable displacement of atoms. Conduction heat transfer can be experienced by holding one end of a metal bar and placing the other end over a source of heat. Soon, the heat will transfer to the other end of the bar.

Convection—Convection is the transfer of heat from one object to another through air or water. Examples of this would be in a room heated with a furnace not equipped with a blower. The air heated by the furnace would rise to the ceiling and the cold air would drop to the floor, where it would then become heated. On the other hand, in gas-fired crucible melting the heat from the gas flame is conducted through the wall of the crucible to the melt close to the crucible wall. From this point on in the melt, the heat is transferred through convection currents.

Radiation—The earth is heated by the sun using radiation as the method of heat transfer. Radiation of heat can occur in air as well as in a vacuum. All bodies radiate heat, and the intensity of the radiation depends on the temperature and the nature of the surface. This latter property is known as the *emissivity* of the body. An example of poor emissivity can be found in aluminum alloy melting. One can almost touch an aluminum alloy casting just removed from the mold before feeling any heat. Aluminum alloys have poor emissivity. Just the opposite of this would be true for a carbon steel casting that has risers open to the atmosphere. One can feel the radiant heat from some distance away. This would be an example of good emissivity. Radiant heat transfer in the metalcasting process is most important at high temperatures.

In addition, it should be remembered that radiation takes place in a straight line. This can be seen in a foundry, or elsewhere, when people will place a cooling fan between themselves and a source of radiant heat. In reality, they are cooling themselves by convection and the evaporation of sweat by the convection currents set up by the fan. If the movement of air could push radiation around, the sun could never warm the earth due to the "jet streams" in the atmosphere.

In the metalcasting process, all three methods of heat transfer—*conduction, convection* and *radiation*—occur as the casting solidifies.

Areas of Heat Transfer

This section will look at where heat transfer by conduction, radiation and convection takes place after melting the charge materials in a melting furnace.

Tapping

The first location the molten metal loses heat is during tapping out of the furnace. As the stream of molten metal exits the furnace, it passes through the air and gives off heat due to *radiation*. It also loses heat to the air currents that may be passing around the stream due to *conduction* and *convection*.

In foundries using continuous tap cupolas, a covered launder is used to transfer the molten cast iron to a holding furnace or forehearth. A launder is a channel for conducting molten metal. The refractory covering helps to reduce the amount of heat loss in the flowing stream of molten metal.

For cast iron alloys, it is estimated that the temperature loss in molten cast iron will be approximately 125F (52C) every time the molten metal stream passes through the air. This would hold true for other alloys especially those with good emissivity. The temperature loss will vary, although there would be heat transfer due to *conduction* and *convection*. Higher molten metal temperatures will increase the amount of heat due to radiation.

Pouring

The next area of concern in heat transfer is within the ladles, both transfer and pouring types. **Figure 16-3** shows three different scenarios in which heat can be lost in a ladle.

In the first situation, the ladle is not preheated and is not covered. Note that there will be heat transfer into the ladle refractory by *conduction,* and into the atmosphere through the open ladle by radiation.

The next situation shows a properly preheated ladle being used to reduce the amount of heat lost by *conduction* into the refractory of the ladle. However, there will still be a temperature reduction due to *radiation* losses. The only way to totally reduce the temperature loss due to *conduction* is to have the ladle lining at the same temperature as the molten metal when poured into the ladle.

Finally, when using a preheated and covered ladle, the temperature loss can be appreciably reduced. As the sketch indicates, the temperature lost by *radiation* will be radiated back into the molten metal. Using covered ladles also makes the work environment around the pouring crew more comfortable. Using a preheated and covered ladle is more energy efficient and cost effective.

Table 16-2 summarizes temperature losses during the pouring and holding of aluminum, copper and iron. Note that aluminum loses a greater amount of temperature through *conduction* than it does through *radiation* and *convection*. Remember that aluminum has poor emissivity and thus loses very little heat due to *radiation*. On the other hand, iron has excellent emissivity and loses most of its heat due to *radiation*. Copper falls between the values for aluminum and iron.

Table 16-2. Temperature Losses During Pouring and Holding

	Aluminum	Copper	Iron
Heat loss from pouring stream for 20 lb casting			
(a) Radiation	1.6	7.0	46.0
(b) Convection	4.8	5.2	6.8
Holding 2000 lb in cold ladle for 10 min			
(a) Radiation	3.0	50.0	250.0
(b) Convection	0.6	0.6	3.6
(c) Conduction	12.4	22.6	31.0
Holding 200 lb in cold ladle for 10 min			
(a) Radiation	6.0	75.0	375.0
(b) Convection	17.0	54.0	62.3
(c) Conduction	27.2	34.0	34.2

Notes: Units = °F
Idealized conditions assumed

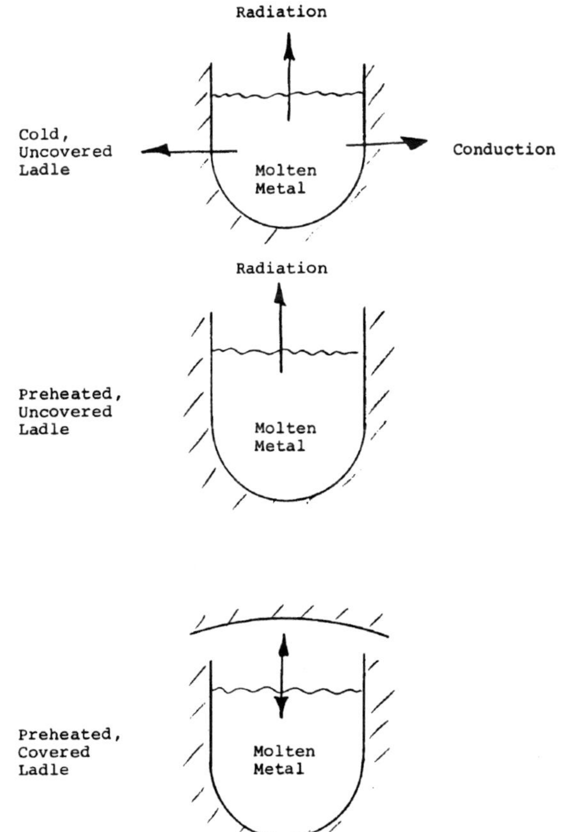

Fig. 16-3. Various ways heat is lost from ladles.

Molds

Depending upon the molding process, when the molten metal is poured into the mold, all or some of the methods of heat transfer occur.

Green Sand Molds—In a green sand mold, all three methods of heat transfer take place. As the molten metal enters the mold, it first *radiates* heat to the internal mold surfaces. Those metal alloys with higher pouring temperature and good emissivity will *radiate* more heat to the mold surfaces than those poured at lower temperatures and having poor emissivity.

When the molten metal touches the mold surfaces, heat transfer by *conduction* begins. Since all of the metal alloys poured in foundries have pouring temperatures above 212F (100C), the water present on and near the surface of the mold turns to steam, which absorbs heat. The steam returns to the mold through the permeability of the sand vapor transport or *convection*. This continues into the mold until the temperature reaches 212F (100C), where the steam condenses back to water. It should be pointed out that the conversion of water to steam would speed up the solidification of thin-walled castings. However, in heavier section metal castings, this conversion of water to steam has little effect on the speed of solidification.

With no water at the mold surface and, theoretically, a dry sand mold, heat transfer takes place by *conduction* and *radiation*. *Conduction* takes place through the grains of sand, individually and collectively, when in contact. *Radiation* takes place from grain to grain through air pores (permeability) of the sand mold. **Figure 16-4** is rough sketch of how these methods of heat transfer take place in a green sand mold.

Nobake Molds—In the case of nobake molds, heat transfer takes place by *conduction* and *radiation*. If there is moisture present in the mold, then *convection* can also become a player in heat transfer. As the binders volatilize they will absorb heat, and the gases formed will move through the mold in the form of *convection*. Heat will also be transferred due to *radiation* between sand grains because of the permeability of the mold.

Heat transfer research has shown that, as the molding sand is heated throughout, its heat *conduction* improves. It has also been shown that the *conductivity* of sand increases with temperature, since most of the heat transfer is by *radiation* from grain to grain...the higher the temperature, the greater the transfer of *radiant* heat.

Metal Molds—In the case of metal molds, such as those used in permanent molding and diecasting, heat is transferred due to *conduction* and *radiation*. In these cases, when the molten metal contacts the mold surface, a very steep thermal gradient is set up and maintained for a period of time, depending upon the ability of the mold material to *conduct* the heat away from the surface. Almost instantaneously, the casting forms a skin of solid metal; as this skin forms, it contracts or shrinks away from the mold surface, leaving an air gap. After this air gap is formed, the only way to transfer the remaining heat from the solidifying casting is by *radiation* to the mold surface.

In permanent molding, the mold face or surface is coated with a refractory. The coating can be used for several reasons. It is primarily used to protect the surface of the mold, but can also be used to control heat transfer. This coating can act as an insulator to slow heat transfer or as a chill to speed up heat transfer to the mold.

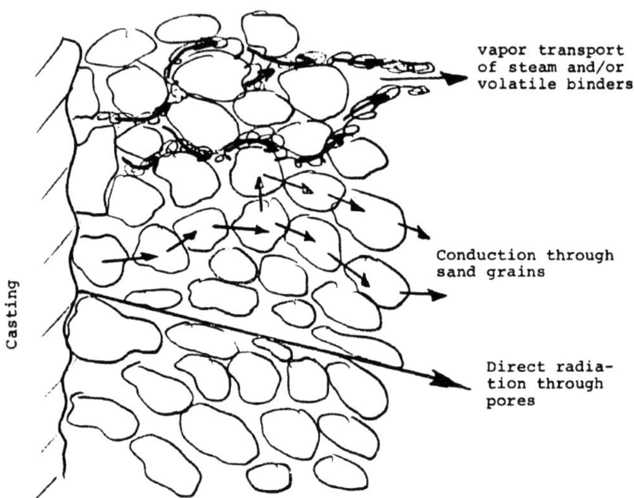

Fig. 16-4. Heat transfer mechanisms in sand molds.

Sand Mold Air Gaps—In reality, the same phenomenon of air gap formation between the solidifying casting and the mold can occur in sand molds. The molten metal-mold interface has many contact points between the molten metal and the mold surface. It is during this period that numerous nucleation sites are set up. However, as the molten metal continues to solidify at this interface, the solidified skin of metal begins to contract or shrink away from the mold surface.

At this time, the sand mold is also beginning to deform. If this deformation of the mold maintains contact with the solidifying metal skin, heat transferred by *conduction* can continue. In most cases, this will not happen, and an air gap will develop between the solidifying skin of metal and the mold surface. *Air gap* may be a misnomer in that the gap is filled with mold gases, mainly hydrogen. This gap will be composed of a mixture of air and hydrogen, which is conducive to heat *conductivity*. **Figure 16-5** illustrates the point.

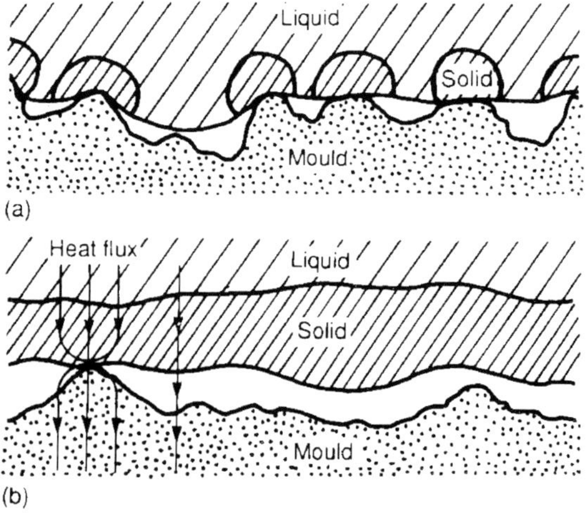

Fig. 16-5. Metal-mold interface at early stage when solid is nucleating at points of good thermal contact. Overall macroscopic contact is good at this state (a). Later (b), the casting gains strength, and casting and mold both deform, reducing contact to isolated points at greater separations on nonconforming rigid surfaces.

Gates/Runners—Heat is lost in the gating system primarily by *conduction.* In order to prevent excessive heat loss, consider the following:

- Area of contact between the molten metal and gating system surfaces.
- Time of contact, which is determined by the pouring rate.
- Thermal properties of the molten metal and mold materials.
- Temperature difference between the molten metal and the mold materials.

Keep in mind that the greater the surface area of the gating system, such as with shallow wide runners, the greater will be the heat loss from the gating system. For this reason, most runners are trapezoidal in shape and not shallow wide rectangles.

Control of Heat Transfer

The casting designer can greatly influence the transfer of heat by the design of the casting. The use of very thin small cores can lead to problems with heat transfer. Pockets of small sections of molding sand enclosed by thick sections of molten metal will also hinder heat transfer. When designing a casting, the designer should think of the thermal design of the casting as well as its physical design.

A simple example of the thermal design of a casting is shown in **Fig. 16-6.** If the temperature of the molding sand was measured in the pocket area between the flanges, around the center body of the casting and in the core sand shortly after pouring, it is highly likely that the temperatures would be almost as high as the solidifying molten metal. This would occur because the heat transferred from the molten metal is trapped within these pockets. Thus, with this design, one would expect to find defects in these areas of the casting. However, computer programs analyzing casting solidification, can help to eliminate these problems. If the design must remain the same, the metalcaster can work around some of these issues to help overcome design-related faults.

Thermal Deception

There are many times when the metalcaster will want to slow or speed up the rate of heat transfer within the solidifying casting. This can be done by thermal deception.

An example of thermal deception involves the casting sketch seen in **Fig. 16-6.** There will be a hot spot in the pocket where the flange and the valve body meet. If the thermal gradient can be increased at this junction, removing much of the heat very quickly, the chances of having heat-related casting defects in this location can be reduced.

How can this be done? Back in Chapter 4 on Molding Sands, specialty sands that could be used as a chill were discussed. A chill is used to quickly remove heat from an area in the mold or casting, when necessary. Two specialty sands that can be used for this purpose are zircon and chromite sand.

Table 16-3 shows a comparison of some physical properties of silica, zircon and chromite sands. Notice the difference in thermal conductivity between the three sands. Chromite has the highest thermal conductivity, followed by zircon and, finally, silica. Next, compare the density and specific heat of these sands and find that zircon and chromite excel in these physical properties. It is for these two reasons—density and specific heat—that zircon and chromite have higher heat diffusivity. Heat diffusivity of a material is its ability to remove heat from the heat source and transfer it to a cooler area. Thus, if production time and costs would allow, a zircon or chromite molding sand mixture could be placed in this pocket, then be backed up with silica molding sand mixture. This is what is meant by thermal deception or trickery.

Many times, cores, which are subject to extreme temperatures over a long period, can be made using zircon or chromite sand. In less extreme cases, a zircon- or chromite-based wash can be applied to the core surface. Permanent mold metalcasters use often use different types of coatings on the mold's internal surfaces. They can also use insulating materials or position water lines around the outside of the mold to control the amount of heat transfer in the mold. All these are forms of thermal deception or trickery.

SOLIDIFICATION (FREEZING)

Now that there is a basic understanding of heat transfer in a casting, the solidification mechanism of the molten metal can be investigated. In this section, solidification will also be referred to as *freezing,* since it is a common term used by metalcasters when discussing solidification.

Depending on the size and section thickness of a casting, solidification can take a few seconds to hours to complete. This is an important phase of the metalcasting process. It is during this time that many important characteristics affecting the casting's quality are determined. The original crystalline structure established dur-

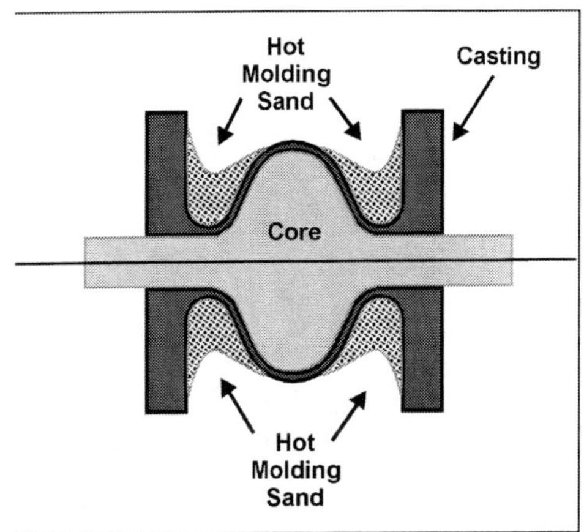

Fig. 16-6. An example of the thermal design of a metal casting.

Table 16-3.
Some Physical Properties of Foundry Sands

	Silica	Zircon	Chromite
Thermal Conductivity (K-Btu/hr ft °F)	0.51	0.57	0.63
Density (ρ-lb/ft³)	93.2	176.0	150.0
Specific Heat (C-Btu/lb °F)	0.280	0.198	0.23
Thermal Diffusivity (K/ρC)	0.0199	0.0164	0.0183
Heat Diffusivity $(K\rho C)^{1/2}$	3.68	4.46	4.67

ing this time frame determines the many physical and mechanical properties of the casting. In addition, during this period, many major casting defects can be formed, i.e., hot tears and shrinkage. Therefore, it is important to consider the mechanisms by which a metal solidifies, and the techniques that are available to control the solidification process.

The subject of solidification can become quite involved, and a detailed discussion is beyond the scope of this text. However, every attempt will be made to keep the subject at a level the reader can readily understand.

For a binary alloy system (a system of two metallic elements), the solidification proceeds in the following manners:

- At a constant temperature for eutectic alloys.
- Over a temperature range in the case of solid solutions.
- By a combination of solidification over a temperature range followed by constant-temperature solidification for proeutectic- plus eutectic-type solidification.

Solidification is primarily a process of nucleation and growth. When the molten metal makes contact with the mold surface, the sand grains or mold coating act as nucleation sites. This means that many small grains of solid metal are formed, almost instantaneously. This is also due to the very steep thermal gradient that is formed at this point. These minute grains or crystals grow under the influence of the crystallographic and thermal conditions that prevail.

Some variables that add to the complexities of solidification include the mold material and its thickness, mold and casting geometry, and thickness of the metal. Other variables to be considered include the metal's properties (such as conductivity), solidification temperature range and control of grain size (such as inoculation and grain refinement).

Heat transfer is the driving force in solidification that causes the heat to leave the molten metal, which, in turn, causes the solid state to become more stable than the liquid state. The two quantities of heat that are important to the solidification of the casting are superheat and the latent heat of fusion. Both were discussed in the cooling curves section of Chapter 13, Microstructure of Ferrous Alloys.

First, superheat must be removed to reduce the molten metal's pouring temperature to the point where solidification can begin. The latent heat of fusion is the heat given off as the molten metal goes from the liquid state to the solid state.

It should be remembered that, as the molten metal begins to cool down to ambient temperature, it will shrink, changing the dimensions of the casting. Shrinkage will be covered in Chapter 17, Risering. There are three major points to be considered when a metal casting solidifies, and they are:

1. Growth of the solid grains.
2. Heat evolution and transfer.
3. Dimensional changes.

Heat transfer and dimensional changes (or contraction) have already been discussed in this chapter. The next section will focus on grain growth.

Grain Growth/Solidification (Freezing) Ranges

Once the minute grains are formed, each of three freezing ranges follows a different mechanism toward the completion of solidification.

When discussing solidification or freezing ranges, there are normally three types: 1) narrow, 2) medium and 3) wide. (In the foundry world, these ranges are also called "short," "intermediate" and "long," as in Chapter 14 on Microstructure.)

The distinguishing factor between the three types is the temperature range at which the molten metal begins to solidify and the temperature at which it is completely solid. Phase diagrams indicate the width of the "mushy" or liquid-solid zone. There is no firm data as to what temperature range separates the three solidification ranges from one another. Generally speaking, the narrow (or short) solidification range would be from 0–75F (–27C–564C), medium (or intermediate) solidification range from 76–250F (24–121C) and finally the wide (or long) solidification range from 251F > (122C >). These temperature ranges can vary with each alloy family.

Narrow (Short) Freezing Range

First, look at the alloys that fall into the narrow freezing range. Among these metals would be the pure metals, eutectic and eutectoid alloys, some of the aluminum and copper alloys, and the carbon steels. The solidification ranges of these alloys can be found in tables that include the melting and solidification temperatures.

The basic reason that a metal or metal alloy solidifies is that the arrangement of atoms in a solid crystal is at a lower state of energy than the that in the molten metal. Above the solidification point, the liquid is more stable; below the solidification point, the solid is more stable. At the precise solidification point, there is no driving force in either direction and, thus, there is equilibrium. The molten metal will begin to solidify at some temperature below the solidification point. The greater the amount of metal cooled below the equilibrium solidification temperature, the greater the driving force for solidification to begin. The degree to which the metal is cooled below the equilibrium solidification temperature is called the *degree of supercooling.*

Because of the steep thermal gradient at the mold surface, the molten metal near the mold surface is supercooled and solidifies as small equiaxed grains. **Figure 16-7** shows the solidification mechanism for a pure metal. Note the fine, equiaxed grains that have formed at the mold surface. To spontaneously form a nucleus like this, the amount of supercooling must be great in order for groups of atoms to form that are large enough to be stable. This spontaneous type of nucleation is called *homogeneous nucleation.*

However, most metal castings solidify by *heterogeneous nucleation,* which means that the atoms attach themselves to foreign particles found in the molten metal. In this case, only a small amount of supercooling is required. These minute particles in the molten metal can be oxide inclusions, etc. Again, the mold surface also provides numerous nucleation sites. The best nucleus is a fine particle of the metal itself.

Once the nuclei or fine equiaxed grains are formed, they continue to grow by acquiring atoms from the molten metal. The latent heat of fusion is being released as the equiaxed grains are formed, which tends to reduce the amount of undercooling in the remaining molten metal; as a result, there is a tendency to stop nucleation. However, as can be seen in **Fig. 16-7,** growth continues on those grains that are favorably oriented and will continue to grow with the less-favorably oriented grains being pinched off. This is called *columnar growth* because the grains are columnar in shape.

This growth continues and is controlled by heat transfer. The greater the thermal gradient, the faster the grains will grow, or vice versa. Since the thermal gradient is perpendicular to the mold surface, the columnar growth continues parallel to, but in the opposite direction of, the heat flow. Thus, the columnar grains continue to grow into the molten metal. This type of solidification mechanism is identifiable by the "smooth" solidification front moving into the molten metal. **Figure 16-8** is another depiction of this mechanism of solidification of pure metals, eutectic or eutectoid alloys and those alloys having a narrow solidification range.

Note in **Fig. 16-8** that solidification begins at the molten metal-mold interface in the form of equiaxed grains. Although not shown, those grains that are favorably oriented begin to grow as columnar grains toward the center of the casting section. The black solid in this figure denotes the columnar grains. There is no "mushy" material in this type of solidification mechanism; there is either solid or molten metal. The solidification front proceeds toward the center of the casting section with a smooth front. The narrow solidification range alloys will solidify last in the center of the casting section. Normally, if shrinkage is to be found, it will be located at the centerline of the casting section.

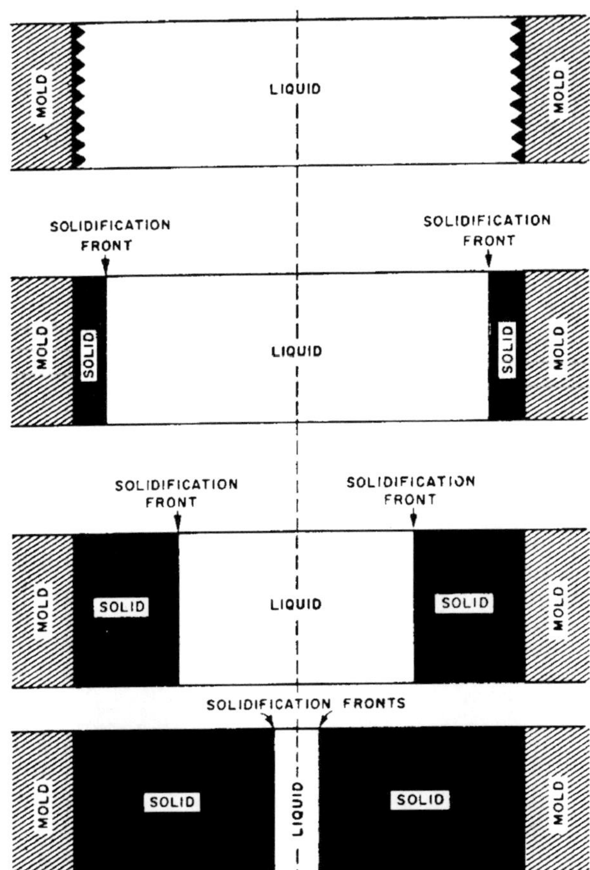

Fig. 16-8. Stages in the freezing of high-purity copper.

Wide (Long) Freezing Range

The majority of the common wide freezing-range alloys fall into the copper-base alloy family and a few in the aluminum alloy family. **Figure 16-9** illustrates the freezing mechanism of an alloy that freezes in the wide freezing-range group. It is obvious that this mechanism of freezing is very different from that of the narrow freezing range metals and alloys.

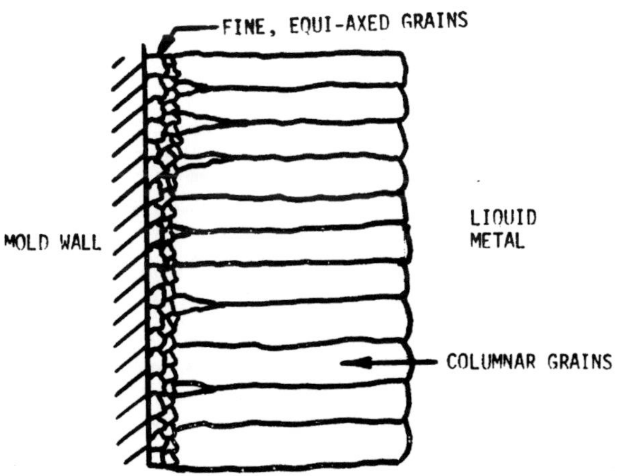

Fig. 16-7. Development of columnar grains from initial fine-grained surface layer during freezing of pure metal. [After Chalmers]

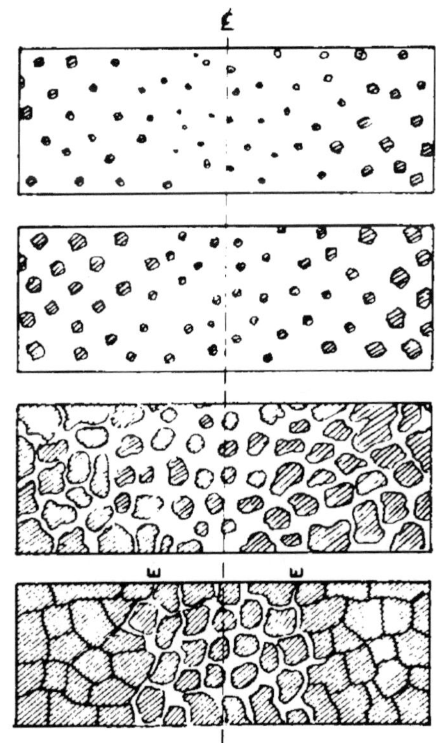

Fig. 16-9. Stages in the freezing of high-alloy copper.

Just as in the narrow freezing-range metals and alloys, solidification begins with the formation of numerous equiaxed grains at the molten metal-mold interface. Growth of these grains is almost immediately retarded or even halted temporarily. The reason for this is that the grains formed are considerably poorer in alloying elements than the molten metal from which they were formed. When these grains are formed, atoms of the alloying elements are rejected into the surrounding molten metal, thus greatly enriching the molten metal in these elements. This substantially lowers the solidification point of the molten metal so that grain growth is temporarily stopped.

Heat extraction by the mold continues and slightly lowers the temperature of the molten metal in this zone, as well as that in the unaffected zone nearer the interior of the casting or casting section. With this drop in temperature, a second group of grains are formed just outside of the enriched area. **Figure 16-10** shows this enriched low freezing-range alloy around the grains. This, in turn, restricts the growth of the second group of grains, and a third batch of grains quickly forms further toward the interior of the casting section, just beyond the enriched region around the second batch of grains.

Freezing then continues by the gradual enlargement of all of the grains. This mechanism of solidification takes place simultaneously throughout the entire casting. Those grains closest to the mold wall grow slightly faster than the grains near the interior of the casting, but the differences are not significant.

With this mode of freezing it is easy to understand why the macrostructures of these metal castings show equiaxial crystallization. This mechanism of solidification is often called "mushy freezing" and can be compared to the setting of cement. The cement is first a liquid, then changes to mushy and finally becomes rigid. In Chapter 17, Risering, there will be an explanation as to why the solidification shrinkage in this mode of freezing becomes difficult to feed.

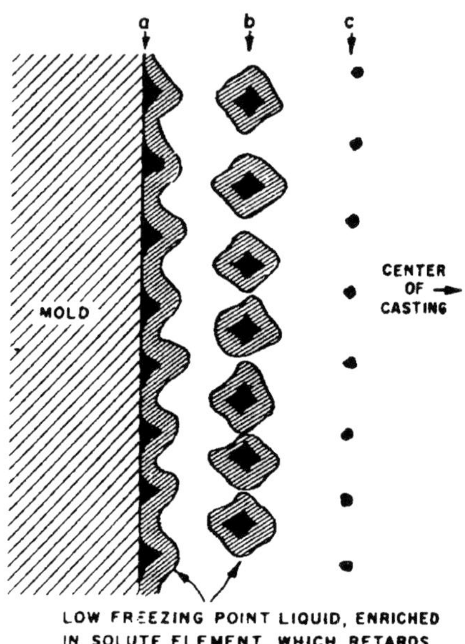

Fig. 16-10. Diagrammatic representation of restriction of crystal growth in the freezing of high-alloy copper.

Medium (Intermediate) Freezing Range

The medium freezing-range alloys solidify using a combination of the narrow and wide freezing ranges. **Figure 16-11** illustrates the medium freezing-range mode of freezing. Again, due to the steep thermal gradient at the molten metal-mold interface, small equiaxed grains of solid metal begin to form. The favorably oriented grains then begin to grow out into the molten metal. This growth is not the same as that of the columnar growth found in the narrow freezing-range metals and alloys, where there was a solid, smooth front of solid metal moving toward the center of the casting section. In this case, the growth is "spiked," with molten metal found between the spikes. These tree-like shapes are given the name *dendrites,* after the Greek word dendron, which means tree-like.

Because the molten metal between the dendrites becomes enriched in alloying elements, the dendrites cannot link up until very late in the freezing process. It can be seen in **Fig. 16-11** that small grains of metal are beginning to form in the molten metal between the two dendrite fronts. These spindly dendrites will sometimes continue to grow until they reach the center of the casting section. When this occurs the macrostructure of the entire casting will be columnar.

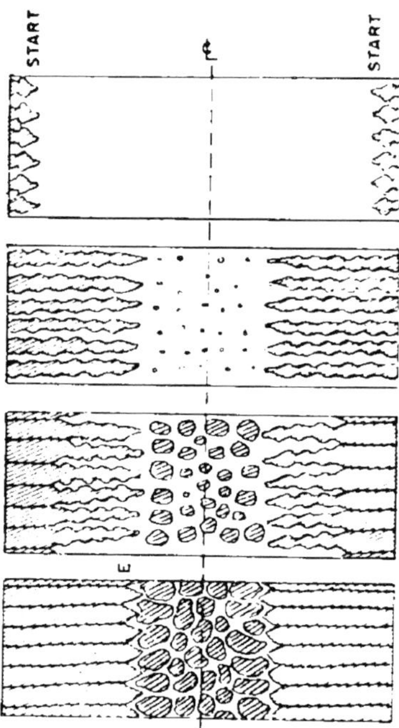

Fig. 16-11. Stages in the freezing of a copper alloy in the medium range.

Frequently, the growth of the dendrites diminishes before reaching the center of the casting section. When this occurs, the central section of the casting will freeze in a mushy manner. In this case, the central section of the casting will have an equiaxed grain structure, and the outer sections will have a columnar structure. It should be noted that, in this type of freezing mechanism, there can be a molten metal pathway that extends through the solidifying crystals to the mold wall itself, almost until the end of freezing.

Progressive and Directional Solidification

Figure 16-12 illustrates both progressive and directional solidification. These two types of solidification will play a big role in preventing shrinkage defects.

When the molten metal begins to solidify at the mold wall and move more or less evenly toward the center of the casting, it is called *progressive solidification.* When the solidification fronts meet at the thermal center of the casting, shrinkage accompanies the solidification of the remaining molten metal. This action eliminates any chance of using feed metal to prevent shrinkage voids. Feed metal is the molten metal used to feed shrinkage in the casting. As explained in Chapter 17, risers will provide this feed metal.

When solidification begins and ends at a point in the mold farthest from the source of feed metal and moves uniformly in the direction of this source, it is called *directional solidification.* Because of the shape of this solidification front, feed metal can reach this area and shrinkage cavities can be avoided. Another term used to describe directional solidification is "end effect."

Now it can be seen that there is a big race occurring between *progressive* and *directional* solidification within the casting. The winner of this race *has* to be directional solidification in order to produce sound castings (from a shrinkage standpoint). The situation can be helped by designing gating systems that will help set up the proper thermal gradients between the mold and the metal casting, which will help promote directional solidification. The casting designer should also be aware of the fact that the design will also determine the winner of the "race."

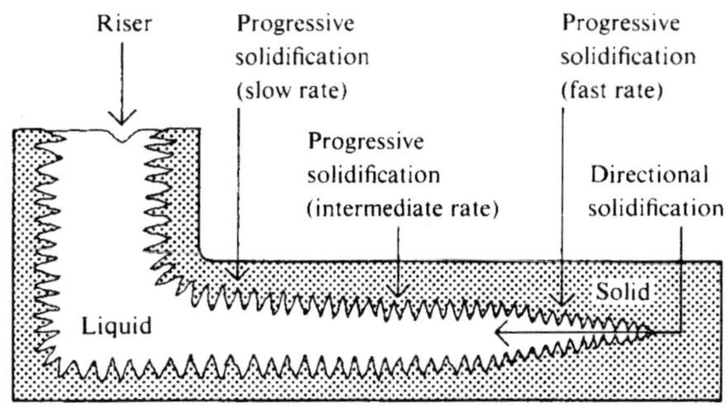

Fig. 16-12. Schematic of wall growth showing progressive and directional solidification.

Solidification of Corners

The freezing rate of a casting and its sections is determined by how fast heat can be removed from the molten metal and transferred away from the mold surface. Again, this involves basic heat transfer.

Back in 1940, a Russian by the name of Chvorinov developed a formula known as "Chvorinov's Rule." Simply put, the rule says: The rate at which a molten casting will solidify is determined by its modulus. The modulus is determined by dividing the Volume of the metal casting by its Surface Area: Modulus = V/A. (In the United States, this rule is sometimes written as Modulus = A/V.) It should be emphasized that the surface area must be cooling surfaces.

Chvorinov's work was later backed by the work of Wlodawer in 1966, whose notable volume *Directional Solidification of Steel Castings* remains an essential source of information used in the steel foundry industry today. However, regardless of the alloy poured, the modulus principle still holds true, as it did back in Chvorinov's day.

Figure 16-13 shows two basic types of corners found on castings: internal and external. The internal corner is a corner of sand that is surrounded by molten metal. If we were to study this internal corner and divide both the sand and molten metal into equal cubes, we would find that the volume of molten metal would be three times that of the sand.

To make it simple, say that each cube is one cubic inch. In the case of the sand, there is one cubic inch of sand having to absorb heat from three cubic inches of molten metal. The surface area available to transfer heat would be the same for the sand as for the molten metal. The solid metal skin shows that, at this corner, there could still be molten metal, while the solid skin has already begun to form to a given thickness, away from the corner.

This type of corner design, as the saying goes, "is waiting for an accident to happen." This type of corner should never be designed, unless it is very crucial to the application of the metal casting. As learned in the Patternmaking Chapter, this corner should have a radius or fillet. One precaution—the fillet can become too large, so the patternmaker design literature should be consulted as to what size the fillet should be.

Contrary to the internal corner, an external corner with the same conditions would have a completely different solidification pattern. Again, using the same cubes of sand and molten metal, it can be seen that the one cube of molten metal has three cubes of sand to give its heat to. This would set up a very steep thermal gradient, as the amount of solid material indicates. In addition, with the amount of

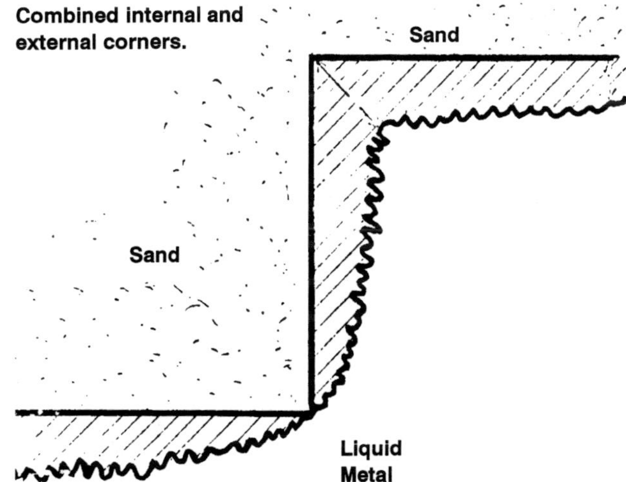

Fig. 16-13. Freezing patterns at mold corners. [After Pellini]

sand volume surrounding this corner of molten metal, a steeper thermal gradient can be maintained between the molten metal and the sand, over a longer period. Thus, there will be a thicker section of solid metal in this corner of the metalcasting.

This may sound well and good, but it can and does cause problems. First, the cast irons have a problem with this type of corner due to the rapid solidification that can cause "chill" or "white" iron, also known as iron carbide, to form. As learned earlier in the microstructure discussions, this material is very hard and can play havoc with cutting tools. Another problem that can arise, in all alloys, is the formation of chill cracks in these edges. These chill cracks can be so minute that they cannot be seen by the human eye, but will become stress raisers and could lead to metalcasting failure later in time. Sharp external corners should be rounded off or "broken" to help reduce or alleviate this problem.

Referring again to **Fig. 16-13,** the application of heat transfer and solidification can be seen in a metal casting having both internal and external corners. The same solidification patterns exist in this case as they did with the individual corners. In many casting designs, there are junctions of sections, pockets of molding sand, that will be surrounded by molten metal on three sides, and, in some cases, there will be small sand cores surrounded by molten metal. All of these situations should be looked at, just as with internal and external corners, with this in mind: How will the heat get out of the molten metal?

SOLIDIFICATION MODELING

In recent years, significant time, energy and monetary resources have gone into removing the tedious and time-consuming labor from predicting how a given casting design will solidify in a given mold when poured using a given metal alloy. This use of computer programs, or software, is called solidification modeling. There are many different types of software available, and the reader is urged to check into the availability and capabilities. A good reference source for this information is the *Modern Casting Buyer's Guide,* which is published each November and is available from the American Foundry Society.

Figure 16-14 shows computer screens from solidification modeling programs. A person draws the design of the casting—either two dimensionally or three dimensionally—and includes the cored areas. The next step is to plug in the necessary data that is used in the process of making the casting. This data usually includes alloy, pouring temperature, mold and core material. With this data, or more data if required, the computer then develops a visual animated model of how the casting will solidify.

Today, these programs have been expanded to include the proposed gating and risering systems, thus showing how the mold cavity(s) will fill and the metal casting(s) solidify. These solidification-modeling programs can be very complex or basic, and it is up to the individual prospective user to determine which one will best meet his or her needs. These solidification modeling programs have been of great help to the designer, patternmaker, manufacturing engineer and foundry team to produce a high-quality casting without a lot of guesswork and wasted time.

SUMMARY

As stated at the beginning of this chapter, the subject of heat transfer and subsequent solidification of the molten metal can be considered two of the most important facets in the process of producing high-quality metal castings. To expand on this information in more depth would require a whole new text. This chapter should provide a basic understanding of how heat transfer and solidification work together to produce castings that will meet the mechanical and, somewhat, the physical properties required of the casting. The material presented in this chapter will also help one to better understand the subject of Chapter 17, Risering Practice.

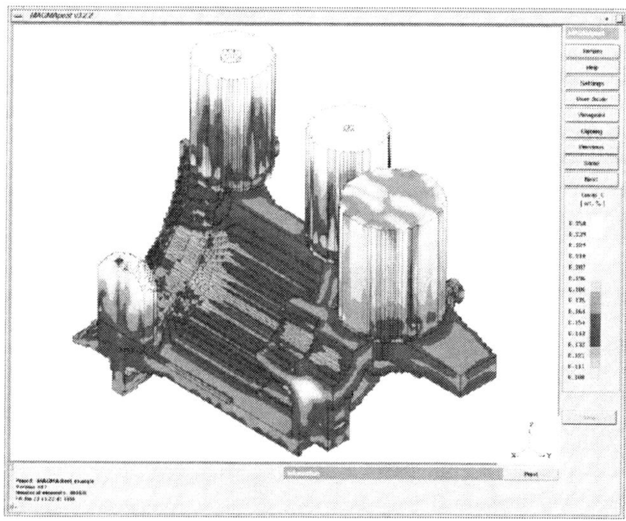

Fig. 16-14. Two examples of how the computer screen appears during modeling of solidification.

BIBLIOGRAPHY

Aluminum Casting Technology, 2nd. Ed., American Foundrymen's Society, Des Plaines, IL (1991).

Bishop, H.F., Pellini, W.S., "Solidification of Metals," *Foundry 80,* p 86 (Feb 1952).

Briggs, C.W., "Solidification of Steel Castings," *AFS Transactions,* pp 157 (1960).

Campbell, J., *Castings,* Butterworth-Heinemann Ltd., Oxford, England (1991).

Casting Copper-Base Alloys, American Foundrymen's Society, Des Plaines, IL (1984).

Chvorinov, N., "The Theory of the Solidification of Castings," *Giesseri,* vol 27, pp 177-225, British Iron and Steel Institute, Translation No. 117 (1940).

Heine, R.W., Loper, Jr., C.R., Roesenthal, C.R., *Principles of Metalcasting,* 2nd ed., McGraw-Hill Book Company, New York (1967).

Henzel, Jr., J.G., Keverian, J., Heat Transfer as Applied to Gating and Risering, Class Notes Prepared for AFS/Cast Metals Institute Courses.

Pellini, W.S., "Practical Heat Transfer, An Interpretive Report," *AFS Transactions,* pp 603 (1953).

Ruddle, R.W., *The Solidification of Castings,* Institute of Metals, London (1957).

Schleg, F.P., Personal class notes from American Foundrymen's Society/ Cast Metals Institute Gating and Risering Courses 1965–1992.

Svoboda, J.M., *Basic Principles of Gating & Risering,* AFS/Cast Metals Institute Course Textbook (1976).

Sylvia, J.G., *Cast Metals Technology,* American Foundrymen's Society/ Cast Metals Institute, Des Plaines, IL (1965).

Taylor, H.F., Flemings, M.C., Wulff, J., *Foundry Engineering,* John Wiley, New York (1959).

Wlowader, R., *Directional Solidification of Steel Castings,* English Translation by Hewitt, L.D. and Riley, R., Pergamon Press (1966).

Risering Practice

17

The primary function of a riser is to serve as a reservoir of molten metal to feed the shrinkage that occurs as the casting solidifies. However, to perform this function, the riser must be designed so that the molten metal remains liquid (or molten) long enough to feed the shrinkage and that it is located in the proper place.

The ability of the riser to meet these requirements depends considerably on the type of metal being cast. For instance, gray cast iron requires less feeding than do other alloys because of the graphitization that occurs during the final stages of solidification. In fact, this graphitization causes an expansion that tends to counteract metal shrinkage. On the other hand, alloys with wide freezing ranges require large and sometimes elaborate risering or feeding systems to produce sound castings.

SHRINKAGE

The term *shrinkage* has been used many times throughout this book. What is shrinkage? As learned earlier, when a solid metal is heated to above room temperature, it expands. This expansion is due to the atoms within the metal becoming excited and moving about. If more heat is added to this metal, the atoms get even more excited, and the solid metal will turn to liquid metal. The liquid metal will be larger in volume than it was in the solid state. If even more heat (superheat) is added to the liquid metal, the atoms increasingly get more excited, and the liquid metal volume continues to increase.

Once the superheat is stopped, the temperature of the liquid metal begins to drop, the atoms begin to relax and the volume decreases or shrinks. When heat was added, the metal expanded; now, as the liquid metal begins to cool, it contracts (shrinkage). As the liquid metal cools, this reduction in volume will continue, even as the liquid metal begins to solidify and finally cools to room temperature. The more technical term for the shrinkage that takes place as the metal changes from liquid to solid is called "volumetric shrinkage."

An example of shrinkage can be seen in **Fig. 17-1**, which depicts the slow cooling of molten steel alloy in a very simple mold. In the liquid state, this molten metal has a volume of 1000 in.^{3} (16,387 cm^3). As the liquid begins to cool, it begins to shrink or contract. This first change in volume is called *liquid* shrinkage or liquid contraction. The next shrinkage to take place begins as the liquid metal begins to solidify. This shrinkage is called *solidification* shrinkage or solidification contraction. Finally, when the liquid metal has completely changed to solid, *solid* shrinkage or solid contraction takes place. After solid shrinkage, the casting volume is 880 in.^{3} (14,421 cm^3).

Figure 17-2 represents the volumetric shrinkage of a steel alloy, by using a simple cooling curve and showing the various amounts of shrinkage that occur. Of the three types of shrinkage, the riser is involved with only the first two: liquid shrinkage and solidification shrinkage. The amount of the third type, solid shrinkage, is built into the pattern by the patternmaker. **Table 3-1** in Chapter 3 provides the approximate amounts of solid shrinkage for various alloys.

Table 17-1 lists approximate total volumetric shrinkages for some commonly poured casting alloys. The literature also contains this information for other casting alloys.

Notice in **Table 17-1** that gray cast iron shows a possible negative volumetric shrinkage. As discussed in earlier chapters, some gray cast irons will actually expand upon solidification when graphitization takes place. If this expansion is great enough and can be contained in the casting, it can eliminate shrinkage. There is a simple formula used to help determine whether this graphite expansion is equal to the volume of shrinkage in the casting.

Graphite expansion = % carbon + 1 / 7% silicon

If the graphite expansion is equal to 3.9% or greater, the expansion will be equal to the volumetric shrinkage in the cast iron casting. The technique in using this type of expansion successfully is to contain the expansion within the casting without distorting the mold. If the mold into which the cast iron is poured is not strong enough to resist or constrain this expansion, the mold wall can become distorted. This action is also called "mold wall movement."

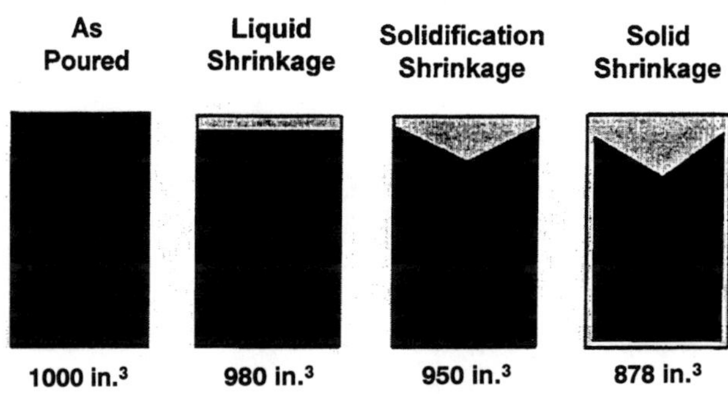

Fig. 17-1. Three phases of shrinkage: liquid shrinkage; solidification shrinkage; solid shrinkage.

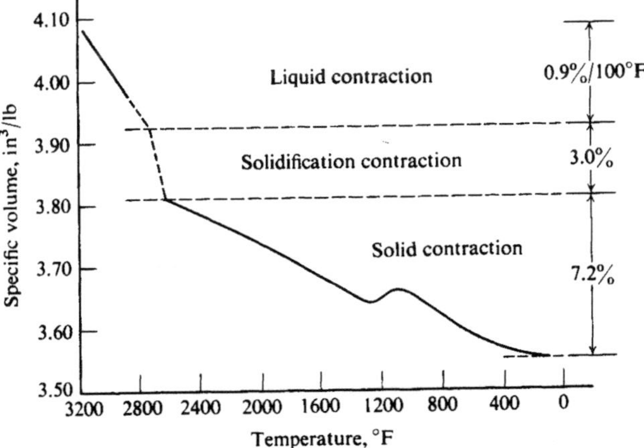

Fig. 17-2. Change in specific volume of solidifying and cooling steel.

Table 17-1.
Approximate Total Volumetric Shrinkages for Some Common Metalcasting Alloys

Metal or Alloy	Total Volumetric Shrinkage (Approx. %)
Pure aluminum	8.0
Al-Mg-Si (5%)	6.7
Al-Si (12%)	3.5
Pure copper	4.0
Brass	6.5
Bronze	7.5
Al bronze	4.0
Medium carbon steel	2.75
1% carbon steel	4.0
Gray cast iron	1.9 to neg.
White cast iron	4.5
Ductile cast iron	
Up to 1-in. section thickness	2.0
Over 1-in. section thickness	4.0

RISER DESIGN

As the casting moves through the stages of shrinkage, so does the riser. Not only is the riser going to be held responsible for feeding the liquid shrinkage and solidification shrinkage in the casting, but it must feed its own shrinkage. The placement, shape and size of risers are very significant elements of the casting design.

Determining Riser Placement

Figure 17-3 shows how and where risers can be attached to a casting. Riser names correspond to their location on the casting. For instance, a riser attached to the top of a casting is known as a top riser. Risers attached to the sides of a casting are called side risers. There is also another matter that determines the name of the riser and that is the path the molten metal travels to fill the risers. **Figure 17-3** shows the molten metal traveling from the gate directly into a riser. This riser is called a hot or live riser. If the molten metal has to pass through the casting before filling the riser, the riser is called a cold or dead riser.

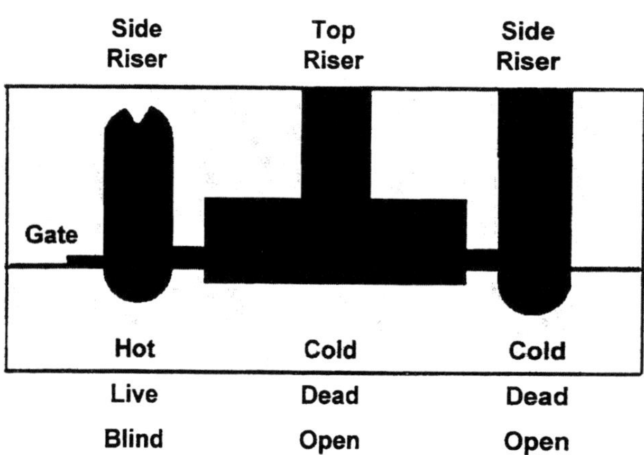

Fig. 17-3. Various ways risers can be attached to a metal casting.

It is desirable, whenever possible, to use a "hot side riser," because the last molten metal, and the hottest, coming out of the gating system will be found in the riser. Using a hot side riser increases the chances of the riser remaining liquid long enough to feed the shrinkage taking place in the casting and in the riser.

There are times when a hot side riser is not practical, and a "cold top" or "cold side riser" must be used. This could occur when the molten metal passes through the mold cavity, where it begins to lose heat and has cooled slightly before reaching the riser. There are methods to help this type of riser stay liquid long enough to be effective, and will be discussed later.

There is also another determining factor to help decide the location of the riser. Again, looking at **Fig. 17-3** (specifically the two side risers), it can be seen that the riser on the left does not extend to the cope surface, whereas, the one on the right does. The riser on the left is called a "blind riser" and the one on the right an "open riser." Simply put, the riser on the right is "open" to the atmosphere, and the riser on the left is "encapsulated" within the mold. Later, it will be explained why a blind riser may be preferred for some alloys.

Grouping Castings

A common practice to more efficiently use risers is to group a number of castings around one riser. The number of castings to be fed from a common riser depends upon the casting size and the layout of the pattern plate. Normally, the riser will feed into the same area of each casting. Using this practice can be more economical, because using one common riser requires less molten metal than if each casting were fed by its own riser. This grouping of castings around the common riser will also reduce the rate of heat loss from the riser and help it to remain molten.

Determining Riser Shape

A riser must be designed and sized so that there is enough molten metal available to compensate for the shrinkage in the casting and that it solidifies after completing its task. For example, if the expected 6% shrinkage will equal 1 lb (0.45 kg) of aluminum, then the riser has to have that amount of aluminum available to feed shrinkage. The key is, however, that the riser has to remain liquid longer than the aluminum casting to accomplish this task.

Remember that the riser is solidifying at the same time as the casting is, but still must have the proper amount of molten metal available to feed the shrinkage occurring in the casting. One way to help the riser meet this challenge is to consider its geometric shape.

The geometric shape should be one that, for a given volume of molten metal, has the least amount of surface area through which it can lose heat. **Figure 17-4** shows four common geometric shapes, all having a volume of 1000 in.^{3} (16,387 cm^3). It will be assumed that the molten metal temperature is the same in all four cases.

When calculating the surface area-to-volume ratio, it is found that the sphere has the smallest ratio and, thus, would be the best choice because the metal would remain molten longer when compared to the other three. Therefore, in theory, spherical shaped risers should be used.

There is a problem with spherical risers, when it comes to horizontally parted molds. A spherical riser will need to have its diameter on the parting line of the mold to facilitate drawing the mold off the pattern. In this case, the riser may not have sufficient metallostatic head above the casting to allow the feed metal to adequately flow into the casting, and shrinkage defects may occur. Spherical risers can be used in the full mold process and, in some cases, vertically parted molds.

The next best shape to use is a cylinder. **Figure 17-5** illustrates the use of a cylindrical blind, hot, side riser. Several alterations can be made to the geometry of this riser to improve its surface area-to-volume ratio:

1. A hemispherical bottom can be placed on the riser.
2. The height-to-diameter ratio of the riser should be maintained at close to 2:1. This is true, provided this ratio makes the riser taller than the casting or casting section it is to feed.
3. If a blind riser is used, as shown, the top of the riser should also be hemispherical.

Following these three steps can help the cylindrical riser approximate the surface area-to-volume ratio of the sphere.

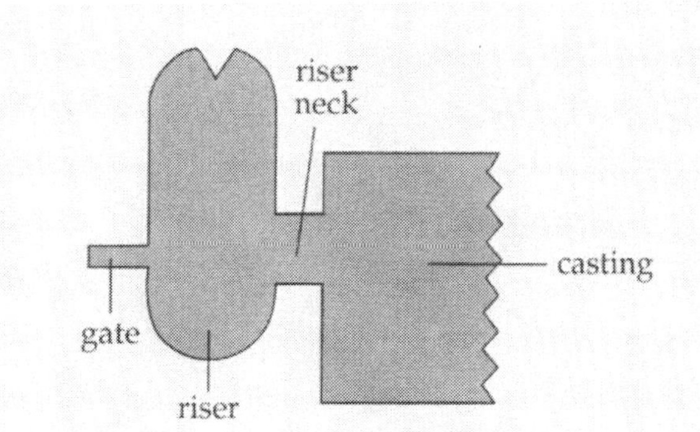

Fig. 17-5. Volume-to-surface area ratio of a cylindrical riser can be improved by using a hemispherical bottom. For blind risers, a hemispherical top also can be incorporated.

Determining Riser Size

For many years, the sizing of risers was done by the trial-and-error method. When Chvorinov published his work on the time it takes a given volume of molten metal in a given shape to solidify, it brought a more technical aspect to determining riser size. Since that time, many new and more accurate methods of sizing risers have been developed.

Very seldom does a riser feed an entire casting. Before determining the riser size, it must be determined where the risers will be required to feed the shrinkage in the casting. Today's solid modeling computer programs prove to be invaluable for determining how the casting will solidify, which also helps to determine where the risers will have to be placed to feed the shrinkage. Through these programs, one can see how sections of castings will feed other sections, etc.

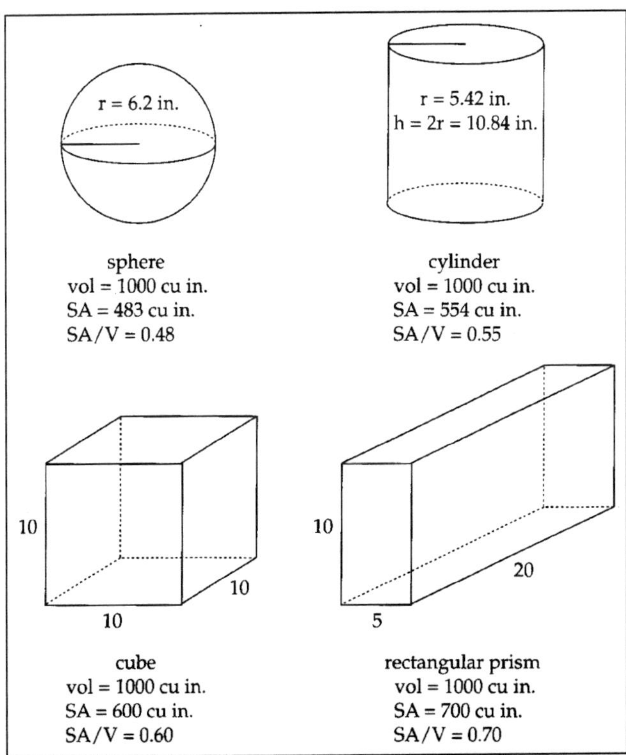

Fig. 17-4. Surface area-to-volume relationships for risers of constant volume and different geometric shapes.

The casting is divided into light and heavy sections. The light sections, of course, solidify rapidly enough; it is only the heavy sections that may require individual risers. The reason is that, when the thin sections solidify, they will draw feed metal from the heavier sections. As discussed in the previous chapter, the aim is to promote directional solidification back to casting sections to which risers can be attached.

It is important to identify the shrinkage characteristics of a given alloy during solidification. For instance, what is the alloy's solidification range, and will the volumetric contraction be large or small?

Pure metals and narrow freezing-range alloys will exhibit a deep "pipe" in the riser. The molten metal that was in the area of the pipe has been fed to the casting. In some cases, this pipe can be very deep and actually extend into the casting, rendering it defective. Thus, when sizing risers for this solidification range of alloys, the riser has to be tall enough to prevent this from occurring. The other two types of solidification (freezing) ranges will normally exhibit a less severe pipe and, in some cases, only have a slight "dishing" of the top surface of the riser. **Figure 17-6** shows an example of piping and dishing in three different aluminum alloys.

The main job of the riser is to provide a given volume of molten metal to the casting when needed. In order to ensure that this volume of molten metal will be available, it has to be kept from solidifying before it is needed. By building an "insulation" of molten metal around this volume, it is kept from solidifying before it is time to do its job.

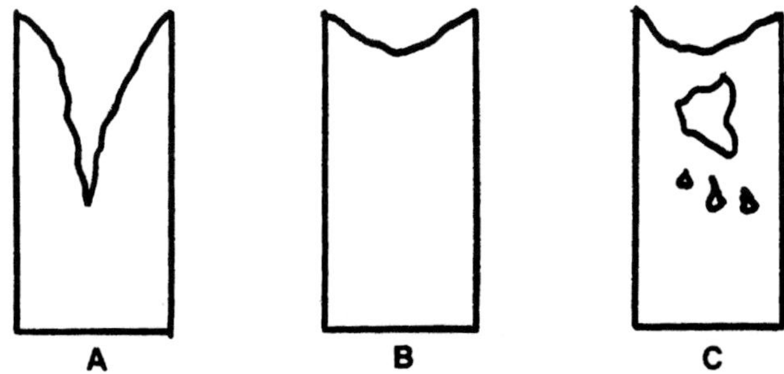

Fig. 17-6. Shrinkage patterns in aluminum castings: (A) pure Al; (B) 195 alloy; (C) A132 alloy. [After Flemmings et al.]

Actually, risering is not the most efficient and economical way to feed shrinkage in a casting, since a very small volume of the molten metal in the riser is used to feed the shrinkage. However, until a better method is found, it is the best there is. Thus, it is important to design and size risers that will meet their task and remain economical.

Next, one must take into consideration the rate of dissipation of heat by the molten metal as it solidifies, and the rate of heat transfer across the mold-metal interface. The problem, in essence, is to determine the smallest riser that will deliver the necessary volume of molten metal to the casting as it solidifies, and to keep this metal molten until the casting or the section to be fed has completely solidified. These two requirements may be referred to as the *freezing time* and the *volume feed capacity criteria*.

Calculation Methods

It is not the intent of this book to actually calculate the riser size. However, several methods that can be used to determine riser size will be presented. There is considerable literature available on the subject. Almost all of the guesswork has been taken out of determining how large a riser has to be to meet its requirements. Computer solid modeling and riser sizing programs have been a very big help in placement and sizing of risers.

The three methods that will be discussed in the following text are: Shape Factor, Mold Dilation and Modulus. Of the three methods discussed, the mold dilation and modulus methods are the more commonly used for determining riser size.

Shape Factor Method—One of the earliest methods of determining the riser size was developed at the U.S. Naval Research Laboratories and is called the "Shape Factor." The shape factor was determined by the following formula:

$$SF = (L + W) / T$$

where, L = length, W = width and T = thickness of a casting.

Figure 17-7 shows an example of how this method would be used to determine the size of a riser needed to feed a plate casting. The first step would be to determine the shape factor, which, in this case, is 20. Next, look at graph (A), locate 20 on the bottom line of the graph, then go up to the curve. From that intersection, go to the left to the vertical line and find a value of 0.25. This number is the riser volume-to-casting volume ratio.

The next step is to calculate the volume of the casting and then multiply this volume by the R_V:C_V ratio to find the riser volume. Then go to graph (B) and locate the riser volume on the bottom line. Reading directly up from this point, look for the places where this line intersects the two curved lines labeled H/D. You will note that the lower of the curved lines is a riser ratio of H/D = 0.5 and the upper curved line is a riser ratio of H/D = 1. H/D is the height-to-diameter ratio of the riser.

In this example, three different sized risers have been chosen. Two of these risers fall within the curved lines, but one is above this zone. A riser of these dimensions, placed as seen in the drawing in **Fig. 17-7,** should feed the shrinkage in this plate casting.

The shape factor method was developed primarily for narrow freezing-range alloys. In fact, the work done at the U.S. Naval Research Laboratories was with carbon steel.

In order to use this method, simple shapes, such as the plate or bar, have to be used. The majority of castings produced today can be broken down into these simple shapes. An example of this would be a cylindrical-shaped casting having a core through the center and length of the casting. This would look like a piece of pipe. This shape, when sliced lengthwise and then flattened out, would be a plate. This same principle would have to be used with the sections of castings.

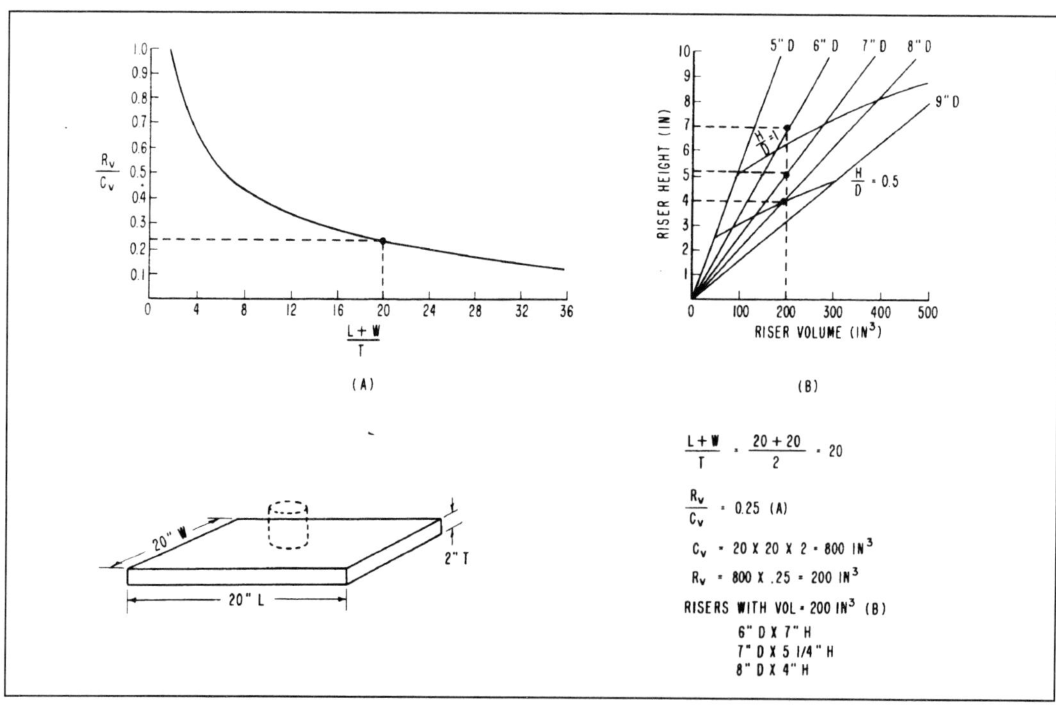

Fig. 17-7. Procedure for determination of minimum riser for plate casting.

Mold Dilation Method—This method is also known as the "Heine Method," so named after Professor R.W. Heine at the University of Wisconsin, who developed this method of calculating riser size.

This method worked very successfully for Heine with malleable iron castings. However, this method has been adapted for use with other alloy castings, both ferrous and nonferrous. It is based on the actual measurement of the pipe in a riser system that will produce a sound casting. In other words, the first step is to determine the volume of feed metal needed in the riser to feed the shrinkage in the casting or section of casting to be fed.

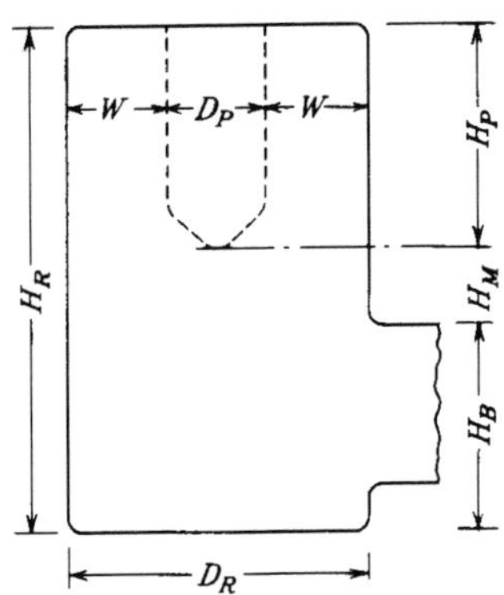

Fig. 17-8a. Cross section of piping type of side riser. [From R.W. Heine]

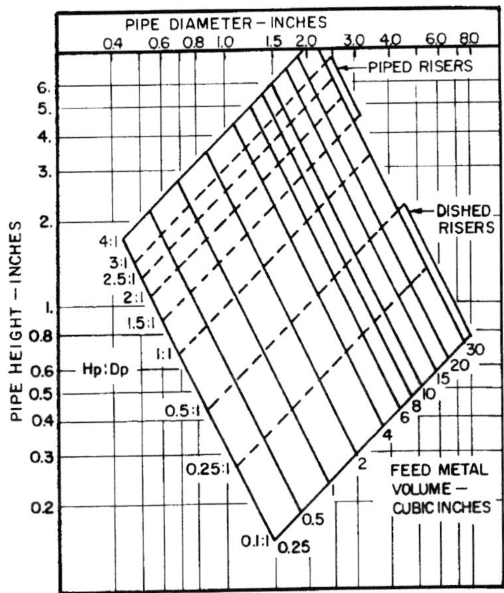

Fig. 17-8b. Graph for finding dimensions of the riser shown in Fig. 17-8a. [From R.W. Heine]

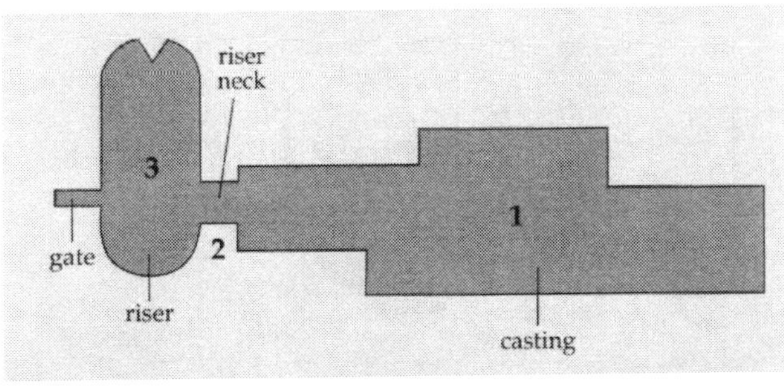

Fig. 17-9. The sequence in which solidification should occur.

Figure 17-8a shows a cross section of a riser sized using the mold dilation method. The diameter of the pipe is D_P and the height is H_P. The dimensions of this pipe are determined by the amount of feed metal required to feed the shrinkage in the casting or the section of casting to be fed. To find the volume of feed metal (V_F) required, the following formula can be used:

$$\frac{\text{casting weight} \times \text{feed metal \%}}{\text{density of alloy}} = V_F$$

Casting weight: calculated
Feed-metal %: See **Table 17-1**
Density of alloy in lb/in.^{3}

Once the V_F is determined, the graph in **Fig. 17-8b** aids in finding the dimensions of a pipe whose volume will be sufficient to feed the shrinkage occurring in the casting. These dimensions are also based on risers having a true pipe and those with a dished upper surface.

As seen in **Fig. 17-8a**, the next thing to determine is the thickness of the molten metal insulating wall (W) around the feed metal, to keep it molten long enough to do it's job. A method is provided to determine this dimension: the diameter of the riser (D_R) is the total of the diameter of the pipe (D_P), plus 2W.

One of the things that helps the riser to feed is the effective head. In order to obtain this, Heine added a pressure section (H_M) to the riser. The effective head would be the sum of the height of the pipe (H_P) and the pressure section (H_M). The minimum dimension for H_M is 0.5 in. (1.27 cm) with 1 in. (2.54 cm) recommended. The bottom section, (H_B) is equal to 1.5W. In some cases, H_B may have to be increased to 2W.

Modulus Method—The modulus method of determining riser size is based on Chvorinov's Rule and later work by Wladower, which calculates how fast a molten metal shape will solidify based on its volume and surface area. This rule is written as:

$$V / A = \text{Modulus}$$

where V = the volume of the casting or casting section to be fed; A = the cooling surface area of the casting or casting section to be fed; and Modulus = the ratio of volume-to-surface area.

Figure 17-9 shows the sequence in which solidification should occur between the casting, riser connection (neck) and the riser. In order for this sequence to take place, the modulus of the neck should be larger than that of the casting, and the riser's modulus should be bigger than that of the riser neck. Most authorities agree that the riser neck's modulus should be 10% larger and the riser's modulus should be 20% larger than that of the casting or casting section to be fed. In this case: $Mc = Vc / SAc$; $Mn = 1.1\ Mc$; and $Mr = 1.2\ Mc$.

As an example, say a determination has been made that a given casting section will be the last to solidify and, during this time, it has been feeding other sections of the casting. The modulus of this casting section is 1.5 (Mc = 1.5). The modulus of the riser neck to this section should have a modulus of 1.65 (Mn = 1.1 x 1.5). The modulus of the riser would be 1.8 (Mr = 1.2 x 1.5). The next step would be to determine the dimensions of a riser neck and riser that would yield the calculated modulus. With these in place, the preferred sequence of solidification takes place.

Determining Riser Connection

Figure 17-9 shows that the riser is attached to the casting by a connection called a neck or riser neck. The riser neck plays a very important role in helping the riser meet its requirement: to feed the shrinkage occurring in the casting. Should the riser neck solidify too soon, it can cut off the path of the feed metal from the riser to the casting.

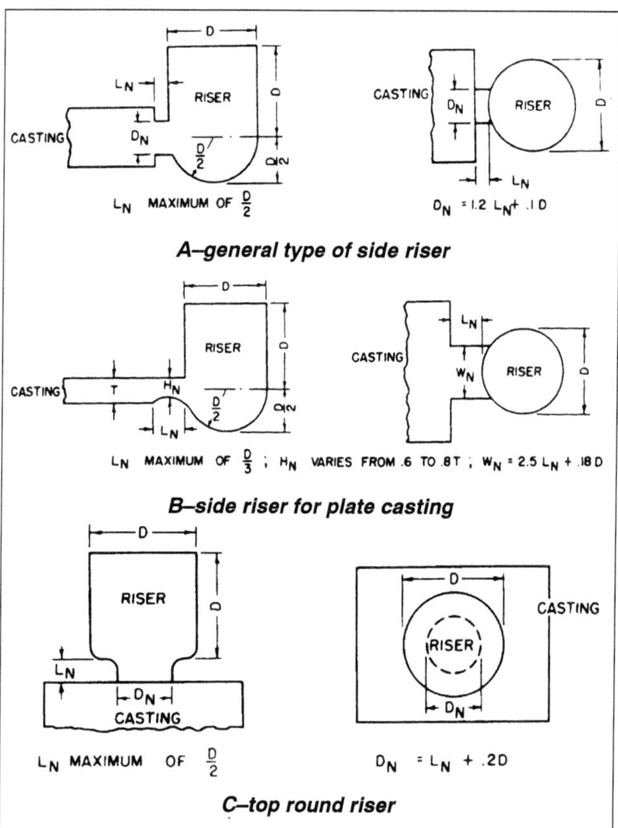

Fig. 17-10. Schematic riser neck dimensions.

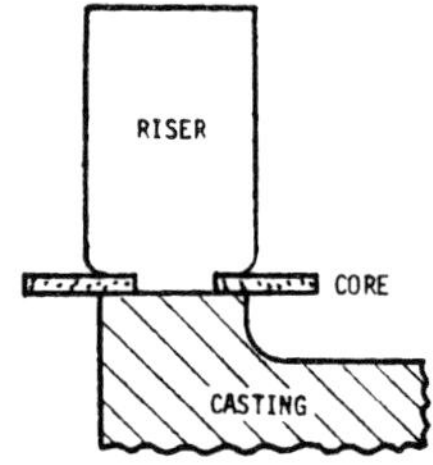

Fig. 17-11. Use of a Washburn core.

In addition to using the modulus method of calculating the size of the riser neck, **Fig. 17-10** shows other formulas that can be used to determine riser neck sizes. In illustration A and C, the riser neck is round, and in B it is rectangular. It should be noted that these formulas could only be used where the mold material surrounding the riser neck has the same thermal properties as the balance of the mold.

Not only should the size of the riser neck be considered, but also the ease with which the riser neck can be removed. Returning to **Fig. 17-3**, note that the top riser, in this case, is a full contact riser. In other words, there is no riser neck. When it comes time to remove this riser, considerable time will be spent removing it and touching up the riser contact area. However, by using basic heat transfer principles, a special core, called a Washburn core (also known as a necked down core or a breaker core) can be used.

Figure 17-11 shows the use of a Washburn core. This core is normally made of a core-sand mixture. The theory here is that the Washburn core becomes so hot that it will help to keep the smaller riser neck liquid, long enough to allow the feed metal to pass from the riser into the casting. **Figure 17-12** shows how the Washburn core can be dimensioned.

In some cases, the metalcaster will design the Washburn core so that there is a chamfer (beveled edge) from the bottom surface and another from the top surface. In this case, the two chamfers will meet at the mid point of the thickness of the core, forming a point. The hole diameter will remain the same as given in **Fig. 17-12.** This notch in the riser neck sets up a stressed area that can allow the riser to break off at this point, leaving minimal riser neck material to be removed from the surface of the casting.

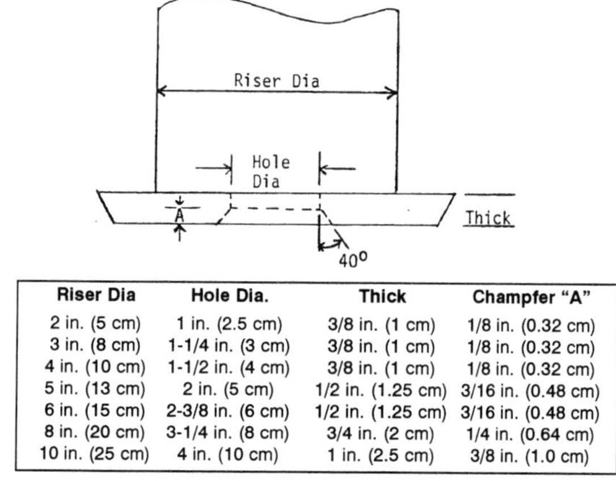

Riser Dia	Hole Dia.	Thick	Champfer "A"
2 in. (5 cm)	1 in. (2.5 cm)	3/8 in. (1 cm)	1/8 in. (0.32 cm)
3 in. (8 cm)	1-1/4 in. (3 cm)	3/8 in. (1 cm)	1/8 in. (0.32 cm)
4 in. (10 cm)	1-1/2 in. (4 cm)	3/8 in. (1 cm)	1/8 in. (0.32 cm)
5 in. (13 cm)	2 in. (5 cm)	1/2 in. (1.25 cm)	3/16 in. (0.48 cm)
6 in. (15 cm)	2-3/8 in. (6 cm)	1/2 in. (1.25 cm)	3/16 in. (0.48 cm)
8 in. (20 cm)	3-1/4 in. (8 cm)	3/4 in. (2 cm)	1/4 in. (0.64 cm)
10 in. (25 cm)	4 in. (10 cm)	1 in. (2.5 cm)	3/8 in. (1.0 cm)

Fig. 17-12. Suggested dimensions for a Washburn core.

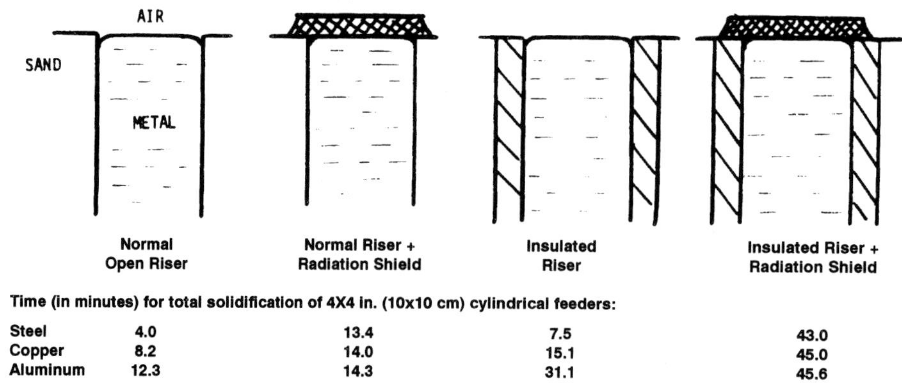

Time (in minutes) for total solidification of 4X4 in. (10x10 cm) cylindrical feeders:

	Normal Open Riser	Normal Riser + Radiation Shield	Insulated Riser	Insulated Riser + Radiation Shield
Steel	4.0	13.4	7.5	43.0
Copper	8.2	14.0	15.1	45.0
Aluminum	12.3	14.3	31.1	45.6

Fig. 17-13. Effect of insulation materials and radiation shields.

RISER-ALTERING AIDS

At times, it may be impossible to use hot or live risers; thus, the molten metal in the riser will be cooler than the molten metal in the casting. In cases like this, it is possible that the riser will solidify before the casting or section of the casting it is to supply feed metal to. In such instances, the riser will be very sound, but there will be shrinkage in the casting.

There are various riser-altering accessories and techniques to aid the metalcaster in reducing the speed at which the riser will give up its heat. **Figure 17-13** shows the results of an experiment that illustrates how these riser aids were used to control the heat loss from risers of various metals. The metals used were steel, copper and aluminum. In this particular experiment, riser sleeves and radiation shields were used.

Riser Sleeves

Riser sleeves can be of three different types: insulating, exothermic or exothermic-insulating. The differences will be explained.

An *insulating sleeve* is just as the name implies; it serves as an insulator that reduces the rate of heat loss from the side of the riser. In other words, it reduces heat loss by conduction, which occurs between the riser and the mold material. Insulating riser sleeves tend to be used on risers of small-to-intermediate size and for alloys having a low pouring temperature range, such as aluminum alloys. Insulating sleeves are less likely to cause problems with metal contamination near the riser in the casting.

The *exothermic sleeve* generates some heat. The sleeve is composed of a chemical material that reacts with the molten metal in the riser. This reaction generates heat. The heat generated does not necessarily increase the amount of heat in the riser; however, it replaces some of the heat lost by the riser.

The exothermic reaction occurs for a short period and can cause some of the solidified metal in the riser to remelt. After the exothermic reaction has ended, the sleeve then has thermal properties similar to the surrounding mold material. Exothermic sleeves are also used on small-to-intermediate sized risers. They are not recommended for large risers used with long freezing-range alloys. The one precaution that must be taken when using this type of sleeve is to be certain that the exothermic material is compatible with the alloy being poured. The exothermic sleeves have the greatest chance of contaminating the molten metal that is closest to the riser.

Exothermic-insulating sleeves are made of exothermic materials surrounded by an insulating material. This type of sleeve is very versatile in that it has an extended exothermic reaction, which, when completed, is backed by an insulating material. Another advantage of this type of riser sleeve is that it can be used for the full range of riser sizes. In addition, the chances of having molten metal contamination near the riser lies between the other two types of riser sleeves.

Radiation Shields

The *radiation shields*, as seen in **Fig. 17-13,** can be solid or granular. The solid shields are usually made of insulating material that can withstand the heat being radiated from the top of the riser. The granular shields can be made of an insulating material or an exothermic material. An insulating material placed on the top of a riser reduces heat loss by radiation; and in addition, the exothermic material generates heat due to a chemical reaction. For large, open risers, metalcasters have been known to toss dry silica sand over the top of the risers to reduce heat loss due to radiation.

The chemical composition of exothermic materials is based on the thermite reaction, which, when it takes place, will generate enough heat to raise the temperature of the molten metal in the top part of the riser. When this thermite reaction takes place, it is very noticeable by the reaction at the top of the riser. These materials are normally used on large, open risers for the ferrous alloys. As with all exothermic materials, the chances of having contamination of the molten metal near the riser must be taken into consideration. The molten metal produced by the thermite reaction must not be allowed to feed into the casting.

Practical Considerations—When it comes to reducing radiation heat loss from open risers, blind risers are recommended. The one exception to this would be the family of aluminum alloys, which lose very little heat due to radiation.

When considering the use of any of these riser aids, it is important to consider the economics. These riser aids will add to the cost of producing the casting. From this standpoint, it is possible that the size of the riser could be reduced, and the cost of the riser aids would be covered by a reduction in the riser size. It has been said that, if the riser aids do not allow a reduction in riser size, then their use is questionable. However, in the case of cold risers, their use may be the only way to help the riser meet its requirement.

Look back at **Fig. 17-13**. All of the risers contain the same volume of molten metal and were poured at their normal pouring temperature.

First, look at the normal open riser. No riser aids are used and, although aluminum is poured at a lower temperature than the other two metals, it takes longer to completely solidify than does steel or copper. The reason is that aluminum, as learned earlier, has poor emissivity and, thus, loses very little heat due to radiation. On the other hand, copper and steel have very good emissivity and will lose more heat due to radiation.

This fact is backed up by the second example, in which a radiation shield was used for all three metals. For all practical purposes, all three risers took the same amount of time to solidify.

In the third example, an insulating sleeve was used on all three risers. The steel riser solidified almost as quickly as it did without the insulation sleeve and radiation shield. However, the copper riser took a little less than twice as long to solidify, indicating that it loses most of its heat due to conduction into the mold material. Heat loss due to conduction is the greatest method of heat loss when it comes to aluminum alloys. With the use of an insulating sleeve, this riser stayed molten over twice as long as one without an insulating sleeve. This is one reason why aluminum foundries will tend to make more use of insulating sleeves than do other alloy foundries.

Finally, in the fourth example, both an insulating sleeve and radiation shield were used. Note the closeness of the solidifying time for all three metals. The loss of heat due to radiation and through conduction into the mold material have been greatly reduced. In this particular example, it may be possible to reduce the size of the riser and still have the riser meet its requirements: to be the last to solidify and feed shrinkage in the casting.

Backpouring

Another form of riser-altering aid is the practice of backpouring into an open cold riser. This method is normally used for top risers but can be used with cold side risers under certain conditions. **Figure 17-14** shows two methods of using the backpour technique. In this example, it shows the backpour technique with a top riser. When using this technique, the pouring of the mold is stopped short of completely filling the riser. After a short period, the pouring ladle is brought back over the open riser and molten metal is poured into the riser.

The key precaution to take when using the backpour technique is to be certain that there is a sufficient amount of molten metal in the riser to not allow the falling stream of molten metal from pushing through the riser into the casting. If this should occur, there is the possibility of the falling stream pushing slag or dross into the casting proper. In addition, air can be entrained in the falling stream due to the turbulence as it enters the molten metal in the riser, and this too can be carried into the casting. To help reduce the amount of turbulence, a diffuser can be used, as seen in **Fig. 17-14** on the right. This technique has proven to be very helpful in aluminum alloys.

The backpour technique is primarily used with very large castings having large open risers. When observing the pouring of a large carbon steel roll (weighing in the tons) for a steel rolling machine, one could see the liquid shrinkage occurring in the riser. After a few minutes, a ladle of fresh hot metal was poured into the top riser to fill it. This was done several times, until most of the liquid shrinkage stopped. At this time, three carbon electrodes, similar to those used in the electric arc furnace, were placed over the top of the riser, and electric arcs were struck between the electrodes and the molten metal in the riser. This was done to help keep the riser liquid long enough to feed the shrinkage taking place in the casting.

RISER FEEDING

What makes a riser feed? Alternatively, what makes the feed metal in the riser want to leave the riser and feed the shrinkage in the casting? Many times, there is more concern about riser shapes and sizes and very little thought given to these questions. Not paying attention to these questions can cause a riser to fail in its assigned task.

There are three essential components that help make the riser fulfill its requirement of feeding shrinkage in the casting: a partial vacuum, metallostatic pressure and atmospheric pressure. All three of these components must be present. If not, chances are that the riser will not do its job. Of these three, the one that is overlooked the most is atmospheric pressure.

Partial Vacuum—**Figure 17-15** shows a cutaway view of a riser and casting during solidification. Remember that the remaining molten metal is encapsulated in a solid skin of solid metal, which includes the casting, riser connection and riser. As solidification continues, a partial vacuum or low-pressure area is formed in the casting. This partial vacuum will try to pull the molten metal out of the riser.

Metallostatic Head—Also, note that the riser is taller than the section of the casting itself. This extra metallostatic head of molten metal will help push the molten metal out of the riser into the casting. Thus far, there is a pulling and pushing action occurring within this system. However, for the pulling action to work, one more ingredient has to be added: atmospheric pressure.

Atmospheric Pressure—To find out more about atmospheric pressure, try this experiment. In the top of an unopened can of soda, drill a hole the size of a soda straw. Place the soda straw through the hole and part way into the can. Next, seal the opening around the soda straw with tape or some other sealing material. When finished with the sealing operation, suck on the straw to get a mouthful of soda. If everything has been done correctly, very little, if any, soda will come out of the can, although a vacuum was formed with the soda straw. The final step in the experiment is to poke a very small hole elsewhere in the top of the can. Again, suck on the soda straw. A mouthful of soda will come out of the can. The difference is atmospheric pressure!

This same "push-pull" situation exists in the riser and casting. The casting tries to pull the molten metal out of the riser; however, the solid skin of metal is impermeable and does not allow atmospheric pressure to enter this enclosed system. In the case of blind risers, a wedge of molding sand is formed in the top of the riser, which will get very hot and prevent a solid metal skin from forming, or at least the skin will be very weak. The sand mold and wedge being permeable will allow atmospheric pressure to pass through and into the riser to help push the molten feed metal into the casting as the partial vacuum is being formed.

In the case of open risers, insulating or exothermic materials can be placed on top of the riser to help prevent a strong skin of solid metal from forming. In this case, the atmospheric pressure can act on the top surface of the riser.

Another practice to use with the skin-forming alloys, such as steel, is to place a William's core (sometimes called atmospheric core, pencil core or firecracker core) in the riser. **Figure 17-16** shows the use of a William's core.

The William's core is normally made of a core sand mixture or graphite. These cores are relatively small and are permeable. When surrounded by molten metal in the riser, they will become very hot. A very weak skin or no skin of solid metal is formed around the core. With the mold and core being permeable, the atmospheric pressure can get into the riser to help push the molten feed metal into the casting as the partial vacuum is being formed. In fact, if the skin of solid metal formed is strong enough, and allows the atmospheric pressure to exert its influence, the riser can feed a casting of greater height than itself. This has been done in carbon steel castings.

Feeding Distance

The distance that a riser can deliver feed metal into the casting section to which it is attached is called its feeding distance. This feeding distance is affected by the alloy being poured and the thickness of the casting section to be fed. One reason a riser may seem ineffective when shrinkage occurs is that the feeding distance of the riser has probably been exceeded.

An example of this happening is seen in **Fig. 17-17**. In this test, an aluminum casting with a length of 14 in. (36 cm) and thickness of 4 in. (10 cm) was used. The feeding distance for aluminum alloys, using a hot riser, is said to be 2T, where T is the thickness of the section. Thus, the feeding distance of this hot riser is 8 in. (20 cm) leaving 6 in. (15 cm) out of reach of the riser. In this example, a shrinkage cavity would occur as seen in the example.

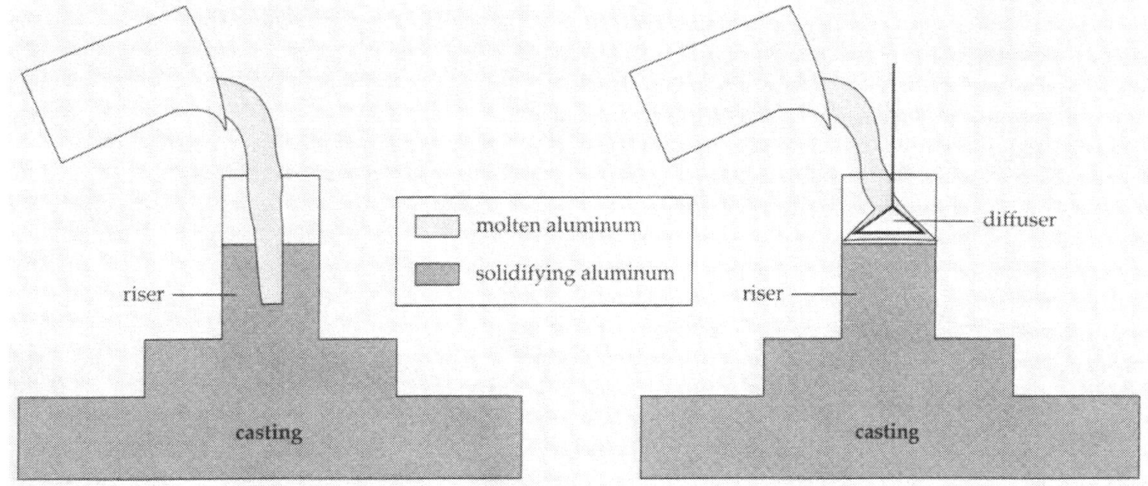

Fig. 17-14. Two methods for backpouring into an open riser. The method on the right is the more desirable way to do this because dross from the top is not forced down the riser toward the casting. As molten metal fills the riser, the diffuser is drawn upward. Although the figures shows aluminum, any metal can apply.

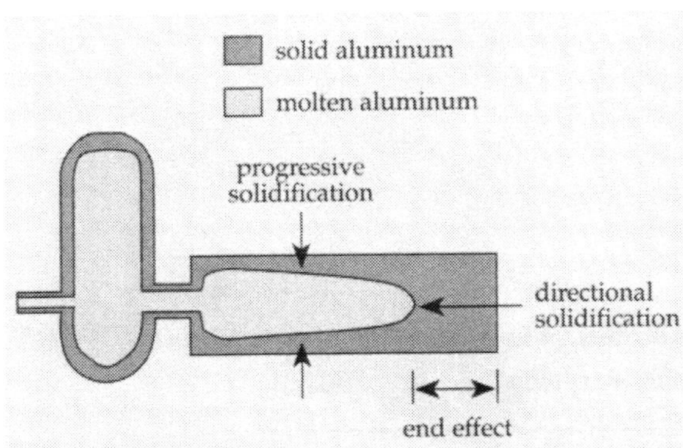

Fig. 17-15. Illustration of a casting during solidification.

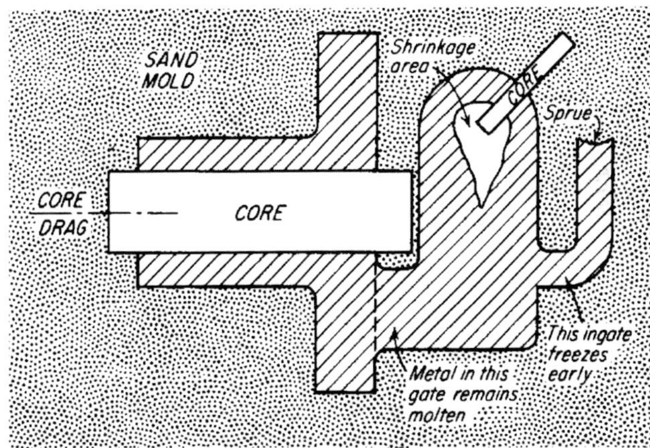

Fig. 17-16. Cross-sectional diagram of a casting fed by a blind riser with atmospheric vent produced by a pencil core.

Returning to the study of solidification, it was found that there are two types of solidification occurring: progressive and directional. The example shown in **Fig. 17-17** shows that progressive solidification did not allow the riser to get its feed metal to the area where the shrinkage is found. It can also be seen that directional solidification did occur but was not great enough to prevent the shrinkage cavity from forming. In this case, simply making the riser larger would not eliminate the shrinkage problem. The answer to this problem is to either slow the rate of progressive solidification or increase the rate of directional solidification or a combination of both.

Table 17-2 lists some approximate feeding distances for various alloys. It is worth mentioning that most of the work done on feeding distance was done at the U.S. Naval Laboratories on steel bars and plates. The data for the other alloys were suggested by researchers specializing in determining feeding distances. The values listed in **Table 17-2** can be used to approximate the feeding distances when multiplied by T.

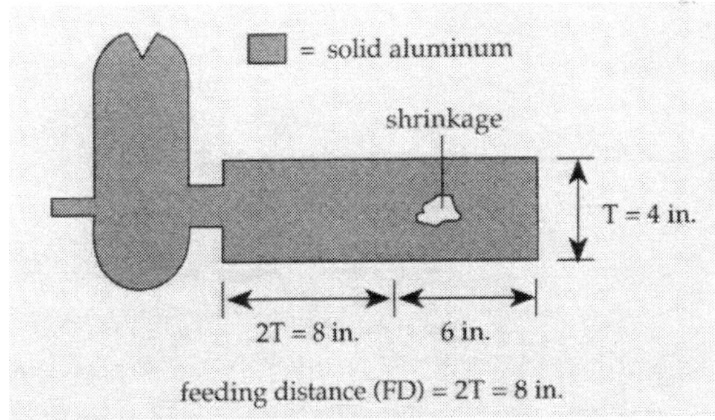

Fig. 17-17. Too large a feeding distance would cause shrinkage in this casting.

Figure 17-18 shows examples of feeding distances for a carbon steel plate. In drawing A, the feeding distance of the alloy is stated as approximately 4.5T. However, in reality, the actual feeding distance of the riser is 2T and the end effect, or directional solidification, is 2.5T. This means that the directional solidification is responsible for the shrinkage-free 2.5T area and the riser took care of the remaining 2T area. It should also be pointed out that 4.5T is the same on either side of the riser.

In drawing B of **Fig. 17-18**, there is a different situation. In this case, the riser can still feed 2T, with some variability. In other words, there may be some centerline shrinkage within 2T from the riser. However, there is a large area in which centerline shrinkage will be found. The end effect is still worth 2.5T and will help keep this area shrinkage-free. This means that the centerline shrinkage is beyond the feeding distance of the riser and the end effect. If the size of the riser is increased, the same situation will occur. Why? Because nothing has been done to increase the feeding distance in this particular case. Shortly, it will be seen how to help the riser meet its requirements in cases such as this.

Table 17-2.
Approximate Feeding Distance Calculations

Alloy	D_F
Carbon steel	4–5
Alloyed steel	4–5
High-Alloy steel	4–5
Malleable iron	5
Gray iron:	
CE 3.0	6.5
CE 3.6	6.5
CE 4.3	10
CE 4.4	9
Ductile iron	6–9
Aluminum alloys	6–10
Pure copper	8
Brass	5
Bronze	2–5
Aluminum bronze	5
Tin bronze	3
Copper nickel (30% Ni)	2–5.8

$F_D = D_F \times T$

where F_D = Feeding Distance
D_F = Distance Factor
T = Thickness (of casting)

Drawing C of **Fig. 17-18** shows the same plate casting with two risers. In this particular case, the total feeding distance between the two risers is 4T, with each riser taking care of 2T of their actual feeding distance. Although not shown, the end effect would still be 2.5T if the edges of the plate were surrounded by a green sand mold.

Finally, drawing D shows an example where the distance between the risers is too great for the feeding distance of the risers. Each riser will still feed a distance of 2T; however, as shown, this will leave an area of centerline shrinkage. Again, increasing the size of the risers will not solve this problem.

Figure 17-19 shows the same results as seen in **Fig. 17-18,** only in this case the castings were carbon steel bars with the thickness, T, equal to the width. Although the alloy used in this case was carbon steel, the same results can be expected with the narrow and medium freezing-range copper-base alloys.

As has been stated several times, simply making the risers larger will not appreciably increase the feeding distance of the riser. Remember that the feeding distance is always stated by "distance factor, D_F," times "T," with "T" being the thickness of the section. The "distance factor" can be changed, based on the alloy being poured, and that explains the different feeding distances for different alloys, although "T" remains the same. It means that if the feeding distance is changed, for a given alloy, T has to be physically increased or something has to be done to change the thermal gradients in the section being fed.

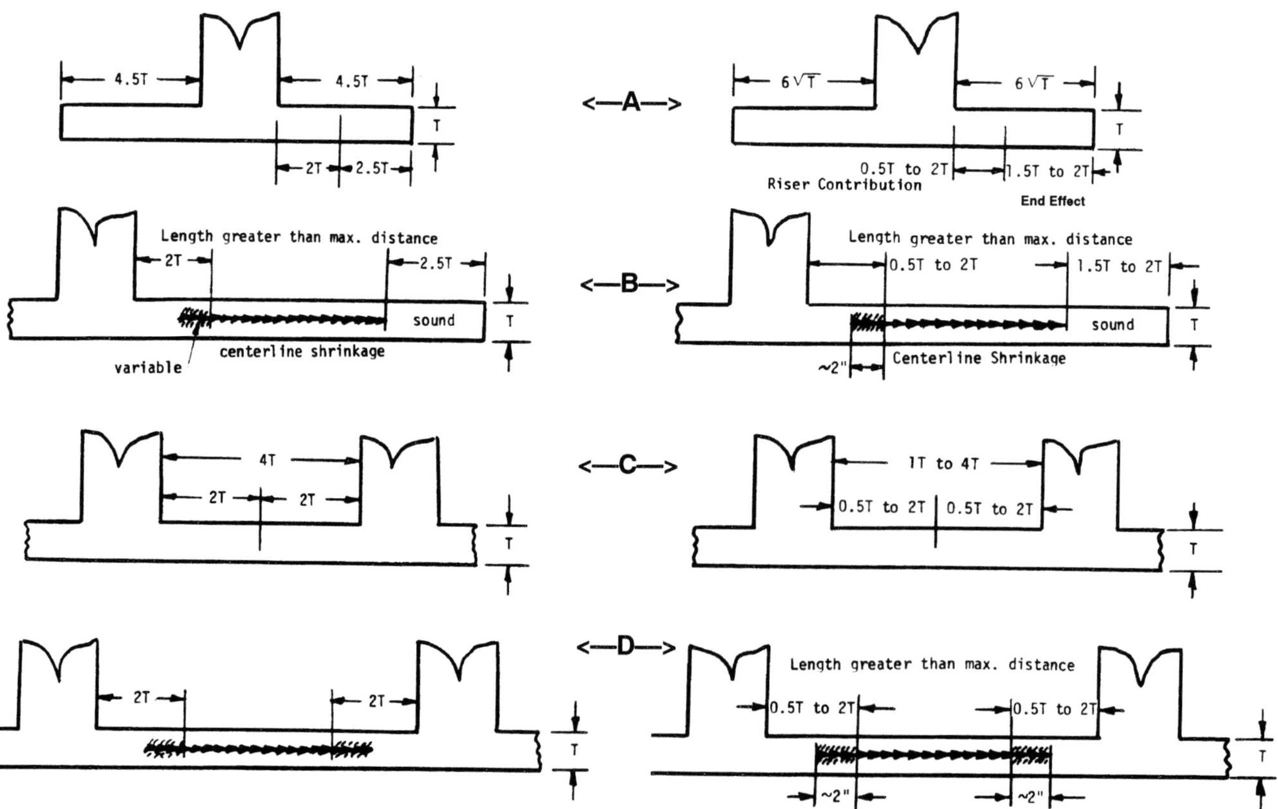

Fig. 17-18. Feeding relationships in plates.

Fig. 17-19. Feeding relationships in bars.

CONTROL OF END EFFECT

Remember that one of the main reasons the riser cannot complete its task is because progressive solidification has restricted the flow of feed metal. If there is some way to change the distance that progressive solidification has to travel toward the centerline, the riser can be helped to meet its requirements. One way to do this is to use padding. Padding can make use of extra metal added to the surface of the casting as shown in **Fig. 17-20.** This padding can also be in the form of insulating or exothermic "boards" placed on the mold surface.

Figure 17-20 shows an example of how metal padding helped cure a centerline shrinkage problem. (As a side point, if all castings were wedge-shaped, risering to get rid of shrinkage would be much easier.) Note that T was changed. In essence, what has been done is to increase the distance that progressive solidification has to travel to reach the centerline of the section. The use of metal padding is not always practical because, in many cases, it will have to be removed to get to the desired section thickness.

Another technique is to use exothermic or insulating material to form a barrier between the casting section and the mold material. **Figure 17-21** shows the use of a tapered exothermic material used as the padding. In this case, the exothermic material is available in board form, can be cut to size and nailed into prints that have been formed in the sand mold. Insulating board is used for aluminum castings. The insulating board is placed in the mold in the same manner as the exothermic material.

When the required feeding distance exceeds that attained under normal conditions, chills may be effectively used to make the end effect greater (in other words, increase the rate of directional solidification). **Figures 17-22 and 17-23** illustrate the use of chills for this purpose. The chill has a higher thermal diffusivity, which results in a greater rate of heat extraction and thermal gradient, which, in turn, promotes directional solidification.

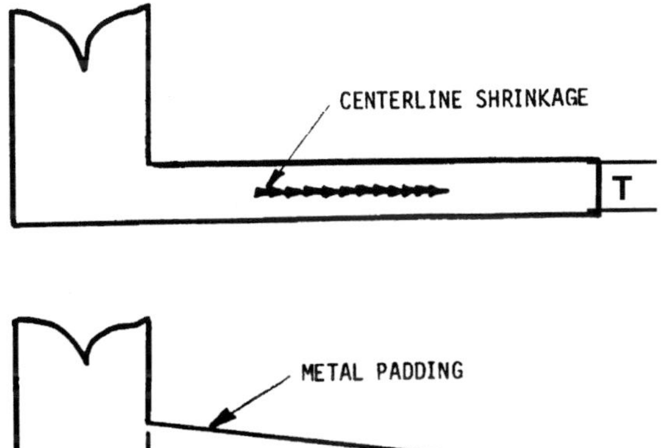

Fig. 17-20. Plate section fed with metal padding.

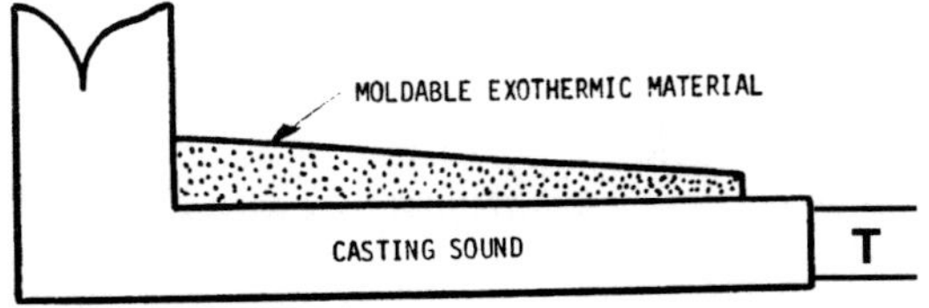

Fig. 17-21. Use of exothermic padding.

First, look at **Fig. 17-22**, where the studies were made using carbon steel *plate* castings. In drawing A, the end effect for this casting is 2.5T when the end is surrounded by molding sand. On the other hand, in drawing B, when a chill is placed against the edge of the casting, the end effect is increased to 2.5T + 2 in. (5 cm). With the aid of the chill, the end effect will proceed 2 in. (5 cm) further into the casting. In this case, the chill was made of metal and molded into the mold when the mold was made. (Chills will be discussed in more detail in the next section.)

In drawing C, when a top riser was used on the plate casting, the feeding distance to the edge of the plate was determined to be 4.5T + 2 in. (5 cm) when a chill was used. The effect of the chill was the same as in drawing B.

Drawing D in **Fig. 17-22** is interesting from two standpoints. First, the chill again increased the distance that each riser can feed, by the same amount: 2 in. (5 cm). Second, if the chill were not used, there would be no end effect between the two risers since this is one continuous section of the casting. In effect, what the chill has done is placed an "artificial end" in this section. In this particular case, the end effect is the same in both directions toward the risers.

Figure 17-23 shows the results of using chills to increase the end effect in carbon steel *bars,* where T is the same as the width of the bar. In other words, the bar is square in geometry. The results are the same as that in **Fig. 17-22**. The only change is that, in this case, $6\sqrt{T}$ was used instead of 2.5T.

An interesting point about creating an artificial end is shown in **Fig. 17-23,** drawing D. Some foundries pour gear blanks, which are nothing more than a bar of metal that is circular in geometry. Given the fact that a circle has no end, if nothing were done to the thermal design of the casting, there would be no end effect or directional solidification. To simulate that the circle has an end, from a thermal standpoint, a chill can be placed in the mold where it can touch the molten metal. In fact, more than one chill can be used between the multiple risers to feed shrinkage in the casting. In many cases, less risers are required because of the increased end effects in the casting caused by the chills.

The use of chills to increase the rate of directional solidification or end effect in castings can work in all alloys cast. However, the results of the work done in carbon steel can be applied to both the narrow and intermediate or medium freezing-range copper-base alloys. On the other hand, the results may vary when chills are used for this purpose in other alloys.

Chills

A chill can be defined as a highly conductive material normally embedded into the surface of a sand mold or core to increase the thermal gradient at that point or in that area of the casting. In other words, if there is an area in the casting or mold where there might be problems caused by excessive heat entrapment, or there is a need for improved directional solidification, the use of chills could improve the situation. Following are several common reasons for using chills:

1. To promote directional solidification.
2. To help risers feed shrinkage.
3. To refine the microstructure of a casting in specific areas.
4. To reduce dendrite arm spacing.
5. To help reduce macro gas cavities in the casting.

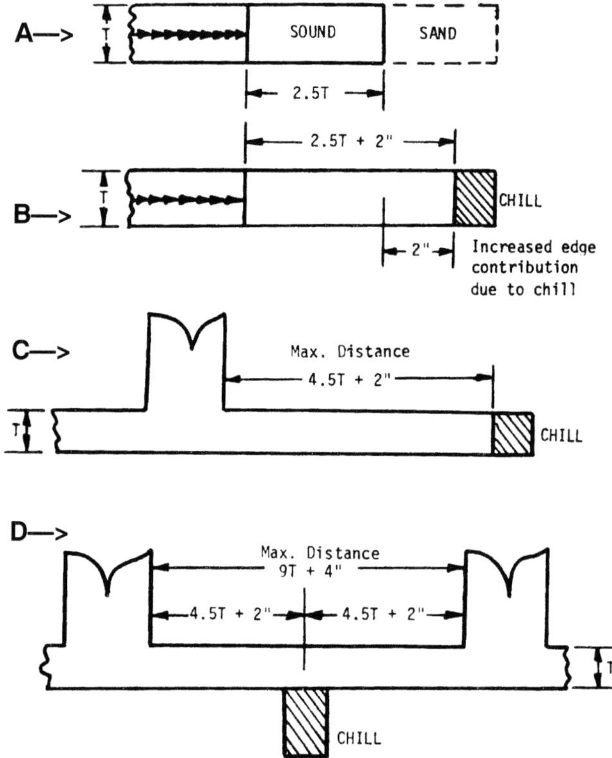

Fig. 17-22. Combination of chills and risers for maximum feeding distance in plates. [After Flinn]

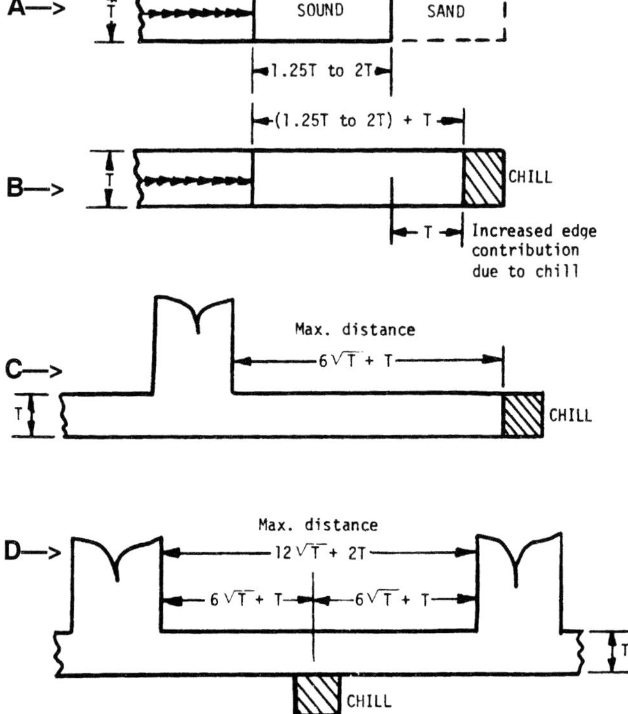

Fig. 17-23. Combination of chills and risers for maximum feeding distance in bars. [After Flinn]

When metal is solidified through a steep thermal gradient, solidification occurs very rapidly and, thus, produces a very fine grain structure. This fine grain structure will also produce higher mechanical strengths and hardnesses. For example, a boss is a protuberance on the surface of a casting offering improved wear resistance against a sliding part. Because a harder surface is more wear resistant, often a chill can be located in this area to improve hardness properties. Here, the microstructure has been "refined" in the area of wear properties.

When discussing dendrite arm spacing in aluminum alloys, it was found that the closer the dendrites are kept together, the better the mechanical properties of the aluminum alloy casting. This spacing can be controlled by the judicious use of chills.

Finally, entrapped gas in the molten metal will remain there, in solution, until the molten metal begins to solidify. Once solidification begins, the solidifying metal begins to expel the gas, which, in turn, seeks more molten metal. The gas stays just ahead of the solidifying front of the solid metal. When solidification is complete, and the gas has not had a chance to escape from the molten metal, it will be trapped in "large" quantities causing macro or large gas cavities to form.

If steep thermal gradients are set up causing the molten metal to solidify very rapidly, this gas will be trapped as micro gas cavities over a larger area in the casting. Although these micro gas cavities may not be harmful to the end use of the casting, chilling should not be used as a cure for improper control of the generation and entrapment of gas in the molten metal or the lack of proper venting of molds and cores.

Types and Purposes of Chills

The three basic types of chills are:

- External Chills
- Internal Chills
- Chill Aggregates

External Chill Purposes—External chills **(Fig. 17-24)** can be used in most any location of a casting where increased cooling is desired. They can be placed on either the external or the internal surfaces of the casting. An external chill may extend directional solidification toward a riser and may reduce the number of risers required to obtain a sound casting.

Pseudo-Chills—In some cases cracking brackets are used as strengtheners on castings. These cracking brackets are usually thin and solidify very rapidly. When completely solidified, they can act as "pseudo-chills." In addition, some large flasks will contain strengthening bars, especially flasks used in floor molding. If the strengthening bars do not have enough molding sand between them and the casting, they can act as pseudo-chills. Even worse, if these bars get to close to a riser, they can actually chill the riser and make it ineffective. Another form of pseudo-chill is the metal flash that is found on the parting line of some castings.

Internal Chill Purposes—Before beginning the discussion on internal chills, it should be mentioned that internal chills are seldom used when making high-quality castings, such as pressure-tight castings. The reason is that, to combine casting soundness with the good "welding-in" of the chill, the temperature of the molten metal surrounding the chill must be known and well controlled. This type of control can be very difficult to maintain on a consistent basis.

An internal chill is one that extends into the mold cavity, where the incoming metal will make contact with the chill.

Internal chills are used primarily in deep recesses where external chilling may be difficult to use. They are also used in locations where they can be removed during machining operations such as drilling holes.

Chill Aggregate Purposes—Chill aggregates are special molding sands, core sands, or steel shot sand mixtures. These can be used selectively in fillets or pockets of castings, to create a more rapid skin formation than regular molding or core sand can provide. These selective chill aggregates can also be used as chill coatings on sand molds and cores, and as coatings for metal molds that are used in permanent mold and diecasting operations.

Chill Material Selection

Chill materials can be metal, or they can be an aggregate used independently, or used in a chill coating or wash. The following materials are used:

Metal: Cast iron, copper, aluminum and carbon steel.
Aggregates: Zircon, chromite, steel shot and sand mixture.
Coatings: Zircon, chromite and graphite.

Solid graphite blocks can also be used as external and internal chills and, in some cases, may replace metal chills.

External Chill Materials—The selection of external chill materials is based on three essential factors.

1. The chill material should have a high volumetric heat capacity. Volumetric heat capacity is determined by the following formula:
   Volume of material × material density × material's specific heat
   Based on this formula and through experimentation, cast iron was found to be the best material.
2. The chill material should have relatively high thermal conductivity.
3. The chill material should have the ability to distribute the absorbed heat.

Internal Chill Materials—The selection of internal chill materials are based on the following:

1. The chill material must be able to fuse with the casting.
2. The chemical composition of the internal chill must be compatible with the alloy being poured. For instance, a cast iron internal chill would not be used for a carbon steel or aluminum alloy casting.

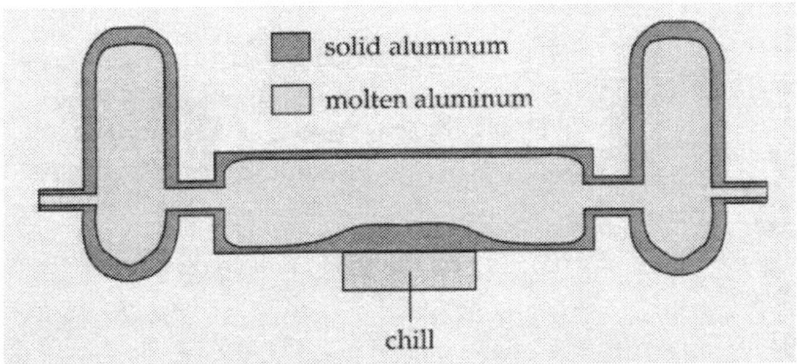

Fig. 17-24. Use of an external chill with an aluminum casting.

Choosing Chill Size

External Chill Size—Chill thickness is determined by the thickness of the casting section to be chilled. Chills that are too thin soon become overcome with heat before the casting is completely solidified. Therefore, these chills will become rapidly ineffective and remove the heat at about the same rate as the molding sand.

Chill Thickness—Chill thickness should equal 0.5T or 1T, where T equals the section thickness. In the case of carbon steel castings, chills of 0.5T are adequate for bar-type castings. For plate-type carbon steel castings, the thickness of the chill should be 1T. Making chills thicker than this will not increase the cooling action of the chill.

Chill Width—The width of the chill should equal from 1T to 2T. Again, T equals the casting section thickness. When the chill width becomes greater than 2T, solidification gradients become adversely affected.

Chill Length—The chill length can be 2T or 3T, but not greater than 3T. If chills are very long, they will bow or distort and produce thickness changes in the casting during solidification.

Internal Chill Size—The sizing of internal chills is markedly affected by the pouring temperature. Because of this, only general guidelines can be offered. Internal chills are sized primarily by weight.

Weight of Chill—In the case of carbon steel castings, the chill should weigh between 5% and 8% of the weight of section to be chilled. In other alloys, chills weighing 2–5% of the weight of the section to be chilled are recommended.

Surface Area-to-Volume Ratio—The recommended surface area-to-volume ratio of an internal chill should be 14:1.

Chill Preparation

External Chill Preparation—Many problems traced to the use of external chills can be blamed on the lack of proper chill preparation.

First, the effectiveness of an external chill lies in the amount of surface contact between the chill and the casting section to be chilled. To increase this surface contact area, metal external chills should be knurled. Knurled surfaces possess a series of protuberances or ridges that increase the surface area. Knurling the contact surface of the external chill will increase the surface area contacting the casting. Knurling will also act as a vent to allow air or mold gases to escape. One drawback of knurling is that the external chill will leave a pronounced mark in the surface of the casting.

Second, it is essential that the surface area of the external chill be clean! This means there should be no oil, moisture or oxidation on the surface of the external chill. A good practice is to sand blast the surface of the external chill before use.

Third, after sand blasting, a chill coating can be placed on the surface of the external chill, so that it will be in contact with the casting. A chill coating can also help prevent the chill from "welding" to the casting. Should this welding occur, the removal of the chill from the casting can be impaired.

Sometimes, aluminum paint is used to coat the external chill. This practice is not recommended because the paint is porous and can hold moisture, which, in turn, can cause gas defects on the casting surface. These paints have a poor binder system and, in most cases, will rub off and become virtually useless.

Fourth, the temperature of the external chill should be at or slightly higher than the mold material onto which it will be placed. If the external chill is cooler than the mold material, there is a good chance that the chill will "sweat." This moisture on the surface of the external chill will cause defects on the chilled surface of the casting. This problem is accentuated in hot, humid sand molds.

Internal Chill Preparation—Just as for external chills, the surface of internal chills must also be clean and dry. Since these chills are often not very large, it is difficult to sand blast them to ensure that their surface is clean and free of oxidation. Thus, internal chills should be handled with care.

Again, the temperature of the internal chill should be at or slightly higher than that of the mold material into which it is placed. If this practice is not adhered to, the internal chill can sweat causing gas defects in the casting. In addition, in regard to moisture, if powdered internal chill materials are to be used, they too must be free of moisture.

Solid internal chills can be coated with a chill coating. The thing to remember, however, is that the coating material has to be compatible with the alloy in which the internal chill is to be used.

Internal chills should have a tapered or pyramid top to reduce entrapped air or gases. Internal chills should be used in the vertical position in the mold to facilitate the passage of air and gases from the top of the chill.

Chill Placement

It is vital that chills be placed in the correct position, if they are expected to do their job.

External Chill Placement—The incorrect placement of external chills can cause premature freezing of the section to be fed, which can result in shrinkage defects. Chilling must not envelop and constrict the feeding section, and feeding must always be directed toward the chills. Generally, it is not possible to feed soundly through chilled areas. In other words, do not place chills in the feed paths of risers. One way to find out if chills are properly placed is to use a solidification-modeling program. The location of chills can become part of the data included in the program.

Chills are normally placed in the drag of the mold. One of the reasons for this is to ensure that the external chill remains in contact with the casting until solidification has been completed. **Figure 17-24** shows the use of an external chill with an aluminum casting. Note the extra amount of solidified metal at the chilled area, compared to other areas of the casting.

During solidification, when the solid metal skin is formed, it begins to draw away from the mold surface; thus, an air gap is formed. External chills should be placed where the solidifying casting can contract on to them, and this area is in the drag of the mold and at the inner surfaces of ring-type castings.

External chills can also be placed in the vertical and cope surfaces of the mold; however, it can be difficult to hold them in place. (External chills can be placed in the drag on the drag pattern plate and become part of the drag mold.) Another problem with external chills placed in the vertical and cope surface of the mold is that heat radiated from the rising surface of molten metal can preheat the chills and reduce their effectiveness. This is especially true for those alloys that have good emissivity.

External chills should not be covered by the molding material. Covering the exposed surface of the external chill that will contact the molten metal will greatly reduce the effectiveness of the chill. Some people will cover the chill with a thin layer of molding sand, so that the casting surface will not have "chill marks." In essence, they may have a smoother casting surface, but they have reduced the ability of the chill to do its job properly.

Distance Between Chills—Where multiple chills are placed around the section to be chilled, they should be spaced equal in distance to the length of the chills. In some cases, half the chill length apart is sufficient.

External chills should not be placed near ingate areas or where considerable molten metal will flow over them. If this is done, the chills will become preheated and lose their effectiveness.

Internal Chill Placement—Internal chills should stand vertically in the mold. Use of horizontal surface placement should be restricted to avoid slag, dross or gas entrapment.

The internal chill should be positioned such that it is not strongly preheated before being enclosed by the molten metal. If this should occur, the effectiveness of the internal chill is reduced. For this reason, do not place them in front of ingates or the flow paths of the molten metal as it fills the mold.

Problems Associated With Chills

Most of the problems associated with the use of chills are caused by not following the rules, some of which were just discussed.

External Chill Problems—Common problems most often encountered with external chills are:

- *Welding of the chill to the casting.* Welding means that the chill sticks to the casting and is difficult to remove. Most often, this problem is caused by poor chill surface preparation. A chill coating can be used to help alleviate this problem.
- *Cracks around the chilled surface of the casting.* This problem can be caused by improperly sized chills (normally, too large a chill). Even with properly sized chills, there is the chance of cracking occurring around the chill. The external chill, at times, will extract heat from this localized area of the casting so quickly that the area will contract too rapidly for the skin of the casting next to the chill to stretch fast enough to keep from tearing. To help prevent this from happening, the chill can be tapered (not on the chilling surface), or better still, a layer of chill aggregate, such as chromite or zircon,

can be placed around the chill. Since these aggregates have a lower conductivity than the metal chill, the cooling effect will be "blended" at the sand-metal interfaces.

- *Cracks directly under the chill.* In most cases, the chill is too large and, thus, the chilling affect is too severe, causing these cracks to occur.
- *Gas cavities in the chilled area.* Again, poor chill surface preparation can be the culprit. In addition, placing cool chills in a hot sand mold can cause the chill to sweat, thus causing gas problems when the molten metal reaches the chill.
- *Chill not working effectively.* If the chills are properly sized and prepared, then the problem can be caused by poor placement in the mold. Remember that chills are only effective over a given distance. For instance, mechanical property improvements in aluminum castings have not been found at a distance greater than 4T from the center of the chill.
- *Chill does not work at all.* In this case, normally the chill is undersized. Also, make certain that no one has covered the chill with a layer of molding material.

Internal Chill Problems—Some problems that can be faced with internal chills are:

- *Improper or incomplete fusion of the chill and the casting.* Over-chilling with internal chills causes unfused chills in the casting. This is normally caused by over-sized chills. Incomplete fusion can also be caused by having the wrong composition coating on the chill.
- *Gas cavities around the chill.* Lack of proper chill preparation before being placed in the mold can cause this problem. Remember that internal chills represent the internal surfaces of the casting and can entrap any air, mold or core gases, slag or dross that may be entrained in the molten metal. Horizontal surfaces on internal chills are more prone to the latter situation.
- *Shrinkage cavities.* Undersized internal chills cause insufficient chilling action, which will result in shrinkage cavities.

Chill Aggregates

So far, not much has been mentioned about the use of chill aggregates, Some may be wondering why the molds and cores aren't just made out of chromite or zircon sand. Then there wouldn't be the hassle of using chills.

If only these mold aggregates were used, there would be reduced feeding distances because of improper directional solidification. The use of these materials should be restricted to selected areas of the casting and can be used in conjunction with external chills. Exceptions to this practice may be in core surfaces and shell molds.

Chill aggregates are very useful materials, since they can be used more easily than metallic external and internal chills. Often, they can be used in areas where metallic chills cannot be used. The chill aggregates can be used in fillets or pockets or as partial chills on casting surfaces to promote directional solidification. These materials are also very useful to help control sand-related defects in areas of the casting, such as reentrant angles, since they can cause a rapid skin formation and they are more refractory than silica sand.

BIBLIOGRAPHY

As with the other material in this book, an attempt has been made to provide a basic understanding of riser practice. For a more in-depth study of risering, the reader is urged to check the literature listed. Although much of this literature may appear to be outdated, it is still as valid today as the day it was written.

Aluminum Casting Technology, 2nd. Edition, American Foundrymen's Society, Inc., Des Plaines, IL, 1991.

Bishop, H.F., "Theoretical Aspects of Moldable Exothermics Applied to Steel Castings," *AFS Transactions,* vol 71, 1963.

Bishop, H.F., Pellini, W.S., "The Contribution of Riser and Chill-Edge Effects to Soundness of Cast Steel Plates," *AFS Transactions,* vol 58, p 185, 1950.

Briggs, C.W., "Risering of Commercial Castings," *AFS Transactions,* vol 63, p 287, 1955.

Caine, J.B., "A Theoretical Approach to the Problem of Dimensioning Risers," *AFS Transactions,* vol 56, p 492, 1948.

Caine, J.B., "Risering Castings," *AFS Transactions,* vol 57, p 66, 1949.

Campbell, J., *Castings,* Butterworth-Heinemann Ltd., Oxford, England, 1991.

Casting Copper-Base Alloys, American Foundrymen's Society, Inc., Des Plaines, IL, 1984.

Creese, R.C., "Generalized Riser Design by Geometric Programming," *AFS Transactions,* vol 87, 1979.

Flinn, R.A., *Fundamentals of Casting,* Addison-Wesley Publishing Co., Reading, MA, 1953.

Gotheridge, J.E., Morgan, J., Ruddle, R.W., "Application of Moldable Exothermic Materials to Steel Castings," *Modern Casting,* Nov 1963.

Heine, R.W., "Design Method for Tapered Riser Feeding of Ductile Iron Castings in Green Sand Molds," *AFS Transactions,* vol 90, 1982.

Heine, R.W., "Riser Design for Mold Dilation," *AFS Transactions,* vol 73, p 34, 1965.

Heine, R.W., Amrhein, R.F., "Experience with Riser Design," *AFS Transactions,* vol 75, 1967.

Heine, R.W., Loper, C.R., Rosenthal, P.C., *Principles of Metal Casting,* Second Edition, McGraw-Hill, New York, 1967.

Introduction to Metalcasting, Handouts and Reference Materials, AFS/Cast Metals Institute, Des Plaines, IL.

John, R.A., "Risering Steel Castings Easily and Efficiently," *AFS Transactions,* vol 88, 1980.

Merchant, D., "Dimensioning of Sand Casting Risers," *AFS Transactions,* vol 67, p 93, 1959.

Meredith, J., "Fundamentals of Gating and Feeding Nonferrous Alloys," *Modern Casting,* Feb 1999.

Myskowski, E.T., Bishop, H.F., and Pellini, W.S., "Application of Chills to Increasing the Feeding Range of Risers," *AFS Transactions,* vol 60, p 389, 1952.

Pellini, W.S., "Factors Which Determine Riser Adequacy and Feeding Range," *AFS Transactions,* vol 61, p 61, 1953.

Plutshack, L.A., Suschil, A.L., *Riser Design,* Foseco, Inc., Cleveland, OH.

Ruddle, R.W., "Risering Copper-Base Castings: Part 2," *Foundry,* Dec 1970.

Sandell, P.K., "Breakability of Knock-Off Risers," *Foundry,* Nov 1960.

Schleg, F.P., Personal class notes from American Foundrymen's Society/Cast Metals Institute, Inc., Gating & Risering Courses, 1965–1992.

Svoboda, J.M., *Basic Principles of Gating & Risering,* AFS/Cast Metals Institute, Course Textbook, 1976.

Sylvia, J.G., *Cast Metals Technology,* American Foundrymen's Society/Cast Metals Institute, Des Plaines, IL, 1965.

Wallace, J.F., Evans, E.B., "Risering of Gray Iron Castings," *AFS Transactions,* vol 66, p 49, 1958.

Wukovich, N., "A Chilling Experience, (2 parts), *Modern Casting,* Jan and Feb, 1986.

Cleaning and Inspection 18

Traditionally, the cleaning or finishing room is the final step in the production of metal castings. In many foundries, the cleaning room often is the last place to be considered for improvement expenditures.

It is estimated that 20% of total labor costs in the average foundry is expended after the casting leaves the mold, and this is primarily in the area of casting cleaning. Not only are labor, equipment and materials costly in the cleaning department, but absenteeism can be disproportionately high, increasing the financial overhead.

Beginning with an optimum casting design can reduce cleaning costs. Careful design of the gating and risering system and attention to the molding and coremaking operations can all lead to reduced cleaning costs.

Casting cleaning must be considered in the early design stage. The need for varying casting wall thickness can result in additional gating and risering and metal removal. These things must be taken into consideration. Risers, larger ingates, riser pads, etc., all contribute greatly, and often unnecessarily, to high cleaning costs. It is true that casting is often the most economical way of making a metal component, but this becomes untrue when costly and involved procedures must be undertaken to remove unwanted appendages, such as flashings, fins and unnecessary gating and risers.

Important, but often forgotten, is the proper maintenance of the equipment used in cleaning castings. Whether all of this is feasible depends on the individual casting, which, in turn, is influenced more and more by the high cost of labor and materials. The environmental controls needed to improve the cleaning room air can be costly; however, it is necessary to attract qualified cleaning room workers, as casting defects can occur here due to poor working conditions.

Today, efforts are being made to modernize the cleaning room and the operations that take place in it. This includes installing improved lighting and ventilation, compliance with environmental, health and safety regulations, installation of newer machinery, and reducing manual labor by utilizing lifting devices and casting conveyor systems.

Grinding machines are located next to conveyor lines, strategically positioned to grind specific areas of the casting. Automatic grinding machines grind castings with CNC (computer numeric control) accuracy and speed. Robots are often used to perform grinding operations.

More and more automation is used in the cleaning room. Blasting cabinets and barrels are becoming more automated. In addition, computers are used extensively throughout the foundry as well as in the cleaning room. Computers control production, track castings as they pass through the cleaning room and increase efficiency.

The ideal situation would be to not have a cleaning room; however, how many customers would accept as-cast castings right out of the mold? Would they be willing to remove the gating and risering system, molding sand, or portions of cores remaining in passageways? Until a new process is developed that can produce castings ready for use as-is, right out of the mold, cleaning rooms will be essential. If more attention was paid to the casting design and the processes used to produce it, much of the time and effort spent in the cleaning room could be reduced considerably!

COOLING

After the mold has been filled with molten metal and before it has completely solidified, the mold is moved to the next production steps: cooling and shakeout. (Shakeout will be discussed in the next section.) **Figure 18-1** shows a mechanized cooling line. The poured molds are conveyed, while cooling, to the shakeout area. The cooling line is environmentally equipped to collect smoke and off-gases from the molds. This type of operation is normally found where a range of smaller castings is produced. In the case of large floor and pit molds, the molds remain in place after pouring.

The rate at which castings solidify has a great impact on the casting's mechanical properties. In order for a casting to offer specific mechanical properties (i.e., hardness, tensile strength), a specific chemistry and cooling rate must be adhered to. If chemistry and/or cooling rate and shakeout are changed, there will be a change in mechanical properties.

For instance, if the cooling line should break down and castings are not shaken out at the correct time, they can become mold annealed. In other words, the casting will have lower hardness readings and lower tensile strength because of the incorrect cooling

Fig. 18-1. During pouring operations, most of the combustible materials in the mold are burned off and are captured in the pouring hoods. Cooling is controlled by the length of the conveyor line.

process. One foundry experienced this problem after the cooling line was enclosed. Before the line was enclosed, the molds could lose heat by radiation and convection. After enclosing the cooling line, the heat lost by radiation was being radiated back into the mold due to the radiating characteristics of the material used to enclose the line. This action caused the castings to have a slower cooling rate and thus the casting's hardness dropped.

Another foundry had a shakeout situation where the shakeout time had been sped up to keep up with production demands. The quicker shakeout times led to castings with higher hardness readings. Why did this occur? The main reason was that the castings were being air quenched at higher temperatures causing them to be chilled.

A foundry in Wisconsin had two cooling lines. One line ran outside of the foundry during the summer months. Another cooling line ran indoors during the winter. The outdoor temperature determined which cooling line was to be used. Running this type of operation needs to be continuously monitored to ensure consistent casting properties. Fluctuations in the casting cooling rate will greatly affect mechanical properties.

Premature shakeout, while the castings are still hot, can cause surface imperfections if the castings are allowed to knock into one another. If castings cool too rapidly, internal stresses build up Subsequent heat treatment would then be required to relieve the internal stress, adding considerable cost to the operation..

Cooling rate and shakeout time play a critical role in the successful production of quality castings having specified mechanical properties.

SHAKEOUT

Shakeout is the term used for removal of the casting from the mold, whether it is a sand mold or a metal mold. **Figure 18-2** shows a casting immediately after shakeout. Note the temperature differences in the varying sections of the casting.

Shakeout Methods

Dump

There are several basic methods to shake out castings from sand molds. The dump method of shakeout is normally used in flaskless molding operations. The mold, along with the casting still inside, is dumped or tipped over, causing the mold and casting to fall onto or into some type of oscillating or vibratory conveyor. The conveyor is normally enclosed to keep the surrounding environment clean.

Another dump method makes use of the rotary drum shakeout device. This particular machine can perform five distinct cleaning operations on a casting. First, the mold and casting fall into the rotating drum. Second, the molding sand falls away from the casting and is blended, conditioned and screened. Third, the casting moves into a zone where cleaning media begins surface cleaning, and fourth, the core sand removal takes place. The cleaning media can be cast iron "stars" or any other hard material shape that will abrade the adhering sand from the casting surface. Any tramp metal is then discharged along with the core sand. The fifth operation is final cooling of the casting before discharge from the rotary drum.

Additional surface cleaning may be needed later in the cleaning operation. This type of equipment is normally used in high-production foundries producing ferrous castings that can withstand tumbling into other castings. Casting size does place some limitations on the use of this type of shakeout device.

Punch/Blow

Another shakeout method uses a mechanical punch to push the mold and casting out of the flask. A cast iron foundry has developed a shakeout method where the mold containing the solidified casting(s) is placed into a chamber. A measured amount of natural gas is injected into the chamber and ignited. The resultant explosion "blows" the mold and casting out of the flask. In this case, the molding sand and casting usually fall onto some type of conveying device to separate the molding sand from the casting. The sand is then returned to the molding sand preparation area for reuse. The conveying device movement causes the core sand to be removed from the internal passageways in the casting. The casting then moves to the next cleaning stage.

In the case of permanent molding and diecasting, the castings are ejected from the dies or molds, manually or automatically placed on conveyors or into "tote boxes" and moved along to the next cleaning operation. In the case of semi-permanent molding (where sand cores are used), core removal from the casting is usually the next step. However, there should be little, if any, surface cleaning done at this point in either process.

Hammering

In investment casting, the investment shell is removed from the casting by a type of force such as hammering or vibration. The cored passageways are cleaned out by ultrasonic methods or etching (with an acidic mixture), if necessary, after the shell has been removed,

Flask Removal

When large floor molds are used, shakeout is done by removing the flasks. The molding sand then falls away from the casting. Molding sand adhering to the casting is removed during subsequent shotblasting operations. In pit molding, the casting is pulled out of a pit with overhead cranes.

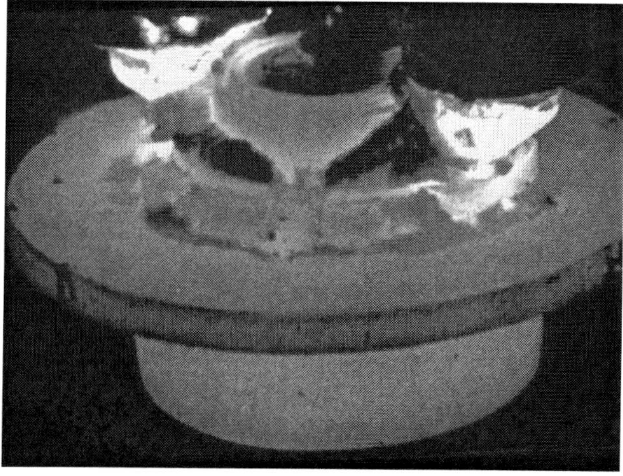

Fig. 18-2. Casting shown immediately after shakeout.

CLEANING AND FINISHING OPERATIONS

The main purposes of the cleaning and finishing department are:

- Remove excess metal from the casting, such as gating and risering system, fins and flash.
- Remove molding sand from the external surfaces and core sand from the internal passages of the casting.
- Remove any surface imperfections found on or within the castings.
- Prepare castings for repair welding and subsequent weldment smoothing.
- Apply any special finishes or treatments to the casting, as specified.

The following text covers the many processes of cutting, blasting, grinding and post-cleaning operations involved in the post-casting operation. The casting design and geometry will determine what process or processes are necessary for the individual casting application and specification.

Cutting

Unwanted appendages, i.e., gates and risers or any other superfluous metal attached to the casting, must be removed. The most efficient way to remove this excess metal is to sever it from the casting in the largest pieces possible. A few of the most common ways of doing this are by knockoff (degating), abrasive cutoff, flame cutting, bandsawing, friction sawing, shearing, water jet cutting and chipping. Each of these methods has its plusses and minuses that will dictate its most effective areas of application. There really is no one method for all situations. The selection of a removal method depends on such factors as casting alloy, mechanical properties, size and shape of the casting.

Knockoff (Flogging)

Previously, when discussing gating and risering, it was mentioned that if the ingates and riser connections were properly sized they would simply break off during shakeout, if the metal has poor ductility. This is especially true for gray and malleable cast iron. In some cases, this may also work for ductile iron.

If this is not the case, the simplest method of removing the gating and risering system (from most cast iron castings) is to knock the excess the metal off with a hammer (small castings) or a sledge (large castings). If the gating or risering system connections are too thick, there is a chance of cracking the casting. Although knockoff is quick, the surface finish where the gating and riser systems have been removed will be rough. However, it can be touched up during the grinding operation.

Abrasive Cutoff

Another method of removing appendages of excess metal from the casting is the use of abrasive cutoff wheels or blades. The wheels or blades are made of the same material as grinding wheels but are much thinner in width. Cutoff wheels cut or slice through the material to be removed. An abrasive wheel can be used on a swing frame, chop-stroke or locked-head push-through cutoff machine. **Figures 18-3 and 18-4** show the use of swing frame and cutoff machines.

Generally, the swing frame grinding machines are used on large castings that are too big to be lifted manually onto the table of the chop-stroke machine. Because of their weight and size, these large castings will remain on the floor during the cutoff operation, but occasionally portable cutoff tools can be used for certain castings. **Figure 18-5** shows an abrasive cutoff wheel on a portable vertical grinder being used to remove fins or flash on a large casting.

The abrasive cutoff method can be used on virtually any metal alloy and is the only method of severing excess metal on certain ultra-hard alloys. Abrasive cutoff is also economical, in terms of the amount of metal removed. Due to the wide range of abrasive wheel sizes and thicknesses, the kerf (notch made by the saw) and consequent loss of metal is smaller than many methods, other than bandsawing, while still obtaining the desired finish. This feature is especially attractive on copper-base alloys where the cost per pound warrants such savings.

There are other advantages to abrasive cutoff. In addition to being one of the fastest methods of removing excess metal from a casting, an abrasive cut is always straight. If necessary, contouring can be done by repeated cuts on a swing frame or portable grinder. This straight cut ability of an abrasive cutoff wheel is extremely effective when, for instance, a number of castings must be cut from the same gating system (such as a tree configuration often found in investment casting).

Another advantage is that when the proper cutoff wheel has been selected, there will be very little burr left on the casting surface. This, in turn, leaves very little touchup around the edges.

Fig. 18-3. Photo of swing frame abrasion cutoff machine.

Fig. 18-4. Photo of abrasion cutoff machine.

Fig. 18-5. Air-powered vertical grinder, ideal for slicing off fins and risers.

Flame Cutting

Flame cutting or torch cutting is commonly used to remove gates and risers from steel castings. An oxy-fuel gas system is used to produce the heat necessary to melt the metal in the area where the cut is being made. As this melting takes place, the melted metal is blown away and the appendage being cut off falls away from the casting. Two gases commonly used are oxygen and acetylene, thus the name oxy-acetylene cutting. When this method is used in conjunction with iron powder (ferro-jet or powder-cutting process), it is effective on oxidation-resistant alloys such as stainless steel. Both methods are less noisy than chipping, for instance, but do produce fumes that must be controlled. **Figure 18-6** shows a riser that has been removed with the oxy-acetylene cutting method.

The tip used on the oxy-acetylene cutting torch has preheating jets and a main oxygen orifice. The preheating jets have holes that surround the oxygen orifice. The oxygen-acetylene gas mixture is adjusted to preheat the cutting area. A trigger mechanism is then used to open the oxygen orifice. This oxygen then combusts with the preheated metal and the cutting action takes place. This cutting action takes place as a result of melting and oxidation of the metal into slag that is blown away from the cutting area by the force of the oxygen. The oxy-acetylene method is capable of cutting considerable depths of penetration in carbon and low-alloy steels, but also depends on the oxidation capability of the metal alloy to be cut.

Flame cutting is not adaptable for use with copper-base alloys, because these alloys do not oxidize enough to facilitate the action of the flame. The thermal conductivity of these alloys is also so high that it is difficult, or impossible, to begin a cut. Aluminum alloys also have extremely high heat conductivity and form a refractory oxide skin that resists smooth burning.

Cast irons are not normally cut using the flame cutting method because carbon levels are such that the metal is embrittled by the rapid cooling that takes place after cutting. In addition, the high carbon content makes cutting slow and inefficient.

Extremely large risers that have contacts between the riser and casting that exceed the capacity of the blowpipe may require the assistance of an oxygen lance to complete the severing operation. In this case, two operators are required: one for the blowpipe and the other for the oxygen lance.

Thermal torches can be used for extra-heavy riser removal. This cutting method uses a steel pipe that encloses rods of steel, aluminum and magnesium alloys and oxygen that burns (oxidizes) at temperatures of 10,000F (6093C) or more. The operator must wear a nonflammable suit and full-face shield, as well as eye protection.

Plasma Arc Cutting

Another method of gate and riser removal is the use of plasma arc cutting. This method requires the cutting torch to have an ultra-high-temperature ceramic nosecone and does not rely on a chemical reaction. This cutting method uses heat generated by applying electrical energy to a gas. Plasma arc cutting can be applied to any electrically conductive metal alloy whose thickness and shape permit full penetration of the plasma jet. One advantage of this method is that it leaves a very clean cut with very little slag on the surface of the casting.

Gouging and Pad Reduction

Other surface cleaning operations that can be included with these flame-cutting methods include gouging and pad reduction. Gouging is used mainly with carbon steel castings. It is used to remove surface imperfections that will later be repaired with arc welding. Pad reduction is the removal of riser pads. Riser pads are the amount of the riser connection remaining on the surface of the casting after flame cutting.

Powder Cutting

Powder cutting is another type of oxygen cutting in foundries. Powder cutting adds iron powder to the oxygen and preheat gases of the cutting blowpipe. This cutting method can be used with oxidation-resistant ferrous alloys, such as cast iron, chrome irons, stainless steel and high-temperature alloys. The powder is also effective in overcoming sand incrustations and layers or inclusions of slag.

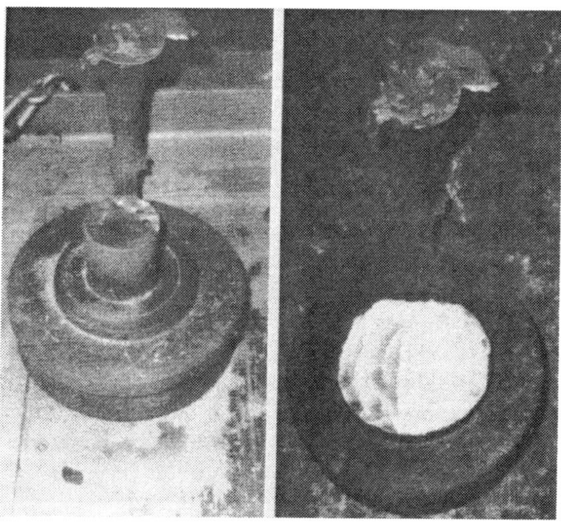

Fig. 18-6. (Left) Spinner head casting with riser. (Right) After removal by cutting torch.

Air-Carbon Arc Process

In this process, a torch, as seen in **Fig. 18-7** (similar to a welding electrode holder) is employed. The torch holds a copper-clad electrode and graphite electrode through which compressed air is blown. A welding power source supplies the electrical energy. The metal is melted by the electric arc and simultaneously the high velocity jets of compressed air (parallel to the electrode) blow thc resultant molten metal away.

The air-carbon arc process is also known as the arc-air process and is used as a means for cutting ingates and risers from stainless steel, ductile iron and nonferrous castings, and small carbon steel castings where ingates and risers are small. **Figure 18-8** shows such an operation performed on a casting.

Pads can also be removed using the arc-air cutting process. High metal-removal rates are possible using this process. This process is also used for the removal of defects, sand inclusions, shrink, etc., in preparation for repair welding.

The arc-air and oxy-acetylene methods are the most common and most frequently used methods for cutting appendages from castings. They allow the placement of ingates and risers in practically any position in order to pour and riser a better casting. This advantage is due to the almost unlimited maneuverability of this equipment, which also makes it a valuable tool for casting cleaning.

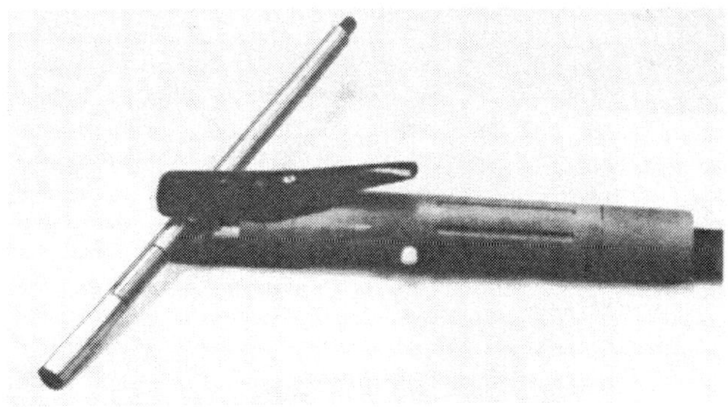

Fig. 18-7. Double-head carbon-arc torch. Equal air flow is directed to each side of the molten puddle for fast, easy metal removal.

Fig. 18-8. Air-carbon arc and oxy-fuel method is used for cutting ingates and risers from castings.

Bandsawing

Bandsawing is another method used to remove excess metal from castings. This method permits the bandsaw operator to follow the contours of the casting more closely than most cutting methods, and can be used effectively on many alloys. With the proper saw, this method will work well, even on certain nickel and chromium alloy castings. Supplementary cleanup of the kerf (notch made by the saw) may be necessary. In most cases, the saw cut will be smooth, but not if the blade is worn or damaged. **Figure 18-9** shows this cutting method being used on an aluminum alloy casting.

When considering the cutting speed in bandsawing, it is important to consider three factors: finish, production and saw life. A good finish requires a faster saw speed, a finer pitched saw and a slower rate of feed. On the other hand, high production calls for faster saw speeds, a coarser pitched saw and a fast rate of speed. In order to maximize saw life, a slower speed, finer pitch and a medium-to-fast rate of feed are needed. Unfortunately, not all three results can be obtained with one technique.

The alloy chemistry, casting cleanliness and metal thickness all determine the correct saw speed. The higher the percent of wear elements, such as nickel, silicon, sand, etc., the higher the hardness and the lower the saw speed to gain maximum saw life. Nonferrous castings can be cut at either high or low saw speeds. The best bandsaw machines are those with infinitely variable speeds, from 40 to 10,000 fpm or higher.

Tooth spacing on the saw band plays an essential role in cutting efficiency and saw life. To form chips fast enough in high-production cutting, some space between the teeth must be provided for chip clearance. Coarse-toothed saws remain sharp longer because the chips are not forced to crowd together and dull the teeth points on the saw. For faster cutting, a ratio of 2-3-4 or 6 teeth per inch (coarse toothed) is recommended. This ratio enables a speed fast enough to keep the teeth from straddling thin sections and being stripped.

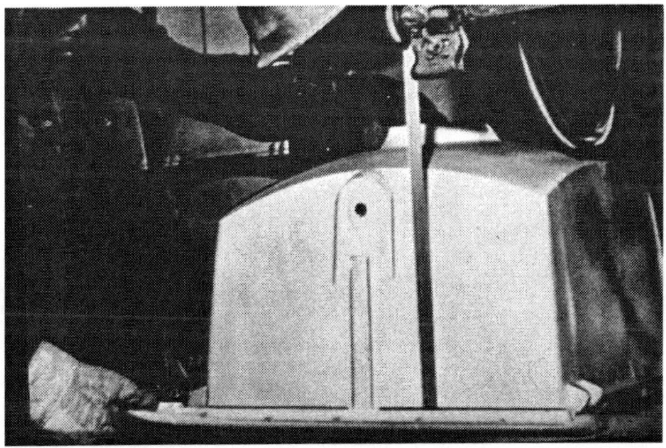

Fig. 18-9. Bandsawing being performed on aluminum alloy casting.

Sand adhering to the castings can have a negative effect on bandsaw blade life. The importance of some form of lubrication and cooling at the point of cut, when cutting castings, is important to achieve prolonged tool life. Some dependable form of drip lubrication or spray system should be used. An excellent lubricant is a high-grade soluble oil added to water. Lubrication can produce considerable increase in band saw life on the light metal alloys.

Friction Sawing

For cuts on ferrous castings up to 1-in. (2.54-cm) deep, friction sawing is effective. It uses the heat of friction to soften the metal so it can be easily removed. Slight pressure of the casting on the points of the band generates high heat immediately in front of those points, faster than the casting can absorb it. This action raises the temperature in this area of the casting, high enough to soften the metal.

In this process, friction caused by saw teeth moveing at 3000–15,000 fpm causes the casting to build up heat and soften the metal prior to the cut. The temperature at this point ranges from 1600–2600F (871–1426C), approaching the forging temperature of most alloys. However, on ferrous castings where the thickness of the ingate or riser exceeds 1 in. (2.54 cm), low speed (40–1500 fpm) should be used.

Although friction sawing (not to be confused with high-velocity sawing) is fast on thin sections, it cannot build up and maintain sufficient heat on thick sections.

The saw band itself is not softened, because the duration of the blade/casting contact is minimal in comparison to the time it is cooling. This same principle works when sliding down a rope, though it may burn the hands, the rope remains the same temperature. **Figure 18-10** shows a ferrous casting with ingates and risers cutoff by friction sawing.

Following are some advantages of friction sawing:

- Low tool cost, as distributed over the amount of cutting before the band is discarded.
- The blade will follow contours.
- Stainless steel, armor plate, steel alloys of almost any hardness and all cast ferrous alloys are easily cut.
- The heat (unlike torch cutting) does not penetrate far enough into the casting, so casting temper is not a consideration.
- Unlike knockoff or flogging, shearing, etc., it is effective on ductile iron.

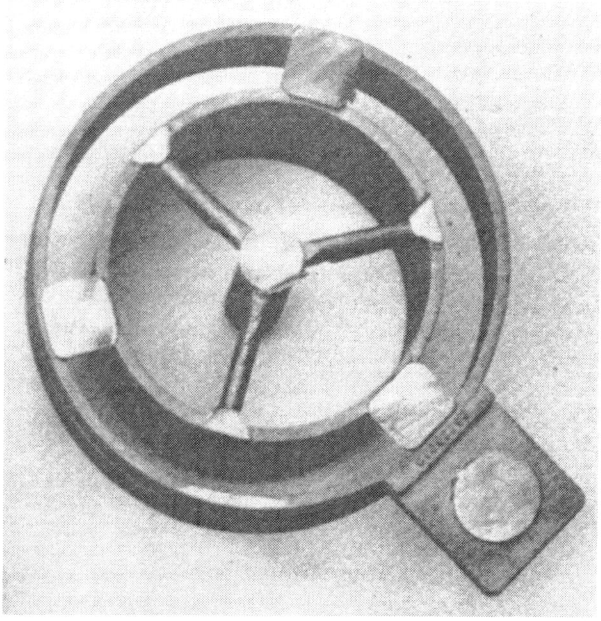

Fig. 18-10. Ferrous casting with ingates and risers cut off by friction sawing.

Friction sawing does have some limitations:

- It can be used only on section thicknesses less than 2 in. (5 cm) and preferably under 1 in. (2.54 cm) since anything over this requires special techniques.
- Steel alloys containing tungsten cannot be friction sawed.
- Most free-machining alloys, such as aluminum and brass being good heat conductors, fail to concentrate the heat immediately ahead of the band and melt close to their softening temperature. Certain aluminum alloys with high silicon content can be friction sawed.
- It cannot be used with stack cutting, since the layers tend to stick together.

Trim Press Operations

Trim press operations use hydraulic and other cutting press equipment to shear off gating and riser connections and casting flash. Many different forms of trim press operations are used in the foundry cleaning room, primarily in high volume operations that produce large numbers of the same casting geometry. The castings are fixtured in a die **(Fig. 18-11).** Pictured in **Fig. 18-12** is a ductile iron casting. The gate connection and all parting line flash have been removed in a single cutting action. The die cast aluminum gear box and motor block casting **(Fig. 18-13)** shows an example of a trim press using simultaneous horizontal and vertical trimming on two planes.

Shearing

Shears are heavy, sharpened matching steel jaws that are worked past one another by a fly wheel and cam assembly. In other words, shearing works on the same principle as cutting paper with a scissors. Shearing can only be used on materials that are softer than the shear blades, which limits them to malleable iron, soft and medium-hard steel, brass, bronze, aluminum and magnesium. They are also limited to small jobs but, when adapted, are fast and economical, as in removing riser pads from malleable iron. **Figure 18-14** shows a press shearing off a crankshaft ingate.

Fig. 18-11. Castings fixtured in a die.

Coining and Broaching

There are two other cleaning processes used that employ a press: coining and broaching. Coining is the method used to straighten and size castings by die pressing. **Figure 18-15** shows a die used for this operation. During coining, some flash around the parting lines can be removed. This process is used for malleable iron castings that can become warped during the malleablizing heat treat cycle. Some alloy castings do have to be preheated to prevent cracking during the coining operation.

Figure 18-16 shows a type of broaching operation. Normally, broaching is done to smooth or slightly enlarge holes such as boltholes. In this case, a series of sharp, hardened blades are brought down under extreme pressures to cut off excess metal, such as ingate and riser pads, from the casting. In order to do this, the casting is held in a fixture.

Fig. 18-14. A press shearing off a crankshaft ingate.

Waterjet Cutting

A high-pressure, abrasive-waterjet stream has been developed for various cutting applications. It is used for quickly and accurately cutting steel superalloys, titanium and other advanced metal alloys, as well as glass, ceramics, Kevlar and other composites. Combining this high-pressure, abrasive-waterjet stream with robotics technology works well for cutting off ingates and risers, especially on steel castings.

Fig. 18-15. Air lines are used on this cold coin die, used to eject castings after flash and pads have been pressed off.

Fig. 18-12. Gate connection and parting line flash removed in a single cutting action.

Fig. 18.13. Castings have been trim pressed using horizontal and vertical trimming on two planes.[Courtesy Renault, Cleon France]

Fig. 18-16. Broaching operation showing close-up of blades, cleaned casting and chips in foreground.

For this operation, a high-pressure unit with a 60-hp pump linked with dual intensifier pumps to pressurize the water to 55,000 psi is needed. The intensifier pump uses a microprocessor-based control system to simplify control and provide a wide range of control features. Garnet particles (#60 grit) provide the preferred abrasive material. Silica sand, silicon carbide and aluminum oxide may be used but should be sharp-edged grains to cut and shear away the target material.

The pressurized water forms a jet stream at a 0.005–0.020-in. (0.13–0.51-mm) sapphire orifice. Then the abrasive particles are added to the jet to become entrained in the stream of water to form the abrasive-waterjet stream that is then expelled through a tungsten carbide nozzle. The nozzle focuses the abrasive-waterjet stream, which exits the nozzle at a tip speed of over 3000 ft/sec (925m/min), on the object to be removed.

The operating life of the synthetic sapphire orifice is 250–500 hours. The operating life of a tungsten carbide abrasive nozzle is approximately only six hours, due to the erosive effects of the accelerated abrasive-waterjet stream.

The abrasive-waterjet system requires a catcher to collect the spent fluid after it passes through the material being cut. This can be a simple tank lined with ceramic pieces, concrete block, brick or white iron scrap to suppress the piercing or cutting of the tank. The accumulated water can be drawn off through a valve placed low in the tank wall. The collected sludge at the bottom of the tank requires periodic removal.

A greatly improved method of collecting and separating the water and abrasive material involves the use of a funnel-shaped steel container containing steel shot and having small water release holes in the bottom. This type of catcher-collector lends itself to following a moving nozzle, while the collector mentioned earlier is stationary and nozzle movement is limited.

Chipping

Chipping hammers are air- or electrically-driven, portable tools that use a mounted chisel. The most commonly used method of powering these tools in the foundry is with pneumatic air systems. Chipping hammers are very versatile but can be quite noisy. The sharpness of the cutting bit (chisel), the way it is ground to shape, the alloy from which it is made and, in particular, the way it is held against the work by the operator affects cutting efficiency greatly. Keeping the cutting tip wet with oil can improve the operation.

Chipping hammers are useful in the removal of fins, gate and riser padding, and to knock out sand cores. **Figure 18-17** shows the use of chipping hammers to punch out boltholes in a gray iron engine block. Most brasses and bronzes tend to work-harden, so that extremely light cuts are generally advised to avoid stalling of the chisel in the cut.

Chipping is one of the noisiest operations in the foundry and the operator must wear the proper safety and hearing protection when performing this operation.

Proper chipping begins with selection of the proper type of chipping chisel for the job to be done. If chisels are used until they are mere stubs and the shank no longer approximates the standards, the piston bore and piston life for the chipping hammer will be shortened and there will be an increase in operator fatigue due to excessive vibration.

Blasting

Immediately after shakeout, much of the sand that adheres to the casting is removed on a cleaning-cooling table by vibration or other mechanical methods and, when feasible, by wirebrushing. At this same time, the core sand should also be draining out of the cored passageways. Sand that continues to adhere, along with scale and other surface protrusions, as well as core sand, can be quickly and uniformly removed by abrasive impacting, commonly known as blasting.

In some foundries, this operation is done immediately after shakeout. In other foundries, abrasive impacting or blasting is done after the gating and risering systems have been removed from the casting. Ideally, the gating system and risers should be blasted at the same time the castings are blasted so that they too are cleaned of any adhering sand. This will result in a much cleaner melting operation, since the cleaned gating system and risers will be remelted.

Not all castings are blasted. Castings such as those produced in permanent molds, investment molds and diecastings will normally have a surface that does not need blast cleaning. Impact cleaning or blasting is often referred to as the most important operation in the cleaning room; certainly, if done properly, it can minimize the cost of subsequent operations.

There are three types of abrasive impact cleaning methods available today: *air, water* and *mechanical.* Perhaps the most efficient of all methods of cleaning castings used in production foundries today is the mechanical or airless-blast method, both small and large castings are cleaned in this manner. Small castings are rotated or tumbled under blasts of grit or shot. Large castings are placed on stationary or rotating tables inside a cabinet, and particles of shot or grit are thrown at them by means of centrifugal force derived from a rapidly rotating wheel, as illustrated in **Fig. 18-18.**

Control of the blast operations and proper maintenance of the machines are essential because:

- The blasting abrasive wears equipment parts, so blast cleaning equipment is self-destructive.
- Unlike machine tools or other production equipment that must be carefully serviced and maintained, blast equipment

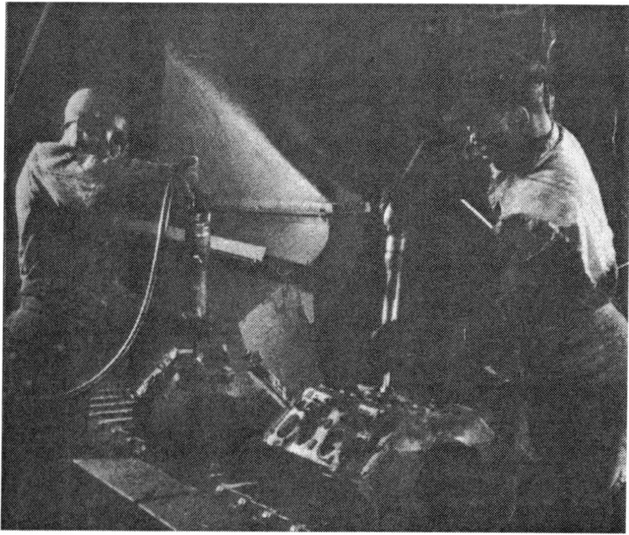

Fig. 18-17. Punching out boltholes in engine castings.

will often continue to produce satisfactory work when its efficiency is at an extremely low and costly level.

- All too often, blast equipment maintenance consists only of repairing breakdowns, or it is confined to keeping the machine running without regard to efficiency. In short, blast equipment maintenance is usually restricted to emergency repair of machine damage, not aimed at correcting the reason for the repairs.
- Due to the generally poor condition of blast equipment, it can be an unpleasant area to work in.

Grit and Shot Blasting

The shot or grit used may be made of white iron, malleable iron or steel. Shot blasting has a peening effect (the surface becomes flattened by the hammering of the steel shot) on the casting. This can be compared to the use of a ball peen hammer, when the peen is used to hammer the surface of metal. Grit has a much harsher effect on the casting surface than shot, sometimes removing small particles of metal from the surface. Shot blasting produces a shiny surface, while grit dulls the surface.

When a nonferrous casting is treated by shot blasting, the force propelling the shot may severely damage the surface of the casting. For treating nonferrous castings, soft steel, copper or bronze shot, or glass beads are quite effective. Sand blasting is used for some softer alloys, but the dust created can present a hazard to the operator of the equipment.

Wire brushing is often sufficient to clean the surface sand from aluminum, brass or bronze alloy castings and produces a shiny surface. More information about the selection of abrasives and their importance will be covered later in this chapter.

Abrasive Blasting Equipment

There are two common types of blasting in use today: the *fixed-cabinet* and the *barrel*. These can be further classified according to whether they are batch-type or continuous. In the case of the fixed-cabinet mills, the means in which the castings are conveyed through the machine is also a consideration.

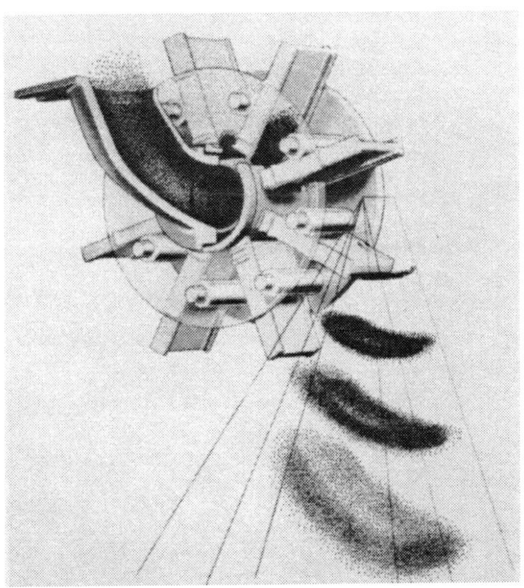

Fig. 18-18. Abrasive blasting equipment uses a rotating wheel to mechanically "throw" metal shot at the casting surface.

No matter which type is selected, the principle of blast cleaning is simple. An abrasive metallic shot or grit or similar abrasive is propelled against the casting either by air, water blast (nozzle), or mechanically by a rotating centrifugal blast wheel **(Fig. 18-18)**, also known as airless blast. The purpose of propelling the abrasive against the casting is to remove sand, core sand, firescale and, to a certain extent, flash from the external and internal surfaces of the casting. The objective of the blast equipment design is to enable the blast stream of abrasive to strike all surfaces with sufficient force to efficiently and thoroughly remove contaminants, without etching the casting surface. To get the most benefit, the stream of abrasive is preset by targeting it against a board or metal plate.

All blast cleaning machines consist of a materials-handling system for positioning the work (castings) for blasting, a cabinet to house the operation and contain the blast abrasive and dust, an abrasive throwing mechanism (usually a centrifugal blasting wheel) and an abrasive cycling system for returning the abrasive and cleaning it so that it can be reused. The materials-handling system must position the work properly for blast and then carry it away after cleaning.

Besides the proper maintenance of the abrasive throwing mechanism, the abrasive cycling system also warrants careful attention. The separator found in the cycling system is one of the most important elements in controlling blast-cleaning costs. Contaminants removed from the workpieces are highly abrasive to the machine, and even a nominal increase in these particles into the abrasive stream can double or triple maintenance costs. Excessive amounts of sand or metal particles left over from previous cleanings retard the effectiveness of the abrasive used in airless blasting. It must be removed from the shot during the recycle process.

The abrasive cycling system controls the efficiency of the machine by establishing the range of particle size in the abrasive mixture. For optimum cleaning, the abrasive mixture must contain large-sized abrasives to loosen sand and scale, and small sized abrasives to scour the surface. Too much of one or the other can seriously restrict cleaning efficiency. Ideally, a maximum efficiency results from using the smallest size of abrasive that has sufficient mass to remove the contaminants on the casting.

Blast Equipment Selection

While most blast machine manufacturers offer engineering services, each foundry should perform a preliminary investigation to evaluate the type of equipment that is appropriate for the operation. In undertaking such an investigation these factors should be considered:

- The quantity and variety of castings and their condition as received in the cleaning room.
- The production rate/hour, in pieces and/or cubic feet. Consider also if the flow is regular or intermittent.
- Special factors, large amounts of sand, binder type, cores and rods to be removed.
- The floor space and height factors pertinent to the equipment location.
- Standard and special electrical control specifications.
- Limitations on pits for foundations or for material handling equipment.
- Environmental, health and safety requirements.

After the investigation has been completed, a decision on whether to purchase standard or specialized equipment must be made. Standard equipment is usually the first consideration because of its availability. Generally, high-production, specialized foundries will require tailor-made or special blasting equipment.

This equipment, by type of process, can be divided into three categories: *airless centrifugal wheel* (most favored), *direct air blast* and *suction air blast*. This equipment, as mentioned earlier, can further be divided into continuous or batch-type in terms of material handling.

The direct air blast and suction air blast equipment will be confined to low production, manual blast cabinets and to touch-up cleaning operations. In addition, where manual flexibility in directing the blast stream is needed, these two types find their application. On the other hand, those foundries engaged in high-volume casting production use the airless centrifugal wheel type of equipment.

How the castings are presented to the abrasive blast stream is also a consideration. Earlier, the tumble blast method was mentioned. In this particular machine, the castings are tumbled as they are blasted. A schematic on how this occurs is shown in **Fig. 18-19**. This machine is usually used for batch-type operations and smaller castings. **Figure 18-20** shows a tumble blast machine being emptied.

Tumbling of the castings takes place in the blast barrel. The castings tumble over one another, similar to clothes tumbling in a clothes dryer. Again, smaller castings are blasted in the blast barrel. In either the tumble blast or blast barrel, it is not advisable to use them with softer alloy castings since the tumbling action will cause the castings to "slam" against each other, which can cause surface imperfections. The blast barrel is also a batch-type operation.

Figure 18-21 shows a continuous blast cabinet in which the castings pass through on a continuous moving slat conveyor belt. The space between the slats allows the abrasive shot to pass through and be collected for reuse. There is no tumbling action taking place in this particular machine.

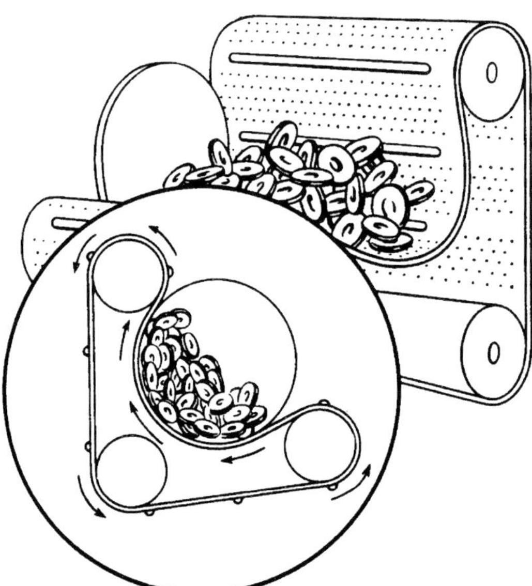

Fig. 18-19. Airless blast cleaning, combined with tumbling and using an endless conveyor of heavy-duty rubber belting to expose all surfaces of the castings to the full effect of a blast of centrifugally thrown abrasive.

Fig. 18-20. Tumble-blast machine being emptied.

Fig. 18-21. Blast cleaning cabinet using a slat belt conveyor capable of cleaning over 200 tons of hard iron castings daily.

In some abrasive blast cabinets, castings are placed on a table that can be swung into and out of the cabinet. The castings are positioned on the table so that the most effective targeting of the blast abrasive is achieved. At times, it may be necessary to reposition the castings to achieve the quality of surface cleaning desired. **Figure 18-22** shows a swing table blast machine with two rear-axle housings on the swing table after blasting. Note that this table is raised up off the floor. This type of blast cabinet is offered in a number of sizes and can be used for various sized castings.

Instead of being called a blast cabinet, the piece of blast equipment seen in **Fig. 18-23** is called a blast room because of the size. This blast room is 30 ft (9 m) long, 20 ft (6 m) wide and 16 ft (5 m) high. The turntable is 12 ft (4 m) in diameter. This particular blast room has three centrifugal wheels propelled by 30-hp motors that throw 138,000 lb (62,100 kg) of shot per hour. The turntable revolves as the blasting is taking place, enabling more of the casting surface area to be exposed to the blast abrasive. If necessary, manual spot-blast touch up can be made with special hookups in the blast room.

Blast rooms can be built to accommodate castings weighing as much as 250 tons (225 metric tons), with special designs available for even larger and heavier loads, if necessary. These blast rooms also use airless centrifugal blast cleaning.

Figure 18-24 shows a barrel type of continuous airless blasting equipment. As the barrel rotates, the castings tumble and are continuously exposing new surface area to the abrasive blast. At the same time, the castings are moving from the charging end to the discharge end of the barrel.

Another type of continuous airless blast cleaning equipment is the spinner monorail as seen in **Fig. 18-25**. In this piece of blast equipment, the castings are placed on hooks suspended from the monorail. As the castings proceed through the blast cabinet, the hooks spin, thus exposing the greatest portion of the casting surface to the abrasive blast stream. Monorail speeds can go as high as 600 hooks per hour, at which speeds the prime consideration is the loading and unloading operations. Hooks can be used from which more than one casting can be hung for blasting.

Monorail-type cleaning can be used in a modular cabinet with a continuous traveling roof seal and blast-tight doors. This design provides for good abrasive retention and control over noise and air pollution. Many types of monorail systems use rotary, self-sealing doors that reduce wear surfaces, simplify maintenance and minimize floor space requirements.

Fig. 18-22. Swing table blast machine.

Fig. 18-23. Abrasive blast machine shown in a blast room.

Fig. 18-24. Hurling 86 tons of steel shot every hour, this continuous flow blast machine handles 25 tons of heat-treated malleable iron castings per hour.

Fig. 18-25. Spinner monorail operation.

Air Blasting

The next few figures will show the air-blast method used to carry the abrasive to the casting surface. **Figure 18-26** shows a person manually air-blasting the surfaces of a casting. In this particular case, sand is the abrasive, and it is called sand blasting.

One precaution should be mentioned regarding aluminum alloy castings. If the castings are to be anodized, silica sand should not be used as the blast abrasive. The silica sand can be imbedded in the surface of the casting and react with the electrolyte used in the anodizing process, causing a milky gray surface appearance.

Figure 18-27 shows a modular blast room in which the operator manually directs the air-blast abrasive material at the casting surfaces. Note the independent air supply hood being used and the well-lighted space allowing the operator full view of the blasting operation.

In **Fig. 18-28**, an individual airblast cabinet is in use. A pair of abrasion-resistant gloves protects the worker's hands. One hand directs the abrasive blast and the other hand can either hold or manipulate the casting. This type of cabinet is usually called a sand blast cabinet.

Abrasives Selection

Blast cleaning abrasives are a major portion of the cost of blast cleaning castings, from 30–75%, depending on the surface finish required. The smoother the finish, the higher the abrasive costs.

Metallic Abrasives—The predominant abrasive in blast cleaning is steel shot and grit with some aluminum oxide added. Ordinary chilled iron, which has great cleaning ability because of its hardness, can be used. However, when using chilled iron shot, higher maintenance costs are usually encountered than when using annealed cast iron or steel abrasives. For certain fine finishes on nonferrous castings, glass beads are used. The best practice to follow when purchasing abrasives is to purchase high-quality abrasives.

It is probable that four independent factors should be considered when selecting a metal abrasive. These factors include:

1) time involved for the cleaning job,
2) equipment wear (maintenance),
3) abrasive consumption (cost),
4) surface finish required.

These factors are all interdependent and no one factor can be changed without affecting the others.

The size of the shot also plays a role in the cleaning ability of the shot abrasive. Larger shot has greater impact, while smaller sizes provide greater coverage. The key is to use a mixture of different sized shot to balance impact with coverage, to obtain the desired results. Unfortunately, there is no easy answer as to what this mixture should be. This test usually begins using large shot, and then gradually adding smaller shot until the best job is done in the shortest time.

Fig. 18-26. Well-protected sand blaster using independent air supply. Floor grates prevent sand buildup underfoot.

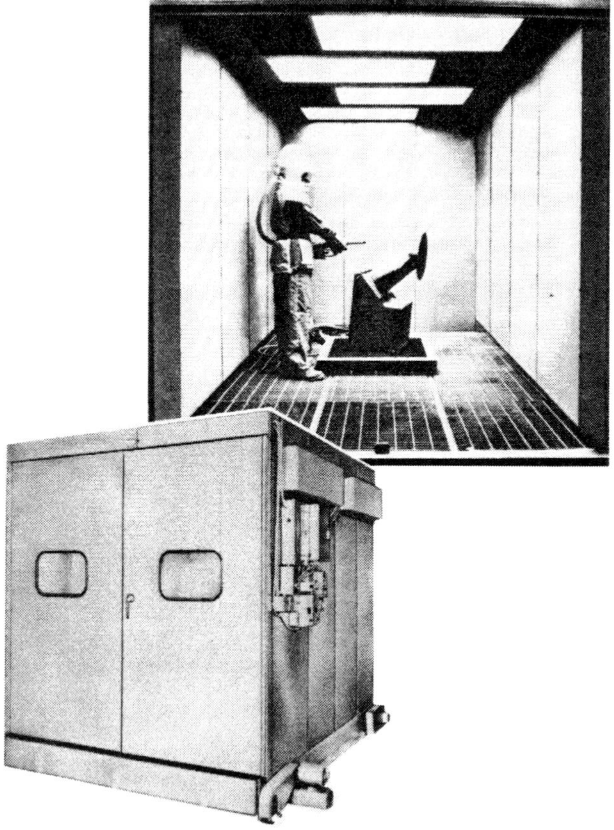

Fig. 18-27. Modular blast room requires no pit excavation. Floor vibrates and shakes used media into bucket elevators for air washing and reuse.

Once the correct shot size mixture has been determined, a testing procedure should be set up to keep the mixture constant. A test similar to the sieve analysis (done with mold and core sand) can be used for this purpose. How often the shot abrasive analysis should be repeated is a matter of trial and error.

Nonmetallic Abrasives—For blast cleaning of castings, nonmetallic abrasives find their greatest use in situations where iron contamination is not acceptable, such as in stainless steel and nonferrous alloys. In any event, foundries are not a major market for aluminum oxide. Sand or glass beads are the most common nonmetallic abrasives.

Manufactured aluminum oxide is fast cutting, uniform in size, readily available and reasonably durable. Although it has been reported that there is difficulty in screening out fines for recycling this abrasive, it should not be a major problem. On airless or wheel-type blasting equipment, the distance between the wheel and the casting should be kept to a practical minimum. In addition, a good dust collection unit should be an essential part of the system. With fast-cutting aluminum oxide, throwing-wheel wear is a problem, and aluminum oxide is a comparatively expensive blasting abrasive.

Silica sand, though inexpensive, is not very efficient and causes considerable wear on equipment. A major problem using silica sand is the added OSHA regulations; thus, it does not find wide usage. When sand is used in the blasting operation, it should be used in a dust-tight room with doors that remain closed during the sand blasting operation. Castings that have been sand blasted must be dusted off prior to their removal from the room, and operators must wear approved dust respirators.

Glass beads are used more for surface blending than for cutting and are used mostly for producing a matte finish on aluminum and similar soft metal alloys.

Fig. 18-28. Individual air blast cabinet in operation.

Water (Hydraulic) Blasting

Water blast cleaning units use a very high-pressure, high-velocity fine stream of water to carry the abrasive to the casting surfaces. This water stream, along with the abrasive, wets, scours and removes sand scale and dust from the casting surfaces. One such piece of blasting equipment has been developed to remove sand cores from hard-to-reach areas in the casting.

Water blasting should be done in an enclosure that will effectively prevent water and/or abrasives from splashing or ricocheting outside the enclosure. **Figure 18-29** shows a water blast unit in operation. In most cases, the water streams are targeted differently for various shaped castings.

The greatest advantage of this blasting process is the complete absence of dust problems. No special ventilation equipment is required. Its biggest disadvantage is the dirtiness it creates by using such a large a quantity of water, and the difficulty in separating it from the abrasive material. In addition, water blasting requires larger amounts of electrical energy, as opposed to air or impeller blasting. It is mainly used in the area of heavy sand core removal.

Mechanical Blasting

Tumbling Mills (Rattlers)—One of the earliest methods used to remove firescale, sand and some fins from castings was to place them in large barrels (tumble mills) along with abrasive media of irregular pieces of hard metal. Rotating the mill causes the casting and the abrasive media to tumble and abrade against one another. This tumbling causes a burnishing action, which, if it becomes excessive, can cause rounded corners, and, with certain configuration castings, can cause deformation. Tumbling is used with copper-base castings for deburring and brightening of the surface.

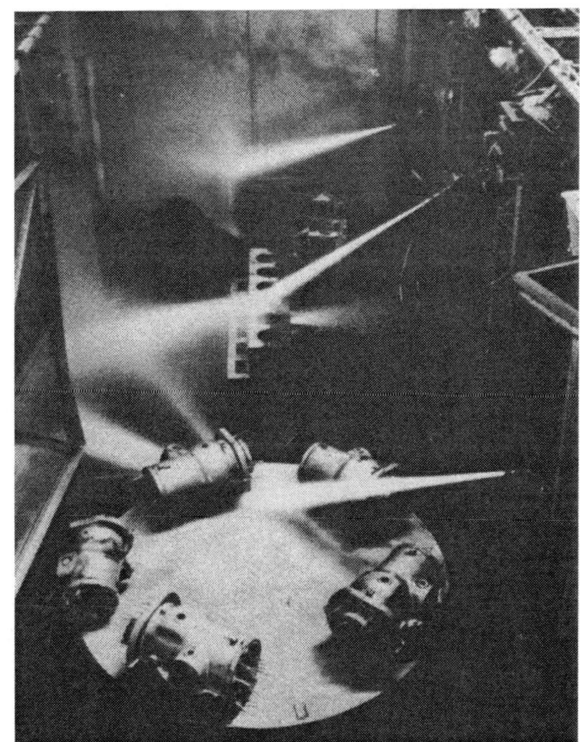

Fig. 18-29. Hydraulic blasting.

Figures 18-30 through 18-32 show different means to cause the tumbling action. **Figure 18-30** is an illustration of vibratory cleaning. A vibratory unit is either a tub machine or a horizontal octagonal barrel type that maintains a constant movement in a mass of abrasives, castings, compound and water. The frequency of vibrations is normally from 600 to 3600 vpm (vibrations per minute). As the mass vibrates, the entire workload is scrubbed and scoured. Because of its scrubbing action, as well as its orbiting, it is well suited for deburring hard-to-reach areas. This type of equipment is most often found in machine shops and diecasting foundries. It is sometimes used on small-to-medium sized castings produced in sand molds.

Spindle Finishing—In spindle finishing, the castings are fixtured or chucked to spindles and lowered into a rotating or vibrating tub containing the abrasive slurry. As the spindles revolve through the slurry, a combination of rotating and vibrating plus a plowing action takes place, resulting in a very rapid finishing action. An illustration of spindle finishing is seen in **Fig. 18-31**.

Rotary Barrel Cleaning—Rotary barrel cleaning is illustrated in **Fig. 18-32**. In this method of cleaning, the upper layer slides downward as the barrel rotates. Up to 95% of the abrading, deburring or polishing takes place at this time. The size of the upper layer, which consists of the abrasive and castings, varies in thickness, depending on the size and speed of the barrel and other conditions such as the amount of water and cleaning compound.

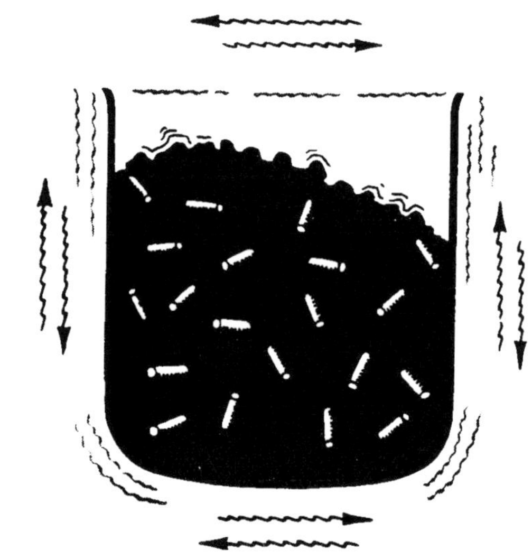

Fig. 18-30. Schematic of vibratory cleaning action.

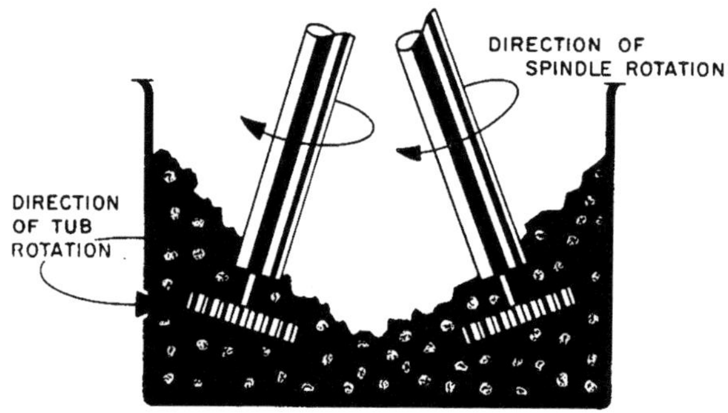

Fig. 18-31. Schematic of spindle finishing action.

Centrifugal Cleaning—Centrifugal cleaning combines the rotating motion of the barrel and a centrifugal force, developing pressures 25 times greater than the weight of the abrasives and castings combined. This high force pushes the abrasive and castings against the forward section, creating a caterpillar action of the castings against the wall of the cylinder that results in fast cutting. This process is well suited for finishing areas of castings inaccessible to other cleaning techniques.

Grinding

Grinding operations are used primarily to remove excess metal from the casting, such as ingates and riser pads. It can also be used to remove surface blemishes. The grinding process removes excess metal in the form of minute chips, but is unlike the cut-off methods, discussed earlier, which remove chunks of metal.

The principal area of grinding concerning the foundry industry is called off-hand grinding. Off-hand grinding includes snagging (which is the rough cleaning of castings), weld grinding, tool sharpening and miscellaneous rough grinding.

In the case of manual grinding, the casting, abrasive wheel, machine and operator are collectively referred to as the metal removal system. A change in any one of these components affects one or more of the others. For example, in a certain instance, it may be necessary to increase the surface speed of the abrasive wheel. Such a change involves more than just changing pulley sizes on the machine. It requires a machine capable of operating at these increased speeds and a change of pressure applied by the operator.

Grinding Wheels

There are no perfect formulas to specify which abrasive wheel will work optimally in a given situation, but there are basic guidelines to follow. (From this point on "wheel" pertains to the abrasive or grinding wheel.) For instance, the diameter of the wheel is determined by the size of the grinding machine, and the thickness is governed by the diameter of the wheel. The wheel is only but one

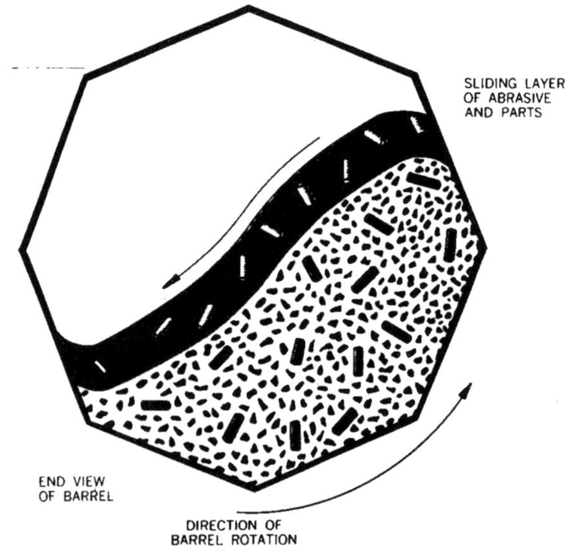

Fig. 18-32. Schematic of rotary barrel cleaning action.

element in this abrasive cutting system. The other two factors that need to be considered are the machine horsepower and the pressure exerted by the operator.

There are some points to be observed. The first is that, the greater the pressure employed on the wheel, the greater are the chances of the casting slipping off the wheel. In this case, the use of fixtures or jigs to hold the casting is advised. The second point is that the grinding machine should be well maintained so that it can deliver the horsepower necessary.

When selecting snagging wheels, the usual compromises must be made in the average jobbing foundry. A large wheel inventory is generally undesirable, but it is possible to group categories of work and establish (by testing) the best wheel for each job. Wheel cost and labor naturally enters into such considerations, and much of this information is available from wheel manufacturers.

Wheel Composition—Most wheels are made from a mixture of bond and abrasive grain, along with some miscellaneous additives. The arrangement of these materials is known as structure. On the blotter of the wheel is a "shorthand" designation of the wheel's composition and other specifications vital to its proper use. The entire composition of the wheel is expressed in a six-position symbol that defines, in order from left to right, the exact type of abrasive, grit or grain size, grade or strength of bonding, structure or grain spacing, type of bond and, finally, any modifications that the manufacturer wishes to include. To delve deeper into this area, the reader is advised to refer to the AFS publication, *Cleaning Castings.*

The larger the grains the faster the rate of metal removal, with certain exceptions. Hard, brittle castings will not always allow the coarse grains to penetrate full depth, and the chips are small and relatively few (coarse grains naturally being wider spaced). In this case, the remedy is to use smaller grains, spaced closer together, to get more cuts per revolution from the wheel.

The grade indicates the relative strength of the bond posts that hold the grains together. The chief determinator as to grade is the nature of the material to be ground. In the case of hard castings, such as cast iron, a soft grade wheel is used. For softer castings, such as aluminum alloys, harder wheels are used. Dullness of the grains raises the pressure and tears dull grains out of the soft grade wheel, letting new sharp grains come into play. With softer castings, it is poor economy to use soft wheels because it will release the grains before the work has dulled them.

The relative spacing of the abrasive grains and bond in the wheel is known as the wheel's structure. The spaces or openings between the abrasive grains provide space for the chips of metal cut by the grain. It might seem better to have as many grains or cutting edges as one could cram into the wheel, but such a wheel would load up fast with metal chips and become smooth and noncutting. The more open the structure, the more space there is for chips. In addition, the cutting surface of the wheel will remain cooler.

Resinoid or synthetic resin (plastic) bond makes up the bulk of grinding and high-speed work done in the foundry. The vitrified bond (glassy) is favored where heavy stock removal and good finish are required. Resinoid bond can be run at considerably higher speeds and is essential on modern, high-speed snagging machines. Their resilience makes them excellent for use where wheels are subject to lateral strains; also where wheels are thin, as in cutting off ingates and risers.

Resinoid wheels are a logical choice where large amounts of metal must be removed, as in the cleaning room. Their resilience helps them to withstand a lot of abuse from well-meaning but heavy-handed operators of floorstand and swing frame grinders. Abrasive/grinding wheels come in a variety of shapes and sizes and are selected by the type of grinding operations to be accomplished.

Grinding Machine Types

Grinding machines used in the foundry can be divided into three main types: *floor stand, swing frame* and *portable.* Portable machines can be further divided into electrically powered or pneumatically powered, with the latter being the most commonly used.

Figure 18-33 shows a swing frame grinder being used to grind a riser pad from a large ferrous casting. The operation of snagging on a floorstand grinder is seen in **Fig. 18-34**.

Portable machine grinding can be done with very small, handheld tools, as is seen in **Fig. 18-35**. This operator is performing the grinding on a workbench, and the array of portable grinding tools is within view. Fine detail grinding is done on this particular casting. In **Fig. 18-36**, two operators are using portable grinding tools to finish-grind a large ferrous casting. The operator on the left is cutting off flash on this casting.

Figure 18-37 shows a cleaning-room grinding operation in which the operator is using a portable grinding tool in a booth. Note how the swarf from the cutting operation is contained within the booth.

Robotics—Robotics are often used in the cleaning room, especially for high volume casting operations. Hand tools can be attached to end-of-arm tooling to perform many of the functions that have been traditionally hand-tool functions **(Figs. 18-38 and 18-39).**

Abrasive Belts and Discs

Coated abrasives are used on belts and discs. They can be any form of abrasive in which the grain is coated onto or affixed to paper (sandpaper) or, usually, cloth-backed belt, with resin, hide glue or a combination of the two. Most coated abrasives are electrocoated because this method aligns the abrasive grains onto a belt in the best position for cutting. Electrocoating also provides control over the density of the coating.

Fig. 18-33. Swing frame grinder.

Traditionally, coated abrasive belts and discs were primarly used in the nonferrous foundries. Through technological improvements, they are now used for many alloy applications. Coated abrasive backing materials have become stronger, joint technology has been improved and new abrasive grains have become available. This has created stronger and more aggressive abrasives than are regularly used in ferrous alloys, even in the high-alloy steels and other applications where grinding wheels and milling machines have been used. Under the right circumstances, coated abrasives can offer higher metal removal rates with less heat generated, resulting in greater energy efficiency and less operator fatigue.

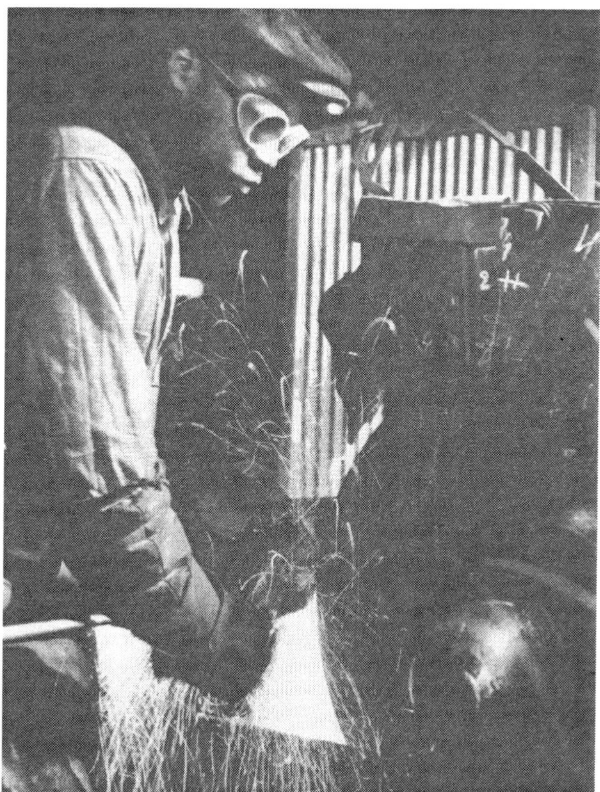

Fig. 18-34. Snagging on a floorstand grinder.

Fig. 18-35. Portable grinding tool in use.

Figure 18-40 shows a coated abrasive belt. These are most often driven on a drive pulley and backstand idler, though very short belts can be mounted on an expanding drum. The advantage of the backstand design is that it gives the grains a longer period to cool between cuts than if the drum were used. **Figure 18-41** shows an aluminum alloy casting being ground using a coated abrasive belt. When coated abrasive discs are used in the cleaning room, they are usually used with portable tools.

The abrasives most commonly used on coated abrasive belts are silicon carbide, aluminum oxide and blended zirconia aluminas. For rough cutting of aluminum alloys, a relatively coarse grit (60–36 grain size) is used.

Fig. 18-36. Portable grinding tools used on large ferrous casting.

Fig. 18-37. Use of portable grinding tool in a booth.

Pneumatic Tools

Probably the largest user of compressed air in the foundry is the cleaning room, especially if air-driven portable grinding tools are used. Having a properly sized and maintained compressed air system is very important for obtaining the maximum casting throughput in the cleaning room.

Compressed air should always be cooled immediately after leaving the air compressor, to remove water vapor. The degree of cooling and separation of condensed water (vapor) determines the dryness of the air. Dry air, in turn, prevents water vapor from condensing in the distribution lines. Water vapor in the air will destroy tool lubrication, resulting in unnecessary wear and repair costs of portable tools. In lieu of immediate cooling, an aftercooler and air dryer can be installed.

Operators and maintenance personnel should be trained to maintain the standards set by the manufacturers and the foundry management. These standards include: 1) never using fittings smaller than those supplied by the manufacturer; 2) keep all air hoses above the floor (provide brackets for tools when not in use); and 3) have a shutoff valve for each hose connection.

When the cleaning room is shut down with only the air compressor running and all other equipment shut off, listen for the sound emitted by the leaking compressed air. These compressed air leaks can be costly in terms of the electricity required to run the air compressor. When the cleaning room is in full production, the air leaks are wasting additional money by forcing the air compressor to work harder to keep up with the demand for compressed air. Therefore, all air leaks should be attended to immediately.

Post-Cleaning Options

Impregnation

Impregnation has been used for many years as a method to seal castings—ferrous and nonferrous—where microporosity is present. Three types of microporosity generally associated with impregnation (enclosed, blind and through) are illustrated in **Fig. 18-42.**

Impregnation is generally thought of as a repair procedure for castings; however, in processes such as diecasting and permanent molding, where porosity can be a problem, it is sometimes used as part of the process.

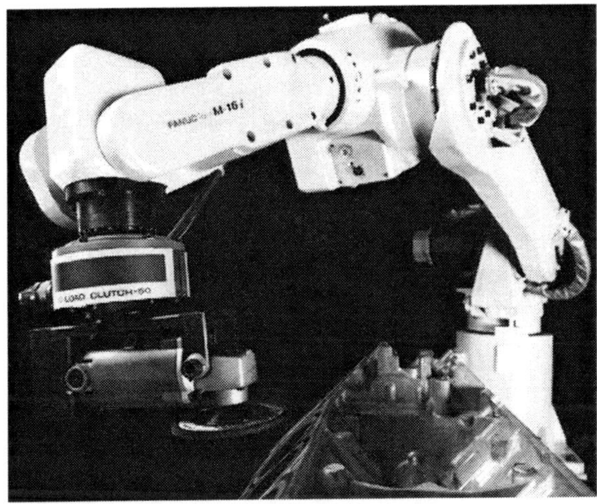

Fig. 18-38. Deflashing of an aluminum engine block performed by robotics.

Fig. 18-39. Deflashing of valve body performed by robotics.

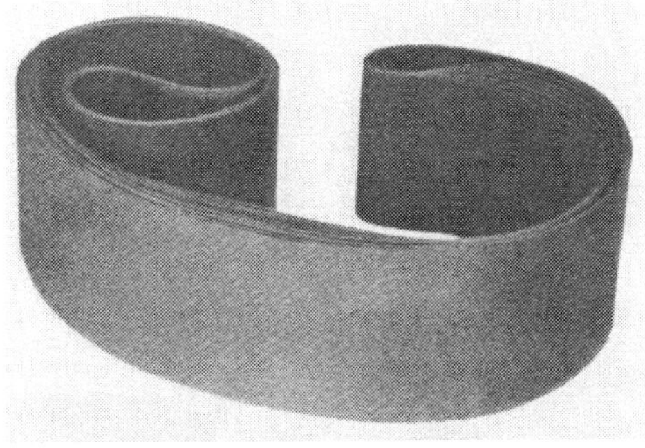

Fig. 18-40. Photo of coated abrasive belt.

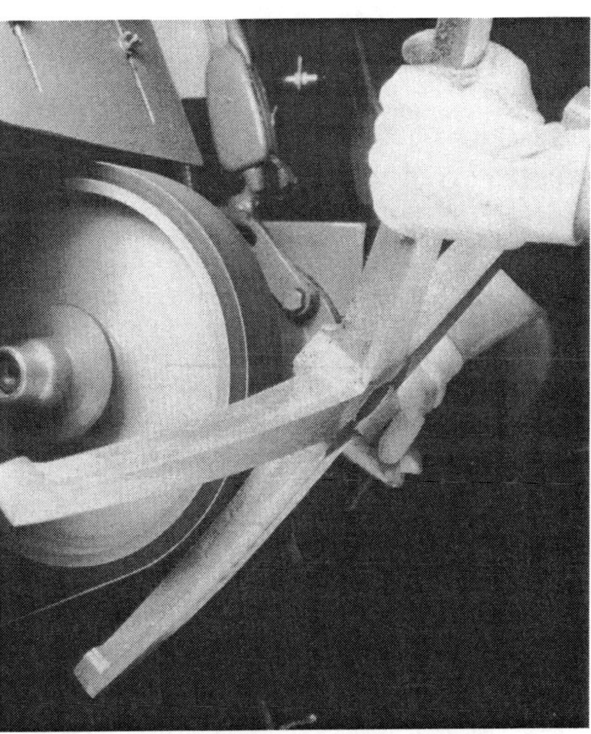

Fig. 18-41. An aluminum casting is ground using a coated abrasive belt. [Photo courtesy of Norton Co.]

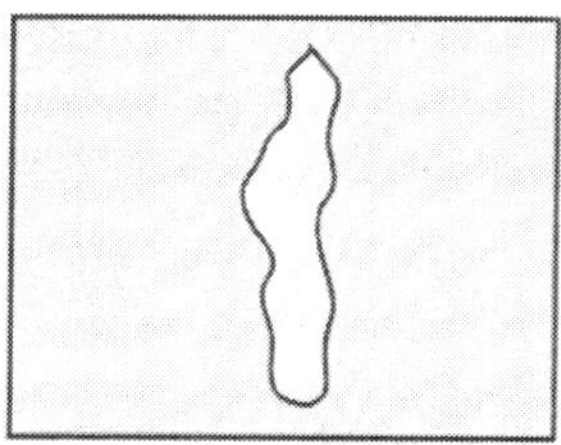

Enclosed porosity is not a problem unless it is uncovered by machining operations.

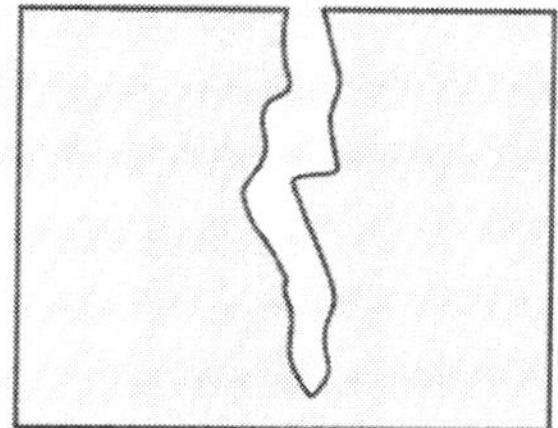

Blind porosity encourages internal corrosion, leading to "spotting out" of plating and "blow out" of paintwork.

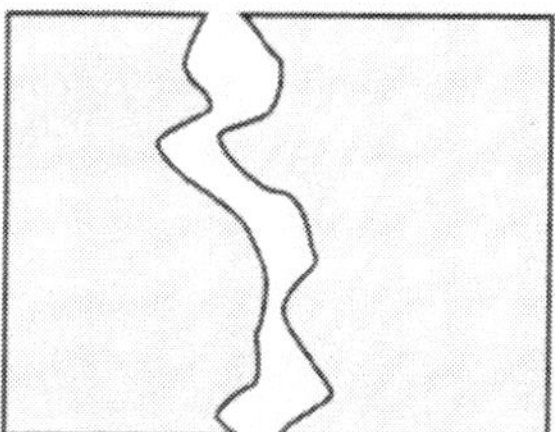

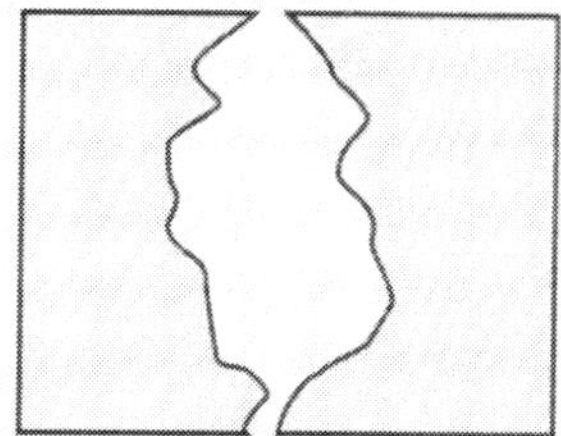

Through porosity allows gas or liquid to leak through the casting. The size of porosity need not be the same as what is seen on the surface. Porosity often is an irregular phenomenon within the casting

Fig. 18-42. Three types of microporosity.

The purpose of impregnation is to fill the voids caused by the microporosity with a material that is chemically resistant to and can withstand the casting application conditions (such as temperature, pressure and stress). Pressure impregnation techniques used in the past resulted in incomplete filling of the voids.

Vacuum impregnation techniques have overcome most of this problem by removing air from the voids before introducing the sealant into the voids. There are two vacuum impregnation processes in use: wet-vacuum and dry-vacuum. In the *wet-vacuum* process, the castings are immersed in the sealant in a pressure vessel and subjected to a high vacuum to remove the air. The vacuum is then released (the vessel is sometimes pressurized), forcing sealant into the porosity. In the *dry-vacuum* process, castings are subjected to the vacuum before they are immersed in the sealant. After immersion, the vacuum is released to the atmosphere, or the vessel may be pressurized to allow sealant to enter the porosity.

There are three types of sealants used for impregnation: 1) aqueous sodium silicate; 2) styrene polyesters; and 3) methacrylate sealants. Each sealant has its own distinct advantages and disadvantages.

Impregnation is a cost-effective way of filling microvoids in castings. It can be used to prevent leakage and improve pressure tightness in castings. It also eliminates bleedout and blemishes in plating and surface conversion coatings, white spotting of anodizing, and blistering of paint finishes due to air expelled on curing.

Fig. 18-43. Aluminum castings are subjected to HIP. [Photo courtesy of Howmet Turbine Components Corp.]

Hot Isostatic Pressing (HIP)

Hot Isostatic Pressing is a process that can be used to improve the internal soundness or integrity, increase the density, and improve mechanical and fatigue properties of castings in all alloy families (ferrous and nonferrous). This process is recognized as a viable tool and has become part of process specifications for critical applications such as structural aerospace and automotive castings.

The HIP process (**Fig. 18-43**) requires the castings to be placed in a pressure vessel where the temperature will be raised to near solidus. This is done in an inert atmosphere. The pressure in the vessel is increased to approximately 15,000 psi. The external pressure on the castings and their temperature cause plastic deformation to take place, which collapses the internal porosity. In the case of aluminum alloy castings, the amount of hydrogen removed by out-gassing during the HIP process is insignificant.

HIPing of castings has proved capable of substantially eliminating microporosity formed by gas precipitation and internal shrinkage formed during solidification. The HIP process makes possible the acceptance of castings that might otherwise be scrapped. This is especially true for castings that are subject to radiographic inspection. Some of the properties benefited by HIP are ductility, low- and high-cycle fatigue, tensile and yield strength, and high temperature creep strength.

Value-Added Operations

Many foundries, ferrous and nonferrous, are finding it profitable to perform value-added operations on the castings they produce. Value-added means that the foundry performs an additional operation(s) on the casting, which was previously performed by the customer or outsourced by the customer. This might include a first-stage or first-cut machining operation in areas where ingates and riser pads would have to be removed anyway. Some foundries will do this, even if ingates and riser pads are not involved.

Some foundries fully machine the casting before shipping it to the customer. In other words, the foundry takes the casting from the mold and prepares it so that it is delivered ready for customer use. Also, some foundries perform secondary operations on the casting as well (such as heat treatment, painting or final surface-finishing).

INSPECTION AND TESTING

Inspection, for all practical purposes, begins with the materials entering the foundry. This section will cover post-casting inspections to ensure that the castings meet the customer's requirements or specifications and that they are of high quality.

Post-casting inspection usually falls into three categories:

1. ***Preliminary (Hot) Inspection***—Foundries cannot afford to spend time and money cleaning defective castings; therefore, when castings are removed from the mold, a visual inspection station is justified. Castings with obvious defects can be sent for repair or be scrapped. Those castings that are visually acceptable are sent to the cleaning room. When the casting is clean enough to reveal surface detail, it should receive a closer visual inspection.
2. ***Intermediate Inspection***—Intermediate inspections may follow specific operations, i.e., hardness checks after heat treatment or visual checks after grinding operations.
3. ***Final Inspection***—Radiographic and other critical, nondestructive testing methods are applied, as required by customer specifications, especially on castings designed for severe service performance. Foundry customers may also require radiographic inspection as an assurance of the casting's quality. Foundries today often use radiographic examination and other critical tests (even when not required by customers) as a control indicator to maintain casting quality.

Many foundries depend upon statistical quality control (QC) methods to inspect castings. In a statistical sampling procedure, only a predetermined percentage of the castings are subjected to critical inspection. This method of inspection saves time and money and is often as adequate as checking every casting in the production run. There are many publications that provide detailed information on this subject.

Diecasting inspection usually consists of chemical analysis, dimensional checks, leak (pressure) tests and visual examination to check for surface defects.

Standards and Specifications

Before setting up a workable quality control program for given castings, suitable standards must be established. These standards are based on the type of casting, the engineering specifications and, often, the casting's end use. For many applications, the requirements are very simple: the casting needs to be reasonably free from defects to be acceptable. However, the word "reasonably" can be hotly debated. In other applications, the requirements may be more stringent.

Rigid quality standards are incorporated in various association and government specifications controlling the acceptance of castings for military, government and commercial use—particularly those used in aircraft and other highly stressed equipment. Commonly, castings are divided into three groups, with different inspection standards for each group:

1. ***Highly Stressed***—The first class of standards includes highly stressed castings. The failure of these castings would result in serious consequences, such as loss of life and property. These castings will usually require 100% x-ray and penetrant inspection.
2. ***Moderately Stressed***—The second class of standards includes castings that are moderately stressed. If these casting were to fail, it would not cause loss of life or serious interruption in the operation of equipment. These castings can be tested for quality using x-ray control, where a certain percentage of castings would be x-rayed after initial runs in the foundry.
3. ***Low Stress***—The third class includes normal commercial castings under relatively low stress in nonstructural applications. Their failure would not result in serious consequences.

The goal and purpose of any inspection program is to determine whether the final castings meet the established standards of quality. As has been indicated, the inspection method and the standards of quality depend on the requirements of the casting and its application.

Regardless of the casting process used or the care that is exercised, it is nearly impossible to produce castings without some minor defect. In fact, a foundry would probably go bankrupt trying to achieve 100% defect-free castings. The establishment of quality standards for any given casting and the decision of whether or not a given casting meets these standards require careful judgment and interpretation.

Inspection and Testing Methods

Inspection methods and techniques for testing castings can be classified into two broad groups: *nondestructive* and *destructive.*

In *destructive testing,* the casting is destroyed, permanently damaged or altered in some way. Included in destructive testing is the determination of tensile properties on specimens cut from the casting proper, determination of soundness by sectioning the casting, performance of breakdown tests and simulated service tests such as fatigue testing.

Nondestructive testing methods include visual and dimensional inspections, fluorescent powder and dye penetrant tests, radiographic (x-ray) inspections, leak (pressure) tests, ultrasonic testing, and mechanical and physical property tests from separately cast samples, such as separately cast test bars.

Visual Inspection

Visual inspection uses the human eye as the chief inspection tool. This type of inspection is used to spot surface defects, nonfilling and molding errors. Casting defects, such as sand holes, excessively rough surface, surface shrinkage, blowholes, misruns, cold shuts and surface dross or slag can usually be detected visually. To aid in visual inspection, magnifiers can be used to help detect surface defects. In most instances, a visual inspection can be performed after the castings have been abrasive blasted.

Dimensional Inspection

Dimensional accuracy testing is a good measure of process control. It is also a requirement, because customers will not tolerate the high cost of machining castings that don't meet their tolerance specifications, or accept inconsistent castings. Casting dimensions can be manually checked with the use of dial indicators, vernier calipers, micrometers and height gauges. These checks can also be aided with the use of "go/no-go" gauges, jigs and fixtures.

Another device for performing dimensional inspection is called the coordinate measuring machine (CMM), as seen in **Fig. 18-44.** The CMM has greatly improved the speed and accuracy of measuring the dimensions of castings. Computerization of the CMM has eased repetitive tasks and has allowed its use as a statistical tool. In addition, it significantly reduces the man-hours required for dimensional inspection and inspection of the casting as a whole.

Fig. 18-44. An aluminum cylinder head is held in a fixture for measurement by a CMM.

Dye Penetrant and Fluorescent Powder Testing

Dye penetrant testing methods are used to inspect surfaces for minute cracks, pores and other surface discontinuities that are difficult to detect visually. A colored dye, suspended in penetrating oil, is applied to the surface of the casting. This solution will then find its way into the surface defects. After excess dye is removed, a special developer is applied, and the defects are clearly indicated.

In the fluorescent powder penetrant testing method, the castings are immersed in a suspension of fluorescent powder and warm penetrating oil. The suspension penetrates the defect(s) and the excess suspension is washed away in warm water. The casting is dusted, sprayed or chalked with a drying powder that draws the suspension from the defect(s), where it fluoresces and is visible under ultraviolet light. This method detects only surface cracks and flaws, but does it more effectively and more economically than does radiographic examination.

Defects show up in fluorescent inspection as glowing yellow-green dots or lines against a dark background. In dye penetrants, defects are indicated as red dots or lines against a white background. Interpreting the characteristic patterns of different types of flaws is very important. A line of penetrant indicates a crack or cold shut. Dots of penetrant indicate pits or porosity. A series of dots indicates a tight crack, cold shut or partially welded lap. Accurate interpretation can be learned only by experience.

Broadly speaking, dye-penetrant techniques identify voids that are open to the surface. They do not identify internal porosity, shrinkage or other defects that are not open to the surface. The penetrant processes are used mainly to detect flaws in nonferrous or ferrous materials. The penetrant system is also used with nonmetallic material such as ceramics and plastic.

Magnetic Particle Inspection

Magnetic particle inspection detects small cracks on or near the surface of ferromagnetic objects. Thus, this inspection process is limited to only those alloys that can be magnetized. In a magnetized object, if a crack or void interrupts the magnetic field, the magnetic field is distorted.

In a material that has an induced magnetic field, defects, such as blowholes, cracks and inclusions, produce a distortion. The path of the magnetic flux is distorted because the defects have magnetic properties that are different from those of the surrounding material. The magnetic flux spreads out in order to detour around the interruption, and flux lines extend outside the metal. Small magnetic particles show the path of the flux line, and, hence, the shape of the crack or void. The magnetic field makes an angle of 90 degrees with the suspected crack.

This inspection method is quick and inexpensive to operate, and very sensitive. It can be applied to nearly all ferrous alloy castings, except austenitic material. The basic principle is that, although magnetic flux lines are caused to flow in a metal object, the local disturbance at a flaw is detected by using a magnetic powder (either dry or as a suspension in a liquid). Iron filings may be used, but ferric oxide is preferred.

Ultrasonic Testing

Internal defects that can be detected with radiography may also be detected with sound. Ultrasonic testing uses high-frequency acoustic energy from a sending unit that is introduced into the casting. The energy beam travels through the casting until it hits the opposite surface or an interface or defect. The interface or defect reflects portions of the energy, which are collected in a receiving unit and displayed on a screen. The pattern on the screen can be analyzed to give both the location and the size of the internal defect, and wall thickness. Ultrasonic testing can also be used to check ductile iron for nodularity and nodule count.

In the early days, ultrasonic testing could only be done on flat surfaces. Today, transducers can be used on curved surfaces as well. Fishermen commonly use a type of ultrasonic testing equipment when using a fish finder or depth detector. The U.S. Navy also uses this principle, but they call it sonar.

Radiographic Inspection

Internal shrinkage, gas porosity, internal dross or slag films and particles of foreign inclusions that normally are not detectable by visual inspection can be detected by radiographic inspection. Internal defects can be detected, within certain limitations, depending on the design of the casting, the size, shape and location of the defect, and the radiographic techniques used.

Film radiography involves exposing a casting to radiation from an x-ray tube. The casting absorbs part of this radiation, and the remainder exposes the radiographic film. Very dense material withstands penetration of the radiation and causes the emulsion on the radiographic film to be affected to a lesser degree. Such areas on the radiographic film will have a light appearance, whereas less-dense materials allow more penetration, and the film appears darker. Any hole, crack or inclusion that is less dense than the alloy in the casting is thus revealed as a dark area.

The amount of electrical energy required to operate this equipment is measured in kilovolts (kV). The section thickness to be radiographed is a deciding factor in the kV required. For instance a 100 x-ray tube is adequate for x-raying aluminum alloy castings up to 2 in. (5 cm) thick.

The source of short-wavelength rays, which can penetrate metal in addition to x-ray tubes, are a radium capsule, a colbalt-60 capsule, betatron or another radioactive source.

Correct radiographic techniques are the best nondestructive methods for determining the presence of internal defects, such as shrinkage and foreign inclusions. Experience gained through practice, under initial metallurgical or qualified metal radiological supervision, is necessary to interpret the results of radiography.

Filmless x-ray equipment is also available. Filmless or real-time radiographic inspection techniques allow a foundry to view an x-ray image on a video screen. **Figure 18-45** shows one such unit being used.

Advanced techniques, such as computerized axial tomography (CAT scanning), originally used in the medical field, are being used to develop three-dimensional computer imagery of castings and their soundness.

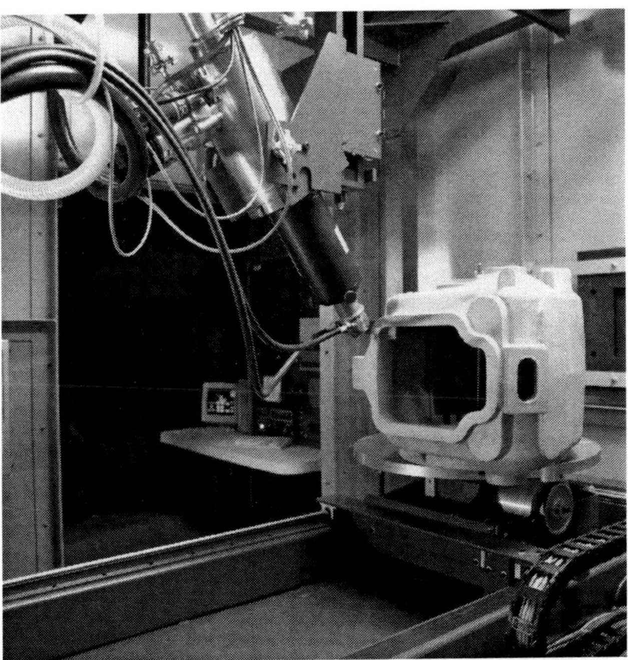

Fig. 18-45. A large aluminum casting is held by a manipulator inside a room-sized lead enclosure. The x-ray tube is shown poised above the casting. The programmable, four-post manipulator can handle castings as heavy as 150 pounds. [Photo courtesy of IRT Corp.]

Eddy Current Inspection

A coil that carries alternating current causes an eddy current to flow in any nearby metal. The eddy current may react on the coil to produce substantial changes in its reactance and resistance, depending on the proximity and conductivity of the metal. If the path of the eddy current is distorted and lengthened by a small crack or defect in the metal, the reactance and resistance of the coil are again modified.

The eddy current is confined to a surface layer whose depth depends inversely on the frequency of the current and on the conductivity and permeability of the metal. Thus, the eddy-current method can be applied only to the detection of cracks at or near the surface.

Pressure (Leak) Testing

A casting that must be pressure-tight or leakproof often is tested by sealing openings in the casting and pressurizing it with air, inert gas or water.

If hydrostatic (water) pressure is used, the seeping of water through the casting wall indicates leaks. If air or gas pressure is used, the pressurized casting is immersed in a tank of clear water, and if bubbles appear, air has penetrated through the casting wall and the casting will leak.

Mechanical and Physical Property Tests

The physical properties of a casting can be determined by metallographic examination: photographs, macrographs and micrographs. The microscope can resolve many casting problems. Grain size, nonmetallic inclusions, submicroscopic pinholes and the effects of thermal treatments can be observed by metallographic examination. Chemical analyses of metal alloys can be made by the wet chemistry method or with an optical emission spectrometer.

Mechanical tests include the following:

- Tensile tests to determine ultimate strength, yield strength and elongation. Both cast test bars and those removed from the casting are used.
- Bend, notch, impact and transverse tests to check ductility and shock resistance.
- Hardness tests to determine ductility, tensile strength and resistance to surface penetration of cast metal alloys.
- Fatigue and fracture toughness tests to determine the durability of the cast metal alloy.

FINAL COMMENTS

The cleaning and inspection of castings is a subject that needs more attention than could be provided in the space allotted. This is especially true of the cleaning of castings. Many new and innovative techniques of cleaning castings have been applied to "old" casting cleaning processes and equipment. The material discussed herein has been gleaned from many publications on this subject. For the latest information on these subjects, please use the AFS Technical Library and refer to manufacturer's technical literature on casting cleaning equipment and materials.

BIBLIOGRAPHY

Aluminum Casting Technology, 2nd. Ed., American Foundrymen's Society, Des Plaines, IL.

Bossi, R.H., Kline, J.L., Georgeson, G.E., "NDT: Computed Tomography of Castings," AFS, *Modern Casting,* pp 19-22 (Mar 1991).

Bralower, P.M., "Nondestructive Testing, Part 1: The New Generation of Radiography," AFS, *Modern Casting,* pp 21-23 (Jul 1986).

Bralower, P.M., "Nondestructive Testing, Part 2: New Applications for Traditional Methods," AFS, *Modern Casting* Magazine, pp 25-27 (Aug 1986).

Cleaning Castings, Reprinted 1992, American Foundrymen's Society, Des Plaines, IL.

Gittens, T.E., "Plasma-Arc Cutting Process Fundamentals," Linde Division, Union Carbide Corp., Danbury, CT.

Heine, R.W., Loper, Jr., C.R., Rosenthal, P.C., *Principles of Casting,* 2nd Ed., McGraw-Hill Book Company, New York (1967).

Introduction to Metalcasting, Handouts and Reference Material, AFS Cast Metals Institute, Des Plaines, IL.

Robison S, Cowden, D., Advances in Coated Abrasives Boost Cleaning Room Optios, AFS, *Modern Casting* (Jan 1997).

Sylvia, J.G., "Abrasive Waterjet Cutting," *ASTM International,* Vol. 14, Ninth Edition, pp 742-755 (1988).

Sylvia, J.G., Pallister, C., James, Jr., C.F., "Cleaning Gray Iron Castings with Robot and Plasma-Arc," Research Report 21, Robotics Research Canter, University Rhode Island (1983).

Environmental Protection and Safety Control 19

The need for environmental protection and safety control is a major concern in the metalcasting industry. Breathing air that contains hazardous contaminants can cause health problems. Water that contains toxic pollutants can contaminate surface and ground water. Coated sand or other solid materials can have adverse impacts on the environment when disposed of in landfills and may have to be cleaned or treated before disposal, or beneficially recycled or reused.

The use of special procedures or equipment to protect workers and the general public from harm is a necessary (and mandatory) part of foundry operations.

Governmental controls on environmental, health and safety issues affects all industries, and the foundry industry is no exception. Larger companies have an entire department devoted to overseeing environmental health and safety programs and government regulations.

There are four important topics covered in this chapter:

1. Air Quality Control, written by J. Radia.
2. Water Pollution Control, written by J. Shifro
3. Solid Waste Management, written by D.E. Oman
4. Safety Control, written by F.H. Kohloff

AIR QUALITY CONTROL

EPA Requirements

What is known today as the Clean Air Act (CAA) was originally passed by Congress in 1955 as Public Law 159. The basic purpose of the legislation was to regulate air emissions and to protect human health and the environment. Several amendments have been made to the Clean Air Act since 1955, including the historic CAA Amendments (CAAA) of 1970, which allowed the U.S. Environmental Protection Agency (EPA) to delegate responsibility to state and local agencies to control ambient air quality.

State Implementation Plans (SIPs) are a product of the 1970 CAA legislation. The SIPs are a vehicle for the state to develop plans to achieve and maintain compliance with the National Ambient Air Quality Standards (NAAQS) set by the CAA. The latest Amendments to the CAA in 1990 broadened coverage of the previous programs to include more commercial and industrial facilities as well as smaller facilities.

The 1990 Amendments also established a new operating permit program, enacted more stringent requirements for air pollution controls and enhanced federal enforcement authority. The potential civil and criminal liabilities for noncompliance were also significantly increased.

Two provisions of the 1990 CAA Amendments that significantly affect foundries are the Title III provisions on hazardous air pollutants (HAPs) and the Title V provisions on operating permits. A major pollutant emitted from most foundry processes is particulate matter (PM). The main sources of the particulate emissions include melting, sand handling, metal pouring, mold cooling, mold/casting shakeout and casting finishing. In addition, volatile organic compounds (VOCs) and carbon monoxide (CO) are emitted from some of these processes.

Emission Control

Equipment needed to clean the air in the foundry is generally located where molds and cores are made, where metal is melted, poured, cooled and shaken out, and where castings are cleaned by blasting, grinding and chipping.

Particulate matter is generally controlled by fabric filter baghouses or wet scrubbers. Gaseous pollutants (VOCs), including carbon monoxide and organic compounds are controlled by thermal oxidation (e.g., afterburners in a cupola) or chemical scrubbing (e.g., TEA or triethylamine removal in coldbox coremaking.)

Particulate Matter Control

Particulate control devices are generally referred to as dust collectors. Wet dust collectors (also known as scrubbers) remove particulates from an exhaust-air stream by wetting the particles with droplets of water and separating the droplets from the exhaust air. Wet scrubbers are basically classified as either low-, medium- or high-energy types, based on the pressure drop across the scrubber. Wet scrubbers operating at low-pressure drops will have a relatively low removal efficiency for fine particles. High-efficiency removal of fine particulates by wet scrubbers requires a high-energy scrubber with a high-pressure drop. **Figure 19-1** shows a typical wet scrubber (venturi). Many foundries use wet scrubbers on sand system emissions with a pressure drop in the range of 5 to 10 inches w.c. (water column). However, for cupolas, a pressure drop of 30 to 80 inches w.c. is typical.

The most common type of dry dust collector is a fabric filter. It removes particulate matter from an exhaust-air stream by passing it through the fabric. The collected particles form a dust cake on the fabric surface. As the dust cake forms, the particulate removal efficiency increases. However, increased cake formation also produces higher pressure drops, which results in decreased airflow. To prevent significant reduction in airflow, the fabric must be cleaned regularly, according to the manufacturer's instructions.

Fabric filters are often referred to as baghouses. The three most common baghouse designs are the pulse-jet type, the reverse-air type and shaker type.

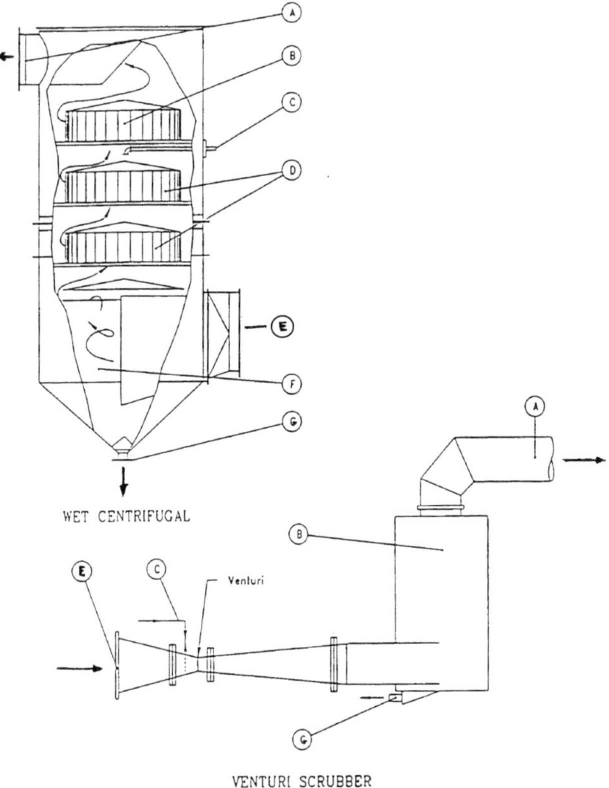

Fig. 19-1. Typical wet type dust collector:
A = Clean air outlet;
B = Entrainment separator;
C = Water inlet;
D = Impingement plates;
E = Dirty air inlet;
F = Wet cyclone for collecting heavy material;
G = Water and sludge drain.

Typical *pulse-jet baghouses* **(Fig. 19-2)** allow collection of dust on the outside surface of the bags. A pulsed jet of air creates an intense pressure wave in the bag, thus knocking off the collected dust. The vigorous nature of this technique can result in reduced bag life if not properly applied.

The *reverse-air baghouse* **(Fig. 19-3)** uses a less vigorous bag cleaning technique. This type of baghouse typically works by collecting the dust inside the bags. The bags are cleaned by periodically reversing the airflow, at a low velocity, causing the dust cake to fall into a bin below.

In the *shaker type baghouse,* dust collection is similar to the pulse jet type, except the bags are shaken using a mechanical device, causing the collected dust to be removed from the fabric.

Typical areas in foundries where baghouses are commonly used are melting, sand system, and casting finishing/cleaning processes.

Cartridge collectors **(Fig. 19-4)** are similar to fabric filters but use cartridges made of pleated filter media. Their design makes it possible to fit a relatively large filter surface area in a small space. This type of collector is commonly used when space is limited or when exhaust air recirculation is needed.

Organic Compound Control

As shown in **Fig. 19-5,** countercurrent packed bed scrubbers are usually vertical columns filled with a plastic packing material to create a large surface area. The exhaust-air stream containing the pollutant moves upward through the packed bed and an absorbing or reacting liquid trickles down through the packed bed. The packing facilitates contact between the scrubbing liquid and gas, and improves the efficiency of the removal of organic compounds from the exhaust-air stream.

This countercurrent method of scrubbing is very efficient and is typically used in two coremaking processes. The first is the phenolic urethane coldbox system, which uses triethylamine (TEA) as the catalyst to cure the cores. The TEA is removed from the exhaust-air stream by a reaction with an acidic solution (generally, dilute sulfuric acid). The second coremaking method uses sulfur dioxide (SO_2) to cure the cores. The SO_2 is removed from the exhaust-air stream by a reaction with a basic solution (generally, dilute sodium hydroxide).

Disposal of collected materials, both solid and liquid, are discussed in the following sections.

WATER POLLUTION CONTROL

Discharged Foundry Waters

The sources of discharged water in foundry operations vary significantly. Major sources of polluted and nonpolluted water are wet collectors (discussed in the previous section on Air Quality Control), and cooling water used in the melting and slag quenching areas. Noncontact cooling water and other nonregulated sources are often mixed with process water. The treatment and discharge of these waters is regulated by various government agencies.

Foundry-specific water pollutants of concern include copper, lead, zinc, total phenols, total toxic organics, and oil and grease.

Water Treatment Options

Foundry water treatment systems can consist of a number of different treatment options, depending on the discharge water quality from in-plant processes. Water may contain large amounts of suspended solids, heavy metals, or oil and grease. Treatment will depend on the nature of the pollutants and may include the use of settling ponds or clarifiers, chemical treatment for heavy metals, pH adjustments, activated carbon beds, and other options to meet specific permit limit conditions.

The method of following guidelines for limiting effluents typically requires a high level of recycle to meet mass loading limits. This may require facilities to recycle process water at rates of 80 to 100%, which in itself may require additional treatment of the recycled water to prevent process problems.

Government Control

Obtaining permits to dispose of waste water is of major importance. Storm water permits will be required where storm water contacts some process-related activities. These permit requirements vary

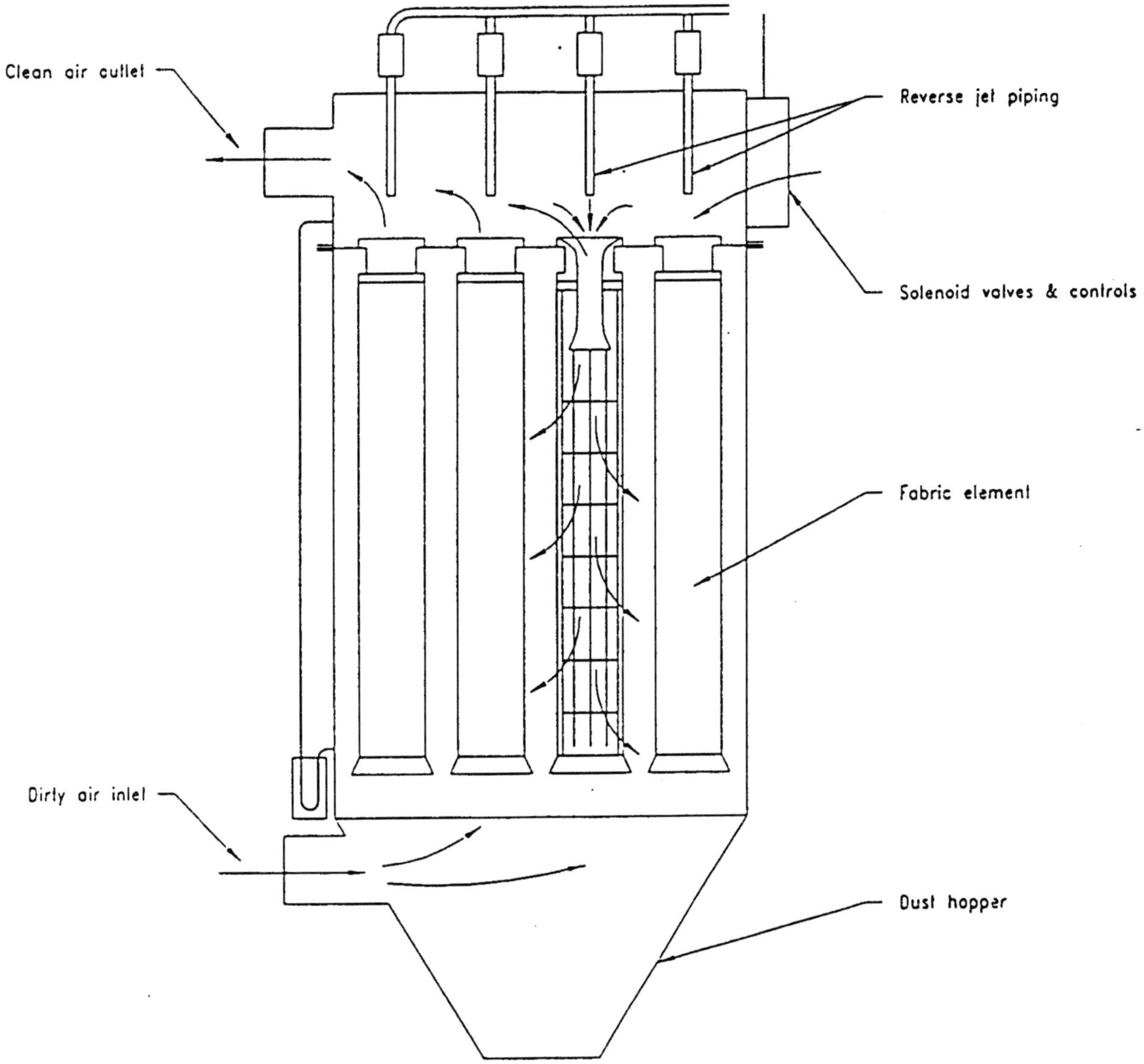

Clean air outlet
Reverse jet piping
Solenoid valves & controls
Fabric element
Dirty air inlet
Dust hopper

Fig. 19-2. Puls-jet type fabric filter collector.

EXHAUST
REPRESSURING VALVE
SIDE VIEW
FILTERING
SIDE VIEW
COLLAPSING
SIDE VIEW
CLEANING

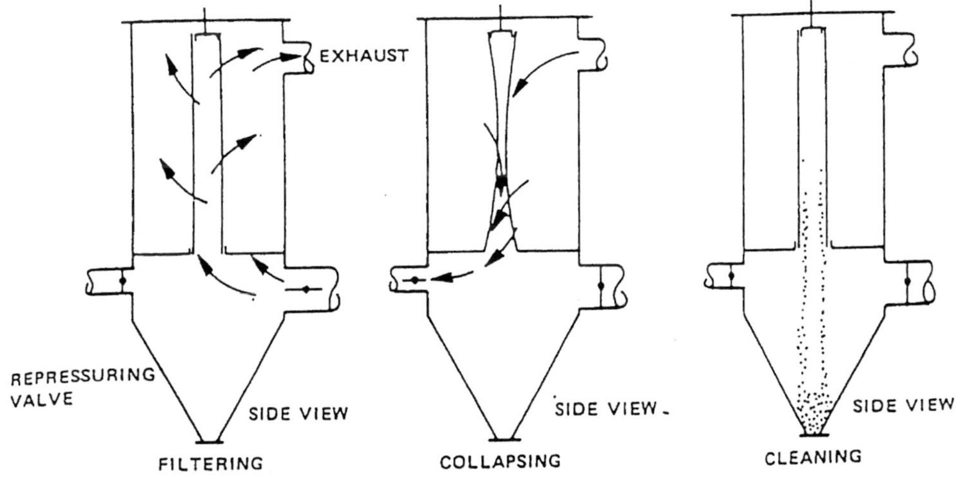

Fig. 19-3. Reverse-air cleaning system.

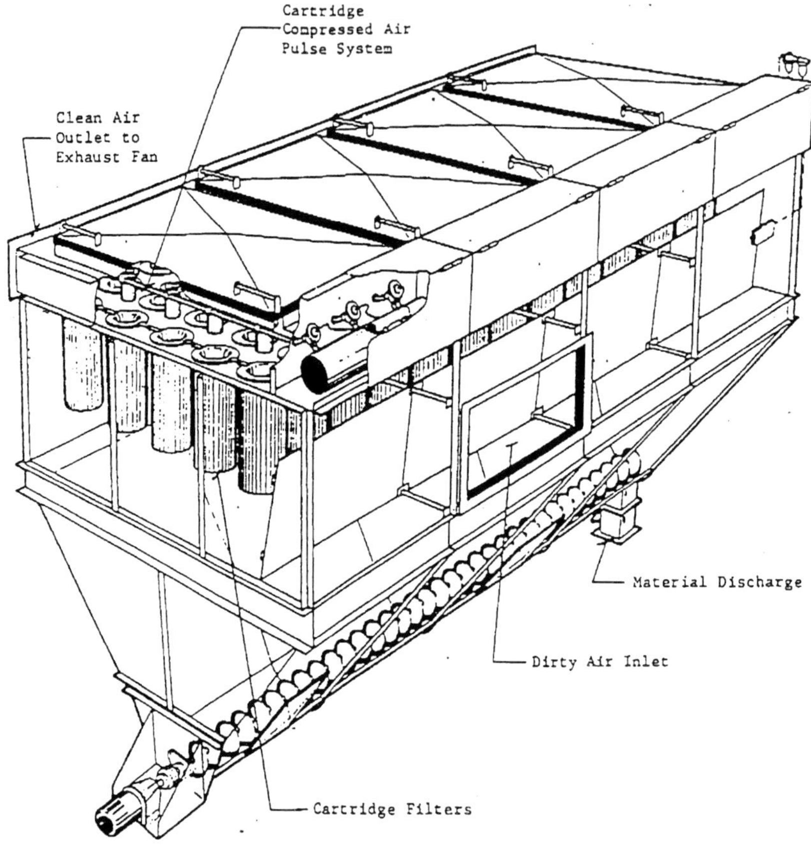

Fig. 19-4. Cartridge collector.

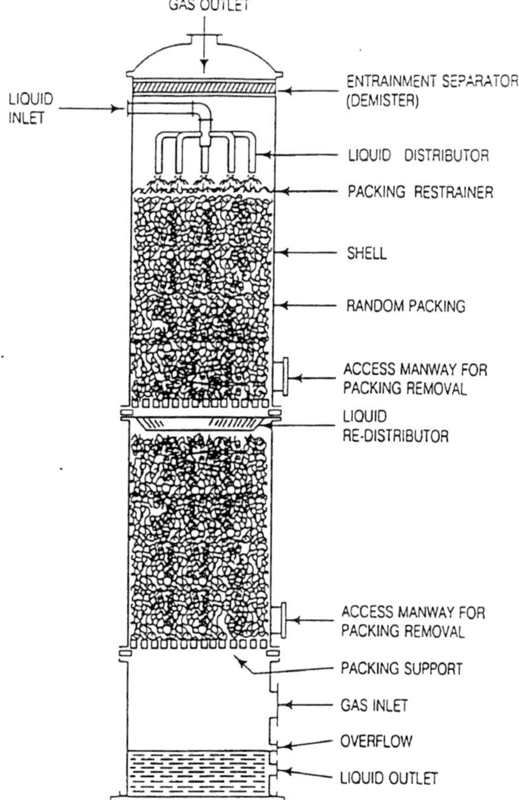

Fig. 19-5. Countercurrent packed bed scrubber column.

significantly, depending on the sources of storm water runoff and the particular requirements of the receiving stream. Failure to adhere to the rules and regulations can result in imprisonment and stiff fines.

Permit requirements vary significantly from foundry to foundry and state to state. Local conditions will control what Water Quality Guidelines apply to foundry National Pollution Discharge Elimination System (NPDES) permitted discharges while local Publicly Owned Treatment Works (POTW) rules or city ordinances govern local pretreatment requirements. In all cases, Categorical Standards or Effluent Limitation Guidelines apply when processes listed in the rules result in processed water discharges. The most stringent of all the rules applies to the pollutant in question.

Water Quality Guidelines will usually result in limits on the concentration of pollutants in the water discharged to waterways, while Categorical Standards or Effluent Limitation Guidelines will result in concentration limits derived from mass loading allowances referenced to average discharge rates. Other types of discharge limits will include pH, biological oxygen demand (BOD), temperature and other requirements to protect the receiving stream or the operation of a POTW for an indirect discharger.

NPDES permits are required for facilities that discharge directly into navigable waters, and/or those facilities that have storm water runoff from industrial activities. Those foundries discharging to a POTW must comply with local discharge requirements in addition to the Categorical Standards. These indirect dischargers may or may not be required to obtain a discharge permit, depending on local POTW requirements. In either case, the POTW is required to have an NPDES permit. Local POTW limits applied to indirect dischargers are, therefore, dependent upon the level of treatment within the POTW as well as its NPDES permit limits.

NPDES permits typically have stringent testing and record keeping requirements. Testing discharge water on a weekly basis, with monthly reporting, is typical, with lesser requirements for water discharged to POTWs.

NPDES permit renewals will need to be submitted a minimum of 180 days prior to the permit expiration date. This time frame will allow the existing permit to remain in affect during permit negotiations or appeals that extend past the current permit's expiration date.

The primary enforcement authority under the Clean Water Act (CWA) is the USEPA, unless the state has received approval to implement the federal program. Enforcement penalties for violations can reach a maximum of 30 years imprisonment and $50,000 per violation up to a $2,000,000 (40 CFR 122 Subpart C).

Foundry water permitting can be a very complicated process. Federal, state and local water quality regulations must be investigated and adequately understood in order to submit required permit applications and insure compliance with applicable rules.

Water permits need to be periodically reviewed and changes need to be submitted when processes or production levels change significantly from current permit submissions. Appropriate training, record keeping and reporting procedures will insure compliance with permit conditions and minimize the risk of enforcement concerns.

Water issues not covered in this section are groundwater and wetland protection. These issues are quite complicated and vary significantly due to differences in local regulations. Regulating agencies might include state and federal Environmental Protection Agencies, cities, counties, the Army Corps of Engineers, and local Boards of Health. Some states may also have requirements such as "water withdrawal" regulations requiring registration and reporting of water withdrawn from surface water or groundwater sources.

Wetland protection is covered under Section 404 of the Clean Water Act. The program is administered by the Army Corps of Engineers and the Environmental Protection Agency.

Legislation

The following text is just a small sample of the large scope of agencies and regulations involved in controlling water pollution.

Groundwater regulations and well registration or monitoring may be covered under different agencies such as state or federal Environmental Protection Agencies (Land or Water Divisions), Boards of Health, cities and counties.

The first major federal water pollution control regulation was the Federal Water Pollution Control Act (FWPCA) of 1972. One of the original goals of the FWPCA was to completely eliminate the discharge of pollutants into surface waters of the United States by 1985. The act made any discharge of pollutants into surface water a violation. The FWPCA set the basic structure for regulating discharges of pollutants into surface waters. The FWPCA has been amended many times and, during this process, became known as the Clean Water Act (CWA). The CWA made it unlawful for any person to discharge any pollutant from a point source into navigable waters of the United States without a National Pollutant Discharge Elimination System (NPDES) permit.

Major revisions to the CWA in 1977 and 1987 focused on toxic pollutants. The 1987 revisions included Storm Water Permitting requirements as well as more stringent discharge limits for NPDES direct dischargers. In addition to these requirements, the Oil Pollution Act of 1990 added additional requirements for oil pollution prevention programs. These requirements included the development and maintenance of a Spill Prevention Control and Countermeasures (SPCC) plan. The major programs under the CWA are covered under 40 CFR (Code of Federal Regulations) Chapter 1, Subchapter D—Water Programs, Part 104–149.

The CWA in 40 CFR, Chapter 1—Part 130–131 also defined water quality goals for specific bodies of water. These Water Quality Standards (WQS) define water quality goals by designating the use or uses to be made of the water and by setting criteria necessary to protect those uses. These uses include protecting public health or welfare, protection and propagation of fish and wildlife, agriculture, and for recreation in and on the water as well as protection of public water supplies. Such standards serve the dual purposes of establishing water-quality-based treatment controls and strategies beyond the technology-based levels of treatment required in other sections of the CWA. The WQS applicable to a facility's discharge will, therefore, be dependent on the uses of receiving streams and the specific requirements of individual state regulations.

Extensions of the Water Quality Standard rules that will not be discussed, but that may affect NPDES permit limits, include the Total Maximum Daily Load (TMDL) program and antidegradation programs. These state programs, responding to the CWA, are being implemented to achieve improvements in the water quality of receiving streams by preventing degradation of the water quality in the stream or limiting the pollutant loading allowed in an impaired watershed.

In addition to general water program requirements, specific standards covering foundry discharges are also covered under the Metal Molding and Casting Point Source category in 40 CFR Subpart N, Part 464 (Categorical Standards/Effluent Limitation Guidelines). These regulations cover all plants that cast metal. The Metal Molding and Casting Point Source category is divided into four subcategories (aluminum, copper, ferrous and zinc casting) and 28 process segments. Some of the segments for these categories consist of Casting Cleaning, Casting Quench, Direct Chill Casting, Die Casting, Dust Collection Scrubbing, Grinding Scrubber, Investment Casting, Melting Furnace Scrubber, Mold Cooling, Slag Quench and Wet Sand Reclamation. These segments are not applicable to every subcategory and can vary depending on individual foundry operations.

The Categorical Standards/Effluent Limitation Guidelines are technology-based discharge guidelines with pollutant allowances dependent upon production levels and specific types of processes or pollution control equipment. These guidelines cover standards for both existing and new sources as well as standards for NPDES permits and indirect dischargers.

SOLID WASTE MANAGEMENT

During the metalcasting process, many of the materials used for moldmaking, coremaking and the final casting are no longer usable in their present state. These solid materials have to be dealt with by disposal, treatment and disposal, recycling or treatment and recycling. Most metallic waste can be remelted and reused. Some of the sand from molds and cores can be treated and added to other sand. Everything else has to be disposed of in the proper manner and according to government regulations.

The foundry industry has always been concerned with the proper management of solid wastes that are generated by the metalcasting process. The industry typically generates fairly large quantities of solid waste. Although the quantities of solid waste generated may be significant, the nature of the vast majority of these materials is such that they pose limited risk to human health and the environment. Prior to the advent of environmental regulations, many foundries found that the solid wastes generated by the industry were ideally suited for reuse as fill material either at or in the general vicinity of the foundry.

Once the Solid Waste Disposal Act was passed in 1965, many states developed their own solid waste regulations. As a result, foundries had to find alternative locations for management of their solid wastes. From the early 1970s through the early 1990s, most foundries either sited their own land disposal facilities, or hauled their solid wastes to permitted landfills in close proximity to the foundry.

Today, many foundries have again realized the potential of beneficial reuse of their solid wastes in many ways. Beneficial reuse or recycling of the solid wastes generated by the industry has become a viable option with the promulgation of new state solid waste regulations, which encourage the reuse of select materials. By employing sound beneficial reuse options, the industry has significantly reduced the quantity of solid wastes that must be managed in disposal facilities. As more options for beneficial reuse are found, the trend toward beneficial reuse will continue in this industry.

Solid Waste Generation

The following is a summary of the major solid wastes generated by foundry operations:

- *Molding Sand or System Sand*—Waste materials produced from molding sand are often in the form of dusts or sludges that are removed by air pollution control devices located over the sand system and shakeout operations. Other wastes generated from molding sands include large clumps of sand that are removed during the screening operations associated with the sand system.
- *Core Sand*—Partially decomposed core sands, which are removed during the shakeout process may be managed as a solid waste. Core sand, which decomposes during the casting process, typically gets recycled back into the molding or system sand process.
- *Melt Emission Dust*—Air pollution control devices such as baghouses or wet scrubbers (discussed in the Air Quality Control section) capture particulates generated over the melting operations at foundries. The resulting dusts or sludges are typically managed as solid wastes.
- *Refractories*—Refractory materials, which provide heat resistant properties, gradually break down as part of normal foundry operations. Furnace and ladle refractory materials are typically managed as solid wastes.
- *Slag*—Metal oxides produced as a result of oxidation of metal during the melting process are typically managed as solid wastes.

Hazardous Wastes

The foundry industry does not generate significant quantities of hazardous wastes. However there are certain waste streams that have the potential to exhibit the characteristic of toxicity, depending on the nature of raw materials used by a given foundry. Foundry wastes that have the potential to be considered as hazardous wastes include:

- *Furnace Emission Control Dusts or Sludges*—Dusts or sludges resulting from the collection of particulates over the melting operations at brass and bronze foundries, especially those that pour leaded-brass alloys, may exhibit the characteristic of toxicity for lead and/or cadmium and thus be considered as a hazardous waste. Certain gray iron foundries have found the emission control dust collected over their melting operations may also exhibit characteristics of toxicity for lead and/or cadmium, although this occurs less frequently than in brass and bronze foundries. Whether or not a gray iron foundry generates hazardous waste from above the melting operation depends on the quality of scrap melted.
- *Molding Sand From Brass or Bronze Foundries*—Foundries that use leaded brass or bronze alloys, have sometimes found that their molding sands or system sands can exhibit the characteristic of toxicity and be classified as hazardous wastes.

Beneficial Reuse

Foundries have historically been major recyclers of discarded material. For centuries, the industry has made metal products by remelting metal scrap that no longer served a useful purpose. Today, the industry has found increasing opportunities to recycle by-products of the manufacturing process such as molding/system sand. This sand is the largest single waste stream generated by foundries. Individual foundries have found the following beneficial reuse options for their molding/system sands:

- Construction Fill/Road Sub-base
- Flowable Fill
- Grouts and Mortars
- Potting and Specialty Soils
- Cement Manufacturing
- Precast Concrete Products
- Highway Barriers
- Pipe Bedding
- Asphalt
- Cemetery Vaults
- Brick Pavers
- Landfill Daily Cover

Sound reuse of foundry by-products, such as molding/system sand, can protect the environment and preserve precious natural resources. As more and more opportunities for reuse of foundry materials are identified, and as more experience with sound recycling options is gained, the industry will significantly reduce its dependence on disposal options.

When foundry waste cannot be beneficially reused or recycled, it must be disposed of properly in accordance with federal, state and/or local requirements.

Legislation

In 1965, Congress passed the Solid Waste Disposal Act (SWDA). The purpose of this act was to address the growing quantity of waste generated in the United States, and to ensure its proper management. Subsequent to passage of the Solid Waste Disposal Act, Congress amended the law when the public became concerned about the safe disposal of huge volumes of municipal and industrial waste. The amendments to the SWDA of 1965 were known as the Resource Conservation and Recovery Act (RCRA) of 1976.

RCRA was established to achieve four specific goals:

1. To protect human health and the environment.
2. To reduce waste and to conserve energy and natural resources.
3. To reduce or eliminate the generation of hazardous waste.
4. To ensure that wastes are managed in an environmentally sound manner.

RCRA is composed of three separate but interrelated programs. Those programs are:

- Subtitle C The Hazardous Waste program
- Subtitle D The Solid Waste program
- Subtitle I The Underground Storage Tank program

Of the three RCRA programs, the Subtitle C and Subtitle D programs most directly impact foundries. The Subtitle C program is probably the most recognized facet of RCRA because it provides for the cradle-to-grave management of hazardous wastes. As a result of the Subtitle C program, generators are responsible for the proper management of hazardous waste from the point at which it is generated until it reaches its ultimate end at some form of treatment or disposal facility.

The Subtitle D program encourages environmentally sound, nonhazardous waste management practices. The program also encourages reuse of recoverable material and fosters resource recovery. Although nonhazardous solid waste management is predominately regulated by the states, the RCRA Subtitle D program sets national goals, provides leadership and technical assistance, and develops guidance and educational materials for use by the states.

SAFETY AND HEALTH PROGRAM

Safety in the foundry should be paramount to all activities within the facility. Methods to minimize injuries and illnesses require support from top management and all levels of supervision.

When safety is given major priority, production and quality will be achieved to expected requirements. No job is so urgent that it cannot be done in a safe and timely manner. At a minimum, the safety of employees should be equal to production and quality within the organization.

To ensure the occupational safety of workers, OSHA (Occupational Safety & Health Administration), under the auspices of the U.S. Government's Department of Labor, enacted the following:

Occupational Safety and Health Act

The Occupational Safety and Health Act of 1970
(Public Law 91-596) effective April 29, 1971
states that the purpose Congress expressed in the Act is
"To assure safe and healthful working conditions for working men and women; by authorizing enforcement of the standards developed under the Act;...."

The Safety and Health Program that follows is suggested, to keep within the OSHA guidelines.

The metalcasting industry presents many possible unsafe situations to employees: molten metal, moving conveyor chains and pulley drives, confined spaces, hazardous atmospheres, harmful noise, heat stress, electrical exposure; and ergonomic hazards associated with melting and pouring of metals, shakeout, degating, grinding, machining and material handling of castings. The list of operations within the facility is numerous. To protect employees and to eliminate identified hazards, a firm commitment by management to a comprehensive safety and health program is needed.

An effective occupational safety and health program will include the following four main elements:

1. Management Commitment and Employee Involvement
2. Worksite Analysis
3. Hazard Prevention and Control
4. Safety and Health Training

Management Commitment and Employee Involvement

The elements of management commitment and employee involvement are complementary and form the core of any occupational safety and health program. Management's commitment provides the motivational force and the resources for organizing and controlling activities within an organization. In an effective program, management regards worker safety and health as a fundamental value of the organization and applies its commitment to safety and health protection with as much vigor as to other organizational goals or operations. Employee involvement provides the means by which workers develop and/or express their own commitment to safety and health protection for themselves and for their fellow workers.

In implementing a safety and health program, there are various ways to provide commitment and support by management and employees. Some recommended actions are described briefly as follows:

- State clearly a worksite policy on safety and healthful work and working conditions so that all personnel with responsibility at the facility fully understand the priority and importance of safety and health protection in the organization. This should be signed by a top ranking official and posted in prominent locations within the facility.
- Establish and communicate a clear goal for the safety and health program and define objectives for meeting that goal so that all members of the organization understand the results desired and measures planned for achieving them.

- Provide visible top management involvement in implementing the program so that all employees understand that management's commitment is serious.
- Arrange for and encourage involvement in the structure and operation of the program and in decisions that affect their safety and health so they will commit their insight and energy to achieving the safety and health program's goals and objectives.
- Assign and communicate responsibility for all aspects of the program so that managers, supervisors and employees in all parts of the organization know what performance is expected of them.
- Provide adequate authority and resources to responsible parties so that assigned responsibilities can be met. This could include the staffing a full time "safety director/engineer/specialist" to manage the program. It is recommended this person report directly to a ranking operations manager.
- Hold managers, supervisors and employees accountable for meeting their responsibilities so those essential tasks will be performed. It is highly recommended that safety be part of the formal job performance appraisal.
- Review program operations, at least annually, to evaluate their success in meeting the goal and objectives, so that deficiencies can be identified and the program and/or the objectives can be revised when they do not meet the goal of effective safety and health promotion.

Worksite Analysis

A practical analysis of the work environment involves a variety of worksite examinations to identify existing hazards, conditions and operations in which changes might occur to create hazards. Unawareness of a hazard, stemming from failure to examine the worksite, is a sign that safety and health policies and/or practices are ineffective. Effective management actively analyzes the work and worksite to anticipate and prevent harmful occurrences. So that all hazards and potential hazards are identified, the following measures are recommended:

- Conduct comprehensive baseline worksite surveys to identify safety and health hazards. Conduct periodic comprehensive surveys to update survey findings. Involve employees in the effort.
- Analyze planned and new facilities, processes, materials and equipment. Include the safety manager in this review.
- Perform routine job hazard analyses for all plant job categories.
- Conduct regular site safety and health inspections so that new or previously missed hazards are identified. Inspections should be conducted monthly to quarterly depending on facility size. Inspection records need to be communicated and maintained. Assign responsibility for abating found hazards.
- Assess risk factors for ergonomic applications to workers tasks.
- Provide a reliable system for employees to notify management about conditions that appear hazardous. Reported hazards need to be addressed and corrected in a timely manner. Employees need not fear reprisal for reporting hazards.
- Investigate accidents and "near miss" incidents so that their causes and means of prevention can be identified.
- Analyze injury and illness trends over time so those patterns with common causes can be identified and prevented.

Hazard Prevention and Control

Where feasible, workplace hazards are prevented by effective design of the job site. Where it is not feasible to eliminate such hazards, they must be controlled to prevent unsafe and unhealthful exposure. Elimination or control must be accomplished in a timely manner once the hazard is recognized. As part of the program, employers should establish procedures to correct or control present or potential hazards in a timely manner. These procedures should include measures such as the following:

- Use engineering techniques to design out the hazard. This approach is the most effective procedure and must be attempted before using administrative or Personal Protective Equipment (PPE).
- Establish, at the earliest time, safe work practices and procedures that are understood and followed by all affected parties. Understanding and compliance are a result of training, positive reinforcement, correction of unsafe performance and, if necessary, enforcement through a clearly communicated disciplinary system.
- Provide PPE only as an interim measure while engineering controls are being developed and installed or are determined not to be feasible.
- Use administrative controls such as reducing the duration of employee exposure. This may be used in combination with PPE.
- Maintain the facility with a preventative maintenance program to avoid machine and equipment breakdowns.
- Plan and prepare for emergencies. Conduct training and emergency drills, as needed, to ensure proper responses to emergencies.
- Establish a medical program that includes first aid, on-site, as well as knowledge of nearby physician and emergency medical care.

Safety and Health Training

Training is an essential component of an effective safety and health program. Training helps identify the safety and health responsibilities of both management and employees at the site. Training is often most effective when incorporated into other education or performance requirements and job practices. The complexity of training depends on the size and organization of the worksite as well as the characteristics of hazards present.

- *Employee Training*—Employee training programs should be designed to ensure that all employees understand and are aware of the hazards to which they may be exposed and the proper methods for avoiding such hazards.
- *Supervisory Training*—Supervisors should be trained to understand the essential roll they play in job-site safety and to carry out their safety and health responsibilities effectively. Training programs for supervisors should include the following objectives:

  1. Analyze the work under the supervisor to anticipate and identify potential hazards.
  2. Maintain physical protection in the work areas.
  3. Reinforce employee training on the nature of potential hazards in their work and on needed protective measures through continual performance feedback, if necessary, through enforcement of safe work practices.
  4. Ensure that all managers understand their safety and health responsibilities.

OSHA Compliance Checklist

OSHA Safety and Health Standards for General Industry can be found in the Code of Federal Regulations 29 CFR Part 1910. These General Industry Standards apply to metalcasting operations. Adhering to these standards is required for OSHA compliance but does not guarantee a safe and healthful work place.

Following are a number of checklists that can be used when setting up programs for following the OSHA guidelines.

Log of Injuries/Illness (OSHA Log 200) 29 CFR 1904

		(Column A)	Incidence Rate		Totals (Columns 2 & 9)	LWDII	
Year	Total Hrs Worked Jan-Dec	Total Recordable Cases		Industry	LostWorkday Injury & Illness Cases		Industry

(Columns A, 2 and 9, referred to in the above chart, are obtained from the OSHA Log 200.)

Lost Workday Injury & Illness (LWDII) Incidence Rate =

[Injury(Col.2) + Illness(Col.9)Cases]
[Total Calendar Hours Worked x 200,000]

- ❑ Log Analysis Completed
- ❑ Posting of Annual Summary
- ❑ Reference: Recordkeeping Guidelines for Occupational Injuries and Illnesses (OSHA "Blue" Recordkeeping Guidelines)

Note: *Effective January 1, 2002 OSHA's revised recordkeeping guidelines went into affect. An OSHA Log 300 has replaced the OSHA Log 200. The same kind of analysis can be made from the new form.*

Hazard Communication – 29 CFR 1910.1200

- ❑ Obtain copy of HAZCOM Rule - 29 CFR 1910.1200
- ❑ Identify Responsible Staff
- ❑ Identify chemicals in the work place (chemical inventory)
- ❑ Ensure containers are labeled
- ❑ Obtain an MSDS for each chemical — dated 1989 or later
- ❑ SARA Title III info on MSDSs
- ❑ Written HAZCOM Program
- ❑ MSDSs readily accessible to employees
- ❑ Employee Training
- ❑ Procedures to Maintain Program Effectiveness
- ❑ Contractor — Notification and Training Procedure

Bloodborne Pathogens – 29 CFR 1910.1030

- ❑ Written Exposure Control Plan
- ❑ HB Vaccination
- ❑ Part of HAZCOM Program
- ❑ Training - Initial/Annual
- ❑ Engineering Controls
- ❑ · Handwashing Facilities
- ❑ Clean-up Kit/Materials
- ❑ · Sharp containers

Emergency Action Plan – 29 CFR 1910.38
(For Fire, Tornado, Earthquake, Explosion, Hazardous Spill)

- ❑ Written Plan
- ❑ Alarm System (loud, distinctive, tested)
- ❑ Evacuation Procedure Route Maps, Designated Gathering Area(s)
- ❑ Training
- ❑ Employee Responsibilities
- ❑ Emergency Lighting Exit Door Identification

HAZardous Waste OPerations & Emergency Response (HAZWOPER) – 29 CFR 1910.120
(For Workers Responding to Hazardous Materials Spills)

- ❑ Written Plan
- ❑ Training
- ❑ Emergency Response Clean Up Materials (Kit)
- ❑ (Or compliance to: 29 CFR 1910.38, whichever is more restrictive)

Lockout / Tagout –29 CFR 1910.147

- ❑ Written Program
- ❑ Copy of Standard 1910.147
- ❑ Training
- ❑ Safety Locks, Safety Lock Identified by Employee
- ❑ Is Procedure Being Followed?
- ❑ List of Equipment and Machines
- ❑ Individual Write-ups for Lockout of each Machine
- ❑ Annual Audit of Lockout/Tagout Procedure Complete

Confined Spaces – 29 CFR 1910.146

- ❑ Identify All Plant Confined Spaces
- ❑ Identify Permit Required Confined Spaces
- ❑ Testing Instrumentation/Procedure for Toxic Gases
- ❑ Entry Permit
- ❑ Written Program
- ❑ Training
- ❑ Confined Space Signs Posted

Personal Protective Equipment (PPE) – 29 CFR 1910.132 -.138

- ❑ Written certification of Workplace Hazard Assessment
- ❑ Selection and use of PPE that will protect affected employee from identified hazards.
- ❑ Training to each employee who is required to use PPE
- ❑ Employer's written training certification containing employee trained, date of training, and subject of certification.

Previous OSHA Citations

Yes___ No___
All Items Abated? Yes___ No___
All Dates Met? Yes___ No___

Safety and Health Programs

- ❑ Management Commitment and Employee Involvement
- ❑ Safety Goals & Objectives
- ❑ Work Site Analysis (Safety Inspections) Inspection Records
- ❑ Training for Employees, Supervisors & Managers / Enforcement
- ❑ Active Employee/Management Safety Committee Meeting minutes kept
- ❑ Rules for Housekeeping and General Maintenance
- ❑ Personal Protective Equipment (PPE) provided and used
- ❑ Accident Investigation Process

(Safety and Health Programs, continued on next page)

(Safety and Health Programs, continued)

- ❑ Are Specific Programs Developed where Required?
- ❑ Emergency Action Plan (1910.38)
- ❑ Fire Prevention Plan (1910.38)
- ❑ Hearing Conservation Program (1910.95)
- ❑ Emergency Response Plan (HAZWOPER) (1910.120)
- ❑ Respirator Program (1910.134)
- ❑ Lockout/Tagout (1910.147)
- ❑ Medical Surveillance Programs:
- ❑ Lead 1910.1025
- ❑ Formaldehyde 1910.1048
- ❑ Cadmium 1910.1027
- ❑ Bloodborne Pathogens (1910.1030)
- ❑ Hazard Communications (1910.1200)
- ❑ Laboratory Chemical Hygiene Plan (1910.1450)
- ❑ Confined Space Entry (1910.146)
- ❑ Ergonomics Program
- ❑ Fork Lift Training Program [1910.178 (l)]

Records

- ❑ OSHA notice (poster) - 29 CFR 1903.2 (a) (1)

Employee Medical Records - 29 CFR 1910.1020

- ❑ Information to employee; location, person responsible, access rights
- ❑ Make copy of this section and appendices available to employees

Preservation of Records

- ❑ Medical Records—Employment plus 30 years
- ❑ Exposure Records— Employment plus 30 years
- ❑ OSHA Log 200/300—Last 5 Years
- ❑ MSDSs—Length of Service plus 30 years

Respirator Program – 29 CFR 1910.134

- ❑ Written Program

Elements of a respirator program:

- ❑ Respirator Selection
- ❑ Training on Proper Use (Beards may not be worn, if they affect face seal)
- ❑ Maintenance, Cleaning and Disinfecting
- ❑ Storage
- ❑ Inspection
- ❑ Workplace Surveillance
- ❑ Medical Review by physician for determining if employee is fit to wear
- ❑ Fit Testing, Required Annually
- ❑ For Voluntary Use Respirators (Appendix D) furnished employees

Overhead Cranes / Hoists – 29 CFR 1910.179

❑ Rated load of crane marked on crane structure and hoist

Inspections (29 CFR 1310.179(j))

Frequent: ❑ Daily Inspection List
❑ Monthly Inspection List

Periodic: ❑ Annual Inspection List

Testing (29CFR 1910.179 (k))

- ❑ Initial and if altered/repaired
- ❑ Load Test
- ❑ Inspection/Test Record Readily Available

Chains / Slings – 29 CFR 1910.184

- ❑ Inspections 1910.184(d)
- ❑ Daily Inspection Check List
- ❑ Annual Inspection Check List
- ❑ Proof Test Certificates on File for New, Repaired or Reconditioned Chains / Slings / Components
- ❑ Records Readily Available

Forklift Operations – 29 CFR 1910.178

- ❑ Training Requirements Paragraph (l)
- ❑ Driver Training Instruction (Class room)
- ❑ Driver Obstacle Test (Practical Demonstration)
- ❑ Daily Forklift Maintenance Checklist— *Is It Used????*
- ❑ Written Program
- ❑ Licenses Issued
- ❑ Performance Re-evaluation Every 3 Years

Abrasive Wheel Guarding – 29 CFR 1910.215

- ❑ Works rests missing or not adjusted to 1/8" from wheel- 29 CFR 1910.215 (a)(4)
- ❑ Tongue guards missing or not adjusted to 1/4" from wheel – 29 CFR 1910.215(b)(9)

Machine Guarding – 29 CFR 1910.212

Look at machines to ensure guarding is present, for example:

- ❑ Point of Operation
- ❑ Nip Points
- ❑ Rotating Parts
- ❑ Flying Chips
- ❑ Sparks
- ❑ Are existing guards in place?

Hearing Conservation – 29 CFR 1910.95

Noise exposures above 85 dBA are common throughout the plant. Elements of a Hearing Conservation program should include:

- ❑ Noise Monitoring
- ❑ Audiometric Tests
- ❑ Engineering Noise Abatement/Administrative Controls
- ❑ Hearing Protection
- ❑ Employee Training (Effects of noise, audiograms, hearing protectors)
- ❑ Recordkeeping/Documentation
  - Audiometric records
  - Noise exposure records
  - Employee exam and noise history records
  - Engineering controls - implementation program
  - Hearing protectors

Industrial Hygiene – Personal Air Monitoring

- ❑ Silica Total Respirable Dust
- ❑ Total Dust
- ❑ Metal Fumes: Pb__ Fe __ Mn __ Cd __ Cu __ Ni __ Cr __
- ❑ Phenol
- ❑ Formaldehyde
- ❑ CO
- ❑ Ammonia
- ❑ Initial Determinations of Employee Air Level Exposures Complete

OSHA Training Requirements

One of the most confusing areas with OSHA compliance is determining which standards require training and how frequently it must be given. To help foundries, the AFS Occupational Health and Safety Committee (10-Q) has put together a list of OSHA training requirements that might apply to metalcasting facilities **(Table 19-1)**.

BIBLIOGRAPHY

Air Pollution Engineering Manual, Air & Waste Management Association, (1992).

American Foundry Society, "The Foundry Industry...Recycling Yesterday, Today & Tomorrow," The Fabric Filter Manual, C.E. Billings, Ed.; McIlvaine Co., Northbrook, IL.

Industrial Ventilation, A Manual of Recommended Practice, 21st Edition, American Conference of Governmental Industrial Hygienists, (1992).

Oman, D. E. (1988), Waste Minimization in the Foundry Industry, *Air Pollution Association Control Journal,* Volume 38, Number 7.

Selecting Baghouse Dust Collectors, S. Moore, J. Ruback and M. Jolin, *Plant Engineering,* p. 58 (Oct 1996).

U.S. Environmental Protection Agency (1998), "EPA Office of Compliance Sector Notebook Project: Profile of the Metal Casting Industry," EPA/310-R-97-004.

Table 19-1.
OSHA Training Frequency

OSHA Std 29 CFR...	Training Requirements	Initially	When workplace changes	Annually	Written program required
1910.38(a)	Emergency Planning & Evacuation	X	X	—	X
1910.38(b)	Fire Prevention Plan	X	—	—	X
1910.95	Occupational Noise Exposure	X	X	X	X
1910.119	Process Safety Management of Highly Hazardous Chemicals	X	X	(1) X	X
1910.120	Hazardous Waste Operations & Emergency Response (HAZWOPER)	X	X	X	X
1910.132-.138	Personal Protective Equipment	X	X	—	X
1910.134	Respiratory Protection	X	X	X	X
1910.146	Permit Required Confined Spaces	X	X	—	X
1910.147	The Control of Hazardous Energy (Lockout/Tagout) - LOTO	X	X	—	X
1910.151	Medical Services & First-Aid	X	—	(2) X	X
1910.157	Portable Fire Extinguishers	X	—	X	X
1910.178	Powered Industrial Trucks:				
	Classroom Training	X	X	—	—
	Performance Training	X	—	(1)X	—
1910.252	General Requirements- Welding, Cutting & Brazing	X	X	—	—
1910.332	Electrical Training	X	X	—	—
5(a)(1)	Ergonomics Exposure Control	(3)X	X	—	—
1910.1020	Access to Employee Exposure and Medical Records	X	—	X	—
1910.1025	Lead-Exposure above the Action Level (AL)				
1910.1027	Cadmium-Exposure in any amount	X	—	X	X
1910.1048	Formaldehyde-If > 0.1ppm	X	—	X	X
1910.1200	Hazard Communication (HAZCOM)	X	X	—	X

(1)Every three years. (2)CPR, annually; First-Aid, every three years. (3)If problem on job

Casting Design 20

Now that the basics of metalcasting have been covered, the casting design parameters should be easier to comprehend. A good understanding of geometry, physics and computer design, along with basic metalcasting, is required to become a casting design engineer.

Structural design engineers commonly design within a narrow group of casting types poured from familiar alloys (i.e., iron alloys or 300 series Al alloys) and molded from familiar foundry processes (i.e., green sand or nobake). Rules of thumb have been developed over the years for common design situations. However, close inspections of these rules reveal a conflict in many cases. For example, the use of *gusseting* instead of *mass* for stiffness might be labeled "recommended" in one set of rules and "poor" in another.

Also, when a design engineer leaves a familiar design realm for an unfamiliar one, unexpected results may occur. For example, when substituting Al bronze for ductile iron while remaining with a familiar foundry process (i.e., nobake molding), there's likely to be trouble with the usual "ductile iron-style" geometry. Good Al-bronze geometry is different from typical ductile iron geometry, and the molding process may need to supplement the different geometry with heat transfer techniques. Not suspecting this, the design engineer's new casting design may suffer from no-quotes, higher-than-expected prices and foundry requests for design changes.

How are design engineers to know that a successful casting geometry for aluminum bronze should be different from ductile iron? Moreover, if the design engineer did know that information, what would be the proper course of design action? The answer lies in a better understanding of the relationship among geometry, metallurgy and physics.

STRUCTURAL GEOMETRY

Because castings can easily apply shape to structural requirements, most casting designs are tailored to manage applied forces. In fact, castings find their way into the most sophisticated applications because they can be so efficient in shape, properties and cost. Examples are turbine blades in jet engines, suspension components (in automobiles, trucks and railroad cars), engine blocks, airframe components, fluid power components, etc.

When designing a component structurally, a design engineer is generally interested in safely controlling forces through choice of allowable stress and deflection (flexibility). Although choice of material affects allowable stress and deflection, the designer's most significant tool is geometry. Geometry directly controls stress and deflection in a structure.

The casting processes are limitless in their combined ability to allow variations in shape. Not many years ago, efficient structural geometry was limited by the designer's inability to visualize in three dimension. Now, computer-generated solid models and rapid prototypes are greatly enhancing the designer's ability to visualize 3D structural shapes. This technology often leads to casting designs of high efficiency and effectiveness. **Figure 20-1** depicts a meshed solid model and a stress analysis via the mesh elements.

Improved efficiency in solid modeling software has led to an interesting design dilemma. Solid models are readily applicable to Finite Element Analysis (FEA) of stress. FEA enables the engineer to quickly evaluate stress levels in the design, and solid models can be tweaked into shape via the software so that geometry can be optimized for allowable, uniform stress.

However, optimum geometry for allowable, uniform stress may not be acceptable geometry for *castability*. When a foundry engineer quotes a design that considers structural geometry only, requests for geometry changes for castability are likely. At this point, the geometry adjustments for castability may be more substantial than the solid model software can handle. The result can be no-quotes, higher-than-expected casting prices or starting over with a new solid model.

A practical solution to this problem is to concurrently engineer the geometry by considering structural, foundry and downstream manufacturing needs. The result can be an optimal casting geometry. The most efficient technique is to make engineering sketches or mark the sections and/or views on blueprints. The idea is to explore overall geometry before locking in to a solid model too quickly. Engineering sketches or mark-ups are easy and quick to change—even dramatically—in the concurrent brainstorming process; solid models are not. A solid model should be the elegant result, not the automatic start.

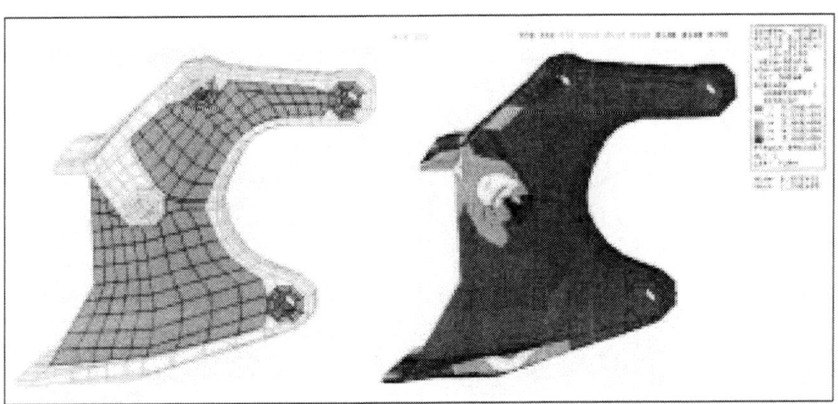

Fig. 20-1. A meshed solid model and a stress analysis.

DESIGN PARAMETERS

There are six parameters (based on physics) that underlie cost-effective casting design. The following six parameters, applied as a system, drive the geometry of casting design:

Casting Properties

1. Fluid Life
2. Solidification Shrinkage
   Type (eutectic, directional and equiaxed)
   Volume (small, medium and large)
4. Pouring Temperature
3. Slag/Dross Formation Tendency

Structural Properties

5. Section Modulus (stiffness of casting geometry)
6. Modulus of Elasticity (stiffness of alloy itself)

This six-parameter system must be considered in order to optimize the geometry for castability, structure, downstream processing (machining and assembly) and process geometry (risering, gating, venting and heat transfer patterns) in the mold. These physics-based design parameters are boundary conditions to optimize the casting geometry.

Optimizing casting geometry using the six-parameter system is not difficult. The four casting and two structural properties influence important variables in designing, producing and using metalcastings. These variables include:

- casting method;
- casting section design;
- junction design between casting sections;
- surface integrity;
- internal integrity;
- dimensional capability;
- cosmetic appearance.

Both the OEM designer and metalcaster must work together to streamline any casting design. Casting geometry is the most powerful tool available to improve castability of the alloy and mechanical stiffness and strength of the casting.

Carefully planned geometry can offset alloy problems in fluid life, solidification shrinkage, pouring temperature and slag/dross forming tendency. Section modulus, an attribute of structural geometry, has the capability to increase stiffness and/or reduce stress—a capability that can be very important when applied to alloys with lower strength and stiffness. Modulus of elasticity, an alloy's inherent stiffness, combined with section modulus and section length, controls deflection in a casting design.

An example of geometry's ability to influence the four casting parameters is given in **Fig. 20-2.** The simple steel fabrication in **Fig. 20-2 (left)** was converted into carbon steel and gray iron casting designs, **Fig. 20-2 (center)** and **20-2 (right),** respectively.

The fabrication is a for a guide block used to constrain low-velocity/low-load sliding motion. It was welded from rectangular bar stock, then milled, drilled and tapped. The geometries in **Figs. 20-2 (center)** and **20-2 (right)** are considerably different because of differences among fluid life, solidification shrinkage type and amount, pouring temperature and tendency to form nonmetallic inclusions (slag/dross).

Fluid Life

Fluid life more accurately defines the alloy's liquid characteristics than does the traditional term fluidity; however, both terms are commonly used. (Fluidity was also discussed in Chapter 15, Gating Practice.) Molten metal's fluidity is a dynamic property, changing as the alloy is delivered from a pouring ladle, diecasting chamber, etc., through the gating system and finally into the mold or die cavity. Heat transfer reduces the metal's temperature, and oxide films form on the metal front as this occurs. Fluidity decreases most rapidly with temperature loss, and it can decrease significantly from the surface tension of oxide films.

The absolute value of temperature is not the test of fluidity at a given moment. For example, some aluminum alloys at 1200-1400F (650-750C) have excellent fluid life. However, some molten steels at 3000F (1650C) have much shorter fluid life. In other words, a molten alloy's fluid life also depends on chemical, metallurgical and surface tension factors.

Fluid life affects the design characteristics of a casting, such as the minimum section thickness that can be cast reliably, the maximum length of a thin section, the fineness of cosmetic detail (i.e., lettering, logos) and the accuracy with which the alloy fills the mold extremities.

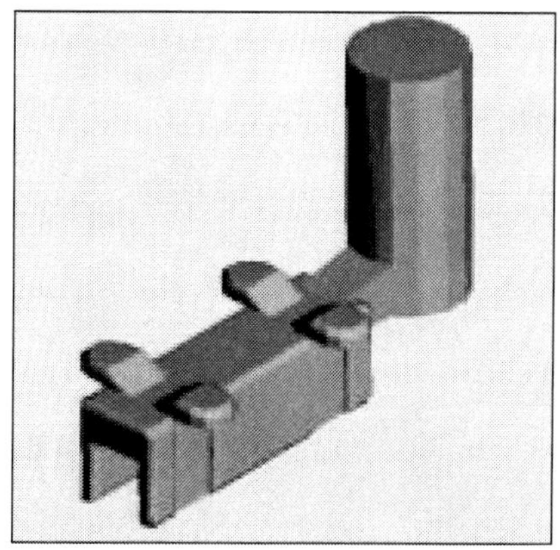

Fig. 20-2. Shown on the left is the original steel fabrication. In the center is a carbon steel casting design featuring geometry that suits the four casting parameters. At the right is a gray iron casting featuring an entirely different geometry that is also based on the four casting parameters.

It is essential to understand that moderate or even poor fluid life does not limit the cost-effectiveness of design. Knowing that an alloy has limited fluid life tells the designer that the part should feature:

- softer shapes and larger lettering;
- finer detail in the bottom portion of the mold, where metal flows first, fastest and generally hottest;
- coarser detail in the upper portions of the mold where the metal is slower to arrive and more affected by oxide films and solidification "skin" formation. Even an alloy with good fluidity, when overexposed to oxygen, may form a high surface tension oxide film that reduces the fluidity, causing a "rounding off" of the leading metal front as it flows;
- more taper toward thin sections.

Some alloys, i.e., 356 aluminum, have been metallurgically designed to enhance fluid life. In the case of 356 aluminum, the addition of high heat capacity silicon retains heat and improves fluid life.

Solidification Shrinkage

There are three distinct stages of shrinkage as molten metals solidify: liquid shrinkage, liquid-to-solid shrinkage and patternmaker's contraction. (Shrinkage was also discussed in Chapter 17, Risering Practice.)

Liquid Shrinkage

Liquid shrinkage is the contraction of the liquid before solidification begins. It is not an important design consideration.

Liquid-to-Solid Shrinkage

Liquid-to-solid shrinkage (also called *solidification shrinkage*) is the shrinkage of the metal mass as it transforms from the liquid's disconnected atoms and molecules into the structured building blocks of solid metal. The amount of solidification shrinkage varies greatly from alloy to alloy. **Table 20-1** provides a guide to the liquid-to-solid shrinkage of common alloys. As shown, shrinkage volumes can vary from low to high.

Table 20-1.
Liquid-to-Solid Shrinkage Characteristics of Common Foundry Alloys

Solidification Shrinkage

Alloy Group	Fluid Life	Type	Amount	Pour Temp.	Slag/Dross
FERROUS:					
Gray Iron	Excellent	Eutectic-Type	Very Small	2500-2600F (1371-1427C)	Little
Ductile Iron	Good	Eutectic/ Directional	Small	2500-2600F (1371-1427C)	Some
Carbon & Low-Alloy Steel	Poor	Directional	Large	2850-3000F (1566-1649C)	Moderate
High Alloy Steels	Fair	[1]Various	[1]Various	[1]Various	Moderate
NONFERROUS:					
Aluminum 356	Excellent	[2] Eutectic-Type	[2] Little	1300-1400F (704-760C)	Moderate
Aluminum 206	Fair/Good	Equiaxed	Moderate/ Large	1300-1400F (704-760C)	Moderate/ Large
Aluminum Bronze	Fair	Equiaxed	Moderate/ Large	2000-2150F (1093-1177C)	Large
Silicon Bronze	Fair	Eutectic-Type	Little	1900-2050F (1038-1121C)	Large
Magnesium ZE43	Excellent	Directional	Moderate	1300-1400F (704-760C)	Little/ Moderate
Yellow Brass	Poor/Fair	Eutectic-Type	Moderate	1800-1950F (982-1066C)	Large
Titanium	Very Good	Eutectic-Type	Little	3200-3300F (1760-1816C)	Very Large
Zirconium	Fair	Eutectic-Type	Little	3300-3400F (1816-1871C)	Very Large

[1] Among martensitic, partly austenitic and fully austenitic grades, solidification shrinkage encompasses all three types. Shrinkage amount and pouring temperature vary also.
[2] For premium structural castings, solidification is more complex. Depending on alloy modifications, section sizes and specifics of liquid-to-solid transformation, directional and/or equiaxed shrinkage may be involved.

Definitions:

Eutectic-Type Solidification: Eutectic alloys or behaving like them. These alloys remain liquid in the mold for a brief period, cool and then solidify very quickly all over. This phenomenon minimizes internal shrinkage and the need for risers.

Directional Solidification: These alloys begin solidifying quickly, perpendicular to molds walls. Solidification "direction" and pathways are predictable from casting geometry and thermal patterns in the mold walls. Without proper pathway geometry, isolated internal shrinkage can result.

Equiaxed Solidification: These alloys not only begin solidifying perpendicular to mold walls, but also solidify in the midst of the liquid, forming equiaxed islands of solid. Solidification pathways are interrupted by the equiaxed islands, making these alloys difficult to feed. Fine, dispersed microporosity is typical.

Alloys are further classified based on their liquid-to-solid solidification type: *eutectic-type, directional* and *equiaxed* **(Table 20-1, bottom).** The type of solidification shrinkage in a casting is just as important as the amount of shrinkage. Specific types of geometry can be chosen to control internal integrity when solidification amount or types are a problem.

Figures 20-3 to 20-5 illustrate what the three solidification shrinkage types defined in **Table 20-1 (bottom)** imply. In each case, a simple plate casting is shown with attached riser. A riser is a reservoir of liquid metal attached to a casting section to feed solidification shrinkage. Cross sections of the plate and riser(s) show conceptually how solidification takes place; metallurgical reality is similar, but microscopic.

Eutectic-Type Solidification—**Figure 20-3** illustrates the eutectic-type alloy solidification, the most forgiving of the three. Such alloys typically have less solidification shrinkage volume. Risers are much smaller and, in special cases, can be eliminated by strategically placed gates. The essential feature with these alloys is the extended time that the metal feed avenue stays open. The plate solidifies more uniformly all over and all at once. Eutectic-type alloys are less sensitive to shrinkage problems attributed to abrupt geometry changes.

Directional Solidification—**Figure 20-4** shows solidification on and perpendicular to the casting surfaces, known as progressive solidification. At the same time, solidification moves at a faster rate from the ends of the section(s) toward the source of feed metal (risers)—this is known as directional solidification. Directional solidification moves faster from the ends of the sections because of the greater amount of surface area through which the solidifying metal can lose its heat. The objective is for directional solidification to beat out progressive solidification before it can "close the door" to the source of the feed metal. As shown, directionally solidifying alloys require extensive risering and tapering, but are capable of providing excellent internal soundness when solidification patterns are designed properly.

Equiaxed Solidification—Alloys that exhibit equiaxed solidification respond the most dramatically to differences in geometry **(Fig. 20-5).** Shrinkage in these alloys tends to be widely distributed as micropores, typically along the center plane of a casting section. The reason is that solidification occurs not only progressively from casting surfaces inward and directionally from high surface area extremities toward lower surface area sections, but also equiaxially via "islands" in the middle of the liquid. These islands of solidification interrupt the liquid pathway of directional solidification. Gradually, the pathways freeze off, leaving micropores of shrinkage around and behind the islands that grew in the middle of the pathway.

Larger risers, thicker sections and tapering **(center, Fig. 20-5)** are counterproductive, causing micropores to coalesce into larger pores across more of the casting cross section. As illustrated at the bottom of **Fig. 20-5,** microporosity is kept small and confined to a narrow mid-plane in the casting section by more "thermally neutral" geometry with smaller, further-spaced risers.

As illustrated in **Figs. 20-3 to 20-5,** there is a significant relationship among solidification shrinkage, geometry and quality. Most simply, *eutectic-type solidification* is tolerant of a variety of geometries; the least reciprocity is required. Most complex, *equiaxed solidification* requires the most engineering foresight in the choice of geometry and may require supplemental heat transfer techniques in the mold process. In the middle lies *directional solidification.* While capable of the worst shrinkage cavities, it is the most capable of very high internal integrity when the geometry is properly designed. Well-planned geometry in a directionally solidifying alloy can eliminate not only shrinkage but also the need for any supplemental heat transfer techniques in the mold.

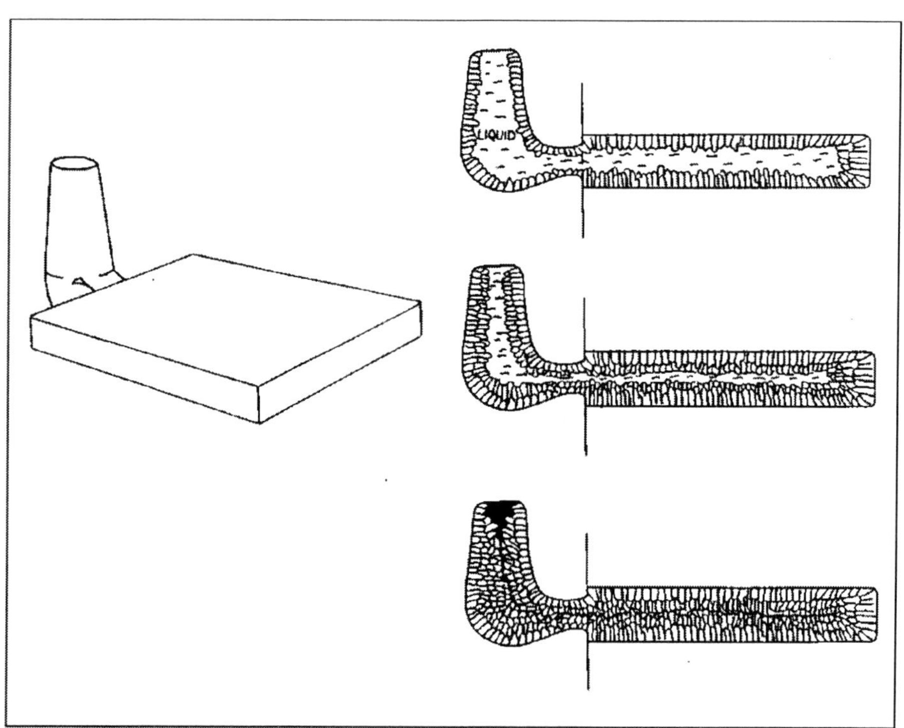

Fig. 20-3. Eutectic-type solidification is the most forgiving of the alloy shrinkage types. Risers may be much smaller with these alloys, as the path of liquid feed metal remains open through solidification.

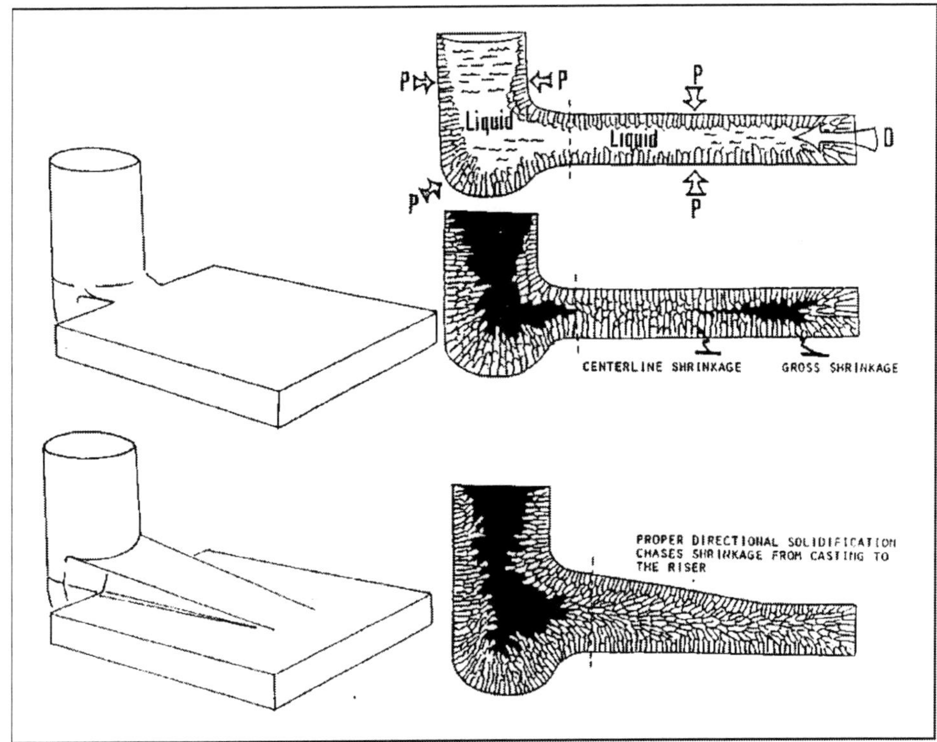

Fig. 20-4. Directional solidification on a plate casting is illustrated. Extensive risering and tapering (bottom) allow for excellent internal casting soundness. (P = progressive solidification and D = directional solidification.)

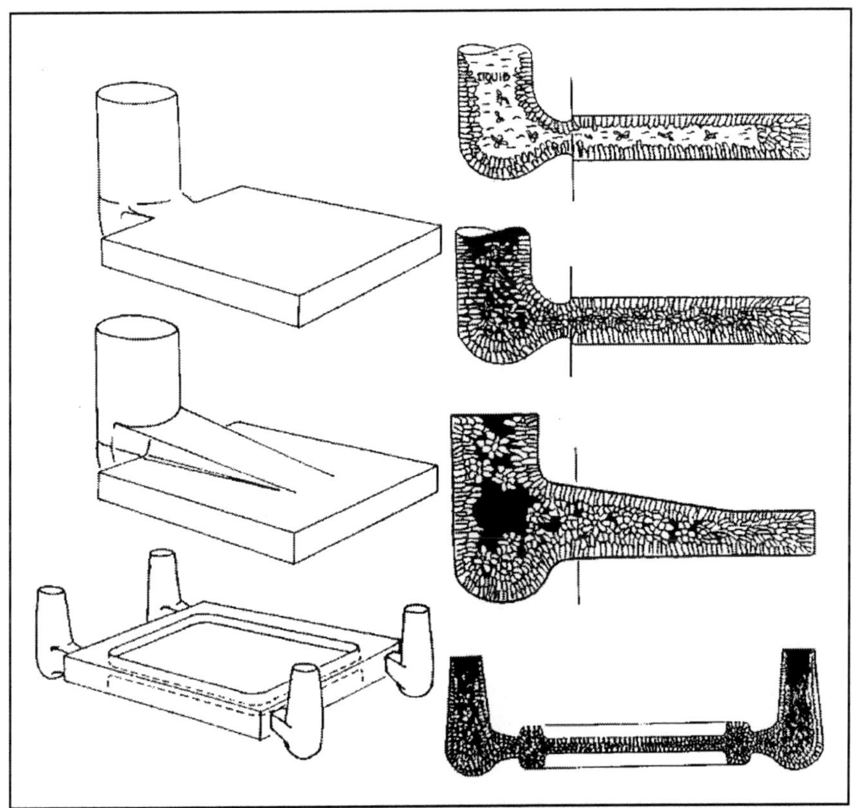

Fig. 20-5. Designs for equiaxed solidifying alloys are shown. The large riser design (second from bottom) illustrates how not to feed a section. While such a taper and large riser worked with directional solidification, using this approach here adds more heat to an area that needs to cool more uniformly and results in larger, coalesced shrinkage. The proper casting and process geometry (smaller risers and a thermally neutral shape) is illustrated at the bottom.

Patternmaker's Contraction

The contraction that occurs after the metal has completely solidified and is cooling to ambient temperature is called patternmaker's contraction (or solid shrinkage). This contraction changes the dimensions of the casting from those of liquid in the mold to those dictated by the alloy's rate of contraction. Therefore, as the solid casting shrinks away from the mold walls, it assumes final dimensions that must be predicted by the pattern- or die-maker. This variability of contraction is another important casting design consideration, and it is critical to dimensional accuracy. Tooling design and construction must compensate for it.

Achieving specified dimensions requires the foundry's patternmaker and/or diemaker to be included in the casting design process. The complex nature of patternmaker's contraction makes tooling adjustments inevitable. For example, a highly recommended practice for critical dimensions and tolerances is to build the patterns/dies/coreboxes with extra material on critical surfaces so that the dimensions can be fine-tuned by removing small amounts of tooling stock after capability castings have been poured and measured.

Pouring Temperature

Although molds must withstand extremely high temperatures of liquid metals, interestingly, there are not many choices of materials with refractory characteristics. When pouring temperature approaches a mold material refractory limit, the heat transfer patterns of the casting geometry become important.

Sand and ceramic materials with refractory limits of 3000-3300F (1650-1820C) are the most common mold materials. Metal molds, such as those used in diecasting and permanent molding, have temperature limitations. Except for special thin designs, all alloys that have pouring temperatures above 2150F (1180C) are beyond the refractory capability of metal molds.

It is also important to recognize the difference between heat and temperature. Temperature is the measure of heat concentration. Lower temperature alloys also can pose problems if heat is too concentrated in a small area. In these cases, better geometry choices allow heat to disperse into the mold. This is discussed later in the Junction Design section.

Slag/Dross Formation

In foundries, the terms slag and dross have slightly different meanings. *Slag* typically refers to high-temperature fluxing of refractory linings of furnaces/ladles and oxidation products from alloying. *Dross* typically refers to oxidation or reoxidation products in liquid metal from reaction with air during melting or pouring, and can be associated with either high or low pouring temperature alloys.

Some molten metal alloys generate more slag/dross than others and are more prone to contain small, round-shaped nonmetallic inclusions trapped in the casting. Unless a specific application is exceedingly critical, a few small rounded inclusions will not affect casting structure significantly. In most commercial applications, nonmetallic inclusions are only a problem if they are encountered during machining or appear in a functional as-cast cosmetic surface.

The best defense against nonmetallic inclusions is to inhibit their formation through good melting, ladling, pouring and gating practices. Ceramic filters, which can be used with alloys that have good fluid life, have advanced the foundry's ability to eliminate nonmetallics. Vacuum melting and pouring are applied in extremely dross-prone alloys, like titanium.

Section Modulus

Experimenting with sketches before building a solid model means that another approach to evaluate stress and deflection will be needed. This alternate approach is the essence of efficient structural evaluation of geometry in casting design.

Geometry, stress and deflection at important cross sections must be quickly evaluated to take full advantage of engineering sketching/print marking. As the design engineer well knows, the classic formulas for bending stress, torsional stress and deflection are relatively simple. Each, however, contains the same parameter, Section Modulus, which is a function of shape and difficult to compute. Therefore, a quick, simple way to compute or estimate Section Modulus (more specifically, its foundational parameter, Area Moment of Inertia) is needed to move from sketch to improved sketch in the casting geometry brainstorming.

This alternate approach for the design engineer is not so simple. The difficulty in computing Area Moment of Inertia for casting shapes is one of the hidden reasons for the design and use of fabrications. Fabrications are made from building blocks of wrought shapes, like I-beams, rectangular bars, angles, channels and tubes. These shapes, which are simple and constant over their length, have Area Moments of Inertia that are easy to calculate or are available in handbooks. Consequently, stress and deflection calculations are relatively easy. Fabricated designs, however, are heavy and nonuniform in stress compared to a casting that is well designed for the same purpose.

Area Moment of Inertia

Although there are five kinds of stress (tension, compression, shear, bending and torsion), the most important stresses for complex structures are bending and torsion. If more than one type of stress is involved in the same section, the Principle of Superposition allows the individual stress types to be analyzed separately and then added together; once again, the larger of the stresses to be combined usually come from bending or torsion. **Figure 20-6** Shows the basic equations for calculating the bending stress, torsional stress and deflection.

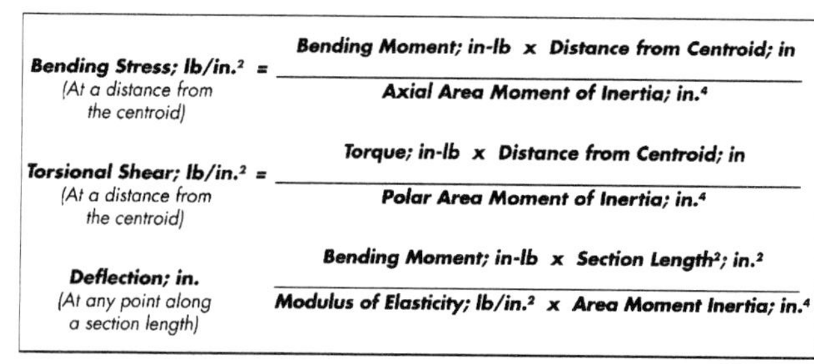

Fig. 20-6. Equations for bending stress, torsional shear and deflection are shown.

In all three cases, the equations use a factor based on the cross section of the geometry called the Moment of Inertia. Since the Area Moment of Inertia is in the denominator in each relationship, increased Area Moment of Inertia reduces stress and deflection. It is easy to draw a scale cross section, whether it is from an engineering isometric sketch or from a marked-up view on a blueprint. If a way to quickly estimate Area Moment of Inertia can be found, stresses in the brainstormed sketches can be estimated, as well as whether deflection will increase or decrease.

Maximum tensile stress in bending is often most critical in structural design. Section Modulus is defined as the Area Moment of Inertia divided by the maximum distance from the center of bending (centroid) to the outermost edge of the casting cross section. Section Modulus is similar to a "stiffness index" because it considers not only magnitude of Area Moment of Inertia, but also maximum section depth. If maximum section depth increases faster than Area Moment of Inertia, a geometry change can actually increase maximum tensile stress, rather than reduce it. This "index," termed Section Modulus, accounts for that potential problem.

The estimation method recommended is based on three principles. One is intuitive and the other two are from the equations of engineering mechanics. The principles are:

- *Learned sense of load magnitudes and component size/shape*—Engineers routinely use this intuitive sense to sketch sized shapes that are close to the final design. Foundry engineers learn this sense through training, thus becoming effective concurrent engineering partners in their customers' casting designs.
- *Equation for Area Moment of Inertia*—Although the calculation for an interesting casting cross section can be very difficult, the relationship expressed between "depth of section" (Y) and "change in cross-sectional area" (δA) is very simple. The position and shape of the two rectangles in **Fig. 20-7,** top demonstrates this simple yet powerful relationship. The change in shape of the inside of the tube **(Fig. 20-7, bottom)** is an even more dramatic illustration. Calculations weren't made in either case, but the qualitative impact of Area Moment of Inertia on stiffness and stress is unmistakable.
- *Parallel Axis Theorem*—Once the engineering sense of structural size and $Y^2\delta A$ have been applied qualitatively to a sketched cross section, the Parallel Axis Theorem can be applied to simple building blocks in the cross section to estimate Area Moment of Inertia quantitatively. A numerical value for Area Moment of Inertia is required to calculate the stress level in the sketched cross section. The Parallel Axis Theorem is illustrated in **Fig. 20-8.**

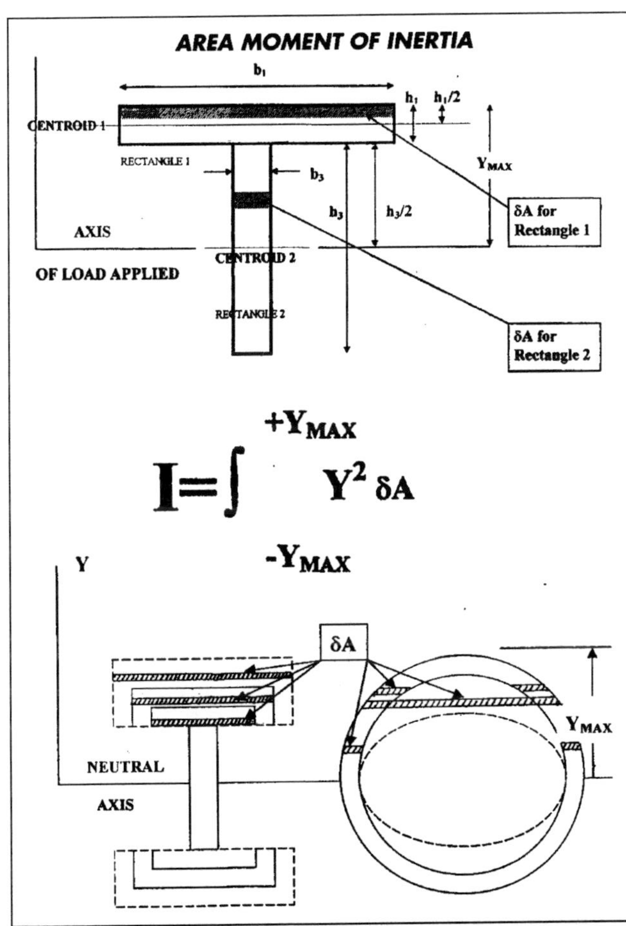

Fig. 20-7. Equation for Area Moment of Inertia for a casting cross section.

Modulus of Elasticity

The measure of a material's stiffness (without regard to material geometry) is known as the Modulus of Elasticity. It is a measure of the interatomic force, and it is a mechanical property of the alloy. Modulus of Elasticity varies widely among materials, and it varies significantly among metals. Alloy groups tend to have the same modulus value. For example, the entire family of steels (carbon, low-alloy and high-alloy) all have the same modulus value of 30×10^6 lb/in.^{2}.

Modulus of Elasticity is an important parameter in structural design, and it is directly involved in the relationship between casting geometry and deflection. A larger Modulus of Elasticity means less deflection. For example, a steel casting would deflect

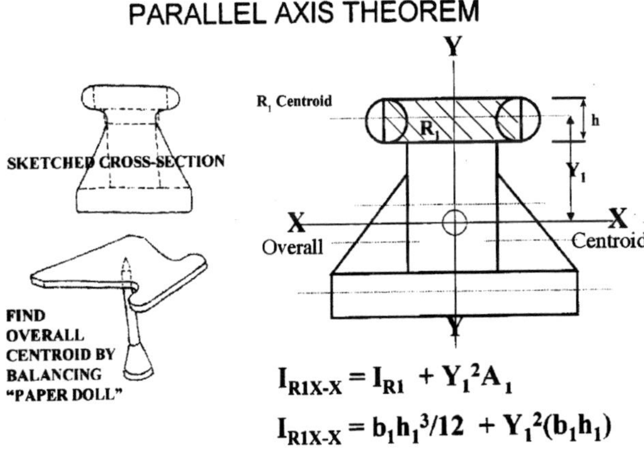

Fig. 20-8. Parallel Axis Theorem is shown.

less than an aluminum casting of identical geometry simply because steel is three times stiffer than aluminum.

The Modulus of Elasticity is simply the elastic slope of the stress versus strain diagram from a test bar. **Figure 20-9** illustrates qualitatively the results of pulled test bars for common groups of foundry alloys. The steepness of the elastic slope of each graph indicates the alloy group's stiffness.

One misunderstanding about Modulus of Elasticity is that it is not affected by heat treatment. However, heat treatment can affect the upper limit of the elastic stress regime. This is very important because the load at which the elastic slope begins to curve is called the metal's "yield stress." This is the stress level at which plastic deformation begins and the metal is permanently distorted. Stresses should be designed below this level so that deflections in the casting under load do not damage it.

For example, consider the family of steels in **Fig. 20-9;** heat treatment can considerably raise the point at which alloy steel yields. Although the steel is no stiffer at higher stress levels, it can withstand the additional stress without permanent deformation. The same is true for heat-treatable aluminum alloys, but the magnitude of heat treatment effect on yield stress in aluminum is considerably less than that for steels.

CASTABILITY VS. GEOMETRY

Castability affects geometry but *well-chosen geometry affects castability.* In other words, a geometry can be chosen that offsets the metallurgical nature of the more difficult-to-cast alloys. Knowing how to choose this "proactive" geometry is the key to consistently good casting designs—in any foundry alloy—that are economical to produce, machine and assemble into a final product.

The focus of the first four design parameters (Casting Properties) is the foundry engineering perspective of geometry for the benefit of design engineers. The focus of the fifth and sixth design parameters (Structural Properties) is the design-engineering perspective of geometry for the benefit of foundry engineers. The geometry found between these two spectrums offers boundless opportunity for improving quality while decreasing costs to add value.

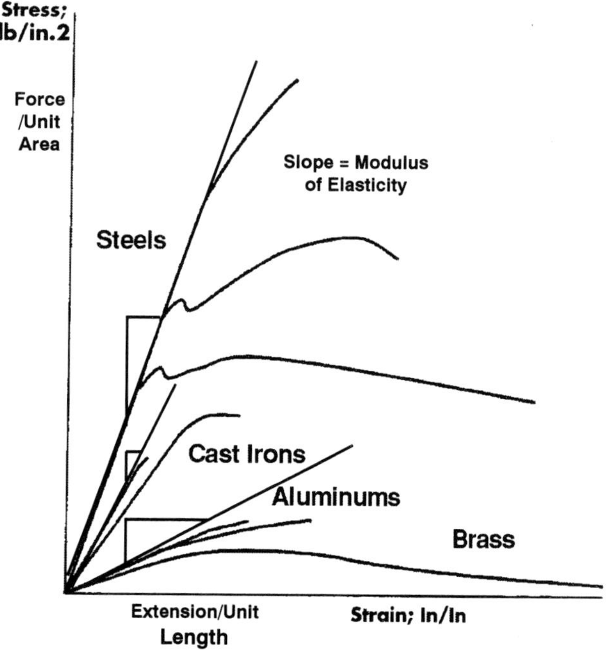

Fig. 20-9. Pulled test bar data for common foundry alloys is shown.

The objective is to explore geometry possibilities, looking for an ideal shape that is both castable in the chosen foundry alloy and allowable in stress and deflection for that alloy. As noted, there is great variety in the four metallurgical parameters that govern alloy castability. Similarly, great variety exists among metals in their allowable stress and deflection. Therefore, an ideal casting shape for all six of the design parameters is not necessarily a trivial exercise. For alloys that have good castability, choosing geometry for allowable stress and deflection is the best place to start. For alloys with less than the best castability, it is better to first find geometry that assists castability, and then modify it for allowable stress and deflection.

Not all alloys are like ductile iron, which is highly castable and has moderate tensile and yield strength. For ductile iron, many geometries may be equally acceptable. Martensitic high-alloy steel has fair-to-poor castability, but can have amazing resistance to stress and can tolerate very large deflections without structural harm. Therefore, structural geometry is easy to develop, but a coincidental castable shape is more difficult to design. Premium A356 aluminum has good castability, but rather weak resistance to stress and low tolerance for deflection. Carefully chosen structural geometry, however, combined with solidification enhancements in the molding process, has resulted in extremely weight-effective A356 structural components for aircraft, cars and trucks.

Junction Design

A junction is a region in which different section shapes come together within an overall casting geometry. Simply stated, junctions are the intersection of two or more casting sections. **Figure 20-10** illustrates both "L" and "T" junctions. The four junction types also include "X" and "Y" designs, which are not discussed here..

Designing junctions is the first step to finding castable geometry via the six-parameter system for casting design. **Figure 20-10** illustrates that there are major differences in allowable junction geometry, depending on alloy solidification. Alloys that have very little solidification shrinkage (Alloy 1) allow abrupt section changes and tight geometry. While the thermally neutral design (third row in **Fig. 20-10**) works in most cases, alloys that have significant solidification shrinkage (Alloy 3) require considerable adjustment of junction geometry, such as radiusing, spacing, dimpling and feeding. **Figure 20-11** illustrates the application of the foregoing principles of junction design in a citical automotive application.

Postcasting Considerations

System-wide thinking also must include the postcasting operations, such as machining, welding and joining, heat treating, painting and plating. One aspect that affects geometry is the use of fixturing to hold the casting during machining. Frequently, the engineers that design machining fixtures for castings are not consulted by either the design engineer or the foundry engineer as a new casting geometry is being developed. Failure to do so can be a significant oversight that adds machining costs. If the casting geometry has been based on the four casting parameters of the alloy, then the designer knows the likely surfaces for riser contacts and may have

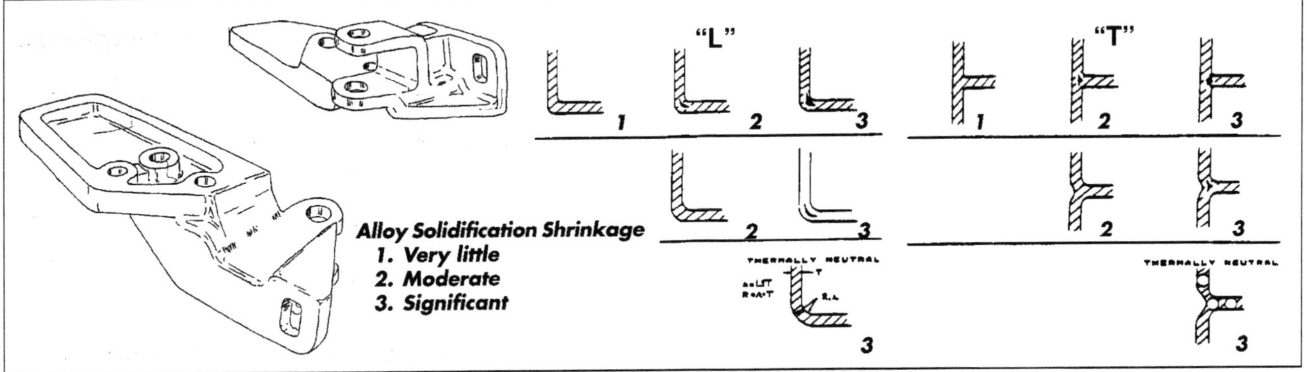

Fig. 20-10. The "L" and "T" junctions and their relation to the degree of solidification shrinkage.

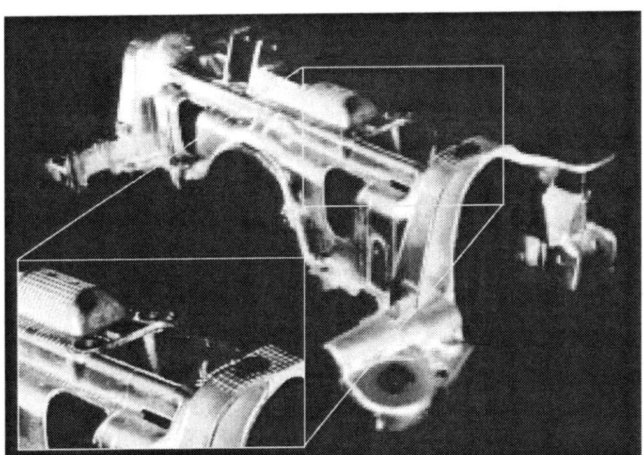

Fig. 20-11. This critical automotive casting application illustrates the importance of good junction design. [Photo courtesy of CMI International]

some idea of likely parting lines and core match lines. These surfaces and lines will be irregularities on the casting geometry and will cause problems if they contact fixturing targets.

Geometric Dimensioning and Tolerancing

The tool (measurement-based system) that has had the most dramatic positive impact on the manufacture of parts that reliably fit together is geometric dimensioning and tolerancing (GD&T), as defined by ANSI Y14.5M-1994 (American National Standards Inst.). When compared to traditional (coordinate) methods, GD&T:

- considers tolerances, feature-by-feature;
- minimizes the use of the "title block" tolerances and maximizes the application of tolerances specific to the requirement of the feature and its function;
- is a contract for inspection, rather than a recipe for manufacture. In other words, GD&T specifies the tolerances required feature-by-feature in a way that does not specify or suggest how the feature should be manufactured. This allows casting processes to be applied more creatively, often reducing costs compared to other modes of manufacture, as well as finish machining costs.

GD&T encourages the manufacturer to be creative in complying with the drawing's dimensional specifications because the issue is in compliance with tolerance, not necessarily in compliance with a manufacturing method. By forcing the designer to consider tolerances feature-by-feature, GD&T often results in broader tolerances in some features, which opens up consideration of lower cost manufacturing methods, like castings. **Figure 20-12** illustrates GD&T principles applied to a design made as a casting. Note use of installation surfaces as datums and use of geometric zones of tolerance.

Tolerance Capabilities

How a cast feature is formed in a mold has a significant effect on the feature's tolerance capability. The following six factors control the tolerance capability of castings. In order of preference, they are as follows:

Molding Process—The type of molding process (such as green sand, shell, investment, etc.) has the greatest single influence on tolerance capability. How a given molding process is mechanized and the sophistication of its pattern or die equipment can refine or coarsen its base tolerance capability.

Casting Weight and Longest Dimension—Logically, heavier castings with longer overall dimensions require more tolerance. These two factors have been defined statistically in tolerance tables for some alloy families.

Mold Degrees of Freedom—This factor is least understood. Just as some *molding processes* have more mold components (mold halves, cores, loose pieces, chills, etc.) than others, some *casting designs* require more mold components. Each mold component has its own tolerances, and tolerances are stacked as the mold is assembled.

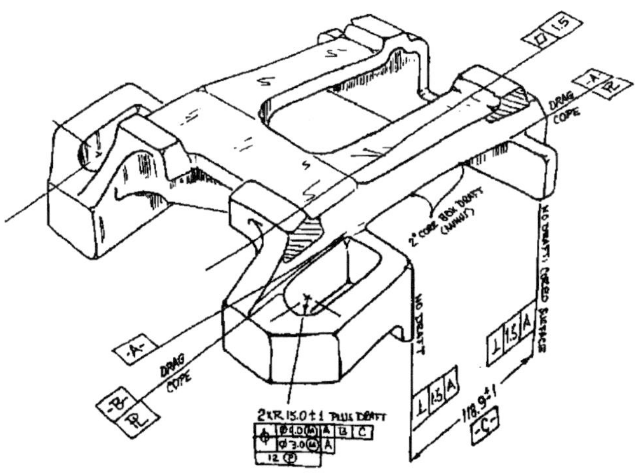

Fig. 20-12. This illustration shows the GD & T principles applied to a design made as a casting.

More mold components mean more degrees of freedom; hence more tolerance. Good design for tolerance capability minimizes degrees of freedom in the mold for features with critical dimensions.

Draft—It is common for casting designs to ignore the certainty of draft, including mold draft, draft on wax and/or Styrofoam patterns made from dies, and core draft. Since one degree of draft angle generates 0.017 in. of offset per in. of draw (about 0.5 mm/30 mm), draft can quickly use up all of a tolerance zone and more.

Patternmaker's Contraction—The uncertainty of patternmaker's contraction is why foundries normally recommend producing first-article and production-process-verification castings (sometimes called "sample" or "capability" castings) to establish what the dimensions really will be before going into production. A common consequence of patternmaker's contraction uncertainty is a casting dimension that is out of tolerance, not because it varies too much, but because its average value is too far from nominal. In other words, the dimension contracted more or less than was expected.

Cleaning and Heat Treating—Many casting dimensions are affected by postcasting processing. At the least, most castings are touched by abrasive cutting wheels and grinding—even precision castings. Many castings are heat treated, which can affect straightness and flatness.

SUMMARY

It is estimated that 85–95% of the cost of a product is fixed in its design. Six design parameters must be considered concurrently by both the design engineer and the foundry engineer. The standards for "optimal" casting geometry are high, but the possibilities for geometry are limitless. Extensive brainstorming geometry is highly recommended through engineering sketching, before committing to a print or solid model.

Casting design is nothing more than an engineering sketch with a sense of size and proportion. Using the "quick method" of sketching cross-sectional areas, Area Moment of Inertia can be estimated with simple building blocks and minimal calculation. Once a value is known, stress can be easily calculated for the chosen cross section. A relative measure for deflection can be easily calculated as well.

Final design would be a solid model, based on at least two or three sketched iterations of combined structural and castable geometry. Detailed structural evaluation could then be done via finite element analysis (FEA). Tweaking the solid model, which is already close to optimal geometry, could easily solve any remaining stress problems. Finally, the solid model could be modified to add risers and a gating system so that computer analysis of solidification and mold filling could verify the geometry chosen for castability.

BIBLIOGRAPHY

Gwyn, M.A., "Cost-Effective Casting Design," *Engineered Casting Solutions,* American Foundry Society, Spring/Summer, 1999, pp 57-66.

Glossary

abrasion—The displacement and/or detachment of metallic particles from a surface from exposure to flowing solids, fluids or gases.

abrasive—Any substance used for abrading, such as grinding, polishing, blasting, etc. Material may be bonded to form wheels, bricks and files or applied to paper and cloth by means of glue. Natural abrasives include emery, corundum, garnet, sand, etc. Main manufactured abrasives are silicon carbide and aluminum oxide. Metallic shot and grit are also used as abrasives in cleaning castings.

acicular structure—A microstructure characterized by needle-shaped constituents.

acid—A chemical term to define a material that gives an acid reaction.

acid bottom or lining (furnace)—The inner bottom or lining of a melting furnace composed of materials having an acid reaction. Materials may be sand, siliceous rock or silica brick.

acid melting—Melting in a furnace with refractory material that has an acid reaction, material may be silica sand, siliceous rock or silica brick.

acid steel—Steel melted in a furnace that has an acid bottom and lining, under a predominantly siliceous slag.

additives—Any material added to molding sand for reasons other than bonding or improvement of bond is considered an additive. Bonds can be of varying types: carbonaceous (seacoal pitch, fuel oil, graphite, Gilsonite): cellulose (wood flour, cereal hulls); fines (silica flour, iron oxide, fly ash); cereals (corn flour, dextrin, sugar); and chemical (boric acid, sulfur, ammonium compounds, diethylene glycol).

aerator—A device for fluffing (or decreasing the density of) and cooling sand by the admixture of air.

AFS grain fineness number (gfn)—The approximate number of meshes per inch in a sieve that allow a sample to pass through if the sand grains were uniform in size. In other words, it is the average of the grains in the sand sample.

aging—A change in properties of metals and alloys that occurs slowly at room temperature and will proceed rapidly at higher temperatures. The change in properties is often, but not always, due to a phase change (precipitation), but never involves a change in chemical composition of the metal or alloy.

air channel—A groove or hole which carries the vent from a core to the outside of a mold.

air control equipment—Any device used to regulate the volume, pressure or weight of air.

air dried (dry)—A core or mold dried in air, without application of heat.

air furnace—A form of reverberatory furnace for melting ferrous and nonferrous metals and alloys. Flame from fuel burning at one end of the hearth passes over the bath to the stack at the opposite end of the furnace.

air hammer—Chipping hammer operated by compressed air.

air hardening—Full hardening of a metal or alloy during cooling in air or other gaseous medium from a temperature above its transformation range.

air hoist—Lifting device operated by compressed air.

air hole—A hole in a casting caused by gas trapped in the metal during solidification.

air injection machine—An early type of diecasting machine in which air pressure acting directly on the surface of molten metal in a closed gooseneck forces the metal into the die.

air setting—The property of some materials to take a permanent set at normal air temperature Examples are gypsum slurry, investment-molding materials, core and mold washes, etc.

airless blast cleaning—A process whereby the abrasive material is applied to the object being cleaned by centrifugal force generated by a rotating-vane-type wheel.

allotropy—Occurrence of an element in two or more modifications. For example, carbon occurs in nature as the hard crystalline diamond, soft flaky crystalline graphite and amorphous coal.

alloy—A substance having metallic properties and composed of two or more chemical elements of which at least one is metal. Usually possesses qualities different from those of the components.

alloying—Procedure of adding elements other than those usually comprising a metal or alloy to change its characteristics and properties.

alloying elements—Elements added to nonferrous and ferrous metals and alloys to change their characteristics and properties.

alpha iron—An allotropic (polymorphic) form of iron, which is magnetic and whose atomic structure is body-centered-cubic lattice. It is soft, ductile, of fair strength and is capable of dissolving a few hundredths of a percent of carbon to form a solid solution.

aluminum—(Al) A metallic element. The atomic weight is 26.98 and melting point is 660C.

amorphous—Having no crystalline form.

anodizing—Forming a conversion coating on a metal surface by electrolytic oxidation with the metal as the anode; most frequently applied to aluminum.

antimony—(Sb) A very brittle metal element, neither malleable nor ductile. Its atomic weight is 121.76 and melting point 630.5C.

Antioch process—Plaster molding process using a mixture of about 50% sand, 40% gypsum, and 8% fibrous talc mixed with water in the proportion of 100 parts by weight material with 50 parts water. After air setting, the mold is placed in an autoclave and subjected to 15 psig steam pressure for 6 to 8 hours. The mold is allowed to rehydrate in air for 14 hours and then baked at 450–475F (232–246C) for 25 to 30 hours.

anti-piping (material)—Usually refers to an insulating material placed on top of a sprue or riser, which keeps the metal in liquid or semi-liquid form for a long period of time and minimizes the formation of the usual conical pipe or shrink in the top of a sprue or riser.

arbor—A metal barrel, frame or plate to support or carry part of a mold or core.

argon—(Ar) An inert gas element with atomic weight of 39.94 and melting point of −189.3C.

arrestor, dust—Equipment for removing dust from air.

as cast—Referring to a casting which has not received finishing (beyond gate removal or sandblasting), or treatment of any kind including heat treatment after casting. (See: finishing)

assembling (assembly) line—Conveyor system where molds or cores are assembled.

atmospheric riser—A riser that uses atmospheric pressure to aid feeding. Essentially, a "blind riser" into which a small core or rod protrudes. The function of the core or rod is to provide an open passage so the molten interior of the riser will not be under a partial vacuum when metal is withdrawn to feed the casting but will always be under atmospheric pressure.

atom—The smallest particle of an element.

austenitic—Usual reference is to an alloy steel or iron with structure at room temperature that is normally composed essentially of austenite. A solid solution of cementite or iron carbide, Fe_3C, in iron.

back draft—Taper or draft that prevents removal of pattern from the mold.

backing board (backing plate)—A second bottom board on which molds are opened.

backing sand—The bulk of the sand in the flask. The sand compacted on top of the facing sand that covers the pattern.

baffle plate—Plate or wall in a firebox or furnace to change direction of the flame.

baked core—A core that has been heated for a sufficient time and temperature to produce the desired physical properties (as opposed to a green-sand core, which is used in the moist state).

baked permeability—Property of a molded mass of sand heated at a temperature above 230F (110C) until dry and cooled to room temperature, to permit passage of gases through it; particularly those generated during pouring of molten metal into a mold.

baked strength—Compressive, shear, tensile or transverse strength of a molded sand mixture when baked at a temperature above 230F (110C) and then cooled to room temperature.

band, inside—A steel frame placed inside a removable flask to reinforce the sand.

banking the cupola—Method of keeping cupola hot and ready for immediate production of hot iron after an unexpected shutdown of several hours. Procedure is to drain all molten iron and slag from the cupola, place extra coke on the top charge and open one or two tuyeres to supply a small natural draft to keep coke combustion going.

bar—A rib in the flask to help hold the sand.

base plate—A plate to which the pattern assemblies are attached and to which a flask is subsequently attached to form the mold container.

basic—A chemical term for a material which gives an alkaline reaction.

basic bottom or lining (furnace)—Inner lining and bottom of a melting furnace composed of materials having a basic reaction. Materials may be crushed burnt dolomite, magnesite, magnesite brick or basic slag.

basic steel—Steel melted in a furnace with a basic bottom and lining under a predominantly basic slag.

basin—A cavity on top of the cope into which metal is poured before it enters the sprue.

batch—Amount or quantity of core or mold sand or other material prepared at one time.

bath—Molten metal on the hearth of a furnace, in a crucible or a ladle.

Baume—Designating or conforming to either of the scales used by the French chemist Antoine Baume in the gradation of his hydrometers for determining the specific gravity of liquids. There are two scales: one for liquids lighter than water and the other for those heavier than water.

bauxite—An ore of aluminum consisting of moderately pure hydrated alumina, $Al_2O_3{\cdot}2H_2O$.

bed charge—The charge of iron placed on the coke bed in a cupola.

bed coke—Coke placed in the cupola well to support the following iron and coke charges.

bedding a core—Placing an irregularly shaped core on a bed of sand for drying.

bed-in—Method of ramming the drag mold without rolling it over.

beehive coke—Coke which is produced in hemispherical ovens about 12 ft in diameter and charged through the top to form a layer of coal 18–24 in. deep. Coke is ignited and air for partial combustion is supplied over the top by doors around the bottom of the ovens. Air burns volatile matter released by coke and during the later stages of carbonization burns some 5–8% of the coke.

bellows—A device operated with both hands, to produce a current of air; some bellows are mechanically operated.

bench—Frame support on which small molds are made.

bench molder—Worker that makes small molds on a molder's bench.

bench rammer—A short rammer used by a bench molder.

bentonite—A widely distributed, peculiar type of clay, which is considered to be the result of devitrification and chemical alteration of the glassy particles of volcanic ash (or tuff). Used in the foundry to bond sand.

Bernoulli's theorem—A theorem which states that in a stream flowing without friction, the total energy in a given amount of the fluid is the same at any point in its path of flow.

Bessemer process—Method of making steel by blowing air through molten pig or carbon-bearing iron contained in a suitable vessel that causes rapid oxidation of silicon, carbon, etc.

binary alloy—An alloy of two metals.

binder—Material to hold the grains of sand together in molds or cores to impart strength or plasticity in a dry state. May be cereal, oil, clay, resin, pitch, etc.

binder, plastic (resin)—Synthetic resin material used to hold grains of sand together in molds or cores; may be phenol formaldehyde or urea formaldehyde thermosetting types.

Blackheart—An American type of malleable iron. The normal fracture shows a velvety black appearance having a mouse-gray rim.

blacking—Carbonaceous material for coating mold or core surfaces.

blast—Air driven into the cupola or furnace for combustion of fuel.

blast cleaning—Removal of sand or oxide scale from castings by the impinging action of sand, metal shot or grit projected under air, water or centrifugal pressure.

blast furnace—Closed-top-shaft furnace for producing pig iron from iron ore.

blast gate—Sliding plate in the cupola blast pipe to regulate the flow of air.

blast meter—Instrument indicates the volume, pressure, or both, of air passing through the blast pipe.

blast pressure—Pressure of air in blast pipe or wind belt of the cupola, depending on location of indicating instrument. Usually given in ounces of water pressure.

blended sand—Mixture of sands of different grain sizes, clay content, etc., to produce one possessing characteristics more suitable for foundry use.

blind riser—An internal riser, which does not reach to the exterior of the mold.

blister—Defect on the surface of a casting appearing as a shallow blow with a thin film of metal over it. In diecasting, it is a surface bubble or eruption caused by expansion of gas (usually results from heating) trapped within the diecasting or beneath the plating on the diecasting.

blow—A casting defect due to trapping of gas in molten or partially molten metal.

blow gun—Valve and nozzle attached to a compressed air line to blow loose sand or dirt from a mold or pattern. Also to apply wet blacking.

blow holes—1) Holes in the head plate or blow plate of a core-blowing machine through which sand is blown from the reservoir into the core box. 2) Irregular shaped cavities with smooth walls produced in a casting when gas is entrapped during mold filling. The gas sources may be air, binder decomposition products or gases dissolved in the molten steel.

blow pipe—A small pipe or tube through which the breath is blown to remove loose sand from small molds.

blower, core or mold—Machine using compressed air to inject sand into a corebox or a flask.

blower—Machine or device for supplying air under pressure to the melting unit.

blowplate—Plate on the bottom of the sand hopper on core or mold blower machines that contains holes through which the sand is blown into the corebox or flask.

bod, bott—A piece of clay or other material to stop the flow of metal from the taphole.

bond—Cohesive material in sand.

bond clay—Any clay suitable for use as a bonding material in molding sand.

bond strength—Resistance of foundry sand to deformation.

boric acid—Inhibitor used in facing sand for magnesium-base and aluminum-base alloys high in magnesium, to prevent reaction with moisture in the sand.

borings—Metal in chip form resulting from machining operations.

boron—(B) One of the metallic elements. The atomic weight is 10.82, melting point is 2300C. In the form of borax and boric oxide, it is used as a flux in nonferrous metallurgy and in the form of an alloy with other elements, as an addition to ferrous alloys.

boron trichloride—A product used for degasification of aluminum alloys.

boss—A projection of circular cross-section on a casting. Usually intended for drilling and tapping for attaching parts.

bottom board—Board supporting the mold.

bottom doors—Doors underneath the cupola.

bottom pour ladle—Ladle in which metal, usually steel, flows through a nozzle in the bottom.

bottom pour mold—Mold gated at the bottom.

bottom sand—Layer of molding sand rammed into place on the doors at the bottom of a cupola.

bracket—Strengthening strip, rib or projection on a casting. Usually used to prevent hot tearing.

branch core—Part of a core assembly.

branch gate—Two or more gates leading into the casting cavity.

brass—Copper-base alloy with zinc as the major alloying element.

brazing—Joining metals and alloys by fusion of nonferrous alloys with melting points above 800F (427C), but lower than those of the materials being joined.

breast—Area surrounding the taphole of a melting furnace.

breeze—Coke or coal screenings.

bridge—Material adhering to the cupola wall that slows or prevents descent of the stock charges.

Brinell hardness—The value of hardness of a metal on an arbitrary scale representing kg/mm^2, determined by measuring the diameter of the impression made by a ball of given diameter applied under a known load. Values are expressed in Brinell hardness numbers, HB or Bhn

briquettes—Compact cylindrical or other shaped blocks formed of finely divided materials by incorporation of a binder, by pressure or by both. Materials may be ferroalloys, metal borings or chips, silicon carbide, etc.

bronze—Copper-base alloy, with tin as the major alloying element.

buckle—1) Bulging of a large flat face of a casting; in investment casting, caused by dip coat peeling from the pattern. 2) An indentation in a casting, resulting from expansion of the sand, may be termed the start of an expansion defect.

built-up plate—A pattern plate with the cope pattern mounted or attached to one side with the drag on the other. (See: matchplate)

bumper—Machine for ramming sand in a flask by repeated jarring or jolting action.

burden—Term used to designate the metal charge for a melting furnace. It is also used in cost accounting to indicate certain additional charges to be included in assessing costs in the different areas.

burner—A device that mixes fuel and air to provide perfect combustion when the mixture is burned. Types include acetylene, oil, gas, powdered coal, stoker, etc.

burnishing—Developing a smooth finish on a metal by tumbling or rubbing with a polishing tool.

burn-on sand—Sand adhering to the surface of the casting that is extremely difficult to remove.

burn-out—Usually refers to removal of the disposable wax or plastic pattern in the investment-molding process by heating the mold gradually to a sufficiently high temperature to consume any carbonaceous residues.

bushing—A sleeve, metallic or nonmetallic, usually removable or replaceable, which is placed in a body to resist wear, erosion, etc.

CAD—Computer Aided Design.

cadmium—(Cd) Soft bluish-gray metal with atomic weight of 112.41 and melting point of 320.9C.

CAE—Computer Aided Engineering.

calcium—(Ca) A metallic element with atomic weight of 40.08 and melting point of 850C.

calcium-aluminum-silicon—An alloy composed of 10–14% calcium, 8–12% aluminum and 50–53% silicon, used for degasifying and deoxidizing steel.

calcium boride—An alloy of calcium and boron corresponding (when pure) to the formula CaB_6 containing about 61% boron and 39% calcium and used in deoxidation and degasification of nonferrous metals and alloys.

calcium carbide—A grayish-black, hard crystalline substance made in the electric furnace by fusing lime and coke. Addition of water to calcium carbide forms acetylene and a residue of slaked lime.

calcium-manganese-silicon—An alloy containing 17–19% calcium, 8–10% manganese, 55–60% silicon and up to 14% iron, used as a scavenger for oxides, gases and nonmetallic impurities in steel.

calcium silicon—An alloy of calcium, silicon and iron, containing 28–35% calcium, 60–65% silicon and 6% maximum iron, used as a deoxidizer and degasifier for steel and cast iron. Sometimes called calcium silicide.

CAM—Computer Aided Manufacturing.

captive foundry—One that is part of a manufacturing plant and whose products (castings) are used in the plant as parts of finished objects.

carbide—A compound of carbon with a more positive element, such as iron. Carbon unites with iron to form iron carbide or cementite, Fe_3C.

carbon—Element occurring as diamond and as graphite. Carbon reduces many metals from their oxides when heated with the latter and small amounts of carbon greatly affect the properties of iron. The symbol is C, atomic weight 12.011.

carbon boil—Refers to the practice of adding oxidizing agents such as iron ore or oxygen to molten steel in the furnace to react with carbon and create a boiling action. In addition to reducing the carbon content, it removes occluded gases such as hydrogen, oxygen and nitrogen.

carbon dioxide process—A process for hardening molds or cores in which carbon dioxide gas is blown through dry clay-free silica sand to precipitate silica in the form of a gel from the sodium silicate binder. Usually expressed as CO_2.

carbon equivalent (CE)—Relationship of total carbon, silicon and phosphorus in gray iron, expressed by the formula: $CE = TC\% + Si\% / 3 + P\% / 3$.

carbon steel—Steel that owes its properties chiefly to various percentages of carbon without substantial amounts of other alloying elements; also known as ordinary steel or straight carbon or plain carbon steel.

carburizing—A form of case hardening that produces a carbon gradient inward from the surface, enabling the surface layer to be hardened by either quenching directly from the carburizing temperature or by cooling to room temperature, then reaustenitizing and quenching.

cast iron—Generic term for a series of alloys of iron, carbon and silicon, in which the carbon is in excess of the amount that can be retained in solid solution in austenite at the eutectic. When cast iron contains a specially added element or elements in amounts sufficient to produce a measurable modification of the physical properties under consideration, it is called alloy cast iron. Silicon, manganese, sulfur and phosphorus, as normally obtained from raw materials, are not considered as alloy additions.

cast plate—Metal plate, usually aluminum, cast with the cope pattern on one side and the drag pattern on the other. (See: matchplate)

cast-weld assembly—Welding one casting to another to form a complete assembly.

casting (noun)—The metal shape, exclusive of gates and risers, that is obtained as a result of pouring molten metal into a mold.

casting (verb)—Act of pouring molten metal into a mold.

casting, machine (verb)—Process of casting by machine.

casting, open sand (noun)—Casting poured into an uncovered mold.

casting layout—A check of dimensions against applicable drawings and specifications.

casting strains—Strains resulting from internal stresses created during cooling of a casting.

cavity, mold or die—Impression or impressions in a mold or die that give the casting its shape.

cement—Mineral substances in finely divided form, which are hardened through chemical reaction or crystallization. A common one is Portland cement.

cement molding—Process in which the sand bonding agent is a type of Portland cement that develops high strength early in the hardening stage. Approximately 13 lb of cement, 6 lb of water and 100 lb of clay-free sand are mixed together. Mixture must be used within 3–4 hours. Molds are air dried for 72 hours before use.

cement, refractory—Highly refractory material in paste or dry form, ready to be mixed with water, which may be used as a mortar, a patching material or to form a complete lining in a furnace or other unit where high temperatures are encountered.

cementation—Process of introducing elements into the outer layer of metal objects by mean of high-temperature diffusion.

cementite—Iron carbide, Fe_3C, a hard, brittle, crystalline compound observed in the microstructure of iron-base alloys.

centrifugal casting—Process of filling molds by pouring the metal into a sand or metal mold revolving about either its horizontal or vertical axis, or pouring the metal into a mold that subsequently is revolved before solidification of the metal is complete. Molten metal is moved from the center of the mold to the periphery by centrifugal action.

ceramic mold—Mold in which the refractory and binder are such that when fired at high temperature, a rigid structure is formed. Mold can be made in a flask or in the form of a shell.

cereal—Substance derived principally from corn flour, which is added to core and molding sands to improve their properties for casting production.

cerium—Metallic element, malleable and ductile, most abundant of rare-earth group. Atomic weight 140.13, specific gravity 7.04, hardness (Moh's) about 2, melting point 640C. Has exceptionally strong affinity for oxygen, sulfur, hydrogen, nitrogen, etc. Readily decomposed silicates, forming cerium oxide and cerium silicide. Nodularizing agent for some cast iron, also said to "neutralize" effect of some subversive elements when producing nodular cast iron with magnesium additions.

chamfer—breaking or beveling the sharp edge or angle formed by two faces of a piece of wood or other material.

chamotte—Coarsely graded refractory material prepared from calcined clay and ground firebrick mulled with raw clay, used in steel foundries.

chaplet—A small metal insert or spacer used in molds to provide core support during the casting process. The molten metal solidifies around the chaplet and fuses it into the finished casting.

charcoal (pig) iron—Pig iron reduced in a blast furnace, using charcoal as the fuel.

charge—A given weight of metal and other primary ingredients (i.e., coke and limestone in a cupola heat) introduced into a furnace to fully utilize its capacity.

charging crane—System for charging the melting furnace with a crane.

charging door—Opening through which the furnace is charged.

charging floor—Floor from which the furnace is charged.

charging machine—Machine for charging the furnace, particularly the open hearth.

Charpy test—A pendulum type of impact test in which a specimen, supported at both ends as a simple beam, is broken by the impact of the falling pendulum. Energy absorbed in breaking the specimen, as determined by the decreased rise of the pendulum, is a measure of the impact strength of the metal.

chill (external)—Metal, graphite or carbon blocks that are incorporated into the mold or core to locally increase the rate of heat removal during solidification and reduce shrinkage defects.

chill (internal)—A metallic device/insert in molds or cores at the surface of a casting or within the mold to increase the rate of heat removal, induce directional solidification and reduce shrinkage defects. The internal chill may then become a part of the casting.

chill coating—A material applied to metal chills to prevent oxidation or other deterioration of the surface that might result in blows when molten metal contacts the chills.

chill coils—Chills made of steel wire formed into helical coils or spirals.

chill nails—Chills in the form of nails.

chill test—Method of determining the suitability of a gray iron for specific castings through its chilling tendency, as measured from the tip of a wedge-shaped test bar.

chill zone—Area of a casting in which chilling occurs, as long sharp edges or exterior corners.

chilled iron—Cast iron poured against a chill to produce a hard, unmachinable surface.

chip (verb)—To remove extraneous metal from a casting with hand or pneumatically operated chisels.

chlorination—A refining or degasification process, wherein dry chlorine gas is passed through molten aluminum-base and magnesium-base alloys to remove entrapped oxides and dissolved gases.

choke—Restriction in a gating system to control the flow of metal beyond that point.

chromium—(Cr) Alloying element used as a carbide stabilizer. (See: ferrochromium) Atomic weight is 51.996, melting point is 1800C.

Chvorinov's rule—A rule that states that solidification time is proportional to the square of the volume of the metal and inversely proportional to the square of the surface area or t (solidification time) = KV^2/SA^2.

clamp—A device for holding parts of a mold, flask, corebox, etc., together.

clamp-off—Indentation on a casting surface due to displacement of sand in the mold.

clay, refractory—A clay which, in addition to its capability of resisting high temperatures, also possesses strong bonding power.

clay wash—Clay and water mixed to a creamy consistency.

cleaning—Process of removing sand, surface blemishes, etc., from the exterior and interior surfaces of castings. Includes degating, tumbling or abrasive blasting, grinding offgate stubs, etc.

CMM—Coordinate Measuring Machine. Used to verify pattern and casting dimensions.

CNC—Computer Numerical Control. Machining is assisted by the aid of computer programmed equipment.

CNC machine tools—Computer Numerical Controlled machine tools.

coalescence—Agglomeration of fine particles into a mass. Also growth of particles of a dispersed phase by solution and reprecipitation. In addition, grain growth by absorption of adjacent undistorted grains.

cobalt—(Co) A blue-white metal. Atomic weight is 58.94, melting point is 1490C.

cobalt 60—Radioisotope of the element cobalt used in radiographic examination of castings and for determining height of molten metal in cupola well.

coke—A porous, gray, infusible product resulting from the dry distillation of bituminous coal that drives off the volatile matter. Used as a fuel in cupola melting. Petroleum coke results from distillation of petroleum and pitch coke from distillation of coal tar pitch. (See: beehive and by-product coke)

coke bed—First layer of coke placed in the cupola. In addition, the coke used as the foundation in constructing a large mold in a flask or pit.

cold blast pig iron—Pig iron produced in a blast furnace without the use of the heated air blast.

cold shortness—Brittleness when metal is at a low temperature.

cold shut—A surface imperfection due to unsatisfactory fusion of metal.

coldbox process—Any core binder process that uses a gas or vaporized catalyst to cure a coated sand while it is in contact with the core box at room temperature.

collapsibility—The requirement that a sand mixture break down under the pressures and temperatures developed during casting, in order to avoid hot tears or facilitate the separation of the sand and the casting.

colloids—Finely divided material, less than 0.5 micron (0.00002 in.) in size; gelatinous, highly absorbent and sticky when moistened.

columnar structure—Coarse structure of parallel columns of grains caused by highly directional solidification resulting from sharp thermal gradients.

combination die—A diecasting die having two or more cavities of dissimilar parts. (See: multiple-cavity die)

combined carbon—The carbon in iron or steel combined with other elements and therefore not in the free state as graphite or temper carbon.

combustibles—Materials capable of combustion; inflammable.

compressive strength (yield)—The maximum stress in compression that can be withstood without plastic deformation or failure.

conduction—The transmission of heat, sound, etc., by the transferring of energy from one particle to another.

conductivity—The quality or power of conducting or transmitting heat, electricity, etc.

conductivity (thermal)—The quantity of heat that flows through a material measured in heat units per unit time per unit of cross-sectioned area per unit of length; **(electrical)** the quantity of electricity that is transferred through a material of known cross-section and length.

continuous annealing furnace—Furnace in which castings are annealed or heat treated by being passed through different zones kept at constant temperatures.

contraction—Act or process of a casting becoming smaller in volume and/or dimensions during the solidification of the metal or alloy that composes the casting.

controlled cooling—Process by which a metal object is cooled from an elevated temperature; a predetermined manner of cooling to avoid hardening, cracking or internal damage.

convection—The motion in a fluid resulting from the differences in density. In heat transmission, this meaning has been extended to include both forced and natural motion or circulation.

converter—Vessel for refining molten metal by blowing air through it. Used in making steel from molten cast iron and in refining copper.

cooling curve—A curve showing the relationship between time and temperature during the solidification and cooling of a metal sample. Since most phase changes involve evolution or absorption of heat, there may be abrupt changes in the slope of the curve.

cope—The upper or topmost section of a flask, mold or pattern.

copper—(Cu) A nonferrous element. Atomic weight is 63.54, melting point is 1083C.

core—Separable part of the mold, usually made of sand and generally baked, to create openings and various shaped cavities in the castings. Also used to designate the interior portion of an iron-base alloy that, after case hardening, is substantially softer than the surface layer or case.

core, Washburn—A thin core constricting the riser where it is attached to the casting. It heats quickly, creating a hot spot that prevents temperature drop in metal passing through and promotes feeding of the casting. In many cases, it speeds riser removal.

core assembly—An assembly made from a number of cores.

core binder—Any material used to hold the grains of core sand together.

core blowing machine—Machine that rams the core by blowing sand into the corebox.

core drier (dryer)—Sand or metal supports to keep cores in shape during baking.

core extruder—A special shell-core-making machine that produces a continuous length of cores, usually of cylindrical cross-section.

core filler—Material used in place of sand in the interiors of large cores—coke, cinder, sawdust, etc.—usually added to aid collapsibility.

core float—A casting defect caused by core movement toward the cope surface of the mold, because of core buoyancy in liquid steel, resulting in a deviation from the intended wall thickness.

core jig (fixture)—Device in which a number of cores are assembled outside the mold, then used to locate the assembly in the proper position in the mold.

core machine—Machine for making cores.

core oil—Linseed-base or other oil used as a core binder.

core oven—An oven for baking cores.

core plate—A plate or board made of metal or heat-resisting material on which certain types of cores are baked.

core rod—A wire or rod of steel (or iron) used to reinforce and stiffen the core internally.

core sand—Sand for making cores.

core sand mixer—Equipment in which cores are made.

core setter—An operator or machine for placing cores in molds.

core setting jig/gage—A device used to help position a core in the mold.

core shift—Defect resulting from movement of the core from its proper position in the mold cavity.

core shooter—A device using low air pressure to fluidize the sand mix that is released quickly in such a way as to force it into a core box.

core vents—1) Holes made in the core for the escape of gas. 2) A metal screen or slotted piece used to form the vent passage in the core box employed in a core-blowing machine. 3) A wax product, round or oval in form, used to form the vent passage in a core.

core wash—A liquid suspension of a refractory material applied to cores and dried (intended to improve surface of casting).

corebox—The wooden, metal or plastic box used to produce cores.

coremaker—A person who makes cores.

coreprint or core print(s)—Portions of a pattern that locate and anchor the core in the proper position in the mold.

coring—Variable composition in solid-solution dendrites: the center of the dendrite is richer in one element, as shown by the pertinent solidus-liquidus lines in a phase diagram.

corrosion—1) Gradual chemical or electrochemical attack on a metal by atmosphere, moisture or other agents. 2) Chemical attack of furnace linings by gases, slag, ash or other fluxes occurring in various melting practices.

cover core—A core set in place during the ramming of a mold to cover and complete a cavity partly formed by the withdrawal of a loose part of the pattern. Also used to form part or all of the cope surface of the mold cavity. A core placed over another core to create a flat parting fine.

crack, cold—Appears in a casting after solidification and cooling due to excessive strain generally resulting from nonuniform cooling.

crack, hot—Developed in a casting before it has cooled completely and usually due to some part of the mold restraining solid contraction of the metal. (See: tear, hot)

creep—Time rate of deformation continuing under stress intensities well within the yield point, proportional limit or the apparent elastic limit for the temperature.

critical cooling rate—minimum rate of continuous cooling just enough to prevent undesired transformations.

critical points (temperatures)—Temperatures at which changes in the phase of a metal take place and are determined by the liberation of heat when the metal is cooled and by the absorption of heat when the metal is heated, resulting in halts or arrests on cooling and heating curves.

crucible—A ceramic pot or receptacle of graphite-clay, clay or other refractory material in which metal is melted. This term is sometimes applied to pots of cast iron, cast or wrought steel.

crucible furnace—Furnace in which metal is melted in crucibles.

crush—Casting defect appearing as an indentation in the surface due to displacement of sand in the mold: usually at the joint surfaces.

crystallization—The formation of crystals by the atoms assuming definite positions in the crystal lattice, e.g. when a metal solidifies.

cupola—A cylindrical, straight shaft furnace (usually lined with refractories) for melting metal in direct contact with coke by forcing air under pressure through openings near its base.

cupola, basic—Cupola with refractory lining that has a basic reaction, usually magnesite and is operated with slags high in lime. Lining may be a neutral material like carbon, used with high lime slags.

cupola, hot blast—Cupola in which the air blast is heated to temperatures from 400–1000F (204–538C).

cupola, water-cooled—Cupola in which melting zone and tuyeres are cooled with water. Cooling of melting zone may be internally through jackets or steel tubing under the refractory lining. Cooling is also accomplished externally by water flowing down the outer shell.

cupola blower—A machine that compresses a large volume of air at low pressure for operation of the cupola.

cure—To harden.

cutoff machines, abrasive—A machine using a thin abrasive wheel and employed in cutting off gates and risers from castings or in similar operations.

cuts—Defects in castings resulting from erosion of the sand by the molten metal pouring over the mold or core surface.

cutter, gate—A piece of sheet metal or other tool for removing a portion of the sand in a mold to form the gate or metal entrance into the casting cavity.

cutter, sprue—A piece of metal tubing or other tool used to remove a portion of the sand from a mold to form the sprue or passage from the exterior of the mold to the gate. In addition, a machine used for shearing sprues and gates from castings.

damping capacity—The ability to absorb vibration. More accurately defined as the amount of work dissipated into heat by a unit volume of material during a reverse cycle of unit stress.

datum points—In layout and machining operations, the reference points that define the datum plane from which dimensions are measured.

decarburization—Loss of carbon from the surface of a ferrous alloy as a result of heating in a medium (usually oxygen) that reacts with carbon.

defect—A discontinuity in the product whose severity is judged unacceptable in accordance with the applicable product specification.

deformation test—An AFS test using an instrument such as the Dietert Universal Sand-Strength Testing machine (with deformation accessory) to determine the amount, in inches, that the sand specimen is compressed before it ruptures.

degasifier—A material employed for removing gases from metals and alloys.

delta iron—An allotropic (polymorphic) form of iron, stable above 2550F (1399C), crystallizing in the body-centered-cubic (BCC) lattice.

dendrite—A crystal formed during the solidification of a metal or alloy characterized by a branching structure like that of a fir tree.

deoxidation—Removal of oxygen from molten metal, usually accomplished by adding materials with a high affinity for oxygen, the oxides of which are either gaseous or readily form slag.

deoxidizer—A material used to remove oxygen or oxides from metals and alloys.

desulfurizer—A material used to remove sulfur from molten metals and alloys.

dezincification—Corrosion of some copper-zinc alloys, involving loss of zinc and the formation of a spongy porous copper.

die—A metal form used as a permanent mold for diecasting or for a wax pattern in investment casting

diecasting (noun)—Casting resulting from diecasting process.

diecasting (verb)—A process in which molten metal is injected at high velocity and pressure into a mold (die) cavity.

diecasting, cold chamber—Type of casting made in a diecasting machine in which the metal injection mechanism is not submerged in the molten metal.

diecasting, hot chamber—Type of casting made in a diecasting machine in which the metal injection mechanism is submerged in the molten metal.

dielectric baking—Baking of cores and molds in a field of high-frequency electric current generated by dielectric equipment; employed with resin-bonded cores.

diffusion—Movements of atoms within a solution. Net movement is usually from regions of high concentration to regions of low concentration to achieve homogeneity of the solution that may be a solid, gas or liquid.

dilatometer—An instrument for measuring the expansion or contraction of a metal sample during heating and cooling.

dimension—A dimension is a numerical value associated with an appropriate unit of measurement. It defines the size, shape and location of features.

dimensional stability—Ability of a casting to remain unchanged in size and shape under ordinary atmospheric conditions.

dimensional tolerance—A tolerance is the permissible variation in size of a feature placed on the specified dimension. It is usually assigned to a dimension based on limitations inherent in production processes and equipment. It can be expressed in three ways: bilateral tolerance, unilateral tolerance and limit dimensions.

direct-arc furnace—Electric furnace in which the material is heated directly by an arc established between the electrodes and the work.

directional solidification—Refers to the arrangement of a solidification pattern in a casting by establishment of high temperature gradients, whereby solidification of the metal begins at the point farthest from the metal entrance or sprue and the metal progressively freezes or solidifies to and including the sprue. (Also see: progressive solidification)

dirt—Indefinite term referring to any extraneous material entering a mold cavity and usually forming a blemish on the casting surface.

dowel—A wooden or metal pin used in the parting surface of patterns and coreboxes to locate and to assure correct registry.

downsprue (sprue, downgate)—The first channel, usually vertical, which the molten metal enters.

draft—Taper on the vertical sides of a pattern or corebox that permits the core or sand mold to be removed without distorting or tearing of the sand.

drag—The lower or bottom section of a mold or pattern.

draw, surface—Appearance of shrink on the upper surface of a casting.

draw bar—A bar used for lifting the pattern from the mold.

draw plate—A plate attached to a pattern to facilitate drawing of a pattern from the mold.

drawback—A part of the mold, made of green sand, which may be drawn back to clear overhanging portions of the pattern to facilitate removal of the pattern.

drawing—Removing the pattern from the sand.

dropping bottom—Removal of the supporting props under the cupola bottom door to permit emptying of the remaining contents.

dross—Metal oxides, etc., on or in a metal or alloy.

dry and baked compression test—A sand test to determine the maximum compressive stress that a baked sand mixture is capable of developing.

dry permeability—The property of a molded mass of sand, bonded or unbonded, dried at 220–230F (105–110C) and cooled to room temperature, that allows passage of gases resulting during pouring of molten metal into a mold.

dry-sand mold—A mold made of prepared molding sand dried thoroughly before being filled with metal.

dry strength—Maximum strength of a sand mixture that has been dried and cooled to room temperature for testing. Value may be in compression, shear, tensile or transverse strength.

ductile iron—Nodular or spheroidal graphite cast iron produced by residual magnesium remaining in the iron after ladle addition of magnesium.

ductility—The property permitting permanent deformation by stress in tension with rupture.

duplexing—Term usually used in reference to melting metals or alloys in one type of furnace and transferring to another for holding, refining, etc. Common in the malleable field, where charges are melted in a cupola and transferred to air or electric furnaces for slight reduction of carbon and an increase in temperature.

dye penetrant—Penetrant used in crack detection, which has a dye added to make it more readily visible under normal or black-lighting conditions. In the case of normal lighting, the dye is usually red and nonfluorescent. With black lighting, the dye is fluorescent and yellow-green in color.

ejector marks—Marks left on diecastings by the ejector pins, which may be raised or depressed from the surface of the casting.

ejector pins—Movable pins in pattern dies that help remove patterns from the die or to eject diecastings from the die.

ejector plate—Movable plate beneath a shell-molding pattern containing the pins for lifting or ejecting the hardened, resin-bonded shell mold from the pattern.

elasticity—The property of recovering original shape and dimensions upon removal of a stress.

elongation—Amount of permanent extension near the fractures in the tensile test; usually expressed as a percentage of original gage length.

embrittlement—Loss of ductility of a metal due to a chemical or physical change.

EPC (Expendable Pattern Casting)—(See: Lost Foam Casting process)

epoxy resin—A plastic resin and hardener that sets or hardens itself at room temperature to form a new chemical, used as an adhesive or workable material.

equilibrium—Dynamic condition of balance between atomic movements where the result is zero, a stable condition.

equilibrium diagram—(See: phase diagram)

erosion—Abrasion of metal or other material by liquid or gas.

erosion scab—Casting defect occurring where the metal has been agitated, boiled or has partially eroded away the sand, leaving a solid mass of sand and metal at that particular spot.

etchant—A solution for the chemical etching of the polished surface of a metal specimen to reveal macro- or microstructures.

ethyl silicate—Light brown liquid consisting predominantly of tetraethyl silicate with some polysilicates that can be hydrolyzed with water to form alcohol and silicic acid; the latter, in turn, dehydrates to an amorphous form of silica extremely resistant to most chemicals and heat. Used as a bonding agent in investment molding.

eutectic—The alloy which has the lowest melting point possible for a given composition.

eutectic reaction—Reaction in which a liquid solution solidifies or transforms at constant temperature to form a solid mass made up of two kinds of crystals.

eutectoid—A solid solution of any series which cools without change to its temperature of final composition.

Evaporative Pattern Casting/Expendable Pattern Casting/EPC—(See: Lost Foam Casting Process)

exothermic reaction—A reaction that produces heat.

expandable polystyrene (EPS)—A foamy plastic composed of bonded beads, density between 1 and 1.2 lb/per ft, compressive strength 13–15 psi.

expanded polyurethane—A lightweight expanded plastic polymer whose foamed texture results from the trappings of CO_2 evolved during production. In resin form, it is used for coating and adhesive.

expansion, sand—Dimensional increase that sand undergoes when subjected to elevated temperature conditions.

expansion scabs—Rough thin layers of metal partially separated from the body of the casting by a thin layer of sand and held in place by a thin vein of metal.

expendable pattern—In investment molding, the wax or plastic pattern that is left in the mold and later melted and burned out.

external chills—Various materials of high heat capacity, such as metals, graphite, etc., forming parts of the walls of the mold cavity to promote rapid heat extraction from molten metal.

fabrication—The joining, usually by welding, of two or more parts to produce a finished assembly The components of the assembly may be a combination of cast and wrought materials.

facing—Refractory material applied to the face of a mold.

facing sand—Specially prepared sand in the mold adjacent to the pattern to produce a smooth casting surface.

false cheek—A cheek used in making a three-part mold in a two-part mold.

Fast Freeform Fabrication (FFF)—(See: rapid prototyping)

fatigue crack—A fracture starting from a nucleus where there is an abnormal concentration of cyclic stress and propagating through the metal. Surface is smooth and frequently shows concentric markings with a nucleus as the center.

feed head—A reservoir of molten metal provided to compensate for contraction of metal as it solidifies, by the feeding down of liquid metal to prevent voids. Also called a riser.

feeder—Sometimes referred to as a riser, it is part of the gating system that forms the reservoir of molten metal necessary to compensate for losses due to shrinkage as the metal solidifies.

feeding—The process of supplying molten metal to compensate for volume shrinkage while a casting is solidifying.

ferric oxide—Red iron oxide, Fe_2O_3, commonly available as hematite ore. Used in ground form in cores and molds to increase hot compressive strength.

ferrite—Iron practically carbon-free. It forms a body-centered-cubic (BCC) lattice and may hold in solution considerable amounts of silicon, nickel or phosphorus; hence, the term is also applied to solid solutions in which alpha or delta iron is the solvent.

ferroalloys—Alloys consisting of certain elements combined with iron and used to increase the amount of such elements in ferrous metals and alloys. In some cases, the ferroalloys may serve as deoxidizers.

ferroaluminum—An alloy of iron and aluminum containing about 20% iron and 80% aluminum.

ferrochromium—An alloy of iron and chromium available in several grades containing 66–72% chromium and 0.06–7% carbon.

ferromanganese—An alloy of iron and manganese containing 78–82% manganese.

ferromolybdenum—An alloy of iron and molybdenum containing 58–64% molybdenum.

ferrophosphorus—An alloy of iron and phosphorus containing about 70% iron and 25% phosphorus.

ferrosilicon—An alloy of iron and silicon available in several grades containing 14–20% silicon, 42.5–52% silicon, 69.5–82% silicon, 82–88% silicon and 88–95% silicon.

ferrostatic pressure—Pressure induced by a head of liquid iron or steel.

fillet—A concave corner piece used on foundry patterns, a radiused joint replacing sharp inside corners.

fin—A thin piece of metal projecting from a casting at the parting line or at the junction of cores or of cores and mold, etc., due to an imperfect joint in the mold.

fines—Sand grain sizes substantially smaller than the predominating grain size in a molding sand; also material remaining on 200- and 270-mesh sieves and pan after tests for grain size and distribution.

finish (machine)—Amount of metal allowed for machining.

finish (verb)—The handwork on a mold after the pattern has been withdrawn.

finish allowance—The amount of stock left on the surface of a casting for machining.

Finite Element Analysis (FEA)—A computerized numerical analysis technique used for solving differential equations to primarily solve mechanical engineering problems relating to stress analysis.

firebrick—Brick made of refractory clay or other material that resists high temperatures.

fireclay—A type of clay that is resistant to high temperatures.

flash—Thin fin or web of metal extending from the casting along the joint line because of poor contact between cope and drag molds.

flask—A metal or wood frame used for making or holding a sand mold. The upper part is the cope and the bottom half is the drag.

flask, slip—A removable flask that can be stripped vertically from the mold.

flask, snap—A hinged flask that can be removed from the mold after completion.

flask, tight—Flask that remains on the mold.

flask clamp—A device for holding together the cope, drag and cheek of a flask.

flask pin(s) (guides)—Guides used to accurately align the matchplate pattern in the flask and flask-to-flask location.

floor molding—Used where the pattern size prohibits the use of molding machines. The pattern is bolted to the floor and the assembled mold is moved by crane.

flowability—Property of a foundry sand mixture that enables it to fill pattern recesses and move in any direction against pattern surfaces under pressure.

fluidity—The ability of molten metal to flow. A common device used to measure fluidity is the fluidity spiral.

fluorescent crack detection—Application of penetrating fluorescent liquid to a part, then removing the excess from the surface, which is then exposed to ultraviolet light. Cracks appear as fluorescent lines.

flux—Any substance used to promote fusion. Also any material that reduces, oxidizes or decomposes impurities so that they are carried off as slags or gases.

foam plastics—Resinoids in spongy form, as polystyrene. The sponge maybe flexible or rigid, the cells closed or interconnected, with a density from that of the parent resin to, in a few cases, 2 lb/ft^3.

followboard—A board shaped to the parting line of the mold.

founding—Art and science of melting and pouring metals and alloys into castings.

foundry returns—Metal in the form of sprues, gates, runners, risers and scrapped castings, with known chemical composition, that are returned to the furnace for remelting. Sometimes referred to as "revert."

freezing—Term used to denote the solidification process.

furans—Generic term for a family of chemical compounds including furfuryl and furfuryl alcohol used as binders for core sands.

fusion—Change from a solid to a fluid state caused by application of heat.

gamma iron—One of the allotropic (polymorphic) forms of iron that crystallizes in the face-centered-cubic (FCC) lattice form. When pure, its range of stability is from 2552 to 1670F (1400–910C).

gas holes—Rounded cavities caused by generation or accumulation of gas or entrapped air in a casting; holes may be spherical, flattened or elongated.

gas porosity—A condition existing in a casting caused by the trapping of gas in the molten metal or by mold gases evolved during the pouring of the casting.

gate—Specifically, the point at which molten metal enters the casting cavity. Sometimes employed as a general term to indicate the entire assembly of connected columns and channels carrying the metal from the top of the mold to that part forming the casting cavity proper. Term also applied to pattern parts that form the passages or to the metal that fills them.

gate (ingate)—The portion of the runner where the molten metal enters the mold cavity.

gated patterns—One or more patterns with gating systems attached.

gating system—A channel, or network of channels, through which the molten metal flows into the mold cavity. It is typically composed of a pouring cup, sprue and runner.

geometry driver—The critical features that may affect or determine the tooling path selection.

Gilsonite—Natural black lustrous asphalt found in the Uinta Mountains in Utah and known as uintaite. It is used as a carbonaceous addition to foundry sands.

grain fineness number (gfn)—A system developed by AFS for expressing the average grain size of a given sand. It approximates the number of meshes per inch of that sieve that would just pass the sample if its grains were of uniform size.

grain refiner—Any material added to a liquid metal or alloy or a treatment that produces a finer grain size in the subsequent solid.

grains—Crystals in metals and alloys.

granular pearlite—A structure formed from ordinary lamellar pearlite by long annealing at a temperature below but near to the critical point, causing the cementite to spheroidize in a ferrite matrix.

graphite—Native carbon in hexagonal crystals, also foliated or granular massive, of black color with metallic luster and soft. Used for crucibles, foundry facings, lubricants, etc. Also made artificially by passing alternating current through a mixture of petroleum coke and coal tar pitch.

graphite, primary—Carbon precipitated as graphite flakes while the iron cools through the freezing eutectic in which austenite, graphite, molten iron and carbide exist together. Usually with reference to white fracture cast iron.

graphite, secondary—Graphite formed by decomposition of austenite during slow cooling of cast iron.

graphitization—The decomposition of carbide to give free carbon as graphite or as temper carbon.

graphitizer—Any substance, such as silicon, titanium, aluminum, etc., which promotes the formation of graphite in cast iron compositions.

green permeability—Property of a molded mass of sand in its tempered condition that is a measure of its ability to permit the passage of gases through it.

green sand—A naturally bonded sand or a compounded molding sand mixture that has been tempered with water for use while still in the damp or wet condition.

green sand core—Core used in the green state; not baked.

green strength—Tenacity (compressive, shear, tensile or transverse) of a tempered sand mixture at room temperature.

grinding—Removing gate stubs, fins and other projections on castings by an abrasive wheel.

growth—With reference to cast iron, permanent increase in volume that results from continued or repeated cyclic heating and cooling at elevated temperatures. For unalloyed iron, temperature is in excess of 900F (482C) and growth is caused by decomposition or graphitization of carbides and by oxidation of the graphite.

guide pin—The pin used to locate the cope in the proper place on the drag.

gypsum cement—Calcined calcium sulfate, commonly called plaster of paris.

hand ladle or shank—A small ladle carried by one worker.

hardness—Resistance of a material to indentation as measured by such methods as Brinell, Rockwell and Vickers. The term hardness also refers to the ability of the metal to resist scratching, abrasion or cutting. It is related to yield strength (YS) and ultimate tensile strength (UTS).

head—Pressure exerted by a fluid such as molten metal. Also used as a term for a riser.

hearth—That portion of a reverberatory furnace on which the molten metal or bath rests.

heat—A single batch of molten metal tapped from a furnace that is the result of melting and refining a charge.

heat transfer—Transmission of heat from one body to another by radiation, convection or conduction.

heat treatment—A combination of heating and cooling operations timed and applied to a metal or alloy in the solid state in a manner that will produce desired mechanical properties.

high pressure mold—A strong high-density mold.

holding furnace—Usually a small furnace for maintaining molten metal at the proper pouring temperature and which is supplied from a large melting unit.

holding ladle—Heavily lined and insulated ladle in which molten metal is placed until it can be used. (See: holding furnace)

hot deformation (sand)—Change of form of a sand specimen that accompanies the determination of hot strength.

hot shortness—Brittleness in metal at elevated temperature.

hot spots—Term applied to gray iron castings to denote chilled areas or inclusions that are harder than the surrounding iron and that cause machining difficulties.

hot strength (sand)—Tenacity (compressive, shear, tensile or transverse) of a sand mixture determined at an elevated temperature.

hot tear—A crack or irregularly-shaped fracture formed prior to completion of metal solidification because of hindered contraction. A hot tear is frequently open to the surface of the casting and is commonly associated with design limitations.

hotbox process—A resin-based process that uses heated metal coreboxes to produce cores.

hydrogen—(H) A nonmetallic element that is the simplest and lightest of all the elements. Atomic weight is 1.008, melting point is –259.2C.

hypereutectic alloy—An alloy containing more than the eutectic amounts of the solutes. Analogous to hypereutectoid.

hypereutectoid—An alloy containing more than the eutectoid composition.

hypoeutectoid—An alloy containing less than the eutectoid composition.

impact strength—The resistance to impact loads; usually expressed as the foot pounds of energy absorbed in breaking a standard specimen.

impregnation—A process for salvaging leaky castings by injecting, under pressure, liquid synthetic resins, tung oil, etc., into the porous area. This material is then solidified in place by heating or baking.

impression—Cavity in a diecasting die or in a mold.

inclusions—Particles of slag, sand or other impurities, such as oxides, sulfides, silicates, etc., trapped mechanically in the casting during solidification.

indirect-arc furnace—Electric furnace in which the arc is struck between two horizontal electrodes, heating the metal charge by radiation.

induction furnace—A melting unit wherein the metal charge is melted electrically by induction.

induction hardening—A surface hardening process involving the localized use of pulsating magnetic currents to achieve heating above the austenite transformation temperature, followed by quenching.

induction heating—Process of heating by electrical resistance and hysteresis losses induced by subjecting a metal to the varying magnetic field surrounding a coil carrying an alternating current.

ingates—(See: gate)

ingot—Commercial pig mold or block in which copper, copper-base, aluminum, aluminum alloys, magnesium, magnesium alloys and other nonferrous materials are made available to the foundry worker.

injection—Forcing molten metal into a diecasting die. Also refers to forcing oxygen, nitrogen and other gases, as well as solids such as calcium carbide and graphite, into molten metal.

inoculation—A process of adding some material to molten metal in the ladle for the purpose of controlling the structure to an extent not possible by control of chemical analysis and other normal variables.

insert—A part usually formed from metal, which is placed in a mold and may become an integral part of the casting.

insulating sleeve—Hollow cylinders or sleeves formed of gypsum, diatomaceous earth, pearlite, vermiculite, etc, placed in the mold at sprue and riser locations to decrease heat loss and rate of solidification of the metal contained in them.

internal chills—Solid pieces of metal or alloy, similar in composition to the casting, placed in the mold prior to filling it with molten metal. Chills increase the rate of solidification in their areas and are employed only where feeding is difficult or impossible.

internal shrinkage—A void or network of voids within a casting caused by inadequate feeding of that section during solidification.

internal stresses (or thermal stresses)—Generally, stresses that occur during the cooling of a part.

inverse chill—A condition in an iron casting section in which the interior is mottled or white while the outer sections are gray. Also called reverse chill, internal chill or inverted chill.

investment casting—A pattern casting process in which a wax or thermoplastic pattern is used. The pattern is invested (surrounded) by a refractory slurry. After the mold is dry, the pattern is melted or burned out of the mold cavity and molten metal is poured into the resulting cavity.

investment molding—Method of molding using a pattern of wax, plastic or other material which is "invested" or surrounded by a molding medium in slurry or liquid form. After the molding medium has solidified, the pattern is removed by subjecting the mold to heat, leaving a cavity for reception of molten metal. Also called lost-wax process or precision molding.

iron—(Fe) A metallic element. Atomic weight is 55.84, melting point is 1535C.

iron, malleable—A mixture of iron and carbon, including smaller amounts of silicon, manganese, phosphorus and sulfur, converted structurally by heat treatment into a matrix of ferrite containing nodules of temper carbon.

iron, white or hard—Iron of suitable composition in which the castings, later to be malleableized, are originally cast. Carbon is in the combined form; hence, its white fracture and name.

isocyanate resin—A basic chemical component of the urethanes.

jacket, mold—A wood or metal form slipped over a mold made in a snap or slip flask, to support the four sides of the mold during pouring. Jackets and mold weights are shifted from one row of molds to another during the pouring period.

jig—A device arranged to expedite a hand or machine operation.

jobbing foundry—Foundry that is not a part of a manufacturing plant and produces castings for sale. Usually pour a variety of castings in small lots or quantities.

jolt machine—Molding machine that packs or rams the sand around the pattern by raising the table on which the flask, sand and pattern are mounted a few inches and allowing it to drop suddenly. The table is raised pneumatically and the operation is repeated until the desired sand density is reached.

jolt-squeeze machine—Combination molding machine on which the sand is rammed into the flask by jolting (see: jolt machine), then compressed further by a mechanism that uses fluid pressure to force the table and contained flask upward against a fixed plate. The plate is slightly smaller in dimension so that it fits inside the flask.

kaolinite—Hydrated silicate of alumina represented by the formula Al_2O_3 SiO_2 $2H_2O$. It is a white, pearly mineral, crystallizing in a monoclinic system in the form of small, hexagonal plates. Constituent of kaolin, white china clay, used for porcelain, etc.

keyhole specimen—A type of notched impact test specimen that has a hole-and-slot notch shaped like a keyhole.

killed steel—Molten steel held in a ladle, furnace or crucible (and usually treated with aluminum, silicon or manganese) until more gas is evolved and the metal is perfectly quiet.

kish—Graphite thrown out by liquid cast iron in cooling.

knock out—To remove sand and casting from a flask. (See: shakeout)

knock-out pins (ejector pins)—Small diameter pins on diecasting machines, permanent molds and shell-molding machines for ejection of castings, etc. (See: ejector pins)

ladle—Metal receptacle lined with refractory for transportation of molten metal. Types include hand, bull, sulky, trolley, crane, bottom-pour and teapot.

ladle, bull—Large ladle for transporting and pouring molten metal.

lance, oxygen—Long steel pipe or tube, usually covered with refractory, used to inject oxygen into molten steel to reduce the carbon content. Also may be used to open frozen tapholes in cupolas, etc.

layout—A full size drawing of a pattern showing its specifications and structural features.

lead—(Pb) Softest of the metal elements. Atomic weight is 207.2, melting point is 327.4C.

leaker—Foundry term for castings that leak under liquid or gaseous pressure.

lining—Inside refractory layer of firebrick, clay, sand or other material in a furnace or ladle

linseed oil—Drying-type oil expressed from flaxseeds and used as a binder for core sand.

liquid contraction—Shrinkage or contraction in molten metal as it cools from one temperature to another while in the liquid state.

liquidus—The temperature at which solidification of metal begins on cooling and the temperature at which the last portion of solid metal becomes liquid on heating.

loam—A coarse, strongly bonded molding sand used for loam and dry-sand molding.

loam molding—A system of molding, especially for large castings, wherein the supporting structure is constructed of brick. Coatings of loam are applied to form the mold face and baked before pouring.

locating pad—A projection on a casting that helps maintain alignment of the casting for machining operations.

locating surface—A casting surface to be used as a basis for measurement in making secondary machining operations.

loose molding—The molding process utilizing unmounted pattern. Gates and runners are usually cut by hand.

loose piece—Part of a pattern that remains in the mold and is taken out after the body of the pattern is removed. It can be indicated as: 1) Corebox part of the corebox that remains embedded in the core and is removed after lifting off the corebox. 2) Pattern laterally projecting part of a pattern so attached that it remains in the mold until the body of the pattern is drawn. Back-draft is avoided by this means. 3) Permanent mold part that remains on the casting and is removed after the casting is ejected from the mold. 4) In diecasting, a type of core that forms undercuts positioned in, but not fastened to, a die and so arranged as to be ejected with the diecasting from which it is removed and used repeatedly for the same purpose.

Lost Foam Casting (LFC) process—A casting process in which a foam pattern is replaced by molten metal in a flask filled with loose sand to form a casting.

machinability—The capability of being cut, turned, broached, etc., by machine tools.

machine allowance—Stock added to the part to permit machining to final dimensions.

machine finish—Allowance of stock on the surface of the pattern in order to permit machining of the casting to the required dimensions.

magnesium—(Mn) Light, ductile, silvery-white metal. Atomic weight is 24.32, melting point is 650C.

magnetic crack detection—Method of locating cracks in materials that can be magnetized; done by applying magnetizing force and applying finely divided iron powder that then collects in the region of the crack.

malleability—The property of being permanently deformed by compression without rupture.

malleableization—Annealing or heat-treating operation performed on white iron castings to transform the combined carbon into temper carbon.

manganese—(Mn) One of the elements. Its atomic weight is 54.93; melting point 1260C. Metallic manganese is used in the nonferrous industry both as a deoxidizing agent and as an essential constituent to improve physical properties of certain alloys.

manganese briquettes—Crushed ferromanganese bonded with a special refractory in briquette form and containing 2 lb metallic manganese and 0.5 lb metallic silicon.

master pattern—The pattern from which the working pattern is cast.

match—A form of wood, plaster of paris, sand or other material on which an irregular pattern is laid or supported while the drag is being rammed.

matchplate—A plate on which patterns and gating systems are attached, split along the parting line. Matchplates are mounted back to back with the gating system to form an integral piece.

mechanical properties—Those properties of a material that reveal the elastic and inelastic properties when force is applied. This term should not be used interchangeably with "physical properties."

melting pot—Metal, graphite-clay or ceramic vessel in which metal is melted.

melting range—Pure metals melt at one definite temperature, but constituents of alloys melt at different temperatures and the variation from the lowest to the highest is called the melting range. Example is copper, which melts at 1981F (1083C), and the 85-5-5-5 alloy, which melts at 1568–1849F (853–1010C).

melting rate—Amount of metal melted in a given period of time, usually one hour.

melting zone—Portion of the cupola above the tuyeres in which the metal melts.

metal penetration—Defect in the casting surface that appears as if the metal has filled the voids between the sand grains without displacing them.

metallurgy—The science and technology of metals. A broad field that includes but is not limited to the study of internal structures and properties of metals and the effects on them of various processing methods.

microporosity—Extremely fine porosity in castings caused by shrinkage or gas evolution and apparent on radiographic film as mottling.

microradiography—Process of passing x-rays through a thin section of an alloy in contact with photographic film and then magnifying the radiograph 50 to 100 diameters to observe the distribution of alloying constituents, of voids and of other microstructural features.

microshrinkage—Very finely divided porosity resulting from interdendritic shrinkage resolved only by use of the microscope; may be visible on radiographic films as mottling. Etching shows they occur at intersections of convergent dendritic directions.

microstructure—The structure and characteristic condition of metals as revealed on a ground and polished (etched or unetched) specimen at magnifications above 10 diameters.

misch metal—Alloy of rare-earth metals containing about 50% cerium and 50% lanthanum, neodymium and similar elements.

misrun—Denotes an irregularity of the casting surface caused by incomplete filling of the mold. A casting not fully formed.

model—A representative of an object; miniature or full size.

Modulus of Elasticity (E)—In tension, it is the ratio of stress to the corresponding strain within the limit of elasticity (yield point) of a material. For carbon and low alloy steels of any composition and treatment, the value is approximately 30,000,000 psi.

mold—Normally consists of a top and bottom form, made of sand, metal or any other investment material. It contains the cavity into which molten metal is poured to produce a casting.

mold blower—Molding equipment for blowing a sand mixture onto the pattern with compressed air; allows for faster production than gravity rollover dump.

mold cavity—The impression in a mold produced by the removal of the pattern. This space is filled with liquid metal to form the casting upon solidification.

mold clamps—Devices used to hold or lock cope and drag flask parts together.

mold coating (mold facing, dressing)—1) Coating to prevent surface defects on permanent mold castings and diecastings. 2) Coating on sand molds to prevent metal penetration and improve metal finish. (See: core wash)

mold conveyor—Power-driven unit on which molds are conveyed from the molding station to pouring station to shakeout.

mold hardener—In sand molds in which sodium silicate is the binder, injection of CO_2 causes a chemical reaction that results in a rigid structure.

mold jacket—A wooden or metal form slipped over a mold to support the sides during pouring.

mold oven—Oven or furnace in which molds are dried.

mold shift—A casting discontinuity resulting from misalignment of the cope and drag halves.

mold wash—Usually an aqueous emulsion, containing various organic or inorganic compounds or both, which is used to coat the face of a mold cavity. Materials include graphite, silica flour, etc.

mold weights—Weights placed on top of molds to offset internal or ferrostatic pressure.

molding, bench—Making sand molds from loose or production patterns at a bench.

molding, floor—Making sand molds from loose or production patterns on the floor. Patterns usually are too large to be handled satisfactorily on the bench or molding machine, the equipment being located on the floor during the entire moldmaking operation.

molding, machine—Making sand molds from production patterns on molding machines.

molding, pit—Molding method in which die mold is made in a pit or hole in the floor.

molding machine—Hand or pneumatically operated machine that rams the sand into the mold by squeezing or jolting or both.

molding sand—A sand that binds strongly without losing its permeability to air or gases.

molybdenum—(Mo) Mainly used as an alloying element. Atomic weight is 95.95, melting point is 2625C.

montmorillonite—A mineral with the formula $(MgCa)O{\cdot}Al_2O_3{\cdot}5\ SiO_2{\cdot}NH_2O$; it is the chief constituent of bentonite, which is used as a sand binder.

mottled—White iron structure interspersed with spots or flecks of gray.

muller—Type of foundry-sand-mixing machine.

mulling and tempering—The thorough mixing of sand with a binder, natural or added, with lubricant of other fluid, as water,

multiple-cavity die—A diecasting die having more than one impression of the same part. (See: combination die)

multiple mold—Composite mold made up of stacked sections. Each section produces a complete gate of castings. All castings are poured from a central downgate.

natural sand—Generic term used to describe clay-bonded sands, suitable for molding operations to produce castings; widely distributed except for the Western section of the U.S.A.

nickel—(Ni) One of the elements used as a base metal or alloying. Its atomic weight is 58.69, melting point 1452C.

nobake process—Molds/cores produced with a resin-bonded air-setting sand. Also known as the airset process because molds are left to harden under atmospheric conditions.

nodular graphite—Graphite or carbon in the form of spheroids.

nodular iron—Cast iron which has the major part of its graphitic carbon in nodular form. (See: ductile iron)

nondestructive testing (NDT)—Testing or inspection that does not destroy the object being tested or inspected.

nucleation—1) Homogeneous—The initiation of solid crystals from the liquid stage, or a new phase within a solid without outside interference-rarely occurs. 2) Heterogeneous—Foreign particles altering the liquid-solid interface energy during phase changes-usually occurs.

nucleus—The first structurally determinate particle of a new phase or structure that may be about to form. Applicable in particular to solidification, recrystallization and transformations in the solid state.

oil core—A core bonded with oil.

oil furnace—Furnaces fired with oil.

oil sand—Sand bonded with oil such as linseed and synthetic oil.

olivine—Magnesium-iron-orthosilicate composed of forsterite and fayalite. Does not contain free silica. Possible molding material.

one-piece pattern—Solid pattern, not necessarily made from one piece of material. May have one or more loose pieces.

open riser—A riser open to the atmosphere. Compare with blind riser.

open sand casting—A casting poured in a mold which has no cope or other covering.

open-hearth furnace—A refractory-lined, shallow-bath, rectangular furnace in which both hearth and charge are subjected to the direct action of the fuel flame. Fuel may be gas, coke-oven gas, powdered coal or oil. Flame is created by mixing preheated air with fuel in ports. Air is preheated in regenerators called checker chambers.

open-hearth steel—Steel made in open-hearth furnace.

optical pyrometer—A temperature-measuring device through which the observer sights the heated object and compares its incandescence with that of an electrically heated filament whose brightness can be regulated.

overflow well—A recess in a diecasting die connected to the die cavity and functioning as a vent.

overheated—A term applied when, after exposure to an excessively high temperature, a metal develops an undesirable coarse grain structure, but is not necessarily damaged permanently. Unlike burned structure, the structure produced by overheating can be corrected by suitable heat treatment, by mechanical work or by a combination of the two.

oxidation—Any reaction of an element with oxygen. In a narrow sense, oxidation means the taking on of oxygen by an element or compound and, based on the electron theory, it is a process in which an element loses electrons.

oxidizing flame—A flame produced with excess oxygen.

oxygen—(O) Odorless, colorless gas element. Atomic weight is 15.9, melting point –218.8C.

pad (padding)—Metal added deliberately to the cross section of a casting wall, usually extending from a riser, to ensure adequate feeding to a localized area in which a shrink might occur without the addition.

parted pattern—A pattern made in two or more parts.

parting—A dividing line at which sections of a mold are separated.

parting compound—Material dusted or sprayed on a pattern or mold to prevent adherence of sand.

parting line—A line on a pattern or casting corresponding to the separation between the cope and drag portions of a sand mold.

parting sand—Unbonded sand that is dusted on the parting surface to prevent the parts of the molds from adhering to each other.

pattern—The wood, metal, plaster, foam or plastic shape used to form the cavity in the sand. A pattern may consist of one or many impressions and would normally be mounted on a board or plate complete with a runner system.

pattern, split—Pattern usually made in two parts, sometimes in more than two parts.

pattern coating—Coating material applied to wood patterns to protect them against moisture and abrasion of molding sand.

pattern draft—The taper allowed on the vertical faces of a pattern to permit easy withdrawal of the pattern from the mold or die. (See: draft)

pattern layout—Full-sized drawing of a pattern showing its arrangement and structural features.

pattern letters—Metal or plastic letters and figures in various sizes that are affixed to patterns for identification purposes.

pattern plates—Straight flat metal or other plates on which patterns are mounted.

patternmaker's contraction—The shrinkage allowance made on all patterns to compensate for the change in dimensions as the solidified casting cools in the mold from freezing temperature of the metal to room temperature. The pattern is made larger by the amount of shrinkage characteristic of the particular metal in the casting and the amount of resulting contraction to be encountered. Also, the final stage of solidification when a casting is cooling, called "solid shrinkage."

pearlite—A microconstituent of iron and steel consisting of alternative layers of ferrite and iron carbide or cementite.

pearlitic malleable iron—Irons made from the same or similar chemical compositions as regular malleable iron, but so alloyed or heat treated so that some of the carbon in the resultant material is in the combined form.

peen—Small end of a molder's rammer.

pencil core—Small cylindrical core used with Williams or atmospheric riser.

permanent mold—A metal mold of two or more parts; not an ingot mold. It is used repeatedly for the production of many castings of the same form.

permeability—The property of a mold material to allow passage of mold/core gases during the pouring of molten metal.

pH—The negative logarithm of the hydrogen ion activity. It denotes the degree of acidity or basicity of a solution. At 25C, the neutral value is 7. Acidity increases with decreasing value below 7 and basicity increases with increasing values above 7.

phase—A constituent that is completely homogeneous and is both physically and chemically separated from the rest of the alloy by definite bounding surfaces; for example, austenite ferrite, cementite. Not all constituents are phases; pearlite, for example.

phase diagram—Graphical representation of the equilibrium temperatures and the composition limits of phase fields and phase reactions in an alloy system.

phosphorus—(P) One of the nonmetallic elements. Its atomic weight is 31.0, melting point 44.1C.

physical properties—Properties of matter such as density, electrical and thermal conductivity, expansion and specific heat. This term should not be used interchangeably with "mechanical properties."

pig iron—Blocks of iron to a known metal chemical analysis that are used for melting (with suitable additions of scrap, etc.) for the production of ferrous castings. Product of the blast furnace by the reduction of iron ore. Also the over-iron in the foundry poured into pig molds.

pilot (or sample) casting—Casting produced prior to the production run to verify correctness of procedures, materials and process to be used in production.

pinhole—Small hole under the surface of a casting.

pins, flask—Hardened steel locating pins used on flasks to ensure proper register of cope and drag molds.

pipe—Cavity formed by contraction in metal during solidification of the last portion of liquid metal, as in a riser.

pit mold—Mold in which the lower portions are made in a suitable pit or excavation in a foundry floor.

pitch—Usually coal-tar pitch obtained in manufacture of coke and distilled off at about 350F (177C). Used as a binder in large cores and molds. Melting range is 285–315F (141–157C).

plaster molding—Molding method wherein gypsum or plaster of paris is mixed with fibrous talc, with or without sand, and with water to form a slurry that is poured around a pattern. In a short period of time, the mass air-sets or hardens sufficiently to permit removal of the pattern. The mold, so formed, is baked at elevated temperature to remove all moisture prior to use. One variation is the Antioch process.

plastic pattern—Pattern made from any of the several thermosetting-type synthetic resins such as phenol formaldehyde, epoxy, etc. Small patterns may be cast solid, but large ones are usually produced by laminating with glass cloth.

plates, bottom—Plates, usually of metal, on which molds are set for pouring.

plates, core drying—Straight, flat plates of metal or heat-resisting composition on which cores are placed for baking.

pneumatic tools—Grinders, rammers, drills, etc., operated by compressed air.

polymer—A compound of high molecular weight, in which the molecules are packed closely, with water, alcohol and the like eliminated.

polymerization—The hardening or setting of plastic materials, as epoxy, urethane, when the resin and hardener are mixed together.

polystyrene—A polymer of styrene used in making molding products. In particular, used in the lost foam casting process.

polyurethane—Synthetic resin polymer used for pattern material, ranging from dense elastomer to expanded, spongy, lightweight.

porosity—Holes in the casting due to gases trapped in the mold, the reaction of molten metal with moisture in the molding sand, or the imperfect fusion of chaplets with molten metal.

postprocessing—The process to remove the extra materials from the prototype, treat, finish and protect the surface.

pour—Discharge of molten metal from the ladle into the mold.

poured short—Casting which lacks completeness due to the cavity not being filled with molten metal.

pouring—Filling the mold with molten metal.

pouring basin—Reservoir on top of the mold to receive molten metal.

pouring cup—Article made of sand or ceramics containing a cup-shaped depression that is placed over a sprue opening and acts as a funnel to receive the metal poured from the ladle (See: pouring basin)

pouring device—Mechanically operated device with a ladle set for controlling the pouring operation.

pouring ladle—Ladle used to pour metal into the mold.

powdered coal—Finely ground, high-volatile coal used for heating furnaces and annealing ovens in the malleable foundry industry.

preheating—General term for heat that is applied preliminary to some further thermal or mechanical treatment.

process capability—The amount of variation in the product of a controlled manufacturing process, the range defined by plus-or-minus three standard deviations.

production foundry—Highly mechanized foundry for manufacturing large quantities of repetitive castings.

progressive solidification—Molten metal starts to solidify at mold wall and moves toward center of casting. (Also see: directional solidification)

purifiers, flux—Various materials added to molten metals and alloys for removing impurities, gases, etc.

push-up—An indentation in the casting surface due to displacement (expansion) of the sand in the mold.

pyrometer—An instrument for determining elevated temperatures.

pyrometry—A method of measuring temperature with any type of temperature-indicating instruments.

quenching—Rapid cooling for hardening; normally achieved by immersion of the object to be hardened in water, oil or solutions of salt or organic compounds in water.

radiographic inspection—Examination of the soundness of a casting by study of radiographs taken in various areas or of the whole casting.

radiographic testing—Use of x-ray or gamma rays to study the internal structure of objects to determine their homogeneity.

ram—To pack the sand in a mold.

rammer—Tool for ramming the sand.

ramming—Packing sand in a mold by raising and dropping the sand, pattern and flask on a table. Jolt squeezers, jarring machines and jolt rammers are machines using this principle. Can also be done by hand with a wooden rammer.

Rapid Prototyping (RP)—RP refers to the physical modeling of component or tooling geometry using layered manufacturing. These technologies make it possible to quickly generate polymer, wax or paper-based prototype parts from three-dimensional solid model computer-aided design (CAD) representations. Parts are typically generated by building up one layer at a time with the thickness of each layer determining the accuracy of the part and the time required to make it.

rapping—Knocking or jarring the pattern to loosen it from the sand in the mold before withdrawing the pattern.

rapping bar—A pointed bar (or rod) made of steel or other metal, which is inserted vertically into a hole in a pattern or driven into it, then struck with a hammer on alternate sides to cause vibration and loosening of the pattern from the sand.

rapping plate—Metal plate attached to a pattern to permit rapping for removal from the sand

rat tail—Minor sand buckle occurring as a small irregular line or series of lines.

rebonding—Term usually employed in reference to adding new bonding material to used molding sand so that it can be used again to produce molds.

recrystallization—A process whereby the distorted grain structure of cold-worked metals is replaced by a new, strain-free grain structure during annealing above a specific minimum temperature.

reducing flame—Flame burning with insufficient oxygen to provide complete combustion resulting in the presence of carbon in the flame.

refractory—Material usually made of ceramics, which is resistant to high temperature, molten metal and slag attack.

resin binder—Any of the thermosetting types of resins used as binders for producing core and shell molds, such as phenol and urea formaldehydes, melamine, furans (fufuryls and furfuryl alcohol), etc.

resin-coated sand—Molding or core sand in which the binder is resin applied to the sand as coating by either cold or hot coating.

resinoid—Solid materials produced by the union (polymerization) of a large number of molecules of one or more relatively simple compounds. Resinoids are classed as thermosetting or thermoplastic.

reverberatory furnace—Melting unit with a roof arranged to deflect the flame and heat toward the hearth on which the metal to be melted rests.

Reynolds number—The ratio of the inertia forces in a flowing fluid to the viscous force. Inertia force is the product of mass and acceleration and viscous force is equal to the shear stress multiplied by the area. Hence, Reynolds number $N_R = vd/\nu$ where v = mean velocity of flow, d = diameter of the channel and ν = kinematic viscosity of the fluid.

riddle—Hand- or power-operated device for removing large particles of sand or foreign material from foundry sand.

rigging—Gates, risers, loose pieces, etc., needed on the pattern to deliver the metal to the mold cavity and produce a sound casting.

riser—Reservoir of molten metal from which casting feeds as it shrinks during solidification. (See: feeder)

riser, blind—A riser that does not extend through the top of the cope and is entirely surrounded by sand. (See: atmospheric riser)

riser, gating—Practice of running metal from the casting through the riser to help in directional solidification.

riser, open—Conventional form of riser usually located at the heaviest section of the casting and extending through the entire height of the cope.

riser gating—Gating system in which molten metal from the sprue enters a riser closest to the mold cavity and flows into the mold cavity.

riser neck—The connecting passage between the riser and casting. Usually only the height and width or diameter of the riser neck are reported, although the shape can be equally important.

riser pad (riser contact)—An enlargement of the riser neck where it joins the casting. The purpose of the pad is to prevent the riser from breaking into the casting when it is struck or cut from the casting.

Rockwell hardness testing—Method of determining the indentation hardness by measuring the depth of residual penetration by a steel ball or a diamond cone.

rolling over—Operation of reversing the position of the mold so that the pattern faces upward in order to be removed.

rollover board—A wood or metal plate on which the pattern is laid top face downward for ramming the drag half mold, the plate and half mold being turned over together before the joint is made.

rollover machine—Molding machine on which the mold is rolled over before the pattern is drawn.

runner—The portion of the gate assembly which connects the downgate or sprue with the casting.

runner extension—In a mold, that part of a runner which extends beyond the farthest ingate as a blind end.

runner system or gating—The set of channels in a mold through which molten metal is poured to fill the mold cavity. The system normally consists of a vertical section (downgate or sprue) to the point where it joins the mold cavity (gate) and leads from the mold cavity through vertical channels (risers or feeders).

runout—A casting defect caused by incomplete filling of the mold due to molten metal draining or leaking out of some part of the mold cavity during pouring; escape of molten metal from a furnace, mold or melting crucible.

sand, backing—Sand in a mold back of the facing.

sand, bank—Sand from a bank or pit.

sand, blast—Sand used in an abrasive blasting machine for cleaning castings.

sand, core—Sand used in making cores.

sand, facing—Prepared sand used next to the pattern.

sand, floor—Sand used in floor molding,

sand, heap—Sand prepared on the foundry floor.

sand, lake—Sharp sand from vicinity of lakes.

sand, molding—Sand used to make molds.

sand, natural—Naturally bonded sand as distinguished from that which is formed synthetically.

sand, open—Sand through which gases can pass freely.

sand, silica—Sand composed of almost pure silica.

sand, synthetic—Molding sand prepared by adding clay or other bond to the sand that is practically free of those materials.

sand blast—Sand driven by a blast of compressed air (or steam). It is used to clean castings, to cut, polish or decorate glass or other hard substances and to clean building facades, etc.

sand castings—Metal castings produced in sand molds.

sand conditioning—Preparation of used molding sand for reuse, which includes additions of bond, additives, moisture, etc.

sand control—Procedure used to adjust various properties of sand such as fineness, permeability, green strength, moisture content, etc., in order to obtain castings free from such defects as blows, scabs, rat tails, veins, etc.

sand control equipment—Testing instruments, such as moisture determinators, permeability air-flow apparatus, etc., for determining the various physical properties of sands.

sand dryer—Apparatus for removing moisture from sand.

sand holes—Cavities of irregular shape and size whose inner surfaces plainly show the imprint of granular material.

sand inclusions—Cavities or surface imperfections on a casting caused by sand washing into the mold cavity.

sand muller—A machine for mixing sand by kneading and squeezing.

sand mulling—A method of evenly distributing the bond around the sand grain by a rubbing action.

sand reclaimer—Equipment for removing extraneous material from used sand and reconditioning it for further use.

sand reclamation—Processing of used foundry sand grains by thermal, attraction or hydraulic methods so that it may be used in place of new sand without substantially changing current foundry sand practice.

sand slinger—Molding machine that throws sand into a flask or corebox by centrifugal action.

sand tempering—Adding sufficient moisture to core or molding sand to make it workable.

sand toughness—Indication of molding sand workability, particularly with reference to rammability, because the tougher the sand, the harder it is to ram tightly against the pattern. It is usually given as a number obtained by multiplying deformation by green compressive strength times 1000.

scab—A blemish on a casting caused by eruption of gas from the mold face.

scale measurement—Measurements taken from a scale drawing or a model to ascertain the true dimensions.

scrap—1) Any scrap metal melted (usually with suitable additions of pig iron or ingots) to produce castings; 2) Reject castings.

seacoal—Finely ground bituminous coal.

seam—Surface defect on a casting similar to a cold shut, but not as severe.

segment—A section of a circle, the ends of which are radial lines.

segregation—Concentration of alloying elements at specific regions, usually because of the primary crystallization of one phase with the subsequent concentration of other elements in the remaining liquid.

SG iron—Term used in Britain and continental Europe for ductile or nodular iron. SG means spherulitic or spheroidal graphite.

shakeout—1) The operation of removing castings from the mold. 2) A mechanical unit for separating the molding materials from the solidified metalcasting.

shakeout machinery—Equipment for mechanical removal of castings from molds.

shank—The handle attached to a small ladle.

sharp sand—Sand that is substantially free of bond; the term does not refer to grain shape.

Shaw (Osborn-Shaw) process—A precision casting technique in ceramic molds that do not require wax or plastic investment.

shear modulus (G)—In a torsion test, the ratio of the unit shear stress to the displacement caused by it per unit length in the elastic range. Units are Pa or psi.

shear strength—Maximum shear stress that a material is capable of withstanding without failure.

shell molding (Croning process)—Process in which clay-free silica sand coated with the thermosetting resin or mixed with the resin is placed on a heated metal pattern for a short period of time to form a partially hardened shell. The unaffected sand mixture is removed for further use. The pattern and the shell are then heated further to harden or polymerize the resin-sand mix and the shell is removed from the pattern.

shift—A casting defect caused by mismatch of cope and drag, or of cores and mold.

shot—Abrasive blast cleaning material. In diecasting, it is the phase of the diecasting cycle when molten metal is forced into the die.

shrink hole—A hole or cavity in a casting resulting from contraction and insufficient feed metal and formed during the time the metal changed from the liquid to the solid state.

shrink rule—Patternmaker's rule graded to allow for metal contraction.

shrinkage—Contraction of metal in the mold during solidification. The term also is used to describe the casting defect, such as shrinkage cavity, which results from poor design, insufficient metal feed or inadequate feeding.

shrinkage, centerline—Shrinkage occurring in the center of casting sections, particularly with platelike or barlike contours, which solidify simultaneously from two faces and cut off feeding in the central portion.

sieve—A device with meshes made from wire or other material for separating fine material from coarse material.

silica—Silicon dioxide, SiO_2, the prime ingredient of sand and acid refractories occurring in nature as quartz, opal, etc. Molding and core sand are impure silica.

silica flour—Silica in finely divided form.

silica wash—Silica flour mixed with water and other materials to form a brushable or sprayable facing material.

silicon—(Si) An abundant element, chemically classed as a nonmetal, metallurgically classed as a metal, used extensively in ferrous and nonferrous alloys; melting point 1423C, atomic weight is 28.06.

silicon-aluminum—An alloy of 50% silicon and 50% aluminum used for making silicon additions to aluminum alloys; also called an intermediate or hardener alloy. Melting point is 1070F (577C).

silicon brass—A series of alloys containing 0.5–6% silicon, 1–19% zinc and a substantial amount of copper.

silicon bronze—A series of alloys containing 1–5% silicon, 0.5–3% iron, under 5% zinc, under 1.5% manganese and the remainder being substantially copper.

silicon carbide briquettes—Silicon carbide in briquette form used as an inoculant and deoxidized in cupola-melted gray iron.

silicon-copper—An alloy of silicon and copper, used as a deoxidizer and hardener in copper-base alloys, which is available in two types containing 10 and 20% silicon. The 10% grade has a melting point of 1500F (816C) while the 20% grade melts at 1650F (899C).

silvery iron—A type of pig iron containing 8–14% silicon, 1.50% carbon max., 0.06% sulfur max. and 0.15% phosphorus max.

sintering—The bonding of adjacent surfaces of particles of a mass of powder or a compact by heating to a suitable temperature and cooling.

sintering point—The temperature at which a molding material begins to adhere to a casting or, in a test, the point when the sand coheres to a heated platinum ribbon under controlled conditions.

skim core (skimmer)—Flat core or tile placed in a runner system to skim the flowing stream of metal. In a pouring basin, it holds back the slag and dross, permitting clean metal to pass underneath.

skim gate—An arrangement that changes the direction of flow of molten metal in the gating system and thereby prevents the passage of slag and other extraneous materials beyond that point.

skin—The surface of a mold or casting.

skin-drying—Drying the surface of the mold by direct application of heat.

slag—A nonmetallic covering on molten metal as the result of the combining of impurities contained in the original charge, such as ash from the fuel and any silica and clay eroded from the refractory lining. Except in bottom pour ladles, it is skimmed off prior to pouring the metal.

slag inclusions—Casting surface imperfections similar to sand inclusions but containing impurities from the charge materials, silica and clay eroded from the refractory lining and ash from the fuel during the melting process. May also originate from metal-refractory reactions occurring in the ladle during pouring of the casting.

slurry—A term loosely applied to any clay-like dispersion. It may be used to wash ladles or other refractory linings to impart a smoother surface. A thin watery mixture such as the gypsum mixture for plaster molding, the molding medium used in investment molding, core dips and mold washes.

slush casting—Casting made by pouring an alloy into a metal mold, allowing it to remain sufficiently long to form a thin solid shell and then pouring out the remaining liquid metal.

smelter—An individual or firm that separates metals from ores or that melts, treats or refines scrap metals and alloys for further use.

snag—Removal of fins and rough places on a casting by means of grinding

snap flask—A flask that has hinges and latches so that it may be removed from the mold prior to the pouring.

sodium—(Na) A ductile element occurring abundantly in nature. Atomic weight is 22.99, melting point is 97.7C.

sodium silicate/CO_2 process—Molding sand is mixed with sodium silicate and the mold is gassed with CO_2 gas to produce a hard mold or core.

soldiers—Wooden pegs used to reinforce a body of sand.

solid contraction—Shrinkage or contraction as a metal cools from the solidifying temperature to room temperature.

solid solution—A single solid homogeneous crystalline phase containing two or more chemical species.

solidification—Process of a metal (or alloy) changing from the liquid to the solid state.

solidification range—Only pure metals solidify or freeze at one definite temperature. Alloys contain different constituents that solidify at different temperatures and the various temperatures from that of the first constituent to solidify to that of the last constituent to freeze is called the solidification range.

solidifying contraction—Shrinkage or contraction as a metal solidifies.

solidus—Temperature at which freezing is completed. Below that temperature, all metals are completely solid.

spectrograph—Optical instrument for determining the concentration of metallic constituent; in a metal (or alloy), by the intensity of specific wavelengths generated when the metal or alloy is thermally or electrically excited.

splash core—A core placed in a mold to prevent erosion of the mold at places where metal impinges with more than normal force. Splash cores are commonly used at the bottom of large rammed pouring basins, at the bottom of long downsprues or at the ingates of large molds.

spline—A thin strip of wood to reinforce butt joints. Also known as "feather" or "tongue."

split pattern—A pattern that is parted for convenience of molding.

sprue (downsprue-downgate)—The vertical portion of the gating system where the molten metal first enters the mold. In diecasting, the metal that fills the conical passage (sprue hole) connecting the nozzle with runners.

sprue base—The lower end of the sprue attached to the runner system. It is usually in the form of an enlargement or reservoir to reduce turbulence.

sprue button—A device attached to the cope pattern to indicate where the sprue should be cut.

sprue cutter—A piece of tubing that cuts the sprue hole through the cope. In addition, a shear-type machine for removing the sprue and gates from the casting.

sprue pin—In diecasting, a tapered pin with a rounded end projecting into a sprue hole acting as a core that deflects the metal and aids in removal of the sprue from the diecasting

sprue plug—A tapered metal or wood pin used to form the sprue opening in a mold. In addition, a metal or other stopper used in a pouring basin to prevent molten metal from flowing into the sprue until a certain level has been reached. It prevents entry of dirt and dross.

spruing—Removing gates and risers from castings after the metal has solidified.

squeeze board—A board used on the cope half of a green sand mold to permit squeezing of the mold.

squeeze pressure—The pressure applied by a molding machine to press the flask and contained sand against the fixed squeeze head or board on a molding machine.

stack molding—Molding method in which the half-mold forms the cope and drag. They are placed one on top of the other and poured through a common sprue. Cavities on the bottom side of one half-mold rest on the flat side of the half-mold beneath. When the cavities are in both sides of the half-molds, the method is called multiple molding.

step gate—A vertical sprue containing a number of side branches or entries at different levels into the casting cavity.

stereolithography—Computerized building of three-dimensional models and patterns. Enables the data representation of a CAD solid model to be directly converted into a plastic model of a casting.

stock core—Core of standard diameter usually made on a core machine and kept on hand, cut to required length.

strained castings—Molten metal, when poured into the mold too fast, raises the cope slightly from the drag and produces an oversize casting with protruding fins; an oversize casting can also be produced from a weak mold.

strainer core—A perforated core placed at the bottom of a sprue or in other locations in the gating system to control the flow of the molten metal. To some extent, it prevents coarse particles of slag and dross from entering the mold cavity.

strains, casting—Strains produced by internal stresses, resulting from unequal contraction of the metal as the casting cools.

stress, residual—Those stresses set up in a metal as a result of nonuniform plastic deformation or the unequal cooling of a casting.

strike off—A straight edge to cut the sand level with the top of the drag or cope flask.

stripping machine—A device for removing the pattern from a mold or a core from the corebox.

stripping plate—A plate, formed to the contour of the pattern, which holds the sand in place while the pattern is drawn through the plate.

strontium—(Sr) A soft white alkaline earth metal. Atomic weight is 87.62, melting point is 770C.

Styrofoam—A proprietary name for expanded polystyrene.

sulfur—(S) A nonmetallic chemical element. Atomic weight 32.06, melting point 120C.

supercooling (undercooling)—Cooling below the temperature at which an equilibrium phase transformation can take place without actually obtaining the transformation.

superheating—Theoretically, the temperature above the liquidus; in practice, it usually means temperature above the usual pouring range.

surface finish—Condition or appearance of the surface of a casting.

sweep—To form a mold or core by scraping the sand with a form sweep having the desired profile.

synthetic molding sand—Any sand compounded from selected individual materials which, when mixed together, produce a mixture of the proper physical and mechanical properties from which to make foundry molds.

tap—To withdraw a molten charge from the melting unit.

taphole—Opening in a furnace through which molten metal is tapped into the forehearth or ladle.

teapot ladle—Ladle with an external spout wherein the molten metal is poured from the bottom rather than from the top.

tear, hot—Same meaning as hot crack, but developing before the casting has solidified completely.

temper carbon—Carbon in nodular form, characteristic of malleable iron.

tempering (sand)—Addition of water to and mixing molding sand to obtain uniform distribution of moisture.

template—A thin piece of material with the edge contour in reverse to the shape to be checked.

tensile strength—The maximum stress in uniaxial tension testing that a material will withstand prior to fracture. The ultimate tensile strength (UTS) is calculated from the maximum load applied during the test divided by the original cross-sectional area.

ternary alloy—An alloy that contains three principal elements.

test bar—Standard specimen bar designed to permit determination of mechanical properties of the metal from which it was poured.

thermal conductivity—The property of matter by which heat energy is transmitted through particles in contact. For engineering purposes, the amount of heat conducted through refractories is usually given in Btu or HB per hour for one square foot of area, for a temperature difference of one degree Fahrenheit and for a thickness of one inch, Btu/hr - ft^2 - F/in.

thermal contraction—Decrease in linear dimensions of a material that accompanies a change in temperature.

thermal expansion—Increase in linear dimensions of a material that accompanies a change in temperature.

thermal shock—Stress developed by rapid and uneven heating of a material.

thermal spalling—Breaking up of refractory from stresses that arise during repeated heating and cooling.

thermal stability—Resistance of a material to drastic changes in temperature.

thermit reaction—Exothermic, self-propagating processes in which finely divided aluminum powder is used to reduce metal oxides to free metals by direct oxidation of aluminum to aluminum oxide, with accompanying reduction of the less stable metal oxide.

thermocouple—A bimetallic device for measuring temperatures by the use of two dissimilar metals in contact; the junction of these metals gives rise to a measurable electrical potential that varies with the temperature of the junction. Thermocouples are used to operate temperature indicators or heat controls.

tight flask—A type of flask that remains on mold during pouring. Lugs are normally provided for clamping cope and drag together for pouring.

tin—(Sn) A soft white metallic element. Atomic weight 118.70, melting point 231.85C.

tin sweat—Beads or exudations of a tin-rich low-melting phase found on the surface of or on risers of bronze castings, which are usually caused from absorption of hydrogen by the molten metal.

titanium—(Ti) A white metallic element. Atomic weight 47.88, melting point 1820C.

tongs—Metal instrument with two legs joined by a hinge for grasping and holding things, e.g., crucible tongs.

tooling path—The procedures from drawing to patternmaking. It generally includes fabrication method, material choosing and tooling approach.

tooling points—The fixed positions on the casting surfaces used for reference during layout and machining.

top board—A wood board on the cope half of the mold to permit squeezing the mold.

transfer ladle—Container used to carry molten metal from the melting furnace to holding furnace or from the furnace to pouring ladles.

trim die—Die for shearing (or shaving) flash from a diecasting.

trimming—Removing fins, gates, etc. from castings.

trowel—Tool for sleeking, patching and finishing a mold.

tucking—Pressing sand with the fingers under the flask bars, around cores and other places to insure firm placement.

tumbling barrel—A revolving metal box, wood box or barrel in which castings are cleaned.

tuyere—Opening through which the air blast enters the cupola.

ultrasonic testing—A nondestructive method of testing metal for flaws based on the fact that ultrasonic waves are reflected and refracted at the boundaries of a solid medium. Use of elastic waves of the same nature as sound, but of shorter wave length and higher frequency than those that affect the human ear (0.5–5 million cycles per sec), for detecting flaws in materials.

undercut—Part of a mold or die requiring a drawback. (See: drawback)

urea formaldehyde resin—A thermosetting product of condensation from urea or thio-urea and formaldehyde, soluble in water and used as a sand binder in core and mold compounds.

vacuum casting—A casting process in which metal is melted and poured under very low atmospheric pressure; a form of permanent mold casting where the mold is inserted into liquid metal, vacuum is applied and metal drawn up into the cavity.

vacuum degassing—Subjecting molten metal to a vacuum to remove deleterious gases such as hydrogen, oxygen and nitrogen.

vacuum melting—Melting, usually by induction heating, in a closed container that is subjected to a vacuum.

vacuum refining—Vacuum melting to remove gaseous metal contaminants.

veining—Surface defect on castings appearing as veins or wrinkles, which results from cracks in the sand due to elevated temperature conditions and occurs mostly in cores.

veins—A discontinuity on the surface of a casting appearing as a raised, narrow, linear ridge that forms upon cracking of the sand mold or core from the sand expansion during filling of die mold with molten metal.

vent—An opening or passage in a mold or core to facilitate escape of gases when the mold is poured.

vibrator—A device that jars or vibrates the pattern (or matchplate) as it is withdrawn from the sand

viscosity—The resistance of fluid substance to flowing, quantitatively characteristic for an individual substance at a given temperature and under other definite external conditions.

wash—Casting defect resulting from erosion of sand by flowing metal. In addition, a term for coating materials applied to molds, cores, etc.

Washburn core—A thin core that constricts the riser at the point of attachment to the casting. The thin core heats quickly and promotes feeding of the casting. Riser removal cost is minimized.

wax—Class of substances of plant, animal or mineral origin, insoluble in water, partly soluble in alcohol, ether, etc. and miscible in all proportions with oils and fats. They consist of esters, free fatty acids, free alcohols and higher hydrocarbons. Common waxes are beeswax, bayberry, paraffin, ozokerite, ceresin and carnauba. Their mixtures are formed into rods and sheets and used for forming vents in cores and molds, repairing patterns, etc.

well (cupola)—Lower portion of a cupola, between the sand bottom and the slaghole, which forms a reservoir for the molten metal.

wetting agent—Surface-active agent that, by reducing surface tension of the wetting liquid, causes a material to be wetted more easily.

white cast iron—Cast iron in which substantially all the carbon is present in the form of iron carbide and which has a white fracture.

wood flour—Finely ground wood, usually hardwood, low in resin.

x-ray—Form of radiant energy with wavelength shorter than that of visible light and with the ability to penetrate materials that absorb or reflect ordinary light. X-rays are usually produced by bombarding a metallic target with electrons in a high vacuum. In nuclear reactions, it is customary to refer to photons originating in the nucleus as gamma rays and to those originating in the extranuclear part of the atom as x-rays.

yield—In production of castings, a value expressed as a percentage indicating the relationship of the weight of a casting to the total composite of the casting and its gating system. For example, if the casting and gating system weigh 125 lb and the casting weighs 100 lb, the yield is 80%.

yield strength (YS)—The stress at which a material exhibits a specified limiting permanent strain.

Young's Modulus (E)—(See: modulus of elasticity)

zinc—(Zn) A metallic element. Atomic weight 65.38, melting point 419.4C.

zircon—1) The mineral zircon silicate, $ZrSiO_4$, a very high melting point acid refractory material used as a molding material in steel foundries. 2) Natural zirconium silicate, $ZrSiO_4$, containing when pure 67.23% zirconium oxide, ZrO_4 and 32.77% silica, SiO_2, is used as a molding medium.

zirconia—ZrO_4, an acid refractory up to 4532F (2500C) having good thermal shock resistance and low electrical resistivity.

zirconium—(Zr) Silvery-white, metallic element. A powerful oxidizer and aluminum stabilizer, when added to molten steel. Atomic weight is 91.22, melting point 1842C.

Index

A

abrasion 35,40,85,87,88,93,94,97,110,138,199,277,286
abrasive blast 283–286,293
abrasive cutoff 277
abrasive impact 282
abrasive-waterjet 281,282
acicular 92,207,208,215
acid demand value (ADV) 41,42,63,64,71,74,75
acid melting practice 160–162
acid slag practice 135,149,164
acid-type binder 65
acrylic binder 68
additives 7,9,42,46–51,53,55,64,65,67,70,76,79,127, 176,185,186,289
adhesive failure 78
ADI See: *austempered ductile iron*
ADV See: *acid demand value*
advanced oxidation (AO) 46,163
AFS gfn 41,49,52,64
age-hardenable 96
aggregate 5,42,44,45,64,68,127,128,152,193,204,270–273
air blasting 286
air pollution 46,149,154,156,160,285,302,307
air-carbon arc 279
air-set 5,10,65,69,74
airless centrifugal wheel 284
airless-blast 282
alloy designation 99,103,106,109
alpha phase 39,191
alpha-quartz 39
alumina oxide 121
aluminum 4,5,11,12,14,16,18,19,21–26,28–32,37,64,67,68, 70,71,77,79,81,82,93,97–107,111–114,120,123, 130,133,135,155,156,161,162,164,171–179,180–186,188, 192,197,199,213–215,217,220,222,223,225,229, 230,237,239,240–249,252,253,257,260,261,264–269, 270–273,278,279,280,282,283,286,287,289,290–292, 294–296,301,309,310,311,316
aluminum alloy ingot 172
aluminum composite 111,218,223,244
aluminum die 11,32
aluminum oxide 12,181,241,282,286,287,290
amine 67
analysis 3,5,49,52,88,92,104,112–114,118,135,154,160–163, 165–167,172,182,185,197,206,207,215,217,223, 287,293 303–305,309,318
angular 39–42,64,77
anneal 3,4,88,89,94,95,109,174,198, 200,201,207,210,217,275,286
anodize 102,286
antimony 189,190,192,194,215
Antioch process 16
AO See: *advanced oxidation*
AOD See: *argon-oxygen decarburization*
arc welding 278
arc-air process 279
argon 18,20,156,163,176,177
argon-oxygen decarburization (AOD) 163
art casting 11
artificial aging 217
as-cast ductile iron 203,206
asphalt 47,77,152,302
aspiration 226,228–230,234,238
ASTM standard grade 92
atmospheric core 266
atmospheric pressure 12,13,142,175,184,226,266
atomic absorption 167
ausferrite 92,93,208
austemper 81,92,96,208,212
austempered ductile iron (ADI) 4,92,93,96,207,212,208
austenite 92,188,192–194,198,208,209
austenite transformation 209
austenitic matrix 208
austenitize 208
automatic ladling 173
automatic pouring 173,204,205,231–233,241,245
automotive industry 12,92,121

B

backpour 266,267
baghouse 46,297,298,302,307
bainite 198,207,208
bale-out furnace 181
bandsaw 277,279
bank sand 39
base block 120
basic melting practice 160,162,165,202
basic slag practice 149
basic-type binder 65
batch melting 113,126
batch-type muller 49,72,73,74
BCC See: *body centered cubic*
bed coke 136,137,140
bench life 9,66,68,69,72,73,75
beneficial reuse 302
bentonite 7,43–45,47,48
Bernoulli's theorem 225,226,227,238
beta phase 191
beta-quartz 39
binary 97,189,252
binder 7–10,16,26,39,40,41,55,63–69,70,71,74–79,109,136,158, 250,272,283
bituminous coal 47
black water 46,53
blackheart 4,88
blast air 115–118,135,138,140–145,147,148,150,152
blast furnace 1,2,105,115
blast room 285,286
blasting 5,271,275–277,282–287,297
blind riser 260,261,265–267

block-and-wedge theory 44
blow method 72
blow slot 67
blowpipe 278
blowtube 35,67
body centered cubic (BCC) 187,188
bonding mechanism 43,64,127
boron 98,154,157,192,214,222
boss 31,270,296
bott-and-tap 115
bottom filling 26
bottom gating 237
bottom sand 135,136
bottom-pour 227,231
brasses 103,105,186,220,222,239,282
bridging 116,137,149,155,182
briquettes 114,185
broaching 281
bronzes 103,105,106,185,186,220,222,239,282
brushing 78,282,283
bubble trap 244
bucket elevator 60,286
buckle 39
bulk density(ies) 12,13,39,41,51
bull's-eye 193,201,203,233
burden 116,117,147,148,151,153
burn-in 70
burn-on 47
burner system 115,133
burnt lime 162,202

C

CAD See: *computer-aided design*
cadmium 190,191,302,306,307
cake filtration 242
calcining 77
calcium 16,39,43–45,48,64,66,93,121,148,156,157,
161–164,197,202,203,205,215
calcium carbide 202
calcium cyanamide 202
calcium silicide 93,197
calcium sulfate 16
calcium sulfide 202
CAM See: *computer-aided manufacturing*
campaign 115,117–119,135,136,150–152,155
carbide slag 163
carbide stabilizer 192,195,199
carbides 90,192,193,198,200,205,208
carbidic matrix 90
carbon 4,5,9,12,18,19,24,26,30,47,49,64,66–68,71,81,85–89,
90,92–97,109,113,115–119,121,130,135,138,141–143,
145,147–149,150,153–159,160–167,169,174,179,
182–184,192–199,201–203,206,210,215,225,230,231,239,
241,245,248,252,259,260,262,266–269,271,278,279,
297,298,310,315
carbon boil 156,161,162
carbon dissolution 154
carbon electrode 115,266
carbon equivalent (CE) 86,89,141,165,166,193,194,199,
201,202,203,206,215,268
carbon injection 163
carbon monoxide 119,145,147,148,161,184,297
carbon pickup 116,118,143,148,184
carbon soot 12
carbon steel 5,12,18,19,24,26,30,47,81,88,90,93–97,113,
121,130,135,142,153,154,156,157,160,162,163,166,
192,198,201,208–210,225,230,231,241,245,248,
252,260,262,266–269,271,278,279,310
carbon-bonded crucible 174,179
carbonaceous 42,47,49,50,165
carrying agent 78
cartridge collector 298,300
cast iron 4,8,9,24,26,28–31,33–35,40,45,47,57,58,64,76,
81,85–89,91–95,97,98,111,113–115,119–121,129,
135,138,141,143–145,147–149,150–156,158,
160,164–167,192–194,198,200,201,205–209,212,214,215,
229–231,239,245,249,256,259,260,271,276–278,286,289
cast steel 4,29,93–96,133,164,165,169,210,273
castability 5,81,90,91,99,101–103,105–107,110,
218,309,310,316,318
casting cleaning 275,295,301
casting design 7,81,88,92,217,237,251,255,256,260,275,277,
309,310,311,313–318
casting geometry 187,252,280,309,310,314–318
casting processes 7,9,11,13,15,17,19,21,23–27,30,36,61,64,
81,97,101,107,309,317
casting properties 68,79,81,83,85,187,217,276,310,316
casting soundness 85,219,270,313
CAT (scan) See: *computerized axial tomography*
catalyst 7,9,10,16,28,35,64–69,70,71,73–76,79,298
caustic soda 202
CE See: *carbon equivalent*
cellulose 42,46
cementite 87,88,193,199,209
centerline shrinkage 267,268,269
centrifugal blast wheel 283
centrifugal blower 141,142
centrifugal casting 24,25,30,34
centrifugal casting mold 25,30,34
centrifugal cleaning 288
ceramic filter 242,244,314
ceramic foam filter 242,243,246
ceramic molding process 16
ceramic slurry 12,17
cereal 42,46,47,48,56,65
cerium 88,91,203,205
CGI See: *compacted graphite iron*
channel furnace 123–126,128,158,168,179,180
channel induction furnace 123,125,126,128,154,157,158,
168,179,180,183,232
charcoal cover 183
charge bucket 136–139,146,147,155,160,165
charge door 118,119,130,136,138,140,143,161
charge makeup 114,140,145–147,154,160,192,201
charge materials 113–116,119–122,124–126,131,132,135–138,
141,145,146–149,153–158,160,162,164–166,
171,172,174,175,179,180,182,183,185,192,194,202,249
charging well 132,180
chemical additives 42,47,185
chemical analysis 104,113,114,135,160,162,167,
172,182,197,206,293
chemical composition 39,41,42,85,87,88,99,
100,103,104,105,107,109,113,114,150,154,161,164,
198,199,202,208,213,265,271
chemical degassing 185
chemical scrubber 73

chill 3,21,24,28,41,64,77,78,87,88,130,143,145,154, 162,163,165–168,184,196,197,199,205,210,214, 248,250,251,256,269,270–273,276,286,301,317
chill aggregates 270,271,273
chill coating 77,271,272
chill placement 272
chill preparation 271–273
chill problem 272,273
chill size 271
chill test 166,167
chipping 5,277,278,282,297
chlorine-base tablets 176
choke 232,233–235,237–239,240,241,243–245
choke calculation 232,243,245
choke core 234,235,240
chrome oxide 121
chromel-alumel thermocouple 166,173
chromite 8,39–41,64,78,128,251,271–273
chromium 81,85,93,94,96,98,109,111, 154,160,163,164,167,187,192,198,200,211,220,221,279
Chvorinov's Rule 255,257,261,263
CLA process 18
CLAS process 14
clay bonded 39,74,127
clay graphite 120,179,183
clay tile 116
clay-bonded crucible 174
Clean Air Act 140,177,297
cleaning department 5,275
cleaning fluxes 176,177
cleaning room 4,40,275,280,282,283,289,290,291,293,296
cluster 33,91,197
CLV process 18
CMM See: *coordinate measuring machine*
CNC See: *computer numeric control*
CO_2 molding 9
coal gasifier 115
coated abrasives 289,290,296
cobalt 4,18,81,109,110,160,171,186,241
coil 117,120,122–127,145,155,179,180,183,295
coining 281
coke 3,5,115,116,118,119,135–138,140,141,143–149, 150–154,158,163,202,204
coke bed 118,138,140,141,145,148,152
coke ignition 136
cold bottom dropping 152
cold chamber diecasting 23
cold coin die 281
cold shortness 95
cold-start batch-type melting 155
coldbox 9,10,65–68,70,71,73,75,78,297,298
coldbox molding 9,10
collapsibility 9,43,46,63,71
columnar growth 253,254
combustible gas 58
combustibles test 53
combustion zone 116,144,145,148
compactability 50,51,53,56
compacted graphite iron (CGI) 4,81,91,92,149,207,212
compaction 9,10,28,45,46,47,51,55,56,57,58,59, 60,61,64,72,74,129
complex gases 184
compliance checklist 305
component solubility 190
composite design 94
compound 7,40,55,56,64,87,97,182,185,189,195, 214,218,219,288,298
compressed air system 291
computer 4,6,27,28,36,51,52,122,148,153, 225,245,251,256,261,262,275,294,295,309,318
computer-aided design (CAD) 27,36
computer-aided manufacturing (CAM) 36
computerized gating 245
computer numerical control (CNC) 4,28,275
computerized axial tomography (CAT) 295
conduction 144,248,249,250,251,265
cone jolt toughness test 53
cone-bottom bucket 137,138
constitution diagram 189
consumable electrode 186
continuous casting 25
continuous mixer 10,74
continuous tapping 115,149,150
continuous-type muller 72
convection 65,144,158,248–250,276
conveyor belt 58,60,284
cooling 9,11,16,21,25,29,34,35,39,47–50,53,60,61,66, 85,87,92,94,117,118,122,124,126,129,148,151,152, 165,166,185–187,189,190–193,196–198,200–203,207, 208,210,215,217,218,220,226,229,247,249,252,253,255,259, 263,270,271,273,275,276,278,280,282,291,297,298,301,314
cooling curve 166,189,190–192,215,247,252,259
cooling cycle 207,217
cooling rate 9,21,85,87,92,185,187,196,198,203,210,220,275,276
coordinate measuring machine (CMM) 37,294
cope 2,3,10–13,16,28–31,47,55–60,67,78,206,232–234,239, 240–245,260,272
cope vent 67
copper 1,3,4,5,18,22,23,25,26,47,48,53,66,81,82,85,94,96–99, 101–107,111–115,117,120,123,125,133,160,163,171, 182–186,188,189,192,194,198,208,213,217,218,220–223, 229,230,239–241,243,245–247,249,252254,257, 260,264,265,268,269,271,273277–279,283,287,298,301
core blower, blowing 71,72,79
core insert 21
core molding 2,10,63
core oil 65,72
core removal 276,287
core sand 4,33,35,47,52,53,63–68,71–75,77, 199,251,266,271,276,277,282,283,287,302
core shooter 72
corebox 4,6,11,27,28,35–37,63–69,70–74,78,314
cored passageway 11,12,77,199,276,282
coreless induction furnace 125–129,153–156,158,168,169, 174,180,183,185
coreprint 30,33,36
coring 218,219,220
corrosion resistant, -ance 81,82,85,87–90,94,96,97, 101,102,105,106,109–111,198,217,218
Cosworth process 12,14
countercurrent packed bed scrubber 298,300
countergravity 12,14,15,17,18,19,20
cover gas 156,182
covered ladle 203,204,249
covering fluxes 177
cracking bracket 270
Croning shell process 14,65

crucible 2,4,5,115,120,121,124–127,129,130,157,174–176,178, 180–183,185,186,189,248
crystal, crystallize, etc. 44,109,187–189,192,193,214,216,220–222, 251,252,254
cupola 3,5,105,113–121,133,135–139,140–149,150–153, 156,158,169,199,202,246,249,297
cupola blast air 141–143,145
cupola bottom installation 135,136
cupola combustion 147
cupola melting 116,118,135,141,143,145,147,149,150,202
cupola zone 147,148
cure time 69
cures 36,65,66,68
cuts 11,17,30,45,46,277,280,282,289,290
cutting tool 97,110,256

D

dam 149–151,233
damping capacity 87,91,93
DAS See: *dendrite arm spacing*
dead clay 45,46,51
deep bed filtration 242
defect 5,8,10,26,29,31,32,39–41,43,45,46,47,53,64,69, 71,75,77–79,116,140,157,160,161,172,173, 182,183,206,226,231,232,236,241,243–245,247, 251,252,255,261,272,273,275,279,293–295
deflashing 291
deflection 76,309,310,314–316,318
deformation 11,12,44,46,94,219,250,287,292,316
degassers 172,176,185
degassing 114,132,157,172,175–178,180–182,184–186,215,219
degassing fluxes 178
degassing salts 177
degassing tubes 172
degating 277,303
degree of supercooling 252
dehumidification 143,145
dehumidifying 116
dehydration 67,74
dendrite arm spacing (DAS) 22,215,223,269,270
dendritic structure 187,210,211
density 7,12,13,32,39–41,50,51,57,60,61,63,64,66,68,74,82,97, 107,111,123,126,127,138,158,161,175,177,181,197, 205,223,229,239,245,247,251,263271,289,292
density measurement 175
deoxidation 114,161,163,183,186
deoxidizer, -ing 161,163,164,203,206
design parameter 309,310,316,318
desulfurization 114,157,164,165,202
desulfurizer 203,206
determinator 167,241,289
die pressing 281
diecasting alloy 98,101,217
diecasting die 24,232
diecasting mold 30,34
diecasting process 23,217
diffuser 145,266,267
dimensional accuracy 7,9–12,16,31,294,314
dimpling 316
dipole 44
dipping 17,18,78,120,132,180,181
dipping well 132,180
direct air blast 284
direct shell production casting 36
direct-fired reverberatory fFurnace 130
direct-pour system fFiltering 243
directional solidification 34,184,230, 236,244,255,257,261,267,269,270,273,312,313
discharge water 298,301
discharged foundry water 298
divided-blast cupola 118
dog bone 75
dolomite 116
double-leaf hinged-bottom 137
double-slag practice 162
down runner 236,238
downcomer 141,142
draft 1,10,12,28,29,31–33,36,234,318
drag 2,8,10–14,16,28–31,55–60,78,180,232,240,272
draw pin 30
drawback 31,66,71,125,271
dried sand molding 8
dropping bottom 117,119,135,138,150,152
dross 21,22,102,114,125,127,128,130–132,158, 172,173,177,178,181,183,185,186, 202,206,225,229,230,233,234,239–241, 266,267,272,273,293,294,310,314
drossing 102,178
drossing-off fluxes 178
drum-type channel furnace 124,125,158
dry slag collection 151
dry-bottom cupola 118,119,150
dry-bottom tapping 150
dryer 65,72,79,284,291
dual-energy resistance 115
dual-energy reverberatory furnace 133
ductile cast iron 81,88,89,201,206,260
ductile iron 4,5,12,25,29,47,49,88–93,96, 133,149,154,164,165,169,193,201–208,212,237,239, 241,243,246,268,273,277,279,280,294,309,316
ductility 88,90,–95,97,101,102,105,107,111,187,193, 199–202,210,215,217,219,277,292,295
dump method 72,276
duplex 96,121,124,128,158
dust collector 297,298,307
dwell time 66,72
dye penetrant test 293,294

E

eddy current inspection 295
ejector die half 23
ejector pins 19,21,22,24,33
electric furnace 4,5,121,135,153,161,163,164,165,168,169
electric globar furnace 130,132
electric radiant reverberatory furnace 132
electric resistance furnace 115,130,180
electric/direct-arc furnace 115,121,158
electrical conductivity 82,97,99,107
electrical resistance furnace 87,111,130,158
electrode 115,121,122,156,158–163,165,186,266,279
electromagnetic pump 12,173
electrostatic bonding 43
elongation 90,95,97,99,101,102,105,109,295

emissivity 167,168,173,248,249,250,265,272
end effect 255,267,268,269
endothermic 145,147,148
engineer, -ing 5,6,36,53,76,85,86,88,90,92,93,95–97,
136,208,212,226,228,246,256,257,283,293,
301,304–307,309,312,314–316,318
entry ports 67
environment 6,9,63,66,76,79,82,96,97,102,110,119,151,155,
167,204,205,215,249,275,276,283,297,301–304,307
environmental 9,63,66,76,79,119,151,155,167,
204,205,215,275,283,297,301–303,307
environmental control 119,275
Environmental Protection Agency (EPA) 177,297,307
EPC See: *expendable pattern casting*
epoxy binder 68
EPS See: *expandable polystyrene*
equiaxed grain 214,252–254
equiaxed solidification 312
equiaxial crystallization 221,254
equilibrium 44,157,184,185,189–192,213,218,252
equilibrium diagram 189–192,213
erosion 47,71,77,106,110,128,136,149,156,157,
161,229,230,236,238,239,241
etching 188,193,201,211,219,220,276,283
eutectic 191,193,194,196,197,199,201,203,208,
213–217,220,221,230,247,252,253,310,312
eutectic-type solidification 252,312
eutectoid 192,194,203,208,209,252,253
eutectometer 166,215
evaporative heat exchanger 126
exhaust port 120,175
exothermic material 265,266,269,273
exothermic reaction 66,147,265
exothermic sleeve 265
expandable polystyrene (EPS) 11,12,27,28,30,32
expansion 32,39–41,46,47,71,79,82,97,99,101,107,117,
128,129,172,175,259
expendable pattern casting (EPC) 11,173
external chill 270,271,272,273
external corner 28,32,255,256
externally fired 116,143
extruded filter 242

F

fabric filter 297–299,307
fabrication 6,32,36,103,106,310,314
face centered cubic (FCC) 187,188,192
fade 68,157,197,198,205
fatigue limit 82
fatigue strength 91,93,97
fayalite 40
FCC See: *face centered cubic*
feeder charging 136,138
feeding distance 266–270,273
ferrite 88–90,92,187,192–195,198,200–202,205,207–211
ferritic ductile iron 90,206
ferroalloy 121,143,154,162,164,199
ferromanganese 3,154,162,163,164
ferrosilicon 9,154,161,162,163,164,197,205
ferrosilicon fines 9
ferrostatic pressure 158
ferrous 4,5,8,12,32,45,53,58,70,78,81,85,88,96,
103,114,116,121,126,133,135,145,147,158,164,
174,179,187,192,194,202,212,213,238,242,263,276,
280,289–292,294,301
ferrous alloys 8,11,18,25,48,67,81,85–87,89–91,93–95,
113,115,116,135,145,146,153,165,168,171,173,187,189,
191–193,195,197,199,201,203,205,207,209,
211–213,239,252265,278,280,290
fillet(s) 28,29,255,271,273
filter 50,126,135,165,228,241–244,246,297–299,307,314
fineness 52,64,74,310
fines 9,40,46,47,52,63,64,77,128,129,138,287
finishing room 275
finite element analysis (FEA) 309,318
firebrick 116,120,131,181,236
fireclay 2,43–45,48,127,135,136,149
firecracker core 266
first-stage graphitization 200
fixed die half 23
fixturing 316,317
flame cutting 277,278
flash 17,49,203,270,277,280,281,283,289
flask 10–13,16,17,24,31,44,55–60,74,236,276
flaskless mold 60,276
flogging 277,280
floor mold 74,233,236,270,276
flotation 202
flowability 28,46,60,65,66,71,74
Flowtret 204
fluid flow 6,36225–228,231,236,238
fluid life 30,225,229,230,241,242,244,310,311,314
fluidity 85,97,99,101,102,110,111,113,161,162,167,168,
182,185,195,229,230,310,311
fluidity spiral 167,168,229,230
fluorescent powder test 294
fluorspar 162,163
flux 39,81,96,114–116,128,132,135,137,145,149,151,163,
172,176–178,181–183,187,191,238,240,241,243,294,314
flux cover 181
fluxing 115,116,132,137,163,172,177,181,182,314
fly cutter 24
FM process 25,26
foam filter 242,243,246
follow board 30,31
forged steel 93
forging 1,2,90,92–94,111,160,280
formaldehyde 66,306,307
forsterite 40
foundry returns 145,151,154,159,160,161,162,164,172,183,186
fracture test 184
free water 46
freeform fabrication (fast, solid) 6,36
freeze 128,145,167,187,199,229,242,253,254,312
freezing 102,105,126,145,166,182,184,185,187,188,
190,193,194,198,220–222,230,242,251–255,259,
261,262,265,268,269,272
friability test 53
friable 45,46,74
friction 43,44,87,106,198,225–228,245,277,280
friction sawing 277,280
frictional force 44,228
fritting 128

front-charging reverberatory furnace 132
front-slagging 116
full mold 11,28,32,261
furnace 1–5,14,17,19,22,23,26,75,77,82,85,96,
105,110,113–115,117,119,120–129,130–133,135,136,
147,150,153–159,160–169,171–179,180–186,192,194,
200,202,204,205,210,231,232,248,249,266,301,302314
furnace lining 122,128,129,130,153,155,174
fused deposition modeling 36
fusion point 41,42,47

G

garnet particle 282
gas bubbles 176,185,237,244
gas contamination 114
gas cover 156,181,182
gas entrainment 180,236
gas evolution test 75
gas impact 58
gas pickup 130,132,157,162,171,177,179,182,183,185,186
gas-fired crucible 180,181,183,248
gating 11,15,17–19,21,22,27,31–36,55,56,60,99,
122,160,173,182,183,198,199,205,206,225–230,231–239,
240–246,248,251,255–257,260,273,275,277,280,282,
310,314,318
gating ratio 238,239,241,244,245
gelation 67
geometric dimensioning and tolerancing (GD&T) 317
Gilsonite 47
glass beads 283,286,287
gouging and pad reduction 278
grain boundaries 188,195,209,219,223
grain distribution 49,52
grain fineness 41,50,51,52,64,74
grain growth 252,254
grain refinement 114,214,215,217,219,220,222,223,252
grain refining 178,182,214,219,222,223
grain shape 40,41,42,49,64,187
grain size 14,18,52,64,98,187,197,
214,215,217,222,252,289,290,295
grain size chart 215
grain structure 22,188,210,214,215,219,220,222,254,270
graphite block 24,34,115,271
graphite crucible 174,179,183
graphite electrode 121,159,121,158,159,279
graphite flakes 85,86,89,91,165,193–195,197,203,207,215
graphite fluxing tube 172
graphite lance 176
graphite molding process 24
graphite seats 185
graphite spheroids 88,89,201,203
graphitization 192–195,197–200,205,214,259
graphitizing 88,158,193,197–199
gravity permanent mold 21,246
gravity pouring 14,17,111,232
gray cast iron 31,33–35,81,85,89,91,92,167,192–194,198,
201,205,207,212,215,229,230,245,259,260
gray iron 18,19,21,29,81,85–90,102,111,130,154,164,165,167,
192–195,197–199,201–203,205–207,227,239,
241,243,245,246,268,273,282,296,302,310
green compressive strength 47,50,51
green sand core 64
green sand mold, -ing 7,8,16,27,31,42,43,45–49,50–53,55–61,
78,88,173,174,185,250,268,273
green strength 43,45,50,51,65,74,127
Griffin process 24,34
grinding 5,28,87,275,277,288–291,293,297,301,303,318
grit 176,282,283,286,289,290
groundwater regulation 301
gypsum 16

H

H grade 96
hammering 1,200,276,283
hand ladle 23,172,178
hand molding 7,30,41,55,225
hand rammer 56
HAP See: *hazardous air pollutants*
hardness 12,52,53,56–58,60,85–88,90,92–97,99,
101,110,111,194,195,198,205,210,218,
270,275,276,279,280,286,293,295
hazard 77,131,150,152–154,158,160,172,183,
202,283,297,302–307
hazard prevention 303,304
hazardous air pollutants (HAP) 297
heap sand 10,11
heat cured 65
heat diffusivity 251
heat exchanger 116,124,126,133,143
heat of fusion 148,171,189,247,252,253
heat sinks 185
heat transfer 77,114,119,120,132,133,144,145,148,187,
247–250,251–253,255–257,262,264,309,310,312,314
heat treatment 77,88,90,92–96,99,101–103,107,109,110,
172,193,198–200,203,206–208,210,211,213,
217–219,276,293,316
heel 113,126,130,132,154,233
Heine method 263
hematite 64
heterogeneous nucleation 252
hexachloroethane 177,182
high-alloy steel 18,81,94,96,113,156,160,166,
211,229,241,268,290,316
high-carbon steel 81,94
high-density green sand mold 60
high-pressure mold 60
HIP See: *hot isostatic pressing*
holding furnace 130,132,150,155,180,181,231,232,249
homogeneous 190,191,205,210,217,252
homogenization 210,217
horizontal gating 226,232,233,236,237
hot blast 116,140,141,142,143,144
hot bottom dropping 152
hot chamber diecasting 22,23
hot coating 66
hot isostatic pressing (HIP) 109,292
hot return sand 47
hot sand 46,48,49,61,271,273
hot shortness 97,102
hot strength 43,45,68,70,71,79,90,135
hot tear 9,10,43,45,46,63,102,214,252
hotbox 5,65,66,71,72,78
HSLA steel 194
humidity degradation 71,78,79

hycon test 175
hydraulic 6,23,33,56,57,59,92,121,124,126,
179,225,229,280,287
hydrogen degassing 175,180
hydrogen porosity 98,175
hydrolyzed ethyl silicate 16
hydroscopic 9,67,145
hydroxide 148,298
hypereutectic 88,89,101,191,193,194,197,201,203,206,209,214
hypereutectoid 208,209
hypoeutectic 191,193,199,206,214,215
hypoeutectoid 208,209

I

ignition 49,50,53,58,63,74,76,136,140,175
immersion thermocouple 167,168
immiscible 189
impact resistance 102,219
impact strength 82,97,99
impact/impulse mold 58–60
impeller 12,58,96,101,102,106,141,142,176,178,181,287
impregnation 158,291,292
in-mold inoculation 198,204,205
incandescent coke 116,145
inclusions 14,21,24,25,29,45,46,114,157,171,177,
180,182,195,208,211,220,229,236,237,241,243,
252,278,279,294,295,310,314
Inconel 111,245
induction furnace 115,120,122–129,147,153–158,168,169,
174,179,180,183,185,232
inductor 124,125,127,128,158
inert gas 114,156,163,176,186,202,295
inert gas cover 156
inert gas melt 114
ingate 47,228,230,232,236,237,240,241,
272,275,277,279–281,288,289,292,293
ingot 92,99,100,114,131,132,164,172,175,180,182,183,186
initial bubble test 175
inoculation 154,182,196,197,198,205,206,212,213,214,252
inorganic additives 46
inorganic binder system 64,66,70
inserts 5,24,32,33,34,198
insoluble solution 247
inspection 4,129,160,275,277,279,281,283,285,287,289,
291–295,304–306,309,317
insulating board 269
insulating sleeve 243,244,265
insulation 130,261,264,265
insulator 21,39,214,250,265
interdendritic region 210,220,221
intermetallic compound 218
intermittent tapping 115,149
internal chill 270–273
internal corner 28,29,32,255
internal soundness 292,312
internal stress 11,43,46,63,276
interparticle friction 43,44
invest cycle 9
investment casting 12,16–18,28,32,33,36,
105,107,109,156,223,237,276,277,301
iron carbide 87,192,193,195,199,203,205,208,209,256
iron oxide 39,42,47,64,129,138,148,161
iron-iron carbide 192,208
isocyanate 67,69,70
isomorphous 190
isotropic 94

J

jolt mold 56
jolt-squeeze mold 58
junction design 310,314,316,317
just-in-time 9

K

kaolinite 43,45
kerf 277,279
kinematic viscosity 229
kinetic energy 44,226,233,235
kish 47,194,197
knockoff 277,280

L

ladle slag control 164
lake sand 39
lamellar 201,209
laminar flow 228,229,241,244
laminated object manufacturing (LOM) 36
laminating 136
lance 161,165,176,177,178,181,278
landfill 77,153,297,302
latent heat 247,252,253
lattice structure 187,188
launder 124,150,155,173,249
law of continuity 226,227,239,241
lead 6,18,23,24,28,36,40,41,46,47,49,68,71,75,81,88,97,
103–106,112,114,120,126,128–130,138,155,159,160,173,
181–183,186,188,189,210,220,221,236,239,251,
275,295,298,302,306,307,311
legislation 297,301,303
lever arm principle 190,191,192
lever rule 190
LFC See: *lost foam casting*
lift-coil 120,126,127,179,183
lignite 47
limestone 115,116,137,145,146,148,149,161,162,202
liquid quench and temper 210
liquid shrinkage 259,260,266,311
liquid-solid zone 252
liquid-to-solid shrinkage 311
liquidus 26,190,197,208,213,220,221
lithium carbonate 64
loam mold 2,5
LOI See: *loss-on-ignition*
LOM See: *laminated object manufacturing*
loose pattern 30,31
loose sand vacuum assisted casting (LSVAC) 14
loss-on-ignition (LOI) 49,50,53,63,74,76
lost foam casting (LFC) 11,12,32,173
lost wax process 2,16
low-alloy steel 81,9496,129,157,163,194,211,278
low-carbon steel 26,81,94,148
low-pressure permanent mold (LPPM) 22,24,33,34,130
LSVAC See: *loose sand vacuum assisted casting*
lustrous carbon 47,64,71

M

machinability 81,85,87,88,90,91,93,96,97,99, 101–103,105,106,157,198,217,218
machine molding 28,56
machine-finish allowance 29
magnesium, Mg alloy 4,23,26,29,40,81,89,91,93,97–99, 102,106,107,111–114,120,121,156,161,165, 171,181,182,186,203–207,217–219, 229,239,240,241,245,278,280
magnesium recovery 203,204
magnesium treatment 165,203,204,206
magnetic particle inspection 294
magnetic properties 88,96,97,294
magnetic stirring 164,185
magnetite 64
maintenance 5,6,12,21,28,49,55,72,114,120,121,127,129, 131,138,141,145–147,154,158,159,167–169,172, 204,231,232,275,282,283,285,286,291,301,304–306
malleable iron 4,81,88,102,154,193,199–201,239,240, 263,268,277,280,281,283,285
malleablization 88,200
malleablizing 88,281
manganese 4,40,81,85,94,95,98,103–107,111,141,148, 154,156,161–164,167,192,194,195,198–201,219,220,222,239
manganese sulfide 195
manual pour 230,231
martempering 207
martensitic 96,198,210,211,316
masteralloy 215
matchplate 16,30,31,56,57,58,111,232
matchplate pattern 16,31,57,58,232
matrix 88–93,111,193,194,197–201,203, 206–208,215,217–219,244
mechanical blasting 287
mechanical degassing 185
mechanical feeder 204
mechanical punch 276
mechanical scrubbing 76
medium-carbon steel 81,94
Meehanite 93
melt, melting 1,2,4,5,12,14–20,22,23,25,39–41,47,66,76,82,85, 94,103,109,110,113–119,120–123,125–129,130–133,135, 137–139,140,141,143–149,150–159,160–165,167–169, 171–179,180–188,192–194,197,201,202,207,212,214,215, 218,219,222,231,232,239,241,242,247–249,252,265, 278–280,282,297,298,301–303,314
melt cover 183,186
melt point 39,75,76,184
melt rate 116,118,140,141,143,144,148,154
meltdown 157,159,160–163,165
melting point 39,40,82,110,171,184,187
melting zone 148
meniscus 125,155
mesh size 197
metal alloy 7–9,26,33,34,81–83,85,90,113,122,123,125–127, 145,166,187,189,229,230,237,239,250,252,256, 277–279,281,287,295,314
metal chemistry 141,166
metal core 21,33,34,35
metal flash 270
metal loss 113,119,132,177,183,185,186
metal matrix composite (MMC) 111
metal mold 4,5,19,21,22,24,34,88,111,250,271,276,301,314
metal oxide 64,69,70,74,302
metal pattern 10,11,12,28,32,35
metal-mold interface 45,187,250,253,254
metal-to-mold ratio 45
metallic abrasive 286
metallics 115,116,135,143,147,152,177,180,314
metallostatic pressure 51,78,229,266
methylene blue clay 50,52
mica screen 241
micelle 43,44
micropores 312
microporosity 291,292,312
microshrinkage 107,208,223
microvoid 292
misrun 173,241,293
mixing 10,40,47–49,53,60,68,71,73,74,121,163,164
MMC See: *metal matrix composite*
modification 28,98,101,114,118,176,178,183, 215–217,230,246,289
modulus 82,87,91,93,97,255,262–264,310,314–316
modulus of elasticity 82,87,97,310,315,316
moisture content 42,50,63,127
mold coating 22,24,252
mold dilation 262,263,273
mold gases 78,229,230,234,238,250,271
mold hardness 52,53,57,58,60
mold test 74,75
mold weight 41
molding process 2,4,5,7–12,14,16,18,21,24–28,33,34,39, 42,53,55,57,59,60,82,229,250,309,316,317
molding sand 4,7–12,28,32,39,41–49,50,51,53,55,57–60, 63,64,74,77,94,128,135,136,167,228,229,236,238,239, 250,251,256,266,269,270–272,275–277,302
molybdenum 85,90,94,96,154,160,163,164,192,198,200,208
moment of inertia 314,315,318
momentum 228,233,236
Monel 111,113,145,245
montmorillonite 43
mottled iron 87,195
mulling 40,45,47,48,49,60,74
multiple cavity matchplate 31
multiple-cavity die 23
mushy 26,131,190,220,221,230,252–254

N

natural gas 4,8,9,72,115,120,131,140,143,276
natural-bonded sand 42
NC See: *numerical control*
neoprene wiper 118
Ni-hard 93
Ni-resist 198
Nichrome 111
nickel 18,26,81,85,88,90,93,94,96–98,102–104,106,107,109, 111,130,156,160,162,163,169,171,183,184,188,189, 192,194,198,199,201,207,208,211,220,221,239,268,279
nitric acid 111,211,219
nitrogen 64,96,119,144,157,161,165,176,177,185,202
nobake 10–12,14,26,39,42,60,64,65,68–71,74,75,78, 109,229,250,309
nobake mold 10,12,14,39,42,64,109,229,250,309
nodular iron 81,89,201
nodularity 91,92,205,207,208,294
nodule 88–92,200,201,205–209,241,294
nondestructive testing 5,293,296

nonferrous 4,5,23,32,40,42,45,81,85,97,103,106,115, 145,154,155,171,186,223,238,242,263, 279,283,286,290–292,294
nonferrous alloys 4,5,10,24,81,97,99,101,103,105,107,109, 111,114,126,130,133,171,178,187,191,213–215, 217,219,221,223,273,287
nonmagnetic 211
nonmetallic 114,149,150,154,157,164,177,180, 205,208,287,294,295,310,314
nonmetallic abrasive 287
nonmetallic inclusion 114,157,177,208,295,310,314
nonpermanent molding 7
nonpressurized gating 238–241,244
normalize, normalizing 90,94,95,198,207,210
nozzle 22,23,109,110,227,231,232,239,282,283
nucleation 153,195,197,200,203,205,250,252,253
nuclei 43,44,187,188,197,205,214,222,253
numerical control (NC) 28,36,88,223,246

O

Occupational Safety and Health Adm./Act (OSHA) 176,287,303,305–307
oil pollution 301
olivine 39–42,64,67,71,78
open-hearth furnace 4,130
orange-peel cone bucket 138
organic additives 46
Osborn-Shaw process 16
OSHA See: *Occupational Safety and Health Adm./Act*
overhead crane 58,146,276,306
oxalic acid 211
oxidation 46,70,81,96,106,110,111,113,114,119,121, 143,147–149,155–157,159–162,164,165,177,178, 181,185,186,198,199,230,234,241, 271,272,278,297,302,314
oxidation loss 114,121,143,159,160,185,186
oxide 4,12,24,39,42,47,64,66–70,74,98, 114,121,127,129,132,135,138,145,147,148,150,156,157, 161,162,171,172,174–176,179–186,203,206,229,232,241, 252,278,282,286,287,290,294,302,310,311
oxide formation 179,180
oxide inclusion 24,98,180,252
oxide removal 180
oxide skin 175,180–182,241,278
oxidizing atmosphere 174,183
oxy-acetylene 278,279
oxygen 47,70,114,116,122,140–145,147,148,156,157, 161–164,171,181,183–186,203,205,278,300,311
oxygen enrichment 116,143,144,145,148
oxygen lance 278

P

padding 269,282
parallel axis theorem 315
partial vacuum 59,266
particulate matter 119,297
parting compound 55,56
parting line 8,9,12,16,21,22,28,30–32,34,36,55,59, 78,233,260,270,280,281,317
pattern 4,6–13,16,17,19,23,27–29,30–33,35–37,43,47, 53,55–60,67,69,70–74,81,118,129,173,184,187,188,194, 210,220,221,225,228,230,234,237,240,255,259, 260,261,272,294,314,317
pattern assembly 11,17
patternmaker's contraction 311,314,318
patternmaking 4–6,27,29,31,33,35–37,255
pearlite 88,89,192–195,198–201,207–210
pearlitic ductile iron 206
pearlitic malleable iron 88,200,201
peel 47,138,165
pencil core 266,267
penetration 45,47,52,64,78,86,117,140,155,160,278,295
permanent mold(s), -ing 4,5,7,12,14,16,18,19,21,22,24–26, 30,33,34,77,98,99,101,102,105,107,111,112, 130,132,172,176,178,214,232,237,246,250,251, 271,276,282,291,314
permeability 12,16,40,42,47,50,51,58,59,74,76, 138,149,155,250,295
permeable oxide 181
petroleum binder 8
pH 41–43,47,49,63–65,156,160,298,300
phase diagram 26,188,189,190,191,192,213,220,221,252
phenolic urethane/amine (PUA) 67,68,71,73
phenolic urethane coldbox (PUCB) 75
phenolic urethane nobake (PUNB) 69,71,74
phos-copper shot 185
phosphide 150,194,195
phosphorus 85,86,94,95,98,149,160,162,163,164, 185,194,195,199,201,202,230
picric acid 211
pig iron 85,114–116,143,162–164,205
pinholing 64
pit mold, -ing 3,10,11,233,236,237,275,276
pitting resistance 96
plain carbon steel 81,94,162,163
plasma arc cutting 278
plasma-fired cupola 119,133
plaster molding 16
plastic deformation 292,316
plastic film 12,13
plastic pattern 7,9,10,16,17,28,32,36
plasticity 43,45,46,127
platelets 46–48,215
plunger(s) -ing 23,172,204
pneumatic air system 282
pneumatic tools 291
pollution 46,121,126,138,149,151,154,156,160, 285,297,298,300–302,307
pollution control 46,121,138,149,151,297,298,301,302
polymerized 65
polymers 42,47
polytetrafluoroethylene (Teflon) 185
polyurethane elastomer 28
porosity 26,98,129,175–178,182,184,208,214,291,292,294
porous plug 176,181,202,204
Portland cement 9
postinoculation 205,206
potassium fluoroborate 64
potential energy 225,226
pouring 1,2,7–14,16,21–25,32–34,39,41–48,51,55–61,67,69–71, 77,79,94,113,114,120–122,124,126,129,130,132, 150,155,157,158,161,163,166–169,172,173,175,178, 180–182,184,187,194,198,199,204–206,214,215,218, 220,222,225–227,230–239,241–245,247, 249–252,256,265,267,271,275,297,303,310,314
powder cutting 278
powder injection 164

powdered coal 115
preblends 48
precipitation hardening 217
preheat, -ing 16,17,21,32,75,116,118,122,131,133, 140,143,147,148,154,155,158,161,162,164, 172,175,180,183,185,186,202,231,249,272,278,281
preheating zone 148
pressed filter 242
pressure (leak) test 295
pressure diecasting 23,178
pressure energy 225,226
pressure head 226
pressure pour 174,232
pressure tightness 99,101,102,107,214,223,292
pressurized gating 238,239,240,241,243
primary ingot 99,172
primary melter 130,158,231,232
principle of superposition 314
process control 173,206,207,208,294
proeutectic 252
progressive solidification 25,255,267,269,312,313
prototyping 4,6,12,17,33,36
pseudo-chill 270
PUA See: *phenolic urethane/amine*
PUCB See: *phenolic urethane coldbox*
pullback 32
PUNB See: *phenolic urethane nobake*
purge air 67,68,73
purging gas 185
push-out 120,126
push-up 179,183
pyrometer 167,168,173,175

Q

quantitative analysis 185
quantitative reduced pressure test 175
quartz, -ite 39,43,44,128,129
quasi-flake 207
quench, -ed, -ing 66,90,95,110,143,151–153,198, 201,207,210,217,276,298,301

R

radiation 122,130,144,160,165,167,168,173,248–250, 264,265,276,295
radiographic inspection 292–295
radius, radii 28,29,229,233,236,255,316
ram half 58
rapid prototyping 4,6,17,33,36
rare earth 88,91,106,107,164,219
rattail 39
reactive gas 176
reagent 114
rear-slagging 116
recarburization 161,163
recirculating gas test 175
reclamation 63,67,70,76,77,133,301
recuperative 115,116,133,143,144
recuperative hot blast system 116
recuperator 133
recycle 46,103,283,297,298,302,303
reduced pressure test 175,176184,185
reducing atmosphere 47
reducing slag 162,163
reduction zone 148
refining 95,114,153,156,160–165,178,182,214,219,222,223
refractory , -ies 4,5,7–9,11,12,16,17,21,24,32,39,40,42,77,78, 94,109,114–119,120–122,124–130,132,135,136, 143,149–152,154,156–158,160–163,165,169, 172,174,176,178,181,183,197,202,204,231,249,250, 273,278,302,314
refractory coating 8,12,21,24,32,77,78,172,176
refractory wash 11,24,176
regenerative 4,115,133
regenerator 133
relative humidity 69,78
residual magnesium 204,206
resin 5,8–10,14,15,46,65–69,70–72,74–79,289
resinoid bond 289
reuse 7,28,31,34,40,43,45,60,76,77, 152,276,283,284,286,297,302,303
reverberatory furnace 115,130–133,172,176,180–182
reverse sprue 237
Reynolds number 228,229
rheocasting 26
ring test 174
riser, -ing 11,15,19,27,31–33,35,36,55,56,122, 230,236–238,241,243–246,252,254,256,257,259, 260–271,273,275,277–282,288,289,293,310–313,316
riser feeding 266,273
riser pad 275,278,280–282,288,289,293
riser volume-to-casting volume ratio 262
robot, -ic(s) 18,173,275,281,289,291,296
rolling 59,87,94,102,266
roof 47,121,122,124,132,159,160,163,165,174,175,285,295
rootile 39
rotary barrel cleaning 288
rotary drum shakeout 276
rotary impeller 176,178,181
rotating mold 24
rounded 39,40,42,64,77,88,91,106,107,207,256,287,314
runner 23,116,121,226–230,232,235–241,243,244,251
runner extension 232,236,237,239,241
runouts 45,75
rust 73,145,148,150,154,155,157,160,161,164

S

safety 125,126,129,132,152–155,158,172,178,231,275, 282,283,297,303,304,306,307
sampling 50,175,293
sand blast 271,272,283,286,287
sand erosion 229,238
sand handling 60,64,297
sand mold air gap 250
sand molding 7,8,10,16,31,33,39,41–43,45,47,48, 50,53,55,59–61,174
sand reclamation 76,301
sand slinger 56,58,59
sandcasting 7,12,178
sandwich 203,204
scab 39,155
scaling resist 198
scrap 4,99,114,121,122,131,132,136,137,145,148,153–156,159, 160–164,172,178,180,182,183,185,194, 225,282,292,293,302
screen analysis 52
screen distribution 41,64,74,77
screening 242,287,302

seacoal 46–48,50
SeBiLOY 103,105,112
second-stage graphitization 200
secondary ingot 99,172,182,183
section modulus 310,314,315
section sensitivity 92,197
section size 45,90–92,206
segregation 64,137,210,211
selective laser sintering 36
semi-continuous 113
semipermanent mold 21
separator 96,118,119,150,283,298
set time 9,64
shakeout 9–11,15,41,43,45–51,53,55,60,63,64,66–71,77,79,88,
199,240,275–277,282,297,302,303
shaking ladle 202
shape factor 262
Shaw process 16
shear strength 53,87
shearing 48,97,277,280,281
shelf life 66,67
shell 1,5,8,9,12,14,16–20,26,28,30–32,36,39,46,
60,65,66,72–77,90,107,115–118,120,121,123,124,
135,137,138,140,142,219,232,237,246,273,276,317
shell investment casting 17,18,36
shell molding 8,9,28,31,39,72,73,232
shell molding pattern 232
shot blast 283
shrink rule 29
shrinkage 29,33,34,36,43,63,68,85,94,99,
101,102,127,182,184,205,214,252–255,259,260–269,
272,273,292–295,310–314,316,317
shrinkage allowance 29,36
shrinkage porosity 182
side-dump skip charge 136,137
silica 5,7–9,16,17,39–42,46,51,53,63,64,66,67,69–71,
73,74,76–78,116,121,128,129,135,136,149,156,157,
160–162,164,186,214,225,239,251,265,
273,282,286,287,292,306
silica sand 7–9,17,39–41,51,53,63,64,76,77,
135,136,161,239,265,273,282,286,287
silicomanganese 162
silicon 4,14,24,39,81,85–88,94,95,98,99,101–106,111,120,126,
130,132,135,141,148–150,154,156,157,
161–167,169,174,183,186,192–195,197–199,201,202,205,
213–218,230,239,244,259,279,280,282,290,311
silicon carbide 14,24,120,126,132,174,183,218,244,282,290
silicon dioxide 39,135,161
simple cubic 187,188
simple gases 184
single-cavity die 23
single-slag practice 162,163
sintering 36,128,129,169
skim core 233
skimmer(s), -ing 132,172,181,241
skin dried molding 8
skull. skull melting 157,159,186
slag 2,14,47,114–118,121,122,124–129,135,
141,145,147–149,150–153,155–158,160–165,
202,229,230,232–234,239–241,266,272,273,278,
293,294,298,301,302,310,314
slag collection/disposal 150–152
slag skimmer 150
slagging 116,124,126,149–151,155,161,162,241
sloping dry-hearth reverberatory furnace 131
sludge 87,181,182,282,298,302
sluice-head 151
smoke 47,68,70,71,155,275
snagging 288,289,290
SO_2 process 10,68
soda ash 202
sodium 9,43–45,48,66,67,70,73,98,164,176,215,225,292,298
sodium silicate 9,66,67,70,73,225,292
solid free-form fabrication 36
solid ground curing 36
solid investment casting 17
solid model, -ing 261,262,309,314,318
solid solution 88,189–191,193,208,211,219,247,252
solid waste disposal 302,303
solid-shrinkage allowance 29
solidification front 220,221,253,255
solidification modeling 6,36,256
solidification process 85,189,195,213,230,252
solidification shrinkage 34,94,99,101,102,
254,259,260,310–312,316,317
solidus 26,190,208,213,220,221,292
solution treatment 217
southern bentonite 45
sows 172
sparging gas 176
specialty sand 40
specific gravity 41,107,154,181
specific heat 39,40,41,82,171,247,251,271
specifications 42,49,77,92,100,103,114,138,
146,149,150,156,158,160,172,177,
208,217,243,283,289,292–294,317
specimen weight 50,51
spectrograph 114
spectrometer 167,295
spheroidal graphite 89,91,92,201,203,206
spindle finishing 288
spinner monorail 285
spraying, spray ring 78,118,153
sprue 13,14,22–24,33,185,226–228,230–239,240–245
sprue basin, sprue cup 13,231,232,233
sprue pin 234
sprue well 235,236,239,240
squeeze casting 25,26
squeeze mold 7,56,57,58,59
stack mold 60
stack reverberatory furnace 181
stack-melting reverberatory furnace 131
stainless steel 14,94,96,145,156,162,163,169,
211,245,278,279,280.287
standards 5,77,103,104,112,177,282,291,293,297,
300,301,303,305,307,317,318
starter block 155,158
static pour process 21
stationary (bowl-type) crucible 120
stationary lift-out crucible 120
statistical quality control 293
steadite 194,195,202
steel lance 176
steel mold 19,23
steel shot 271,282,283,285,286
stereolithography 6,17,36
stick point 75,76
stirrers 172

stopper rod 231,232
storage area 5,114,145,146,151,172
strainer core 234,235
Straube-Pfeiffer test 175
strengthening bar 270
stress relief 105,198,210,217
stress-relief annealing 210
strip time 10,69,71,74
strontium 98,176,215
strontium modification 215
structural design 309,315
structural geometry 309,316
structural properties 310,316
subangular 40,42,64
suction air blast 284
sulfide 150,195,202,203,206
sulfur 10,47,49,68,71,81,85,88,94,95, 138,149,154,157,160,162–165,167,182,195,198,199, 201–203,205,207,298
supercooling 252
superheat 113,148,157,164,185,186,197,198, 205,229,247,252,259
surface area-to-volume 114,260,261,271
surface cleaning 5,276,278,284
surface finish 5,7,9,10,12,14,16,18,21,41,47, 55,64,77,97,185,277,286
surface hardening 95
surface imperfection 276–278,284
surface oxide 229
surface tension 43,44,47,229,242,310,311
suspension agent 78
sweep mold 2
swing cover 120
swing frame grinding 277
swing half 58
swing table blast 284,285
swing-away furnace 179,183
synthetic sand 42
system sand 45,71,74,77,302

T

tap-and-charge 113,114,126,154,155,158
tap hole 115,116,118,136,140,149,150,152,153
tap temperature 122,148
taper 28,118,234,236,240,311,313
tapping 52,115,116,118,121,122,124,129,149,150,156,157, 161,162,163,165,168,181,249
temperature control 69,156,157,162,173,180
tempering water 46
tensile strength 67,69,74–76,82,86,87,90–92,94,95,97, 102,105,199,210,215,219,275,295
tensile strength test 75
ternary 97,189,213
testing 4,5,39,50,51,52,53,60,74,75,135,166,175,184, 206,225,229,287,289,293,294,295,296,301,305,306
thermal arrest 166,215
thermal conductivity 39,82,91,92,97,105,107,120,128,181,251,271,278
thermal emissivity 167
thermal expansion 40,41,47,97,99,101,107,128
thermal gradient 77,99,185,187,230,236,248,250–256, 268,269,270
thermal reclamation 67,77
thermal shock resistance 90
thermal torch 278
thermite reaction 265
thermoset 5,8,9,28,65
thermosetting polyurethane 28
thermosetting resin-coated sand 5,9
thin-wall 14,18,25,101,168,244,250
thixomolding 26
thorium 106,107,219
tilt pour process 21,22
tilting 22,120,121,126,146,161,179,180,183
tilting crucible 120
tin 1,23,81,97,98,99,103,104,106,154,155,160,164, 184,186,192,194,213,221,223,239,268
titanium 4,8,9,10,18,24,81,92,97,98,109,111,113,114,117,122,156, 171,185,186,192,199,207,212,214,241,246,281,314
titanium electrode 186
tolerance capability 317,318
tongs 175
tool 3,4,6,16,23,24,28,30–34,36,52,53,65,85,87,97, 101,102,110,119,121,136,163,164,171,172, 215,247,256,277,279,280,282,288–294,309,310,314,317
tool steel 157
tooling 27,30
tooth spacing 279
torch 8,119,140,278–280
torsion strength 87
total maximum daily load 301
toughness 53,81,88,90,92,94–96,99,109,110, 200,202,210,217,295
toxic 176,177,215,297,298,301,302,305
transverse strength test 75
tree assembly 17
trim press operation 280
tumble blast 284
tumbling mills 287
tundish 25,203
turbo blower 141
turbo-head mixer 74
turbulence 22,24,77,131,157,168,176,178,179,181, 206,225,226,228–231,233–236,239–241,243,244,266
turbulent flow 21,228,229
turnings 114,119,160,164,172,183,185
tuyere injection 145
tuyeres 1,105,115–119,137,140,145,147–150,152,163
two-chamber coreless induction furnace 180

U

ultrasonic test 293,294
undercooling 165,197,198,253
Unicast process 16
unit cell 187,188
unit die 23

V

V-process 12,13
vacuum assisted casting (VAC) 14
vacuum chamber 14,18,20,122,186,188
vacuum melt 114,129,161,314
vacuum-squeeze mold 59
vapor loss 183,186
VCM See: *volatile combustible matter*

velocity 22,24,101,117,118,128,151,158,
225–229,232,233,235,239–241,244,279,280,287,298,310
venting 21,32,33,51,59,66,72,73,77,78,270,310
vents 13,21,35,67,69,72,73,77,78
venturi 297
vermicular 91
vertical gating 237,238,241,242,246
vertical skip hoist 136
vertical-shaft 115
vibratory compaction 10
vibratory conveyor 276
vibratory feeder 138
visual inspection 293,294
volatile 8,47,49,50,77,138,172,297
volatile combustible matter (VCM) 49
volume feed capacity 262
volumetric contraction 261
volumetric shrinkage 259,260
vortex 152,233–235

W

wall movement 39,45,46,259
warm coating 66
warmbox 65,66,71,72,78
wash 11,24,77,127,151,157,176,183,251,271
wash heat 157,183
washes 39,40,45,46,127
waste stream 302
wastewater 77,151
water blast cleaning 287
water cooled shell 117
water cooling 25
water jacket 25,101,117
water jet cutting 277
water pollutant 298
water-cooled coils 127
water-cooled crucible 186
water-cooled inductors 127
water-to-clay ratio 46
wax 2,16,17,27,28,30,32,33,36,318
wear resist 27,81,85,87,90,93,94,97,
101,106,110,111,198,200,270
web 26,237,242
weighing hopper 146
weir 118
weldability 96,97,99,105,109
welding 5,94,97,105,107,160,270,272,277,278,279,307,316
well 116–118,131–133,140,145,147,148,150,152,176,180,
203,233,235–241,301
well injection 145
well registration 301
well zone 147,148
western bentonite 45
wet chemistry 167,295
wet reclamation 67,77
wet scrubber 297,302
wet slag collection 151
wet tensile test 53
wet-bath reverberatory furnace 132
wetted 39,40,48
wetting agent 47,78
wheel composition 289
white cast iron 81,87,88,89,260
white iron 29,87,88,93,167,193,196–200,205,282,283
white-bronze 111
whiteheart 4
William's core 266
wind drum 141,142
windbox 116,117,118,142,150
wire injection 164,198
wirebrushing 282
wood pattern 12,27,28
work time 10,69,70,73,74
worksite analysis 303,304
wrought shapes 25,314
wrought steel 210

Y

yield 5,7,22,25,56,58,59,85,93,95,97,99,
101,102,105,107,109,160,187,
210,215,230,241,244,263,292,295,316
yield strength 93,95,97,99,101,102,107,109,210,292,295,316
yttrium 106,203

Z

zinc 4,18,19,23,24,26,81,97,99,102,103,105–107,111,113,114,
155,171,183,186,189,213,217,219,298,301
zinc-aluminum alloy 18,19,24,111
zircon 8,17,39–41,64,67,78,106,107,113,128,181,182,185,197,
214,219,222,251,271–273,290